AF522208

ENCYCLOPEDIA OF BIOCHEMISTRY

Vol. 1

By

Sananda Chatterjee

DAE, DCA, CMCNet

Scientific Consultant

Institute for Natural Sciences

Kolkata

(West Bengal)

DISCOVERY PUBLISHING HOUSE PVT. LTD.

NEW DELHI-110 002

Published by:
Namit Wasan
DISCOVERY PUBLISHING HOUSE PVT. LTD.
4383/4B, Ansari Road, Darya Ganj
New Delhi-110 002 (India)
Phone : +91-11-23279245; 23253475; 43596065
E-mail : discoverybooksindia@gmail.com
discoverypublishinghouse@gmail.com
namitwasan9@gmail.com
web : www.discoverypublishinggroup.com

Reprinted: 2020
First Edition: 2012

ISBN: 978-81-8356-876-0 (Set)

Encyclopedia of Biochemistry

Printed at:
Infinity Imaging Systems
Delhi

Preface

It is always an ultimate outlook to give students the best information in a good package. So I took up the job of writing a series of books in chemistry starting from a very introductory level to this advance level of biochemistry.

Biochemistry, to express something about it, in one word, would be difficult for me. It should be ones part of life; the reactions of protein, amino-acids, the DNA RNA—double helix pattern. The nynhydrin reactions are all the wonders of biochemistry—man's anato-biochemical linkages.

Students who want to become doctors or medical practitioners must be a good biochemist first because biochemistry is the theory behind and the medicine is the practical application.

The effects and side effects of the particular medicine, whether it is allergic or non-allergic, to the patients are well understood by the doctors but are taught in biochemistry.

The chemical functions of the different human organs and systems; the kidney, the respiratory system, the blood, the chemical functions of hormones, vitamins, the liver etc are all well covered by this book.

The subject of this book is therefore deeply intended for those upcoming doctors who keep themselves for the real service of the mankind.

The object of this book is not only meant for the doctors but it is the last bangle of that long garland which has started all along from the introduction chapter of The Architecture of Chemistry, The Mans Scientific Attitude.

Biochemistry also comes in use in the field of agriculture, the biochemical configuration of the soil. The biochemical configuration of the plants to be cultivated, the application of the pesticides, and the relation of the pesticides with the human population are well studied under biochemistry.

The Agricultural Scientist, the Pathological Scientist, Forensic Scientist, is definitely inter-joined with each other by a single buffer of the Biochemistry.

This book is dedicated to those promising scientific readers and interested students of biochemistry and advanced chemical science, who wants to make a fruitful attempt in the field of modern scientific world.

The author will be grateful and surely thankful to see the best utilization of this book in any of the field of application.

Sananda Chatterjee

Contents

Enzymatic—Chemical—Evolutionary Conservation—Other Uses/information—Reduction of Galactose Produces Dulcitol—Lipids—Fatty Acids—Soaps and Detergents—Fats and Oils—4. Waxes—5. Phospholipids—Terpenes—Lipid Soluble Vitamins—Biosynthetic Pathways—Condensations—Alkylations—An Alternative Isoprenoid Synthesis—Prostaglandins—Blood Clotting—Prostaglandins—Inflammatory Response and Aspirin—Sphinogomyelin—Disorders—Lipoproteins—Cholesterol Regulation—Arteriosclerosis—Triacylglycerols from Seed Oils—Triacylglycerols from Animal Fats—Adipose Tissue—Triacylglycerols from Animal Fats - Milk Fat—Triacylglycerols from Animal Fats - Other Tissues—Triacylglycerols from Fish Oils—Triacylglycerols of Bacteria and Yeast—SN-1,2-Diacylglycerols in Tissues - Biological Importance—Plant Sterols—Sterol Esters—Sterol Glycosides—Sterols in Yeasts and Fungi—Bacterial Sterols—Analysis—Structure, Composition, Biology and Analysis—Structure and Biosynthesis—Tocopherols Metabolism in Animals—Tocopherols as Antioxidants—Other Biological Functions of Tocopherols—Analysis—Analysis—Glycerophospholipids—Phosphatidic Acid—Occurrence and Biosynthesis—Phosphatidic Acid-Biological Functions—Lysophosphatidic Acid—Cyclic Phosphatidic Acid—Pyrophosphatidic Acid—Analysis—Phosphatidylglycerol in Bacteria and Plants—Phosphatidylglycerol in Animal Tissues—Biosynthesis of Phosphatidylglycerol—Acylphosphatidylglycerol—Complex Lipoamino Acids—Analysis—Phosphatidylcholine-Biosynthesis and Biological Function—3. Lysophosphatidylcholine—Platelet-Activating Factor—Analysis—Glycoglycero-lipids—Sphingomyelin and Glycosphingolipids—Fatty Acids—Eicosanoids and Related Lipids—Nomenclature Of Saturated And Unsaturated Fatty Acids—Melting Points of Saturated vs. Unsaturated Fatty Acids—Saturated Fatty Acids —Solution Properties—Branched-chain Fatty Acids —Mono or multibranched chain fatty acids—Branched methoxy fatty acids and Branched hydroxy fatty acids (Mycolic acids)—Isoprenoid Fatty Acids—Branched Methoxy Fatty Acids—Essential Fatty Acids—w-3 Fatty Acids or Omega-3 Fatty Acids)—List of n-3 fatty acids—Biological Significances—Health Risks—Cardiac Risk—Fish—Eggs—Meat—Other Sources—The n-36 to n-33 ratio—Conversion Efficiency of ALA to EPA and DHA—Prostaglandins : Biosynthesis and Functions—Prostaglandin—Biosynthesis—Release of Prostaglandins from the Cell—Cyclooxygenases—Prostaglandin E synthase—Other Terminal Prostaglandin Synthases—Function—Types—Role in Pharmacology—Inhibition—Clinical Uses—Leucotriens and Thromboxanes—Biochemistry and Synthesis—Function—Leukotrienes in Asthma—Role of Cysteinyl Leukotrienes—Production—Mechanism—Functions—Role of A2 in Platelet Aggregation—Pathology—Suppression—Classification And Structure of Phospholipids—a Surfactant Glycolipids—Types of Glycolipids—Derived Lipids—Cholesterol—Clinical Significance (Hypercholesterolemia)—Steroid Hormones—Steroid Hormone Biosynthesis Reactions—Steroids of the Adrenal Cortex—Regulation of Adrenal Steroid Synthesis—Functions of the Adrenal Steroid Hormones—Gonadal Steroid Hormones—Steroid Hormone Receptors—Bile Acids—Production and Distribution—Types—Regulation—Clinical Significance—Potential Applications—Characterisation of Lipids—Saponification Number, Iodine Number, Acid Number,

Brain Nonsynaptic Glutamatergic Signaling Circuits—GABA Precursor—Flavor Enhancer—Nutrient—Plant Growth—Production—Pharmacology—Role in Sickle-cell Disease—Production—Biosynthesis—Degradation—Physiological Function—As a Biosynthetic Intermediate—As a Neurotransmitter—As a Potential Antipsychotic—Industrial Uses—Presence in the Interstellar Medium—Chemical Properties—Metabolism—Supplementation—Structure—Production and Function—Clinical Significance—Biosynthesis—Isomers of Isoleucine—Synthesis—Biosynthesis—Biological Uses—Chemical Properties—Food Additive—Biosynthesis—Metabolism—Synthesis—Dietary Sources—Properties—Modifications—Clinical Significance—In Popular Culture—Biosynthesis—Other Biochemical Pathways—Propionyl-COA, Pathway—Synthesis—Dietary Aspects—Role in Urea Cycle—Lactamization—Other Biological Roles—Phenylketonuria—In Non-Humans—D- and DL-phenylalanine—History—History—Biosynthesis—Structural Properties—Cis-trans Isomerization—Uses—Specialities—Structure—Biology—Occurrence—Biosynthesis—Chemical Synthesis—Metabolic Function—Structural Role—Signaling—Allo-threonine—Biosynthesis—Metabolism—Sources—Isolation—Biosynthesis and Industrial Production—Function of Tryptophan—Dietary Sources—Use as a Dietary Supplement—Metabolites—Turkey Meat and Drowsiness—Fluorescence—Functions—Dietary Sources—Biosynthesis—Metabolism—Phosphorylation and Sulphation—Precursor to Hormones—Precursor to Alkaloids—Precursor to Pigments—Degradation—Ortho- and Meta-tyrosine—Medical Use—Drug Detoxification—Biosynthesis—Synthesis—Dietary Aspects—a Chemical Reactions of Amino acid—Isoelectric Point—The Isoelectric Point—pH—Acid Base Properties of Amino acids—Why isn't the Isoelectric Point of an Amino Acid at pH 7?—Gel Filtration —Nutshell Overview of Gel Filtration —Principles of Gel Filtration —Notes on Use of Gel Filtration —Part I: Column Pouring and Calibration —Materials—Procedure—Problems—Electrophoresis—Theory—The Electrophoresis System: The Electrophoresis Chamber and the Complete Unit—Applications—Factors Affecting Migration—Visualisation: Ethidium Bromide (EtBr) and Dyes—Percent Agarose and Resolution Limits—Buffers—Analysis—Materials and Method—Preparation—Procedure—Ultracentrifugation—Analytical Ultracentrifuge—Preparative Ultracentrifuge—Hazards—Ninhydrin Reaction—Uses—Reactivity—Determination fo protein Structure with Ninhydrin—Separations of Amino Acids—Paper Chromatography—Thin Layer Chromatography—Problems in TLC—High Performance Liquid Chromatography—Polyacrylamide gel electrophoresis—Determination of Molecular Weight—Continuous and Discontinuous Buffer Systems—Protocol—Assembling Gel Apparatus—Resolving Gels—Stacking Gels—Electrophoresis Buffer—Samples—Staining of Gels—Classification and Structure of Immunoglobins and Plasma Proteins—Antibody forms—Isotypes—Structure—Immunoglobulin Domains—Heavy Chain—Light Chain—Fab and Fc Regions—Function—Activation of Complement—Activation of Effector Cells—Immunoglobulin Diversity—Domain Variability—V(D)J Recombination—Somatic Hypermutation and Affinity Maturation—Class Switching—Medical Applications and Disease Diagnosis—Disease Therapy—Prenatal Therapy—Heamoglobin and Myoglobin—Meat Colour—Role in Disease—Structure and Bonding—Role of 2, 3 DPG—Effects of Binding—Fetal Hemoglobin—

Thalassemia — Genetic Classification of the Thalassemias — Alpha Thalassemia — Beta Thalassemia — Clinical Classification of the Thalassemias — Alphathalassemia — Betathalassemia — Relationship of the Genetic and Clinical Classifications of Thalassemia — collagen — Uses — Industrial Uses — Medical Uses — Conformation and Structure — History and Background — Molecular Structure — Fibrillar Structure — Types and Associated Disorders — Staining — Synthesis — Amino Acids — Collagen I Formation — Synthetic Pathogenesis — Variety of Animal Uses — Cornification — Molecular Biology and Biochemistry — Glycine and Alanine — Disulphide Bridges — Silk — Pairing — Clinical Significance — Composition — Locations in Body — Purines — Pyrimidines — Nucleosides — Pharmacological Effects — Adenosine Receptors — Anti-inflammatory Properties — Action on the Heart — Action in CNS — Dosage — Drug Interactions — Side Effects — Metabolism — Dietary Sources of Uridine — Dietary Sources of Cytidine — Nucleotides — Polynucleotides — Hydrolysis of Polynucleotides — Purine Catabolism — Nucleotides to Bases — Bases to Uric Acid — Gouts and Hyperuricemia — Summary — Pyrimidine Catabolism — Ring Cleavage — General Comments — De Novo Synthesis of Purine Nucleotides — PRPP — Commitment Step — Formation of IMP — Formation of AMP and GMP — Control of De Novo Synthesis — Possible Scenario — De Novo Synthesis of Pyrimidine Nucleotides — Carbamoyl Phosphate — Formation of Orotic Acid — Formation of the Nucleotides — Control — Interconversion of Nucleotides — Salvage of Bases — Salvaging Purines — Lesch-Nyhan Syndrome — Salvaging Pyrimidines — Formation of Deoxyribonucleotides — Synthesis of dTMP — Chemotherapeutic Agents — structures and properties of DNA in different forms — Properties — Grooves — Base Pairing — Sense and Antisense — Supercoiling — Alternative Structures — Quadruplex Structures — Further Information: G-quadruplex — Branched DNA — Chemical Modifications — Base Modifications — Damage — Biological Functions — Genes and Genomes — Transcription and Translation — Replication — Interactions with Proteins — DNA-binding Proteins — DNA-modifying Enzymes — Nucleases and Ligases — Topoisomerases and Helicases — Polymerases — Genetic Recombination — Evolution — Uses in Technology — Genetic Engineering — Forensics — Bioinformatics — DNA Nanotechnology — History and Anthropology — Primary DNA Molecular Structure — Basic Properties of Primary DNA Molecular Structure — Fundamental Properties of DNA — Sugar-Phosphate Backbone — Molecular Structure of DNA — Complimentary Core of Bases — Hydrostatic Interactions — Antiparallel — Direction and Polarity in DNA — DNA is Antiparallel — Structure and properties of m-RNA, t-RNA, r-RNA and hn-RNA — Synthesis, Processing, and Function — Transcription — Eukaryotic pre-mRNA Processing — 5' cap Addition — Splicing — Editing — Polyadenylation — Transport — Translation — Structure — 5′ Cap — Coding Regions — Untranslated Regions — Poly(A) Tail — Monocistronic versus Polycistronic mRNA — mRNA Circularization — Degradation — Prokaryotic mRNA degredation — Eukaryotic mRNA Turnover — AU-rich Element Decay — Nonsense mediated decay — Small interfering RNA (siRNA) — Micro RNA (miRNA) — Main article: microRNA — Other Decay Mechanisms — Transfer RNA — Anticodon — Aminoacylation — Binding to Ribosome — tRNA Genes — History — Processing.

CHAPTER 1

General Review

ORIGIN OF LIFE

Molecular Logic of Living Matters

> *Marco Polo describes a bridge, stone by stone*
> *"But which is the stone that supports the arch?" Kublai Khan asks*
> *"This bridge is not supported by one stone or another,"—Marco Polo answers*
> *"But by the line of the arch that they form."*
> *Kublai Khan remains silent, reflecting. Then he adds, "Why do you speak*
> *to me of the stones? It is only the arch that matters to me."*
> *Polo answers, "Without stone there is no arch."*
>
> ***—Italo Calvino***

When biochemists set out to tackle a problem, our first step is commonly grind the intricate fabric of cells and tissues into a pulp (a homogenate). This is a significant action, representing a drastic reduction in the level of organization. It allows us to treat living matter as a mixture of chemical and encourages us to isolate and also purify individual constituents. Every student of biochemistry is warned not to misuse clean thinking on dirty enzymes. To be sure, something is sacrificed by this destructive procedure—not only life itself, but all the special order that impresses any one who inspects a photomicrograph.

But never mind: biochemists still cherish the premise that nothing irretrievable is lost by homogenization; and that given the macromolecules, all the essentials are present and accounted for.

We know quite well that this cannot be true, but the focus on the molecules defines that layer of knowledge that we designate as biochemistry molecular biology, and undergirds our professional identity.

Let us for the present set aside the levels of order lost to the tissue grinder, and celebrate the astonishing achievements that grew from the meticulous examination of life's fragments. All the activities of living things are carried out by molecules, and are thus ultimately rooted in molecular structures and interactions. Without subscribing to the lop sided view that, therefore, all of biology is molecular biology, it must still be said that one cannot reflect usefully on the phenomenon of life without taking account of its material basis. There can be no arch without stones. Besides the molecular level provides an excellent introduction to the exploration of biological order in general. Regularity, purpose, and complexity pervade the field. We can observe individual molecules coming together to perform novel and emergent functions, each of them a whole larger than the sum of its parts. We can discern generalizations that apply to all organisms on earth; they qualify locally as laws of biology, and assure us that all life is of one kind. We also find variations on all but the most basic themes; these report the roles of mutation and selection, constraints and contingency, in the genesis of living order. And we can ponder the deep question, what might we expect to find, if and when we see life beyond the solar system?

Biochemists and molecular biologists revel in the details of their subject; the key to finding order in the profusion is the concept of *function.* "Living organisms are composed of lifeless molecules" the late Albert Lehninger's proclaims on the opening page of his classic textbook, but those molecules are special. The molecules of life differ from those encountered in the inorganic world, not in their chemical qualities, but in their biological ones: with, few exceptions such as waste products each perform a job in the service of the organism as a whole. The notion of function is meaningless when applied to the constituents of clay or of petroleum, for those molecules are the products of physical and chemical forces alone but function becomes crucial when we ask why leaves are green and blood is red. Function implies purpose, and therefore, order. "The molecules of which living things are composed conform to all the familiar laws of chemistry, but they also interact with each other in accordance with another set of principles, which we shall refer to collectively as the *molecular logic of living matter*. In this chapter we will be walking in Lehninger's trail, most of its factual content is drawn from the textbook of biochemistry and also from the students of biochemistry to the general reader by underscoring the principles and omitting the details; some of the latter will be found in Boyce Rensberger's useful book, which is specifically addressed to a lay audience.

For the purposes of this chapter, we may think of a cell as an intricate and sophisticated chemical factory (Fig. 1.1). Matter, energy and information enter the cell from the environment while waste products and heat are discharged. The object of the entire exercise is to replicate the chemical composition and organization of the original cell, making two cells grow where there was one before. Even in the simplest cells, this calls for the collaborative interactions of many thousands of molecules large and small, and requires hundreds of concurrent chemical reactions. These breakdown foodstuffs extract energy manufacture precursors. Assemble constituents note and execute generic instructions and keep all this frantic activity coordinated. The term "metabolism" designates the sum total of all these chemical processes derived from the Greek word for "exchange". Biochemistry, then, is the study of the chemical basis of all biological activity. My purpose here is of necessity much more modest: to introduce the reader to terms, discoveries and ideas that are indispensable for closer reflection on the nature of life.

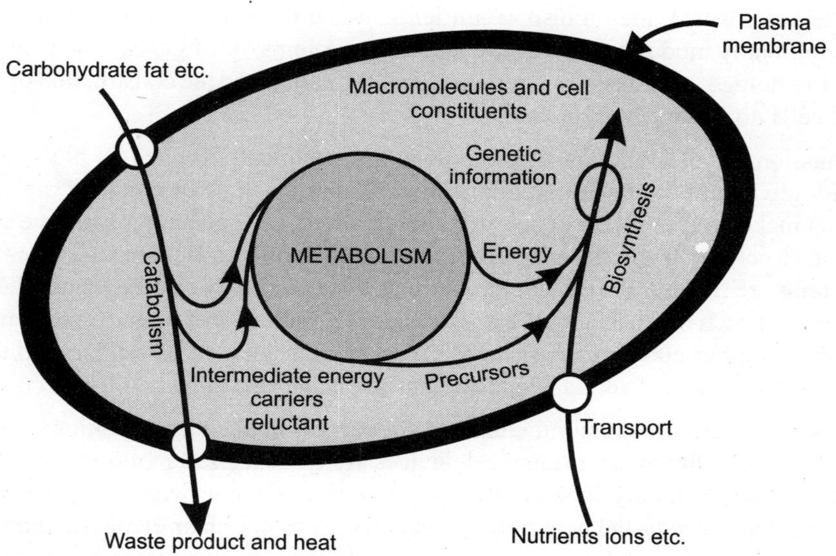

Fig. 1.1 : The Cell as a Chemical Factory

A beginner, on first encountering the profusion of chemical reactions that take place in every cell is apt to suspect that any reaction that can possibly go, does go. Closer inspection do correct the false impression: in reality, cellular metabolism is highly selective and quite purposeful. Each reaction is mediated by a particular enzyme whose function is to enable that reaction to proceed at a high rate, often with extreme specificity and with minimal formation of useless by products. Step by simple step, the cell's complement of enzymes breaks down foodstuffs turns them into metabolites and then into cell constituents, and harnesses the energy of some reactions drive others (Fig. 1.2). Enzymes select the channels through which matter and energy flow. They can be studied as single molecules and often are, but they derive meaning from being parts of a larger whole, the metabolic web.

How enzymes perform their catalytic feats, greater by many orders of magnitude than those of inorganic catalysts, has long been one of the central questions in biochemistry. The heart of the matter is the specific, intimate, and tight binding of the substrate (or substrates) to the enzyme proteins (and virtually all enzymes are proteins) are not commonly entails changes in the configuration of both substrate and enzyme, inducing stresses and strains that contribute to the mechanism of catalysis. Besides, the catalytic site supplies chemically active groups in the form of amino acid side chains that actually participate in the reaction. The catalytic site is tailored, as it were to its particular task, linking its structure to its function.

The genome of *E. coli* encodes approximately 4.000 proteins that of yeast 6.000; it takes 50.000 proteins or more to make a man. What do they all do? Many proteins are enzymes but by no means all. Some proteins serve as the building blocks of structural scaffolding. Some make tracks for the movement of organelles itself mediated by motor proteins. Proteins act as receptors for signals from within the cell or from the outer world; they transport nutrients, waste products and viruses across membranes. Proteins also commonly modulate the activities of other proteins, or of genes. The general principle is that except for the storage and transmission of generic information and the construction of compartments, almost all that cells do is done by proteins.

The explanation for the function of proteins is not chemical so much as physical. Amino acid chains can fold into a variety of shapes globular and fibrous, each determined by the sequence of the amino acids that make up the protein in question. As they fold, each generates a unique contour with its own pattern of structural features: rods and hinges, platforms and channels, holes and crevices. Moreover, proteins are flexible and dynamic constructs that commonly change shape when they interact with legends or with each other. The range of stable configurations that amino acid chains can assume is wider than that of other classes of macromolecules, nucleic acids in particular; and their flexibility permits all sorts of mechanical actions demanded of molecular machines.

Proteins, as catalyses and structural elements are part of biochemical tradition; more recently we have come to see many of them as mechanical devices are on energized motion to perform their tasks. Even enzymes can be profitably looked at from this point of view: with the growing catalogue of enzyme structures has come to the recognition that active sites and their elements commonly undergo rearrangement as part of the catalytic cycle and its regulation. Other proteins are there to bring about overt movement, either of molecules or of larger objects. Transport carriers reorient the binding site from one membrane surface to the other, and back again; sometimes the mechanical cycle is coupled to an energy source, turning the carrier into a pump. Students of eukaryotic cells are finding ever more motor proteins that translocate vesicles, chromosomes, or elements of the cytoskeleton fıom one place to another. The most familiar example is myosin whose cyclic change of conformations underlies muscle contraction and some instances of cell motility. And bear in mind ribosome and polymerizes that transcribe and replicate genetic information: energized movements are central to their operations. As we unravel the molecular workings of life, the cell presents itself an assemblage of tiny machines; mundane mechanical engineering looms as large as the subtle flow of the energy and information.

Few generalizations in biology get by without qualification, and that applies to the status of proteins as the tools for all tasks: surprisingly, some tasks have to ribonucleic acids. RNAs make up almost two□ thirds of the ribosome's mass and play a catalytic role in linking up the amino acids during protein synthesis. They contribute to the structure and function of several less familiar organelles, such as the particles involved in the translocations of proteins across membranes. Most remarkably, RNA alone sometimes serves as a catalyst. Catalytic RNAs or ribozymes participate chiefly in the manipulation of RNA, and they do not challenge the predominance of proteins in the catalysis line. But their discovery in the eighties overthrew the universal consensus that all enzymes must be proteins, and it suggested an altogether novel line of enquiry to the students of origin of life.

The composition of living cells is grossly different from that of their environment. That is possible because each cell is enclosed by a membrane; extremely thin and flexible, that is essentially impermeable to molecules large and small. Biological membranes are composed of phospholipids (or in Archæa, of the related ether□lipids), that form closed bilayered structures spontaneously (Fig. 1.2). The oily core of the bilayer excludes water□soluble molecules, and thus imposes a barrier to the diffusion of most substances; water itself, oxygen and other gases are among the exceptions. Membranes are as essential to life as genes and proteins: "To stay alive you have to be able to hold out against equilibrium, maintain imbalance, bank against entropy, and you can only transact this business with membranes in our kind of world".

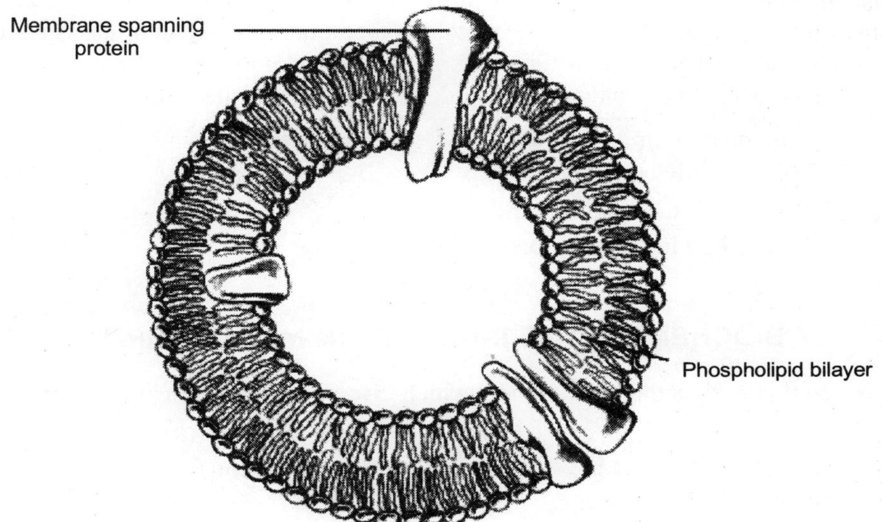

Fig. 1.2 : The Architecture of Biological Membrane. A Phospholipid Bilayer makes up the Basic Structure.

The corollary is that cells require special means to transport nutrients in and waste products out. With very few exceptions, this is accomplished with the aid of proteins that span the membrane, linking the aqueous phases inside and out. Transport proteins are akin to enzymes in recognizing and binding particular substrates but most of them do not catalyze any chemical reaction, they merely translocate the substrate from one side to the other by shifting the orientation of the binding site from one surface to the other. In some cases this reorientation is linked to an energy source, allowing the transport protein to "pump" its substrate against a concentration gradient, others merely facilitate downhill diffusion. The plasma membrane of a bacterial cell may contain as many as a hundred different transport catalysts, each more or less specific for a particular substrate.

Transport proteins carry matter; receptors deal in information. Bacteria cells sense the presence of nutrients in the medium by virtue of specialized proteins that span the plasma membranes. When a potential nutrient binds to the receipt of protein, the news is carried across the membrane by a conformational change; this, in turn activate an enzyme or an enzyme cascade. What passes across the membrane is neither matter nor energy, but a signal: "now", or sometimes "here".

Bacteria, as a rule feature but a single membrane, the plasma membrane that defines the cell; but in eukaryotic cells, intercellular membranes are a conspicuous feature. Each intercellular membrane enclose a defined space whose composition and function is more or less distinct and which communicates cytoplasmic fluid by means of specialized transport catalysts. These compartments are also dynamic, fusing with one another and detaching a controlled and functional manner.

Traffic across membranes is not confined to small nutrients and waste products. In a bacterial cell many constituents are situated external to the plasma membrane: the cell wall is an array of enzymes proteins that serve as receptors for environmental signals, flagella and they must meanwhile maintain the integrity of their bounding membrane. The role of membranes, like that of the sea in human history, is dual; they separate compartments while channeling the flow of matter and of information. When one thinks about membranes, one stands on the border between molecules and cells.

> The man behind the microscope
> Has this advice for you
> Never ask what some thing is
> Just ask, what does it Do?
>
> —***Hilaire Belloc***

WHY BIOCHEMISTRY IS INCLUDED IN MEDAL CURRICULUM?

Biochemistry, as it is said, is the basic subject which discusses the chemical constitution of the human body.

The chemical constitution of the body that means the different function of different organs are dealt in biochemistry.

The action of vitamins, enzymes co□enzymes, hormones, carbohydrates etc are all preprogrammed by biochemistry. The function of kidney, liver, the respiratory system (digestive system is a total biochemical programmed) work on complicated chemical reactions. These chemical reactions are all preprogrammed, as said earlier, any disturbance in them can cause a fatal disaster in human metabolism. To set right these disturbances—"diseases" as they are commonly called, some correcting agents called "medicines" are admixture.

So in order to give or admixture these medicines doctors had to know biochemistry very well otherwise a big incorrigible problem can occur which could be more fatal. Some patients can be allergic to many drugs, the study of these problems is carried in the advances of medical biochemistry.

The hypo□and hyper□activity of these drugs precisely can cause what kind of metabolic disorder can only be understood in biochemistry. We can take a very common example, a very common disease the diabetes, the inconsiderable increase of sugar level in blood; to decrease this sugar level increment in blood insulin is administered.

Apart from this the different pathological tests are all been developed in biochemistry, without which a medical practitioner would not be able to diagnose the problem of the patient.

Biochemistry also deals with the mechanism of drugs, which makes a doctor well updated to treat a patient. The steroids, the cholesterol',

So in the end it can be translated that biochemistry must become a useful tool for those doctors who can offer a good treatment to the patients.

CELLULAR STRUCTURE AND FUNCTION IN CHEMICAL TERMS

Introduction

In this section we discuss the cell, the smallest unit of the human organism capable of reproducing itself. An understanding of cellular structure and function forms the basis for understanding malignant growth and anticancer therapy because (1) all malignancies are composed of abnormal cells, and (2) chemotherapy attacks cancer on the cellular level. In Section 5 the cell life cycle, i.e., the various phases through which a cell passes during the process of cell division, will be reviewed. A knowledge of the life cycle of the cell helps us to better understand how chemotherapeutic agents work.

Functional Anatomy of the Cell

Because cells are so different, there is no *typical* cell, but for purposes of illustration we can construct a hypothetical *composite* cell having all the component structures observed in many cells. For our purposes, we need only to study the cell parts and "organelles" shown in Figure 1.3a cell membrane, cytoplasm, nucleus, nucleolus, nuclear membrane, ribosomes, centrioles, and endoplasmic reticulum.

Cell Membrane

Separating the internal structures of the cell from the intercellular fluid is the *cell membrane*. Also called the plasma or cytoplasmic membrane, it is a very thin, flexible, elastic membrane composed of a lipid (fat) bilayer with protein molecules embedded within the bilayer. Passing through the lipid layers are channels, which allow the controlled passage of chemicals into and out of the cell. Proteins embedded in the membrane perform these specialized transport functions. These special transport structures give the cell membrane its *selective permeability*. One that is *selectively permeable* permits only certain sizes and types of molecules to pass through it, while excluding others. Because of its selective permeability, the cytoplasmic membrane regulates the amount and type of material entering and leaving the cell, thus helping to maintain the cell's homeostasis.

Homeostasis is the condition whereby the internal and external environments of a cell remain relatively constant and in balance. Any disruption of homeostasis can damage the cell and cause it to malfunction. An understanding of *homeostasis* has helped develop cellular poisons (anticancer agents) capable of disrupting homeostasis of cells and killing them. Because the conditions of homeostasis differ slightly from one type of cell to another, it is possible to exercise some selectivity in the types of cells affected by a poison.

The *cytoplasm* surrounds the nucleus and consists of the cytosol, or fluid portion, and the various cellular organelles suspended in it. Most intermediary metabolism takes place in the cytosol, which represents about 55 per cent of total cell volume. In the cytosol, one finds thousands of enzymes that catalyze reactions involving (1) energy substrate production (gluconeogenesis and glycolysis);

(2) biosynthesis of sugars, fatty acids, nucleotides, and amino acids; and (3) different cytoskeletal proteins, which give a structural framework to the cell. Within the cytoplasm are various organelles that perform specific functions:

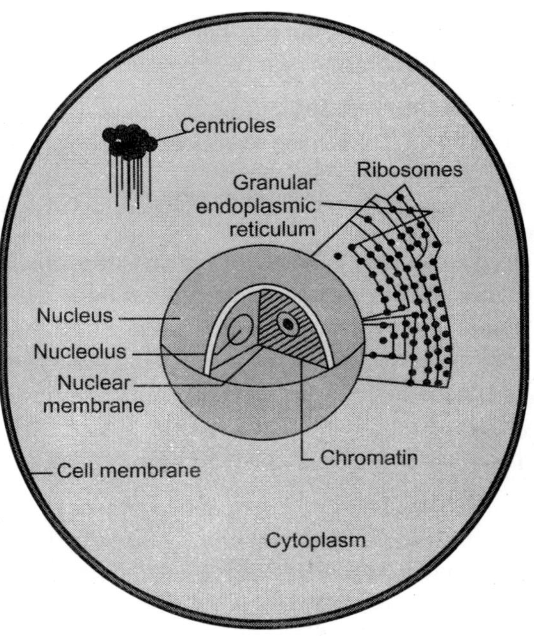

Fig. 1.3 : Diagrammatic representation of the living cell illustrating the major organelles.

Diffusion is the process by which molecules or ions spread from an area of high concentration to an area of low concentration until the concentration is uniform throughout, whereupon a state of equilibrium is said to exist. Most chemotherapeutic agents enter the cell through the cell membrane by *diffusion.* Also called *passive transport,* diffusion requires no expenditure of energy while transporting a substance across the cell membrane. Molecules and ions in body fluids are constantly in motion and they are thus available for diffusion through membranes. Diffusion can move substances either into or out of a cell—oxygen diffuses into cells while carbon dioxide diffuses out of cells is the process by which molecules or ions are moved across the cell membrane, sometimes in a direction opposite to that of diffusion. It is a process that allows a cell to maintain a state of non equilibrium at the cost of using energy in the transport process. Most chemotherapeutic agents enter the cell by active transport. Active transport systems act on *groups* of similar.

Within the cytoplasm are various organelles that perform specific functions which are shown in table 1.1.

Table 1.1 : The Different Functions of Cell

Organelle	% Cell Volume	Function
Free ribosomes	<1%	Protein synthesis
Endoplasmic reticulum (with membrane-bound ribosomes) and Golgi apparatus	9%	Synthesis of membrane is lipid and proteins to be secreted intracellular transport of macromolecules
Mitochondria (about 1,700/cell)	22%	ATP (energy) production
Lysosomes (about 300/cell)	1%	Digestive enzyme packets to degrade foreign macro-molecules
Peroxisomes (about 400/cell)	1%	Enzyme packets involved in oxidative reactions

The *nucleus* is usually located near the center of the cell. Relatively large and often spherical in shape, the nucleus is the control center of the cell. Separating the nucleus from the cytoplasm is the clear nuclear membrane, whose porous nature permits most substances to move back and forth between the nucleus and the cytoplasm. Unlike the wide variety of organelles found in the cytoplasm, the only prominent structural feature in the *nucleus* is the nucleolus, or "little nucleus", which is a minute body composed primarily of proteins and RNA. The nucleolus functions in the processing of the RNA that moves to the ribosomes in the cytoplasm. The nucleus of a cell contains the hereditary material DNA. The DNA in the nucleus is tightly coiled around specialized proteins called *histones*. The double-helical DNA strands packed about the histone proteins are called *chromatin fibers* (DNA plus histone proteins). The bulk of DNA, wrapped around histone proteins, is called a nucleosome. It is attached to other nucleosomes by a short "linker" section of DNA. The chromatin is organized into massive, short, rod-like bodies called *chromosomes*. The normal human cell contains 46 chromosomes. This is the diploid state. In contrast, the sperm and egg each contain only 23 chromosomes (haploid). Cancer cells may be aneuploid (unusual number of chromosomes) or tetraploid (twice the normal number of chromosomes).

Nucleic acids (DNA) are huge organic molecules, so named because they were first discovered in the nuclei of cells. The two main kinds of nucleic acids are DNA and RNA.

Deoxyribonucleic acid (DNA) is a very long helical structure that associates with a complementary strand via hydrogen-bonding reactions. The single strands are made up of an unbranched linear polymer composed of millions of nucleotides (a nucleic acid base, a phosphate, and the sugar deoxyribose) (Figure 1.4). The sequence of nucleotide bases represents the genetic code, which carries the cells' hereditary information. A gene is a section of DNA that codes for a single cellular trait or function. One

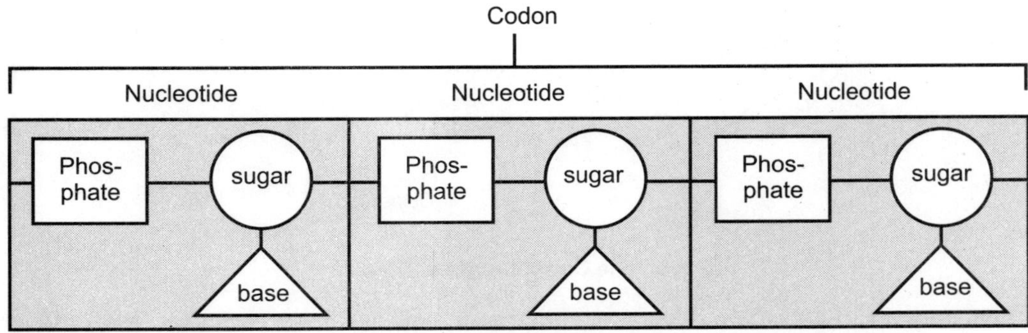

Fig. 1.4 : Linear organization of nucleotides of the DNA molecule. A "codon" consists of three sequential nucleotides

chromosome may contain more than 20,000 genes. The four *bases* in DNA are adenine, thymine, guanine, and cytosine (Figure 1.5). Each genetic unit in the DNA code is composed of three nucleotides in a specific order. This is called a codon. When a strand of DNA acts as a template for the synthesis of RNA (transcription), there is a "start" codon and a "stop" codon at the beginning and the end of the sequence that codes for the RNA that carries the message to the ribosome, where the protein is

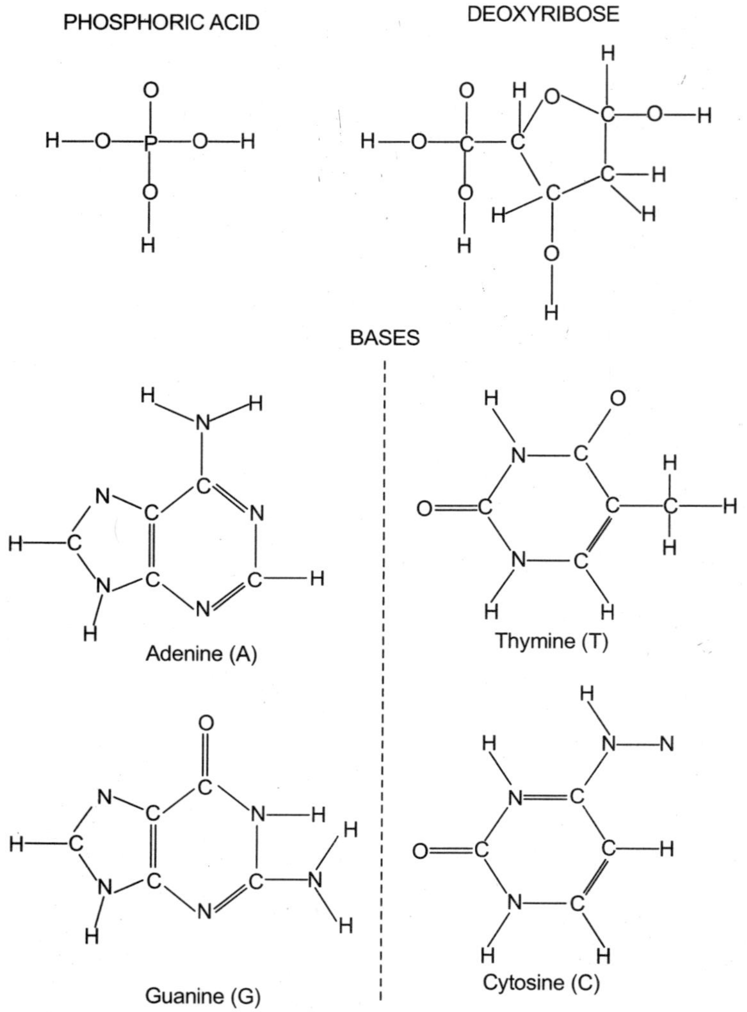

Fig. 1.5 : The Constituents of DNA

formed. Some portions of the DNA code for an amino acid sequence in the protein (exons), and other portions of the DNA strand do not (introns). The non-coding portions of the RNA message are removed when the RNA is processed in the nucleus before it migrates to the ribosome. The whole DNA molecule consists of two strands—two chains of nucleotides—that wind around each other in the form of a double helix. This is illustrated in Figure 1.6(*a*). The two strands are held together by hydrogen bonding between the bases. There is a very specific order in the bonding process. Adenine bonds only to thymine, and guanine bonds only to cytosine. This is called *complementary base pairing,* as illustrated in Figure 1.6(*b*). Complementary base pairing makes it possible for the DNA molecule to engage in self-replication.

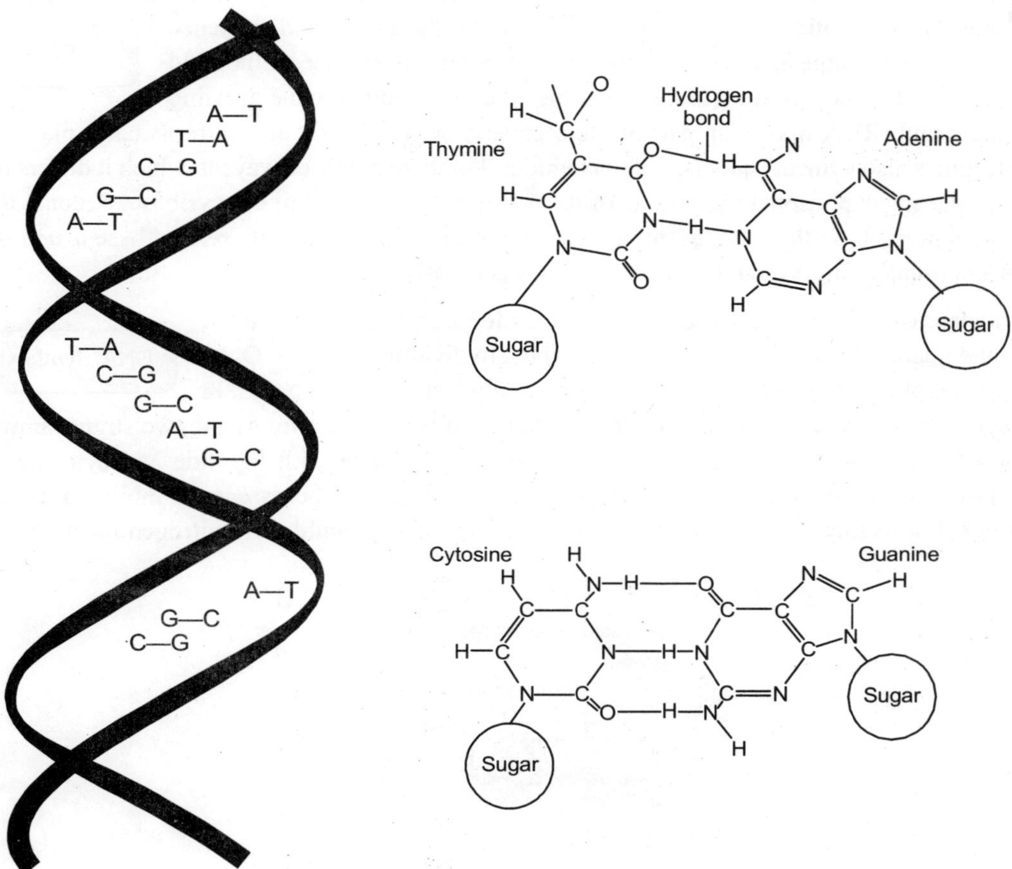

Fig. 1.6(*a*) : The DNA double helix. Note that adenine always pairs with thymine, while ganine always pairs with cytosine

Fig. 1.6(*b*) : The Complementary base pairing. The bases are united by hydrogen bonds, holding the double helix together

Coded information stored in DNA directs the production of RNA (transcription), which in turn directs the chemical synthesis of proteins (translation) (see Figure 1.7). *Enzymes* are proteins that act as catalysts to induce or speed up chemical reactions inside or outside the cell. The unique characteristic of a catalyst is that it remains unchanged by the chemical reaction it engenders. Enzymes govern all of a cell's metabolic reactions, and the enzymes themselves are proteins synthesized under the control of DNA. Some antineoplastic agents act by interfering with the activity of enzymes involved in the synthesis of DNA.

Ribonucleic acid (RNA) is part of the messenger system through which DNA controls protein production within the cell. There are at least three different kinds of RNA: messenger RNA (mRNA), transfer RNA (tRNA), and ribosomal RNA (rRNA). mRNA carries the genetic message for the gene to the ribosome. The nucleotide sequence of mRNA specifies the amino acid sequence in the protein to be synthesized at the ribosomes. tRNA transfers the amino acid from the enzyme that recognizes it to the ribosome where it is added to the growing peptide chain, and rRNA makes up part of the organelle on which protein synthesis takes place. The RNA molecule is also a linear sequence of nucleotides, but there are three ways in which it differs from DNA. First, the sugar-phosphate backbone of RNA uses ribose instead of deoxyribose. Second, RNA does not contain the base thymine, but in its place RNA contains the closely related base *uracil.* And third, RNA is usually single stranded, whereas DNA is usually double stranded.

The nucleus contains both DNA and free nucleotide bases (adenine, thymine, cytosine, and guanine) that are not bound to the DNA. In dividing cells, the DNA component is replicated during S phase, which normally lasts about 12 hours. In order to synthesize new DNA, the two opposing strands of DNA separate. As the two strands unwind, the nucleotide bases that were once paired together (i.e., adenine with thymine and cytosine with guanine) now become exposed to the nuclear environment. This process can be inhibited by some cancer drugs. For example, alkylating agents such as cyclophosphamide and nitrogen mustards form

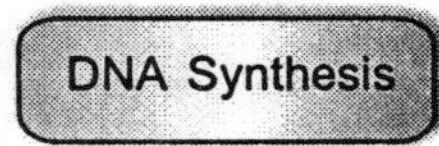

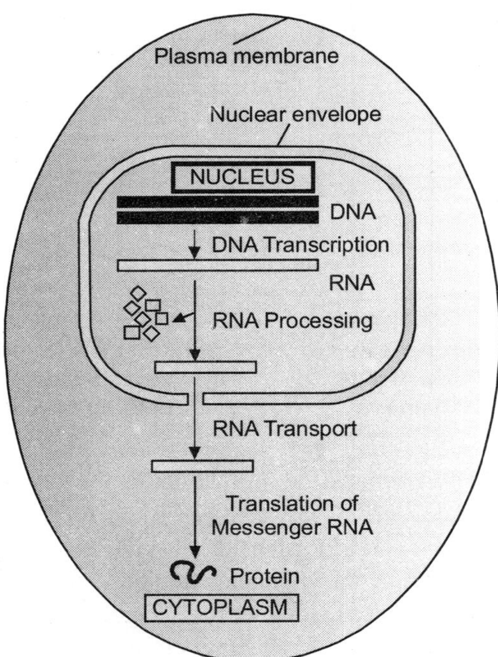

Fig. 1.7 : A schematic view of the expanded process of protein synthesis ("DNA—RNA—protein") in eukaryotes. Because of the nuclear envelope, RNA processing and RNA transport steps are interposed between transcription and the translation of mesenger RNA into protein

covalent crosslinks between the two strands of DNA so that they cannot unwind and separate for replication. If unperturbed, the enzyme DNA polymerase matches free (unbound) nucleotides with the exposed nucleotide bases along one strand of DNA to form a complementary copy. The antimetabolite cytosine arabinoside is known to inhibit this enzyme. As noted previously, the chemical structure of the bases requires that adenine pair with thymine and cytosine with guanine. When the DNA component is doubled, replication ends and the two new double stranded DNA molecules coil up to form helix just like the parent molecule. The cell now has twice the normal amount of DNA. Thus, each strand of the parent DNA molecule serves as the template for its own replication (Figure 1.8). The two resulting DNA molecules each have one old DNA strand and one new DNA strand. At this point, S phase ends and the cell moves into the G2 phase of division. Because of the way the enzyme DNA synthetase works, it cannot complete the DNA chain all the way to the end. So with each cell division, a few bases are lost at each end of the chromosome. The bases lost make up the telomere, which is a special section of DNA at the end that protects the DNA. This "end replication problem" leads to unstable chromosomes after 50 to 100 doublings, and this is thought to be a biologic clock that accounts for cell aging. Germ cells and stem cells have an enzyme, telomerase, which completes the telomere. Cancer cells also express this enzyme, which explains their immortality.

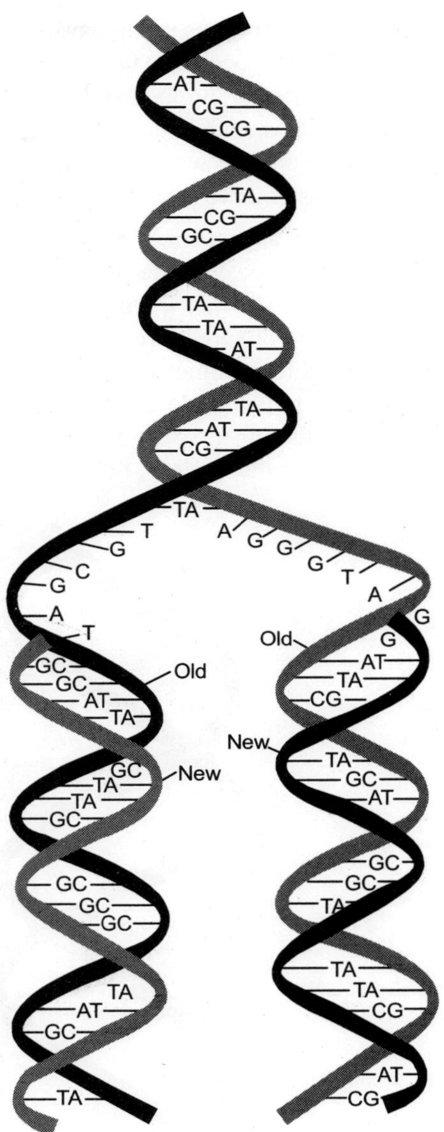

Fig. 1.8 : The DNA replication

Proteins are composed of long chains of amino acids linked to each other by peptide bonds. Each protein has a unique three-dimensional structure designed to perform a specific function. It is the sequence of amino acids in a protein that determines the details of the three-dimensional structure.

Proteins typically contain hundreds of amino acids and the chain has two distinct ends: an amino terminus and a carboxyl terminus. The functions performed by these molecules include all the functions necessary for life: enzymatic reactions, active transport, structural functions as for bone and muscle, etc. Any errors in the sequence of amino acids in a protein (a mutation) will cause an error in the three-dimensional structure and will cause a malfunction of that protein. For example, a single incorrect amino acid in hemoglobin results in sickle cell hemoglobin and a serious defect in red blood cell function, leading to anemia. When new proteins or enzymes are needed, DNA begins unraveling at a specific point along the chain. Unraveling occurs at the specific point that bears the code for the particular enzyme that is needed. Only a specific number of sites are exposed. As in DNA synthesis, the exposed nucleotides along the sugar-phosphate backbone are acted upon by an enzyme, RNA synthetase, to pair the free nucleotides in the nucleus with the nucleotide backbone. The result is that mRNA is formed (see Figure 1.9). This process is called transcription.

An example of an anticancer drug that inhibits RNA transcription is actinomycin D (dactinomycin). The mRNA is processed in the nucleus and then goes to the cytoplasm through small holes in the nuclear membrane. The mRNA migrates to the ribosome in the endoplasmic reticulum. Figure 1.10 is a highly schematic picture of protein synthesis. The ribosome moves along the strand of mRNA in assembly-line fashion; amino acids are "delivered" by the carrier molecule, tRNA, and are attached one by one until the polypeptide chain is complete. Each codon, or sequence of three bases, such as adenine-adenine-thymine, "spells out" a particular amino acid. Using the same process of complementary base pairing used in the nucleus, tRNA molecules

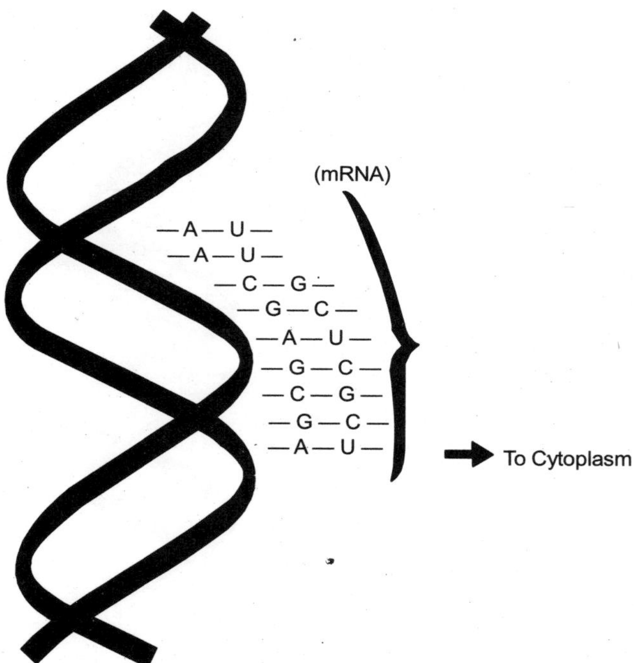

Fig. 1.9. Messenger RNA (mRNA) synthesis.
(Note that uracil pairs with adenine on the DNA strand.)

"know" exactly where to deliver amino acids along the mRNA molecule, thus insuring that the polypeptide chain of the protein molecule will be synthesized according to the DNA "blueprint." Figure 1.11 is an overview showing how DNA controls cellular homeostasis.

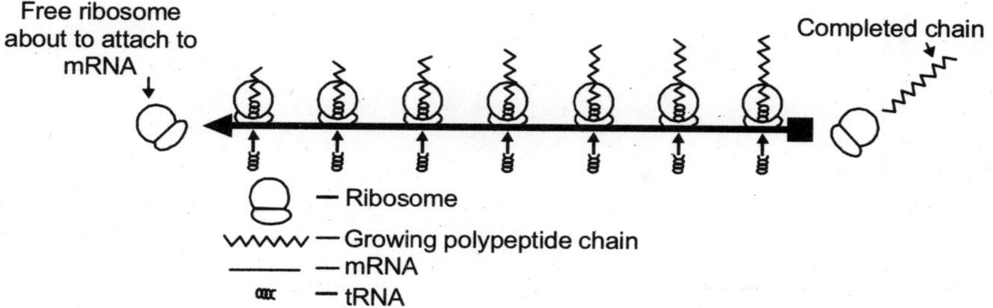

Fig. 1.10 : The ribosomal "protein factory". mRNA moves along the ribosome and amino acids attach one at a time until the chain is synthesized

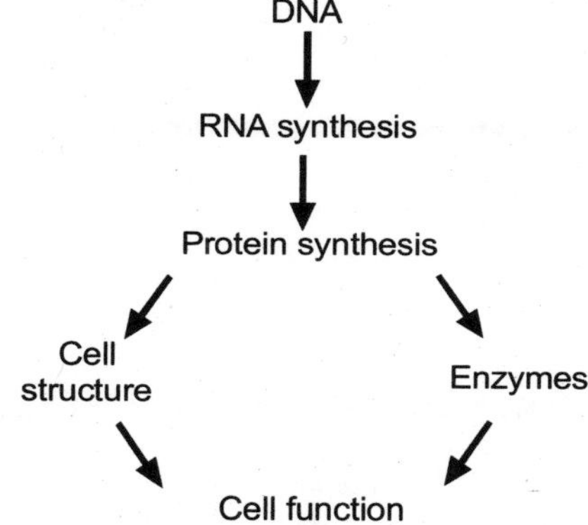

Fig. 1.11 : Overview: the role of DNA in homeostasis

Chemistry of Living Matters

CARBOHYDRATES AND ITS NOMENCLATURE

Carbohydrates, as the name implies believed to consist solely of the elements carbon and water in a ration 1:1 but our recent understanding is that a number of other atoms and functional groups can be present in a carbohydrate and the original formula of $C_xH_{2x}O_x$ is no long is no longer rigidly adhered to. Carbohydrates are traditionally considered to three broad groups :

1. Monosaccharide
2. Oligosaccharides
3. Polysaccharide

With monosaccharide being the simplest one, which cannot be hydrolyzed into smaller ones and from which the two other groups are obtained by formation of glycosidic linkages. This simple classification system defines oligosaccharides as simple polymers of monosachharides containing between two and 10 monosachharides unites or residues. Polysaccharides are higher molecular weight polymers of monosachharides which contain in excess of 10 residues, whilst the division between oligosaccharides and polysaccharides is arbitrarily set at 10 residues, nature has made the division much easier to observe since carbohydrates containing between five and 15 residues rarely exist in nature while only a few consist of 25 to 75 residues. The majority of naturally occurring polysaccharides contain 80 to 100 residues although a few negative cellulose comprising a series of polymers with an average molecular weight distribution equivalent to 3000 residues. Such weight but a range of macromolecules having a distribution of molecular weights. This phenomenon is known as mocroheterogneity.

Nomenclature

The members of these groups of compounds are denoted by the term – *ose*, the hydroxy – aldehyde (%CHO group) being called *aldoses* and the hydroxy ketones (%CO) are called *ketoses*. Whilst the number of carbon atoms in the compound is indicated by prefixing the great numerical to the syllable – ose. Thus glycolic aldehyde is an aldo – triose; where as dihydroxy – acetone is a keto – triose. The necessity for this system of nomenclature is apparent when dealing with the higher members of the series, each of which can exist, in several (glyceric aldehyde) 4 possible aldo – ketoses, 8 – possible aldopentoses and 16 possible aldo – hexoses. There are also heptoses, octoses, nonoses and decose with 7, 8, 9 and 10 atoms.

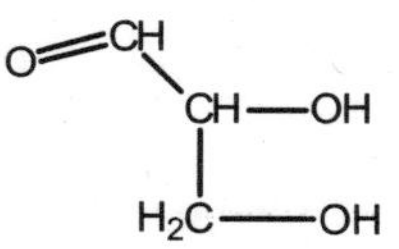

O=CH
CH—OH
H_2C—OH

Aldotriose

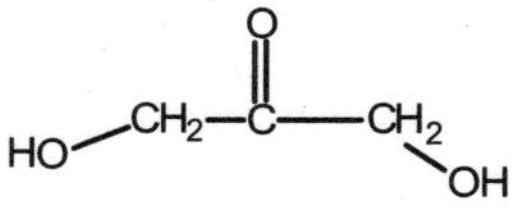

HO—CH_2—C(=O)—CH_2—OH

keto triose

The existence of the large number of isomers in the example given above is due to the difference in space arrangement or configuration of the atoms, which is discussed in a later part of this chapter.

CLASSIFICATION WITH EXAMPLES

A. Monosaccharides (simple sugars)

1. Hexoses

(a) ***Glucose:*** honey, fruits, corn syrup, sweet grapes, sweet corn; hydrolysis of starch and cane sugar. Physiologically the most important sugar; the "sugar" carried by the blood and the principal one used for tissues.

Classification of Carbohydrates

(b) ***Fructose:*** honey, ripe fruits, some vegetables; hydrolysis of sucrose insulin. Can be changed to glucose in the liver and intestine and is an intermediate metabolite in glycogen breakdown.

(c) ***Galactose:*** not found free in nature; digestive end product of lactose hydrolysis. Can be changed to glucose in the liver; synthesized in body to make lactose and is a constituent of glycolipids.

(d) ***Mannose:*** found in legumes; hydrolysis of plant mannosans and gums. A constituent of polysaccharide of albumins, globulins, and mucoids.

2. Pentoses

(a) ***Arabinose:*** derived from gum arabic plus plum and cherry gums; not found free in nature. Has no known physiologic function in man; used in metabolism studies of bacteria.

(b) ***Ribose:*** derived from nucleic acid of meats and seafoods. Structural element of nucleic acids, ATP, and coenzymes (NAD and FAD).

(c) ***Ribulose:*** formed in metabolic processes. Intermediate in direct oxidative pathway of glucose breakdown.

(d) ***Xylose:*** wood gums, corncobs, and peanut shells; not found free in nature. Very poorly digested and has no known physiologic function; used medicinally as a diabetic food.

B. Oligosaccharides (2-10 sugar units)

1. Disaccharides

(a) ***Sucrose:*** cane and beet sugar, maple syrup, molasses, and sorghum. Hydrolyzed to glucose and fructose; a non-reducing sugar.

(b) ***Maltose:*** malted products and germinating cereals; an intermediate product of starch digestion. Hydrolyzed to two molecules of glucose; a reducing sugar; does not occur free in tissues.

(c) ***Lactose:*** milk and milk products; formed in the body from glucose nature. Hydrolyzed to glucose and galactose; may occur in urine during pregnancy; a reducing sugar.

2. Trisaccharides

(a) ***Raffinose:*** cottonseed meal, molasses, sugar beets and stems. Only partially digestible but can be hydrolyzed by enzymes of intestinal bacteria to glucose, fructose, and galactose.

(b) ***Melizitose:*** honey, poplars, and conifers. Composed of one fructose unit and two glucose units.

C. Polysaccharides (more than 10 sugar units)

1. Digestible

(a) ***Glycogen:*** meat products and seafoods; polysaccharides of the animal body, often called animal starch; storage form of carbohydrates in body, mainly in liver and muscle.

(b) ***Starch:*** cereal grains, unripe fruits, vegetables, legumes, and tubers. Most important food source of carbohydrates; storage form of carbohydrates in plants. Composed chiefly of amylose and amylopectin; hydrolyzed to glucose.

(c) ***Dextrin:*** toasted bread, intermediate product of starch digestion. Formed in course of hydrolytic breakdown of starch.

2. Partially Digestible

(a) ***Inulin:*** tubers and roots of dahlias, artichokes, dandelions, onions, and garlic. Hydrolizable to fructose; used in physiologic investigation for determination of glomerular filtration rate.

(b) ***Mannosan:*** legumes and plant gums. Hydrolyzable to mannose but digestion incomplete; further splitting by bacteria may occur in large bowel.

3. Indigestible

(a) ***Cellulose:*** skins of fruits, outercoverings of seeds, plus stalks and leaves of vegetables. Not subject to attack of digestive enzymes in man, thus an important source of "bulk" in diet; may be partially split to glucose by bacterial action in large bowel.

(b) ***Hemicellulose and pectin:*** woody fibers and leaves. Less polymerized than cellulose; may be digested to some extent by microbial enzymes, yielding xylose.

Indigestible is another name for indigestable polysaccharides is dietary fiber which is comprised of 2 groups:

(c) ***Insoluble dietary fibers*** (cellulose, lignin, and cutin) which are the most abundant organic compounds in the world. They help prevent constipation, colon cancer, and diverticulosis, but not hypercholesterolemia.

(d) ***Soluble dietary fibers*** (hemicellulose, pectins, gums, and algal polysaccharides) which are useful in decreasing serum cholesterol and in regulating blood glucose levels.

CHEMICAL STRUCTURES OF CARBOHYDRATES IN PYRANOSE AND FURANOSE FORM

Pyranose is a collective term for carbohydrates which have a chemical structure that includes a six-membered ring consisting of five carbons and one oxygen. The pyranose ring is formed by the reaction of the C-5 alcohol group of a sugar with its C-1 aldehyde forming an intramolecular hemiacetal. The name derives from its similarity to the oxygen heterocycle pyran.

Furanose ring is a cyclic hemiacetal of an aldopentose or a cyclic hemiketal of a ketohexose.

A furanose ring structure consists of four carbon and one oxygen atom with the anomeric carbon to the right of the oxygen. The highest numbered chiral carbon (typically to the left of the oxygen in a Haworth projection) determines whether or not the structure has a D-configuration or L-configuration. In an L-configuration furanose, the substituent on the highest numbered chiral carbon is pointed downwards out of the plane, and in a D-configuration furanose, the highest numbered chiral carbon is facing upwards.

The furanose ring will have either alpha or beta configuration, depending on which direction the anomeric hydroxy group is pointing. In a D-configuration furanose, alpha configuration has the hydroxy pointing down, and beta has the hydroxy pointing up. It is the opposite in an L-configuration furanose. Typically, the anomeric carbon undergoes mutarotation in solution, and the alpha-beta configuration switches constantly—thus it is in an equilibrium state.

Monosaccharides

The number of monosaccharides known in the neighbourhood of fifty, of which 10 occur in nature and the remaining are synthetic. The existence of such a large number of comnpounds is due to the presence of presence of asymmetric carbon atoms within the molecules. Aldohexoses, for example which includes glucose, a sugar of great theoretical and practical interest, contain no less than four asymmetric atoms, each of which may be present in either the *d* and *l* – configuration. It has already been shown that how rapidly stereo isomerides increases with each additional asymmetrical atom.

A list of best monocaccharides is given below in detail in the table.

An explanation is required as to the way in which the optical isomerides are named. In accordance with a suggestion of Fischer, the structure of the sugar, where ever possible, referred back to that of

Names & Chemical Formula of Glucose

Name	Formula	Chemical name
Biose	$CH_2OH.CHO$	Glycolic aldehyde
Trioses	1. $CH_2OH.CHOH.CHO$	Glyceric aldehyde
	2. $CH_2OH.CO.\ CH_2OH$	Dihydroxy acetone
Tetrose	$CH_2OH.(CHOH)_2.CHO$	Erithrose
Pentose	$CH_2OH(CHOH)_3CHO$	Binose, xylose, ribose
Methyl pentose	$CH_3(CHOH)_4CHO$	Rhamnose
Hexose	1. $CH_2OH(CHOH)_4CHO$	Glucose, gulose, talose mannose
	2. $CH_2OH(CHOH)_4CHO$	Fructose , sorbose
Heptose	$CH_2OH(CHOH)_3CO\ CH_2OH$	Monoheptose, glucoheptose, galaheptose
Octose	$C_8H_{16}O_8C_8H_{16}O_9$	
Nonose	$C_9H_{18}O_9C_9H_{18}O_9$	

the active glucoses. Any other monosaccharide standing in close structural relationship to one or other of the glucose is generally labeled with the corresponding letter *d* and *l*-irrespective of the actual sign of its rotation. Ordinarily fructose, or example, which is lævorotatory, but termed as *d*-fructose, owing to its spatial relationship to *d*-fructose. Naturally occurring arbinose, which is dexorotatory, is termed as 1-arbinose to indicate its relationship to 1-glucose. This is illustrated in the following formulæ, in which the terminal aldehydic or ketonic group is written uppermost links blinding the H and OH addenda are assumed to be inclined from the chain of central carbon atoms towards the observer. According to the usual convention a CHOH group having the hydroxyl on the right of the formula employed.

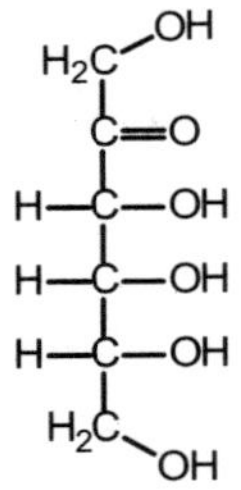

d-fructose 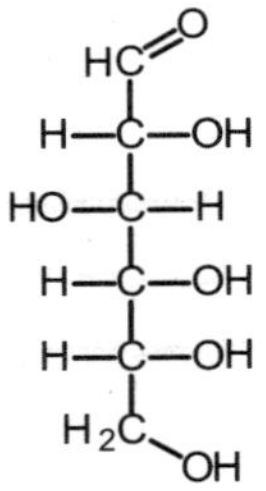 *d*-glucose 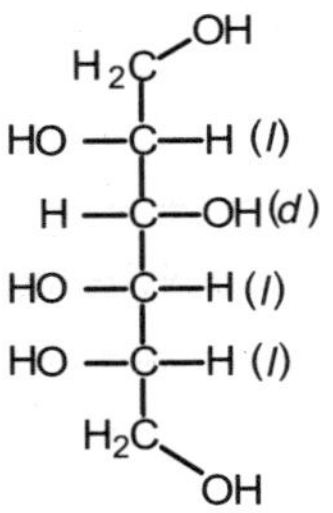 *l*-glucose 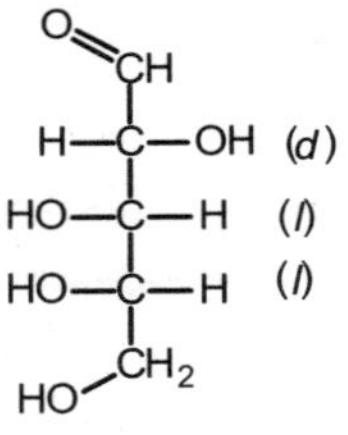 *l*-arbinose

In some cases the actual sign of the rotation given by a compound is indicated in the following pages by the use of the sign + and %, eg d(%) fructose for ordinary fructose. In addition, the prefix dl – indicates a racemic and mesoform.

Wohl's work on glycerose now enables the family relationship to be carried back to the *d* and *l* glyceroses from which all other sugars can be derived by extending the molecule on the side of the aldehyde group by reaction with. The nomenclature fortunately remains unchanged as *d*-glycerose is *d*-rotatory and is genetically related to *d*- glucose. Hence it will be seen that the family of a sugar is determined in each case by the spatial arrangement of the CHOH group adjustment to the terminal CH_2OH, no matter what the disposition of the rest of the molecule may be. If this group has a dexo configuration, as indicated by writing it with OH to the right the sugar is classified as belonging to the *d*-family. The same system is applied to other derivatives of monosaccharides.

d glecerose L glecerose

The monosaccharides possess strong reducing properties, causing the separation of the silver form ammonia-silver nitrate solution and precipitating cuprous oxide from an alkaline solution of copper sulphate. They thus resembles aldehydes, moreover, like aldehydes and ketones, they form cyanohydrines with hydrocyanic acid, oximes with hydroxylamine, and phenylhydrazones with phenylhy-drazones. The phenylhydrazones are as rule very soluble in water but by the further action of phenylhydrazine, they are converted into a yellow insoluble compound called osazones. The osazones are nearly insoluble in water, and readily separate from a solution of monosaccharide. They have, more over a definite melting point, and seen under microscope possess a characteristic appearance.

General Properties or Monosaccharides

The reaction is therefore regarded as the most important process in detecting and identifying certain sugars. The monosaccharides are readily oxidized. The aldoses yield mono and di basic acids containing the same number of carbon atoms; the ketones break up into acids with a fewer carbon atoms. They all yield oxalic acid when warmed with strong nitric acid, with strong hydrochloric acid they form levulinic acid. Finally they undergo alcoholic fermentations with yeast. These reactions are going to be discussed under glucose.

GLUCOSE

Introduction

Glucose a monosaccharide (or simple sugar), is an important carbohydrate in biology. The living cell uses it as a source of energy and metabolic intermediate. Glucose is one of the main products of photosynthesis and starts cellular respiration in both prokaryotes and eukaryotes. The name comes from the Greek word *glykys*, which means "sweet", plus the suffix "-ose" which denotes a sugar.

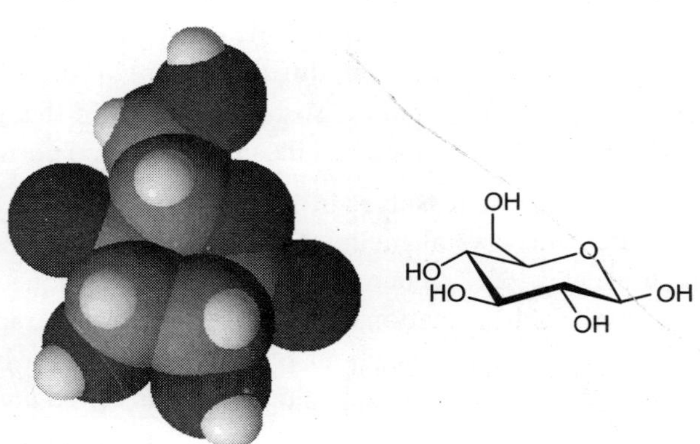

Two stereoisomers of the aldohexose sugars are known as glucose, only one of which (D-glucose) is biologically active. This form (D-glucose) is often referred to as **dextrose monohydrate**, or, especially in the food industry, simply **dextrose** (from *dextrorotatory glucose*. The mirror-image of the molecule, L-glucose, cannot be metabolized by cells in the biochemical process known as glycolysis. Glucose is commonly available in the form of a white substance or as a solid crystal. It can also be commonly found as an aqueous solution.

Synthesis

In small quantities pure glucose is most readily obtained from cane sugar. Cane sugar is dissolved in 80% alcohol and a little strong hydrochloric acid added. On gently warming the mixture, the cane sugar is hydrolysed, and breaks up into glucose and fructose

$$\underset{\text{Cane sugar}}{C_{12}H_{22}O_{11}} + H_2O \rightarrow \underset{\text{Glucose}}{C_6H_{12}O_6} + \underset{\text{Fructose}}{C_6H_{12}O_6}$$

Glucose, being less soluble in alcohol separates out as crystals.

Glucose is manufactured by boiling starch with dilute sulphuric acid; the starch is hydrolysed and converted into glucose. The liquid neutralized with soda ash and filtered and filtrate decolourised by filtration through animal charcoal. (See fig. 2.1)

The solution is evaporated to the requisite consistency in vacuum pans. The product solidifies on cooling and forms an amorphous looking mass, which always contains dextrin.

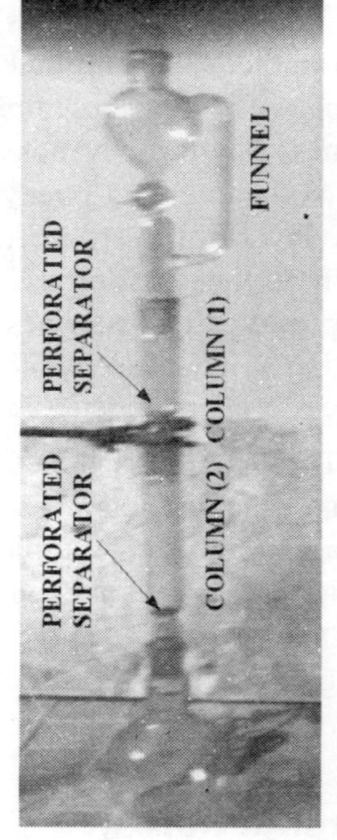

Fig. 2.1 : The picture of the column filter with animal charcoal used in Author's lab

The Vacuum Pan (See Fig. 2.2)

It consists of an iron pan which is heated by vertical steam coils placed in the lower part of the vessel; two or three pans are connected so that the steam rising from the evaporation of the liquid in the first pan is utilized for heating the next and its vapour passes on to the third. Between the pans a small cylindrical vessel is interposed, which serves to collect any juice which "primes" or collects and carried over during boiling. The evaporation is continued until the liquid is so far concentrated that it shows "grains" or commences to crystallize. It is then run out and cooled and the un crystallized portion or molasses separate in a centrifugal extractor.

Pure glucose dissolves in 1.2 parts of water. It crystallizes from aqueous solution with 1 molecule of water, the crystal melts at 66°C, and whilst from alcohol the anhydrous compound separates, melting at 146°C. Glucose is dexorotatory in aqueous solution, when freshly dissolved in water it shows a specific rotation of +110.2°, which falls rapidly, and then gradually more and more slowly until after the lapse of about 6 hours it reaches a constant value of + 52.5°. This change of rotatory behaviour is called *mutarotation*, and this is due to structural changes which occur in solution. Many other sugars

Properties of Glucose

under go muta rotation. Another variety of glucose obtained by crystallization from alcohol has a specific rotation of + 19°, a solution of which on standing rises to + 52.5°. The two sugars are called α and β glucose (see constitution) and the value + 52.5°C is the specific rotatory power of an equilibrium mixture. In genral the α-sugars are characterized by a *fall* and the β-sugars a *rise* in the initial value of the rotation.

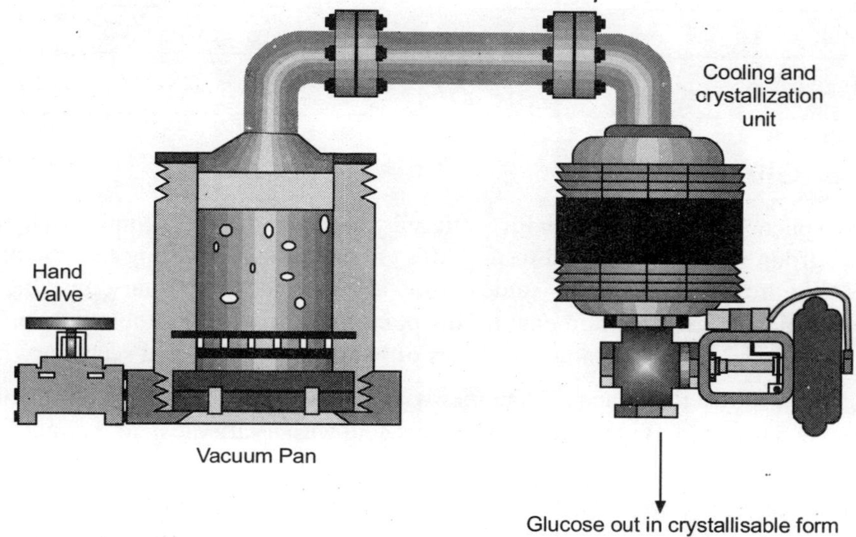

Fig. 2.2 : The Manufacturing process of Glucose from starch

When lime or baryta solution is added to a solution of glucose and then alcohol, glucose, glucosates of calcium or barium are precipitated. These compounds are soluble in water and are decomposed by carbon dioxide into the original sugar and the carbonate of the metal. Calcium glucostate has the formula $C_6H_{12}O_6CaO$

Some Detail Physical Properties of Glucose

Molecular Formula	$= C_6H_{12}O_6$
Formula Weight	≐ 180.15588
Composition	= C(40.00%) H(6.71%) O(53.29%)
Melting point	= 146°C
Molar Refractivity	= 37.25 cm^3
Molar Volume	= 104.0 cm^3
Parachor	= 312.7 cm^3
Index of Refraction	= 1.635

Surface Tension	= 81.7 dyne/cm
Density	= 1.732 g/cm^3
Polarizability	= 14.76 10^{-24}cm^3
Monoisotopic Mass	= 180.063388 Da
Nominal Mass	= 180 Da
Average Mass	= 180.1559 Da

Structure of Glucose

Glucose forms a phenoacetyl derivative with acetic anhydride and therefore contains a hydroxyl groups. Each hydroxyl group probably attached to the different carbon atom, seeing that the arrangement of attachment of 2 hydroxyl groups to the same carbon atom would form a very unstable arrangement. The only point which is left uncertain chain. This point is determined by the reduction of glucose to sorbitol, and the conversion of the lätter by means of hydriodic into normal secondary hexyl iodide.

It has also been shown that glucose combines with hydrocyanic acid and forms cyanhydrin which, on hydrolysis, yields an acid, By the reduction of this acid with hydriodic acid, normal heptylic acid is produced. These changes are represented by the following formulæ.

glucose glucose cyanohydrin hydroxy heptic acid heptylic acid

The formula of glucose would therefore appear to be a penta hydroxy aldehyde

$$CH_2(OH).CH(OH).CH(OH).CH(OH).CH(OH).CHO$$

The formula does not, however strictly confirm to its behaviour for a study of the two methyl glucosides has revealed the fact that both glucose and the glucosides exist in cyclic form. Since the terminal (aldehyde) group of glucose is involved in the ring formation, this carbon atom becomes asymmetric, and in this way gives rise to the stereoisomers α and β-glucose and α and β glucosides. Further investigations has shown that the rings are 6 membered, with oxygen atom linked to the first and fifth group were present, it can presumably isomerise easily to an open chain structure. The a and β ring structure and the open chain formula for glucose are as follows:

It is convenient to be able to refer to the different carbon atoms by means of numbers, which are indicated on the right hand side. The two ring structures differ only in the disposition of the groups attached to No. 1 carbon atom. The configuration of the other four asymmetric carbon atoms (no 2, 3,

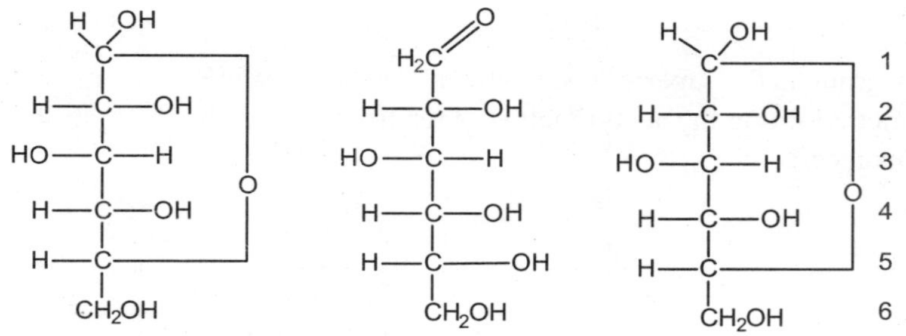

4, and 6) has been accurately determined, and is here indicated by means of projection formulæ (see chap organic mechanism). Although these projection formulæ are quite satisfactory for open chain compounds, some confusion is at to arise whenever the chain is closed to a ring, so that Howorth has adopted the use of perspective formulæ, which reveal the ring as a hexagon the groups attached to the carbon atoms being shown either above or below the plane of the hexagon as the case may be.

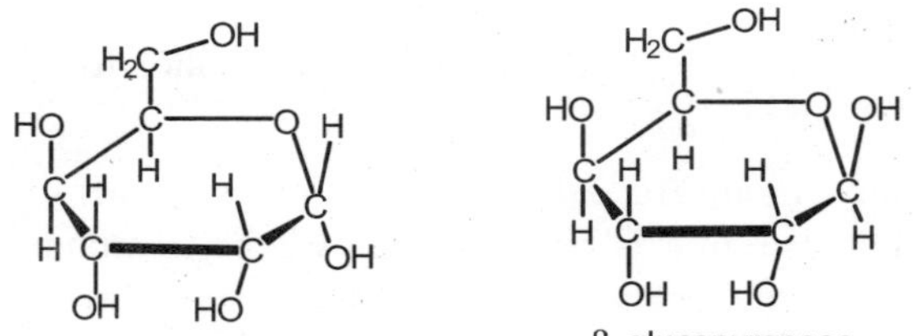

α-glucopyranose β-glucopyranose

These are called pyranose forms. Other rings called furanose rings are formed with only four carbon atoms and one of oxygen. These terms are derived from *pyron* and *furon*.

Since every asymmetric atom gives rise to two stereoisomers every additional one of them must produce two more form each of these. Thus the aldopentoses with three asymmetric carbon atoms exist in eight active form, whilst the aldo hexoses with 4 asymmetric atoms give rise to 2 × 8 or 16, all of which are known, the most important of these are *d*-glucose, *d*-galactose and *d*-mannose.

Chemical Properties

In considering simple sugars or monosaccharides, it is useful to discuss in detain the chemical properties of the glucose. The fact is that it may be isolated from the acid -catalyzed hydrolysis of starch, suggests that the carbon atoms in glucose are attached to each other by carbon-carbon bonds rather than more labile ether bonds (although the existence) of the later is addition to the carbon-carbon bonds not excluded).

Quantitative Acylation of glucose (or equivalent benzoylation) suggests (in the equation below) that there are five hydroxy groups per molecule.

$$(i)\ C_6H_7O(OH)_5 + 5(CH_3CO)_2O \xrightarrow{Na^+OCOCH_3^{\%}} C_6H_2O(OCOCH_3)_3 + 5CH_3CO_2H$$

Treatment of glucose with anhydrous methanol in presence of hydrogen chloride replaces only one hydroxy group with dilute aqueous acid, so that the product assumed be acetal.

$$C_6H_{11}O_8(OH) + CH_3OH \rightleftharpoons C_6H_{11}O_5(OCH_3) + H_2O$$

All the hydroxy group in the sugar can be methylated by the treatment with methyl iodide and silver hydroxide. Of the methyl group the acetyl methyl is readily removed by the hydrolysis, others are not because they are linkages.

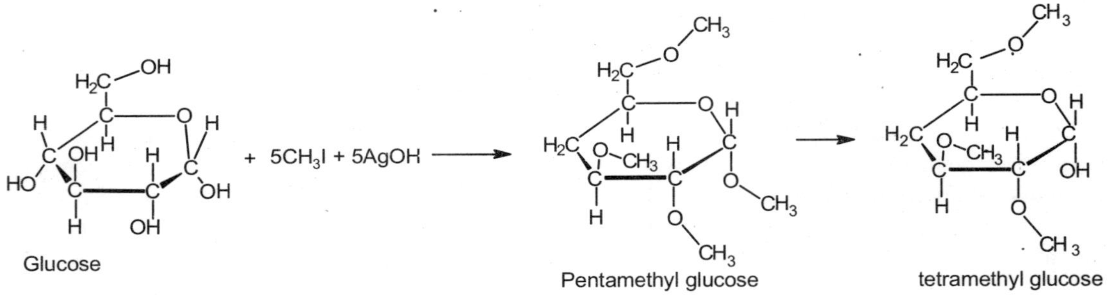

Glucose + $5CH_3I$ + $5AgOH$ → Pentamethyl glucose → tetramethyl glucose

These results show that four of the five hydroxy groups are alcohol groups, while one is a hemiacetal group.

The presence of a carbonyl group (or equivalent) in glucose is demonstrable by treatment with hydroxylamine. The anticipated oxime is produced. However treatment of monosaccharide with phenylhydrazine produce an osazone reaction similar to that observed with benzoin. Oxidation occurs only at the hydroxy group adjustment to the original carboxyl group, so that a phenyl osazone result.

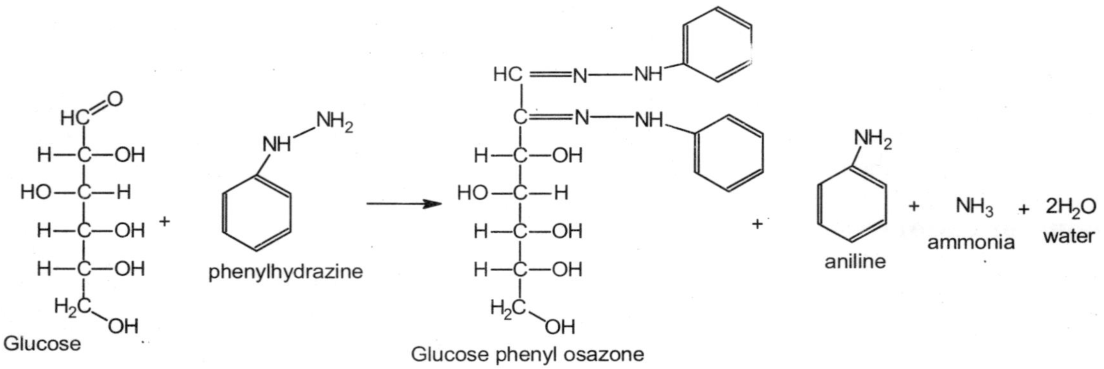

Glucose + phenylhydrazine → Glucose phenyl osazone + aniline + NH_3 (ammonia) + $2H_2O$ (water)

Both aldoses and ketoses are readily oxidized by Fehling's solution and Benedict's reagents these reagents are important for the determination of free available carboxyl group. Both consist of alkaline complexes of cupric ion. Fehling's solution is a mixture of citrate cupric ion and sodium carbonate. These reagents are reduced to cuprous oxide by α-hydroxy aldehydes and α-hydroxyketones. A positive test is indicated by complete disappearance of colour from the solution and the formation of a coral to red-brown precipitate.

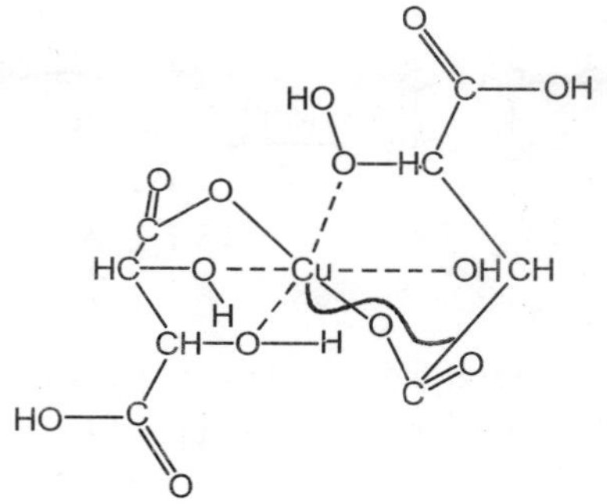

One Possible Tatatocupric Complex

Bromine and dilute acid are reagents useful for oxidation of aldehyde groups of aldoses. The products are termed aldonic acids. Oxidation of a aldoses with boiling concentrated nitric acid attack both the aldehyde group and the primary alcohol group to give a dicarboxylic acid formerly called saccharic acid, but now given the generic name glyceric acid. The aldehydic acids produced by oxidation of the primary alcohol group are called glycuronic acid. This oxidation involves procedures where the other groups susceptible to oxidation are suitable protected and then the protecting substituents are removed later.

$$RCO + Br_2 + H_2O \rightarrow RCO_2H + 2HBr$$
$$HCCH_2(CHOH)_nCHO + 6HNO_3 \rightarrow HOCO(CHOH)_nCO_2H + 6NO_2 + 4H_2O$$

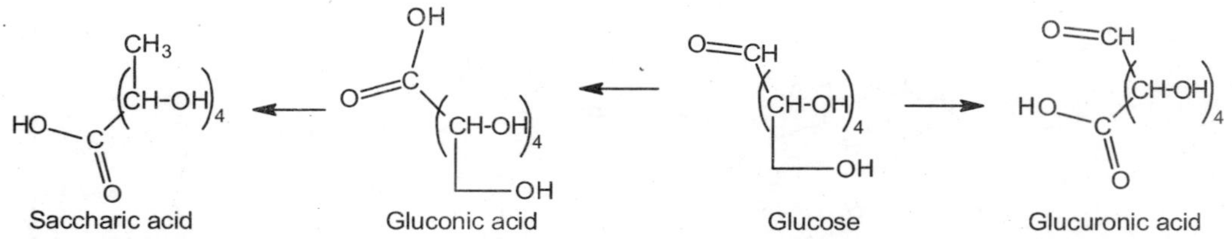

Saccharic acid Gluconic acid Glucose Glucuronic acid

Like similar aldehydes and ketones monosaccharides react with hydrogen producing cyanohydrines. This fact provides a tool effectual in the structural analysis of mono-saccharides as well as a means of synthesis of aldoses of one more carbon atom per molecule than the original aldose.

Interconversions: Heinrich Killani was first to apply cyanohydrin formation to the synthesis of higher aldoses. The reactions of his method (Killani synthesis) are summerised in outline. The new asymmetric center marked with an asterisk makes possible two product dia stereoisomers, which are formed in unequal amounts and differ in configuration at only one carbon atom (the 2-position). These are called *epimers.*[1]

[1] In chemistry, an **epimer** is a stereoisomer of another compound that has a different configuration at only one of several stereogenic centers. Stereoisomers include enantiomers and diastereomers, both which contain a stereogenic center (excluding geometric isomers, which is a class of diastereomers).

...(Contd.)

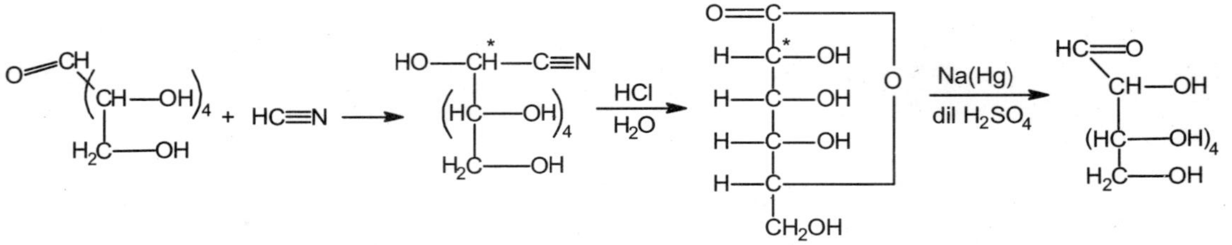

Vigorous reduction of the seven carbon lactone (I) with hydriodic acid results in the formation of heptanoic acid, This demonstrates the straight – chain nature of the carbon skeleton of glucose, as well as aldehyde (as opposed to keto) structure,

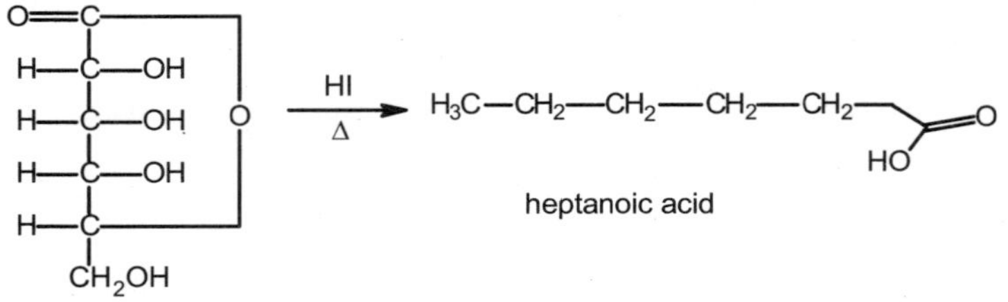

heptanoic acid

Degradation of aldoses to aldoses with one less carbon atom can be accomplished by either, in retrograde order of the Killani synthesis. This method is summerised in equation; the second degradation is the Ruff degradation.

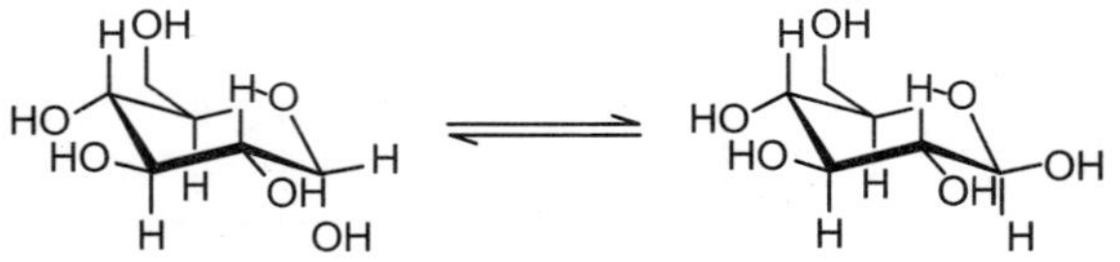

α-D-glucopyranose β-D-glucopyranose

For example, the sugars α-glucose and β-glucose are epimers. In α-glucose, the -OH group on the first (anomeric) carbon is in the direction opposite the methylene group (in the axial position). In β-glucose, the -OH group is oriented in the same direction as the methylene group (in the equatorial position). These two molecules are both epimers and anomers.

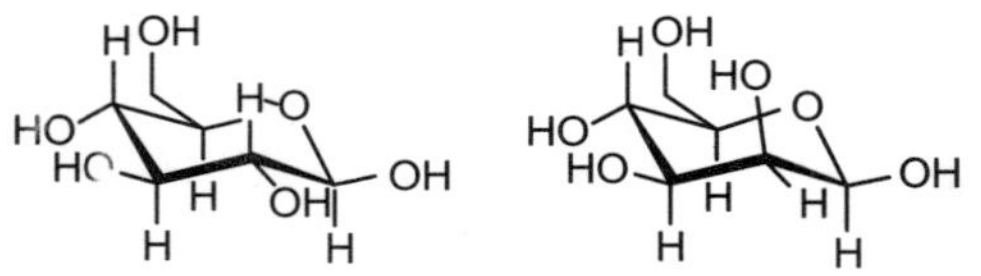

β-D-glucopyranose β-D-mannopyranose

In this case, β-D-glucopyranose and β-D-mannopyranose are epimers because they differ only in the stereochemistry at the 2 position. The hydroxyl group in β-D-glucopyranose is equatorial (in the "plane" of the ring) while in β-D-mannopyranose the 2 hydroxyl group is axial (up from the "plane" of the ring). These two molecules are epimers but not anomers.In chemical nomenclature, one of the epimeric pairs is given the prefix **epi-** for example in quinine and *epi-quinine.* When the pairs are enantiomers, the prefix becomes **ent**.

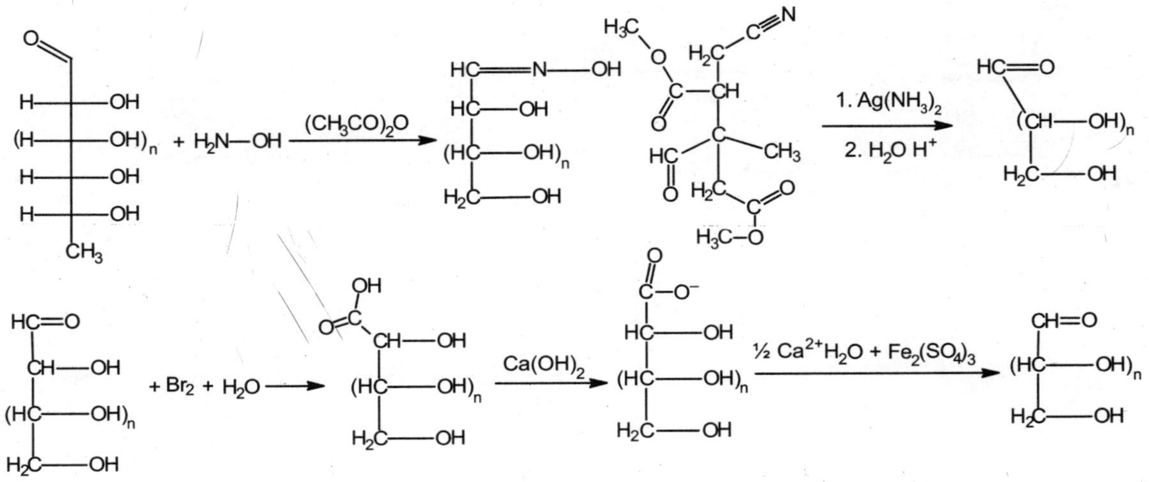

Conversion of aldose to a 2 ketose involves formation of the 1.2-dicarboxyl compounds either by osazone synthesis and hydrolysis or by oxidation of the 2-hydroxy group with hydrogen peroxide and ferric sulphate. The aldehyde group can be reduced preferentially with powdered zinc and acetic acid.

The reverse, conversion of a ketose to an aldose, is more equivocal. The first step is reduction of the ketose to the two corresponding stereo-isomeric polyols. Either of the two primary alcohols groups of a molecule of each polyol is then oxidized to each polyol is then oxidized to yield a lactone mixture. The lactone are then reduced to aldoses. Either the mixture of intermediate lactones or that of final aldoses must be separated by fractional crystallization.

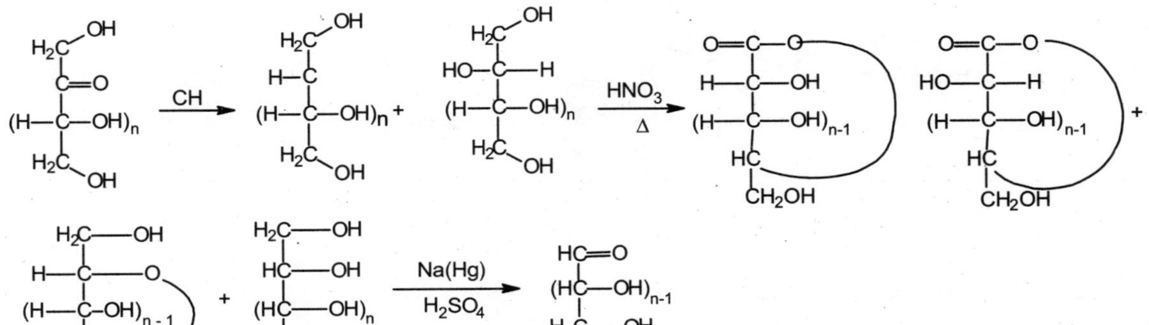

possible separation

Conversion of an aldose to its epimer at C_2 is readily accomplished. In practice, epimerization depends upon formation of equilibrium between the two epimeric aldonic acid in boiling pyridine. Consequently the same equilibrium results by starting with either epimer.

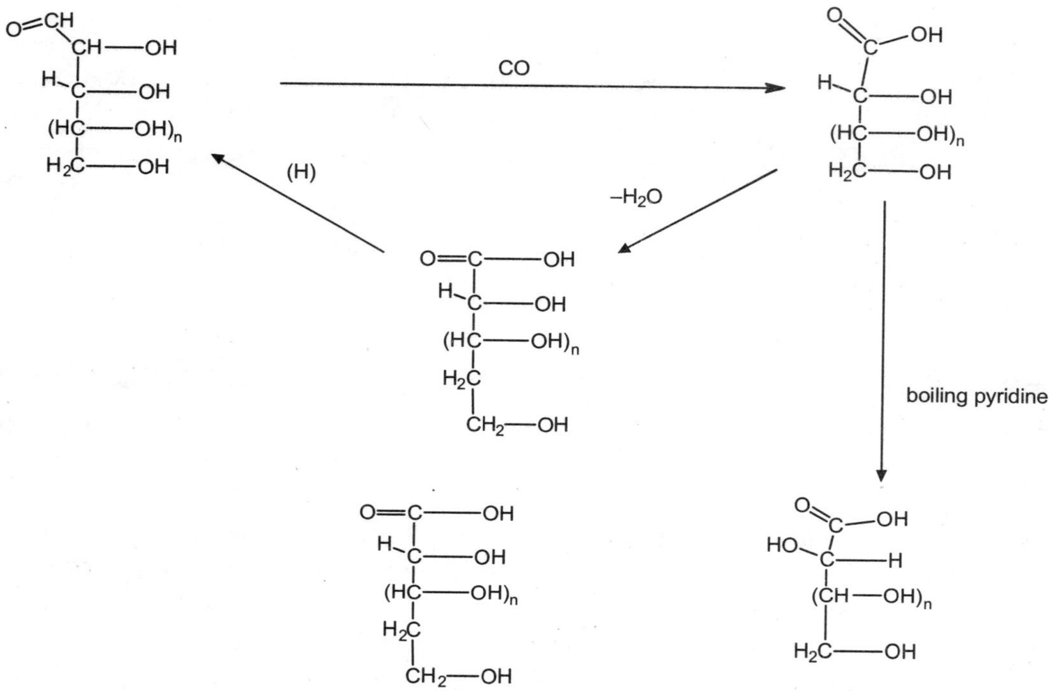

O=CH
CH—OH
H-C—OH
(HC—OH)$_n$
H_2C—OH

CO

(H)

$-H_2O$

boiling pyridine

Another method of epimerization introduces the ketoses into the equilibrium. Treatment with dilute alkali operates through enedol forms to give the equilibrium mixture shown in the equation below.

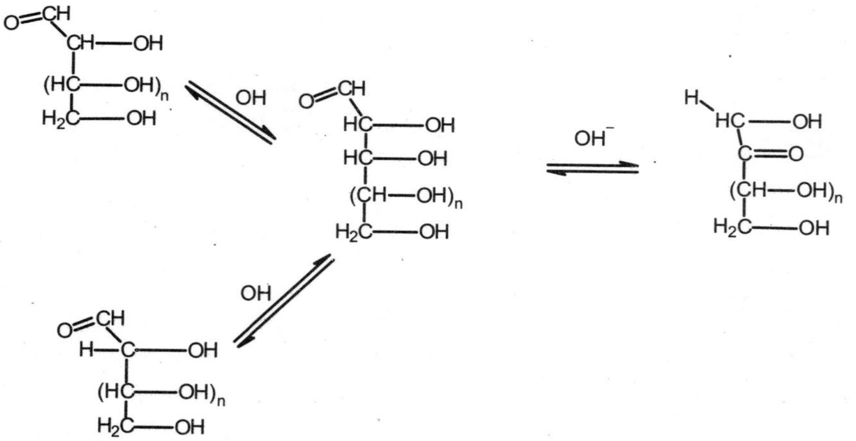

OH

OH^-

OH

Osazone and the Conversion of Aldoses And Ketoses

As aldehydes or ketones the glucose also reacts with hydrazines and hydroxylamine. Phenyl – hydrazine $C_6H_5NHNH_2$ has provided of the greatest value in the separation identification and interconversion of the various other monosaccharides, without the aid of this reagents the brilliant researches of Fischer in the sugar group would hardly have been possible.

When I mol of phenylhydrazine reacts with I mol of an aldose or ketoses, the first product is a normal hydrazone.

$$CH_2OH[CH.OH]_4.CHO{:}O + H_2N - NHC_6H_5 \rightarrow CH_2.OH\,[CH.OH]_4\,NH - N\,C_6H_5 + H_2O$$

Glucose phenyl hydrazone

On warming with excess of phenylhydrazine, however, the hydrazone first formed is oxidized in such way that the CHOH group adjacent to the original aldehydic or ketonic group is converted into a CO group. The latter than combines with more phenyl-hydrazine to give a di-hydrazone containing the group

I. These compounds are termed as osazones.

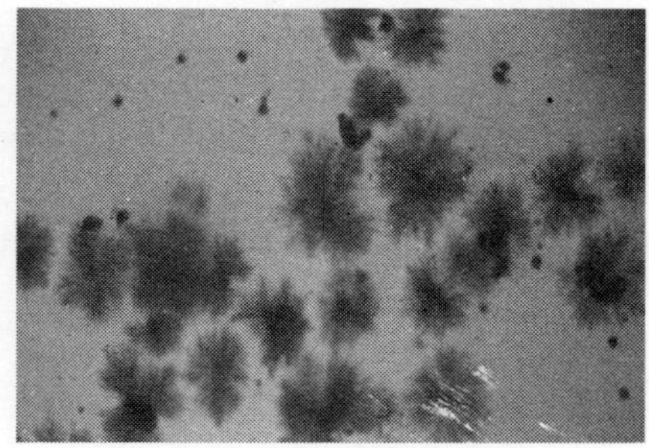

Fig. 2.3 : The Osazones under microscope

Prior to Fischer's researches one of the greatest barriers to a wider knowledge of the monosachharides lay in the difficulty of spreading mixtures of these sugars by crystallization, owing to their high solubility in water and tendency to form syrups. The value depends on the fact that they are sparingly soluble and easily separable by crystallization, and in addition, from their characteristic melting – points and crystalline forms as seen under a microscope it is possible to identify.

The recognition and isolation of ketoses is a matter of some difficulty owing to their lack characteristic reactions, with secondary hydrazine, however, the ketoses give phenyl – methyl osazones by which they may be identified. Aldodses usually react with this base to form colourless hydrazones, which in all cases are readily distinguished or separated from the highly coloured osazones.

Osazones like hydrazones, are hydrolyzed on being heated with hydrochloric acid, when phenyl hydrazine is regenerated, The sugar originally employed, however, is not regained, as the group

$C(:N.NH.C_6H_6) - CH(:N.NH.C_6H_5)$ is converted into the group —CO, —CHO. The compound so formed is thus and oxidation product of the original sugar, and is termed as *osone*

$CH_2OH.(CHOH)_3CO.CHO$ Glucosone.

On mild reduction of this compound with zinc dust and dilute acetic acid, the aldehydic group remaining unchanged. In this case, therefore the sugar finally obtained is fructose in place of the glucose used as starting material.

In these reactions we have a general method of transforming an aldose into a ketose, according to the scheme.

$$\text{Aldose} \xrightarrow[\text{hydrazine}]{\text{Phenyl}} \text{Osazone} \xrightarrow{\text{hydrolysis}} \text{Osone} \xrightarrow{\text{rediction}} \text{Ketose}$$

Further Structural Details of the Glucose Molecule

The data given by the idea that glucose is a mixture of tautomers involving equilibrium between free hydroxyaldehyde (2, 3, 4, 5, 6-pentahydroxyhexanal) and one or more internal hemiacetal forms.

Although there must be small amount of free aldehyde present, the principal substances in a solution of glucose are the hemiacetals and we must now enquire into their number and their structure. Experimentally, two hemiacetals α and β termed anomers are obtainable by special procedures. Thus, when glucose is crystallized from solution at room temperature, the α anomers is obtained. It has a specific rotation. + 111 when initially dissolved in water, but the rotation gradually falls to + 52.5. Crystallization of glucose at about 100 results in the β-anomers. The rotation of the solution of the β-anomers = + 19. The change in rotation with time, which is catalyzed by both acids and bases is called mutarotation and is caused by equilibrium mixture of the two containing 63% of β and 37% of an β anomers.

The mutarotation of glucose and the existence of two crystalline forms are accountable on the basis of the formation of cyclic hemiacetals (with introduction of an additional asymmetric center at carbon I) we have noted may times before that both five and six membered ring are readily formed and therefore sugar chemists had to understate a study of the ring size in sugars.

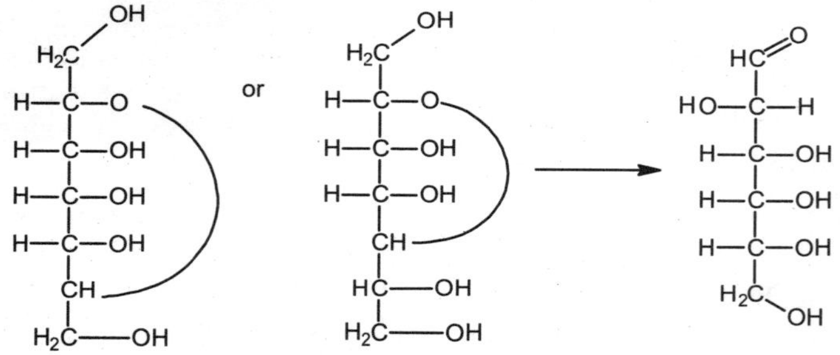

To ascertain the structure of the cyclic form of the glucose, the molecular configuration must be secured so that the molecule cannot change structure during a chemical reaction utilized form analysis. This apparently is achieved by complete methylation of glucose withy methyl sulphate and sodium hydroxide.

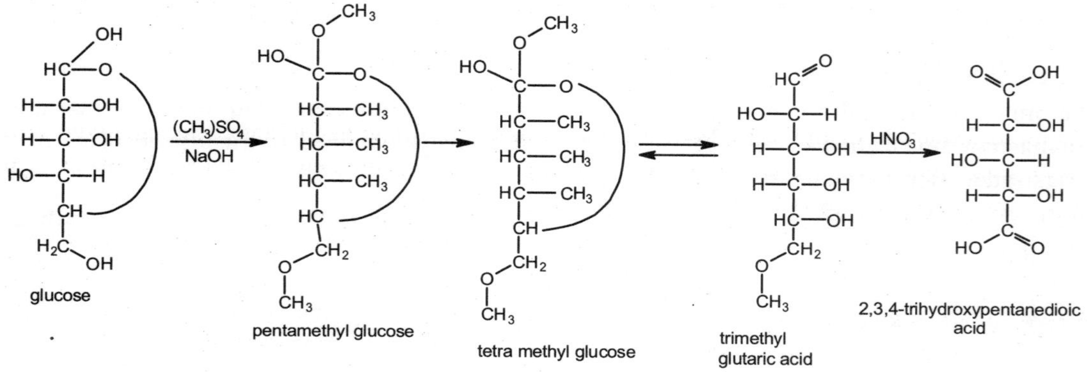

glucose

$(CH_3)SO_4$ / NaOH

pentamethyl glucose

tetra methyl glucose

trimethyl glutaric acid

HNO_3

2,3,4-trihydroxypentanedioic acid

Hydrolysis of phentamethyl glucose in acid then frees the aldehyde group and the hydroxy group involved in the acetal linkage. Oxidation of the tetramethyl glucose occurs at the aldehyde group and the free hydroxy group. Upon oxidation the tetramethyl glucose forms a trimethoxy glutaric acid. These reactions are summerised in the above equation. Those sugar or sugar derivatives having a five membered ring lead to dimethoxy succcinic acids by the Haworth Procedure. Since the six membered ring is structurally related to pyran a six membered acetal ring in a sugar is called a pyranose ring. A furanose ring because of its relation to furan.

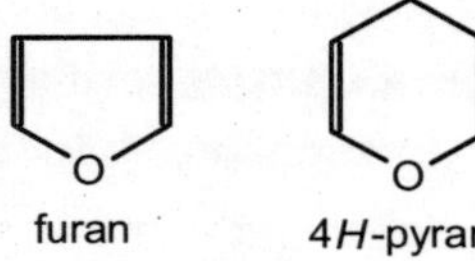

Structural analysis of the aldoses and the ketoses follows a pattern much like that outline d for glucose.

Cyclic Formulas For Glucose

After this work determining the relative configurations of the aldohesoses and aldopentoses. Emil Fischer devises a projection formation to represent the open(aldehydo) forms of these monosaccharides. In the Fischer projection, the carbon chain is represented vertically and the bonds from each carbon atom to its neighbouring carbon atoms are thought of a lying behind the carbon atom under consideration. The hydrogen atoms and hydroxy groups are considered to project up in front of the carbon atom to which they are attached. Thus the configurationally formulas, II, represents a molecule in which all hydroxy groups are on the right side and all hydrogen atoms on the left side and the middle of the carbon chain loops upon towards the observer III the top carbon atom is the Ist position.

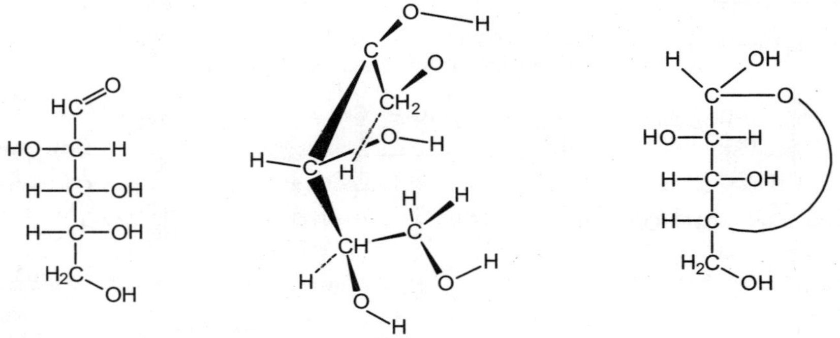

(II) (III) (IV)

Although the Fischer projection serves well enough for the open forms of the monosaccharides it is afterward when used to represent cyclic forms such as IV, The long bent "bond" between the oxygen atom and one of the carbon atoms does not represent the situation well; well hence it use is misleading in a formula intending to be a spatial representation of a molecule. To correct this, Haworth devised a modified projection formula which represent carbohydrates in conventionalized cyclic forms. The ring oxygen is usually placed at the upper right, with the carbon atoms placed in clockwise order of numbering after the oxygen atom. Hydroxy group placed to the right if Fischer projection of α-D fructofuranose and of the anomeric β-D fructofuranose.

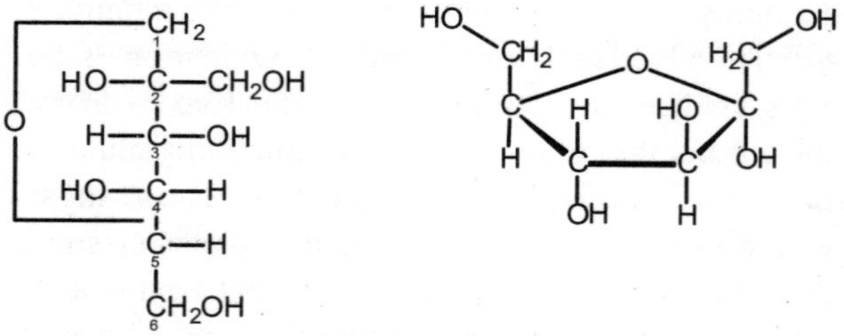

α D fructofuranose

In the oligosaccharides and polysaccharides the virtues of the Haworth projections are even more striking since, the links between units as well as the rings oxygens must be distorted in Fischer projections, where as Haworth projections cope well with this feature.

Sometimes the Haworth projection is inverted or reversed so as to avoid other obstacles to clear formulation. The projection is a picture of the ring with the top always farthest from the viewer.

Compare the formulas for β-D fructofuranose below with the given below:

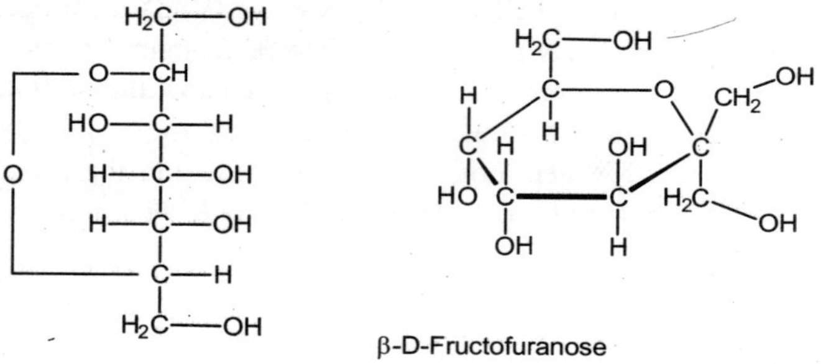

β-D-Fructofuranose

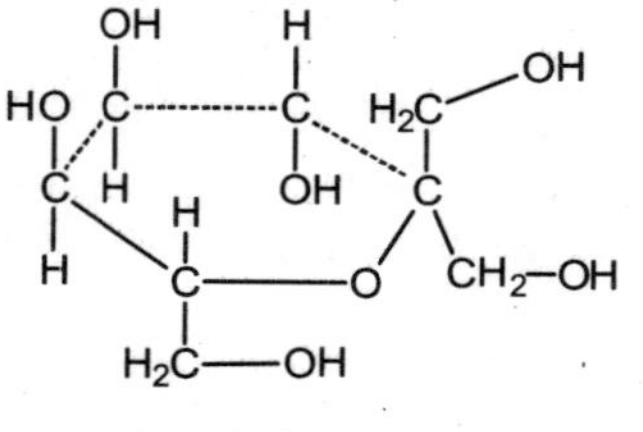

Inverted

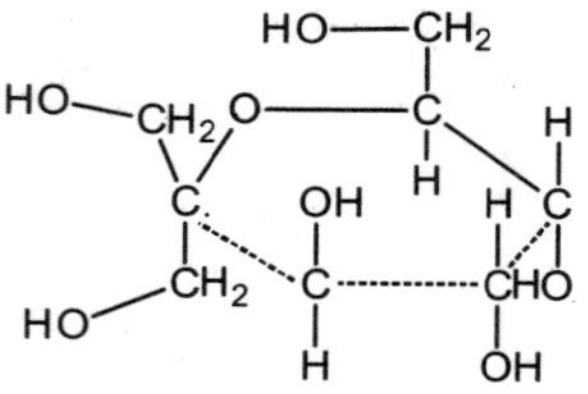

reversed

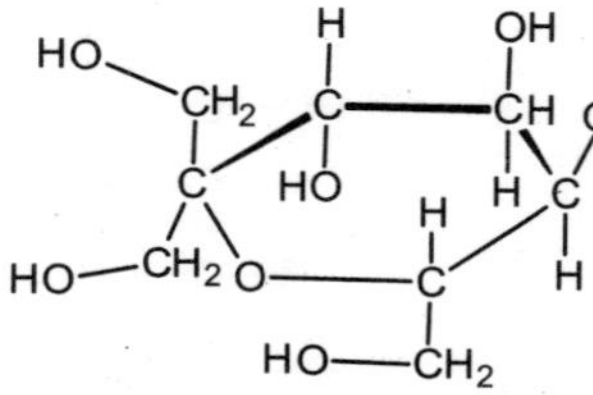

rotated

Still more realistic configuration at formula are often used in the consideration of spatial influences on reactive of equilibrium positions Compare the Haworth projection of α-D glucopyranose and the formational formula below.

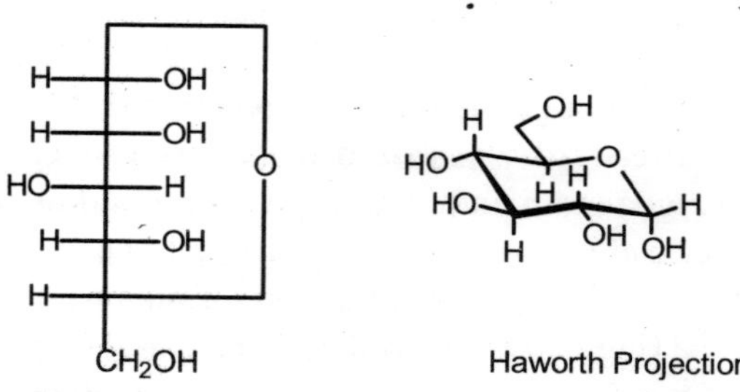

Fischer projection

Haworth Projection

α-D-Glucopyranose

Optical Configuration of Aldoses

As late as 1885, although the tetrahedral theory of carbon valences required sixteen isomeric aldohexoses and eight aldopentoses, and although several of the required isomers were recognized, the configurations of the monosaccharides remained unknown. It was the great structural chemist, Emil Fischer, who basing much of his work on the studies of Killani, Wöhl and Ruff, solved the problem required.

The point of departure is the establishment of epimeric relationship and relationships among the aldoses via Killani synthesis and the Wohl and Ruff degradation. These relationships are given in the table. Corresponding to the epimeric and series relationships are the configuration relationship of aldoses below. In this figure the molecular configuration are indicated by abbreviated Fisher projections using triangles to represent the orientation of hydroxy groups relative to the backbone of the molecule the orientation of hydroxy groups relative to the back of the molecule, which is the vertical line connecting the circle and triangle. Since the optically opposite series, but in minor image position, the configurational relationships of only one series need be considered. The D-series aldoses are aldoses discussed in the analysis. These are the aldoses that can be built up D-glyceraldehydes by successive Killani synthesis and, therefore, have the lowest asymmetric carbon in the Fisher convention attached to hydroxy on the right hydrogen on the left.

The Table of Killani Synthesis Series

Triose	Tetroses	Pentoses	Hexoses
D-Glyceraldehyde	→ D-threose	→ D-lyxose	→ D-galactose
			→ D-talose
		→ D-xylose	→ D-idose
			→ D-gulose
	→ D-erythrose	→ D-arabinose	→ D-glucose
			→ D-mannose
		→ D-ribose	→ D-allose
			→ D-altrose

It should be apparent at first glance the whereas D-glyceraldehydes can be assigned by convention the configuration, the relationship between the other numbered formulæ and the aldoses listed in the table above are by no means self-evident. Tetrose 2 might be either D-threose or D-erithrose. To ascertain the configurational identity of the various aldoses. Fisher resorted to oxidations leading to glyceric (saccharic)acids. These yield structurally symmetrical compound of which a certain number should be meso acids. For example oxidation of ribose and xylose by nutric acid yields optically inactive hence, these two aldopentoses must be 4 and 6 (not necessary in order. To check this the carbonyl groups can be indicated by another sign like X and the configuration of the molecule checked for a plane of symmetry.

Lyoxose and arbinose, which yield optically active glyceric acids, are 5 and 7 (not necessarily in order). Tallose and galactose are epimers derivable from lyxose glucose and mannose epimers derivable from arbinose. Of these hexode only galactose, yield a meso glyceric acid hence galactose is 14 and

Tallose 15. Lyxose must by 7, arbinose 5 Xylose, the epimer of lyxode is 6 and ribose, the epimer of lyxose is 6 and ribose, the epimer of arbinose, is 4. Another inactive acid is produced by allose. Hence, allose is 8. Altrose which yields an optically active acid and is the epimer of allose is 9.

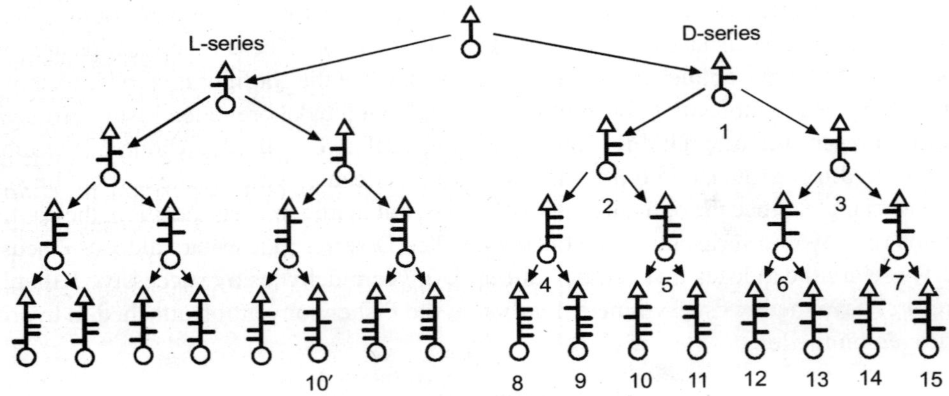

Fig. 2.4 : The Configurational Relationship of Aldoses

To establish the configuration of glucose recourse was made to Killani synthesis to the aldoheptoses. Oxidation of the two aldoheptoses yields one optically active glyceric acid and one meso acid. Consideration of formula 10 and 11, the epimeric hexoses related to arbinose, shows that the result substance which produces this result must have the configuration 10. Hence the glucose has the configuration 10. Mannose the epimer of glucose, which yields two optically active heptaglyceric acid is 11.

Upon oxidation, glucose gives glyceric acid specifically named v saccharic acid. Glucose yields enantimorph L-saccharic acid. Since D-glucose cannot belong to the L-series, this means that when the aldehyde group and the primary alcohol group lose their identify by transformation to carboxyl groups, the same result unuses as when L-glucose is oxidized. To see the relationship, one rotates a model of glucose 180°C so that its aldehyde group coincides with the primary alcohol group of L-glucose 10 and its primary alcohol group coincides with the aldehyde group of L glucose. The formula which fits this description is 12. The remaining hexode iodose is 13.

Occasionally one wishes to indicate the direction of optional rotation od an optically active rotation of an optically active material. This is done by placing (+) or (%) between the series indication and the remainder of the name, Thus D (+) glucose D(%) ribose D(+) glyceraldehydes and D(%) tartaric acid are examples.

Analysis of Glucose

Glucose gives the following series of reactions during analysis; caustic alkalies added to a solution of glucose and warmed, produce a brown solution.

Exp: Add a few drops of sodium hydroxide solution to a dilute solution of glucose, and warm gently. The colour of this liquid turns from yellow to brown.

Glucose reduces an Ammonical – Silver oxide solution to metallic silver which gets deposited on heating.

Exp: Add a few drops of Tollen's Reagent (see A chemical Analyser's Guide) glucose solution and heat the mixture in a water bath, for few minutes the metallic silver is deposited.

The following reaction which is given by all soluble carbohydrate is Molisch's Test.

Exp: Add 1 ml of alcoholic α-naphthol solution (Molisch's reagent) to a solution of glucose, then 1 ml of conc. Sulphuric acid is allowed to run down by the side of the tube so that a junction of two layers are formed which slowly turns blue.

An alkaline copper sulphate solution is reduced and cupric oxide is precipitated.

Exp: Add a few drops of copper sulphate solution to a solution of glucose and add a few drops of sodium hydroxide when the solution is heated the colour changes from blue to red and a red precipitate of cuprous oxide is produced.

Quantitative Test of Glucose

The above reaction is utilized for the quantitative estimation of glucose as well as other sugars. A standard solution of copper sulphate is prepared by dissolving 34.64 gm in 500 ml (.434mol/L) A second solution is made by dissolving 60 gms of sodium hydroxide in 500 ml water. This solution when mixed and run into a flask and boiled in a measured amount, and the sugar solution is added gradually from a burette into the hot solution mixture until all the cuprous oxide is precipitated. The quantity of sugar solution taken is a measure of the amount of glucose present 10cc of the alkaline copper solution corresponds to 0.05 gm of glucose.

Use

We can speculate on the reasons why glucose, and not another monosaccharide such as fructose (Fru), is so widely used in evolution, the ecosystem, and metabolism. Glucose can form from formaldehyde under abiotic conditions, so it may well have been available to primitive biochemical systems. Probably more important to advanced life is the low tendency of glucose, by comparison to other hexose sugars, to non-specifically react with the amino groups of proteins. This reaction (glycation) reduces or destroys the function of many enzymes. The low rate of glycation is due to glucose's preference for the less reactive cyclic isomer. Nevertheless, many of the long-term complications of diabetes (e.g., blindness, kidney failure, and peripheral neuropathy) are probably due to the glycation of proteins or lipids. In contrast, enzyme-regulated addition of glucose to proteins by glycosylation is often essential to their function.

As an Energy Source

Glucose is a ubiquitous fuel in biology. It is used as an energy source in most organisms, from bacteria to humans. Use of glucose may be by either aerobic or anaerobic respiration (fermentation). Carbohydrates are the human body's key source of energy, through aerobic respiration, providing approximately 3.75 kilocalories (16 kilojoules) of food energy per gram. Breakdown of carbohydrates (e.g. starch) yields mono- and disaccharides, most of which is glucose. Through glycolysis and later in the reactions of the Citric acid cycle (TCAC), glucose is oxidized to eventually form CO_2 and water, yielding energy, mostly in the form of ATP. The insulin reaction, and other mechanisms, regulate the concentration of glucose in the blood. A high fasting blood sugar level is an indication of prediabetic and diabetic conditions.Glucose is a primary source of energy for the brain, and hence its availability influences psychological processes. When glucose is low, psychological processes requiring mental effort (e.g., self-control) are impaired.

FRUCTOSE

Introduction

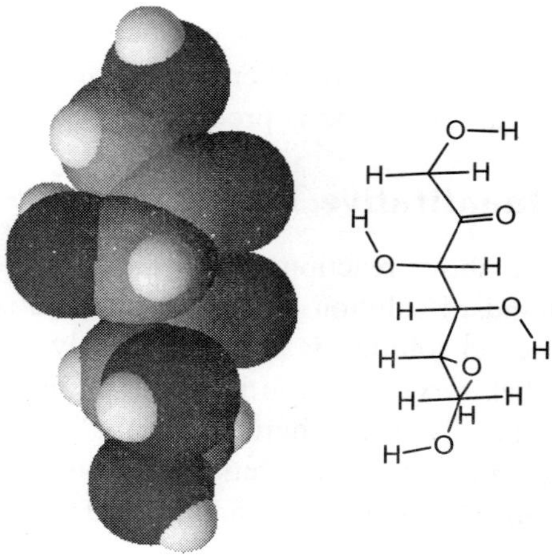

Fructose (also **levulose** or **laevulose**) is a simple reducing sugar (monosaccharide) found in many foods and is one of the three most important blood sugars along with glucose and galactose. Honey, tree fruits, berries, melons, and some root vegetables, such as beets, sweet potatoes, parsnips, and onions, contain fructose, usually in combination with sucrose and glucose. Fructose is also derived from the digestion of sucrose, a disaccharide consisting of glucose and fructose that is broken down by glycoside hydrolase enzymes during digestion. Fructose is the sweetest naturally occurring sugar, estimated to be twice as sweet as sucrose. Fructose is often recommended for, and consumed by, people with diabetes mellitus or hyperglycemia, because it has a very low glycemic index (GI) relative to cane sugar (sucrose). However, this benefit is tempered by concern that fructose may have an adverse effect on plasma lipid and uric acid levels, and the resulting higher blood levels of fructose can be damaging to proteins. The low GI is due to the unique and lengthy metabolic pathway of fructose, which involves phosphorylation and a multi-step enzymatic process in the liver. See health effects and glycation for further information.

Synthesis

Fructose can be obtained from cane sugar by hydrolysis with sulphuric acid in dilute state. The acid is removed by the precipitation with barium carbonate and the filtrate is concentrated. Milk of lime is then added, when the lime compound or the calcium fructosate (corresponding to calcium glucosate which is only slightly soluble separates out and is filtered and washed. The calcium compound is then suspended in water and decomposed by carbon dioxide. The solution is again filtered from calcium carbonate and evaporated. On introducing a crystal of fructose into the syrup, the later crystallizes.

Fructose is also prepared from inulin which turns to fructose inulin which turns to fructose on hydrolysis with sulphuric acid. After removing the acid the liquid is evaporated preferably under reduced pressure. When a syrup is left which solidifies if a crystal of the solid is added.

Physical Properties

Molecular Formula	= $C_6H_{12}O_6$
Formula Weight	= 180.15588
Composition	= C(40.00%) H(6.71%) O(53.29%)
Melting point	= D-fructose: 103°C
Molar Refractivity	= 37.42 cm^3
Molar Volume	= 113.3 cm^3
Parachor	= 351.7 cm^3
Index of Refraction	= 1.574
Surface Tension	= 92.6 dyne/cm
Density	= 1.589 g/cm^3
Polarizability	= 14.83 $10^{-24}cm^3$
Monoisotopic Mass	= 180.063388 Da
Nominal Mass	= 180 Da
Average Mass	= 180.1559 Da
Log P	= -1.63

Chemical Properties and Constitution

Fructose is more soluble than glucose but crystallizes from rhombic prisms which melt at 103°C. it has a sweet taste and gives many of the reactions of glucose. Although fructose is not an aldehyde but a ketone, it nevertheless reduces alkaline copper solution. This is due to the presence of the easily oxidisable group $-CO_2-CH_2(OH)$ with phenylhydrazine, fructose yields an osazone, which is identical with glucosazone. Fructose also undergoes fermentation with yeast, through less readily than glucose, glucose being first removed when a solution of the two sugars is fermented. Like glucose fructose undergoes mutarotation in solution the specific rotation rising from –135.5 to –92.0°

Fructose forms a pentacetyl derivative like glucose. On reduction it is converted into a mixture of sorbitol and mannitol, since an additional asymmetric combination has been introduced. On oxidation it does not like glucose form an acid with same number of carbon atom but breaks up into formic acid and trihydroxybutaric acid. This decomposition points to the presence of a ketone group in the molecule, which is further confirmed by the following reactions.

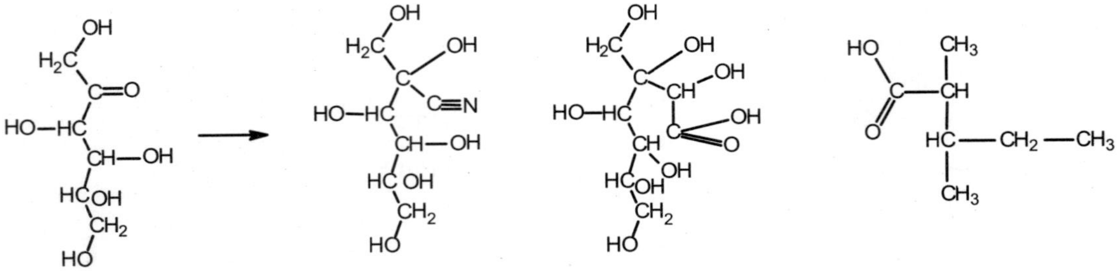

fructose → Cyanohydrin of fructose → hydroxy acid

The above formula for fructose agrees moreover, with the synthesis of inactive fructose from the mixture glyceric aldehyde and dihydroacetone and with the production of the same osazone as that obtained from glucose, a reaction in which the two end carbon atoms of the chain are involved.

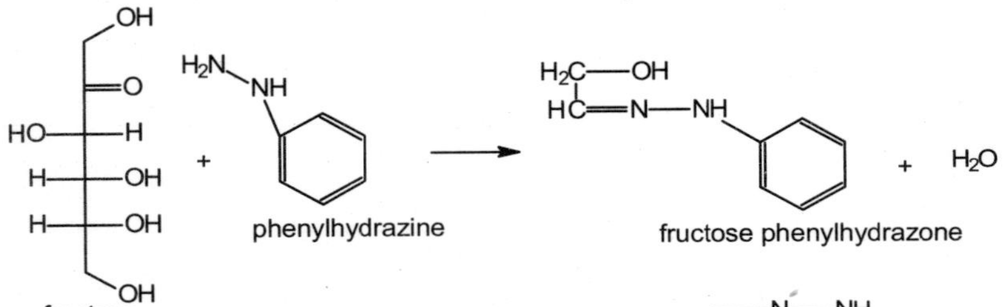

fructose + phenylhydrazine → fructose phenylhydrazone + H_2O

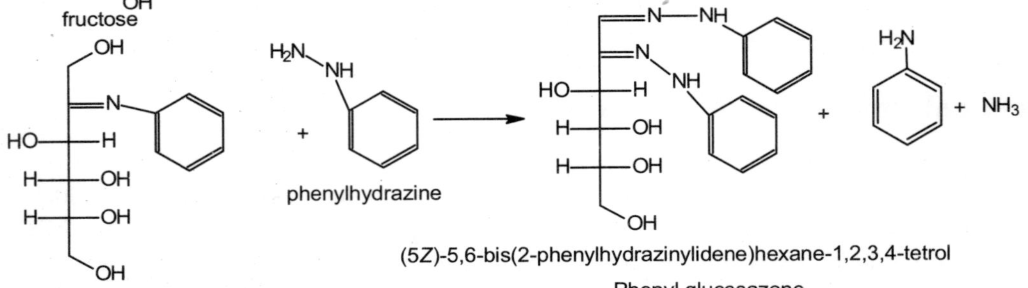

fructose phenylhydrazone + phenylhydrazine → (5Z)-5,6-bis(2-phenylhydrazinylidene)hexane-1,2,3,4-tetrol Phenyl glucosazone + H_2N-phenyl + NH_3

Although fructose in many of its reactions a ketone its exists like glucose in a and stereoisomeric pyranose forms and also in γ or furanose rings.

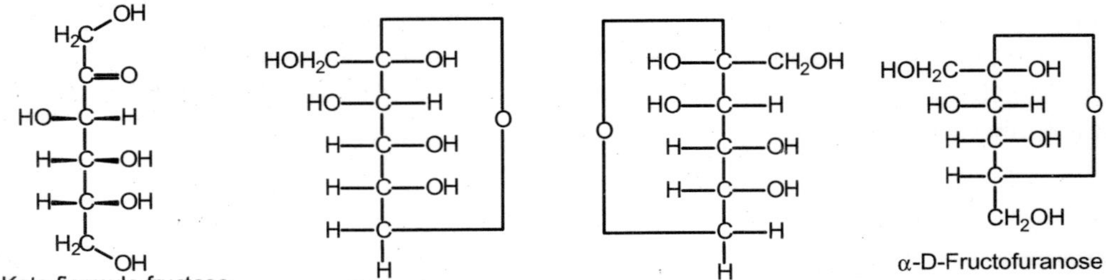

Keto fiormula fructose

α-D-Fructopyranose

β-D-Fructopyranose

α-D-Fructofuranose

The perspective formulæ of α and β fructopyranose are γ - fructose is a constituent of cane sugar but when liberated from its union with glucose passes into the pyranose form.

α-D-Fructofuranose

β-D-Fructopyranose

Uses

Fructose is often recommended for diabetics due to its glycemic index being significantly lower than both glucose, sucrose and starches. The medical profession thinks fructose is better for diabetics than sugar, but every cell in the body can metabolize glucose. However, all fructose must be metabolized in the liver. The livers of the rats on the high fructose diet looked like the livers of alcoholics, plugged with fat and cirrhotic This is not entirely true as certain other tissues do use fructose directly, notably the cells of the intestine, and sperm cells (for which fructose is the main energy source).

Fructose is a reducing sugar, as are all monosaccharides. The spontaneous addition of single sugar molecules to proteins, known as glycation, is a significant cause of damage in diabetics. Fructose appears to be as dangerous as glucose in this regard and so does not seem to be a better answer for diabetes for this reason alone. This may be an important contribution to senescence and many age-related chronic diseases. Fructose is used as a substitute for sucrose (composed of one unit each of fructose and glucose linked together with a relatively weak glycosidic bond) because it is less expensive and has little effect on measured blood glucose levels. Often, fructose is consumed as high fructose corn syrup, which is corn syrup (glucose) that has been enzymatically treated by the enzyme glucose isomerase. This enzyme converts a portion of the glucose into fructose thus making it sweeter. This is done to such a degree as to yield corn syrup with an equivalent sweetness to sucrose by weight. While most carbohydrates have around the same amount of calories, fructose is sweeter and manufacturers can use less of it to get the same result. The free fructose present in fruits, their juice, and honey is responsible for the greater sweetness of these natural sugar sources. Some studies point to fructose as key factors in hyperactivity and tooth decay in children.

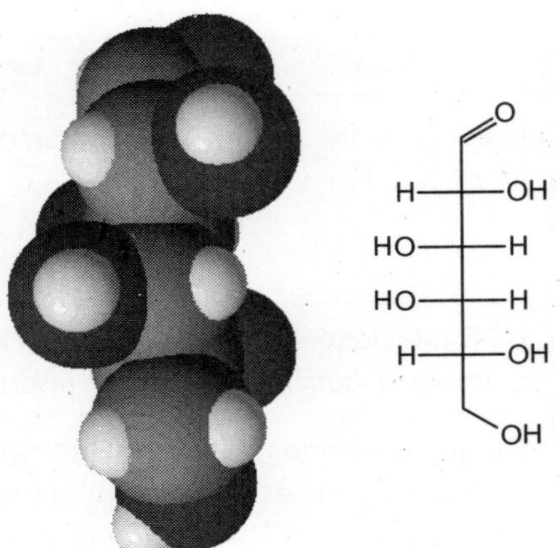

GALACTOSE

Galactose (Gal) (also called *brain sugar* is a type of sugar which is less sweet than glucose and not very water-soluble. It is considered a nutritive sweetener because it has food energy. Galactan is a polymer of the sugar galactose. It is found in hemicellulose and can be converted to galactose by hydrolysis.

It is found in dairy products, in sugar beets and other gums and mucilages. It is also synthesized by the body, where it forms part of glycolipids and glycoproteins in several tissues.

Relationship to Lactose

Galactose is a monosaccharide constituent, together with glucose, of the disaccharide lactose. The hydrolysis of lactose to glucose and galactose is catalyzed by the enzyme lactase, a β-galactosidase. In the human body, glucose is changed into galactose in order to enable the mammary glands to secrete lactose. Galactose and glucose are produced by hydrolysis of lactose by β-galactosidase. This enzyme is produced by the *lac* operon in *Escherichia coli* (*E. coli*).

Clinical Significance

Two studies have suggested a possible link between galactose in milk and ovarian cancer. Other studies show no correlation, even in the presence of defective galactose metabolism. More recently, pooled analysis done by the Harvard School of Public Health showed no specific correlation between lactose containing foods and ovarian cancer, and showed statistically insignificant increases in risk for consumption of lactose at >30 g/d. More research is necessary to ascertain possible risks. There are some ongoing studies which suggest that galactose may have a role in treatment of focal segmental glomerulosclerosis (a kidney disease resulting in kidney failure and proteinuria). This effect is likely to be a result of binding of galactose to FSGS factor.

Structure and Isomerism

The first and last –OH groups point the same way and the second and third –OH groups point the other way. D-Galactose has the same configuration at its penultimate carbon as D-glyceraldehyde. Galactose is a diastereomer of glucose.

Metabolic Disorders

There are 3 important disorders involving galactose:

Name	Enzyme	Description
Galactosemia	Galactokinase	Causes cataracts and mental retardation. If a galactose-free diet starts sufficiently early, the cataracts will regress without complications however neurological damage is permanent
UDPgalactose-4-epimerase deficiency	UDPgalactose-4-epimerase	Is extremely rare (only 2 reported cases). It causes nerve deafness
Galactose-1-phosphate uridyl transferase deficiency	Galactose-1-phosphate uridyl transferase	Is the most problematic, as galactose-free diets do not have considerable long-term effects

MANNOSE

Mannose is a sugar monomer of the hexose series of carbohydrates.

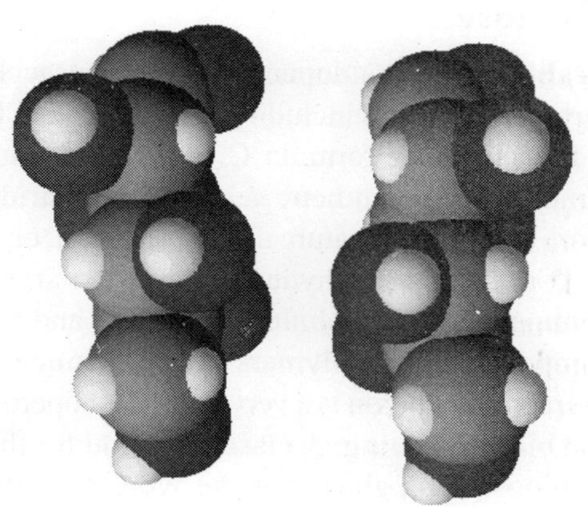

Formation

Mannose can be formed by the oxidation of mannitol.

It can also be formed from D-glucose in the Lobry-de Bruyn-van Ekenstein transformation

Etymology

The root of both "mannose" and "manitol" is manna, which the Bible records as the food supplied to the Israelites during their journey through the Sinai Peninsula. Manna is a sweet secretion of several trees and shrubs, such as *Fraxinus ornus*.

Cranberry juice

D-Mannose, which appears in some fruits including cranberry, has been postulated to prevent the adhesion of bacteria to the epithelium of the urinary tract and Urinary bladder through a mechanism presumed to be competitive in nature with the polysaccharide coating of the cystic epithelium. While no studies have objectively verified this mechanism, anecdotal reports from patients show some improvement in symptoms.

D-Mannose L-mannose

Configuration

The fact that D-mannose has the same configuration at its penultimate carbon as D-glyceraldehyde is unsurprising as that is what defines the dextro classification. However, mannose differs from D-glucose by inversion of the C2 chiral centre. This apparently simple change leads to the drastically different chemistry of the two hexoses, as it does the remaining six hexoses.

Metabolism

Mannose enters the carbohydrate metabolism stream in two steps:

- First it undergoes phosphorylation to mannose-6-phosphate by hexokinase.
- Then, it is converted to fructose-6-phosphate by mannose phosphate isomerase.

Mannose is present in numerous glycoconjugates including N-linked glycosylation of proteins. C-mannosylation is also abundant and can be found in collage-like regions. Mannose is a C2 epimer of glucose and displays a ^2C_4 pucker in the solution ring form.

PENTOSES

Arbinose

Arabinose is an aldopentose—a monosaccharide containing five carbon atoms, and including an aldehyde (CHO) functional group. It has chemical formula $C_5H_{10}O_5$ and a molar mass of 150.13 g/mol. For biosynthetic reasons, saccharides are almost always more abundant in nature as the "D" form, or structurally analogous to D-(+)-glyceraldehyde. However, L-arabinose is in fact more common than D-arabinose in nature and is found in nature as a component of biopolymers such as hemicellulose and pectin. The L-arabinose operon is a very important operon in molecular biology and bioengineering. A classic method for the organic synthesis of arabinose from glucose is the Wohl degradation.

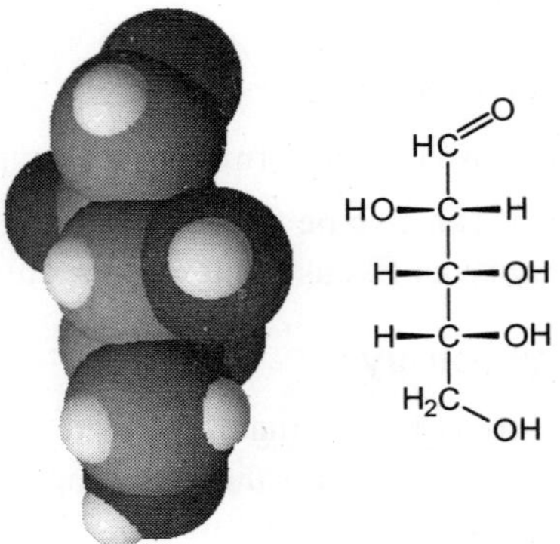

Ribose

Ribose, primarily seen as **D-ribose**, is an aldopentose—a monosaccharide containing five carbon atoms, and including an aldehyde functional group in its linear form. It has the chemical formula $C_5H_{10}O_5$, and was discovered in 1905 by Phoebus Levene.As a component of the RNA that is used for genetic transcription, ribose is critical to living creatures. It is related to deoxyribose, which is a component of DNA. It is also a component of ATP, NADH, and several other chemicals that are critical to metabolism.Refer to the article on deoxyribose for more information on both sugars, how they relate to each other, and how they relate to genetic material.

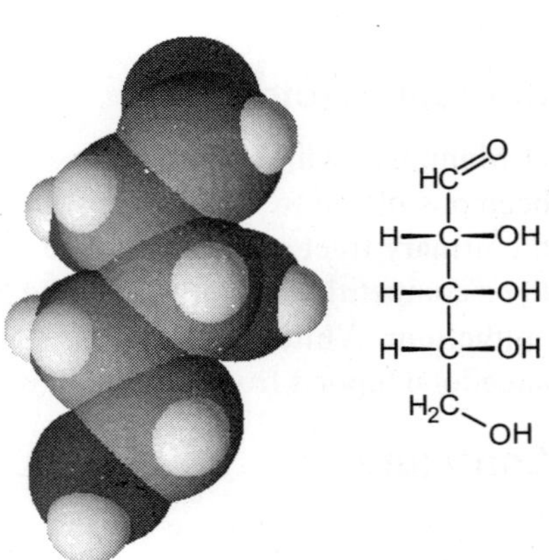

Physical Properties

Molecular Formula	= $C_5H_{10}O_5$
Formula Weight	= 150.1299
Composition	= C(40.00%) H(6.71%) O(53.29%)
Melting Point	= 99 °C
Molar Refractivity	= 31.41 cm^3
Molar Volume	= 99.5 cm^3
Parachor	= 299.0 cm^3
Index of Refraction	= 1.543

Surface Tension	= 81.4 dyne/cm
Density	= 1.508 g/cm^3
Polarizability	= 12.45 10^{-24}cm^3
Monoisotopic Mass	= 150.052823 Da
Nominal Mass	= 150 Da
Average Mass	= 150.1299 Da

Ribulose

Ribulose is a ketopentose—a monosaccharide containing five carbon atoms, and including a ketone functional group. It has chemical formula $C_5H_{10}O_5$. Two enantiomers are possible, D-ribulose (D-erythro-pentulose) and L-ribulose (L-erythro-pentulose). D-Ribulose is the diastereomer of D-xylulose.Ribulose sugars are composed in the pentose phosphate pathway. They are important in the formation of many bioactive substances. For example, D-ribulose is an intermediate in the fungal pathway for D-arabitol production. Also, as the 1,5-bisphosphate, D-ribulose combines with carbon dioxide at the start of the photosynthetic process in green plants (carbon dioxide trap).

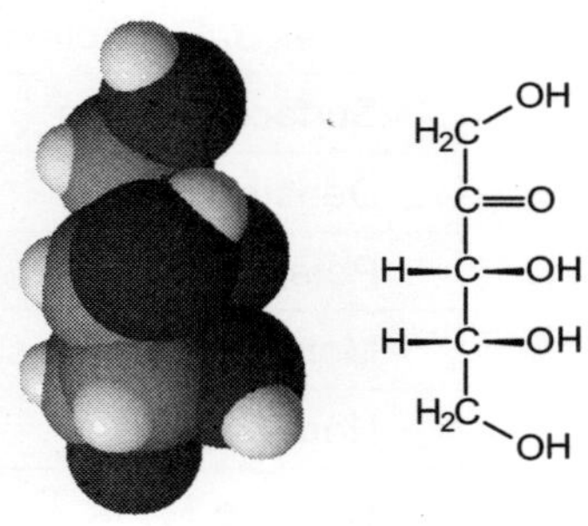

A synthetic form of ribulose known as sucroribulose is found in many brands of artificial sweeteners.

Xylose

Xylose or wood sugar is an aldopentose—a monosaccharide containing five carbon atoms and including an aldehyde functional group. It has chemical formula $C_5H_{10}O_5$. Xylose is found in the embryos of most edible plants.

In animal medicine, xylose is used to test for malabsorption by administering to the patient in water after fasting. If xylose is detected in blood and/or urine within the next few hours, it has been absorbed by the intestines.

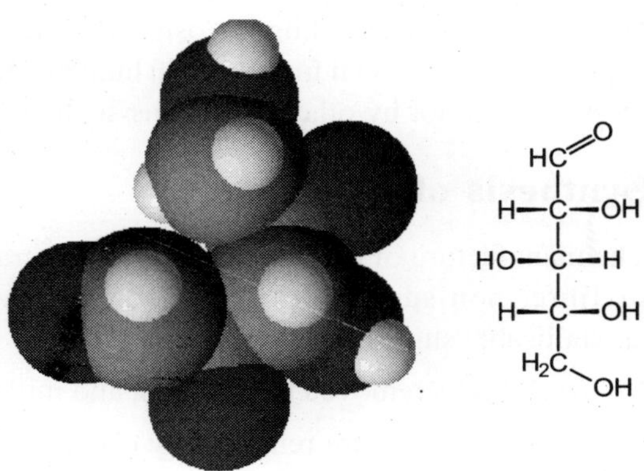

Physical Properties

Molecular Formula	= $C_5H_{10}O_5$
Formula Weight	= 150.1299
Composition	= C(40.00%) H(6.71%) O(53.29%)
Melting Point	=144-145 °C
Molar Refractivity	= 31.41 cm^3
Molar Volume	= 99.5 cm^3
Parachor	= 299.0 cm^3
Index of Refraction	= 1.543
Surface Tension	= 81.4 dyne/cm
Density	= 1.508 g/cm^3
Polarizability	= 12.45 $10^{-24}cm^3$
Monoisotopic Mass	= 150.052823 Da
Nominal Mass	= 150 Da
Average Mass	= 150.1299 Da

OLIGOSACCHARIDES

Disaccharides

Sucrose

Sucrose (common name: **table sugar**, also called **saccharose**) is a disaccharide (glucose + fructose) with the molecular formula $C_{12}H_{22}O_{11}$. Its systematic name is α-D-glucopyranosyl-(1→2)-β-D-fructofuranoside (ending in "oside", because it's not a reducing sugar). It is best known for its role in human nutrition and is formed by plants but not by other organisms such as animals.

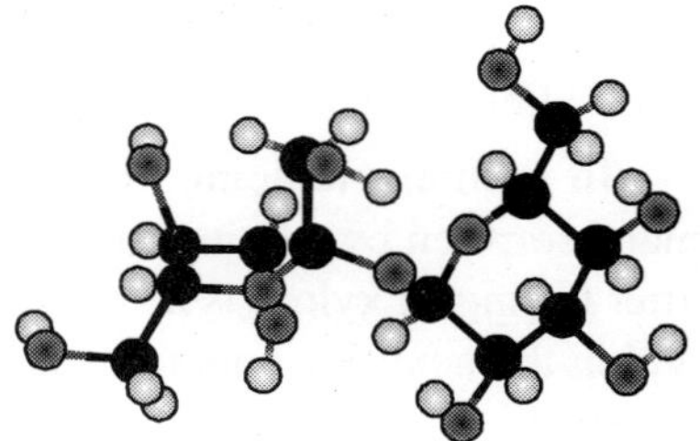

Synthesis of Sucrose

The manufacture of cane sugar is based on gradual removal of fibre, non-sugars, water, colouring matter and non crystallisable sugars.

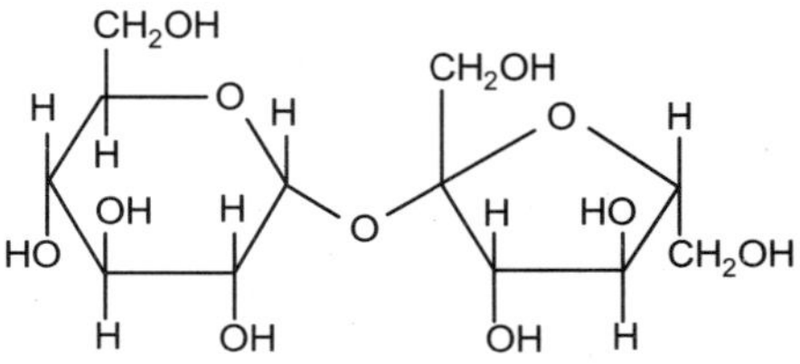

(*a*) Fibre is removed by crushing and milling;

(*b*) Non-sugars are removed by clarification;

(*c*) Water is removed by evaporation and centrifuging;

(d) Colouring matter is removed by bleaching *or* absorption by bone charcoal *or* activated carbon;

(e) Non-crystallisable sugars are eliminated during crystallisation. White sugar is produced either in two steps *or* in one step.

(a) In two steps method, first raw *or* un-refined crystalline sugar is manufactured, and then white sugar is manufactured by refining raw sugar.

(b) In one step process in which chrification is done by sulphitation *or* double carbonation method. In this process the syrup is bleached by sulphur dioxide and the thin film *of* molasses on the crystals is finally removed by double centrifuging.

Extraction of the Juice or Separation of the Fibre from the Juice (Fig. 2.5)

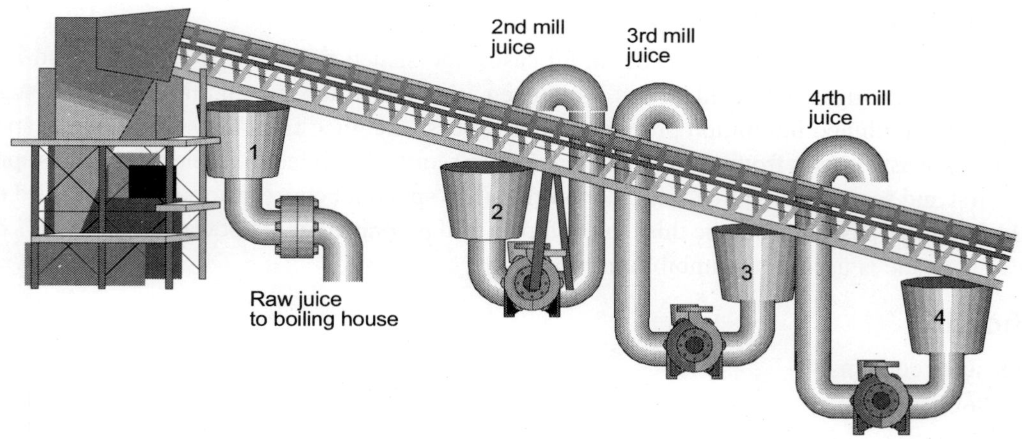

Fig. 2.5 : The Pressing of Sugar Cane

The Process of Extraction Consists

(*a*) Cutting of the stalk and leaves.

(*b*) Cutting of the canes by means of sharp rapidly revolving knives (550-600 r.p.m.) fitted on a horizontal shaft 2" inches apart. The canes move on a carrier under the knives which make diagonal cuts, converting the canes into chips in which the soft portion is exposed to the maximum.

(*c*) The chips are then brought to the extractor consisting of one two-rolled toothed crusher and four three-rolled grooved mills. (In India, the crusher and the mills are mostly made of grey cast iron; in other places it is made of cast steel)

The three rollers of the mill are placed at the crusher of a triangle; only the top roller of the crusher or of the mill moves, the bottom rollers in the milling train remain stationery during operation. A pressure of 3 to 6 tons per sq. in. is maintained on each of the top roller. There are intermediate carriers between the crusher and the first mill and between the two subsequent mills to carry the bagasse to the next mill or out of the milling train. The crusher and the first mill are provided with vibrating screens and arrangement for transferring the juice to the raw juice tank. The 2nd, 3rd and 4th mills are provided with containers to hold unstrained juice which is used for combined imbibition process (Fig. 2.5).

When the chips are fed into the crusher, the chips during their passage through the rollers are pressed to extract about 50% of the juice in the cane. The bagasse (the fibre) containing 50% of the juice is fed into the first mill. The juice from these two are strained on the strainer below and disposed as mentioned .above. The bagasse still containing some juice pass successively through-the second, third and fourth mills and then to the storage or to the kiln in which it is fed for steam raising. Even in recent times, most of the Indian mills used bagasse as fuel, but the present tendency is to replace it by coal or coke.

Compound Imbibition Process

After the crusher, the baggasse has to be, in all cases, moistened for efficient pressing. Had it been done by water the sugar extracted would have been less; hence the maximum amount of sugar is pressed out by combined imbibition process. In this process the juice from the second mill is sprayed all the baggasse as it passes from' the crusher and the first 'milt. The juice from the third mill is sprayed between first and second mill, juice from the fourth mill is sprayed between the second and third mills, while the baggasse passing from the third to the fourth mill is sprayed with hot water. About 25% of the weight of the cane is used in the imbibition process.

Clarification

Cane juice contains:

78 - 86% water

10% - 20% sucrose

0.5 - 1% organic matter

0.5 - 2.5% reducing sugar

0.3 - 0.77% ash

Besides these the cane juice contains finely divided pith and fibre and also some colloidal matters. These substances together with gums, pectin's, proteins and other non-sugars definitely hamper crystallisation of sucrose. Some of these are also responsible for the colouring of the crystallized sugar.

The process of clarification by which the above impurities are eliminated is, therefore, an important step in the production of maximum quantity of white crystalline sugar. In simple defecation by lime the impurities are not completely eliminated; hence by crystallising the syrup (the concentrated form of juice) a coloured crystalline sugar known as raw sugar is obtained. This raw sugar is then converted into white crystalline sugar by a refining process.

In most of the countries, white sugar is produced by the above-mentioned process. In India"Java and some other parts of the world white sucrose is produced by one step. In these places clarification is done either by sulphitation or carbonation process. In the sulphitation and sometimes in the carbonation process also the concentrated juice is. treated with sulphur dioxide. In carbonation process, more. efficient clarification is done by double carbonation process. Hence these processes are known as double sulphitation and double carbonation processes.

Lime Defecation Process: The cane juice is acid having pH from 5.1 to 5.7. This acidity helps the inversion of sucrose to reducing non-crystallisable sugars. For this reason the juice after extraction should be made alkaline without delay. If acidity is removed by liming, i.e. by adding adequate amount of milk of lime, then insol1ible salts of organic acids such as oxalic and' propionic are precipitated. Lime reacts with phosphoric' acid or insoluble, phosphates to form precipitate of. tricalcium phosphate Now 'when the alkalinity is increased, gums, pectins, and proteins, etc. which are in colloidalform are precipitated due to coagulation in presence of alkali. These colloidal matters are also precipitated by being entrained in the precipitates of phosphates etc. Along with colloidal matter finely divided pith and fibre and also clay are precipitated.

The raw cane juice is heated in juice heaters to 60°C and then limed to a pH of about 8. The mixing is helped by stirring. The limed juice is then passed through brass tubular heater, where it is heated to boiling. The hot limed juice is then pumped into settling tank for subsidization of the scum and the mud. By heating the protein matter.is coagulated and settling of the precipitates is accelerated. In the settling tank the hot limed juice separates into three layers. The dear juice in the middle lies between the top layer of scum (consisting of fine fibre particles) and the bottom layer of mud (consisting of the heavy precipitates. The scum is skimmed of and the clear juice is allowed to drain off through a pipe which is gradually lowered to a level above the mud layer. The clear juice is sent to the boiling house for evaporation to syrup. The mud and the scum are filtered in a filter press. The cakes are washed during filtering (the wash water is sent back to the clarifier), and then dried. These tilter cakes contain sucrose and phosphate. Hence, the added cakes are used as manure after rotting for a year. This removes the sucrose which attracts white ants.

Clarification by Sulphitation: To prepare white sugar by One step clarification is done in India by passing sulphur dioxide gas through limed juice. The production of white is further helped by bleaching the syrup by sulphur dioxide. Better clarification takes place due to two causes:

(1) By adding more lime, more collidal matters and non-sugars are precipitated, because increased alkalinity increases the coagulation of colloids;

(2) By addition of precipitates of calcium sulphite formed by the action of sulphur dioxide on lime as well as on the tricalcium sucrate ($C_{11}H_{22}O_{11}\cdot 3GO + 3SO_2 = 3CaSO_3 + C_{12}H_{22}O_{11}$) there takes place the precipitation of more colloidal matter by entrainment. The heavy precipitate of calcium sulphite also brings about quick and better subsidization. Sulphur dioxide also helps to bleach the juice.

Sulphur dioxide is passed from the bottom of a tank (with conical bottom) fitted with perforated baffles (at different levels) down which heated (temp. 140°F) limed cane juice comes down in a shower. The gas is passed until the pH of the resultant solution is 7 (neutral). The discharge from the sulphitation tank is heated almost to boiling point in a second juice heater. The amount of lime added is more than that added in simple defecation process (2-4 lbs of lime per ton of cane). From the second heater the hot fluid containing suspended precipitates is sent to the settling tank, where, after subsidization, the discharge separates into dear juice and mud. The subsequent operations have been already mentioned.

Clarification by Carbonation: In India a *few* factories have substituted the sulphitation process by double carbonation process. In this process, the limed juice is treated with carbon dioxide gas to precipitate the excess of lime as $CaCO_3$. This $CaCO_3$ subsequcntly acts like $CaSO_3$in the sulphitation process. CaCOs is formed by the action of CO: on $Ca(OH)_2$ as well as on $C_{12}H_{22}O_{11}$ 3CaO (tricalcium sucrate). In this process much more lime is added than in sulphitation process. (1-1.5 p.c. of lime on the weight of cane). This amount of lime works out to be 16-24 Ibs of lime per ton of ca.ne. Due to the large alkalinity prodliced by such excessive amount of lime, the removal of non-sugar and colouring matter become more complete. Moreover, due to the formation of much larger quantity of insoluble precipitate in the juice, the subsidization becomes quicker and clarification becomes more effective. Due to removal of solu.ble impurities more completely the resultant syrup becomes more mobile and more sucrose crystallises cut. Due to the mobility of the syrup the movement of the syrup up and down the calendria in the striking pan becomes easier. This ensures uniform heating and better crystallization. Had sulphur dioxide been used for the CO_2, very large amount of valuable sulphur would have been necessary. Hence, the tenuency in Indian mills is to take to carbonation process. In carbonation process, the excessive amount of lime precipitates the gums, pectins, etc. much more completely when liming takes place below 60°C. But when the alkalinity is neutraliscd by CO_2 ,these-precipitates pass into solution. Also these precipitates become soluble near boiling point of the juice. Hence, instead of single carbonation, double carbonation is adopted. (Described below).

First, cane juice is heated to 55°C and limed by adding milk of lime to the heated juice. Temperature during liming is not allowed to rise above 55°C. Keeping the limed juice at this temperature in ajacketed tank carbon dioxide obtained by heating lime stone in a kiln, is pumped.through the limed juice until the alkalinity is represented by 0.05 p.c. of lime.

The juice along with the precipitate is filtered en-masse (without settling). The juice is then raised to boiling point, saturated with CO_2, again boiled to break the bicarbonate formed, and then filtered en-masse. The filter cake is washed; washings may be added to the juice or sent back for clarification. The juice is sent to the boiling house to make syrup.

Evaporation to make Syrup (Fig. 2.6)

Crystalline sugar is obtained from clarified cane juice in two steps. First, the juke is concentrated to syrup containing 50% water; next the syrup is concentrated to massecuite A, containing 9-11% water. Crystalline sugar—coloured or white-is obtained from the massecuite by removing the molasses.

Before the introduction of the present day method of evaporation under vacuum, the juice was concentrated in open cauldron 'heated by solid fuels. In India, 'Gur' (crystalline sugar mixed with molasses) is still made by this process. Formerly, crystalline sugar was manufactured from 'gur; by removing the molasses by an indigenous process. By this process much sucrose is lost as caramel, and portion of sucrose is converted into glucose'. Even concentration by steam coils under atmospheric pressure produces loss. Hence, the present-day method of evaporation at a lower temperature under reduced pressure has been adopted.

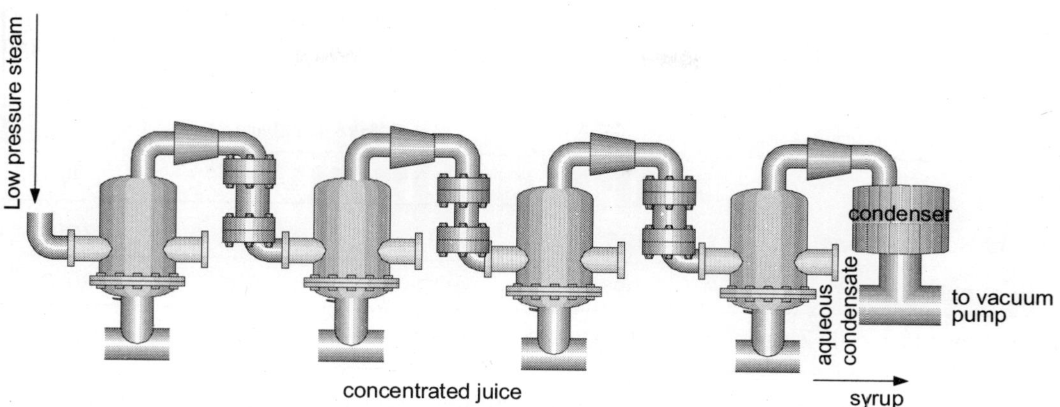

Fig. 2.6 : The Quadruple Effect Vacuum Evaporator

The clarified juice obtained from any of the above methods is concentrated into the thick syrup by concentrating in multiple-effect evaporators. Usually quadruple-effect or tripple-effect evaporators are used.

An evaporator, is a tall cylindrical vessel made of riveted mild steel plates or cast in sections from cast iron. The top is bottle-necked and the bottom is spheroidal. The evaporators are placed vertically and in a battery of three or four. The cylinder is divided into three chambers by fitting two parallel plates a few feet arart at the lower part of the cylinder; the chambers above the upper plate and below the lower plate are the juice chambers, and the chamber between the two plates form the steam chamber or calendria.

Several narrow vertical pipes join the upper plate to the lower plate (the pipes open in, the two juice chambers). A few wider pipes also vertically connect the upper and the lower pipe. The juice circulates inside the tubes and the steam circulates outside the tubes. There is a steam inlet pipe to each of the calendrias. The bottle-neck is closed at the top, but is fitted with a vapour outlet pipe at one side. The outlet pipe of the first evaporator is joined to the calendria of the second evaporator, and so on up to the last calendria. Arrester baffle plates are fitted to the neck to prevent any juice or syrup to pass through the vapour outlet pipe. A narrow pipe known as "Ammonia pipe" (by analogy with beet juice evaporators) joins the calendria to the main body to remove air and other uncondensible gases accumulated in the calendria of the body, and ultimately delivers these gases to the vacuum and condensers attached to the tast body.

Heating of the juice in the first evaporator is done by passing steam exhaust or line through the calendria. The juice submerges the calendria, and during heating the juice which is admitted from the bottom by a feed pipe circulates up through the narrow tubes, and down through the wider tubes. In this way by rapid circulation and heating at a vacuum, the water is rapidly evaporated. The vapour from the first evaporator enters into the calendria of the second evaporator. This stem, although of the lower temperature, helps rapid evaporation of the partially thickened syrup which passed from the first evaporator to the second by the opening of the valve of the bottom feed pipe: being actuated by difference of pressure. The low temperature steam helps rapid evaporation because of a higher vacuum maintained in the second evaporator. In quadruple-effect evaporators, the vacuum maintained in the 3rd body is greater than in the second, and that maintained in the last body is greater than in the 3rd. The vacuum pump together with a condenser is attached to the vapour outlet pipe of the 4th evaporator. This pump gradually reduces the pressure from the first to the last body. The vacuum and corresponding temperature maintained in each evaporator along with the steam temperature is as follows:

	Evaporators			
	I	II	III	IV
Temp of Heating Steam	125°C	99°C	86°C	76°C
Temp of boiling Juice	99°C	86°C	76°C	60°C
Vacuum maintained (in cm)	2.744	30.48	50.8	62.03

From the data mentioned above, it is found that boiling of the juice take place from 210°F to 140°F in the last evaporator. This means that most of the water has been removed at comparatively lower temperature and in absence of air, to ensure least loss of sugar by evaporation. The evaporation and subsequent transfer of fluid is so adjusted as to keep the level of the liquid in the evaporators above the level of the calendria.

From the last evaporator the pale yellow thick syrup (50-66% solid matter) is discharged into the syrup tank. From this tank it is pumped to the top of the sulphitation tank wherefrom it is allowed to percolate down the cylinder over the perforated baffle plates, in a shower, to meet an ascending stream of sulphur dioxide.

The sulphur dioxide bleaches the syrup and reacts with residual calcium ions. This process is omitted when manufacture of raw sugar is aimed at. This process is, therefore, followed by the Indian factories a few factories where clarification is done by carbonation, bleaching by sulphur dioxide may be omitted if double carbonation has been properly carried out.

Crystallization (Fig. 2.7)

The syrup is next treated for forming crystalline sugar. This is done by removing most of the water by heating in a single-effect vacuum evaporator known as pan. The vacuum pan is either fitted with calendria or steam coils. The calendria fitted pan is similar to the evaporator used for boiling the juice. But it is larger in diameter and the tubes of its calendria are wider. It is fitted with vacuum gauge and a condenser, a thermometer, a sampling stick and safety device to prevent the boiling mass getting into the vapour) outlet pipe and outlet pipe at the bottom to dump the massecuite. There is the usual feed pipe at the bottom.

The crystallization takes place in two steps, viz.;

(*a*) Formation of seed crystals by graining;

(*b*) Allowing the seed crystals to grow into fully developed crystals.

(*a*) The syrup is introduced into the pan until the steam chest is under it. Then by heating under vacuum the syrup is concentrated until, it is supersaturated with sucrose (83.5-84..5%. Sucrose at 80°C). Then the operator adopts either waiting method or shock method to arrive at the graining point, i.e., induce the formation of seed crystals by breaking the supersaturation. In the waiting method the

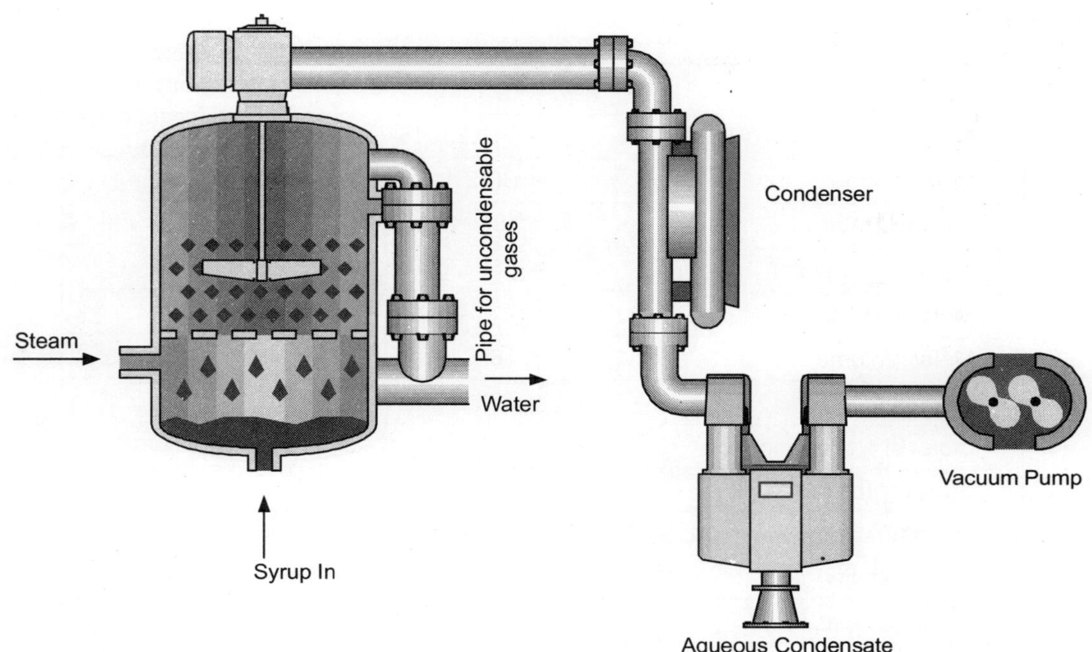

Fig. 2.7 : The Single Effect Vacuum Evaporators

operator gradually cools down to break the supersaturation. In the shock method which is usually adopted, the supersaturation is broken by sudden chilling which is, done either by increasing the vacuum or by decreasing the rate of steam flow or by suddenly introducing large' amount of cold syrup'. Circulation. of the syrup is necessary for producing proper seed crystal.

(*b*) After graining point is reached more and more syrup is introduced in instalments. By proper circulation of the very thick syrup up and down the tubes of the calendria which is helped by injecting dry steam through, the bottom even, boiling is carried out at high vacuum (28.5 t inches) to force out sucrose out of the concentrated syrup which is deposited as crystals on the nuclei formed at the grainfng point. As more and more syrup is introduced the crystats already formed grow more and more, until a semi-solid of mass fills up the pan. When this point is reached the flow of syrup is stopped and Concentration continued up to 92° - 93° Brix (9 to 11% water) at a temperature of 65°C and vacuum of 28.7 inches. The semi-solid mass consisting of sugar crystals and molasses-called massecute A is removed through the bottom outlet in crystalliser tanks. During the formation of crystals in the pan, false grains are formed which hamper the growth of crystals. False grains consist of minute crystals formed between the proper crystals. The object of the operator is to allow the grains to grow and to prevent the formation of fresh grains.

Physical Properties

Molecular Formula	= $C_{12}H_{22}O_{11}$
Formula Weight	= 342.29648
Composition	= C(42.11%) H(6.48%) O(51.42%)
Melting Point	= 186 °C
Solubility in water	= 211.5 g/100 ml (20 °C)
Molar Refractivity	= 71.04 cm^3
Molar Volume	= 189.2 cm^3
Parachor	= 583.9 cm^3
Index of Refraction	= 1.674
Surface Tension	= 90.6 dyne/cm
Density	= 1.808 g/cm^3
Polarizability	= 28.16 $10^{-24}cm^3$
Monoisotopic Mass	= 342.116212 Da
Nominal Mass	= 342 Da
Average Mass	= 342.2965 Da
Log P	= -1.82

Chemical Properties

Pure sucrose is most often prepared as a fine, white, odorless crystalline powder with a pleasing, sweet taste; the common table sugar. Large crystals are sometimes precipitated from water solutions of sucrose onto a string (or other nucleation surface) to form rock candy, a confection. Like other carbohydrates, sucrose has a hydrogen to oxygen ratio of 2:1. It consists of two monosaccharides, α-glucose and fructose, joined by a glycosidic bond between carbon atom 1 of the glucose unit and carbon atom 2 of the fructose unit. What is notable about sucrose is that unlike most polysaccharides, the glycosidic bond is formed between the reducing ends of both glucose and fructose, and not between the reducing end of one and the nonreducing end of the other. The effect of this inhibits further bonding to other saccharide units. Since it contains no free anomeric carbon atom, it is classified as a nonreducing sugar. Sucrose melts and decomposes at 186°C to form caramel, and when combusted produces carbon, carbon dioxide, and water. Water breaks down sucrose by hydrolysis, however the process is so gradual that it could sit in solution for years with negligible change. If the enzyme sucrase is added however, the reaction will proceed rapidly. Reacting sucrose with sulfuric acid dehydrates the sucrose and forms the element carbon, as demonstrated in the following equation:

$$C_{12}H_{22}O_{11} + H_2SO_4 \text{ catalyst} \rightarrow 12\ C + 11\ H_2O$$

Sugar as a Macronutrient

In mammals, sucrose is very readily digested in the stomach into its component sugars, by acidic hydrolysis. This step is performed by a glycoside hydrolase, which catalyzes the hydrolysis of sucrose to the monosaccharides glucose and fructose. Glucose and fructose are rapidly absorbed into the bloodstream in the small intestine. Undigested sucrose passing into the intestine is also broken down by sucrase or isomaltase glycoside hydrolases, which are located in the membrane of the microvilli lining the duodenum. These products are also transferred rapidly into the bloodstream.

Sucrose is digested by the enzyme invertase in bacteria and some animals. Acidic hydrolysis can be used in laboratories to achieve the hydrolysis of sucrose into glucose and fructose.

Use in Human Nutrition

Sucrose is an easily assimilated macronutrient that provides a quick source of energy to the body, provoking a rapid rise in blood glucose upon ingestion. However, pure sucrose is not normally part of a human diet balanced for good nutrition, although it may be included sparingly to make certain foods more palatable. Overconsumption of sucrose has been linked with some adverse health effects. The most common is dental caries or tooth decay, in which oral bacteria convert sugars (including sucrose) from food into acids that attack tooth enamel. Sucrose, as a pure carbohydrate, has an energy content of 3.94 kilocalories per gram (or 17 kilojoules per gram). When a large amount of foods that contain a high percentage of sucrose is consumed, beneficial nutrients can be displaced from the diet, which can contribute to an increased risk for chronic disease. It has been suggested that sucrose-containing drinks may be linked to the development of obesity and insulin resistance. However, most soft drinks in the USA are now made with high-fructose corn syrup, not sucrose. The rapidity with which sucrose raises blood glucose can cause problems for people suffering from defects in glucose metabolism, such as persons with hypoglycemia or diabetes mellitus. Sucrose can contribute to development of the metabolic syndrome. In an experiment with rats that were fed a diet one-third of which was sucrose, the sucrose first elevated blood levels of triglycerides, which induced visceral fat and ultimately resulted in insulin resistance. Another study found that rats fed sucrose-rich diets developed high triglycerides, hyperglycemia, and insulin resistance. Extension of average lifespan can be achieved by good diet, exercise and avoidance of hazards such as smoking and excessive eating of sugar-containing foods.

Maltose

Maltose, or malt sugar, is a disaccharide formed from two units of glucose joined with an $\alpha(1\rightarrow4)$ linkage. It is the second member of an important biochemical series of glucose chains. The addition of another glucose unit yields maltotriose; further additions will produce dextrins (also called maltodextrins) and eventually starch.

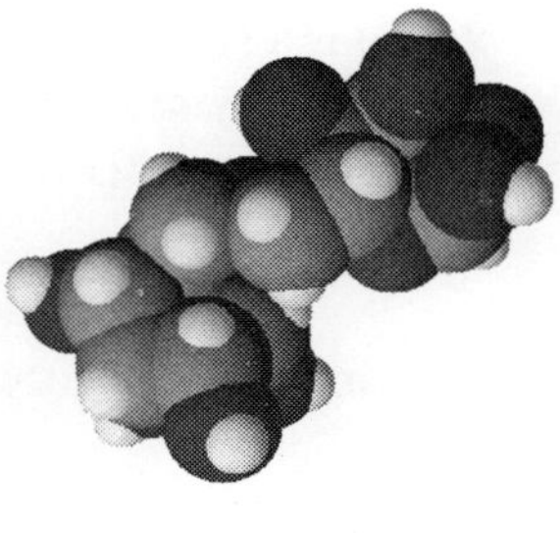

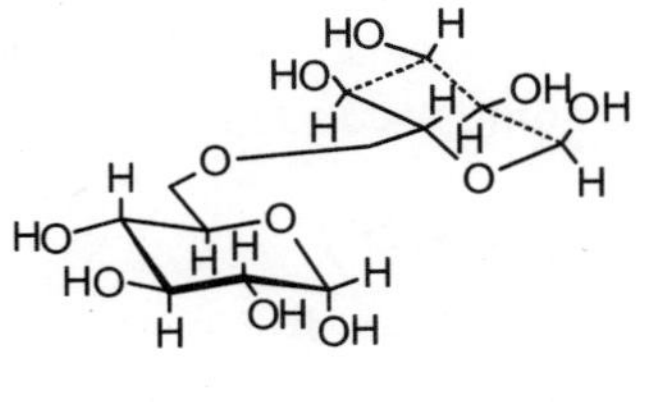

Maltose can be broken down into two glucose molecules by hydrolysis. In living organisms, the enzyme maltase can achieve this very rapidly. In the laboratory, heating with a strong acid for several minutes will produce the same result.

The production of maltose from germinating cereals, such as barley, is an important part of the brewing process. When barley is malted, it is brought into a condition in which the concentration of maltose-producing amylases has been maximized. Mashing is the process by which these amylases convert the cereal's starches into maltose. Metabolism of maltose by yeast during fermentation then leads to the production of ethanol and carbon dioxide

Physical Properties

Molecular Formula	= $C_{12}H_{22}O_{11}$
Formula Weight	= 342.29648
Composition	= C(42.11%) H(6.48%) O(51.42%)
Melting Point	= 102-103 °C
Molar Refractivity	= 71.13 cm^3
Molar Volume	= 190.8 cm^3
Parachor	= 579.5 cm^3
Index of Refraction	= 1.668
Surface Tension	= 85.0 dyne/cm
Density	= 1.793 g/cm^3
Polarizability	= 28.20 $10^{-24}cm^3$
Monoisotopic Mass	= 342.116212 Da
Nominal Mass	= 342 Da
Average Mass	= 342.2965 Da

Lactose

Lactose (also referred to as *milk sugar*) is a sugar which is found most notably in milk. Lactose makes up around 2–8% of milk (by weight). The name comes from the Latin word for milk, plus the-ose ending used to name sugars. Its systematic name is β-D-galactopyranosyl-(1→4)α-D-glucopyranose.

Chemistry

Lactose is a disaccharide that consists of β-D-galactose and β-D-glucose molecules bonded through a β1→4 glycosidic linkage.

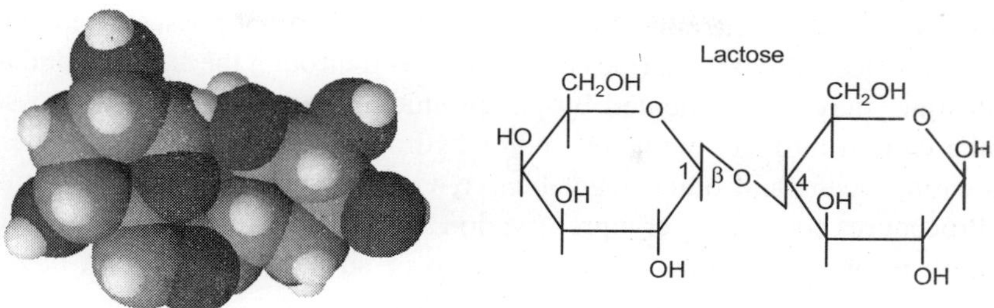

Solubility

Lactose has a solubility of 1 in 4.63%w/v measured . This translates to 0.216 g of lactose dissolving readily in 1 mL of water. The solubility of lactose in water is 18.9049 g at 25°C, 25.1484 g at 40°C and 37.2149 g at 60°C per 100 g solution. Its solubility in ethanol is 0.0111 g at 40°C and 0.0270 g at 60°C per 100 g solution.

Digestion of Lactose

Infant mammals are fed on milk by their mothers. To digest it an enzyme called lactase (β-D-galactosidase) is secreted by the intestinal villi, and this enzyme cleaves the molecule into its two subunits glucose and galactose for absorption.Since lactose occurs mostly in milk, in most mammals the production of lactase gradually decreases with maturity. Many people with ancestry in Europe, the Middle East, India, or parts of East Africa, maintain normal lactase production into adulthood. In many of these cultures, mammals such as cattle, goats, and sheep are milked for food. Hence, it was in these regions that genes for lifelong lactase production first evolved.

TRISACCHARIDES

Raffinose

Raffinose is a trisaccharide composed of galactose, fructose, and glucose. It can be found in beans, cabbage, brussels sprouts, broccoli, asparagus, other vegetables, and whole grains. Raffinose can be hydrolyzed to D-galactose and sucrose by the enzyme α-galactosidase (α-GAL), an enzyme not found in humans. α-GAL also hydrolyzes other α-galactosides such as stachyose, verbascose, and galactinol, if present. The enzyme does not cleave β-linked galactose, as in lactose.The raffinose family of oligosaccharides (RFOs) are alpha-galactosyl derivatives of sucrose, and the most common are the trisaccharide raffinose, the tetrasaccharide stachyose, and the pentasaccharide verbascose. RFOs are almost ubiquitous in the plant kingdom, being found in a large variety of seeds from many different families, and they rank second only to sucrose in abundance as soluble carbohydrates.

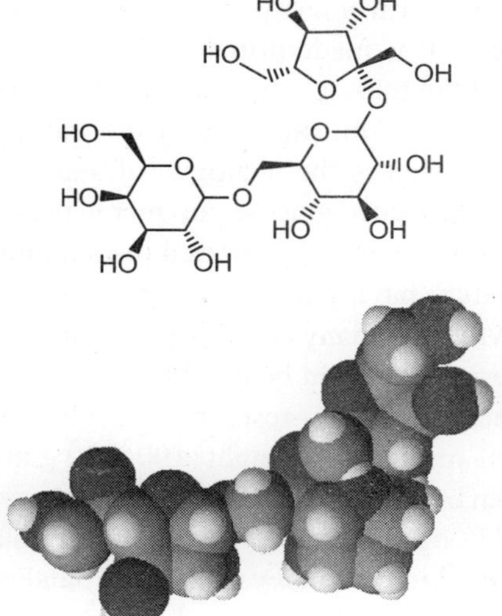

Humans and other monogastric animals (pigs and poultry) do not possess the α-GAL enzyme to break down RFOs and these oligosaccharides pass undigested through the stomach and upper intestine. In the lower intestine, they are fermented by gas-producing bacteria which do posses the α-GAL enzyme and make carbon dioxide, methane, and/or hydrogen—leading to the flatulence commonly associated with eating beans and other vegetables. α-GAL is present in digestive aids such as the product Beano. Procedures concerning cryopreservation have utilized raffinose to provide hypertonicity for cell desiccation prior to freezing. Either raffinose or sucrose is used as a base substance for sucralose.

Melezitose

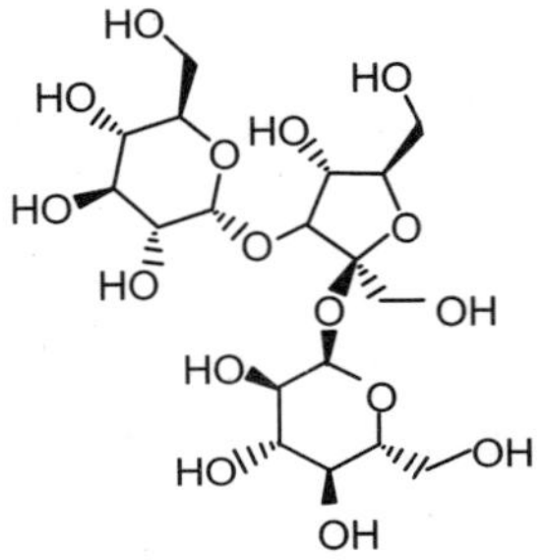

Melezitose, also spelled **melicitose**, is a nonreducing trisaccharide sugar that is produced from lice such as *Cinara pilicornis* by an enzyme reaction. This is beneficial to the insects, as it reduces the stress of osmosis by reducing their own water potential. The melezitose is part of the honeydew of the honey produced by bees, and acts as an attractant for ants. This is useful to the lice as they have a symbiotic relationship with ants. Melezitose can be partially hydrolyzed to glucose and turanose the latter of which is an isomer of sucrose

POLYSACHHARIDE

Digestible

Glycogen

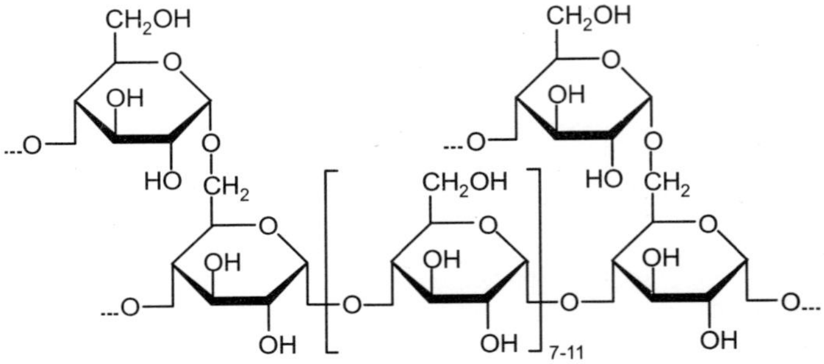

Glycogen is a polysaccharide of glucose (Glc) which functions as the primary short term energy storage in animal cells. It is made primarily by the liver and the muscles, but can also be made by the brain, uterus, and the vagina. Glycogen is the analogue of starch, a less branched glucose polymer in plants, and is commonly referred to as animal starch, having a similar structure to amylopectin. Glycogen is found in the form of granules in the cytosol in many cell types, and plays an important role in the glucose cycle. Glycogen forms an energy reserve that can be quickly mobilized to meet a sudden need for glucose, but one that is less compact than the energy reserves of triglycerides (fat). In the liver hepatocytes, glycogen can compose up to 8% of the fresh weight (100–120 g in an adult) soon after a meal. Only the glycogen stored in the liver can be made accessible to other organs. In the muscles, glycogen is found in a much lower concentration (1% of the muscle mass), but the total amount exceeds that in liver. Small amounts of glycogen are found in the kidneys, and even smaller amounts in certain glial cells in the brain and white blood cells. The uterus also stores glycogen during pregnancy to nourish the embryo.

Synthesis

Glycogen synthesis differs from glycogen breakdown. Unlike breakdown, synthesis is endergonic, meaning that glycogen is not synthesized without the input of energy. Energy for glycogen synthesis comes from UTP, which reacts with glucose-1-phosphate, forming UDP-glucose, in reaction catalysed by UDP-glucose pyrophosphorylase. Glycogen is synthesized from monomers of UDP-glucose by the enzyme Glycogen synthase, which progressively lengthens the glycogen chain. As glycogen synthase can only lengthen an existing chain, the protein glycogenin is needed to initiate the synthesis of glycogen.

Breakdown

Glycogen is cleaved from the nonreducing ends of the chain by the enzyme glycogen phosphorylase to produce monomers of glucose-1-phosphate that is then converted to Glucose 6-phosphate. A special debranching enzyme is needed to remove the alpha(1-6) branches in branched glycogen and reshape the chain into linear polymer. The G6P monomers produced have three possible facts:

- G6P can continue on the glycolysis pathway and be used as fuel.
- G6P can enter the pentose phosphate pathway via the enzyme Glucose-6-phosphate dehydrogenase to produce NADPH and 5-carbon sugars.
- In the liver and kidney, G6P can be dephosphorylated back to Glucose by the enzyme Glucose 6-phosphatase. This is the final step in the gluconeogenesis pathway.

Structure and Biochemistry

Glycogen is a highly branched polymer that is better described as a dendrimer of about 60,000 glucose residues and has a molecular weight between 10^6 and 10^7 daltons (~4.8 million) Most of Glc units are linked by α-1,4 glycosidic bonds, approximately 1 in 12 Glc residues also makes -1,6 glycosidic bond with a second Glc, which results in the creation of a branch. Glycogen does not possess a reducing end: the 'reducing end' glucose residue is not free but is covalently bound to a protein termed glycogenin as a beta-linkage to a surface tyrosine residue. Glycogenin is a glycosyltransferase and occurs as a dimer in the core of glycogen.

The glycogen granules contain both glycogen and the enzymes of glycogen synthesis (glycogenesis) and degradation (glycogenolysis). The enzymes are nested between the outer branches of the glycogen molecules and act on the non-reducing ends. Therefore, the many non-reducing end-branches of glycogen facilitate its rapid synthesis and catabolism.

Function and Regulation of Liver Glycogen

As a meal containing carbohydrates is eaten and digested, blood glucose levels rise, and the pancreas secretes insulin. Glucose from the portal vein enters the liver cells (hepatocytes). Insulin acts on the hepatocytes to stimulate the action of several enzymes, including glycogen synthase. Glucose molecules are added to the chains of glycogen as long as both insulin and glucose remain plentiful. In this postprandial or "fed" state, the liver takes in more glucose from the blood than it releases. After a meal has been digested and glucose levels begin to fall, insulin secretion is reduced, and glycogen synthesis stops. About four hours after a mealglycogen begins to be broken down and converted again to glucose. Glycogen phosphorylase is the primary enzyme of glycogen breakdown. For the next 8–12 hours,

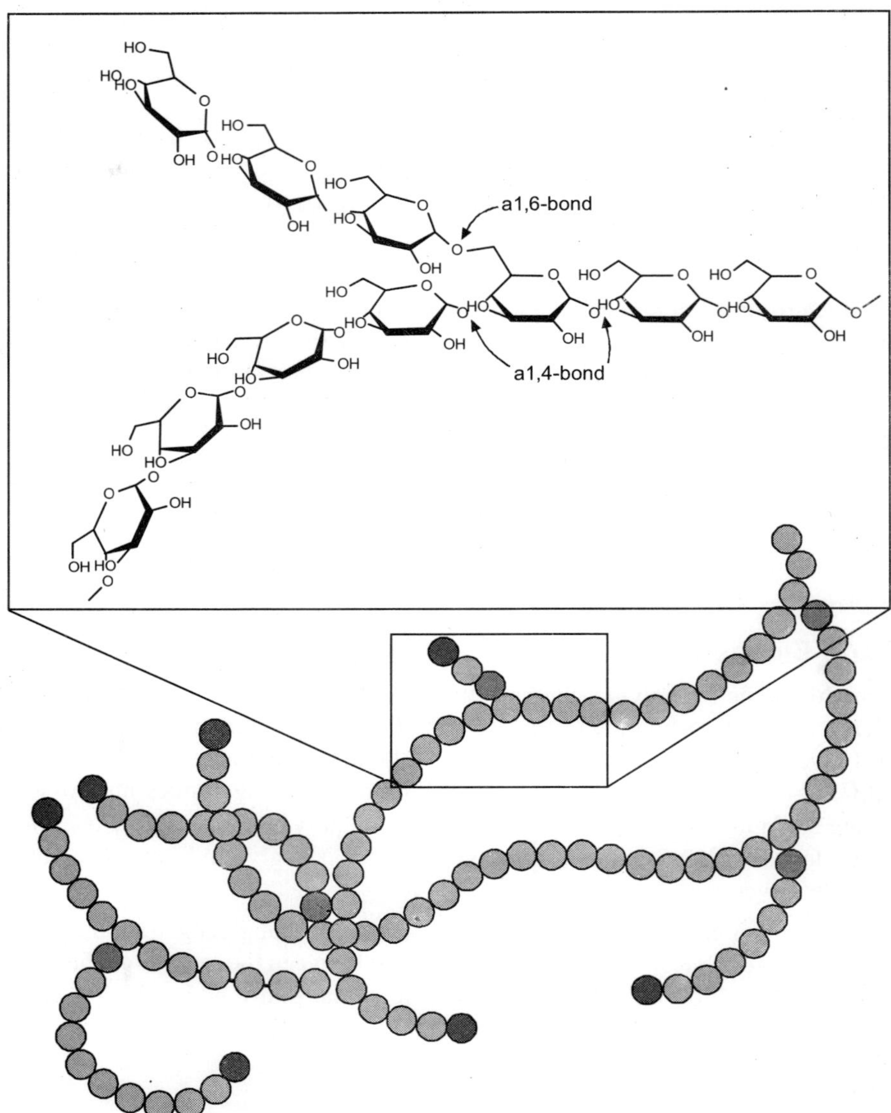

F'g. 2.8 : Sharing the Glycogen Structure

glucose derived from liver glycogen will be the primary source of blood glucose to be used by the rest of the body for fuel. Glucagon is another hormone produced by the pancreas, which in many respects serves as a counter-signal to insulin. When the blood sugar begins to fall below normal, glucagon is secreted in increasing amounts. It stimulates glycogen breakdown into glucose even when insulin levels are abnormally high.

In Muscle and other Cells

Muscle cell glycogen appears to function as an immediate reserve source of available glucose for muscle cells. Other cells that contain small amounts use it locally as well. Muscle cells lack glucose-6-phosphatase enzyme, so they lack the ability to pass glucose into the blood, so the glycogen they store internally is destined for internal use and is not shared with other cells, unlike liver cells.

Glycogen Debt and Endurance Exercise

Due to the body's inability to hold more than around 2,000 kcal of glycogenlong-distance athletes such as marathon runners, cross-country skiers, and cyclists go into glycogen debt, where almost all of the athlete's glycogen stores are depleted after long periods of exertion without enough energy consumption. This phenomenon is referred to as "hitting the wall" or "bonking". In marathon runners it normally happens around the 20 mile (32 km) point of a marathon, where around 100 kcal are spent per mile depending on the size of the runner and the race course. However, it can be delayed by a carbohydrate loading before the task. When experiencing glycogen debt, athletes often experience extreme fatigue to the point that it is difficult to move.

Disorders of Glycogen Metabolism

The most common disease in which glycogen metabolism becomes abnormal is diabetes, in which, because of abnormal amounts of insulin, liver glycogen can be abnormally accumulated or depleted. Restoration of normal glucose metabolism usually normalizes glycogen metabolism as well. In hypoglycemia caused by excessive insulin, liver glycogen levels are high, but the high insulin level prevents the glycogenolysis necessary to maintain normal blood sugar levels. Glucagon is a common treatment for this type of hypoglycemia. Various inborn errors of metabolism are caused by deficiencies of enzymes necessary for glycogen synthesis or breakdown. These are collectively referred to as glycogen storage diseases.

Starch

Introduction

These are both complex carbohydrate polymers of glucose, making starch a glucose polymer as well, as seen by the chemical formula for starch, regardless of the ratio of amylose:amylopectin. The word is derived from Middle English *sterchen*, meaning to stiffen, which is appropriate since it can be used as a thickening agent when dissolved in water and heated.

Synthesis

The starch is located in the crown and glutinous regions of the endosperm. Large, round granules are loosely held in the crown region, associated with a relatively small amount of protein. The separation of these granules in the milling process offers no serious problem. Smaller, polygonally shaped granules are tightly held in a water- insoluble protein matrix in the glutinous region. The hydration and disintegration

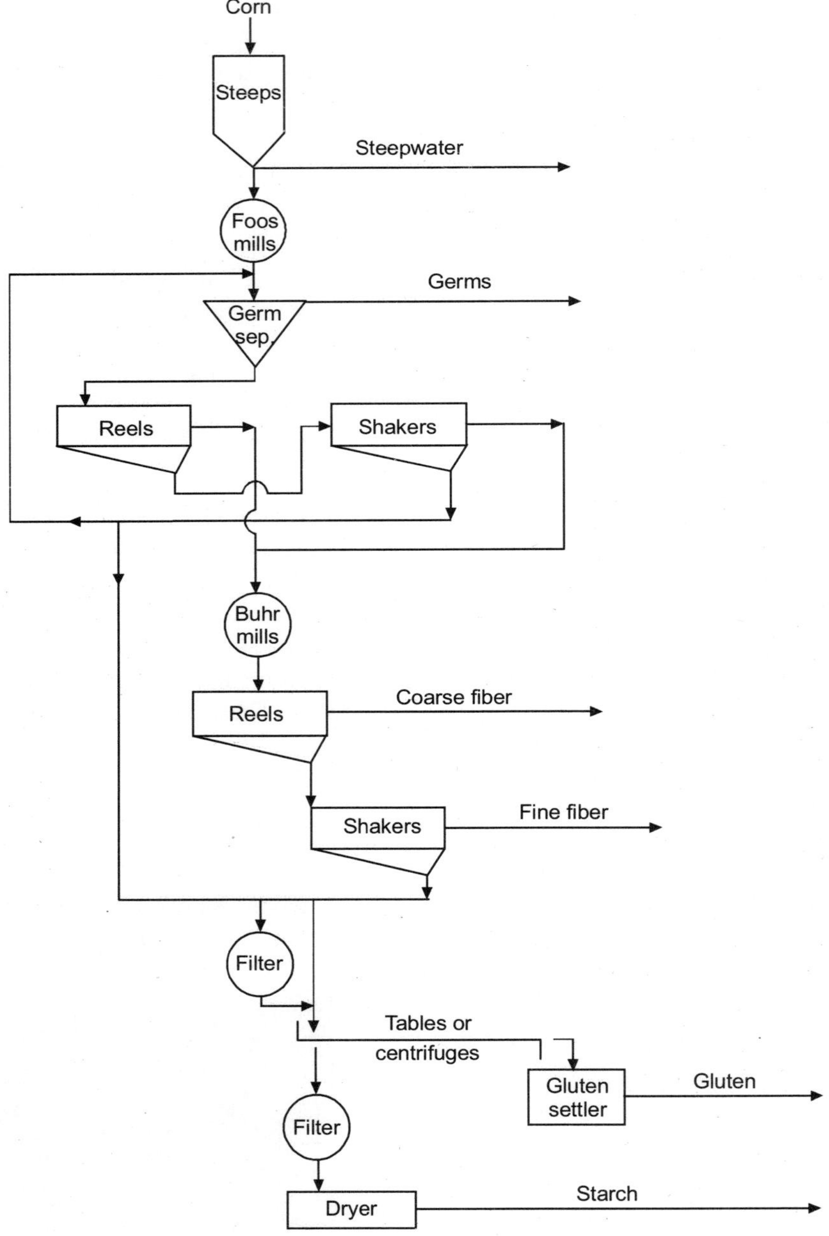

Fig. 2.9 : The Industrial Prepartion of Starch

of this protein in order to loosen the starch, the separation of the starch from the finely dispersed protein, and the accomplishment of these ends without a serious deterioration of the more fragile and less resistant granules in the crown region constitute the chief problems in milling. The steps of the *wet milling process* are indicated diagrammatically in Figure 2.9.

In order to facilitate the separation of the constituents of the grain, the corn is hydrated by soaking *(steeping)* in water at about 50°C for 35-45 hours. Usually, about 0.15 – 0.20% of sulfur dioxide is added to the water (pH about 4.0) in order to control biological activity and to facilitate the disintegration of the protein structures. The moisture content of the corn increases during steeping to about 45%. The steeped corn is passed to coarse grinding mills (such as the Foos mill) called degenninators, which loosen the hulls and separate the tough, rubbery germs with very little disintegration of these constituents. The grist, as a water slurry with a density corresponding to 8-10.5°Be. at 90-95°F, is fed continuously to V-shaped tanks. Here the germs tend to float because of their high oil content and are removed from the surface by overflowing into a weir box. The heavier part of the grist is continuously removed from the bottom of the tank.

The coarse portion of the degerminated grist is finely ground in buhrstone mills. Hulls and fiber are removed by passage over a series of metal screens and finally over a series of screens made of bolting silk. The reslilting slurry contains only finely dispersed protein, starch, and some water-soluble material. The density is adjusted to betwem 6 and 12°Be., the sulphur dioxide concentration to between 0.06 and 0.08%, and the temperature to about 85-90 of. Adjustment of density by concentration is usually required in mills with table feed liquors of the order of 10 to 12° Be. Concentration may be affected by passing a portion of the liquors from the shakers over filters and then adding the filter cake to the balance of the unfiltered slurry.

Starch is *sedimented* from the gluten (water-insoluble proteins, zein and zeanin, together with minor amounts of corn fiber, endosperm lipides, and unseparated starch) either by gravity, by the use of centrifuges, or by a combination of both methods. In separation by gravity, the slurry is allowed to run down inclined troughs called tables. These are about 2 ft. wide and 120 ft. long. For feed rates of 1.5-2.0 gaL/minute of mill liquor, the slope of these tables is about 3>16 in.jft. Variables affecting the separation are: (1) rate of feed to the tables, (2) slope of the table compared to its dimensions, (3) density of the slurry, (4) temperature, and (5) pH and ionic strength of the liquor. Starch sediments on the tables, while the gluten passes off the lower end. Gluten particles which are not sufficiently hydrated and which contain a relatively large number of starch granules will obviously settle with the starch and increase the protein content of the latter. A smaller number of starch granules may not cause the gluten to sediment, but this starch will increase the carbohydrate content of the gluten fraction and result in a reduced yield of starch. Centrifuges of the Merco (bowl) type may be used to separate the starch, or these centrifuges may be used after tabling in order further to purify either the starch or the gluten fractions. Starch is washed from the tables by high-pressure water jets. After filtration, the starch is washed with water and dried. American (drum-type) and Oliver (disk-type) filters are usually used in these operations followed by drying either in kilns, rotary dryers, or preferably endless belt dryers of the type manufactured by Proctor and Schwartz

The wet milling of corn employs the countercurrent principle throughout in the usn of water. Aside from the water which enters the system as moisture in the corn grain, the balance, making a total of

10-12 gal. per bushel of corn processed, is added in the final washing of the starch before filtration. This wash water is then used to flush the starch from the tables and is used in turn in various cycles of the milling process, until as gluten overflow water from the tables it contains about 1-2 grams of dissolved material per 100 ml. After sedimentation of the gluten in tanks called gluten settlers, the water is finally used in the steeping operation. After steeping and before final dewatering, the pH of the starch is usually adjusted to a range between 5 and 7. The water is removed from the corn and evaporated to recover the extracted corn solids. Steep water may contain as much as 8 grams of nonvolatile solids per 100 ml. and has proved to be a practical commercial source of meso-inositol, which is of interest as a member of the vitamin-B complex. The various by-product constituents of the grain such as the germs, the hulls, and fiber are in turn washed free of adhering starch by the process waters in the system and are then further processed

Physical Properties

Molecular Formula	= $C_{48}H_{76}O_{38}$
Formula Weight	= 1261.09424
Composition	= C(45.72%) H(6.07%) O(48.21%)
Molar Refractivity	= 262.69 cm^3
Molar Volume	= 774.8 cm^3
Parachor	= 2243.7 cm^3
Index of Refraction	= 1.593
Surface Tension	= 70.3 dyne/cm
Density	= 1.627 g/cm^3
Polarizability	= 104.14 $10^{-24}cm^3$
Monoisotopic Mass	= 1260.401458 Da
Nominal Mass	= 1260 Da
Average Mass	= 1261.0942 Da

Chemical Properties

Most varieties of starch contain two types of polymers which differ from each other in molecular weight and in chemical structure. The linear polymer, amylose, consists of 200-1000 glucopyranose units joined to each other through a-1,4-glucosidic linkages, whereas the branched or ramified polymer, amylopectin, is made up of 1500 or more glucopyranose units (6,10,15,20) (see Fig. 2.10). In addition to the normal or predominating a-1,4 linkages, an anomalous a-1,6 linkage is present in the ramified structure at the origin or point of branching in a ratio of about 1:25.

The linear polymer and the longer branches of the nonlinear polymer show a pronounced tendency to orient and associate with other linear members. This property is characteristic of many thread-like molecules which contain a large number of hydroxyl or other hydrogen-bond-forming groups distributed along their chain length. Starch molecules are highly associated in the granule.

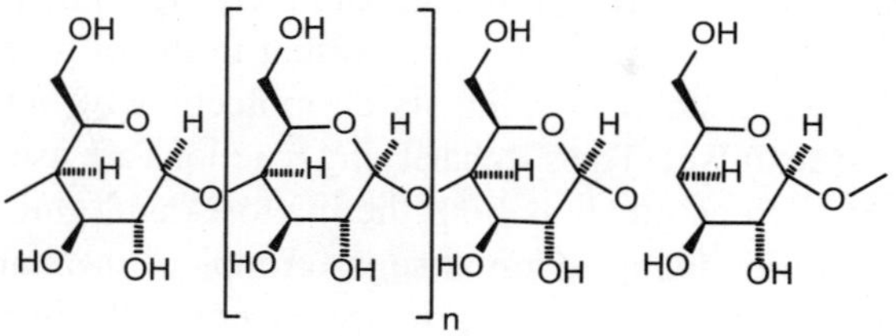

Amylose Molecule

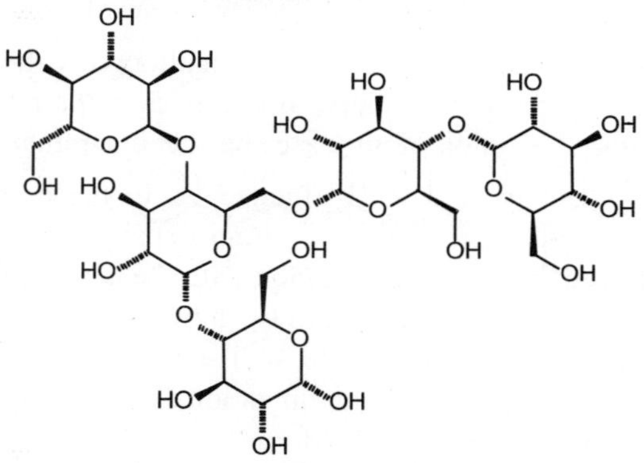

Amylopectine Molecule

When starch is dispersed in water, the sol may be fractionated into its polymeric components by the addition of polar substances, particularly the higher alcohols. The amylose forms an insoluble complex with butyl alcohol which may be sedimented by centrifugation. The amount of amylose obtained varies with the origin of the starch. Starches from so-called waxy grains, such as waxy corn and sorghum, contain no amylose; in contrast, the starch from certain varieties of peas contains 75% by weight of amylose. The common industrial starches contain about 15-30% by weight of amylose.

The chemical property of starch which is at present of greatest industrial importance is the comparative ease with which the plymeric bond is hydrolyzed by both enzymes and acids. When starch is gelatinized by heating in water or when it is dissolved by suitable solvating agent, for example, sodium hydroxide, followed by neutralization, its glucoside linkages (the polymeric bonds) are readily hydrolyzed by amylases. These enzymes are obtained from a large number of sources such as bacteria, fungi, vegetables, and animals. The amylases present in these preparations are grouped into three broad classifications or types since the course of the hydrolysis varies in a characteristic manner with the type of amylase employed.

α-Amylase splits the high-polymer starch molecules into molecules of progressively smaller size. Eventually, sugar molecules are produced. Characteristic of this reaction is the rapid decrease in viscosity of the starch medium with very little production of reducing sugars. Accordingly, the reaction is used industrially to produce low-viscosity sizes and adhesives from native starch. {3-*Amylase* splits the disaccharide *maltose* directly from starch molecules. Amylose molecules are almost quantitatively hydrolyzed to maltose in this manner. Amylopectin molecules are hydrolyzed 50-60%. The residue which remains ("beta-amylase limit. dextrin") is of comparatively high molecular weight and still contains all of the anomalous glucoside linkages which were present in the original starch molecules. A third type of amylase, called *amyloglucosidase*, splits starch molecules of both structures directly to the monosaccharide, D-glucose (dextrose). The saccharifying reactions are used by fermentation industries employing starch as substrate. In as much as both the maltose- and glucose-producing enzymes are believed to act only on the non-reducing terminal sugar groups of the starch polymer, it follows that one of the principal functions of a-amylase, as it occurs normally in nature in conjunction with saccharifying enzymes, is to facilitate the action of the latter by creating more terminal groups per unit weight. Thus, the complete hydrolysis of starch to sugars is accelerated.

Although amylases of anyone class act on starch in qualitatively the same manner, members of a class, depending on source, may show quantitative differences in their mode of action. The relatively complex chemistry involved in the action of amylases has been summarized by Kerr and Gehman.

All starches are hydrolyzed by acid to D-glucose, as in the commercial production at high temperatures of dextrose *(q.v.)* and starch sirups (mixtures consisting of dextrose, maltose, and higher polysaccharides generally called dextrins). In common with the action of enzymes, acid at first causes a reduction in paste viscosity. In contrast to enzymes, however, acid readily penetrates at least the amorphous regions of the native un swollen granule, so that hydrolysis may take place even at temperatures materially lower than those required for gelatinization. This fact is utilized in the manufacture of modified, "thin-boiling" starches. When these acid-modified starches are gelatinized, the consistency of their pastes at a given concentration is less than that of the native starch. The intermediate product of acid degradation is known as **amylodextrin.**

The alcohol groups in starch may be oxidized to form carbonyl or carboxyl groups. The course of oxidation varies with the oxidant and conditions employed. Specific oxidants such as nitrogen dioxide act principally to convert primary alcoholic groups to carboxyls, whereas others such as periodate act on the two adjacent secondary hydroxyl groups, converting them first into ketone groups and subsequently into aldehydes with a rupture of the bond between carbons 2 and 3. Supplementary oxidants such as bromine water may then convert them further to diacids. Less specific agents used industrially, such as hypochlorite, very likely create many carboxyl groups along the starch molecule, thus increasing its solubility in water particularly in neutralized or dilute alkaline solution. The alcohol groups in starch react with aldehydes, under acidic conditions, by a condensation reaction, in which cross-linkages are very likely established between two adjacent starch molecules through the carbonyl carbon of the aldehyde. The end product is completely nondispersible in water. Starch also reacts to form a product which is nondispersible in water with agents such as epichlorohydrin, dibasic acids or their acid chlorides, potassium pyroantimonate, and allyl bromide. Where the reaction involves the uniting of two starch molecules, it may be viewed as a polymerization. Starch, or more particularly the linear polymer of starch, reacts characteristically with iodine to form a blue-colored complex. The iodine test is used in food analysis as a test for unconverted starch.

The aldehyde group in the starch molecule is reducible. Polyhydric alcohols such as sorbitol are formed by hydrogenation (involving also hydrolysis) and hydrogenolysis

Tests

Iodine solution is used to test for Starch. A bluish-black color indicates the presence of iodine in the starch solution. It is thought that the iodine fits inside the coils of amylose. A 0.3% w/w solution is the standard concentration for a dilute starch indicator solution. It is made by adding 4 grams of soluble starch to 1 litre of heated water; the solution is cooled before use (starch-iodine complex becomes unstable at temperatures above 35 °C). This complex is often used in redox titrations: in presence of an oxidizing agent the solution turns blue, in the presence of reducing agent, the blue color disappears because triiodide (I_3^-) ions break up into three iodide ions, disassembling the complex.

Under the microscope, starch grains show a distinctive Maltese cross effect (also known as 'extinction cross' and birefringence) under polarized light.

USES

Starch in Food

In terms of human nutrition, starch is by far the most consumed polysaccharide in the human diet. It constitutes more than half of the carbohydrates even in many affluent diets, and much more in poorer diets. Traditional staple foods such as cereals, roots and tubers are the main source of dietary starch.

Starch (in particular cornstarch) is used in cooking for thickening foods such as sauces. In industry, it is used in the manufacturing of adhesives, paper, textiles and as a mold in the manufacture of sweets such as wine gums and jelly beans. It is a white powder, and depending on the source, may be tasteless and odorless.

Fruit, seeds, rhizomes or tubers of plants and is the major source of energy in these food items. The major resources for starch production and consumption worldwide are rice, wheat, corn, and potatoes. Cooked foods containing starches include boiled rice, various forms of bread and noodles (including pasta).

As an additive for food processing, arrowroot and tapioca are commonly used as well. Commonly used starches around the world are: arracacha, buckwheat, banana, barley, cassava, kudzu, oca, sago, sorghum, regular household potatoes, sweet potato, taro and yams. Edible beans, such as favas, lentils and peas, are also rich in starch.

When a starch is pre-cooked, it can then be used to thicken cold foods. This is referred to as a pregelatinized starch. Otherwise starch requires heat to thicken, or "gelatinize." The actual temperature depends on the type of starch.

A modified food starch undergoes one or more chemical modifications, which allow it to function properly under high heat and/or shear frequently encountered during food processing. Food starches are typically used as thickeners and stabilizers in foods such as puddings, custards, soups, sauces, gravies, pie fillings, and salad dressings, but have many other uses.

Resistant starch is starch that escapes digestion in the small intestine of healthy individuals.

Plants use starch as a way to store excess glucose, and thus also use starch as food during mitochondrial oxidative phosphorylation.

Commercial Applications

Papermaking is the largest non-food application for starches globally, consuming millions of metric tons annually. In a typical sheet of copy paper for instance, the starch content may be as high as 8%. Both chemically modified and unmodified starches are used in papermaking. In the wet part of the papermaking process, generally called the "wet-end", starches are chemically modified to contain a cationic or positive charge bound to the starch polymer, and are utilized to associate with the anionic or negatively charged paper fibers and inorganic fillers. Starch also helps get out cleaning stains from dirty washing.

These cationic starches impart the necessary strength properties for the paper web to be formed in the papermaking process (wet strength), and to provide strength to the final paper sheet (dry strength). In the dry end of the papermaking process the paper web is rewetted with a solution of starch paste that has been chemically, or enzymatically depolymerized. The starch paste solutions are applied to the paper web by means of various mechanical presses (size press). The dry end starches impart additional strength to the paper web and additionally provide water hold out or "size" for superior printing properties.

Corrugating glues are the next largest consumer of non-food starches globally. These glues are used in the production of corrugated fiberboard (sometimes called corrugated cardboard), and generally contain a mixture of chemically modified and unmodified starches that have been partially gelatinized to form an opaque paste. This paste is applied to the flute tips of the interior fluted paper to glue the fluted paper to the outside paper in the construction of cardboard boxes. This is then dried under high heat, which provides the box board strength and rigidity.

Another large non-food starch application is in the construction industry where starch is used in the or wall board manufacturing process. Chemically modified or unmodified starches are added to the rock mud containing primarily gypsum. Top and bottom heavyweight sheets of paper are applied to the mud formulation and the process is allowed to heat and cure to form the eventual rigid wall board. The starches act as a glue for the cured gypsum rock with the paper covering and also provide rigidity to the board.

Clothing starch or **laundry starch:** is a liquid that is prepared by mixing a vegetable starch in water (earlier preparations also had to be boiled), and is used in the laundering of clothes. Starch was widely used in Europe in the 16th and 17th centuries to stiffen the wide collars and ruffs of fine linen which surrounded the necks of the well-to-do. During the 19th century and early 20th century, it was stylish to stiffen the collars and sleeves of men's shirts and the ruffles of girls' petticoats by applying starch to them as the clean clothes were being ironed.

Aside from the smooth, crisp edges it gave to clothing, it served practical purposes as well. Dirt and sweat from a person's neck and wrists would stick to the starch rather than fibers of the clothing, and would easily wash away along with the starch. After each laundering, the starch would be reapplied.

Starch is also used to make some packing peanuts, and some dropped ceiling tiles.

Printing industry - in the printing industry food grade starch is used in the manufacture of anti-set-off spray powder used to separate printed sheets of paper to avoid wet ink being set off. Starch is also used in the manufacture of glues for book-binding.

Hydrogen production - Starch can be used to produce Hydrogen.

Oil exploration - starch is used as to adjust the viscosity of drilling fluid which is used to lubricate the drill head in (mineral) oil extraction.

Use as a mold: Gummed sweets such as jelly beans and wine gums are not manufactured using a mold in the conventional sense. A tray is filled with starch and leveled. A positive mold is then pressed into the starch leaving an impression of 1000 or so jelly beans. The mix is then poured into the impressions and then put into a stove to set. This method greatly reduces the number of molds that must be manufactured.

Starch can be modified by addition of some chemical forms to be a hard glue for paper work, some of those forms are Borax, Soda Ash, which mixed with the starch solution at 50-70C to gain a very good adhesive, Sodium Silicate can be added to reinforce this formula.

Dextrin

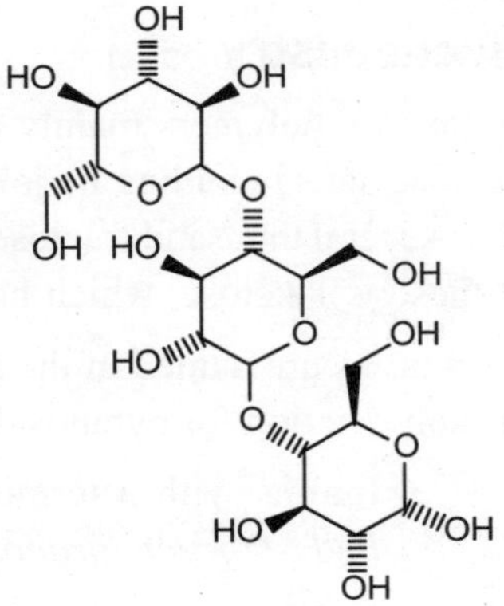

Dextrins are a group of low-molecular-weight carbohydrates produced by the hydrolysis of starch. Dextrins are mixtures of linear α-(1,4)-linked D-glucose polymers starting with an α-(1,6) bond. Because branched amylopectin and glycogen also contain α-(1,6) bonds, which α-amylase cannot hydrolyze in humans, the digest resulting from this action contains a mixture of dextrins. They have the same general formula as carbohydrates but are of shorter chain length. Industrial production is, in general, performed by acidic hydrolysis of potato starch. Dextrins are water-soluble, white to slightly yellow solids that are optically active. Under analysis, dextrins can be detected with iodine solution, giving a red coloration.

The cyclical dextrins are known as cyclodextrins. They are formed by enzymatic degradation of starch by certain bacteria, for example, Bacillus macerans. Cyclodextrins have toroidal structures formed by 6-8 glucose residues.

Dextrins find widespread use in industry, due to their non-toxicity and their low price. They are used as water-soluble glues, as thickening agents in food processing, and as binding agent in pharmaceuticals. In pyrotechnics, they are added to fire formulas, allowing them to solidify as pellets or "stars." Cyclodextrins find additional use in analytical chemistry as a matrix for the separation of hydrophobic substances, and as excipients in pharmaceutical formulations. Not all forms of dextrin are digestible, and indigestible dextrin is sometimes used in fiber supplements.

For example, **maltodextrin** either can be moderately sweet or have hardly any flavor at all. Maltodextrin is a polysaccharide that is used as a food additive. It is produced from starch and is usually found as a creamy-white hygroscopic powder. Maltodextrin is easily digestible, being absorbed as rapidly as glucose.

Maltodextrin can be derived from any starch. In the US, this starch is usually rice, corn or potato; elsewhere, such as in Europe, it is commonly wheat. This is important for coeliacs, since the wheat-derived maltodextrin can contain traces of gluten. There have been recent reports of coeliac reaction to maltodextrin in the United States. This might be a consequence of the shift of corn to ethanol production and its replacement with wheat in the formulation.

Foods containing maltodextrin may contain traces of amino acids, including glutamic acid as a manufacturing by product. The amino acids traces would be too small to have any dietary significance.

PARTLY DIGESTIBLE

Inulin

Inulins are a group of naturally occurring polysaccharides (several simple sugars linked together) produced by many types of plants. They belong to a class of fibers known as fructans. Inulin is used by some plants as a means of storing energy and is typically found in roots or rhizomes. Most plants which synthesize and store inulin do not store other materials such as starch.

Biochemistry

Inulins are polymers mainly comprised of fructose units and typically have a terminal glucose. The fructose units in inulins are joined by a β-(2-1) glycosidic bond. Plant inulins generally contain between 20 to several thousand fructose units. Smaller compounds are called fructooligosaccharides, the simplest of these is 1-ketose, which has 2 fructose units and 1 glucose unit.

Inulins are named in the following manner, where **n** is the number of fructose residues and **py** is the abbreviation for pyranosyl:

- Inulins with a terminal glucose are known as *α-D-glucopyranosyl-[β-D-fructofuranosyl](n-1)-D-fructofuranosides*, abbreviated as **GpyFn**.
- Inulins without glucose are *β-D-fructopyranosyl-[D-fructofuranosyl](n-1)-D-fructofuranosides*, abbreviated as **FpyFn**.

Hydrolysis of inulins may yield fructooligosaccharides, which are oligomers with a degree of polymerization (DP) of $<=$ 10.

USES

Processed Foods

Inulin is used increasingly in foods because it has unusual nutritional characteristics. It ranges from completely bland to subtly sweet and can be used to replace sugar, fat, and flour. This is particularly advantageous because inulin contains a third to a quarter of the food energy of sugar or other carbohydrates and a sixth to a ninth of the food energy of fat. It also increases calcium absorption and possibly magnesium absorption, while promoting intestinal bacteria. Nutritionally, it is considered a form of soluble fiber, and it is important to note that consuming large quantities (particularly for sensitive and/or unaccustomed individuals) can lead to gas and bloating. Inulin has a minimal impact on blood sugar, and—in sharp contrast to fructose—is not insulemic and does not raise triglycerides, making it generally considered suitable for diabetics and potentially helpful in managing blood sugar-related illnesses.

Industrial use

Nonhydrolyzed inulin can also be directly converted to ethanol in a simultaneous saccharification and fermentation process which may have great potential for converting crops high in inulin into ethanol for fuel.

Medical

Inulin is used to help measure kidney function by determining the Glomerular filtration rate (**GFR**). GFR is the volume of fluid filtered from the renal (kidney) glomerular capillaries into the Bowman's capsule per unit time.

Calculation of Glomerular Filtration Rate (GFR)

Inulin is uniquely treated by nephrons in that it is completely filtered at the glomerulus but neither secreted nor reabsorbed by the tubules. This property of inulin allows the clearance of inulin to be used clinically as a highly accurate measure of Glomerular filtration rate (GFR)—the percentage of plasma from the afferent arteriole that is filtered into Bowman's capsule.

It is useful to contrast the properties of inulin with those of para-aminohippuric acid (PAH). PAH is completely filtered from plasma at the glomerulus and not reabsorbed by the tubules, in a manner identical to inulin. PAH is different from inulin in that the fraction of PAH that bypasses the glomerulus and enters the nephron's tubular cells (via the Peritubular capillaries) is completely secreted. Renal Clearance of PAH is thus useful in calculation of renal plasma flow (RPF), which empirically is (1-Hematocrit) times renal blood flow. Of note, the clearance of PAH is reflective only of RPF to portions of the kidney that deal with urine formation, and thus underestimates actually RPF by about 10%.

The measurement of GFR by inulin is still considered the gold-standard. Practically, however, it has now been largely replaced by other, simpler measures that are approximations of GFR. These measures, which involve clearance of such substrates as EDTA and creatinine, have had their utility confirmed in large cohorts of patients with chronic kidney disease.

Health effects

Inulin is indigestible by the human enzymes ptyalin and amylase, which are adapted to digest starch. As a result, inulin passes through much of the digestive system intact. It is only in the colon that bacteria metabolise inulin, with the release of significant quantities of carbon dioxide, hydrogen and/or methane. Inulin-containing foods can be rather gassy, particularly for those unaccustomed to inulin, and these foods should be consumed in moderation at first.

There are two types of dietary fiber, soluble and insoluble. Insoluble fiber increases the movement of materials through the digestive system and increases stool bulk; it is especially helpful for those suffering from constipation or stool irregularity. Soluble fiber dissolves in water to form a gelatinous material. Some soluble fibres may help lower blood cholesterol and glucose levels. Inulin is considered a soluble fiber. Because normal digestion does not break inulin down into monosaccharides, it does not elevate blood sugar levels and may therefore be helpful in the management of diabetes. Inulin also stimulates the growth of bacteria in the gut. Inulin passes through the stomach and duodenum undigested and is highly available to the gut bacterial flora. This contrasts with proprietary probiotic formulations based on Lactic acid bacteria (LAB) in which the bacteria have to survive very challenging conditions through the gastrointestinal tract before they are able to colonize the gut.

Some traditional diets contain up to 20g per day of inulin or fructo-oligosaccharides. Many foods naturally high in inulin or fructo-oligosaccharides, such as chicory, garlic, and leek, have been seen as "stimulants of good health" for centuries. Inulin is also used in medical tests to measure the total amount of extracellular volume and determine the function of the kidneys.

Inulin is generally recognized as safe (GRAS) by the FDA. Contrary to the health benefits as mentioned above, allergic reactions to inulin in foods have been reported in a letter to the New England Journal of Medicine. Between about 30-40% of the population suffers from fructose malabsorption. Since inulin is a fructan, it is problematic for people with fructose malabsorption. It is recommended that fructan intake for people with fructose malabsorption be kept to less than 0.5 grams/serving.

Natural Sources of Inulin

Plants that contain high concentrations of inulin include:

- Elecampane (*Inula helenium*)
- Dandelion (*Taraxacum officinale*)
- Wild Yam (*Dioscorea* spp.)
- Jerusalem artichokes (*Helianthus tuberosus*)
- Chicory (*Cichorium intybus*)
- Jicama (*Pachyrhizus erosus*)
- Burdock (*Arctium lappa*)
- Onion (*Allium cepa*)
- Garlic (*Allium sativum*)
- Agave (*Agave* spp.)
- Yacon (*Smallanthus sonchifolius* spp.)

INDIGESTIBLE

Cellulose

History

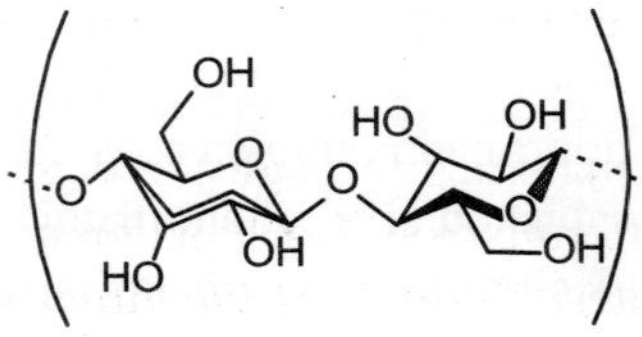

Cellulose was discovered in 1838 by the French chemist Anselme Payen, who isolated it from plant matter and determined its chemical formula. Cellulose was used to produce the first successful thermoplastic polymer, celluloid, by Hyatt Manufacturing Company in 1870. Hermann Staudinger determined the polymer structure of cellulose in 1920. The compound was first chemically synthesized (without the use of any biologically-derived enzymes) in 1992, by Kobayashi and Shoda.

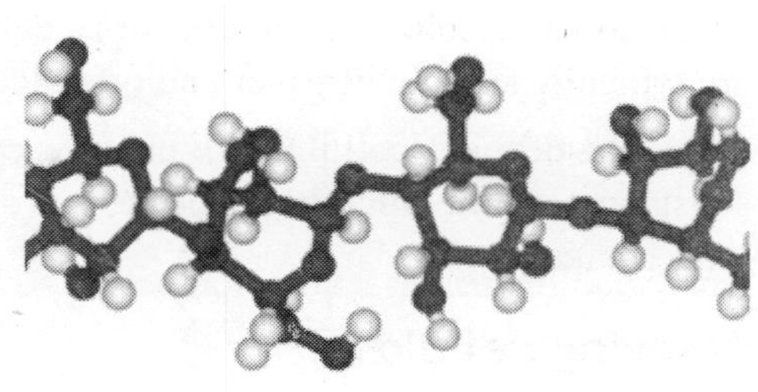

Cellulose is an organic compound with the formula $(C_6H_{10}O_5)_n$, a polysaccharide consisting of a linear chain of several hundred to over ten thousand $\beta(1\rightarrow4)$ linked D-glucose units.

Cellulose is the structural component of the primary cell wall of green plants, many forms of algae and the oomycetes. Some species of bacteria secrete it to form biofilms. Cellulose is the most common organic compound on Earth. About 33 per cent of all plant matter is cellulose (the cellulose content of cotton is 90 per cent and that of wood is 50%).

For industrial use, cellulose is mainly obtained from wood pulp and cotton. It is mainly used to produce cardboard and paper; to a smaller extent it is converted into a wide variety of derivative products such as cellophane and rayon.

Some animals, particularly ruminants and termites, can digest cellulose with the help of symbiotic micro-organisms that live in their guts. Cellulose is not digestible by humans and is often referred to as 'dietary fiber' or 'roughage', acting as a hydrophilic bulking agent for feces.

Structure and Properties

Cellulose is taste- and odorless, hydrophilic, not soluble in water or most organic solvents, chiral and biodegradable. Cellulose is derived from D-glucose units, which condense through $\beta(1\rightarrow4)$-glycosidic bonds. This linkage motif contrasts with that for $\alpha(1\rightarrow4)$-glycosidic bonds present in starch, glycogen, and other carbohydrates. Cellulose is a straight chain polymer: unlike starch, no coiling occurs, and the molecule adopts an extended and rather stiff rod-like conformation. The multiple hydroxyl groups on the glucose residues from one chain form hydrogen bonds with oxygen molecules on another chain, holding the chains firmly together side-by-side and forming *microfibrils* with high tensile strength. This strength is important in cell walls, where they are meshed into a carbohydrate *matrix*, conferring rigidity to plant cells.

Compared to starch, cellulose is also much more crystalline. Whereas starch undergoes a crystalline to amorphous transition when heated beyond 60-70°C in water (as in cooking), cellulose requires a temperature of 320°C and pressure of 25 MPa to become amorphous in water. Chemically, cellulose can be broken down into its glucose units by treating it with concentrated acids at high temperature. Many properties of cellulose depend on its degree of polymerization or chain length, the number of glucose units that make up one polymer molecule. Cellulose from wood pulp has typical chain lengths between 300 and 1700 units; cotton and other plant fibers as well as bacterial celluloses have chain lengths ranging from 800 to 10,000 units. Molecules with very small chain length resulting from the break down of cellulose are known as cellodextrins; in contrast to long-chain cellulose, cellodextrins are typically soluble in water and organic solvents.

Plant-derived cellulose is usually contaminated with hemicellulose, lignin, pectin and other substances, while microbial cellulose is quite pure, has a much higher water content, and consists of long chains.

Assaying cellulose

Given a cellulose-containing material, the portion that does not dissolve in a 17.5% solution of sodium hydroxide at 20°C is *α cellulose*, which is true cellulose. Acidification of the extract precipitates *β cellulose*. The portion that dissolves in base but does not precipitate with acid is *a cellulose.*

Cellulose can be assayed using a method described by Updegraff in 1969, where the fiber is dissolved in acetic and nitric acid, and allowed to react with anthrone in sulfuric acid. The resulting coloured compound is assayed spectrophotometrically at a wavelength of approximately 635 nm.

In addition, cellulose is represented by the difference between acid detergent fiber (ADF) and acid detergent lignin(ADL).

Biosynthesis

In vascular plants cellulose is synthesized at the plasma membrane by rosette terminal complexes (RTC's). The RTC's are hexameric protein structures, approximately 25 nm in diameter, that contain the cellulose synthase enzymes that synthesise the individual cellulose chains. Each RTC floats in the cell's plasma membrane and "spins" a microfibril into the cell wall.

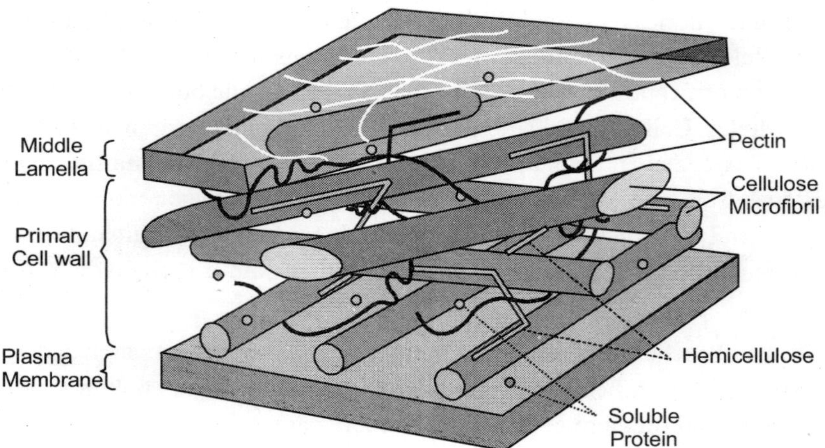

Fig. 2.10 : Location and arrangement of cellulose microfibrils in the plant cell wall.

The RTC's contain at least three different cellulose syntheses, encoded by *CesA* genes, in an unknown stoichiometry. Separate sets of *CesA* genes are involved in primary and secondary cell wall biosynthesis. Cellulose synthase utilizes UDP-D-glucose precursors to generate microcrystalline cellulose.

Cellulose synthesis requires chain initiation and elongation, and the two processes are separate. *CesA* glucosyltransferase initiates cellulose polymerization using a steroid primer, 'sitosterol-beta-glucoside' and UDP-glucose. A cellulase may function to cleave the primer from the mature chain.

Breakdown (cellulolysis)

Cellulolysis is the process of breaking down cellulose into smaller polysaccharides called cellodextrins or completely into glucose units; this is a hydrolysis reaction. Because cellulose molecules bind strongly to each other, cellulolysis is relatively difficult compared to the break down of other polysaccharides.

Mammals do not have the ability to break down cellulose directly. Some ruminants like cows and sheep contain certain symbiotic anaerobic bacteria (like *Cellulomonas*) in the flora of the gut wall, and these bacteria produce enzymes to break down cellulose; the break down products are then used by the mammal. Similarly, lower termites contain in their hindguts certain flagellate protozoa which produce such enzymes; higher termites contain bacteria for the job. Fungi, which in nature are responsible for recycling of nutrients, are also able to break down cellulose.

The enzymes utilized to cleave the glycosidic linkage in cellulose are glycoside hydrolases including endo-acting cellulases and exo-acting glucosidases. Such enzymes are usually secreted as part of multienzyme complexes that may include dockerins and cellulose binding modules; these complexes are in some cases referred to as cellulosomes.

Hemicellulose

Hemicellulose is a polysaccharide related to cellulose that comprises ca. 20% of the biomass of most plants. In contrast to cellulose, hemicellulose is derived from several sugars in addition to glucose, including especially xylose but also mannose, galactose, rhamnose, and arabinose. Hemicellulose consists of shorter chains - around 200 sugar units as opposed to 7,000 - 15,000 glucose molecules in the average cellulose polymer. Furthermore, hemicellulose is branched, whereas cellulose is unbranched.

Commercial products

Cellulose is the major constituent of paper and cardboard and of textiles made from cotton, linen, and other plant fibers.

Cellulose can be converted into cellophane, a thin transparent film, and into rayon, an important fiber that has been used for textiles since the beginning of the 20th century. Both cellophane and rayon are known as "regenerated cellulose fibers"; they are identical to cellulose in chemical structure and are usually made from viscose, a viscous solution made from cellulose. A more recent and environmentally friendly method to produce rayon is the Lyocell process. Cellulose is used in the laboratory as the stationary phase for thin layer chromatography. It is the raw material in the manufacture of nitrocellulose (cellulose nitrate) which was historically used in smokeless gunpowder and as the base material for celluloid used for photographic and movie films until the mid 1930s. Cellulose is used to make hydrophilic and highly absorbent sponges and water-soluble adhesives and binders such as methyl cellulose and carboxymethyl cellulose which are used in wallpaper paste. Microcrystalline cellulose (E460i) and powdered cellulose (E460ii) are used as an inactive filler in tablets and as thickeners and stabilizers in processed foods.

Cellulose Source and Energy Crops

The major combustible component of non-food energy crops is cellulose, with lignin second. Non-food energy crops are more efficient than edible energy crops (which have a large starch component), but still compete with food crops for agricultural land and water resources. Typical non-food energy crops include industrial hemp, switchgrass, *Miscanthus*, *Salix* (willow), and *Populus* (poplar) species.

Some bacteria can convert cellulose into ethanol which can then be used as a fuel; see cellulosic ethanol.

Nitrocellulose (also: cellulose nitrate, flash paper) is a highly flammable compound formed by nitrating cellulose through exposure to nitric acid or another powerful nitrating agent. When used as a propellant or low-order explosive, it is also known as guncotton.

Guncotton and Gunpowder

Various types of smokeless powder, consisting primarily of nitrocellulose

Henri Braconnot discovered in 1832 that nitric acid, when combined with starch or wood fibers, would produce a lightweight combustible explosive material which he named *xyloïdine*. A few years later in 1838 another French chemist Théophile-Jules Pelouze (teacher of Ascanio Sobrero and Alfred Nobel) treated paper and cardboard in the same way. He obtained a similar material he called *nitramidine*. Both of these substances were highly unstable, and were not practical explosives.

However, Christian Friedrich Schönbein, a German-Swiss chemist, discovered a more practical solution around 1846. As he was working in the kitchen of his home in Basle, he spilled a bottle of concentrated nitric acid on the kitchen table. He reached for the nearest cloth, a cotton apron, and wiped it up. He hung the apron on the stove door to dry, and as soon as it was dry there was a flash as the apron exploded. His preparation method was the first to be widely imitated—one part of fine cotton wool to be immersed in fifteen parts of an equal blend of sulfuric and nitric acids. After two minutes the cotton was removed and washed in cold water to set the esterification level and remove all acid residue. It was then slowly dried at a temperature below 100°C. Schönbein collaborated with the Frankfurt professor Rudolf Böttger, who had discovered the process independently in the same year. By a strange coincidence there was even a third chemist, the Braunschweig professor F. J. Otto, who had also produced guncotton in 1846 and was the first to publish the process, much to the disenchantment of Schönbein and Böttger. (Itzehoer Wochenblatt, 29 October 1846, columns 1626 f.)

The process uses the nitric acid to convert the cellulose into cellulose nitrite and water:

$$2HNO_3 + C_6H_{10}O_5 \rightarrow C_6H_8(NO_2)_2O_5 + 2H_2O$$

The sulfuric acid is present to prevent the water produced in the reaction from diluting the concentrated nitric acid.

The power of guncotton made it suitable for blasting. As a projectile driver, it has around six times the gas generation of an equal volume of black powder and produces less smoke and less heating. However the sensitivity of the material during production led the British, Prussians and French to discontinue manufacture within a year.

Jules Verne viewed the development of guncotton with optimism. He referred to the substance several times in his novels. His adventurers carried firearms employing this substance. Most notably, in his *From the Earth to the Moon*, guncotton was used to launch a projectile into space.

Further research indicated that the key was the very careful preparation of the cotton: unless it was very well cleaned and dried, it was likely to explode spontaneously. The British, led by Frederick Augustus Abel, also developed a much lengthier manufacturing process at the Waltham Abbey Royal Gunpowder Mills, patented in 1865, with the washing and drying times each extended to 48 hours and repeated eight times over. The acid mixture was also changed to two parts sulfuric acid to one part nitric. Guncotton remained useful only for limited applications. For firearms, a more stable and slower burning mixture would be needed. Guncotton-like preparations were eventually prepared for this role, known at the time as smokeless powder. Guncotton, dissolved at approximately 25% in acetone, forms a lacquer used in preliminary stages of wood finishing to develop a hard finish with a deep luster. It is normally the first coat applied, sanded, and followed by other coatings that bond to it.

Nitrate Film

Nitrocellulose was used as the first flexible film base, beginning with Eastman Kodak products in August, 1889. Camphor is used as plasticizer for nitrocellulose film. It was used until 1933 for X-ray films (where its flammability hazard was most acute) and for motion picture film until 1951. It was replaced by safety film with an acetate base. German WW2 newsreel film (circa 1940) believed to be nitrocellulose, made by AGFA The use of nitrocellulose film for motion pictures led to a widespread requirement for fireproof projection rooms with wall coverings made of asbestos. Famously, the US Navy shot a training film for projectionists which included footage of a controlled ignition of a reel of nitrate film which continued to burn even when fully submerged in water. Due to public safety precautions, the London Underground forbade transport of nitrate films on its system until well past the introduction of safety film. A cinema fire caused by ignition of nitrocellulose film stock (foreshadowed by an earlier small fire) was a central plot element in the Italian film *Cinema Paradiso*. Today nitrate film projection is usually highly regulated and requires extensive precautionary measures including extra projectionist health and safety training. Additionally, projectors certified to run nitrate films have many containment strategies in effect. Among them, this includes the chambering of both the feed and takeup reels in thick metal covers with small slits to allow the film to run through. Furthermore, the projector is modified to accommodate several fire extinguishers with nozzles all aimed directly at the film gate; the extinguishers automatically trigger if a piece of flammable fabric placed near the gate starts to burn. While this triggering would likely damage or destroy a significant portion of the projection components, it would prevent a devastating fire which almost certainly would cause far greater damage.

It was discovered decades later that nitrocellulose gradually decomposes, releasing nitric acid which further catalyzes the decomposition (usually into a still-flammable powder or goo). Low temperatures can delay these reactions indefinitely. It is estimated that the great majority of films produced during the early twentieth century were lost forever either through this accelerating, self-catalyzed disintegration or studio warehouse fires. Salvaging old films is a major problem for film archivists (see film preservation).

Nitrocellulose film base manufactured by Kodak can be identified by the presence of the word *Nitrate* in dark letters between the perforations. Acetate film manufactured during the era when nitrate films were still in use was marked *Safety* or *Safety Film* between the perforations dark letters. Letters in white or light colors are print-through from the negative.

Color negative film was never manufactured with a nitrate base, nor were 8 mm or 16 mm motion picture film stocks.

Other Uses

Depending on the manufacturing process, nitrocellulose is esterified to varying degrees. Table tennis balls, guitar picks and some photographic films have a fairly low esterification level and burn comparatively slowly with some charred residue. See celluloid.

In professional wrestling, nitrocellulose, in the form of a small piece of flash paper, is used to execute an illegal (and dangerous) move called a Fireball. The wrestler sets the piece of quick-burning paper alight (using a concealed lighter) and throws it at the opponent, giving the impression of a supernatural ball of fire emerging from their hand.

Nitrated cotton: Used as liftcharge for indoor fireworks, for hand flashers and magicians

Nitrated yarn: Used as fuse for indoor fireworks and to make things fall down on command

Nitrated paper: Mainly used by magicians to make paper disapear in a flash, but also indoor fireworks as comets

Nitrated cellulose: Ice fountains, indoor fireworks, making smokeless gunpowder, celluloid, paints

Collodion is a solution of nitrocellulose in ether or acetone, sometimes with the addition of alcohols. Its generic name is pyroxylin solution. It is toxic and highly flammable. As the solvent evaporates, it dries to a celluloid-like film. It was discovered about 1846 by the French chemist and writer Louis Ménard. n 1851, the Englishman Frederick Scott Archer discovered that collodion could be used as an alternative to albumen on glass plates. This also reduced the exposure time when making the image. This became known as the wet plate Collodion or wet collodion method. Collodion was also grainless and colorless, and allowed for one of the first high quality duplication processes, also known as negatives. This process also produced positives, the Ambrotype and the Ferrotype (aka Tintypes).

Celloidin is a pure type of pyroxylin used to embed specimens which will be examined under a microscope.

Uses

- Collodion is widely used to glue electrodes to the head for electroencephalography.
- Pyroxylin with added pigments is used as a nitrocellulose lacquer.
- It was also added to nitroglycerine to stabilise it as blasting gelatine.
- Collodion is also used in theatrical makeup for various effects, such as simulating old-age wrinkles or scars.

- Similarly, it was sometimes used in boxing and other applications to cover up cuts. However, it has been generally replaced by the less toxic Liquid Bandage, which includes pyroxylin as an ingredient.
- Collodion also finds use in the cleaning of optics such as telescope mirrors. The collodion is applied to the surface of the optic, usually in two or more layers. Sometimes a piece of thin cloth is applied between the layers, to hold the collodion together for easy removal. After the collodion dries and forms a solid sheet covering the optic, it is carefully peeled away taking contamination with it.
- Collodion was also used by pathologist Thomas Stoltz Harvey to preserve Einstein's brain in 1955.

ISOMERISM

Aldoses

- The simplest aldosugar (aldose) is glyceraldehyde and the simplest ketosugar (ketose) is dihydroxyacetone.

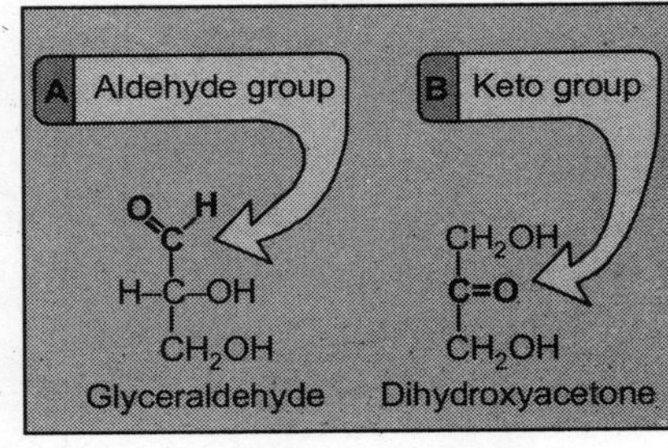

Definition

Carbohydrates are defined as organic polyhydroxyaldehydes or polyhydroxyketones or compounds which produce them on hydrolysis.

Functions of Carbohydrates

They are important due to their multiple roles in all forms of life:

- They serve as energy stores and fuels. Energy from the sun captured by plants during photosynthesis is stored in the form of carbohydrates. Breakdown of carbohydrates provides the energy that maintains animal life. Glucose is main source of energy for the brain.
- They participate in the structure of cell membranes.
- Pentose sugars (ribose and deoxyribose) form the structure of nucleic acids.
- Carbohydrates play important roles in recognition between cell types or recognition of cellular structures by other molecules. Recognition events are important in normal cell growth, fertilization and transformation of cells.
- Carbohydrates form the structure of glycoproteins and glycolipids.

Classification of Carbohydrates

They are classified according to the "number of structural units" into:

- Monosaccharides
- Disaccharides
- Oligosaccharides and
- Polysaccharides

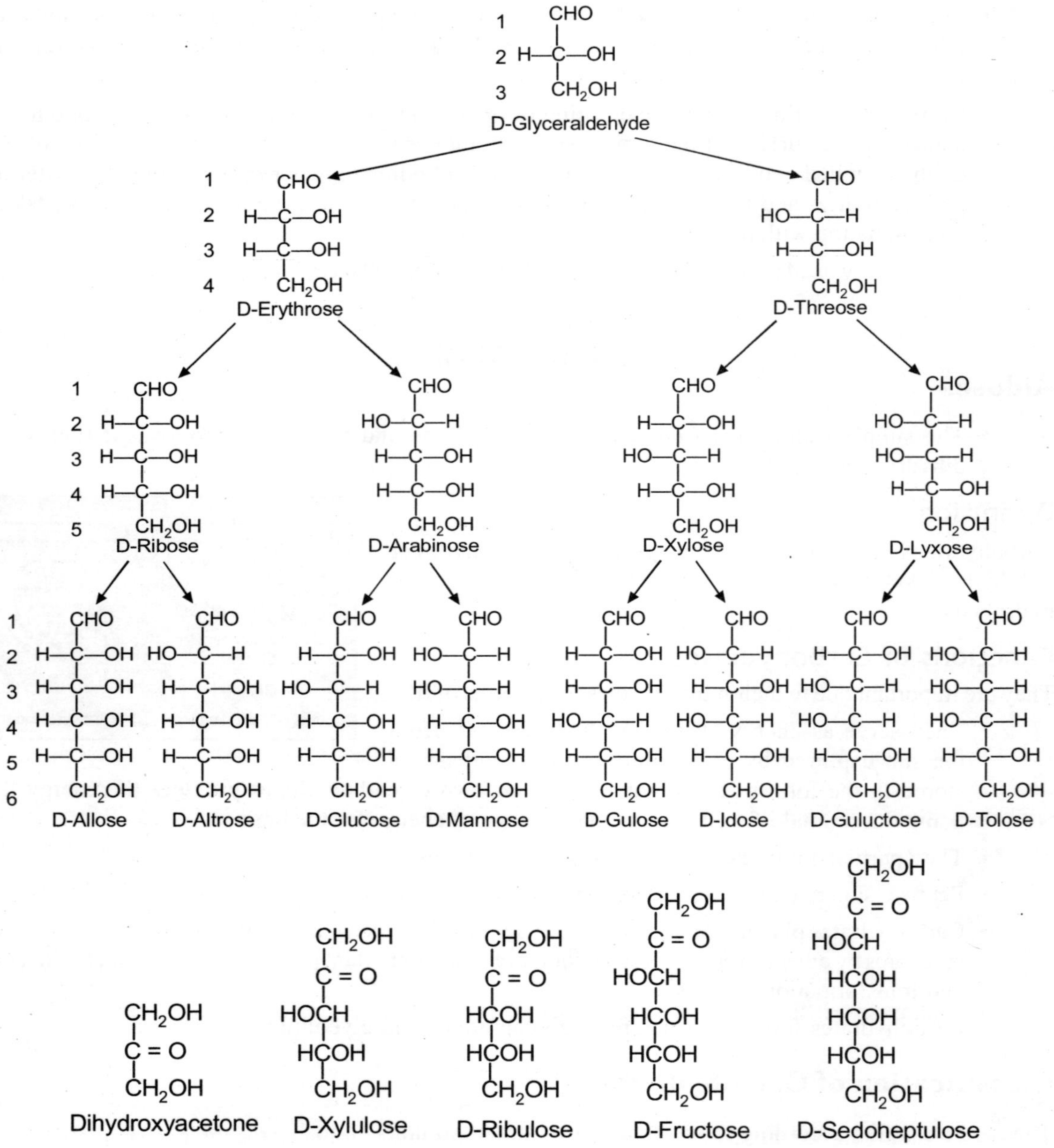

D-Glyceraldehyde

D-Erythrose D-Threose

D-Ribose D-Arabinose D-Xylose D-Lyxose

D-Allose D-Altrose D-Glucose D-Mannose D-Gulose D-Idose D-Guluctose D-Tolose

Dihydroxyacetone D-Xylulose D-Ribulose D-Fructose D-Sedoheptulose

Monosaccharides

- They are carbohydrates that cannot be hydrolysed into simpler units. General formula is $(CH_2O)_n$.

- The smallest ones, for which $n = 3$, are glyceraldehyde and dihydroxyacetone. They are trioses.
- Glyceraldehyde is also an aldose because it contains an aldehyde group; whereas dihydroxyacetone is a ketose because it contains a keto group, (the suffixose means sugar)

Classification of Monosaccharides

They are classified according to :

1. Number of carbon atoms in the molecule: They may be sub-divided into, trioses, tetroses, pentoses, hexoses, heptoses or octoses.

General formula	Aldosugars	Ketosugrs
• Trioses ($C_3H_6O_3$)	Glyceraldehyde	Dihydroxyacetone
• Tetrases ($C_4H_8O_4$)	Erythorse	Ertythrulose
• Pentoses ($C_5H_{10}O_5$)	Ribose	Ribulose
• Hexoses ($C_6H_{12}O_6$)	Glucose	Fructose

2. Type of functional group as aldosugars and ketosugars.
   - Examples of aldoses are glyceraldehyde, erythrose, ribose, glucose and galactose.
   - Examples of ketoses are dihydroxyacetone, Erythrulose, Ribulose, Fructose.

General formula	Aldosugars	Ketosugrs
• Trioses ($C_3H_6O_3$)	Glyceraldehyde	Dihydroxyacetone
• Tetrases ($C_4H_8O_4$)	Erythorse	Ertythrulose
• Pentoses ($C_5H_{10}O_5$)	Ribose	Ribulose
• Hexoses ($C_6H_{12}O_6$)	Glucose	Fructose

Diasaccharides

- These carbohydrates yield two molecules of monosaccharides when hydrolysed.

MALTOSE ⟶ GLUCOSE + GLUCOSE

SUCROSE ——————→ GLUCOSE + FRUCTOSE

LACTOSE ——————→ GLUCOSE + GALACTOSE

Oligosaccharides

- They yield three to ten molecules of monosaccharides on hydrolysis, example is Maltotriose which is a trisaccharide containing three glucose residues.

Polysaccharides

- They yield more than ten molecules of monosaccharides on hydrolysis. They may be either linear or branched polymers and may contain hundreds or even thousands of monosaccharide units. Examples are starch, glycogen, dextran.

Properties of Monosaccharides In Reference to Glucose

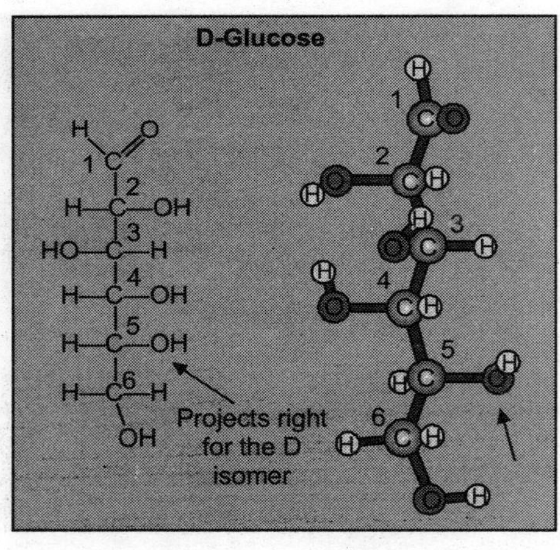

D-glucose

- Glucose is an aldohexose sugar.
- Common names include dextrose, grape sugar, blood sugar.
- Most important sugar in our diet.
- Most abundant organic compound found in nature.
- Level in blood can be as high as 0.1%.

```
        H
        |
        C=O
        |
   H ---+--- OH
        |
  HO ---+--- H
        |
   H ---+--- OH
        |
   H ---+--- OH
        |
      CH2OH
```

1. All monosaccharide are crystals and dissolve in water.
2. All have a Sweet taste.
3. Monosaccharides contain asymmetric carbon atom (which is a carbon atom attached to four different chemical groups) except dihydroxyacetone.

Due to presence of asymmetric carbon atom, monosaccharides have two main physical properties which are:

- Optical Isomerism or Stereoisomerism
- Optical Activity

4. Cyclic structure of Monosaccharides
5. Mutarotation

Asymmetric Carbon

CHO
H—C—OH
OH—C—H
HC—OH
H—C—OH
CH_2OH

Asymmetric carbon

CHO
H—*C—OH
OH—*C—H
H*C—OH
H—*C—OH
CH_2OH

4 asymmetric carbons

CHO
H—C—OH
CH_2OH

Asymmetric carbon atiom

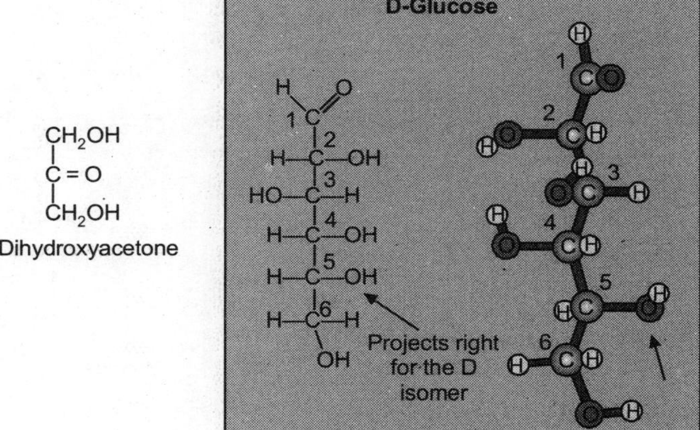

OPTICAL ISOMERISM OR STEREOISOMERISM

- Optical isomerism or stereoisomerism is the ability of substances to be present in more than one form. It is due to the presence of asymmetric carbon atom. These forms are named optical isomers.
- So optical isomers or stereoisomers or isomers are compounds that have same structural formula but differ in their configuration around the asymmetric carbon atom.

- A substance containing more than one asymmetric carbon atom can exist in a number of isomers equal to 2*n*, where *n* equal to the number of asymmetric carbon atoms e.g Glucose has 4 asymmetric carbon atoms, so the number of optical isomers equal to 24=2×2×2×2 = 16.

Enantiomers

- Enantiomers are a type of isomers designated as "D" and "L" isomers that are mirror images of each other and differ in the configuration around the asymmetric carbon before the last (sub-terminal) e.g. D and L glyceradehyde, D and L glucose.

D-glyceraldehyde		L-glyceraldehyde
H—C=O	1	H—C=O
H—C—(OH)	2	(HO)—C—H
CH₂—OH	3	CH₂—OH

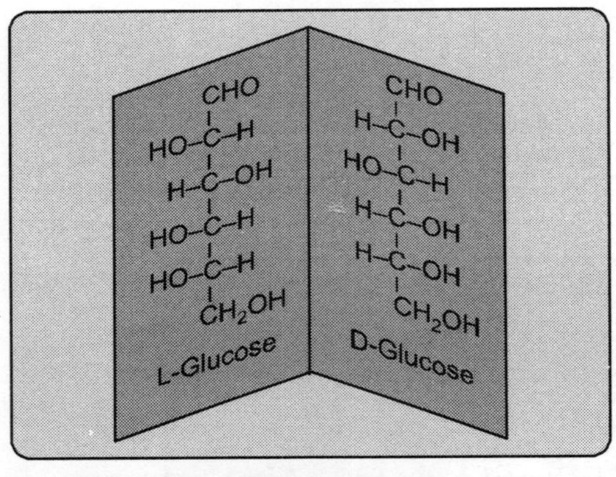

	D-glucose	L-glucose
1	H—C=O	H—C=O
2	H—C—OH	HO—C—H
3	HO—C—H	H—C—OH
4	H—C—OH	HO—C—H
5	H—C—(OH)	(HO)—C—H
6	CH₂— OH	CH₂— OH

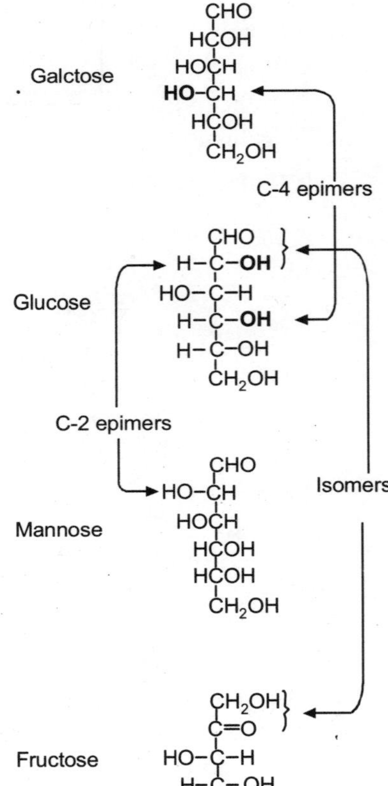

- In D isomer (D-glucose) the OH group attached to the asymmetric carbon (sub-terminal C5) is towards the right while in L isomer (L-glucose) OH group is on the left.

"The majority of the sugars in humans are D-sugars".

Optical Activity

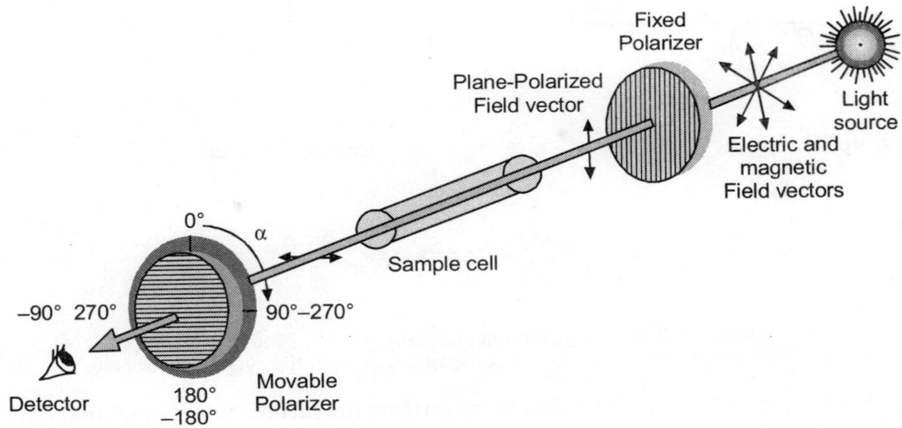

Fig. 2.11 : The Polarimeter arrangement for optical isomerism experiment

- It is the ability of an organic compound containing one or more asymmetric carbon atom to rotate a beam of polarized light either to the right or to the left.
- If the rotation is to the right it is called dextrorotatory (+ or d) and if it is to the left it is known as levorotatory (– or 1).
- Polarized light is the light which vibrates in only one direction. It is obtained by the passage of ordinary light through a prism.

EPIMERS

- Epimers are sterioisomers that differ in the position of (OH) group at only one of their asymmetric carbon atom other than before the last (sub-terminal) carbon atom. e.g.
- D-Glucose and D-Galactose are C4 epimers.
- D-Glucose and D-Mannose are C2 epimers.

MUTAROTATION AND SPECIFIC ROTATION

Mutarotation: interconversion of α- and α- anomers

- The α- and β- anomers of carbohydrates are typically stable solids.
- However, in aqueous solution, they quickly equilibrate to an equilibrium mixture of the two forms.
- For example, in aqueous solution, glucose exists as a mixture of 36% α- and 64% β- (>99% of the pyranose forms exist in solution).
- The equilibration occurs via the ring opening of the cyclic sugar at the anomeric center with the acyclic form as the intermediate.

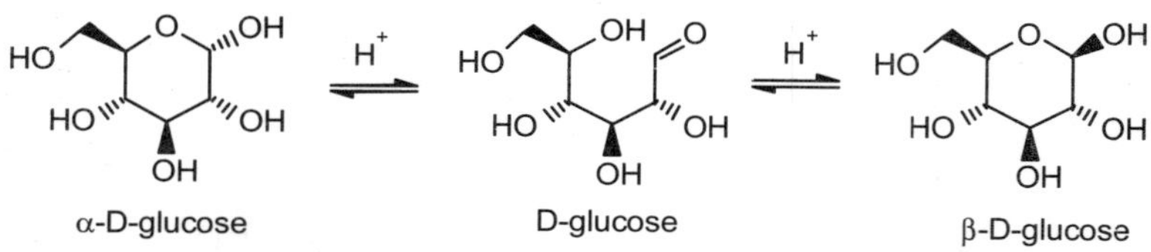

α-D-glucose D-glucose β-D-glucose

Or, in a simplified form with all the "extra" hydroxy groups removed:

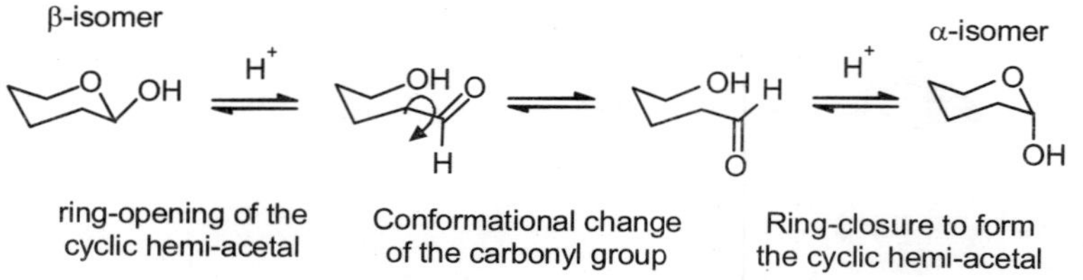

β-isomer α-isomer

ring-opening of the cyclic hemi-acetal

Conformational change of the carbonyl group

Ring-closure to form the cyclic hemi-acetal

Chemists use polarimeters (see Fig. 2.11) to investigate the influence of compounds (in the sample cell) on plane polarized light. Samples composed only of achiral molecules (e.g. water or hexane), have no effect on the polarized light beam. However, if a single enantiomer is examined (all sample molecules being right-handed, or all being left-handed), the plane of polarization is rotated in either a clockwise (positive) or counter-clockwise (negative) direction, and the analyzer must be turned an appropriate matching angle, α, if full light intensity is to reach the detector. In the above illustration, the sample has rotated the polarization plane clockwise by +90°, and the analyzer has been turned this amount to permit maximum light transmission.

The observed rotations (α) of enantiomers are opposite in direction. One enantiomer will rotate polarized light in a clockwise direction, termed **dextrorotatory** or (+), and its mirror-image partner in a counter-clockwise manner, termed **levorotatory** or (–). The prefixes dextro and levo come from the Latin *dexter*, meaning right, and *laevus*, for left, and are abbreviated *d* and *l* respectively. If equal quantities of each enantiomer are examined , using the same sample cell, then the magnitude of the rotations will be the same, with one being positive and the other negative. To be absolutely certain whether an observed rotation is positive or negative it is often necessary to make a second measurement using a different amount or concentration of the sample. In the above illustration, for example, α might be –90° or +270° rather than +90°. If the sample concentration is reduced by 10%, then the positive rotation would change to +81° (or +243°) while the negative rotation would change to –81°, and the correct α would be identified unambiguously.

Since it is not always possible to obtain or use samples of exactly the same size, the observed rotation is usually corrected to compensate for variations in sample quantity and cell length. Thus it is common practice to convert the observed rotation, α, to a **specific rotation, [α]**, by the following formula:

Specific Rotation = where l = cell length in dm, c = concentration in g/ml
D is the 589 nm light from a sodium lamp

Compounds that rotate the plane of polarized light are termed **optically active**. Each enantiomer of a stereoisomeric pair is optically active and has an equal but opposite-in-sign specific rotation. Specific rotations are useful in that they are experimentally determined constants that characterize and identify pure enantiomers. For example, the lactic acid and carvone enantiomers discussed earlier have the following specific rotations.

Carvone from caraway: $[\alpha]_D = +62.5^\circ$ this isomer may be referred to as (+)-carvone or *d*-carvone

Carvone from spearmint: $[\alpha]_D = -62.5^\circ$ this isomer may be referred to as (–)-carvone or *l*-carvone

Lactic acid from muscle tissue: $[\alpha]_D = +2.5^\circ$ this isomer may be referred to as (+)-lactic acid or *d*-lactic acid

Lactic acid from sour milk: $[\alpha]_D = -2.5^\circ$ this isomer may be referred to as (–)-lactic acid or *l*-lactic acid

A 50:50 mixture of enantiomers has no observable optical activity. Such mixtures are called **racemates** or racemic modifications, and are designated (±). When chiral compounds are created from achiral compounds, the products are racemic unless a single enantiomer of a chiral co-reactant or catalyst is involved in the reaction. The addition of HBr to either cis- or trans-2-butene is an example of racemic product formation (the chiral center is colored red in the following equation).

$$CH_3CH{=}CHCH_3 + HBr \rightarrow (\pm)\ CH_3CH_2CHBrCH_3$$

Chiral organic compounds isolated from living organisms are usually optically active, indicating that one of the enantiomers predominates (often it is the only isomer present). This is a result of the action of chiral catalysts we call enzymes, and reflects the inherently chiral nature of life itself. Chiral synthetic compounds, on the other hand, are commonly racemates, unless they have been prepared from enantiomerically pure starting materials.

There are two ways in which the condition of a chiral substance may be changed:

1. A racemate may be separated into its component enantiomers. This process is called resolution.
2. A pure enantiomer may be transformed into its racemate. This process is called racemization.

GLYCOSIDIC LINKAGES

Deoxy and Amino Sugars and Homopolysaccharides

Deoxy sugars are sugars that have had a hydroxyl group replaced with a hydrogen.

HO O OH HO OH

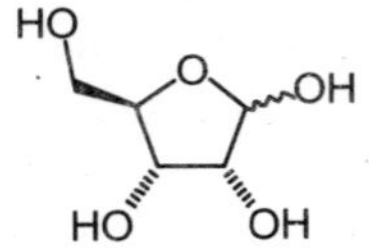

Comparison of the chemical structure of ribose (top) and deoxyribose (bottom)

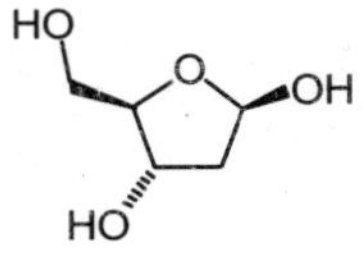

HO O OH HO

Deoxyribose, also known as **D-Deoxyribose** and **2-deoxyribose**, is an aldopentose—a monosaccharide containing five carbon atoms, and including an aldehyde functional group in its linear structure. It is a deoxy sugar derived from the pentose sugar ribose by the replacement of the hydroxyl group at the 2 position with hydrogen, leading to the net loss of an oxygen atom.

Replacement of the hydroxyl group also shifts the conformation of the ring from C3'-endh to C2'-endh. It has a chemical formula of $C_5H_{10}O_4$; it was discovered in 1929 by Phoebus Levene.

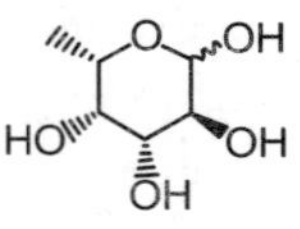

Rannose 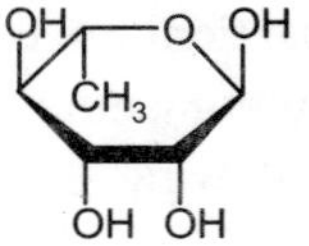 Fucose

Ribose forms a five-member ring composed of four carbon atoms and one oxygen. Hydroxyl groups are attached to three of the carbons. The other carbon and a hydroxyl group are attached to one of the carbon atoms adjacent to the oxygen. In deoxyribose, the carbon furthest from the attached carbon is stripped of the oxygen atom in what would be a hydroxyl group in ribose. Due to the common C3' and C4' stereochemistry of D-ribose and D-arabinose, D-2-deoxyribose is also D-2-deoxyarabinose.

Deoxyribofuranose is an alternative name for the ring structure of deoxyribose. This alternative name merely refers to the fact that deoxyribose has a five membered ring consisting of four carbons and an oxygen and is more a structural description than a name: ribose and 2-deoxyribose derivatives have an important role in biology. Among the most important derivatives are those with phosphate groups attached at the 5 position. Mono-, di-, and triphosphate forms are important, as well as 3-5 cyclic monophosphates. There are also important diphosphate dimers called coenzymes that purines and pyrimidines form an important class of compounds with ribose and deoxyribose. When these purine and pyrimidine derivatives are coupled to a ribose sugar, they are called nucleosides. In these compounds, the convention is to put a 2 (pronounced "prime") after the carbon numbers of the sugar, so that in nucleoside derivatives a name might include, for instance, the term "52-monophosphate", meaning that the phosphate group is attached to the fifth carbon of the sugar, and not to the base. The bases are attached to the 12 ribose carbon in the common nucleosides. Phosphorylated nucleosides are called nucleotides.

Biological importance of deoxyribose

One of the common bases is adenine (a purine derivative); coupled to ribose it is called adenosine; coupled to deoxyribose it is called deoxyadenosine. The 52-triphosphate derivative of adenosine, commonly called ATP, for adenosine triphosphate, is an important energy transport molecule in cells.

See Nucleic acid nomenclature for a diagram showing the numbered positions in a 52-2-Deoxyribose and ribose nucleotides are often found in unbranched 52 -32 polymers. In these structures, the 32 carbon of one monomer unit is linked to a phosphate that is attached to the 52 carbon of the next unit, and so on. These polymer chains often contain many millions of monomer units. Since long polymers have physical properties distinctly different from those of small molecules, they are called macromolecules. The sugar-phosphate-sugar chain is called the backbone of the polymer. One end of the backbone has a free 52 phosphate, and the other end has a free 32 OH group. The backbone structure is independent of which particular bases are attached to the individual sugars.

Genetic material in earthly life often contains poly 52 -32 , 22 -deoxyribose nucleotides, in structures called chromosomes, where each monomer is one of the nucleotides deoxy- adenine, thymine, guanine or cytosine. This material is commonly called deoxyribonucleic acid, or simply DNA for short.

DNA in chromosomes forms very long helical structures containing two molecules with the backbones running in opposite directions on the outside of the helix and held together by hydrogen bonds between complementary nucleotide bases lying between the helical backbones. The lack of the 22 hydroxyl group in DNA appears to allow the backbone the flexibility to assume the full conformation of the long double-helix, which involves not only the basic helix, but additional coiling necessary to fit these very long molecules into the very small volume of a cell nucleus.

In contrast, very similar molecules, containing ribose instead of deoxyribose, and known generically as RNA, are known to form only relatively *short* double-helical complementary base paired structures. These are well known, for instance, in ribosomal RNA molecules and in transfer RNA (tRNA), where so-called *hairpin* structures from palindromic sequences within one molecule.

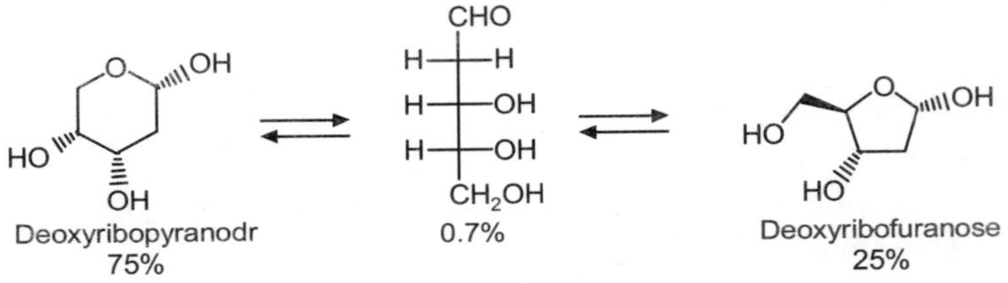

Deoxyribopyranodr 75% ⇌ CHO, H–C–H, H–C–OH, H–C–OH, CH_2OH 0.7% ⇌ Deoxyribofuranose 25%

Fucose is a hexose deoxy sugar with the chemical formula $C_6H_{12}O_5$. It is found on *N*-linked glycans on the mammalian, insect and plant cell surface, and is the fundamental sub-unit of the fucoidan polysaccharide. α→3 linked core fucose is a suspected carbohydrate antigen for IgE-mediated allergy.

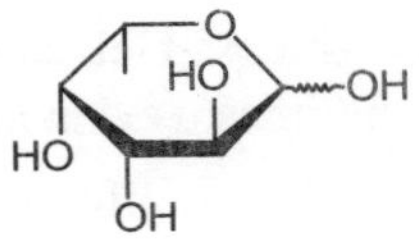

Two structural features distinguish fucose from other six-carbon sugars present in mammals: the lack of a hydroxyl group on the carbon at the 6-position (C-6) and the L-configuration. It is equivalent to 6-deoxy-L-galactose.

In the fucose-containing glycan structures, fucosylated glycans, fucose can exist as a terminal modification or serve as an attachment point for adding other sugars.[2] In human *N*-linked glycans, fucose is most commonly linked α-1,6 to the reducing terminal beta-*N*-acetlyglucosamine. However, fucose at the non-reducing termini linked α-1,2 to galactose forms the H antigen, the substructure of the A and B blood group antigens.

Fucose is metabolized by an enzyme called α-fucosidase. Fucose is abundant in human breast milk and certain mushrooms. It has been shown to influence brain development, act as immune modulator by inhibiting tumour growth and its spread and enhancer of cellular communication. It also guards against respiratory tract infections and inhibits allergic reactions. High concentrations of fucose are found at the junctions of nerves, as well as in the kidneys, testes, and in the outer layer of skin.

Rhamnose is a naturally-occurring deoxy sugar. It can be classified as either a methyl-pentose or a 6-deoxy-hexose. Rhamnose occurs in nature in its L-form as L-rhamnose (6-deoxy-L-mannose). This is unusual, since most of the naturally-occurring sugars are in D-form. Exceptions are the methyl pentoses L-fucose and L-rhamnose and the pentose L-arabinose. L-Rhamnose can be isolated from Buckthorn (Rhamnus) and poison sumac. It is also found as a glycoside in a variety of other plants. Rhamnose is a component of the outer cell membrane of acid-fast bacteria in the Mycobacterium genus, which includes the organsism that causes tuberculosis.

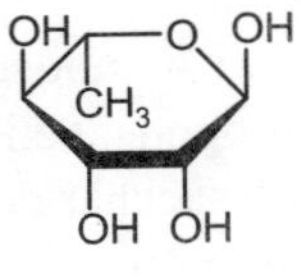

Amino Sugars

In chemistry, an **amino sugar** contains an amine group in place of a hydroxyl group. Derivatives of amine containing sugars, such as N-acetylglucosamine and sialic acid, while not formally containing an amine, are also considered amino sugars.

Aminoglycosides are a class of antimicrobial compounds that inhibit bacterial protein synthesis. These compounds frequently contain amino sugars (but in some cases contain aminocyclitols).

Common examples of amino sugars include:

- Galactosamine
- Glucosamine
- Sialic acid
- N-Acetylglucosamine

Galactosamine is a hexosamine derived from galactose with the molecular formula $C_6H_{13}NO_5$. This amino sugar is a constituent of some glycoprotein hormones such as follicle-stimulating hormone (FSH) and luteinizing hormone (LH). Other sugar constituents of FSH and LH include glucosamine, galactose and glucose.Galactosamine is a hepatotoxic, or liver-damaging, agent that is sometimes used in animal models of liver failure.

Glucosamine was first identified in 1876 by Dr. Georg Ledderhose, but the stereochemistry was not fully defined until 1939 by the work of Walter Haworth. D-Glucosamine is made naturally in the form of glucosamine-6-phosphate, and is the biochemical precursor of all nitrogen-containing sugars. Specifically, glucosamine-6-phosphate is synthesized from fructose-6-phosphate and glutamine as the first step of the hexosamine biosynthesis pathway. The end-product of this pathway is UDP-N-acetylglucosamine (UDP-GlcNAc), which is then used for making glycosaminoglycans, proteoglycans, and glycolipids. As the formation of glucosamine-6-phosphate is the first step for the synthesis of these products, glucosamine may be important in regulating their production. However, the way that the hexosamine biosynthesis pathway is actually regulated, and whether this could be involved in contributing to human disease, remains unclear.

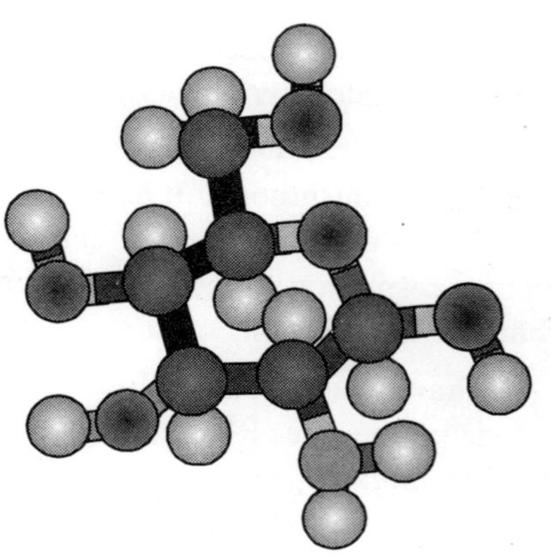

Glucosamine ($C_6H_{13}NO_5$) is an amino sugar and a prominent precursor in the biochemical synthesis of glycosylated proteins and lipids. A type of glucosamine forms chitosan and chitin, which composes the exoskeletons of crustaceans and other arthropods, cell walls in fungi and many higher organisms. Glucosamine is one of the most abundant monosaccharides. It is produced commercially by the hydrolysis of crustacean exoskeletons or, less commonly and more expensive to the consumer, by fermentation of a grain such as corn or wheat.

Glucosamine is commonly used as a treatment for osteoarthritis, although its acceptance as a medical therapy varies.

Sialic acid The term "sialic" (from the Greek (salia) 'saliva') was first introduced by Swedish biochemist, Gunnar Blix, in 1952

Sialic acid is a generic term for the N- or O-substituted derivatives of neuraminic acid, a monosaccharide with a nine-carbon backbone. It is also the name for the most common member of this group, N-acetylneuraminic acid (Neu5Ac or NANA).sialic acids are found widely distributed in animal tissues and in bacteria, especially in glycoproteins and gangliosides. The amino group bears either an acetyl or a glycolyl group. The hydroxyl substituents may vary considerably: acetyl, lactyl, methyl, sulphate, and phosphate groups have been found. Siaiic acid-rich glycoproteins bind selectin in humans and other organisms. Cancer cells that can metastasize often have a lot ofsialic acid-rich glycoproteins. This helps these late-stage cancer cells enter the blood stream.sialic acid also plays an important role in Human Influenza infections. The influenza viruses (orthomyxoviridae) have Hemagglutin Activity (HA) glycoproteins on their surface that bind tosialic acids found on the surface of human erythrocytes and on the cell membranes of the upper respiratory tract. This is the basis of heme-agglutination when viruses are mixed with blood cells, and entry of the virus into cells of the upper respiratory tract. Sialic acid-rich oligosaccharides on the glycoconjugates found on surface membranes help keep water at the surface of cells. Thesialic acid-rich regions contribute to creating a negative charge on the cells' surface. Since water is a polar molecule with partial positive charges on both hydrogen atoms, it is attracted to cell surfaces and membranes. This also contributes to cellular fluid uptake.sialic acid can "hide" mannose antigens on the surface of host cells or bacteria from mannose-binding lectin. This prevents activation of complement.

N-Acetylglucosamine (N-Acetyl-D-Glucosamine, or GlcNAc, or NAG) is a monosaccharide derivative of glucose. Chemically it is an amide between glucosamine and acetic acid. It has a molecular formula of $C_8H_{15}NO_6$, a molar mass of 221.21 g/mol, and it is significant in several biological systems. It is part of a biopolymer in the bacterial cell wall, built from alternating units of GlcNAc and N-acetylmuramic acid (MurNAc), cross-linked with oligopeptides at the lactic acid residue of MurNAc. This layered structure is called peptidoglycan. GlcNAc is the monomeric unit of the polymer chitin, which forms the outer coverings of insects and crustaceans. GlcNAc is also of note in neurotransmission, where it is thought to be an atypical neurotransmitter functioning in nocioceptive (pain) pathways. Polymerized with glucuronic acid it forms hyaluronan. It has been proposed as a treatment for autoimmune diseases.

Homopolysaccharides

Homopolysaccharides are polymers composed of a single type of sugar monomer. For example, cellulose is an unbranched homopolysaccharaide comprised of glucose monomers connected via beta-glycosidic linkages; glycogen is a branched form, where the glucose monomers are joined by alpha-glycosidic linkages

Glycosaminoglycans (GAGs) or **mucopolysaccharides** are long unbranched polysaccharides consisting of a repeating disaccharide unit.

Classification

Members of the glycosaminoglycan family vary in the type of hexosamine, hexose or hexuronic acid unit they contain (e.g. glucuronic acid, iduronic acid, galactose, galactosamine, glucosamine).

Geometry of the glycosidic linkage.Examples of GAGs include:

Name	Hexuronic acid/ Hexose	Hexosamine	Linkage geometry between predominant monomeric units	Unique features
Chondroitin sulphate	GlcUA or GlcUA(2S)	GalNAc or GalNAc(4S) or GalNAc(6S) or GalNAc(4S,6S)	-4GlcUAβ1-3GalNAcβ1-	Most prevalent GAG
Dermatan sulphate	GlcUA or IdoUA or IdoUA(2S)	GalNAc or GalNAc(4S) or GalNAc(6S) or GalNAc(4S,6S)	-4IdoUAβ1-3GalNAcβ1-	Distinguished from chondroitin sulphate by the presence of iduronic acid, although some hexuronic acid monosaccharides may be glucuronic acid.
Keratan sulphate	Gal or Gal(6S)	GlcNAc or GlcNAc(6S)	-3Gal(6S)β1-4GlcNAc(6S)β1-	Keratan sulphate type II may be fucosylated.
Heparin	GlcUA or IdoUA(2S)	GlcNAc or GlcNS or GlcNAc(6S) or GlcNS(6S)	-4IdoUA(2S)β1-4GlcNS(6S)β1-	Highest negative charge density of any known biological molecule
Heparan sulphate	GlcUA or IdoUA or IdoUA(2S)	GlcNAc or GlcNS or GlcNAc(6S) or GlcNS(6S)	-4GlcUAβ1-4GlcNAcβ1-	Highly similar in structure to heparin, however heparan sulphates disaccharide units are organised into distinct sulphated and non-sulphated domains.
Hyaluronan	GlcUA	GlcNAc	-4GlcUAβ1-3GlcNAcβ1-	The only GAG that is exclusively non-sulphated

Abbreviations

- **GlcUA** = β-D-glucuronic acid
- **GlcUA(2S)** = 2-O-sulfo-β-D-glucuronic acid
- **IdoUA** = α-L-iduronic acid

- **IdoUA(2S)** = 2-O-sulfo-α-L-iduronic acid
- **Gal** = β-D-galactose
- **Gal(6S)** = 6-O-sulfo-β-D-galactose
- **GalNAc** = β-D-N-acetylgalactosamine
- **GalNAc(4S)** = β-D-N-acetylgalactosamine-4-O-sulphate
- **GalNAc(6S)** = β-D-N-acetylgalactosamine-6-O-sulphate
- **GalNAc(4S,6S)** = β-D-N-acetylgalactosamine-4-O, 6-O-sulphate
- **GlcNAc** = α-D-N-acetylglucosamine
- **GlcNS** = α-D-N-sulfoglucosamine
- **GlcNS(6S)** = α-D-N-sulfoglucosamine-6-O-sulphate

Glycosaminoglycans

Glycosaminoglycans (GAGs) are a family of sulphated sugar chains that play important roles in neuronal communication by binding to different proteins. Subtle variations in **stereochemistry**, **length**, and **patterns of sulfation** differ between and within GAG families. This structural diversity creates an enormous number of protein binding motifs. However, it also has hampered efforts to define the structural and functional properties of GAGs.

We are developing two general methods for the synthesis of glycosaminoglycans. Chemical and chemoenzymatic approaches will allow us to synthesize sugar chains with precise control over length, stereochemistry, and sulfation patterns. Synthetic access to defined GAG structures will allow us to understand their biological roles in the brain. For example, we will address which GAG structures are important for learning and memory and how changes in sulfation pattern impact aging and neurodegenerative disease.

These molecules are long unbranched polysaccharides containing a repeating disaccharide unit. The disaccharide units contain either of two modified sugars, *N*-acetylgalactosamine (GalNAc) or *N*-acetylglucosamine (GlcNAc), and a uronic acid such as glucuronate or iduronate. GAGs are highly negatively charged molecules, with extended conformation that imparts high viscosity to the solution. GAGs are located primarily on the surface of cells or in the extracellular matrix (ECM). Along with the high viscosity of GAGs comes low compressibility, which makes these molecules ideal for a lubricating fluid in the joints. At the same time, their rigidity provides structural integrity to cells and provides passageways between cells, allowing for cell migration. The specific GAGs of physiological significance are hyaluronic acid, dermatan sulphate, chondroitin sulphate, heparin, heparan sulphate, and keratan sulphate. Although each of these GAGs has a predominant disaccharide component (see Table below), heterogeneity does exist in the sugars present in the make-up of any given class of GAG. Hyaluronic is unique among the GAGs in that it does not contain any sulphate and is not found covalently attached to proteins as a proteoglycan. It is, however, a component of non-covalently formed complexes with proteoglycans in the ECM. Hyaluronic acid polymers are very large (with molecular weights of 100,000 - 10,000,000) and can displace a large volume of water. This property makes them excellent lubricators and shock absorbers.

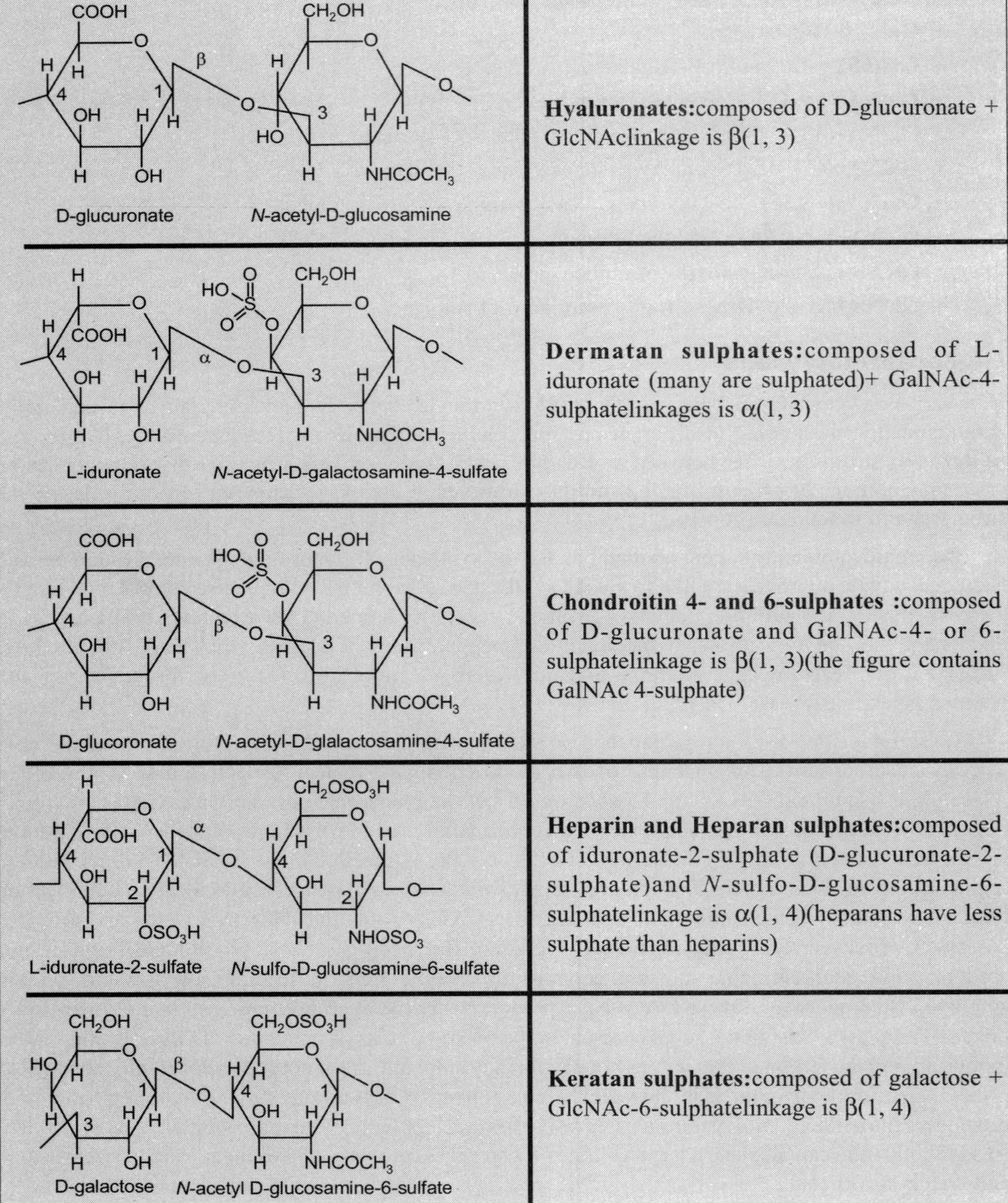

Hyaluronates:composed of D-glucuronate + GlcNAclinkage is β(1, 3)

Dermatan sulphates:composed of L-iduronate (many are sulphated)+ GalNAc-4-sulphatelinkages is α(1, 3)

Chondroitin 4- and 6-sulphates :composed of D-glucuronate and GalNAc-4- or 6-sulphatelinkage is β(1, 3)(the figure contains GalNAc 4-sulphate)

Heparin and Heparan sulphates:composed of iduronate-2-sulphate (D-glucuronate-2-sulphate)and *N*-sulfo-D-glucosamine-6-sulphatelinkage is α(1, 4)(heparans have less sulphate than heparins)

Keratan sulphates:composed of galactose + GlcNAc-6-sulphatelinkage is β(1, 4)

GAG	Localization	Comments
Hyaluronate	synovial fluid, vitreous humor, ECM of loose connective tissue	large polymers, shock absorbing
Chondroitin sulphate	cartilage, bone, heart valves	most abundant GAG
Heparan sulphate	basement membranes, components of cell surfaces	contains higher acetylated glucosamine than heparin
Heparin	component of intracellular granules of mast cellslining the arteries of the lungs, liver and skin	more sulphated than heparan sulphates
Dermatan sulphate	skin, blood vessels, heart valves	
Keratan sulphate	cornea, bone, cartilage aggregated with chondroitin sulphates	

Proteoglycans

The majority of GAGs in the body are linked to core proteins, forming proteoglycans (also called **mucopolysaccharides**). The GAGs extend perpendicularly from the core in a brush-like structure. The linkage of GAGs to the protein core involves a specific trisaccharide composed of two galactose residues and a xylose residue (GAG-GalGalXyl-O-CH_2-protein). The trisaccharide linker is coupled to the protein core through an *O*-glycosidic bond to a S residue in the protein. Some forms of keratan sulphates are linked to the protein core through an *N*-asparaginyl bond. The protein cores of proteoglycans are rich in S and T residues, which allows multiple GAG attachments.

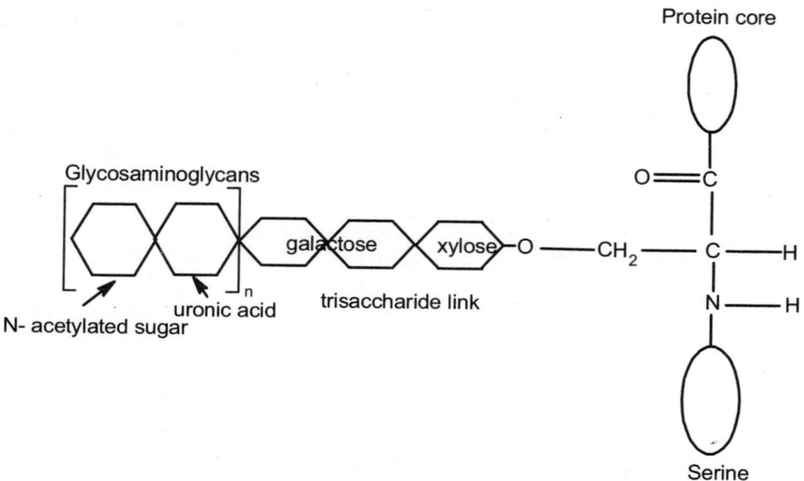

Fig. 2.12 : Structure of the GAG linkage to protein in proteoglycans

Clinical Significance

Proteoglycans and GAGs perform numerous vital functions within the body, some of which still remain to be studied. One well-defined function of the GAG heparin is its role in preventing coagulation of the blood. Heparin is abundant in granules of mast cells that line blood vessels. The release of heparin from these granules, in response to injury, and its subsequent entry into the serum leads to an inhibition of blood clotting, in the following manner. Free heparin complexes with and activates antithrombin III, which in turn inhibits all the serine proteases of the coagulation cascade. This phenomenon has been clinically exploited in the use of heparin injection for anti-coagulation therapies. Several genetically inherited diseases, for example the lysosomal storage diseases, result from defects in the lysosomal enzymes responsible for the metabolism of complex membrane-associated GAGs. These specific diseases, termed **mucopolysaccharidoses (MPS)** (in reference to the earlier term, mucopolysaccharide, used for glycosaminoglycans) lead to an accumulation of GAGs within lysosomes of affected cells. There are at least 14 known types of lysosomal storage diseases that affect GAG catabolism; some of the more commonly encountered examples are indicated in the Table below. All of these disorders, excepting Hunter's syndrome (X-linked), are inherited in an autosomal recessive manner. To see a diagram of the locations of the enzyme defects in GAG degradation go to the Mucopolysaccharidoses page.

Type: Syndrome	Enzyme Defect	Affected GAG	Symptoms
Hurler MPSIH (MPS1H)	α-L-iduronidase	dermatan sulphate, heparan sulphate	corneal clouding, dystosis multiplex, organomegaly, heart disease, dwarfism, mental retardation; early mortality
Scheie MPSIS (MPS1S)	α-L-iduronidase	dermatan sulphate, heparan sulphate	corneal clouding; aortic valve disease; joint stiffening; normal intelligence and life span
Hurler-Scheie MPSIHS (MPS1HS)	α-L-iduronidase	dermatan sulphate, heparan sulphate	intermediate between I H and I S
Hunter MPSII (MPS2)	L-iduronate-2-sulfatase	dermatan sulphate, heparan sulphate	mild and severe forms, only X-linked MPS, dystosis multiplex, organomegaly, facial and physical deformities, no corneal clouding, mental retardation, death before 15 except in mild form then survival to 20 - 60
Sanfilippo A MPSIIIA (MPS3A)	heparan N-sulfatase	heparan sulphate	profound mental deterioration, hyperactivity, skin, brain, lungs, heart and skeletal muscle are affected in all 4 types of MPS-III

Type: Syndrome	Enzyme Defect	Affected GAG	Symptoms
Sanfilippo B MPSIIIB (MPS3B)	α-N-acetyl-D-glucosaminidase	heparan sulphate	phenotype similar to III A
Sanfilippo C MPSIIIC (MPS3C)	acetylCoA:α-glucosaminide-acetyltransferase	heparan sulphate	phenotype similar to III A
Sanfilippo D MPSIIID (MPS3D)	N-acetylglucosamine-6-sulfatase	heparan sulphate	phenotype similar to III A
Morquio A MPSIVA (MPS4A)	galactose-6-sulfatase	keratan sulphate, chondroitin 6-sulphate	corneal clouding, odontoid hypoplasia, aortic valve disease, distinctive skeletal abnormalities
Morquio B MPSIVB (MPS4B)	β-galactosidase	keratan sulphate	severity of disease similar to IV A
MPS V, a designation no longer used			
Maroteaux-Lamy MPSVI (MPS6)	arylsulfatase B also called N-acetyl-galac-tosamine-4-sulfatase	dermatan sulphate	3 distinct forms from mild to severe, aortic valve disease, dystosis multiplex, normal intelligence, corneal clouding, coarse facial features
Sly MPSVII (MPS7)	β-glucuronidase	heparan sulphate, dermatan sulphate, chondroitin 4-, 6-sulphates	hepatosplenomegaly, dystosis multiplex, wide spectrum of severity, hydrops fetalis
MPS VIII, a designation no longer used			

Carbohydrates in Blood Group and Bacterial Cell-Dell

Blood Group Polysaccharides

The so – called blood group polysaccharides are present in erythrocytes; saliva, gastric mucin, cystic fluids, and other body secretions, when combined with proteins, they constitute the A B O (H), Rh and other antigens of the erythrocytes and differentiate the blood groups or types, when red cells containing a specific type polysaccharide antigen are mixed with specific antibodies of serum, agglutinated by antibodies (isoagglutinins)found in type O or B blood group A serum agglutination characteristic of cells and sera, it is possible to determine whether or not a given blood is of the proper type for transfusion into a particular patient so that agglutination and disastrous result will not follow:

Blood type	Serum Isoagglutinins	Agglutination of Red cells			
		A	B	AB	O
A	β	–	+	+	–
B	α	+	–	+	–
AB	None	–	–	–	–
O(H)	α and β	+	+	+	–

The various blood group polysaccharides are generally polymers of D-glucosomanine or D-glactsamine and simple sugar. Both D-galactosamine and D-glucosoamine may be present. Blood group polysaccharides A, B and O (H) from ovarian cyst fluid yield D-galactosamine, D-glucosoamine D-glactose L-fructose, and sialic acid (N-acetyleneuraminic acid). The specific polysaccharisdes from hog gastric mucin are made up to D-galactose N-acetyl-D-glucosomanine and L-fucose

Bacterial Polysaccharides

Complex carbohydrate substances have been isolated from many different species of bacteria. A number of these carbohydrates have been found to be responsible for the specificity of immune reaction. They are also referred as specific polysaccharides.

While the hydrolytic products of the different polysaccharides vary, the following substances have been obtained from the group as a whole; uronic acids especially glucouronic acids and glactouronic; monosachharides, such as glucose, galactose arbinose mannose rhamnose, inosotol, amino sugar, notably glucosomanine galactosamine and acetic acid. The amino sugars present are N-acetylated.

Heidelberger and Avery first demonstrated the presence of specific bacterial polysaccharides by their work on penemococcus types II and III. Where the polysaccharides are found in the bacterial capsules. Type II polysaccharides are compounds consisting D-glucose, L-rhamnose (40%) and D-glucouronic acid units. While type III polysaccharides contains only D-glucose and D-glucouronic acid groups. By careful hydrolysis of type III polysaccharide and aldobiuronic acid units, 3–Oβ-D-

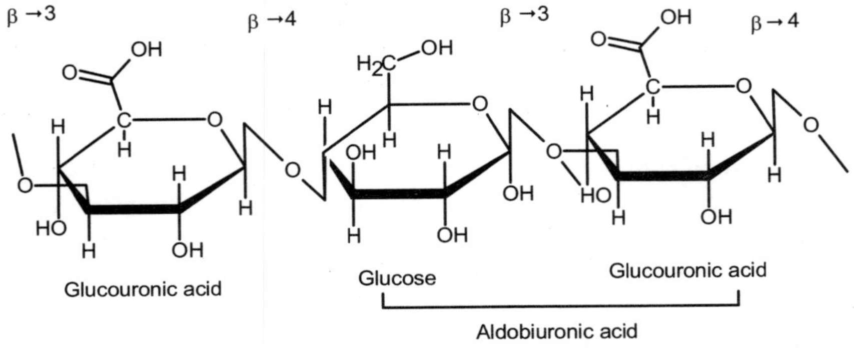

Glucouronic acid Glucose Glucouronic acid

Aldobiuronic acid

glucopyranosyluronic acid. The structure of the polysaccharide is a linear chain with alternative glucose and glucouronic acid groups. It may be considered chain polymer of the above aldobiuronic acid.

When the type III polysaccharide molecule is hydrolyzed at the 1, 4 linkages an aldobiuronic acid with D-glucose linked to D-glucouronic acid by 1,3 linkage is obtained, 3-O-β D-glucopyranosyl-O-glucopyranosyluronic acid is obtained, 3-O-β-D-Glucopyranosyl-O- glucopyranosyluronic acid. If however the 1, 3 linkages are hydrolyzed, leaving the glucouronic acid and glucose joined by 1-4 bonds, a different aldobiuronic acid is formed cello biuronic acid 4-O-β-D-glucopyranosyluronic acid-D-glucopyranose.

The polysaccharide type I penemococcus is composed of D-glactouronic acid (28%) and N-acetyl-D-glucosamine, while type VIII polysaccharide is made up of glucose and D-glucose and D-glucouronic acid (7: 2), and type XIV polysaccharide has D-galactose and N-acetyl-D-glucosoamine as constituent units. Only the type III polysaccharide structure has been definitely established.

The immunologically active polysaccharides to contain D-glactose, D mannose D-arbinose inositol.

The so called V_1 antigen of *Salmonella typhosa* and other bacterial series appears to be largely a polymer of an N-acetyl aminohexuronic acid possibly of N-acetyl-D-glucosaminuronic acid

Dextran is a complex, branched glucan (polysaccharide) made of many glucose molecules joined into chains of varying lengths (from 10 to 150 kilodaltons), used as an antithrombotic (anti-platelet), and to reduce blood viscosity.

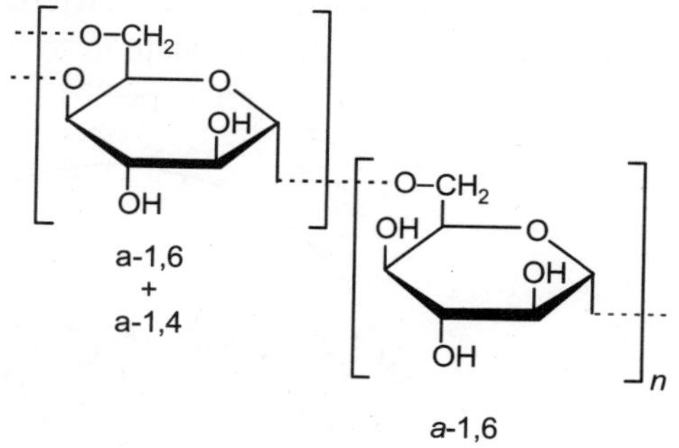

The straight chain consists of α-1,6 glycosidic linkages between glucose molecules, while branches begin from α-1,4 linkages (and in some cases, α-1,2 and α-1,3 linkages as well). (For information on the numbering of carbon atoms in glucose, see the glucose article.) Dextran is synthesized from sucrose by certain lactic-acid bacteria, the best-known being *Leuconostoc mesenteroides* and *Streptococcus mutans*. Dental plaque is rich in dextrans. Dextran is also formed by the lactic acid bacterium *Lactobacillus brevis* to create the crystals of tibicos, or water kefir fermented beverage which supposedly has some health benefits.

O–CH_2
O
O
OH
OH
a-1,6
+
a-1,4
O–CH_2
OH
O
OH
OH
n
a-1,6

CARBOHYDRATES IN CELL WALLS OF BACTERIA

Bacterial Cell Walls

Schematic of typical gram-positive cell wall showing arrangement of N-Acetylglucosamine and N-Acetlymuramic acid.

Around the outside of the cell membrane is the bacterial cell wall. Bacterial cell walls are made of peptidoglycan (also called murein), which is made from polysaccharide chains cross-linked by unusual peptides containing D-amino acids. Bacterial cell walls are different from the cell walls of plants and fungi which are made of cellulose and chitin, respectively. The cell wall of bacteria is also distinct from that of Archaea, which do not contain peptidoglycan. The cell wall is essential to the survival of many bacteria. The antibiotic penicillin is able to kill bacteria by inhibiting a step in the synthesis of peptidoglycan.

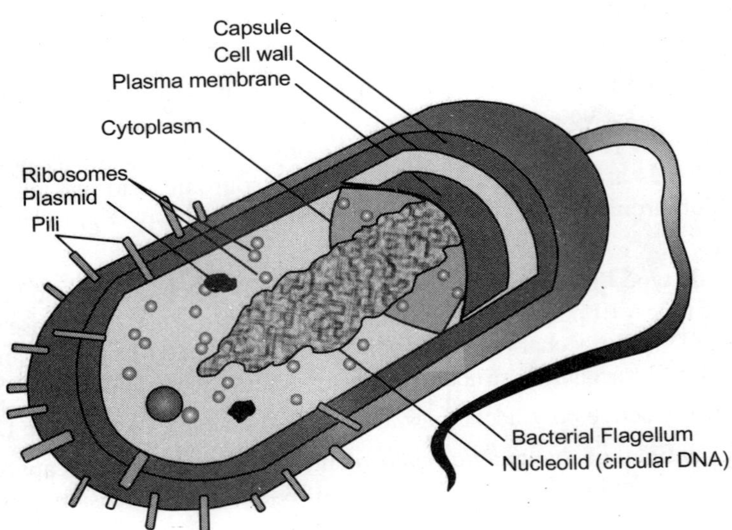

Fig. 2.13 : A typical gram-negative bacterium, with the thin cell wall sandwiched between the red outer membrane and the thin green plasma membrane

There are broadly speaking two different types of cell wall in bacteria, called Gram-positive and Gram-negative. The names originate from the reaction of cells to the Gram stain, a test long-employed for the classification of bacterial species.

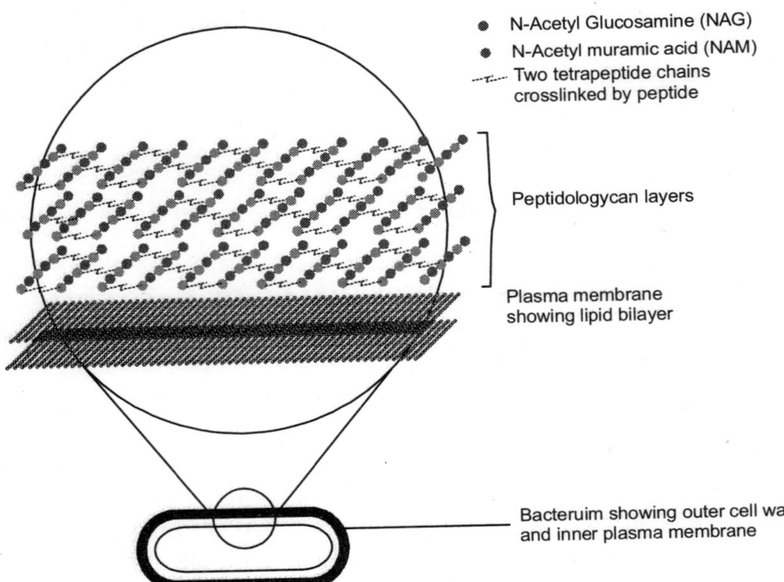

Fig. 2.14 : Showing the Simplified Schematic of cell wall in a gram positive bacterium

Gram-positive bacteria possess a thick cell wall containing many layers of peptidoglycan and teichoic acids. In contrast, Gram-negative bacteria have a relatively thin cell wall consisting of a few layers of peptidoglycan surrounded by a second lipid membrane containing lipopolysaccharides and lipoproteins. Most bacteria have the Gram-negative cell wall and only the Firmicutes and Actinobacteria (previously known as the low G+C and high G+C Gram-positive bacteria, respectively) have the alternative Gram-positive arrangement. These differences in structure can produce differences in antibiotic susceptibility, for instance vancomycin can kill only Gram-positive bacteria and is ineffective against Gram-negative pathogens, such as *Haemophilus influenzae* or *Pseudomonas aeruginosa*.

Archaeal Cell Walls

Although not truly unique, the cell walls of Archaea are unusual. Whereas peptidoglycan is a standard component of all bacterial cell walls, all archaeal cell walls lack peptidoglycan, with the exception of one group of methanogens. In that group, the peptidoglycan is a modified form very different from the kind found in bacteria There are four types of cell wall currently known among the Archaea.

One type of archaeal cell wall is that composed of pseudopeptidoglycan (also called pseudomurein). This type of wall is found in some methanogens, such as *Methanobacterium* and *Methanothermus*. While the overall structure of archaeal *pseudo*peptidoglycan superficially resembles that of bacterial peptidoglycan, there are a number of significant chemical differences. Like the peptidoglycan found in bacterial cell walls, pseudopeptidoglycan consists of polymer chains of glycan cross-linked by short peptide connections. However, unlike peptidoglycan, the sugar N-acetylmuramic acid is replaced by N-acetyltalosaminuronic acid, and the two sugars are bonded with a β,1-3 glycosidic linkage instead of β,1-4. Additionally, the cross-linking peptides are L-amino acids rather than D-amino acids as they are in bacteria.

A second type of archaeal cell wall is found in *Methanosarcina* and *Halococcus*. This type of cell wall is composed entirely of a thick layer of polysaccharides, which may be sulphated in the case of *Halococcus*. Structure in this type of wall is complex and as yet is not fully investigated.

A third type of wall among the Archaea consists of glycoprotein, and occurs in the hyperthermophiles, *Halobacterium*, and some methanogens. In *Halobacterium*, the proteins in the wall have a high content of acidic amino acids, giving the wall an overall negative charge. The result is an unstable structure that is stabilized by the presence of large quantities of positive sodium ions that neutralize the charge. Consequently, *Halobacterium* thrives only under conditions with high salinity.

In other Archaea, such as *Methanomicrobium* and *Desulfurococcus*, the wall may be composed only of surface-layer proteins, known as an *S-layer*. S-layers are common in bacteria, where they serve as either the sole cell-wall component or an outer layer in conjunction with polysaccharides. Most Archaea are Gram-negative, though at least one Gram-positive member is known.

Microsurgery Uses

These agents are used commonly by microsurgeons to decrease vascular thrombosis. The antithrombotic effect of dextran is mediated through its binding of erythrocytes, platelets, and vascular endothelium, increasing their electronegativity and thus reducing erythrocyte aggregation and platelet adhesiveness. Dextrans also reduce factor VIII-Ag Von Willebrand factor, thereby decreasing platelet function. Clots formed after administration of dextrans are more easily lysed due to an altered thrombus structure (more evenly distributed platelets with coarser fibrin). By inhibiting α-2 antiplasmin, dextran serves as a plasminogen activator and therefore possesses thrombolytic features.

Outside from these features, larger dextrans, which do not pass out of the vessels, are potent osmotic agents, and thus have been used urgently to treat hypovolemia. The hemodilution caused by volume expansion with dextran use improves blood flow, thus further improving patency of microanastomoses and reducing thrombosis. Still, no difference has been detected in antithrombotic effectiveness in comparison of intraarterial and intravenous administration of dextran.

Dextrans are available in multiple molecular weights ranging from 10,000 Da to 150,000 Da. The larger dextrans are excreted poorly from the kidney and therefore remain in the blood for as long as weeks until they are metabolized. Subsequently, they have prolonged antithrombotic and colloidal effects. In this family, dextran-40 (MW: 40,000 Da), has been the most popular member for anticoagulation therapy. Close to 70% of dextran-40 is excreted in urine within the first 24 hours after intravenous infusion while the remaining 30% will be retained for several more days.

Other Medical Uses

- It is used in some eye drops as a lubricant, and in certain intravenous fluids to solubilise other factors, e.g. iron (=iron dextran).
- Dextran in intravenous solution provides an osmotically neutral fluid that once in the body is digested by cells into glucose and free water. It is occasionally used to replace lost blood in emergency situations, when replacement blood is not available, but must be used with caution as it does not provide necessary electrolytes and can cause hyponatremia or other electrolyte disturbances.
- It also increases blood sugar levels.

Laboratory Uses

- Dextran is used in the osmotic stress technique for applying osmotic pressure to biological molecules.
- It is also used in some size-exclusion chromatography matrices; an example is Sephadex.
- Dextran has also been used in bead form to aid in bioreactor applications.
- Dextran has been used in immobilization in biosensors.
- Dextran preferentially binds to early endosomes; fluorescently-labelled dextran can be used to visualize these endosomes under a fluorescent microscope.
- Dextran can be used as a stabilising coating to protect metal nanoparticles from oxidation and improve biocompatibility.

Side Effects

Although there are relatively few side-effects associated with dextran use, these side-effects can be very serious. These include anaphylaxis, volume overload, pulmonary edema, cerebral edema, or platelet dysfunction. An uncommon but significant complication of dextran osmotic effect is acute renal failure.

The pathogenesis of this renal failure is the subject of many debates with direct toxic effect on tubules and glomerulus versus intraluminal hyperviscosity being some of the proposed mechanisms. Patients with history of diabetes mellitus, renal insufficiency, or vascular disorders are most at risk. Brooks and others recommend the avoidance of dextran therapy in patients with chronic renal insufficiency and CrCl<40 cc per minute.

Alginic acid (**algin**, **alginate**) is a viscous gum that is abundant in the cell walls of brown algae.

Structure

It is a linear copolymer with homopolymeric blocks of (1-4)-linked β-D-mannuronate (M) and its C-5 epimer α-L-guluronate (G) residues, respectively, covalently linked together in different sequences or blocks.

The monomers can appear in homopolymeric blocks of consecutive G-residues (G-blocks), consecutive M-residues (M-blocks), alternating M and G-residues (MG-blocks), or randomly organized blocks.

Forms

Commercial varieties of alginate are extracted from seaweed, including the giant kelp *Macrocystis pyrifera*, *Ascophyllum nodosum*, and various types of *Laminaria*. It is also produced by two bacterial genera *Pseudomonas* and *Azotobacter*, which played a major role in the unravelling of its biosynthesis pathway. Bacterial alginates are useful for the production of micro- or nanostructures suitable for medical applications.[1]

Alginate absorbs water quickly, which makes it useful as an additive in dehydrated products such as slimming aids, and in the manufacture of paper and textiles. It is also used for waterproofing and fireproofing fabrics, as a gelling agent, for thickening drinks, ice cream and cosmetics, and as a detoxifier that can absorb poisonous metals from the blood. Alginate is also produced by certain bacteria, notably Azotobacter species.

Alginate ranges from white to yellowish-brown, and takes filamentous, grainy, granular, and powdered forms.

Uses

Alginate is used in various pharmaceutical preparations such as Gaviscon, Bisodol, and Asilone. Alginate is used extensively as a mold-making material in dentistry, prosthetics, lifecasting, and textiles. It is also used in the food industry, for thickening soups and jellies. Calcium alginate is used in different types of medical products, including burn dressings that promote healing and can be removed with less pain than conventional dressings.

Also, due to alginate's biocompatibility and simple gelation with divalent cations such as Ca^{2+}, it is widely used for cell immobilization and encapsulation.

Alginic acid (alginato) is also used in culinary arts, most notably in the "Esferificación" (Sphereification) techniques of Ferrαn Adric ᶠEl Bulli in Roses, Girona, where natural juices of fruits and vegetables are encapsulated in bubbles that "explode" on the tongue when consumed. One of the most famous examples of this use of alginic acid was when Ferrαn Adriα used alginic acid to make apple caviar.

Due to its ability to absorb water quickly, alginate can be changed through a lyophilization process to a new structure that has the ability to expand. It is used in the weight loss industry as an appetite suppressant. A new type of diet using alginate is the CM3-Alginate Diet.

Chitin ($C_8H_{13}O_5N)_n$ is a long-chain polymer of a N-acetylglucosamine, a derivative of glucose, and is found in many places throughout the natural world. It is the main component of the cell walls of fungi, the exoskeletons of arthropods, such as crustaceans (like the crab, lobster and shrimp) and the insects, including ants, beetles and butterflies, the radula of mollusks and the beaks of the cephalopods, including squid and octopi. Chitin has also proven useful for several medical and industrial purposes. Chitin is a biological substance which may be compared to the polysaccharide cellulose and to the protein keratin. Although keratin is a protein, and not a carbohydrate like chitin, both keratin and chitin have similar structural functions. Chemistry, physical properties and biological function Chitin is a polysaccharide; it is synthesized from units of N-acetylglucosamine (more completely, N-acetyl-D-glucos-2-amine). These units form covalent β-1,4 linkages (similar to the linkages between glucose units forming cellulose). Chitin may therefore be described as cellulose with one hydroxyl group on each monomer substituted with an acetylamine group. This allows for increased hydrogen bonding between adjacent polymers, giving the chitin-polymer matrix increased strength. In its unmodified form, chitin is translucent, pliable, resilient and quite tough. In arthropods, however, it is often modified, becoming embedded in a hardened proteinaceous matrix, which forms much of the exoskeleton. In its pure form it is leathery, but when encrusted in calcium carbonate it becomes much harder. The difference between the unmodified and modified forms can be seen by comparing the body wall of a caterpillar (unmodified) to a beetle (modified).Chitin is one of many naturally occurring polymers. Its breakdown may be catalyzed by enzymes called chitinases, secreted by microorganisms such as bacteria and fungi, and produced by some plants. Some of these microorganisms have receptors to simple sugars from the decomposition of chitin. If chitin is detected, they then produce enzymes to digest it by cleaving the glycosidic bonds in order to convert it to simple sugars and ammonia. Chemically, chitin is closely related to chitosan (a more water-soluble derivative of chitin). It is also closely related to cellulose in that it is a long unbranched chain of glucose derivatives. Both materials contribute structure and strength, protecting the organism.

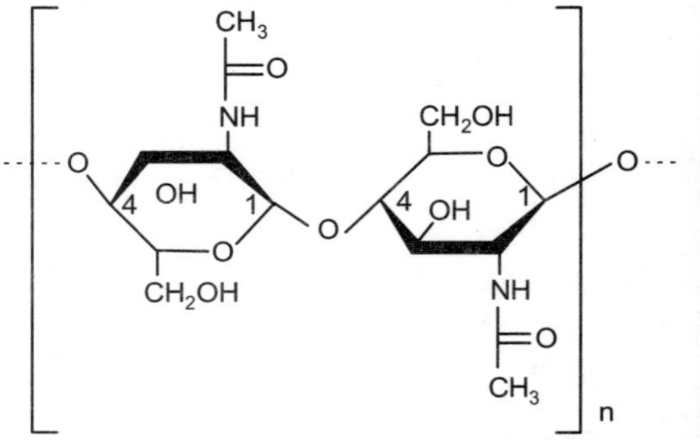

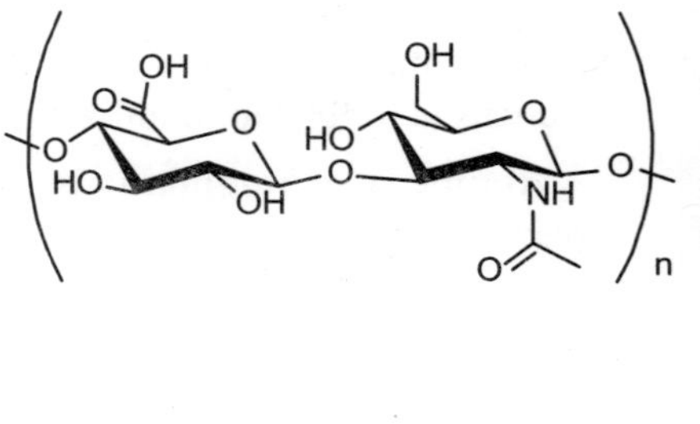

Hyaluronan (also called hyaluronic acid or hyaluronate) is a non-sulphated glycosaminoglycan distributed widely throughout connective, epithelial, and neural tissues. It is one of the chief components of the extracellular matrix, contributes significantly to cell proliferation and migration, and may also be involved in the progression of some malignant tumors. The average 70 kg (154 lbs) person has roughly 15 grams of hyaluronan in his body, one-third of which is turned over (degraded and synthesised) every day. Hyaluronic acid is also a component of the group A streptococcal extracellular capsule, and is believed to play a role in virulence

Functions

Until the late 1970s, hyaluronan was described as an "ectoplasm" molecule, an ubiquitous high energy carbohydrate polymer that is part of the extracellular matrix. For example, hyaluronan is a major component of the synovial fluid and was found to increase the viscosity of the fluid. Along with lubricin, it is one of the fluid's main lubricating components.

Hyaluronan is an important component of articular cartilage, where it is present as a coat around each cell (chondrocyte). When aggrecan monomers bind to hyaluronan in the presence of link protein, large highly negatively-charged aggregates form. These aggregates imbibe water and are responsible for the resilience of cartilage (its resistance to compression). The molecular weight (size) of hyaluronan in cartilage decreases with age, but the amount increases.

Hyaluronan is also a major component of skin, where it is involved in tissue repair. When skin is excessively exposed to UV rays, it becomes inflamed (sunburn) and the cells in the dermis stop producing as much hyaluronan, and increase the rate of its degradation. Hyaluronan degradation products also accumulate in the skin after UV exposure.

While it is abundant in extracellular matrices, hyaluronan also contributes to tissue hydrodynamics, movement and proliferation of cells, and participates in a number of cell surface receptor interactions, notably those including its primary receptor, CD44. Upregulation of CD44 itself is widely accepted as a marker of cell activation in lymphocytes. Hyaluronan's contribution to tumor growth may be due to its interaction with CD44. Receptor CD44 participates in cell adhesion interactions required by tumor cells.

Although hyaluronan binds to receptor CD44, there is evidence that hyaluronan degradation products transduce their inflammatory signal through Toll-like receptor 2 (TLR2), TLR4 or both TLR2, and TLR4 in macrophages and dendritic cells. TLR and hyaluronan play a role in innate immunity.

High concentrations of hyaluronan in the brains of young rats, and reduced concentrations in the brains of adult rats suggest that hyaluronan plays an important role in brain development.

Structure

The chemical structure of hyaluronan was determined in the 1950s in the laboratory of Karl Meyer. Hyaluronan is a polymer of disaccharides, themselves composed of D-glucuronic acid and D-N-acetylglucosamine, linked together via alternating β-1,4 and β-1,3 glycosidic bonds. Hyaluronan can be 25,000 disaccharide repeats in length. Polymers of hyaluronan can range in size from 5,000 to 20,000,000 Da *in vivo*. The average molecular weight in human synovial fluid is 3.4 million Da, and hyaluronan purified from human umbilical cord is 3,140,000 Da.

Hyaluronan is energetically stable in part because of the stereochemistry of its component disaccharides. Bulky groups on each sugar molecule are in sterically favored positions, whereas the smaller hydrogens assume the less-favorable axial positions.

Biological Synthesis

Hyaluronan is synthesized by a class of integral membrane proteins called hyaluronan synthases, of which vertebrates have three types *e.g.,* HAS1, HAS2, and HAS3. These enzymes lengthen hyaluronan by repeatedly adding glucuronic acid and N-acetylglucosamine to the nascent polysaccharide as it is extruded through the cell membrane into the extracellular space.

Hyaluronan synthesis (HAS) has been shown to be inhibited by 4-Methylumbelliferone (hymecromone, heparvit), a 7-Hydroxy-4-methylcoumarin derivative. This selective inhibition (without inhibiting other Glycosaminoglycans) may prove useful in preventing metastasis of malignant tumor cells.

Degradation

Hyaluronan is degraded by a family of enzymes called hyaluronidases. In humans, there are at least seven types of hyaluronidase-like enzymes, several of which are tumor suppressors. The degradation products of hyaluronan, the oligosaccharides and very low-molecular-weight hyaluronan, exhibit pro-angiogenic properties. In addition, recent studies showed that hyaluronan fragments, not the native high-molecular mass of hyaluronan, can induce inflammatory responses in macrophages and dendritic cells in tissue injury and in skin transplant rejection.

Medical Applications

Hyaluronan is naturally found in many tissues of the body, such as skin, cartilage, and the vitreous humour. It is therefore well suited to biomedical applications targeting these tissues. The first hyaluronan biomedical product, Healon, was developed in the 1970s and 1980s by Pharmacia, and is approved for use in eye surgery (i.e., corneal transplantation, cataract surgery, glaucoma surgery and surgery to repair retinal detachment). Other biomedical companies also produce brands of hyaluronan for ophthalmic surgery.

Hyaluronan is also used to treat osteoarthritis of the knee. Such treatments, called *viscosupplementation*, are administered as a course of injections into the knee joint and are believed to supplement the viscosity of the joint fluid, thereby lubricating the joint, cushioning the joint, and producing an analgesic effect. It has also been suggested that hyaluronan has positive biochemical effects on cartilage cells. However, some placebo controlled studies have cast doubt on the efficacy of hyaluronan injections, and hyaluronan is recommended primarily as a last alternative to surgery. Oral use of hyaluronan has been lately suggested, although its effectiveness needs to be demonstrated. At present, there are some preliminary clinical studies that suggest that oral administration of Hyaluronan has a positive effect on osteoarthritis, but it remains to be seen if there is any real benefit to the treatment.

Due to its high biocompatibility and its common presence in the extracellular matrix of tissues, hyaluronan is gaining popularity as a biomaterial scaffold in tissue engineering research.

In some cancers, hyaluronan levels correlate well with malignancy and poor prognosis. Hyaluronan is thus often used as a tumor marker for prostate and breast cancer. It may also be used to monitor the progression of the disease.

Hyaluronan may also be used postoperatively to induce tissue healing, notably after cataract surgery. Current models of wound healing propose that larger polymers of hyaluronic acid appear in the early stages of healing to physically make room for white blood cells, which mediate the immune response.

Hyaluronan has also been used in the synthesis of biological scaffolds for wound healing applications. These scaffolds typically have proteins such as fibronectin attached to the hyaluronan to facilitate cell migration into the wound. This is particularly important for individuals with diabetes who suffer from chronic wounds.

In 2007, the EMEA extended its approval of Hylan GF-20 as a treatment for ankle and shoulder osteoarthritis pain.

Heparin, a highly-sulphated glycosaminoglycan, is widely used as an injectable anticoagulant and has the highest negative charge density of any known biological molecule. It can also be used to form an inner anticoagulant surface on various experimental and medical devices such as test tubes and renal dialysis machines. Pharmaceutical grade heparin is derived from mucosal tissues of slaughtered meat animals such as porcine (pig) intestine or bovine (cow) lung.

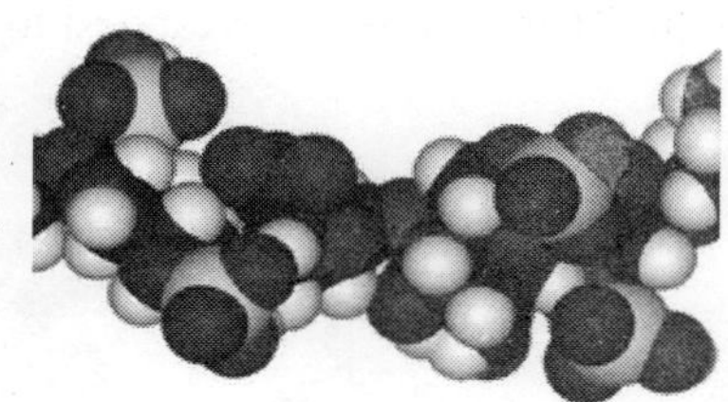

Although used principally in medicine for anticoagulation, the true physiological role in the body remains unclear, because blood anti-coagulation is achieved mostly by endothelial cell-derived heparan sulphate proteoglycans. Heparin is usually stored within the secretory granules of mast cells and released only into the vasculature at sites of tissue injury. It has been proposed that, rather than anticoagulation, the main purpose of heparin is in a defensive mechanism at sites of tissue injury against invading bacteria and other foreign materials.[4] In addition, it is preserved across a number of widely different species, including some invertebrates which lack a similar blood coagulation system. Heparin structure

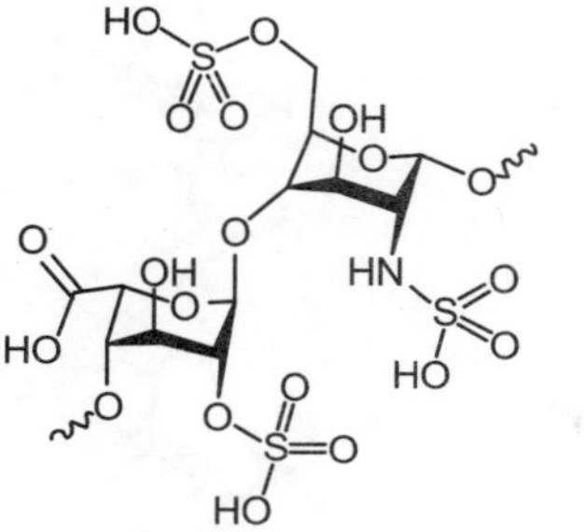

Native heparin is a polymer with a molecular weight ranging from 3 kDa to 50 kDa, although the average molecular weight of most commercial heparin preparations is in the range of 12 kDa to 15 kDa. Heparin is a member of the glycosaminoglycan family of carbohydrates (which includes the closely-related molecule heparan sulphate) and consists of a variably-sulphated repeating disaccharide unit. The main disaccharide units that occur in heparin are shown below. The most common disaccharide unit is composed of a 2-O-sulphated iduronic acid and 6-O-sulphated, N-sulphated glucosamine, IdoA(2S)-GlcNS(6S). For example, this makes up 85% of heparins from beef lung and about 75% of those from porcine intestinal mucosa. Not shown below are the rare disaccharides containing a 3-O-sulphated glucosamine (GlcNS(3S,6S)) or a free amine group ($GlcNH_3^+$). Under physiological conditions, the ester and amide sulphate groups are deprotonated and attract positively-charged counterions to form a heparin salt. It is in this form that heparin is usually administered as an anticoagulant.

1 unit of heparin (the "Howell Unit") is an amount approximately equivalent to 0.002 mg of pure heparin, which is the quantity required to keep 1 mL of cat's blood fluid for 24 hours at 0°C.

Three-dimensional Structure

The three-dimensional structure of heparin is complicated by the fact that iduronic acid may be present in either of two low-energy conformations when internally positioned within an oligosaccharide. The conformational equilibrium being influenced by sulfation state of adjacent glucosamine sugars. Nevertheless, the solution structure of a heparin dodecasacchride composed solely of six GlcNS(6S)-IdoA(2S) repeat units has been determined using a combination of NMR spectroscopy and molecular modeling techniques. Two models were constructed, one in which all IdoA(2S) were in the $^{2}S_{0}$ conformation (**A** and **B** below), and one in which they are in the $^{1}C_{4}$ conformation (**C** and **D** below). However there is no evidence to suggest that changes between these conformations occur in a concerted fashion. These models correspond to the protein data bank code 1HPN.

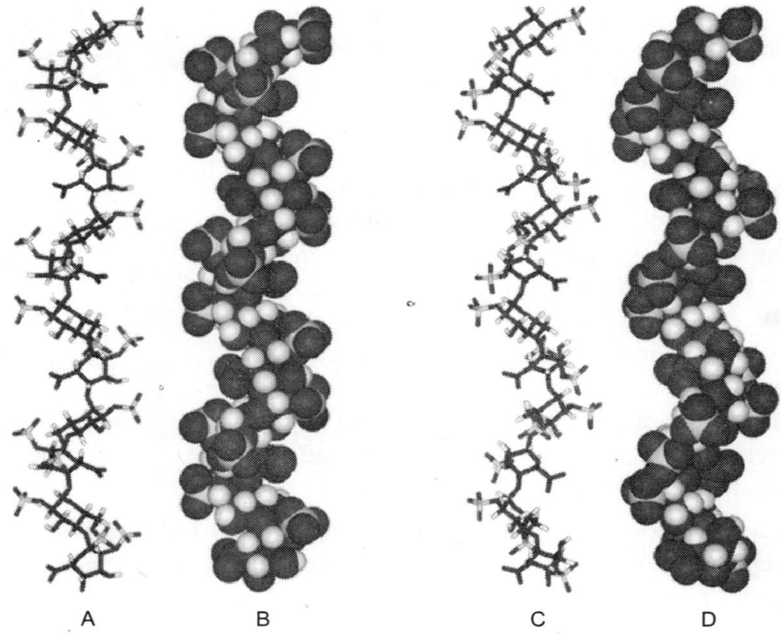

In the image above:

- **A** = 1HPN (all IdoA(2S) residues in $^{2}S_{0}$ conformation) Jmol viewer
- **B** = van der Waals radius space filling model of *A*
- **C** = 1HPN (all IdoA(2S) residues in $^{1}C_{4}$ conformation) Jmol viewer
- **D** = van der Waals radius space filling model of *C*

Fig. 2.15 : Thru-dimensional structure of DNA

In these models, heparin adopts a helical conformation, the rotation of which places clusters of sulphate groups at regular intervals of about 17 angstroms (1.7 nm) on either side of the helical axis.

Medical use

Heparin is a naturally-occurring anticoagulant produced by basophils and mast cells. Heparin acts as an anticoagulant, preventing the formation of clots and extension of existing clots within the blood. While heparin does not break down clots that have already formed (unlike tissue plasminogen activator), it allows the body's natural clot lysis mechanisms to work normally to break down clots that have already formed. Heparin is used for anticoagulation for the following conditions:

- Acute coronary syndrome, e.g., NSTEMI
- Atrial fibrillation
- Deep-vein thrombosis and pulmonary embolism
- Cardiopulmonary bypass for heart surgery.

Heparin and its derivatives (enoxaparin, dalteparin, and so forth) are effective at preventing deep-vein thromboses and pulmonary emboli in patients at risk, but there is no evidence that they are effective at preventing death. Current guidelines recommend aspirin and leg stockings instead.

Heparin binds to the enzyme inhibitor antithrombin (AT) causing a conformational change that results in its activation through an increase in the flexibility of its reactive site loop. The activated AT then inactivates thrombin and other proteases involved in blood clotting, most notably factor Xa. The rate of inactivation of these proteases by AT can increase by up to 1000-fold due to the binding of heparin.

AT binds to a specific pentasaccharide sulfation sequence contained within the heparin polymer

GlcNAc/NS(6S)-GlcA-GlcNS(3S,6S)-IdoA(2S)-GlcNS(6S)

The conformational change in AT on heparin-binding mediates its inhibition of factor Xa. For thrombin inhibition however, thrombin must also bind to the heparin polymer at a site proximal to the pentasaccharide. The highly-negative charge density of heparin contributes to its very strong electrostatic interaction with thrombin. The formation of a ternary complex between AT, thrombin, and heparin results in the inactivation of thrombin. For this reason heparin's activity against thrombin is size-dependent, the ternary complex requiring at least 18 saccharide units for efficient formation. In contrast anti factor Xa activity only requires the pentasaccharide binding site.

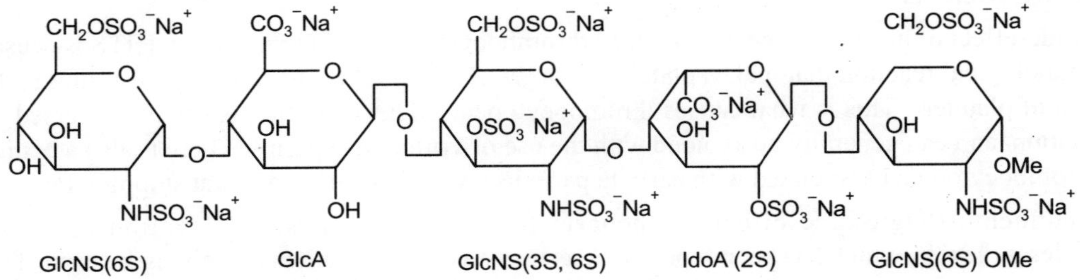

$CH_2OSO_3^-Na^+$ $CO_3^-Na^+$ $CH_2OSO_3^-Na^+$ $CH_2OSO_3^-Na^+$

O O O O O

OH OH $OSO_3^-Na^+$ $CO_3^-Na^+$ OH OH

OH O O O O OMe

$NHSO_3^-Na^+$ OH $NHSO_3^-Na^+$ $OSO_3^-Na^+$ $NHSO_3^-Na^+$

GlcNS(6S) GlcA GlcNS(3S, 6S) IdoA (2S) GlcNS(6S) OMe

Fig. 2.16 : Chemical structure of fondaparinux

This size difference has led to the development of low-molecular-weight heparins (LMWHs) and more recently to fondaparinux as pharmaceutical anticoagulants. Low-molecular-weight heparins and fondaparinux target anti-factor Xa activity rather than anti-thrombin (IIa) activity, with the aim of facilitating a more subtle regulation of coagulation and an improved therapeutic index. The chemical structure of fondaparinux is shown to the left. It is a synthetic pentasaccharide, whose chemical structure is almost identical to the AT binding pentasaccharide sequence that can be found within polymeric heparin and heparan sulphate.

With LMWH and fondaparinux, there is a reduced risk of osteoporosis and heparin-induced thrombocytopenia (HIT). Monitoring of the APTT is also not required and indeed does not reflect the anticoagulant effect, as APTT is insensitive to alterations in factor Xa.

Danaparoid, a mixture of heparan sulphate, dermatan sulphate, and chondroitin sulphate can be used as an anticoagulant in patients who have developed HIT. Because danaparoid does not contain heparin or heparin fragments, cross-reactivity of danaparoid with heparin-induced antibodies is reported as less than 10%.

The effects of heparin are measured in the lab by the partial thromboplastin time (aPTT), (the time it takes the blood plasma to clot).

Administration

Details of administration are available in clinical practice guidelines by the American College of Chest Physicians:

- Non-weight-based heparin dose adjustment
- Weight-based-heparin dose adjustment

Heparin is given parenterally, as it is degraded when taken by mouth. It can be injected intravenously or subcutaneously (under the skin). Intramuscular injections (into muscle) are avoided because of the potential for forming hematomas.

Because of its short biologic half-life of approximately one hour, heparin must be given frequently or as a continuous infusion. However, the use of low molecular weight heparin (LMWH) has allowed once daily dosing, thus not requiring a continuous infusion of the drug. If long-term anticoagulation is required, heparin is often used only to commence anticoagulation therapy until the oral anticoagulant warfarin takes effect.

Adverse Reactions

A serious side-effect of heparin is heparin-induced thrombocytopenia (HIT syndrome). HITS is caused by an immunological reaction that makes platelets a target of immunological response, resulting in the degradation of platelets. This is what causes thrombocytopenia. This condition is usually reversed on discontinuation, and can generally be avoided with the use of synthetic heparins. There is also a benign form of thrombocytopenia associated with early heparin use, which resolves without stopping heparin.

Two nonhemorrhagic side effects of heparin treatment. The first is elevation of serum aminotransferase levels, which has been reported in as many as 80% of patients receiving heparin. This abnormality is not associated with liver dysfunction, and it disappears after the drug is discontinued. The other complication is hyperkalemia, which occurs in 5 to 10% of patients receiving heparin, and is the result of heparin-induced aldosterone suppression. The hyperkalemia can appear within a few days after the onset of heparin therapy.

Rarer side-effects include alopecia and osteoporosis with chronic use.

As with many drugs, overdoses of heparin can be fatal. In September 2006, heparin received worldwide publicity when 3 prematurely-born infants died after they were mistakenly given overdoses of heparin at an Indianapolis hospital.

Treatment of Overdose

In case of overdose, protamine sulphate (1 mg per 100 Units of Heparin that had been given over 4 hours) can be given to counteract the action of heparin.

History

Heparin is one of the oldest drugs currently still in widespread clinical use. Its discovery in 1916 predates the establishment of the United States Food and Drug Administration, although it did not enter clinical trials until 1935. It was originally isolated from canine liver cells, hence its name (*hepar* or Greek "liver"). Heparin's discovery can be attributed to the research activities of two men, Jay McLean and William Henry Howell.

In 1916, McLean, a second-year medical student at Johns Hopkins University, was working under the guidance of Howell investigating pro-coagulant preparations, when he isolated a fat-soluble phosphatide anti-coagulant. It was Howell who coined the term *heparin* for this type of fat-soluble anticoagulant in 1918. In the early 1920s, Howell isolated a water-soluble polysaccharide anticoagulant, which was also termed *heparin*, although it was distinct from the phosphatide preparations previously isolated. It is probable that the work of McLean changed the focus of the Howell group to look for anticoagulants, which eventually led to the polysaccharide discovery.

Between 1933 and 1936, Connaught Medical Research Laboratories, then a part of the University of Toronto, perfected a technique for producing safe, non-toxic heparin that could be administered to patients in a salt solution. The first human trials of heparin began in May 1935, and, by 1937, it was clear that Connaught's heparin was a safe, easily-available, and effective blood anticoagulant. Prior to 1933, heparin was available, but in small amounts, and was extremely expensive, toxic, and, as a consequence, of no medical value.

Novel Drug Development Opportunities for Heparin

As detailed in the table below, there is a great deal of potential for the development of heparin-like structures as drugs to treat a wide range of diseases, in addition to their current use as anticoagulants.

Disease states sensitive to heparin	Heparins effect in experimental models	Clinical status
1	2	3
Adult respiratory distress syndrome	Reduces cell activation and accumulation in airways, neutralizes mediators and cytotoxic cell products, and improves lung function in animal models	Controlled clinical trials

...(Contd.)

1	2	3
Allergic encephalomyelitis	Effective in animal models	—
Allergic rhinitis	Effects as for adult respiratory distress syndrome, although no specific nasal model has been tested	Controlled clinical trial
Arthritis	Inhibits cell accumulation, collagen destruction and angiogenesis	Anecdotal report
Asthma	As for adult respiratory distress syndrome, however it has also been shown to improve lung function in experimental models	Controlled clinical trials
Cancer	Inhibits tumour growth, metastasis and angiogenesis, and increases survival time in animal models	Several anecdotal reports
Delayed type hypersensitivity reactions	Effective in animal models	—
Inflammatory bowel disease	Inhibits inflammatory cell transport in general. No specific model tested	Controlled clinical trials
Interstitial cystitis	Effective in a human experimental model of interstitial cystitis	Related molecule now used clinically
Transplant rejection	Prolongs allograph survival in animal models	

—indicates no information available

As a result of heparin's effect on such a wide variety of disease states a number of drugs are indeed in development whose molecular structures are identical or similar to those found within parts of the polymeric heparin chain.

Drug molecule	**Effect of new drug compared to heparin**	**Biological activities**
Heparin tetrasaccharide	Non-anticoagulant, non-immunogenic, orally active	Anti-allergic
Pentosan polysulphate	Plant derived, little anticoagulant activity, Anti-inflammatory, orally active	Anti-inflammatory, anti-adhesive, anti-metastatic
Phosphomanno-pentanose sulphate	Potent inhibitor of heparanase activity	Anti-metastatic, anti-angiogenic, anti-inflammatory
Selectively chemically O-desulphated heparin	Lacks anticoagulant activity	Anti-inflammatory, anti-allergic, anti-adhesive

De-polymerisation Techniques

Either chemical or enzymatic de-polymerisation techniques or a combination of the two underlie the vast majority of analyses carried out on the structure and function of heparin and heparan sulphate (HS).

Enzymatic

The enzymes traditionally used to digest heparin or HS are naturally produced by the soil bacterium Pedobacter heparinus (formerly named Flavobacterium heparinum). This bacterium is capable of utilizing either heparin or HS as its sole carbon and nitrogen source. In order to do this it produces a range of enzymes such as lyases, glucuronidases, sulfoesterases and sulfamidases. It is the lyases that have mainly been used in heparin/HS studies. The bacterium produces three lyases, heparinases I (EC 4.2.2.7), II (no EC number assigned) and III (EC 4.2.2.8) and each has distinct substrate specificities as detailed below.

Heparinase enzyme	Substrate specificity
Heparinase I	GlcNS(±6S)-IdoA(2S)
Heparinase II	GlcNS/Ac(±6S)-IdoA(±2S) GlcNS/Ac(±6S)-GlcA
Heparinase III	GlcNS/Ac(±6S)-GlcA/IdoA (with a preference for GlcA)

The lyases cleave heparin/HS by a beta elimination mechanism. This action generates an unsaturated double bond between C4 and C5 of the uronate residueThe C4-C5 unsaturated uronate is termed ÄUA or UA. It is a sensitive UV chromaphore (max absorption at 232nm) and allows the rate of an enzyme digest to be followed as well as providing a convenient method for detecting the fragments produced by enzyme digestion.

Chemical

Nitrous acid can be used to chemically de-polymerise heparin/HS. Nitrous acid can be used at pH 1.5 or at a higher pH of 4. Under both conditions nitrous acid effects deaminative cleavage of the chain.

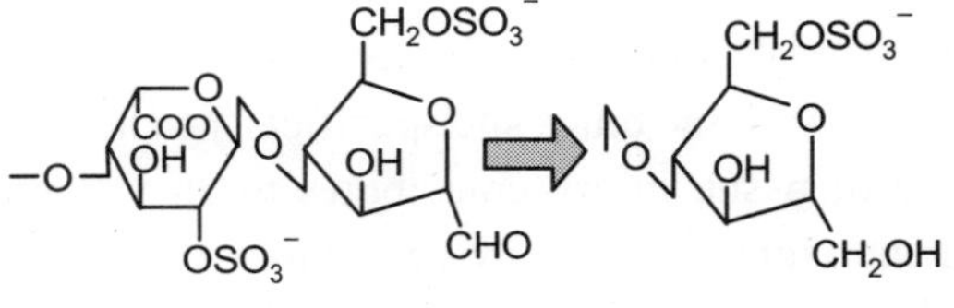

$CH_2OSO_3^-$ $CH_2OSO_3^-$ O COO O O OH OH –O– O OH OSO_3^- CHO CH_2OH

IdoA(2S)-aMan: The anhydromannose can be reduced to an anhydromannitol

At both 'high' (4) and 'low' (1.5) pH, deaminative cleavage occurs between GlcNS-GlcA and GlcNS-IdoA, all be it at a slower rate at the higher pH. The deamination reaction, and therefore chain cleavage, is regardless of O-sulfation carried by either monosaccharide unit.

At low pH deaminative cleavage results in the release of inorganic SO_4, and the conversion of GlcNS into anhydromannose (aMan). Low pH nitrous acid treatment is an excellent method to distinguish N-sulphated polysaccharides such as heparin and HS from non N-sulphated polysacchrides such as chondroitin sulphate and dermatan sulphate; chondroitin sulphate and dermatan sulphate being unsusceptible to nitrous acid cleavage.

Evolutionary Conservation

In addition to the bovine and porcine tissue from which pharmaceutical-grade heparin is commonly extracted, heparin has also been extracted and characterized from the following species:

1. Turkey.
2. Whale.
3. Dromedary camel.
4. Mouse.
5. Humans.
6. Lobster.
7. Fresh water mussel.
8. Clam.
9. Shrimp.
10. Mangrove crab.
11. Sand dollar.

The biological activity of heparin within species 6–11 is unclear and further supports the idea that the main physiological role of heparin is not anticoagulation. These species do not possess any blood coagulation system similar to that present within the species listed 1–5. The above list also demonstrates how heparin has been highly evolutionarily conserved with molecules of a similar structure being produced by a broad range of organisms belonging to many different phyla.

Other Uses/information

- Heparin gel (topical) may sometimes be used to treat sports injuries. It is known that the diprotonated form of histamine binds site specifically to heparin. The release of histamine from mast cells at a site of tissue injury contributes to an inflammatory response. The rationale behind the use of such topical gels may be to block the activity of released histamine, and so help to reduce inflammation.
- Heparin gains the capacity to initiate angiogenesis when its copper salt is formed. Copper-free molecules are non-angiogenic. In contrast heparin may inhibit angiogenesis when it is administered in the presence of corticosteroids. This anti-angiogenic effect is independent of heparins anticoagulant activity.

- Test tubes, Vacutainers, and capillary tubes that use the lithium salt of heparin (lithium heparin) as an anticoagulant are usually marked with green stickers and green tops. Heparin has the advantage over EDTA of not affecting levels of most ions. However, it has been shown that the levels of ionized calcium may be decreased if the concentration of heparin in the blood specimen is too high. Heparin can interfere with some immunoassays, however. As lithium heparin is usually used, a person's lithium levels cannot be obtained from these tubes; for this purpose, royal-blue-topped Vacutainers containing sodium heparin are used.
- Heparin-coated blood oxygenators are available for use in heart-lung machines. Among other things, these specialized oxygenators are thought to improve overall biocompatibility and host homeostasis by providing characteristics similar to native endothelium.
- The DNA binding sites on RNA polymerase can be occupied by heparin, preventing the polymerase binding to promoter DNA. This property is exploited in a range of molecular biological assays.
- Common diagnostic procedures require PCR amplification of a patient's DNA, which is easily extracted from white blood cells treated with heparin. This poses a potential problem, since heparin may be extracted along with the DNA, and it has been found to interfere with the PCR reaction at levels as low as 0.002 U in a 50 iL reaction mixture.
- Immobilized heparin can be used as an affinity ligand in protein purification. The format of immobilized heparin can vary widely from coated plastic surfaces for diagnostic purposes to chromatography resin. Most types of immobilized heparin can be used in three ways. The first of which is to use heparin to select out specific coagulation factors or other types of heparin-binding proteins from a complex mixture of non-heparin-binding proteins. Specific proteins can then be selectively dissociated from heparin with the use of differing salt concentrations or by use of a salt gradient. The second use is to use heparin as a high-capacity cation exchanger. This use takes advantage of heparin's high number of anionic sulphate groups. These groups will capture common cations such as Na^+ or Ca^{2+} in solution. The third use for immobilized heparin is group-specific purification of RNA and DNA binding proteins such as transcription factors and/or virus coat proteins. This methodology takes advantage of heparin's similar properties to RNA and DNA i.e. negatively charged sugar molecule.
- Heparin does not break up fibrin, it only prevents conversion of fibrinogen to fibrin. Only thrombolytics can break up a clot.

THE INTERPRETATION OF THE CHEMICAL REACTION OF CARBOHYDRATES

1. Reaction with hydrazine from hydrazines and osazones react with other monosaccharides and other carbohydrates containing a free sugar group to form hydrazones and osazones. The reactions of glucose with phenyl hydrazine may be used for illustration as commonly give they are:

D-Glucose + phenylhydrazine → D-Glucose phenyl hydrazine → NH_2 NH_3 +

(1E)-3,4,5,6,7-pentahydroxy-1-(2-phenylhydrazinylidene) heptan-2-one intermediate product + phenylhydrazine → (5Z)-5,6-bis(2-phenylhydrazinylidene)hexane-1,2,3,4-tetrol D-Glucose phenyl osazone

However, it appears that although the above sequence of reactions represents the result the conversation of the phenyl hydrazones to the phenyl osazones actually involves a more complex mechanism because phenyl hydrazine is reducing rather than an oxidizing agent. Weygand proposed that osazones formation may involve an Amadori Rearrangement.

phenylhydrazine ⇌ (Amadori rearrangement) ⇌ ⇌

$+ H_2O$ ⇌ ⇌ phenyl osazone

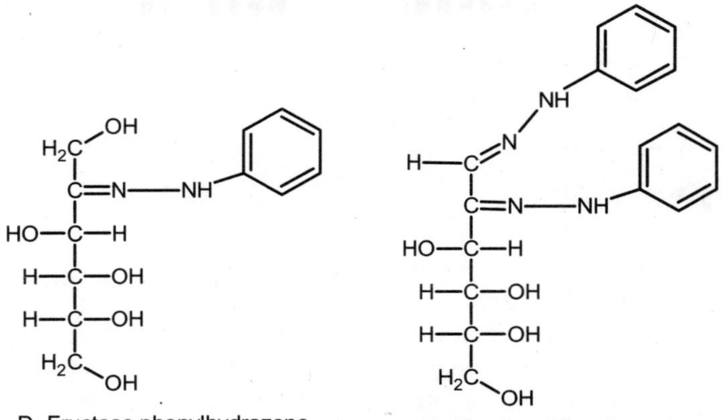

D -Fructose phenylhydrazone

D -Fructose phenylosazone

Fig. 2.17 : Fructose and other ketoses react similarly to form the hydrazones and osazones.

With few exceptions the hydrazones are soluble and difficult to isolate. One the other hand, the osazones are relatively insoluble and crystallize in a beautiful; and characteristic form for different sugars (see fig. 2.17 for glucose) . It must be noted that the osazones of glucose fructose and mannose are identical and by this fact Fischer demonstrated that glucose, fructose and mannose are identical in the lower four carbon atoms. Similar relations were also found in the case of other groups of sugars, and thus the osazones become of prime importance in helping to determine the structural configurations of the sugars.

Osazones when treated with strong hydrochloric acid.

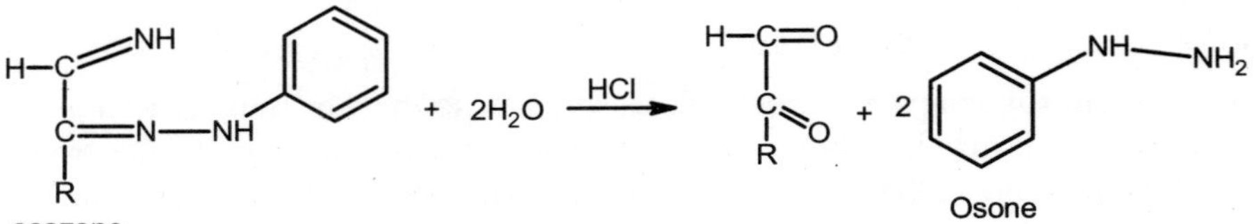

osazone + $2H_2O$ $\xrightarrow{HCl}$ Osone + 2 phenylhydrazine

When osones are treated with zinc and acetic acid, the aldehyde group is preferentially reduced to form the corresponding ketoses.

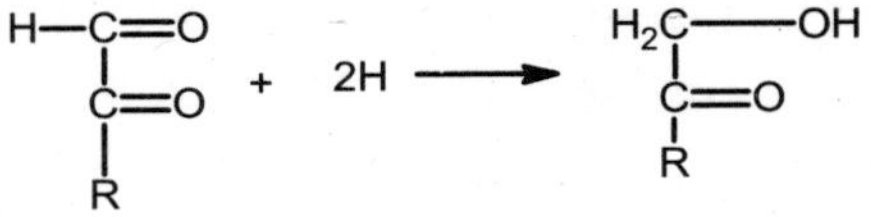

Osone + 2H → Ketoses

These reactions provide a method of converting an aldose into a ketose for example, the conversion of glucose to fructose.

Reaction with Hydroxylamine to form oximes hydroxylamine condenses with aldoses and ketoses as it does with ordinary aldehydes and ketoses, to form oximes.

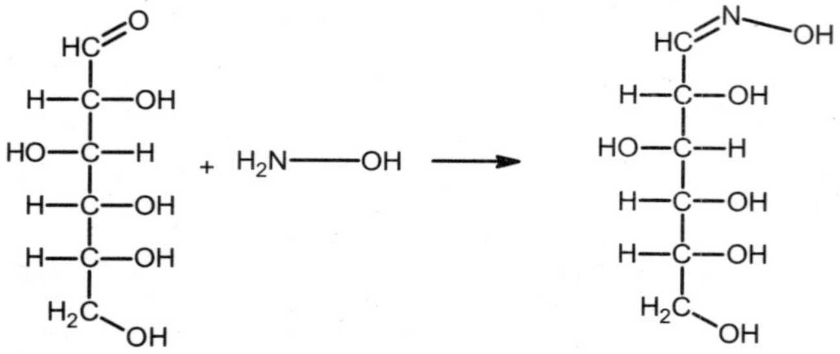

D - Glucose + H_2N—OH → D - Glucose oxime

Whole developed a method based upon use of the sugar oxime, of shortening the sugar chain one carbon atom at a time. In this way a hexone may be degraded to pentose, a pentose to a tetrose etc, In order to do so the oxime is treated with acetic anhydride, which removed a molecule of water and converts the oxime to a cyanohydrine or nitrile ad also acetates the hydroxyl groups. The cyanohydrine then treated with ammonical silver nitrate solution (Tollen's Reagent) which removes the hydrogen cyanide to form an acetylated sugar with one less carbon atom than the original, which can be decomposed into sugar. The process is shown in outline form as follows the acetyl group has been omitted.

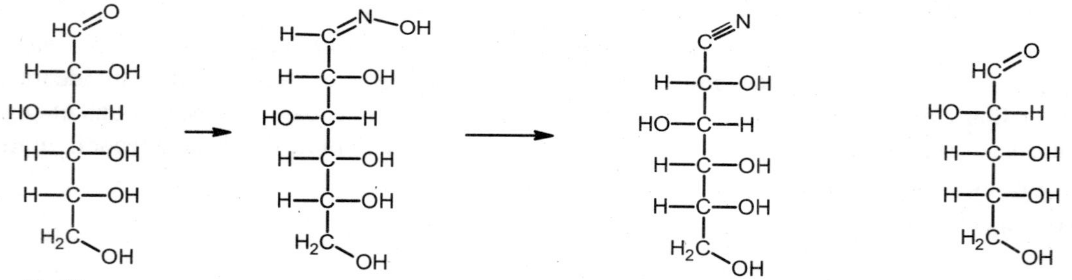

D - Glucose → D- Glucose Oxime → D - Gluconitrile D - Arbinose

Reduction to form sugar alcohols both aldoses and ketoses may be reduced to the corresponding polyhydroxy alcohols. This may be accomplished with sodium amalgam or, better electrolytically or by hydrogen under high pressure in the presence of a catalyst. The alcohols formed from glucose mannose, and fructose is:

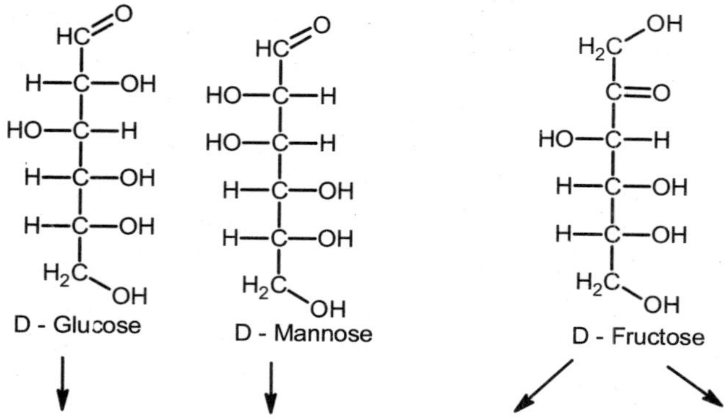

D - Glucose D - Mannose D - Fructose

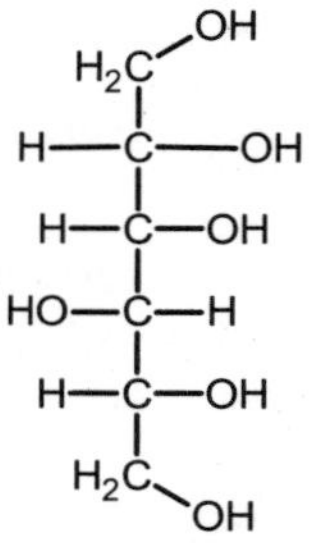

D - Sorbitol 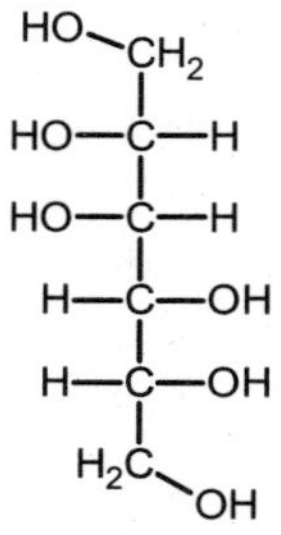 D - Mannitol 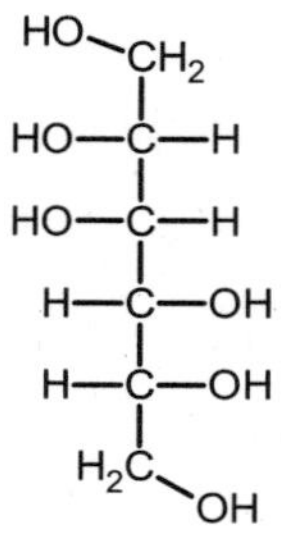 D - Mannitol 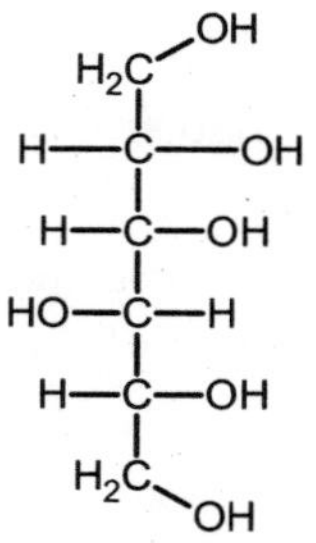 D - Sorbitol

Each aldoses yield the corresponding alcohol upon reduction while a ketose form to alcohols.

Because of the appearance of a new asymmetry carbon atoms in the process. It is interesting to note that reduction of the ketohexose L – sorbose forms D – sorbitol and L – iditol

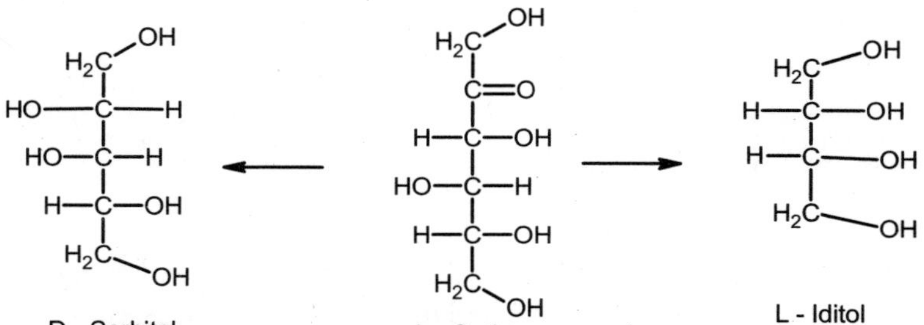

D - Sorbitol L - Sorbose L - Iditol

If the above formula of D Sorbitol is rotated longitudinally through the plane of the molecule 180° it will appear, the conventional formula of D – Sorbitol.

Reduction of glyceric aldehyde and dihydroxyacetone forms the trihydroxy alcohol glycerol.

Erythritol is the reduction product of the aldotetrose erithrose. Ribitol is the alcohol derived from aldopentose ribose.

Reduction of Galactose Produces Dulcitol

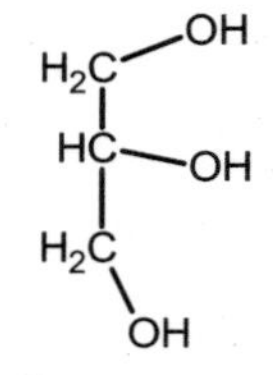

Glycerol 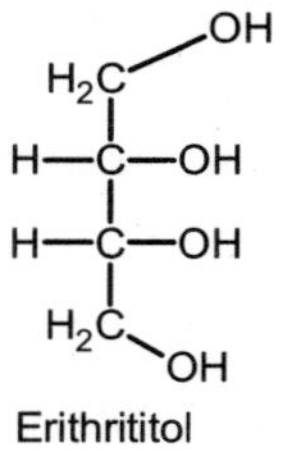 Erithrititol 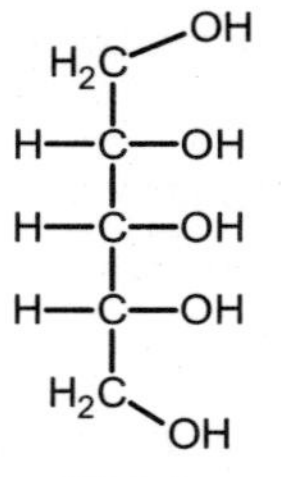 Ribitol 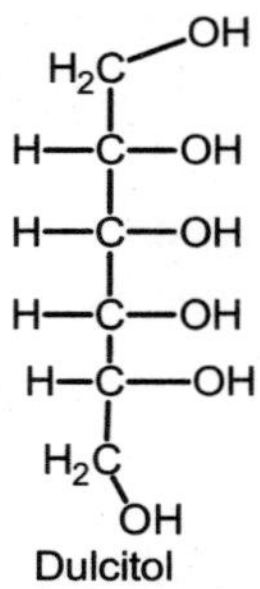 Dulcitol

Each of these sugar alcohols is derived by the reduction of both D and L corresponding sugar, and the alcohols are not prefixed by the either D or L. They are optically inactive, though Erythritol, Ribitol and dulcitol possesses asymmetric carbon atoms. This is due to the fact that the molecules are symmetrical and are internally compensated relative to polarized light just as meso- tartaric acid is .

All the sugar alcohols are well crystallized compounds, soluble in water and alcohol, and they have a sweet taste.

A number of the sugar alcohols may be oxidized to the corresponding ketoses by approximate bacteria in the presence of oxygen. For example D Sorbitol is oxidized as follows by *Acetobactor suboxidans.*

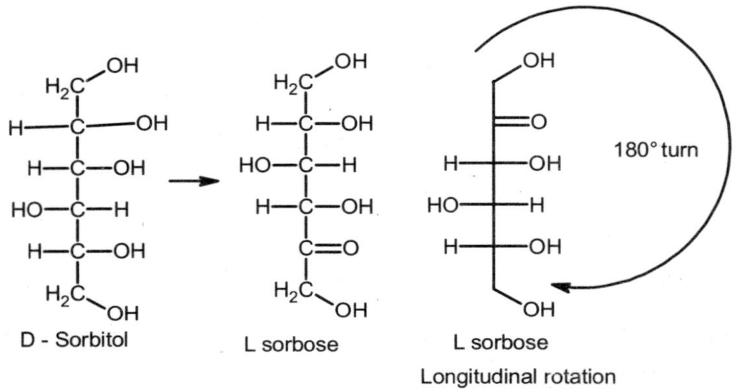

D - Sorbitol → L sorbose; L sorbose (180° turn), Longitudinal rotation

Oxidation occurs on the fifth carbon atom of D – Sorbitol to form a ketone belonging to the l – series, L Arbose. This reaction is used to produce large quantities of L Sorbose for the synthesis of ascorbic acid, or the vitamin C.

Dihydroxyacetone may be prepared similarly by oxidation of glycerol.

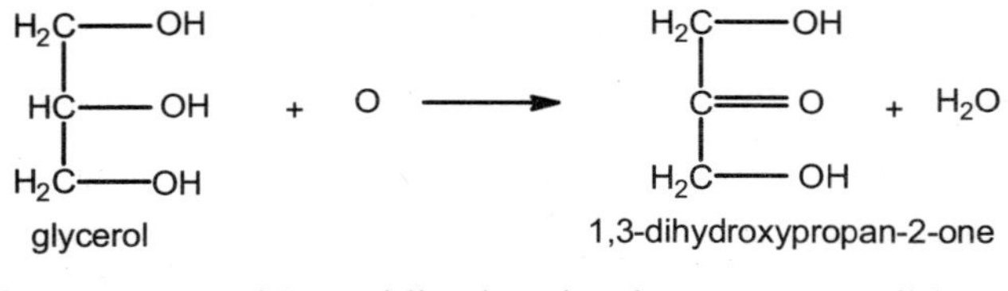

glycerol + O → 1,3-dihydroxypropan-2-one + H_2O

Oxidation to produce sugar acids, oxidized under the proper conditions, the aldoses may form monobasic aldonic acids, or dibasic saccharic acids, or monobasic uronic acid containing the aldehyde group.

Oxidation of a aldoses with bromine water converts the aldehyde group to carboxyl group and thereby forms the corresponding acid which water to form hypobromous acid which acts as the oxidizing agent.

$$Br_2 + HOH \rightleftharpoons HOBr + Br$$

So it is said that the oxidation is avoided in the normal metabolic state the aerated waters are one of the source of the oxidants by taking anti oxidants.

Oxidation of glucose forms glyconic acid:

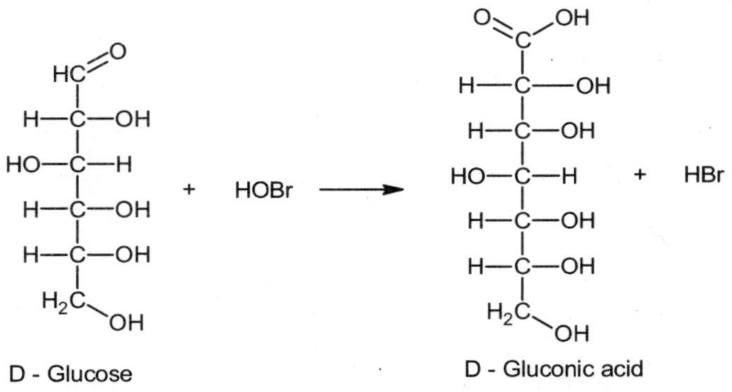

D - Glucose + HOBr → D - Gluconic acid + HBr

Similarly mannose, galactose, and arbinose give mannoic galactonic and arbonic acids, respectively. Other aldoses form the corresponding aldonic acid. Ketoses are not readily oxidized by bromine water.

Aldoses are readily oxidized to aldonic acids by alkaline iodine solution but ketoses are not oxidized. The reactions may be represented.

$$I_2 + 2NaOH \rightarrow NaOI + NaI + H_2O$$

$$R - CHO + NaOI + NaOH \rightarrow \underset{\text{Sodium alconate}}{R - COONa} + NaI + H_2O$$

Willstätter and Schudel developed a method for the quantitative determination of aldoses based upon oxidation.

So it is said that the oxidation is avoided in the normal metabolic state the aerated waters are one of the source of the oxidants by taking anti oxidants

When gluconic acid is heated water is readily lost and it forms a mixture of gamma (γ) and delta (δ) lactones.

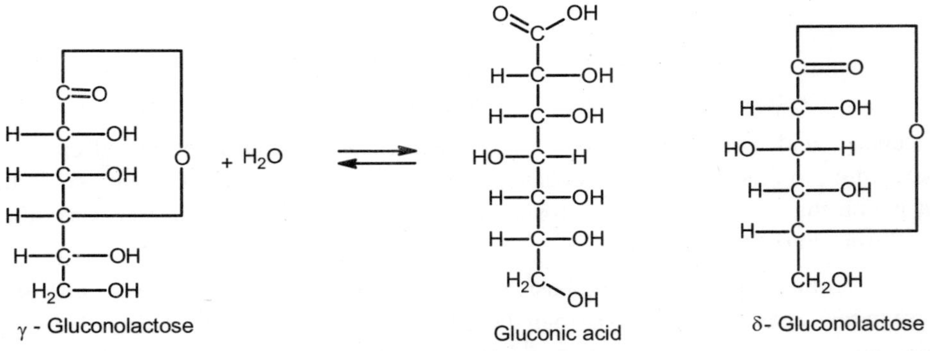

γ - Gluconolactose + H_2O ⇌ Gluconic acid ⇌ δ- Gluconolactose

In aqueous solutions equilibrium is established between gluconic acid its two lactones. Other sugar acids containing five or more carbon atoms similarly form two lactones. Acids with four carbon atoms can form only the γ-lactones.

Sugar acids lactones may be reduced to the corresponding sugars by treatment with sodium amalgam in the presence of dilute sulphuric acid.

When the calcium salts of aldonic acids are oxidized with hydrogen peroxide in the presence of Fe^{+++} ions used as catalyst, carbon dioxide is split out and sugars of one less carbon are formed.

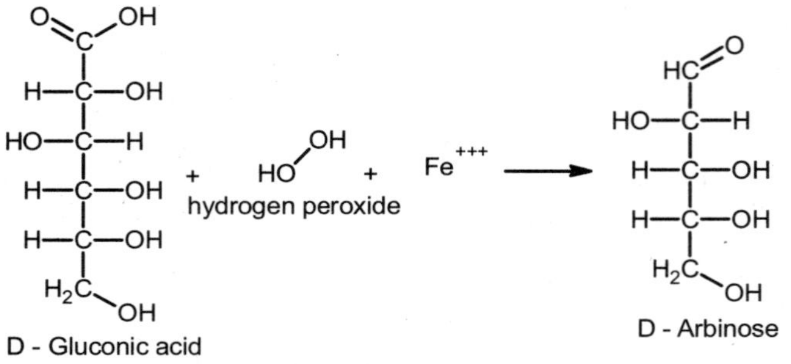

D - Gluconic acid + HO–OH (hydrogen peroxide) + Fe^{+++} ⟶ D - Arbinose

A sugar may be oxidized to the corresponding aldonic acid, and this then converting to the sugar with one less carbon atom.

Calcium gluconate is often administered as a source of calcium. Solutions of it are given intravenously to raise the blood calcium.

Saccharic or Aldaric acids Oxidation of aldoses with nitric acid under the proper conditions converts both aldehyde and primary alcohol groups to carboxyl forming dibasic sugar acids, the saccharic or Aldaric acids.

D - Gluconosaccharic acid D - Mannosaccharic acid Mucic Acid

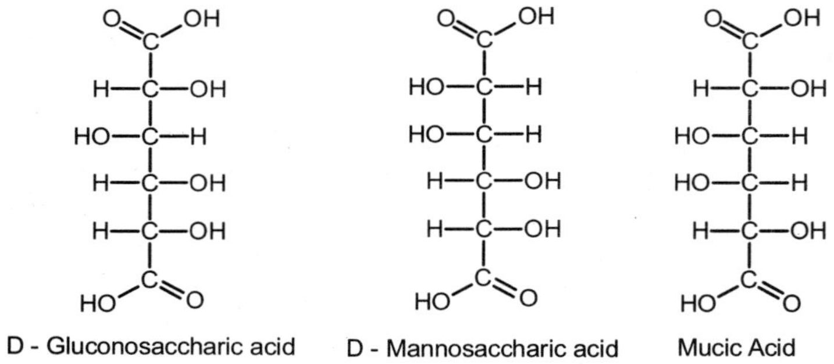

The saccharic acid nomenclature for these dibasic sugar acids has been used exclusively until recently. According to this system, the acids are named by adding to saccharic a prefix indicating the sugar from which the acids is derived – for example glucosaccharic (also called simply saccharic). Similarly "arbosaccharic" xylosaccharic.

The more recent designation of the dibasic sugar acids as aldaric acids appears preferable because the names are in many cases much shorter than those based upon the saccharic nomenclature. According to the aldaric acid system, the name of the acid from a sugar is the name of the sugar with the ending "– aric replacing "– ose for example "glucaricsw" "xylaric" " arbaric" "threaric" "hexaric" (form hexose) "pentaric" from pentose.

The aldaric acid form more complex mixtures of lactose than do the aldaric acid, due to the presence of carboxyl groups.

The acid salts of the alderic acids are often used in the identification of sugars because of their low solubility in water.

Galactaric (galactosaccharic) or mucic acid, produced by oxidation of glucose is relatively insoluble in water and well crystallized and its format is the free and combined states.

The aldaric acids have been of much value in proving the configurations of the aldoses sugars.

Uronic acids are produced when an aldoses is oxidized in such a way that the primary alcohol group is converted to carboxyl without oxidation of the aldehyde or sugar group is protected by conversion to a glucoside or the oxidized by oxygen with an activated platinum carbon catalyst.

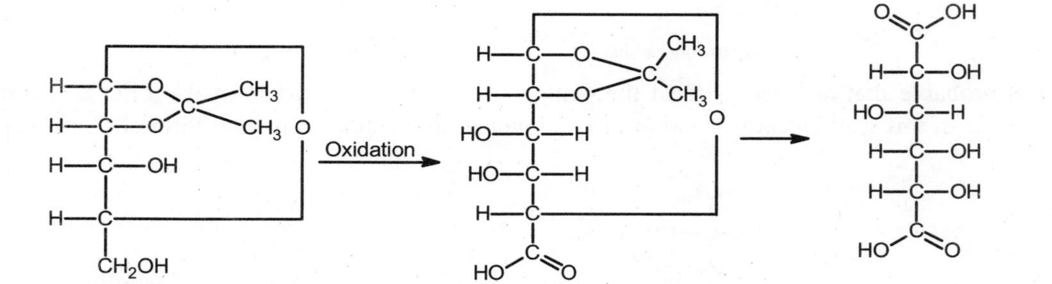

Oxidation

1,2,0 Isopropyledene D - glucose

1,2,0 Isopropyledene glucuronic acid

Glucouronic acid

The isopropylidene group may be removed as acetone by carful acid hydrolysis to yield D-Glucuronic acid.

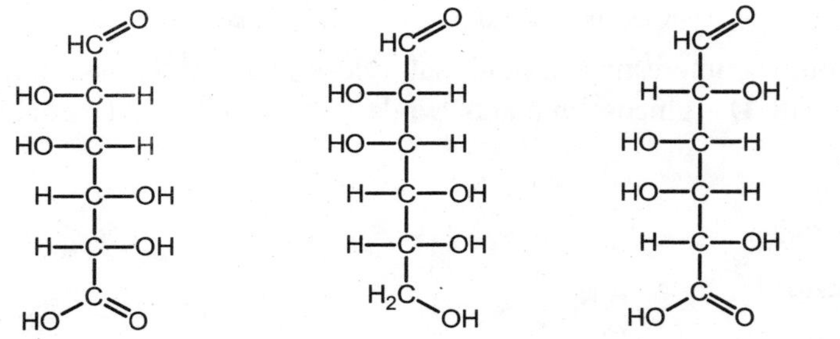

D - Glucouronic acid

D - Mannuronic acid

D - Galactouronic acid

All three of these uronic acids are natural products.

D-Glucouronic acid occurs in combined in plant materials. It is also a constituent of chondoitin mucoitin sulphuric acid sulphuric acids of glycoproteins. Glucouronic acid is formed in the animal body in the process of detoxifying substances such as borneol camphor and benzoic acid; the Glucouronic acid may be prepared in the laboratory by feeding borneol to a dog and hydrolyzing the borneol compound of Glucouronic acid which is excreted in the urine.

D-Galactouronic acid is widely distributed as a constituent of pectin and many plant gums and mucilage. It may me conveniently prepared from pectin.

D – Mannouronic acid occurs as a polymer anhydride in alginic acid and to form the pentose D - xylose.

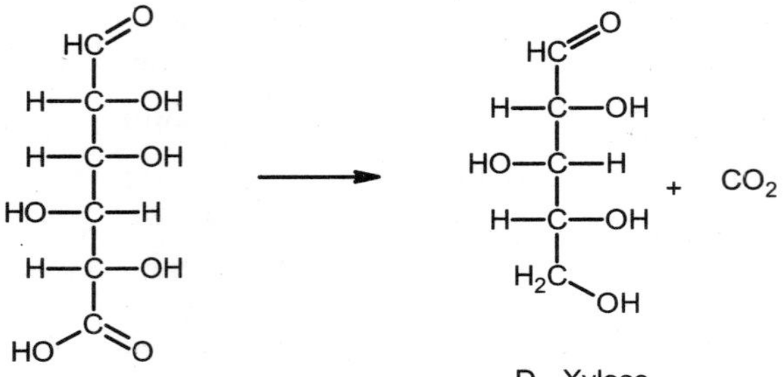

D - Glucouronic acid → D - Xylose + CO_2

It is probable that at least some of the pentose of plants is formed form from hexoses through uronic acids in this way Decarboxylation of D – Galactouronic acid form acid forms L – arbinose.

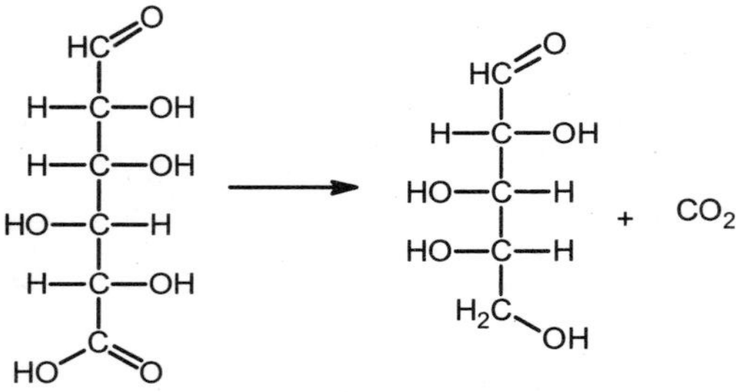

D - Glucouronic acid → L - arbinose + CO_2

In this connection it is interesting to note that xylose is generally associated with D – xylose is generally associated with D – glucose in plants, while L – arbinose and D – galactose is associated.

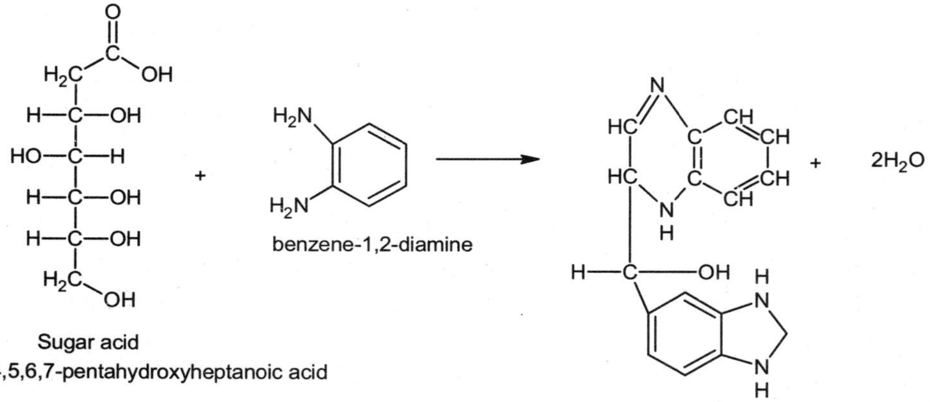

Sugar acid
3,4,5,6,7-pentahydroxyheptanoic acid
+ benzene-1,2-diamine → + $2H_2O$

2,3-dihydro-1*H*-benzimidazol-5-yl(1,2-dihydroquinoxalin-2-yl)methanol
Sugar acid benzimidiazoles

Formation of benzimidiazoles forms sugar acids, aldonic, aldaric (saccharic) uronic and saccharic acids all react with o – phenyl enediamine to form benzimidiazoles, which are useful derivatives for the identification of these acids and the sugars form which they are derived.

The reactions are carried out in the presence of hot acid.

Action of alkalies upon sugars, where sugar behaves like weak acids and form salts high alkalinities. The pK values (pH at which the acid is half converted to salts of four common sugars are given below.

Sugar	pK_1	pk_2
Glucose	12.9	13.85
Fructose	11.68	13.24
Sucrose	12.60	13.52
Lactose	11.92	13.44

Monosaccharides, both aldoses and ketoses and compound carbohydrates containing a free sugar tatutomerise and form the enol salt in alkaline solution.

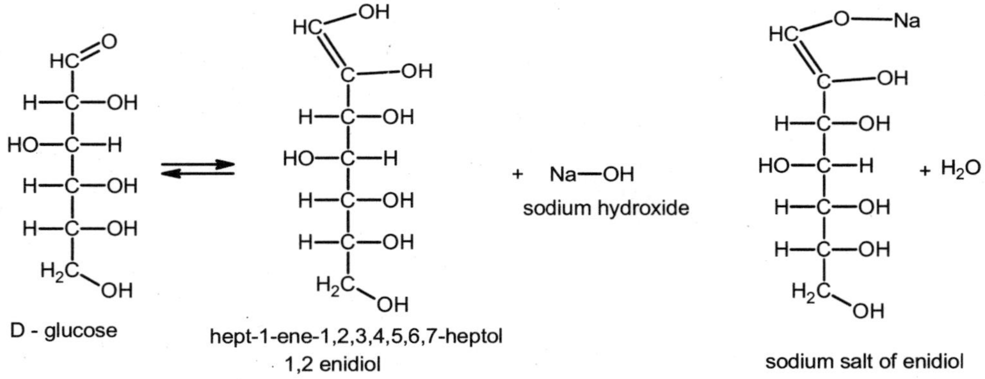

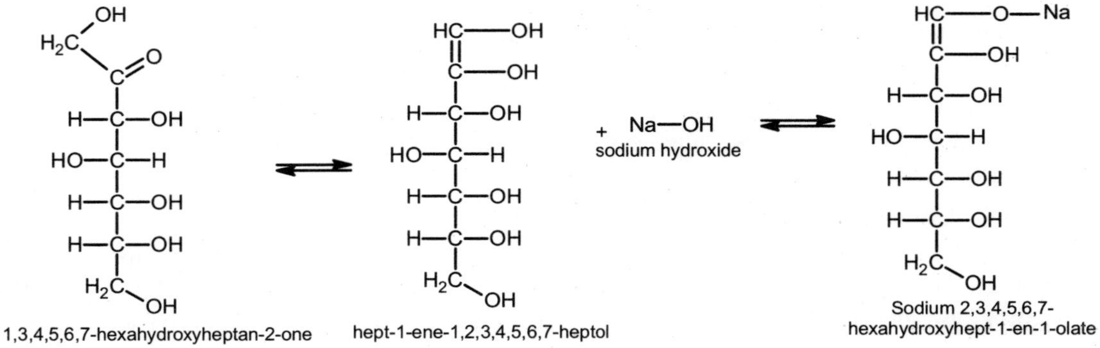

D - glucose ⇌ hept-1-ene-1,2,3,4,5,6,7-heptol 1,2 enidiol + Na—OH sodium hydroxide → sodium salt of enidiol + H_2O

1,3,4,5,6,7-hexahydroxyheptan-2-one ⇌ hept-1-ene-1,2,3,4,5,6,7-heptol + Na—OH sodium hydroxide ⇌ Sodium 2,3,4,5,6,7-hexahydroxyhept-1-en-1-olate

The enol forms of the sugar are enediols because two hydroxyl groups are attached to the double – bonded carbon system. It is uncertain which enolic hydroxyl first will react to form the salt.

It will be noted that glucose, mannose and fructose from the same enediols and enediols salts forms, When a solution of any one of these sugars is allowed to stand in dilute alkali (0.05N) for same time and then acidified a mixture of all these sugars is obtained. This is due to the fact that they all give the same enediols salts, and when this is decomposed to the free enediols by acidified with acetic acid, the enediols tatutomerise into all three sugar as follows:

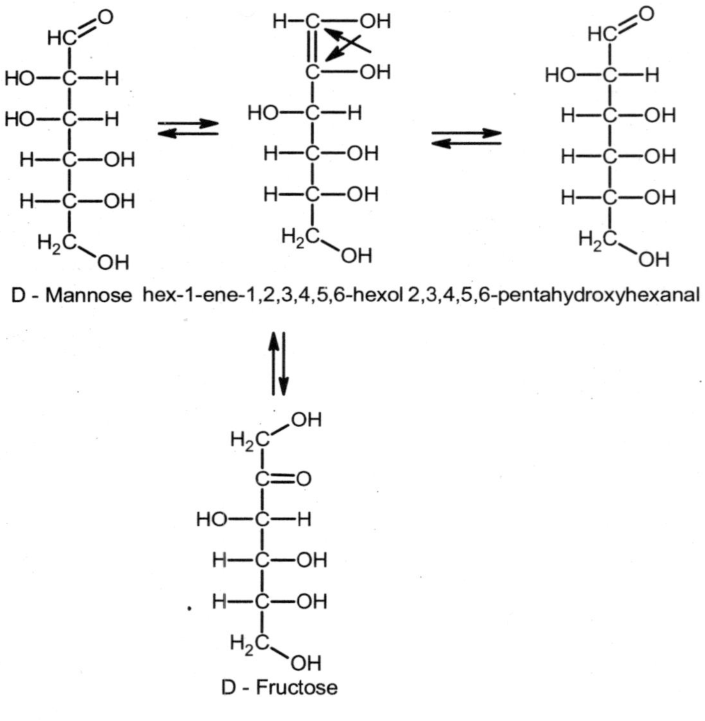

D - Mannose hex-1-ene-1,2,3,4,5,6-hexol 2,3,4,5,6-pentahydroxyhexanal

D - Fructose

This inter conversion of related sugar by the action of dilute alkali is referred to as the **Lobry de Bruyn – Von Ekenstein Equation**; it has been used for the synthesis of certain sugars.

When sugars with free groups are treated with stronger alkali (0.5N and stronger) enolization occurs to form not only 1 – 2 but, also 2 – 3 and 3 – 4 enediols. The 2 – 3 and 3 – 4 enidiols are apparently produced from the ketose sugars present in the mixture. An outline of the probable process in gen below, glucose being used as the initial sugar.

In strong alkali the enediols forms are very unstable and reactive. The enediols break at the double bond forming a complex mixture of products For example, a – 1 – 2 enediol decomposes to give formaldehyde and pentose; a 2 – 3 enediol gives glycolic aldehyde CH_2OH . CHO and a tetrose; and a 3 – 4 enediol form glyceric aldehyde which is a triose. However, since aldoses forms by scission of the parent enediol also enolize and rearrange and therefore an extremely complex sugar is produced.

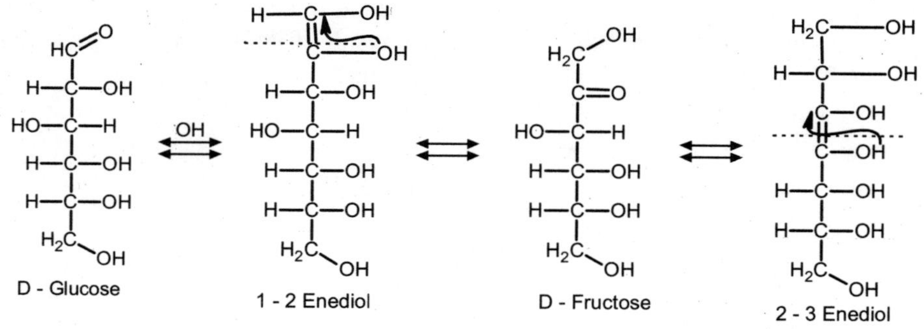

D - Glucose

1 - 2 Enediol

D - Fructose

2 - 3 Enediol

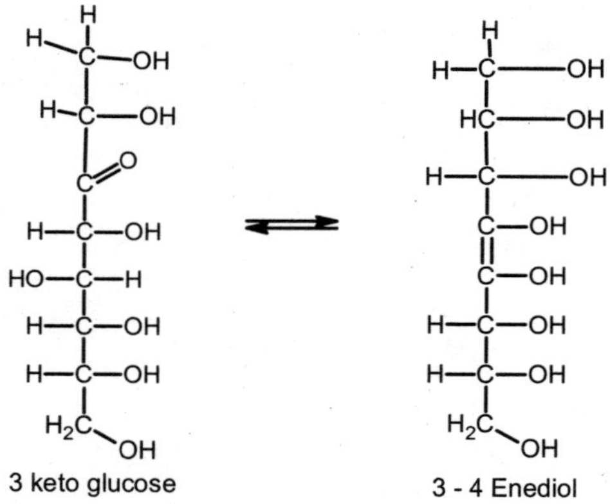

3 keto glucose

3 - 4 Enediol

In the presence of oxygen or other oxidizing agents such as Cu^{++} (Fehling's Solution) the acids corresponding to the various scission products of the alkaline mixture are obtained. For, example, the following acids have been isolated from an oxidized alkaline sugar solution and identified as : formic acid D and L glyceric acid four trihydroxybutyric acid, eight tetrahydroxyvaleric acid, and eight pentahydroxyhexoic acids.

Oxidation of the enediols involves rupture of the double bond with formation of shorter – chain acids.

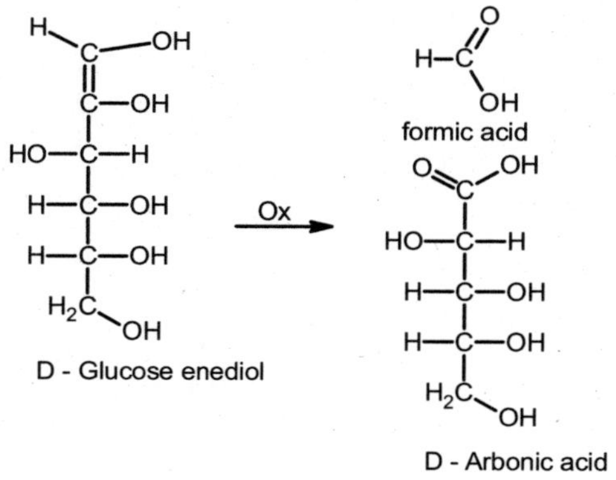

D - Glucose enediol

formic acid

D - Arbonic acid

When strongly alkaline solutions of sugars are permitted to stand in the absence of an oxidizing agent, complex mixture of saccharic acid is formed as a result of intra-molecular oxidation-reduction and arrangement. The formulas of a number of these that have been identified are.

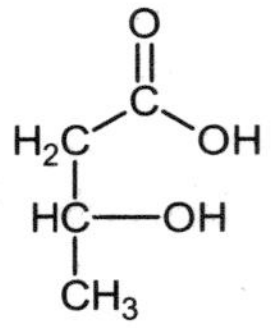

3-hydroxybutanoic acid
D - L Lactic

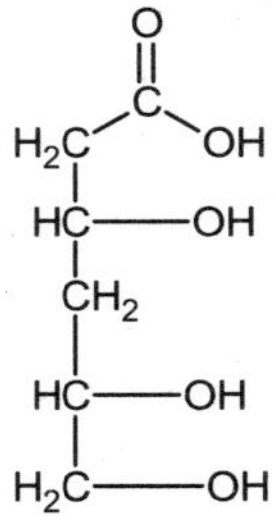

3,5,6-trihydroxyhexanoic acid

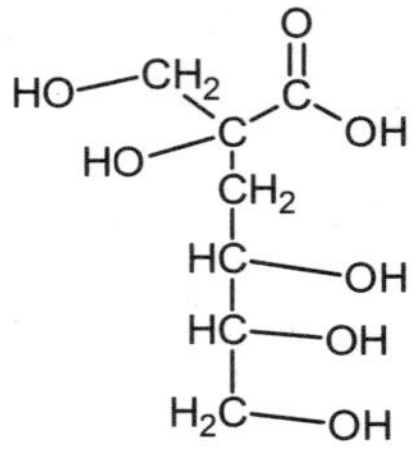

2,4,5,6-tetrahydroxy-2-(hydroxymethyl)hexanoic acid

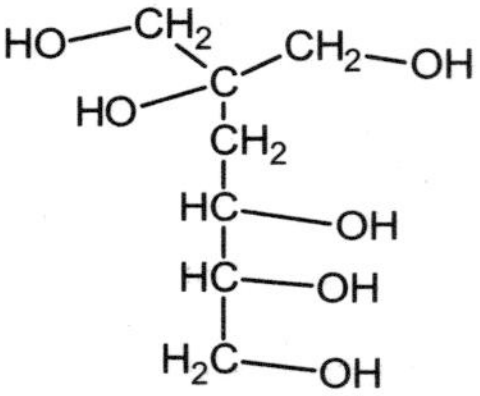

5-(hydroxymethyl)hexane-1,2,3,5,6-pentol

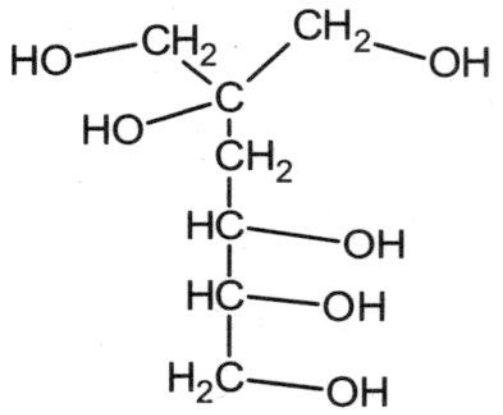

5-(hydroxymethyl)hexane-1,2,3,5,6-pentol

Lactic acid is apparently formed glyceric aldehyde by intermolecular oxidation-reduction involving the first and third carbon atoms. It is generally formed in rather large amounts.

Lime water at ordinary temperature causes formaldehyde to polymerise to a sweet $C_6H_{12}O_6$, which Fischer called "formose" Fischer formose to be a complex mixture, Fischer and Tafel oxidized glycerol to form a mixture dihydroxyacetone and glyceric aldehyde treated the mixture with dilute alkali and obtained a sweet syrup, which is called "α-acrose" D-L Fructose was identified as constituent of α - acrose.

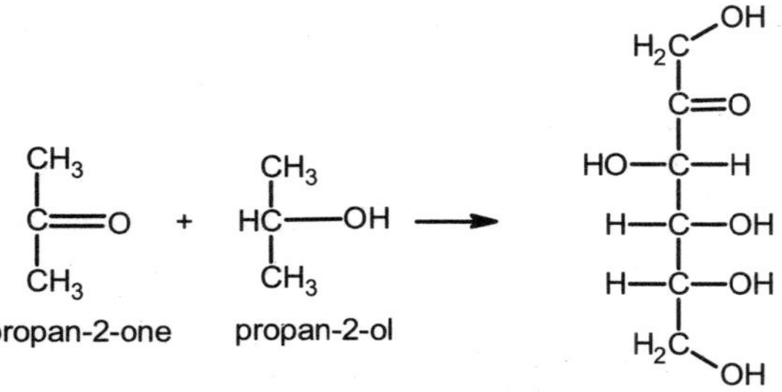

propan-2-one + propan-2-ol → D - L Fructose

Three molecules of glycolic aldehyde $CH_2OH - CHO$ have also been found to polymerise to α – acrose when treated with dilute alkali.

The action of alkali upon formaldehyde and the lower sugars produces a very complex mixture of sugars.

Carbohydrates such as sucrose, which do not contain a free sugar group, are not enolized by alkali are therefore stable in alkaline solution.

When sugars containing free sugar group is heated with alkali, the solutions turn yellow to reddish brown as result of the formation of the complex resinous substances.

This is known as Moor's Test; the sugar enediols exist in the cis and trans configuration about the double bond. Topper has obtained evidence that the enzyme phosphoglucose isomerise acts stereo specifically upon H or C-2 glucose-6-phosphate to form this cis-1, 2-enediol while the enzyme phosphomannose isomeriase acts upon the H on C-2 of Mannose-6-phosphate.

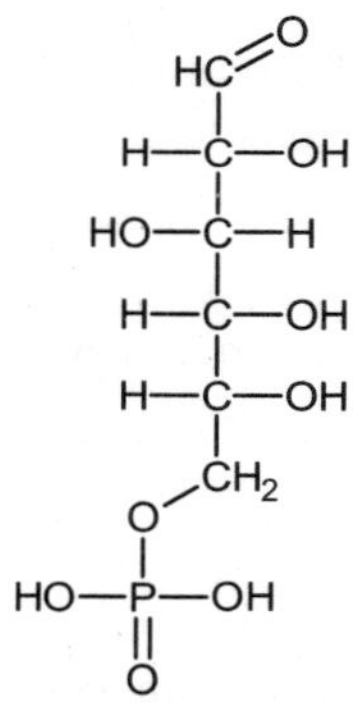

2,3,4,5-tetrahydroxy-6-oxohexyl dihydrogen phosphate
D-Glucose-6-PO_4 chain form

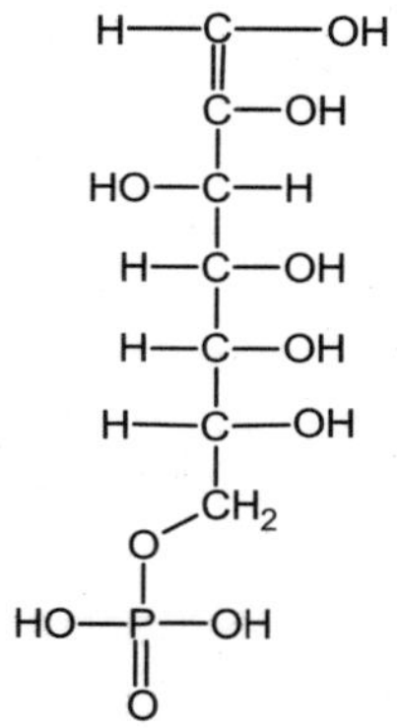

2,3,4,5,6,7-hexahydroxyhept-6-en-1-yl dihydrogen phosphate
D-Glucose-6-PO_4 chain form cis 1,2 enediol

Phosphoglucose isomerise will not acts upon D – mannose 6 – PO_4, and phosphomannose isomerise will not act upon D – glucose 6 – PO_4

Reducing action of sugars in alkaline solution, determinations of sugars. All the sugars which contain the free sugar in alkaline solution, there the enediol forms are very reactive and also are readily oxidized by oxygen and other oxidizing agents which are very strong. This means that these sugars in alkaline solution are very strong reducing agents. As a consequence, they readily reduce oxidizing ions such as Ag^{++}, Hg^{++}, Bi^{+++}, Cu^{++} and $Fe(CN)_6^{+++}$ and the sugar are oxidized to complex mixtures of acids. This reducing action of sugars in alkaline solution is utilized for both the qualitative and the quantitative determination of sugars. Reagents containing Cu^{++} are commonly used; these are generally alkaline solutions of cupric sulphate containing sodium potassium tatrate (Rochelle) Sodium citrate. Sodium hydroxide is used as the alkali in the reagents, such as Fehling's Reagent but weaker alkalies, such as Sodium carbonate and sodium bicarbonate, are used in the more recent reagents like Benedict, Folin and Shaffer and Hatmann.

Citrate and Rochelle salt prevent the preparation of the cupric ions in the form of cupric hydroxide or carbonate by forming soluble, slightly dissociated complex with the Cu^{++} ions. These complexes dissociate sufficiently to provide a continuous supply of readily available Cu^{++} ions for oxidation but at concentrations which prevent the solubility products of cupric hydroxide and carbonate being exceeded.

The alkali of the sugar reagents enolize the sugars and thereby cause them to be strong reducing agents.

When a reducing sugar is heated with one of alkaline copper reagents the process occur accordingly.

Sugar + Alkali ⟶ enediols, reducing agents

\+

Cu^{++} Copper Complex of tartrate and citrate

↓

Cu_2O ⟵ CuOH ⟵ OH^- + Cu^+ + mixture of sugar acids

The Cu^{++} ions take electrons from the enediols and oxidize them to sugar acids and are in turn reduced to cuprous ions Cu^{++}. The cuprous ions combine with hydroxyl ions to form yellow cuprous hydroxide, which upon heating is converted to red cuprous oxide Cu_2O

$$2CuOH \rightarrow Cu_2O + H_2O$$

The appearance of a yellow-to-red precipitate indicates reduction and the quantity of sugar present can be roughly determined from the amount of precipitate. In quantitative estimation the amount of copper reduced is obtained by iodometric titration or colorimetric methods, and from this the amount of sugar is calculated.

Alkaline ferricyanide ($K_3Fe(CN)_6$) solutions are also used in the quantitative determination of sugars the most common reagent being that of Hagendrom and Jenson. These are stronger oxidizing solutions than the copper reagents, which are more liable to oxidize non-sugars present. Their action involves oxidation of the sugar enediols by the ferricyanide, ion $Fe(CN)_6^{+++}$, which is reduced to the ferricyanide ion. The amount of ferricyanide reduced is easily determined by iodine titration and from this the quantity of sugar is determined.

Polysaccharides such as glycogen as starch are generally quantitatively determined by preliminary hydrolysis with acid to the constituent monosaccharide, which are then estimated by one of the above reduction method.

An interesting determination of glucose is based upon its quantitative oxidation to D-glucono-δ-lactone is converted to D-gluconic acid, which is then titrated with alkali

Hydrogen peroxide also is formed in the oxidation with is used as the basis of a very sensitive colorimetric method for glucose estimation.

$$\text{Glucose} + O_2 \xrightarrow[\text{oxidase}]{\text{glucose}} \text{gluconolactone} + \text{Hydrogen Peroxide}$$

$$\Downarrow +H_2O$$

$$\text{D - Gluconicacid}$$

$$\text{Hydrogen Peroxide} + \text{o-dianicidine} \xrightarrow{\text{peroxidase}} \text{yellow compound}$$

---

*Now a days the Follin U method of sugar estimation is used in the modern pathological labs

A very useful method for the colorimetric determination of carbohydrates is based upon the reaction of anthrone with a carbohydrate in the presence of strong hydrochloric acid.

The carbohydrate is decomposed by the acid to give a substance, probably a furfural derivative, which reacts with anthrone to form a blue compound. The reaction is given by both free and combined sugar and is quite sensitive.

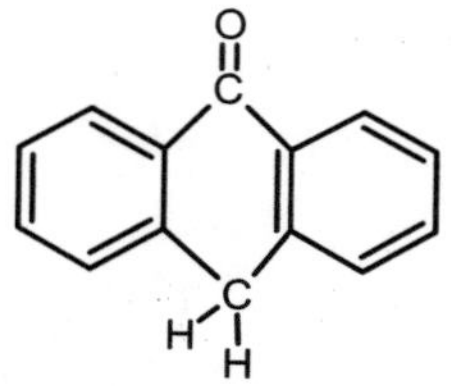

anthracen-9(10H)-one
anthrone

In general the Monosaccharides are relatively stable to that hot dilute hydrochloric acid though the ketoses are apparently decomposed by prolonged action. When the concentration of acid is increased to several moralities, the Monosaccharide molecules are decomposed. Pentoses yield the cyclic aldehyde furfural, as shown in the reaction equation below.

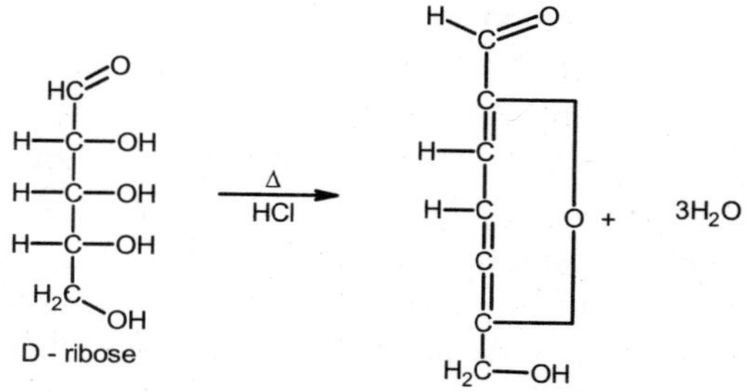

D - ribose $\xrightarrow[HCl]{\Delta}$ hydroxymethylfurfural + $3H_2O$

This reaction is used for the quantitative determination of Pentoses compound carbohydrates containing Pentoses (pentosan etc).12% hydrochloric acid has been found the most satisfactory acid for decomposition. Furfural forms, with phloroglucinol, a relatively insoluble compound, furfural – phoroglucide, which can be used in the estimation of the furfural formed in the reaction as a measure of the Pentoses present. The furfural can also be determined by use of its colour reaction with aniline acetate or by a titrimetric reaction.

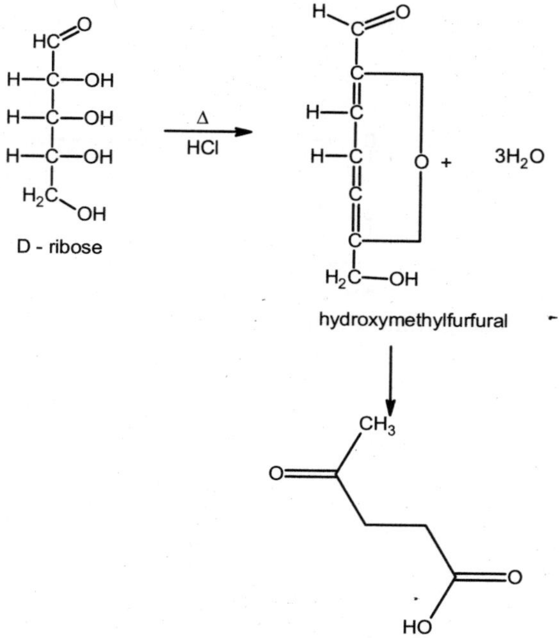

D - ribose $\xrightarrow[HCl]{\Delta}$ hydroxymethylfurfural + $3H_2O$

4-oxopentanoic acid
Levulinic acid

Hexoses are decomposed by hot strong acid to give hydroxymethylfurfural which decomposes into levulinic acid and other products.

Considerable brown or black resinous substances are also formed. When uronic acids such as Glucouronic acid and Galactouronic acids are heated with acid, they loose carbon dioxide and form the corresponding pentose which in turn is converted to furfural.

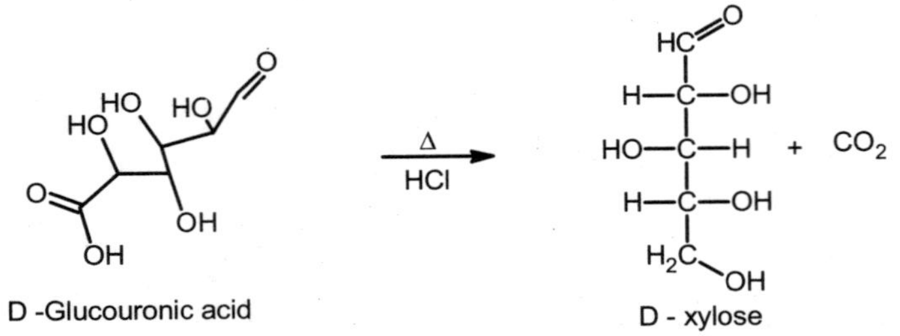

D -Glucouronic acid $\xrightarrow[HCl]{\Delta}$ D - xylose + CO_2

The yield of carbon dioxide is quantitative, and uronic acids are determined by estimating the amount of carbon dioxide liberated in the above reaction. The yield of furfural from the pentose in this reaction is not quantitative.

Reactions of Sugars due to hydroxyl group Most of the chemical properties of the sugars previously discussed were concerned directly or indirectly with the aldehyde or ketone groups, though a few reactions involving hydroxyl groups were considered. The hydroxyl groups of sugars have the properties of ordinary alcoholic groups but, because of the number present, may give rise to derivatives which are impossible with monohydroxy alcohols.

1. Formation of glucosides (glycosides). Production of the glucosides of the simple sugars has been referred to in a previous section. In general, the ring forms of the simple sugars react with alcohols in the presence of hydrogen chloride as catalyst to form the glucosides. Only the hydroxyl on carbon 1 (glucosidic hydroxyl) of the sugar reacts under these conditions. Generally a mixture of the *a*- and, β-glucosides is obtained, and when the reaction is carried out at elevated temperatures, the glucosides contain the pyranose ring. Reaction at room or lower temperatures also may produce glucosides with the furanose ring.

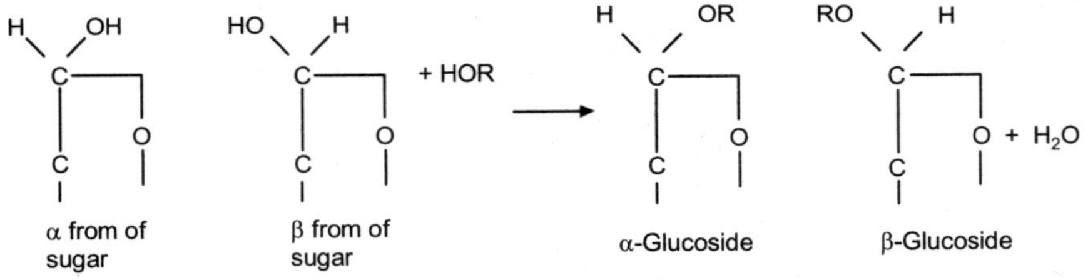

α from of sugar + β from of sugar + HOR → α-Glucoside + β-Glucoside + H_2O

The pyranose glucosides are called "pyranosides" whereas the furanose the glucosides are "furanosides" The formula and names are below will show:

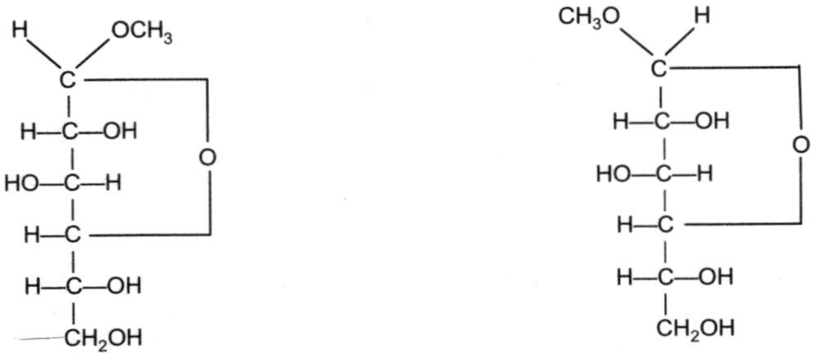

or

or

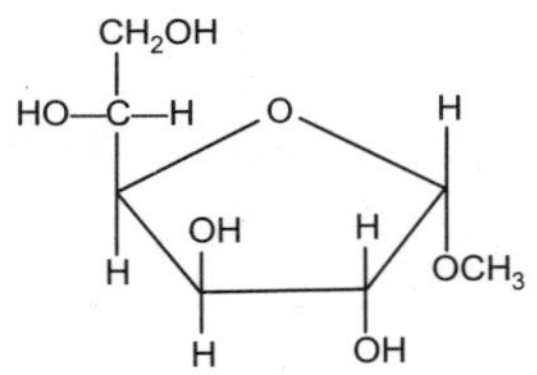

α-Methyl-D-glucofuranoside

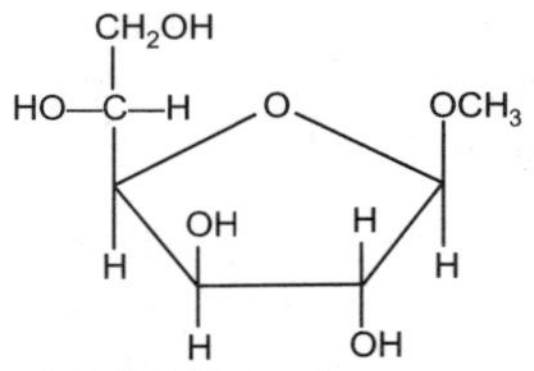

β-Methyl-D-glucofuranoside

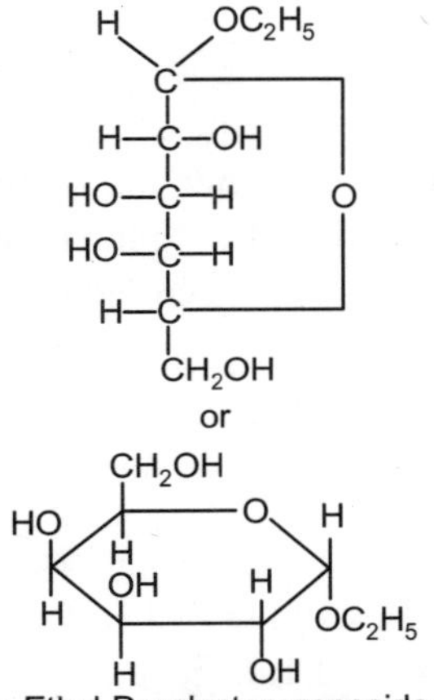

or

α-Ethyl-D-galactopyranoside

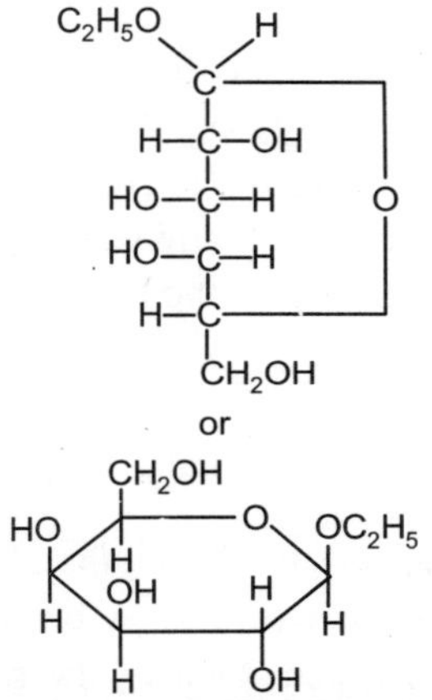

or

β-Ethyl-D-galactopyranoside

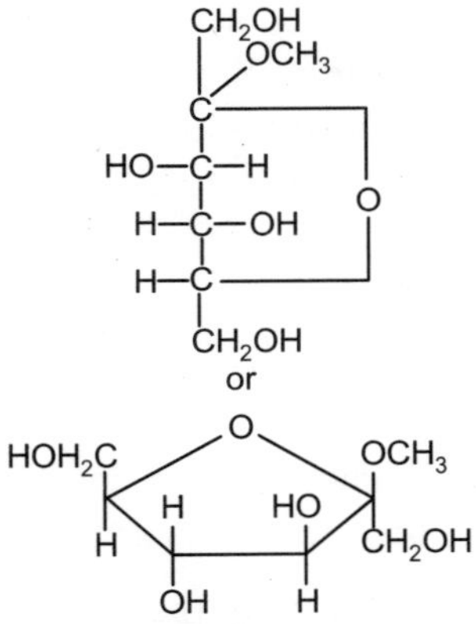

or

α-Methyl-D-fructofuranoside

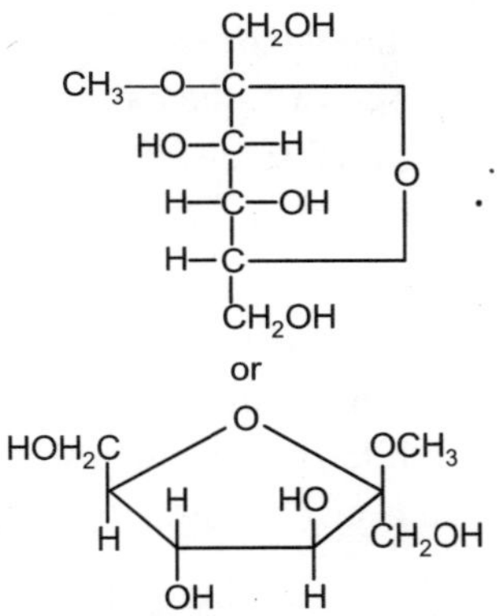

or

β-Methyl-D-fructofuranoside

When the α and β forms of a free sugar or a glucoside are prepared, the problem of determining which shall be designated as α and which as β arises.

Boeseken found that polyhydroxy compounds having two hydroxyl groups on adjacent carbon atoms and on the same side of the chain form complexes with boric acid in solution which ionize more than boric acid alone and consequently increase the electrical conductivity. For example, α-D-glucose increases the conductivity of boric acid solutions, but the conductivity falls as the sugar mutarotates and passes into the *β* isomer. The same thing is true with α-D-galactose. According to these findings the glucosidic hydroxyls of a-D-glucose and a-D-galactose lie on the same side of the carbon chain as the hydroxyls on carbon 2, while the hydroxyls are in the trans position for the *β* forms.

```
 H  OH                 HO   H
  \ /                    \ /
   C-------              C-------
   |       |             |       |
H--C--OH   O          H--C--OH   O
   |       |             |       |
α Forms               β Forms
```

The configurations about carbon 1 of the α and *β* anomers of various sugars have been established by other methods, especially from the rotations of the dialdehydes obtained in the periodate oxidation of glycosides. In general, in the D-series the hydroxyl of the a monomer lies to the right in Fischer's projection ring formulas and below the ring in Haworth's ring formulas. In the L-series the hydroxyl of the a-anomer lies to the left in the Fischer formulas and above the ring in the Haworth formulas.

In the case of a sugar which exists as both α and β anomeric forms, belongs to the v-series, and mutarotates in solution, the a form of the sugar usually is the one which mutarotates to a value less positive than the initial rotation.

Isbell has pointed out that, when the glucosidic hydroxyl of a sugar lies in the same direction as the ring oxygen, the sugar is oxidized by bromine water more slowly than when the hydroxyl is directed away from the ring oxygen. That is, the β forms of glucose and galactose are more readily oxidized than the α forms. Isbell also proposed a general rule for differentiating the α and β pairs of a sugar, or its glucosides, or other derivatives, based upon optical rotation: "When the oxygen ring lies to the right, as in D-glucose, the more dextrorotatory member of the α-β pair shall be designated α, and the less dextrorotatory member β when the oxygen ring lies to the left, as in L-glucose, the more levorotatory member shall be designated α, and the less levorotatory β. This rule is applicable to the furanoses as well as to the pyranoses and to the derivatives of both."

The glucosides do not reduce alkaline copper solutions, because the sugar group is combined. For the same reason, they are resistant to the action of alkali. They may be hydrolyzed to the constituent reducing sugars by boiling with dilute mineral acids. α-Glucosides are hydrolyzed by maltase, an enzyme from yeast, while *β*-glucosides are hydrolyzed by the enzyme emulsin, from bitter almonds. Enzyme hydrolysis thus affords a method of distinguishing between the two forms.

Many glucosides occur in the roots, bark, and fruit, and frequently the leaves of various plants. Glucosides are usually well-crystallized, colourless, bitter solids, soluble in water and alcohol. A number of the natural glucosides are important in medicine or otherwise. The groups attached to the sugars in the natural glucosides are frequently quite complex, but the union is always through condensation of an alcoholic or phenolic hydroxyl with the glucosidic hydroxyl of the sugar.

The group attached to the sugar in a glucoside is often referred to as the "aglucone" or "aglycone"; thus mandelonitrile is the aglucone of amygdalin, and digitogenin is the aglucone of digitonin.

Table below lists a number of the natural glucosides, with their hydrolytic products and source.

Glucoside	Hydrolytic Products	Source
Arbutin	Glucose + Hydroquinone	Arbutus The European species are called Strawberry Trees
Phlorizin	Glucose + Phloretin	Rosewood Bark
Amygdalin	2 Glucose + D-mandelonitrile	Seeds of bitter almonds
Digitonin	4 Galactose + xylose + digitogenin ($C_{27}H_{44}O_5$)	Leaves of Foxglove
Saponin	Sugar + Sapogenin	Soapwort
Indican	Glucose + Indoxyl	Leaves of Indigofera

2. Formation of ethers. The hydrogen of hyoroxyl groups of sugars and of carbohydrates in general may be replaced by alkyl groups to form ethers. The methylated sugars which have already been discussed in some detail are methyl ethers of the sugars. Ethylated sugars, or ethyl ethers of the sugars, may be prepared by reactions analogous to those for the prepara-tion of the methylated sugars.

3. Formation of esters. The hydroxyl groups of the sugars may be esterified to give esters such as the sugar acetates, propionates, stearates, and benzoates. This is generally accomplished by treating the sugar with the appropriate acid anhydride or chloride under the proper conditions.

For example, when glucose is treated with acetic anhydride, pentaacetyl glucose or glucose pentaacetate is formed. By varying the conditions the α or β form may be obtained as the chief product:

In the case of sugar esters rigid systematic nomenclature requires that the acyl groups be designated" -O-acyl" as in "penta-O-acetyl-α-D-glucopyranose."

The acetyl group on the first carbon is readily hydrolyzed off, leaving tetraacetyl glucose with a free sugar group. All the acetyl groups may be removed by mild alkaline hydrolysis to re-form the sugar. The sugar acetates and other esters are used especially in the preparation of other sugar derivatives. They are generally insoluble in water and soluble in organic solvents.

While sugars form various inorganic esters such as sulphates and nitrates, the sugar phosphates are outstanding in biologic importance. The breakdown or metabolism of glucose and other sugars by animal tissues and by yeast and other microorganisms involves a succession of the phosphates of sugars and sugar derivatives. Nucleoproteins of cell nuclei also contain sugar phosphates in combination. Sugar phosphates have been found as intermediate products in the carbohydrate metabolism of plants. Many of these substances have been isolated and identified, and a number have been synthesized. Diagrams below gives the structures of phosphates of glucose and fructose which are of biologic importance.

It will be observed that, whereas glucose phosphates have pyranose rings, the phosphates of fructose are of the furanose type.

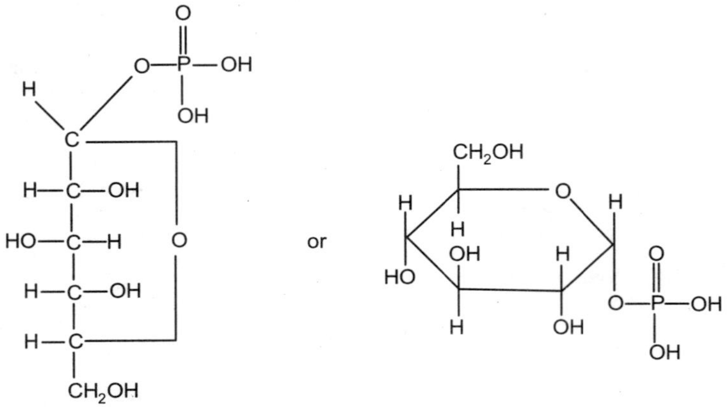

α-D-Glucopyranosyl-I-phosphate
Glucose-I-phosphate (Cori ester)

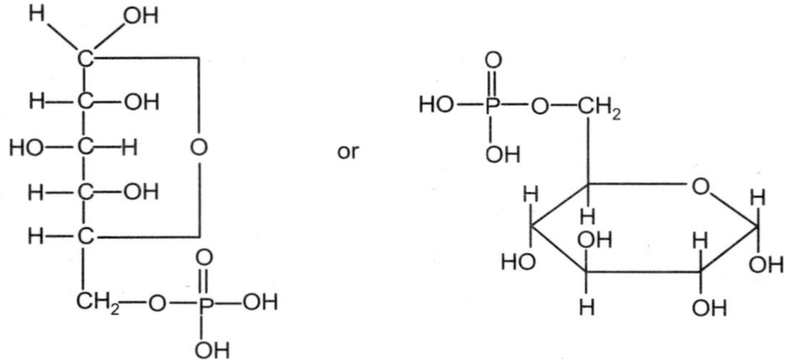

D-Glucopyranosyl-6-phosphate
Glucose-6-phosphate (Robison ester)

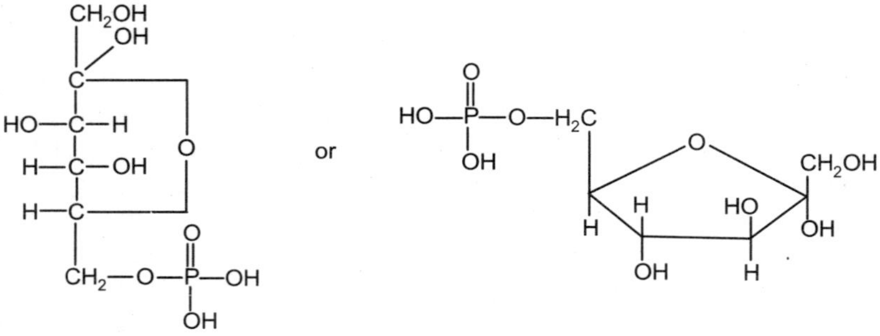

D-Fructofuranosyl-6-phosphate
Fructose-6-phosphate (Neuberg ester)

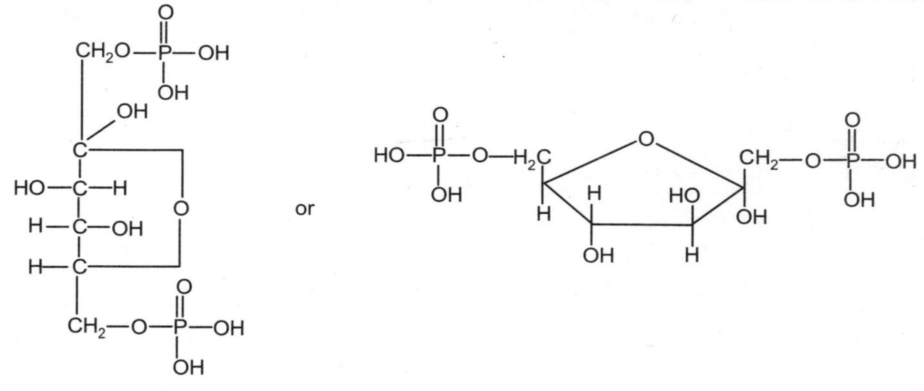

D-Fructofuranosyl-1,6-diphosphate
Fructose diphosphate (Harden-Young ester)

LIPIDS

The lipids are a large and diverse group of naturally occurring organic compounds that are related by their solubility in nonpolar organic solvents (e.g. ether, chloroform, acetone & benzene) and general insolubility in water. There is great structural variety among the lipids, as will be demonstrated in the following sections:

Fats, Oils, Waxes & Phospholipids

Fatty Acids

The common feature of these lipids is that they are all esters of moderate to long chain fatty acids. Acid or base-cayalyzed hydrolysis yields the component fatty acid, some examples of which are given in the following table, together with the alcohol component of the lipid. These long-chain carboxylic acids are generally referred to by their common names, which in most cases reflect their sources. Natural fatty acids may be saturated or unsaturated, and as the following data indicate, the saturated acids have higher melting points than unsaturated acids of corresponding size. The double bonds in the unsaturated compounds listed on the right are all cis (or Z).

Fatty Acids

Saturated Formula	Common Name	Melting Point
$CH_3(CH_2)_{10}CO_2H$	lauric acid	45°C
$CH_3(CH_2)_{12}CO_2H$	myristic acid	55 °C
$CH_3(CH_2)_{14}CO_2H$	palmitic acid	63 °C
$CH_3(CH_2)_{16}CO_2H$	stearic acid	69 °C
$CH_3(CH_2)_{18}CO_2H$	arachidic acid	76 °C

Unsaturated Formula	Common Name	Melting Point
$CH_3(CH_2)_5CH{=}CH(CH_2)_7CO_2H$	palmitoleic acid	0 °C
$CH_3(CH_2)_7CH{=}CH(CH_2)_7CO_2H$	oleic acid	13 °C
$CH_3(CH_2)_4CH{=}CHCH_2CH{=}CH(CH_2)_7CO_2H$	linoleic acid	-5 °C
$CH_3CH_2CH{=}CHCH_2CH{=}CHCH_2CH{=}CH(CH_2)_7CO_2H$	linolenic acid	-11 °C
$CH_3(CH_2)_4(CH{=}CHCH_2)_4(CH_2)_2CO_2H$	arachidonic acid	-49 °C

The higher melting points of the saturated fatty acids reflect the uniform rod-like shape of their molecules. The cis-double bond(s) in the unsaturated fatty acids introduce a kink in their shape, which makes it more difficut to pack their molecules together in a stable repeating array or crystalline lattice. The trans-double bond isomer of oleic acid, known as elaidic acid, has a linear shape and a melting point of 45 °C (32 °C higher than its cis isomer). The shapes of stearic and oleic acids are displayed in the models below.

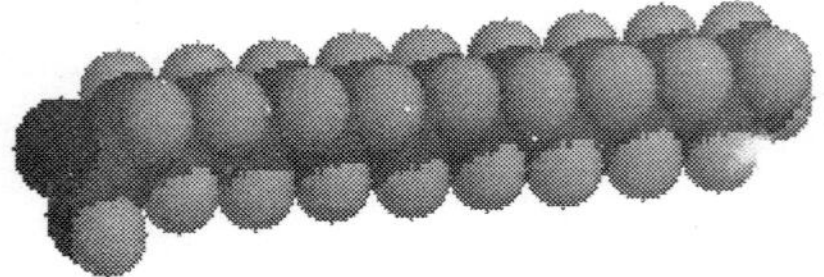

stearic acid

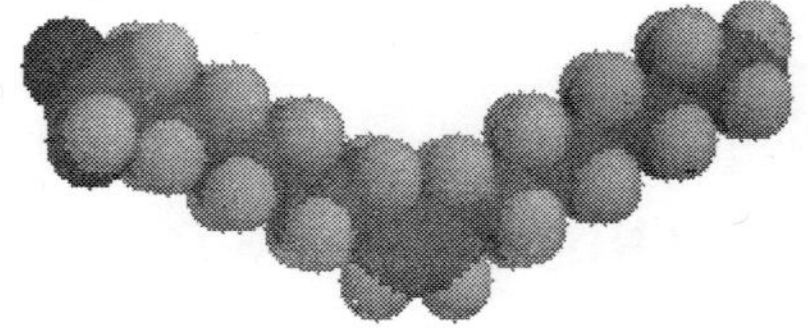

oleic acid

Two polyunsaturated fatty acids, linoleic and linolenic, are designated "essential" because their absence in the human diet has been associated with health problems, such as scaley skin, stunted growth and increased dehydration. These acids are also precursors to the prostaglandins, a family of physiologically potent lipids present in minute amounts in most body tissues.

Because of their enhanced acidity, carboxylic acids react with bases to form ionic salts, as shown in the following equations. In the case of alkali metal hydroxides and simple amines (or ammonia) the resulting salts have pronounced ionic character and are usually soluble in water. Heavy metals such as silver, mercury and lead form salts having more covalent character (3rd example), and the water solubility is reduced, especially for acids composed of four or more carbon atoms.

$$RCO_2H \quad + \quad NaHCO_3 \quad \rightarrow \quad RCO_2^{(-)}\,Na^{(+)} \; + \; CO_2 \; + \; H_2O$$

$$RCO_2H \quad + \quad (CH_3)_3N: \quad \rightarrow \quad RCO_2^{(-)}\,(CH_3)_3NH^{(+)}$$

$$RCO_2H \quad + \quad AgOH \quad \rightarrow \quad RCO_2^{\delta(-)}\,Ag^{\alpha(+)} \; + \; H_2O$$

Soaps and Detergents

Carboxylic acids and salts having alkyl chains longer than eight carbons exhibit unusual behaviour in water due to the presence of both hydrophilic (CO_2) and hydrophobic (alkyl) regions in the same molecule. Such molecules are termed **amphiphilic** (Gk. amphi = both) or **amphipathic**. Fatty acids made up of ten or more carbon atoms are nearly insoluble in water, and because of their lower density, float on the surface when mixed with water. Unlike paraffin or other alkanes, which tend to puddle on the waters surface, these fatty acids spread evenly over an extended water surface, eventially forming a monomolecular layer in which the polar carboxyl groups are hydrogen bonded at the water interface, and the hydrocarbon chains are aligned together away from the water. This behavior is illustrated in the diagram on the right. Substances that accumulate at water surfaces and change the surface properties are called **surfactants**.Alkali metal salts of fatty acids are more soluble in water than

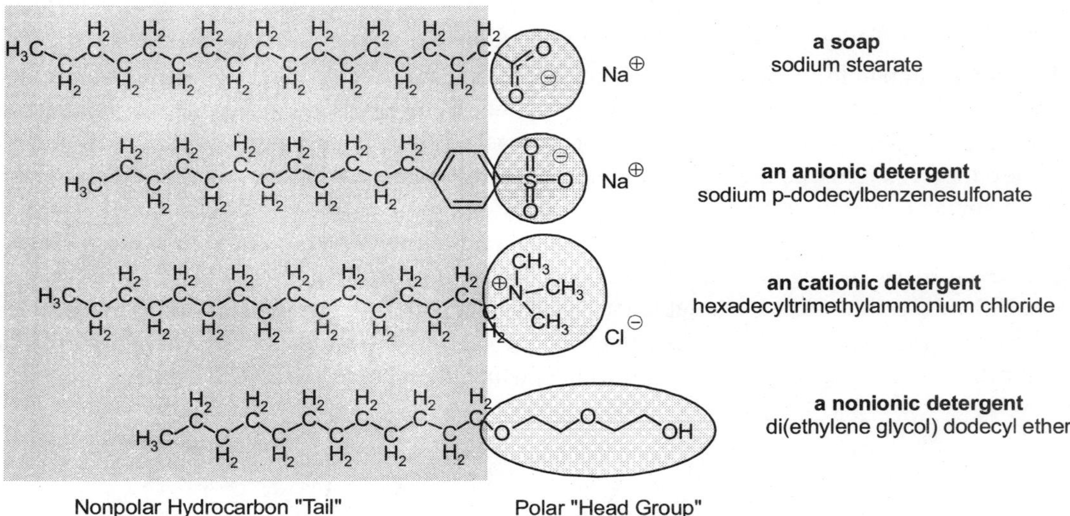

the acids themselves, and the amphiphilic character of these substances also make them strong surfactants. The most common examples of such compounds are soaps and detergents, four of which are shown below. Note that each of these molecules has a nonpolar hydrocarbon chain, the "tail", and a polar (often ionic) "head group". The use of such compounds as cleaning agents is facilitated by their surfactant character, which lowers the surface tension of water, allowing it to penetrate and wet a variety of materials.

Very small amounts of these surfactants dissolve in water to give a random dispersion of solute molecules. However, when the concentration is increased an interesting change occurs. The surfactant molecules reversibly assemble into polymolecular aggregates called **micelles**. By gathering the hydrophobic chains together in the center of the micelle, disruption of the hydrogen bonded structure of liquid water is minimized, and the polar head groups extend into the surrounding water where they participate in hydrogen bonding. These micelles are often spherical in shape, but may also assume cylindrical and branched forms, as illustrated on the right. Here the polar head group is designated by a blue circle, and the nonpolar tail is a zig-zag black line. An animated display of micelle formation is presented below. Notice the brownish material in the center of the three-dimensional drawing on the left. This illustrates a second important factor contributing to the use of these amphiphiles as cleaning agents. Micelles are able to encapsulate nonpolar substances such as grease within their hydrophobic center, and thus solubilize it so it is removed with the wash water. Since the micelles of anionic amphiphiles have a negatively charged surface, they repel one another and the nonpolar dirt is effectively emulsified. To summarize, the presence of a soap or a detergent in water facilitates the wetting of all parts of the object to be cleaned, and removes water-insoluble dirt by incorporation in micelles.

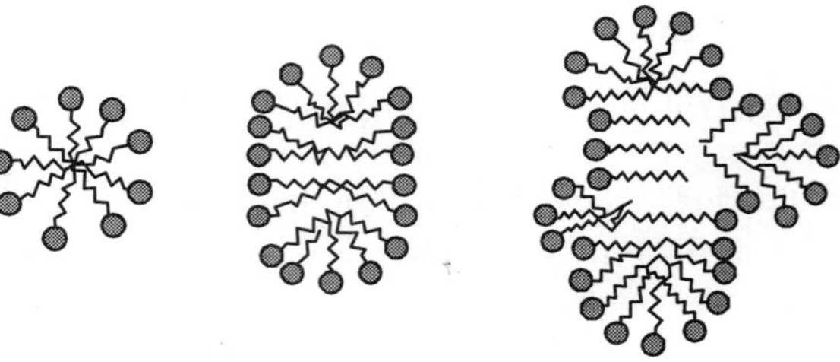
Micelle assemblies of amphiphiles

The oldest amphiphilic cleaning agent known to humans is soap. Soap is manufactured by the base-catalyzed hydrolysis (saponification) of animal fat. Before sodium hydroxide was commercially available, a boiling solution of potassium carbonate leached from wood ashes was used. Soft potassium soaps were then converted to the harder sodium soaps by washing with salt solution. The importance of soap to human civilization is documented by history, but some problems associated with its use have been recognized. One of these is caused by the weak acidity (pK_a ca. 4.9) of the fatty acids. Solutions of alkali metal soaps are slightly alkaline (pH 8 to 9) due to hydrolysis. If the pH of a soap solution is lowered by acidic contaminants, insoluble fatty acids precipitate and form a scum. A second problem is caused by the presence of calcium and magnesium salts in the water supply (hard water). These divalent cations cause aggregation of the micelles, which then deposit as a dirty scum.

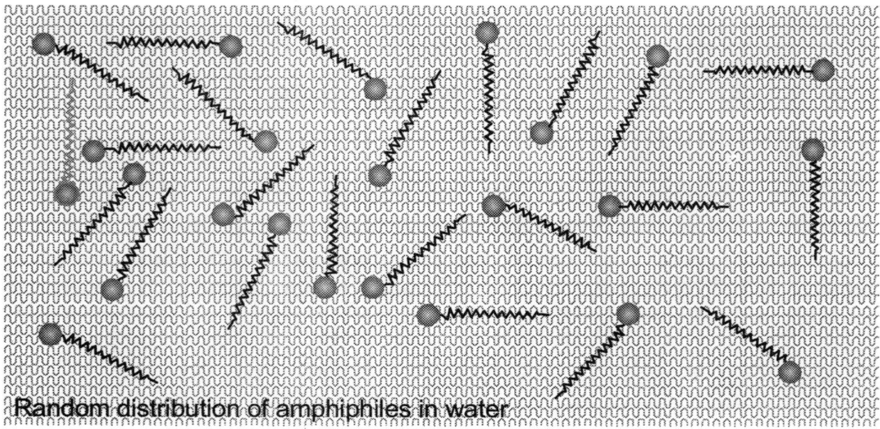
Random distribution of amphiphiles in water

These problems have been alleviated by the development of synthetic amphiphiles called detergents (or syndets). By using a much stronger acid for the polar head group, water solutions of the amphiphile are less sensitive to pH changes. Also the sulfonate functions used for virtually all anionic detergents confer greater solubility on micelles incorporating the alkaline earth cations found in hard water. Variations on the amphiphile theme have led to the development of other classes, such as the cationic and nonionic detergents shown above. Cationic detergents often exhibit germicidal properties, and their ability to change surface pH has made them useful as fabric softners and hair conditioners. These versatile chemical "tools" have dramatically transformed the household and personal care cleaning product markets over the past fifty years.

Fats and Oils

The triesters of fatty acids with glycerol (1,2,3-trihydroxypropane) compose the class of lipids known as fats and oils. These **triglycerides** (or triacylglycerols) are found in both plants and animals, and compose one of the major food groups of our diet. Triglycerides that are solid or semisolid at room temperature are classified as fats, and occur predominantly in animals. Those triglycerides that are liquid are called oils and originate chiefly in plants, although triglycerides from fish are also largely oils. Some examples of the composition of triglycerides from various sources are given in the following table.

Some Examples of the Composition of Triglycerides from Various Sources

Source	Saturated Acids (%)				Unsaturated Acids (%)			
	C_{10} & less	C_{12} lauric	C_{14} myristic	C_{16} palmitic	C_{18} stearic	C_{18} oleic	C_{18} linoleic	C_{18} unsaturated
Animal Fats								
butter	15	2	11	30	9	27	4	1
lard	-	-	1	27	15	48	6	2
human fat	-	1	3	25	8	46	10	3
herring oil	-	-	7	12	1	2	20	52
Plant Oils								
coconut	-	50	18	8	2	6	1	-
corn	-	-	1	10	3	50	34	-
olive	-	-	-	7	2	85	5	-
palm	-	-	2	41	5	43	7	-
peanut	-	-	-	8	3	56	26	7
safflower	-	-	-	3	3	19	76	-

As might be expected from the properties of the fatty acids, fats have a predominance of saturated fatty acids, and oils are composed largely of unsaturated acids. Thus, the melting points of triglycerides reflect their composition, as shown by the following examples. Natural mixed triglycerides have somewhat lower melting points, the melting point of lard being near 30ºC, whereas olive oil melts near -6 ºC. Since fats are valued over oils by some Northern European and North American populations, vegetable oils are extensively converted to solid triglycerides (e.g. Crisco) by partial hydrogenation of their unsaturated components. Some of the remaining double bonds are isomerized (to trans) in this operation. These saturated and trans-fatty acid glycerides in the diet have been linked to long-term health issues such as atherosclerosis.

$H_2C{-}OCO(CH_2)_{10}CH_3$
$HC{-}OCO(CH_2)_{10}CH_3$
$H_2C{-}OCO(CH_2)_{10}CH_3$
trilaurin
mp 45°C

$H_2C{-}OCO(CH_2)_{16}CH_3$
$HC{-}OCO(CH_2)_{16}CH_3$
$H_2C{-}OCO(CH_2)_{16}CH_3$
Tristearin
mp 71°C

$H_2C{-}OCO(CH_2)_7CH\overset{cis}{=}CH(CH_2)_7CH_3$
$HC{-}OCO(CH_2)_7CH\overset{cis}{=}CH(CH_2)_7CH_3$
$H_2C{-}OCO(CH_2)_7CH\overset{cis}{=}CH(CH_2)_7CH_3$
triolein
mp-4°C

Triglycerides having three identical acyl chains, such as tristearin and triolein (above), are called "simple", while those composed of different acyl chains are called "mixed". If the acyl chains at the end hydroxyl groups (1 & 3) of glycerol are different, the center carbon becomes a chiral center and enantiomeric configurations must be recognized. The hydrogenation of vegetable oils to produce semisolid products has had unintended consequences. Although the hydrogenation imparts desirable features such as spreadability, texture, "mouth feel," and increased shelf life to naturally liquid vegetable oils, it introduces some serious health problems. These occur when the cis-double bonds in the fatty acid chains are not completely saturated in the hydrogenation process. The catalysts used to effect the addition of hydrogen isomerize the remaining double bonds to their trans configuration. These unnatural **trans-fats** appear to to be associated with increased heart disease, cancer, diabetes and obesity, as well as immune response and reproductive problems.

4. Waxes

Waxes are esters of fatty acids with long chain monohydric alcohols (one hydroxyl group). Natural waxes are often mixtures of such esters, and may also contain hydrocarbons. The formulas for three well known waxes are given below, with the carboxylic acid moiety colored red and the alcohol colored blue.

spermaceti	beeswax	carnuba wax
$CH_3(CH_2)_{14}CO_2$- $(CH_2)_{15}CH_3$	$CH_3(CH_2)_{30}CO_2$- $(CH_2)_{33}CH_3$	$CH_3(CH_2)_{24}CO_2$- $(CH_2)_{29}CH_3$

Waxes are widely distributed in nature. The leaves and fruits of many plants have waxy coatings, which may protect them from dehydration and small predators. The feathers of birds and the fur of some animals have similar coatings which serve as a water repellent. Carnuba wax is valued for its toughness and water resistance.

5. Phospholipids

Phospholipids are the main constituents of cell membranes. They resemble the triglycerides in being ester or amide derivatives of glycerol or sphingosine with fatty acids and phosphoric acid. The phosphate moiety of the resulting phosphatidic acid is further esterified with ethanolamine, choline or serine in the phospholipid itself. The following diagram shows the structures of some of these components. Clicking on the diagram will change it to display structures for two representative phospholipids. Note that the fatty acid components (R & R') may be saturated or unsaturated.

Phospholipid Components

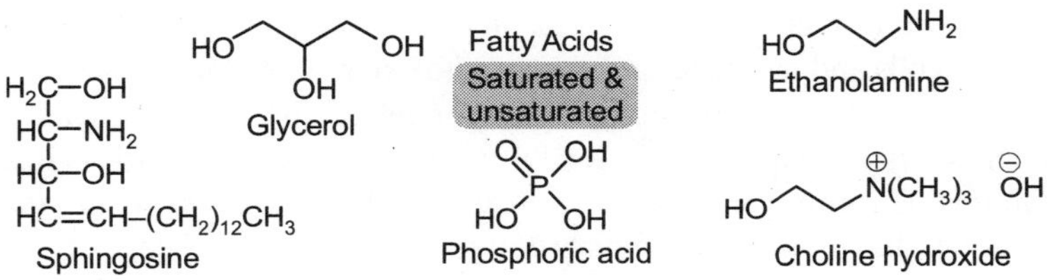

$H_2C{-}OH$
$HC{-}NH_2$
$HC{-}OH$
$HC{=}CH{-}(CH_2)_{12}CH_3$
Sphingosine

HO OH
OH
Glycerol

Fatty Acids
Saturated & unsaturated

O OH
P
HO OH
Phosphoric acid

HO NH_2
Ethanolamine

HO $\overset{\oplus}{N}(CH_3)_3$ $\overset{\ominus}{O}H$
Choline hydroxide

As ionic amphiphiles, phospholipids aggregate or self-assemble when mixed with water, but in a different manner than the soaps and detergents. Because of the two pendant alkyl chains present in phospholipids and the unusual mixed charges in their head groups, micelle formation is unfavorable relative to a bilayer structure. If a phospholipid is smeared over a small hole in a thin piece of plastic immersed in water, a stable planar bilayer of phospholipid molecules is created at the hole. As shown in the following diagram, the polar head groups on the faces of the bilayer contact water, and the hydrophobic alkyl chains form a nonpolar interior. The phospholipid molecules can move about in their half the bilayer, but there is a significant energy barrier preventing migration to the other side of the bilayer.

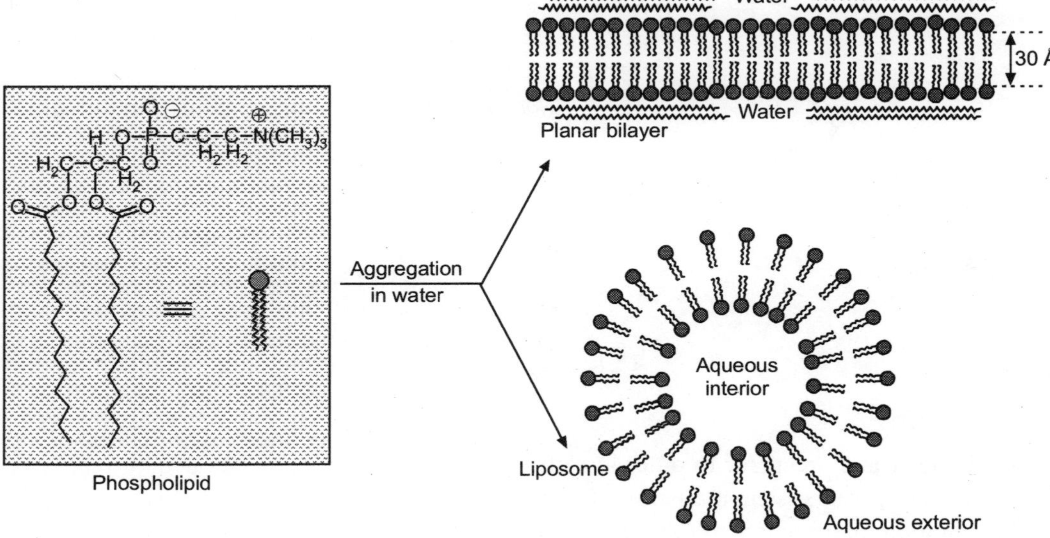

This bilayer membrane structure is also found in aggregate structures called **liposomes**. Liposomes are microscopic vesicles consisting of an aqueous core enclosed in one or more phospholipid layers. They are formed when phospholipids are vigorously mixed with water. Unlike micelles, liposomes have both aqueous interiors and exteriors.

A cell may be considered a very complex liposome. The bilayer membrane that separates the interior of a cell from the surrounding fluids is largely composed of phospholipids, but it incorporates many other components, such as cholesterol, that contribute to its structural integrity. Protein channels that permit the transport of various kinds of chemical species in and out of the cell are also important components of cell membranes. The interior of a cell contains a variety of structures (organelles) that conduct chemical operations vital to the cells existence. Molecules bonded to the surfaces of cells serve to identify specific cells and facilitate interaction with external chemical entities.The sphingomyelins are also membrane lipids. They are the major component of the myelin sheath surrounding nerve fibers. Multiple Sclerosis is a devastating disease in which the myelin sheath is lost, causing eventual paralysis.Prostaglandins Thromboxanes & Leukotrienes. The members of this group of structurally related natural hormones have an extraordinary range of biological effects. They can lower gastric secretions, stimulate uterine contractions, lower blood pressure, influence blood clotting and induce asthma-like allergic responses. Because their genesis in body tissues is tied to the metabolism of the essential fatty acid arachadonic acid (5,8,11,14-eicosatetraenoic acid) they are classified as **eicosanoids**. Many properties of the common drug asprin result from its effect on the cascade of reactions associated with these hormones.

The metabolic pathways by which arachidonic acid is converted to the various eicosanoids are complex and will not be discussed here. A rough outline of some of the transformations that take place is provided below. It is helpful to view arachadonic acid in the coiled conformation shown in the shaded box.

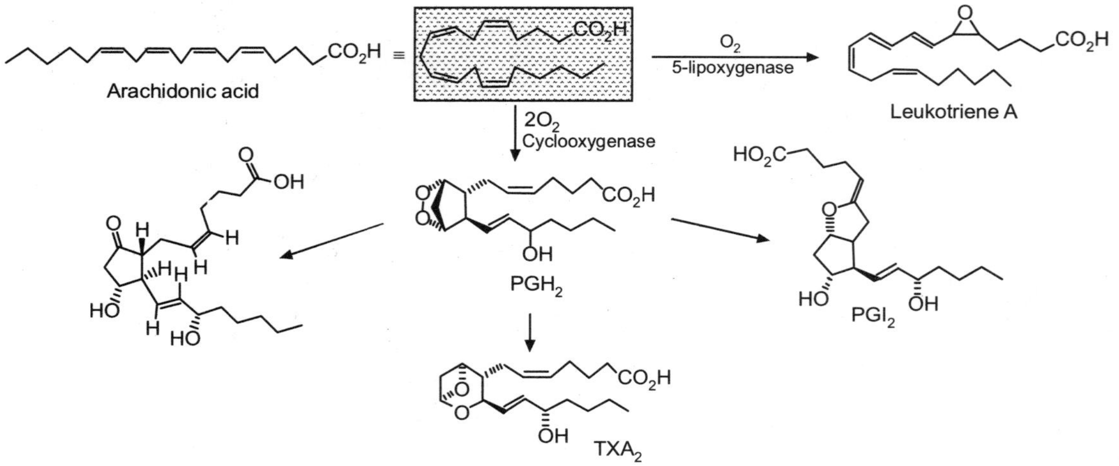

Leukotriene A is a precursor to other leukotriene derivatives by epoxide opening reactions. The prostaglandins are given systematic names that reflect their structure. The initially formed peroxide PGH_2 is a common intermediate to other prostaglandins, as well as thromboxanes such as TXA_2.

Terpenes

Compounds classified as terpenes constitute what is arguably the largest and most diverse class of natural products. A majority of these compounds are found only in plants, but some of the larger and more complex terpenes (e.g. squalene & lanosterol) occur in animals. Terpenes incorporating most of the common functional groups are known, so this does not provide a useful means of classification. Instead, the number and structural organization of carbons is a definitive characteristic. Terpenes may be considered to be made up of isopréne (more accurately isopentane) units, an empirical feature known as the **isoprene rule**. Because of this, terpenes usually have 5n carbon atoms (n is an integer), and are subdivided as follows:

Classification	Isoprene Units	Carbon Atoms
monoterpenes	2	C_{10}
sesquiterpenes	3	C_{15}
diterpenes	4	C_{20}
sesterterpenes	5	C_{25}
triterpenes	6	C_{30}

Isoprene itself, a C_5H_8 gaseous hydrocarbon, is emitted by the leaves of various plants as a natural byproduct of plant metabolism. Next to methane it is the most common volatile organic compound found in the armosphere. Examples of C_{10} and higher terpenes, representing the four most common classes are shown in the following diagram. The initial display is of monoterpenes; larger terpenes will be shown by clicking the "Toggle Structures" button under the diagram. Most terpenes may be structurally dissected into isopentane segments.

Monoterpenes

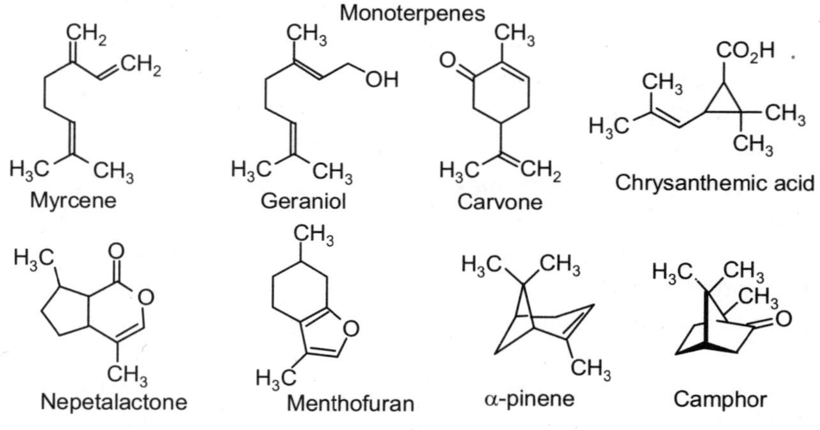

Myrcene · Geraniol · Carvone · Chrysanthemic acid

Nepetalactone · Menthofuran · α-pinene · Camphor

The isopentane units in most of these terpenes are easy to discern, and are defined by the shaded areas. In the case of the monoterpene camphor, the units overlap to such a degree it is easier to distinguish them by coloring the carbon chains. This is also done for alpha-pinene. In the case of the triterpene lanosterol we see an interesting deviation from the isoprene rule. This thirty carbon compound is clearly a terpene, and four of the six isopentane units can be identified. However, the ten carbons in center of the molecule cannot be disected in this manner. Evidence exists that the two methyl groups circled in magenta and light blue have moved from their original isoprenoid locations (marked by small circles of the same color) to their present location. This rearrangement is described in the biosynthesis section. Similar alkyl group rearrangements account for other terpenes that do not strictly follow the isoprene rule.Polymeric isoprenoid hydrocarbons have also been identified. Rubber is undoubtedly the best known and most widely used compound of this kind. It occurs as a colloidal suspension called latex in a number of plants, ranging from the dandelion to the rubber tree (*Hevea brasiliensis*). Rubber is a polyene, and exhibits all the expected reactions of the C=C function. Bromine, hydrogen chloride and hydrogen all add with a stoichiometry of one molar equivalent per isoprene unit. Ozonolysis of rubber generates a mixture of levulinic acid ($CH_3COCH_2CH_2CO_2H$) and the corresponding aldehyde. Pyrolysis of rubber produces the diene isoprene along with other products.

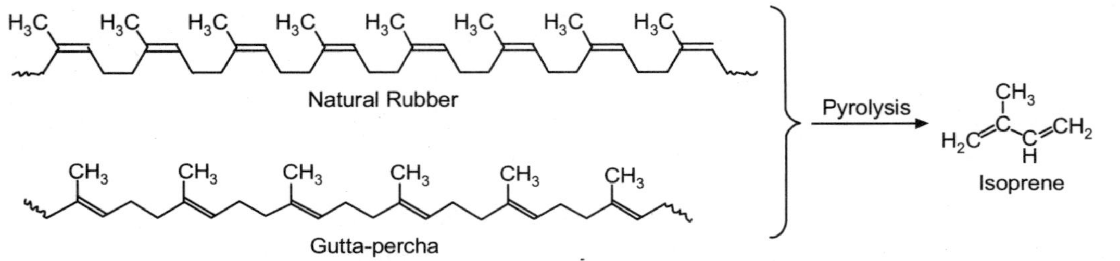

H_3C H_3C H_3C H_3C H_3C H_3C H_3C H_3C

Natural Rubber

CH_3 CH_3 CH_3 CH_3 CH_3 CH_3

Gutta-percha

Pyrolysis

CH_3 H_2C C CH_2 C H

Isoprene

The double bonds in rubber all have a **Z**-configuration, which causes this macromolecule to adopt a kinked or coiled conformation. This is reflected in the physical properties of rubber. Despite its high molecular weight (about one million), crude latex rubber is a soft, sticky, elastic substance. Chemical modification of this material is normal for commercial applications. Gutta-percha (structure above) is a naturally occuring **E**-isomer of rubber. Here the hydrocarbon chains adopt a uniform zig-zag or rod like conformation, which produces a more rigid and tough substance. Uses of gutta-percha include electrical insulation and the covering of golf balls.. Steroids

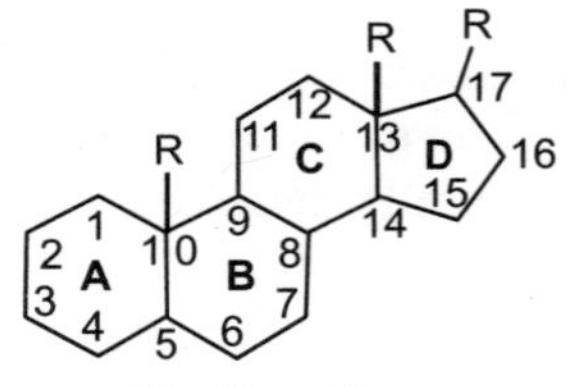

R R 17 12 R 11 C 13 D 16 15 14 1 9 2 10 8 A B 3 7 4 5 6

The Steroid Carbon Skeleton

The important class of lipids called **steroids** are actually metabolic derivatives of terpenes, but they are customarily treated as a separate group. Steroids may be recognized by their tetracyclic skeleton, consisting of three fused six-membered and one five-membered ring, as shown in the diagram to the right. The four rings are designated A, B, C & D as noted, and the peculiar numbering of the ring carbon atoms (shown in red) is the result of an earlier misassignment of the structure. The substituents designated by R are often alkyl groups, but may also have functionality. The R group at the A:B ring fusion is most commonly methyl or hydrogen, that at the C:D fusion is usually methyl. The substituent at C-17 varies considerably, and is usually larger than methyl if it is not a functional group. The most common locations of functional groups are C-3, C-4, C-7, C-11, C-12 & C-17. Ring A is sometimes aromatic. Since a number of tetracyclic triterpenes also have this tetracyclic structure, it cannot be considered a unique identifier.

Steroids are widely distributed in animals, where they are associated with a number of physiological processes. Examples of some important steroids are shown in the following diagram. Different kinds of steroids will be displayed by clicking the "Toggle Structures" button under the diagram. Norethindrone is a synthetic steroid, all the other examples occur naturally. A common strategy in pharmaceutical chemistry is to take a natural compound, having certain desired biological properties together with undesired side effects, and to modify its structure to enhance the desired characteristics and diminish the undesired. This is sometimes accomplished by trial and error.The generic steroid structure drawn above has seven chiral stereocenters (carbons 5, 8, 9, 10, 13, 14 & 17), which means that it may have as many as 128 stereoisomers. With the exception of C-5, natural steroids generally have a single common configuration. This is shown in the last of the toggled displays, along with the preferred conformations of the rings.

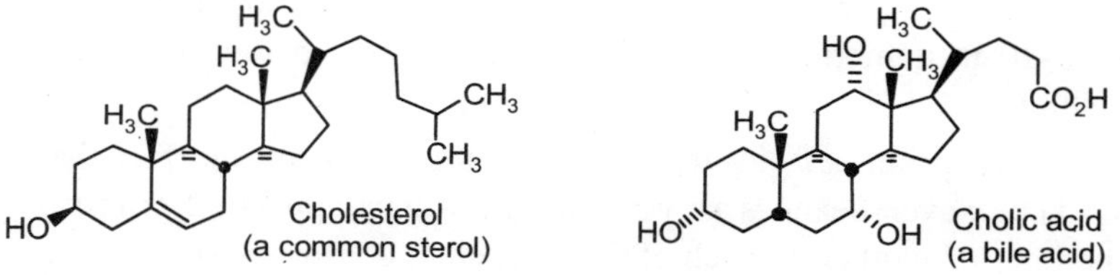

Cholesterol (a common sterol)

Cholic acid (a bile acid)

Typical Animal Steroids

Chemical studies of the steroids were very important to our present understanding of the configurations and conformations of six-membered rings. Substituent groups at different sites on the tetracyclic skeleton will have axial or equatorial orientations that are fixed because of the rigid structure of the trans-fused rings. This fixed orientation influences chemical reactivity, largely due to the greater steric hindrance of axial groups versus their equatorial isomers. Thus an equatorial hydroxyl group is

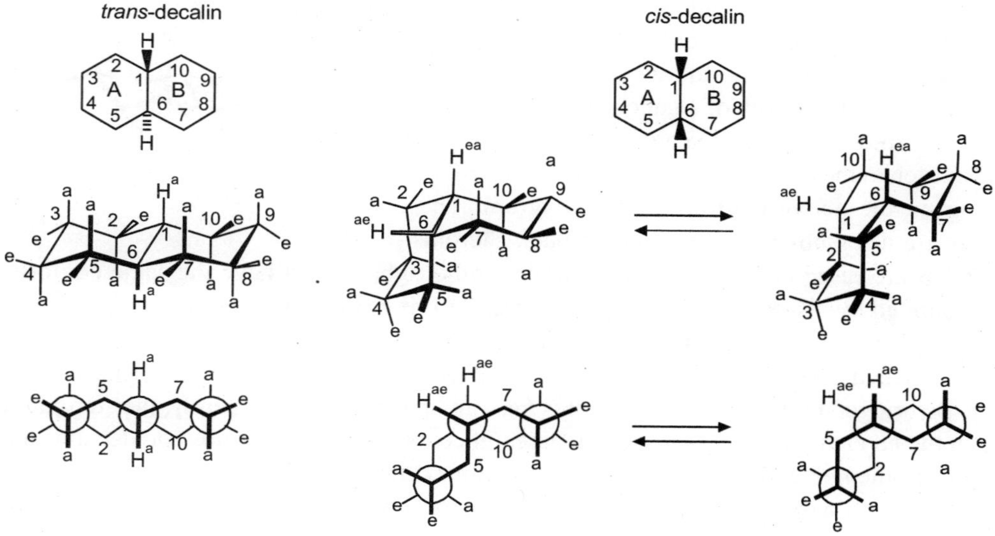

trans-decalin

cis-decalin

esterified more rapidly than its axial isomer. It is instructive to examine a simple bicyclic system as a model for the fused rings of the steroid molecule. Decalin, short for decahydronaphthalene, exists as cis and trans isomers at the ring fusion carbon atoms. Planar representations of these isomers are drawn at the top of the following diagram, with corresponding conformational formulas displayed underneath. The numbering shown for the ring carbons follows IUPAC rules, and is different from the unusual numbering used for steroids. For purposes of discussion, the left ring is labeled A (colored blue) and the right ring B (colored red). In the conformational drawings the ring fusion and the angular hydrogens are black.

The trans-isomer is the easiest to describe because the fusion of the A & B rings creates a rigid, roughly planar, structure made up of two chair conformations. Each chair is fused to the other by equatorial bonds, leaving the angular hydrogens (H^a) axial to both rings. Note that the bonds directed above the plane of the two rings alternate from axial to equatorial and back if we proceed around the rings from C-1 to C-10 in numerical order. The bonds directed below the rings also alternate in a complementary fashion.

Conformational descriptions of cis- decalin are complicated by the fact that two energetically equivalent fusions of chair cyclohexanes are possible, and are in rapid equilibrium as the rings flip from one chair conformation to the other. In each of these all chair conformations the rings are fused by one axial and one equatorial bond, and the overall structure is bent at the ring fusion. In the conformer on the left, the red ring (B) is attached to the blue ring (A) by an axial bond to C-1 and an equatorial bond to C-6 (these terms refer to ring A substituents). In the conformer on the right, the carbon bond to C-1 is equatorial and the bond to C-6 is axial. Each of the angular hydrogens (H^{ae} or H^{ea}) is oriented axial to one of the rings and equatorial to the other. This relationship reverses when double ring flipping converts one cis-conformer into the other. Cis-decalin is less stable than trans-decalin by about 2.7 kcal/mol (from heats of combustion and heats of isomerization data). This is due to steric crowding (hindrance) of the axial hydrogens in the concave region of both cis-conformers, as may be seen in the Chime display activated by the following button. This difference is roughly three times the energy of a gauche butane conformer relative to its anti conformer. Indeed three gauche butane interactions may be identified in each of the cis-decalin conformations, as will be displayed by clicking on the above conformational diagram. These gauche interactions are also shown in the Chime model. Steroids in which rings A and B are fused cis, such as the example on the right, do not have the same conformational mobility exhibited by cis-decalin. The fusion of ring C to ring B in a trans configuration prevents ring B from undergoing a conformational flip to another chair form. If this were to occur, ring C would have to be attached to ring B by two adjacent axial bonds directed 180° apart. This is too great a distance to be bridged by the four carbon atoms making up ring C. Consequently, the steroid molecule is locked in the all chair conformation shown here. Of course, all these steroids and decalins may have one or more six-membered rings in a boat conformation. However the high energy of boat conformers relative to chairs would make such structures minor components in the overall ensemble of conformations available to these molecules.

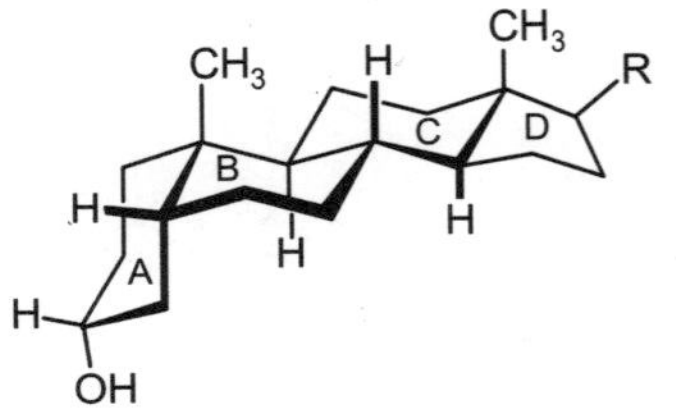

Lipid Soluble Vitamins *(See Fig. 2.18)*

The essential dietary substances called **vitamins** are commonly classified as "water soluble" or "fat soluble". Water soluble vitamins, such as vitamin C, are rapidly eliminated from the body and their dietary levels need to be relatively high. The recommended daily allotment (RDA) of vitamin C is 100 mg, and amounts as large as 2 to 3 g are taken by many people without adverse effects. The lipid soluble vitamins, shown in the diagram below, are not as easily eliminated and may accumulate to toxic levels if consumed in large quantity. The RDA for these vitamins are:

Vitamin A	800 μg (upper limit ca. 3000 μg)
Vitamin D	5 to 10 μg (upper limit ca. 2000 μg)
Vitamin E	15 mg (upper limit ca. 1 g)
Vitamin K	110 μg (upper limit not specified)

From this data it is clear that vitamins A and D, while essential to good health in proper amounts, can be very toxic. Vitamin D, for example, is used as a rat poison, and in equal weight is more than 100 times as poisonous as sodium cyanide.

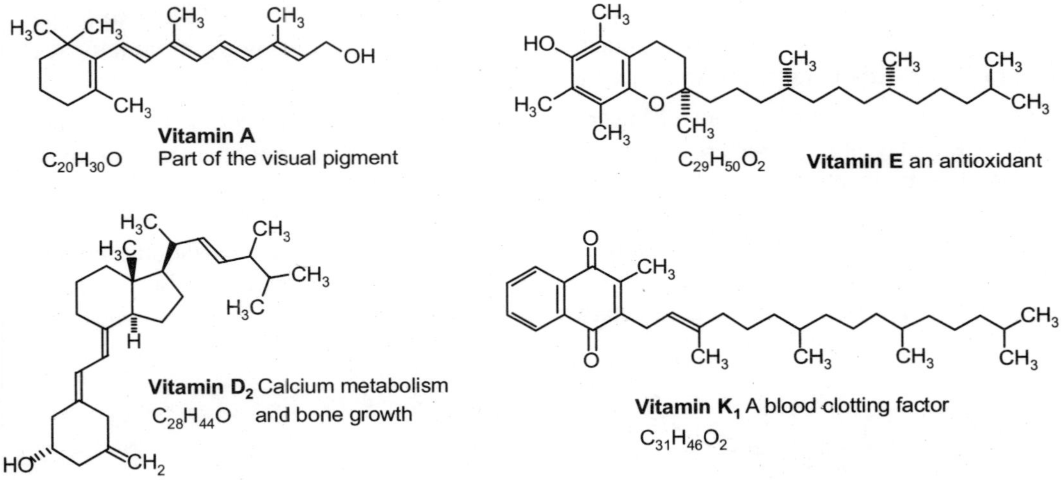

$C_{20}H_{30}O$ **Vitamin A** Part of the visual pigment

$C_{29}H_{50}O_2$ **Vitamin E** an antioxidant

Vitamin D_2 Calcium metabolism and bone growth $C_{28}H_{44}O$

Vitamin K_1 A blood clotting factor $C_{31}H_{46}O_2$

Fig. 2.18 : Lipid soluble Vitamins

From the structures shown here, it should be clear that these compounds have more than a solubility connection with lipids. Vitamins A is a terpene, and vitamins E and K have long terpene chains attached to an aromatic moiety. The structure of vitamin D can be described as a steroid in which ring B is cut open and the remaining three rings remain unchanged. The precursors of vitamins A and D have been identified as the tetraterpene β-carotene and the steroid ergosterol, respectively.

Biosynthetic Pathways

The complex organic compounds found in living organisms on this planet originate from photosynthesis, an endothermic reductive condensation of carbon dioxide requiring light energy and the pigment chlorophyll.

$$x\,CO_2 + x\,H_2O + \text{energy} \rightarrow C_xH_{2x}O_x + x\,O_2$$

The products of photosynthesis are a class of compounds called carbohydrates, the most common and important of which is glucose ($C_6H_{12}O_6$). Subsequent reactions effect an oxidative cleavage of glucose to pyruvic acid (CH_3COCO_2H), and this in turn is transformed to the two-carbon building block, acetate. The multitude of lipid structures described here are constructed from acetate by enzymatic reactions that in many respects correspond to reactions used by chemists for laboratory syntheses of similar compounds. However, an important restriction is that the reagents and conditions must be compatible with the aqueous medium and moderate temperatures found in living cells. Consequently, the condensation, alkylation, oxidation and reduction reactions that accomplish the biosynthesis of lipids will not make use of the very strong bases, alkyl halides, chromate oxidants or metal hydride reducing agents that are employed in laboratory work.

Condensations

Claisen condensation of ethyl acetate (or other acetate esters) forms an acetoacetate ester, as illustrated by the top equation in the following diagram. Reduction, dehydration and further reduction of this product would yield an ester of butyric acid, the overall effect being the elongation of the acetate starting material by two carbons. In principle, repetition of this sequence would lead to longer chain acids, made up of an even number of carbon atoms. Since most of the common natural fatty acids have even numbers of carbon atoms, this is an attractive hypothesis for their biosynthesis.

Nature's solution to carrying out a Claisen-like condensation in a living cell is shown in the bottom equation of the diagram. Thioesters are more reactive as acceptor reactants than are ordinary esters, and preliminary conversion of acetate to malonate increases the donor reactivity of this species. The thiol portion of the thioester is usually a protein of some kind, with efficient acetyl transport occuring by way of acetyl coenzyme A. Depending on the enzymes involved, the condensation product may be reduced and then further elongated so as to produce fatty acids (as shown), or elongated by further condensations to polyketone intermediates that are precursors to a variety of natural phenolic compounds.

Claisen condensation

R = small alkyl group

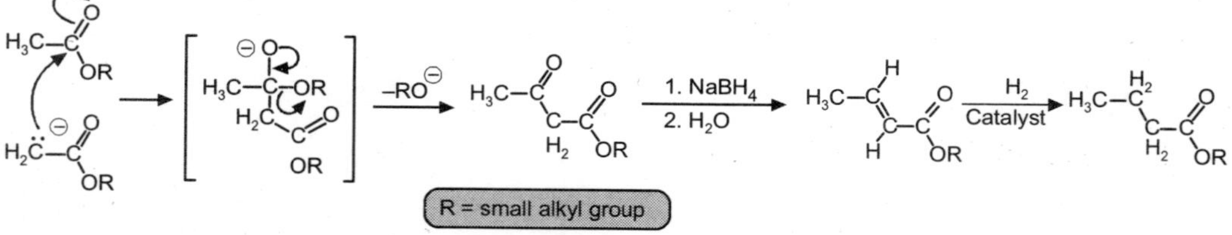

Thioester condensation

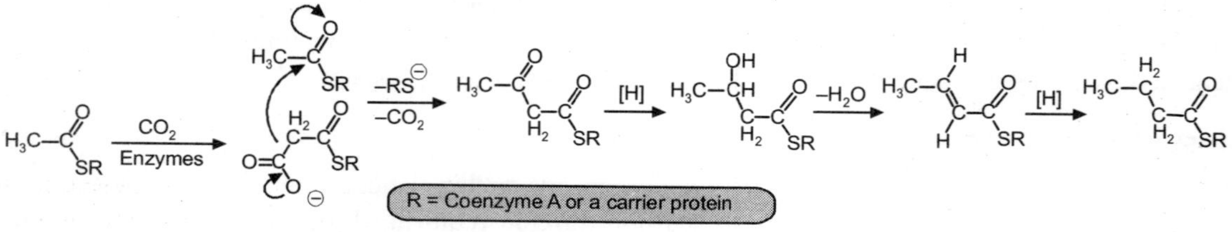

R = Coenzyme A or a carrier protein

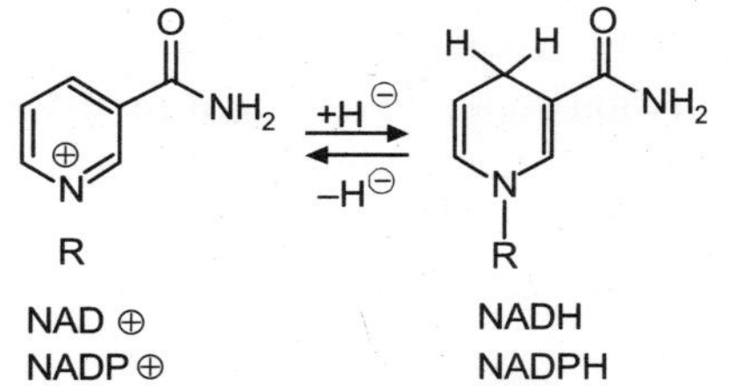

NAD ⊕ NADP ⊕ NADH NADPH

The reduction steps (designated by [H] in the equations) and the intervening dehydrations needed for fatty acid synthesis require unique coenzymes and phosphorylating reagents. The pyridine ring of nicotinamide adenine dinucleotide (NAD) and its 2'-phosphate derivative (NADP) function as hydride acceptors, and the corresponding reduced species (NADH & NADPH) as a hydride donors. Partial structures for these important redox reagents are shown on the right.

As noted earlier, the hydroxyl group is a poor anionic leaving group (hydroxide anion is a strong base). Phosphorylation converts a hydroxyl group into a phosphate (PO_4) or pyrophosphate (P_2O_7) ester, making it a much better leaving group (the pK_as at pH near 7 are 7.2 and 6.6 respectively). The chief biological phosphorylation reagents are phosphate derivatives of adenosine (a ribose compound). The strongest of these is the triphosphate ATP, with the diphosphate and monophosphate being less powerful.

The overall process of fatty acid synthesis is summarized for palmitic acid, $CH_3(CH_2)_{14}CO_2H$, in the following equation:

$$8CH_3CO\text{-}CoA + 14\ NADPH + 14\ H^{(+)} + 7\ ATP + H_2O \rightarrow CH_3(CH_2)_{14}CO_2H + 8\ CoA + 14\ NADP^{(+)} + 7\ ADP + 7\ H_2PO_4^{(-)}$$

Alkylations

The branched chain and cyclic structures of the terpenes and steroids are constructed by sequential alkylation reactions of unsaturated isopentyl pyrophosphate units. As depicted in the following diagram,

Pyrophosphate intermediates in terpene synthesis

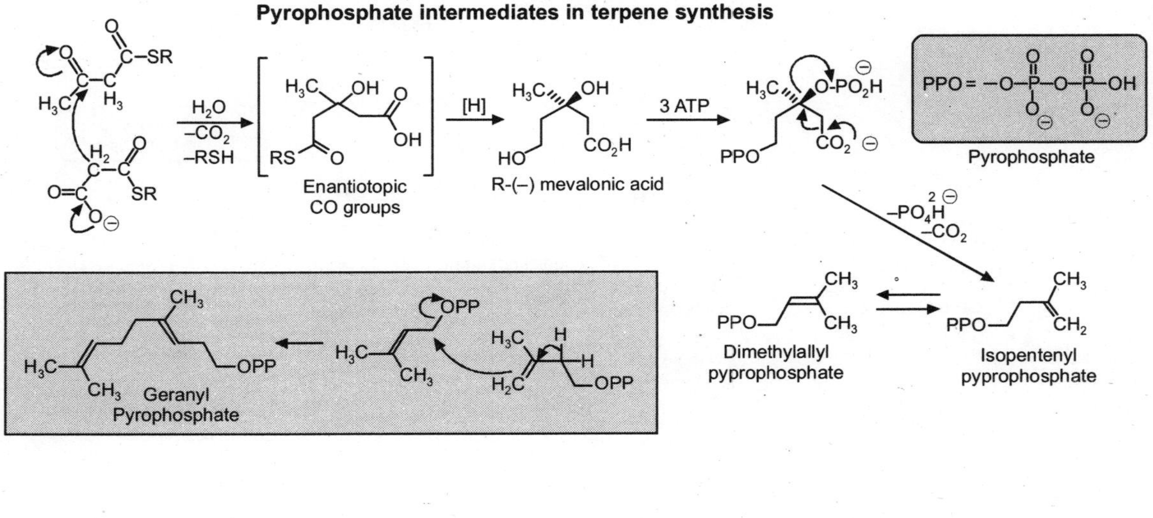

Enantiotopic CO groups

R-(–) mevalonic acid

Pyrophosphate

Dimethylallyl pyprophosphate

Isopentenyl pyprophosphate

Geranyl Pyrophosphate

these 5-carbon reactants are made from three acetate units by way of an aldol-like addition of a malonate intermediate to acetoacetate. Selective hydrolysis and reduction gives a key intermediate called **mevalonic acid**. Phosphorylation and elimination of mevalonic acid then generate isopenenyl pyrophosphate, which is in equilibrium with its double bond isomer, dimethylallyl pyrophosphate. The allylic pyrophosphate group in the latter compound is reactive in enzymatically catalyzed alkylation reactions, such as the one drawn in the green box. This provides support for the empirical isoprene rule.

The simplest fashion in which isopentane units combine is termed "head-to-tail". This is the combination displayed in the green box, and these terms are further defined in the upper equation that will appear above on clicking the "Toggle Examples" button. Non head-to-tail coupling of isopentane units is also observed, as in the chrysanthemic acid construction shown in the second equation. A second click on the diagram displays the series of cation-like cyclizations and rearrangements that have been identified in the biosynthesis of the triterpene lanosterol. Lanosterol is a precursor in the biosynthesis of steroids. This takes place by metabolic removal of three methyl groups and degradation of the side chain.

An Alternative Isoprenoid Synthesis

For many years, the mevalonic acid route to isopentenyl pyrophosphate was considered an exclusive biosynthetic pathway. Recently, an alternative reaction sequence, starting from pyruvic acid and glyceraldehyde-3-phosphate, has been identified (bottom equations in the following diagram). By labeling selective carbon atoms (colored red) these distinct paths are easily distinguished. The new, **DXP** (1-deoxyxylulose-5-phosphate) path is widespread in microorganisms and chloroplast terpenes. The rearrangement to 2-methylerythritol-4-phosphate is an extraordinary transformation.

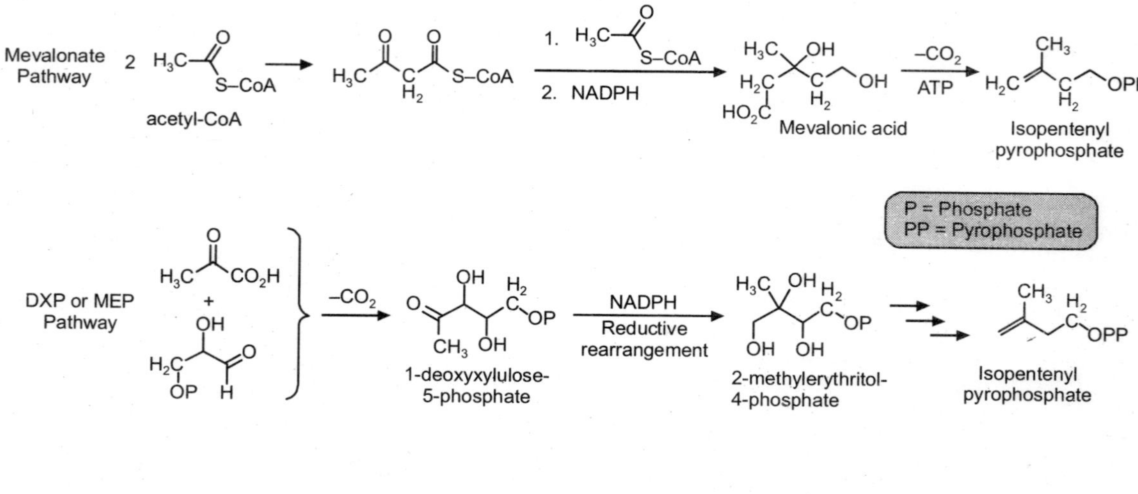

Mevalonate Pathway
acetyl-CoA
1. H_3C–CO–S–CoA
2. NADPH
Mevalonic acid
$-CO_2$
ATP
Isopentenyl pyrophosphate

P = Phosphate
PP = Pyrophosphate

DXP or MEP Pathway
$-CO_2$
1-deoxyxylulose-5-phosphate
NADPH
Reductive rearrangement
2-methylerythritol-4-phosphate
Isopentenyl pyrophosphate

Prostaglandins

or C=O
cis or trans
HO
O
OH
Arachidonic acid
Cyclooxygenase
HO
OH
Prostaglandin

Prostalandins are hormones.

They transmit signals within cells and between cells.

Local activity only.

All cells make prostaglandins.

Different classes made in different tissues.

Action is determined by class produced.

Many different functions.

Blood Clotting

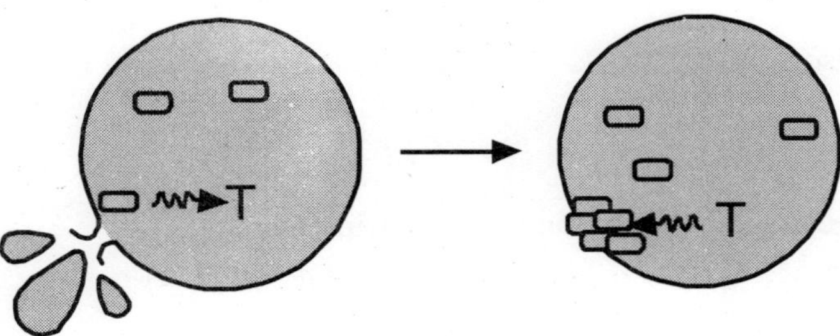

Thromboxin A (T) causes platelet aggregation → clot.

Prostacyclin (PGI_2) inhibits aggregation → no clot.

Prostaglandins

Inflammatory Response and Aspirin

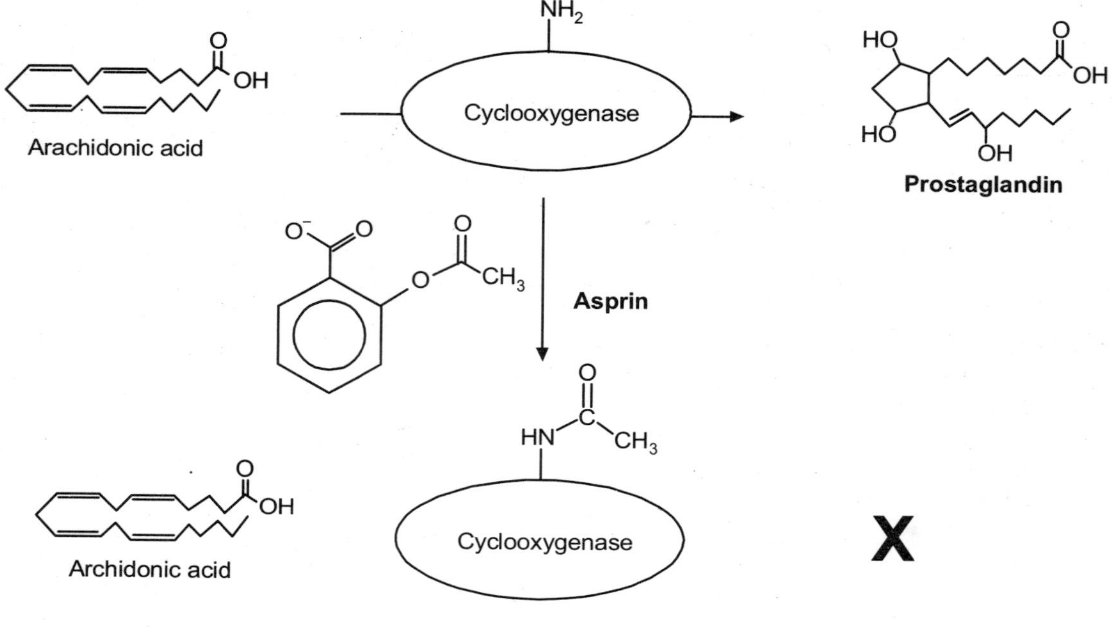

Enzyme is acetylated ⟶ inactivation

But ... Prostaglndins also

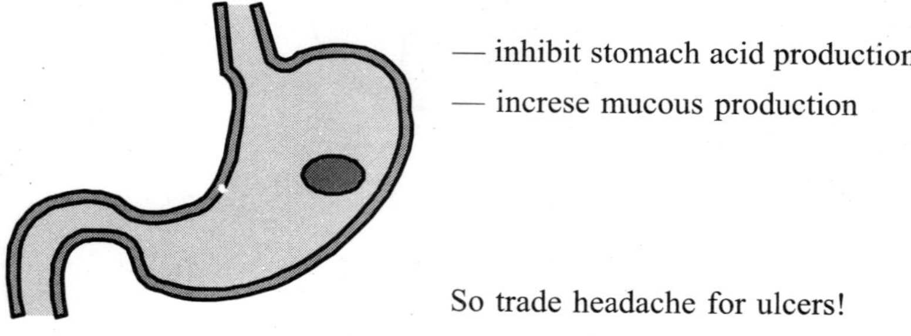

— inhibit stomach acid production

— increse mucous production

So trade headache for ulcers!

Sphinogomyelin

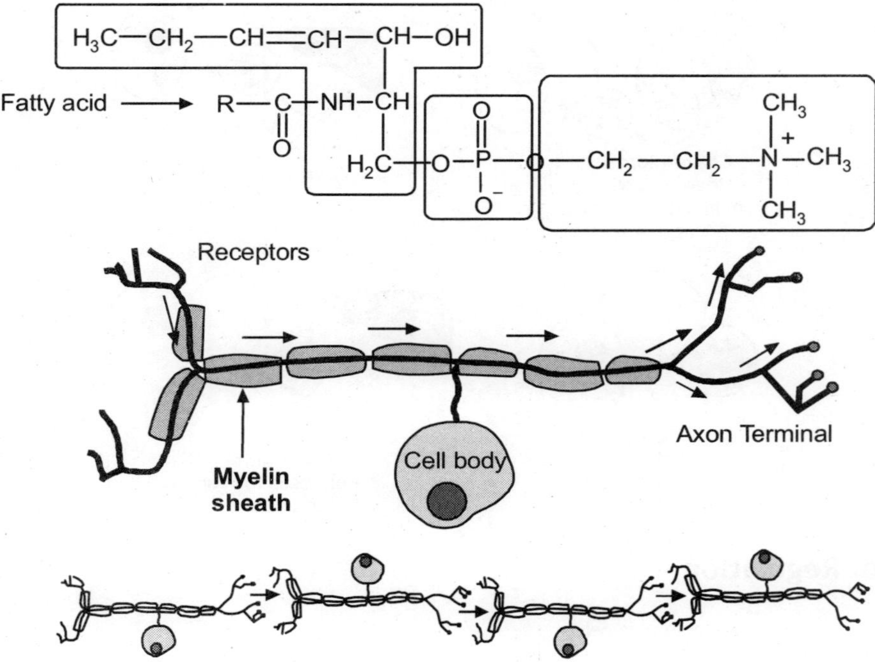

Myelin increases the speed of nerve impulse conduction & protects axon.

Disorders :

Niemann-Pick Disease:

- No Sphingomyelinase (breaks sphingomyelin down)
- accumulates incells (brain, bone marrow, liver)
- Lack of motor skills,muscle strength and tone
- Lose vision and hearing
- Death within a few years

Multiple Sclerosis:

- Believed to be anautoimmune response
- Antibodies attack myelin around nerves
- Different levels of severity

Lipoproteins

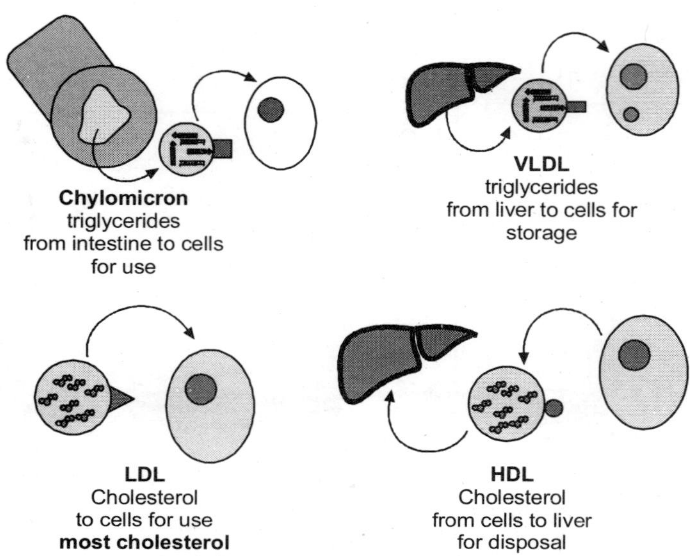

Cholesterol Regulation

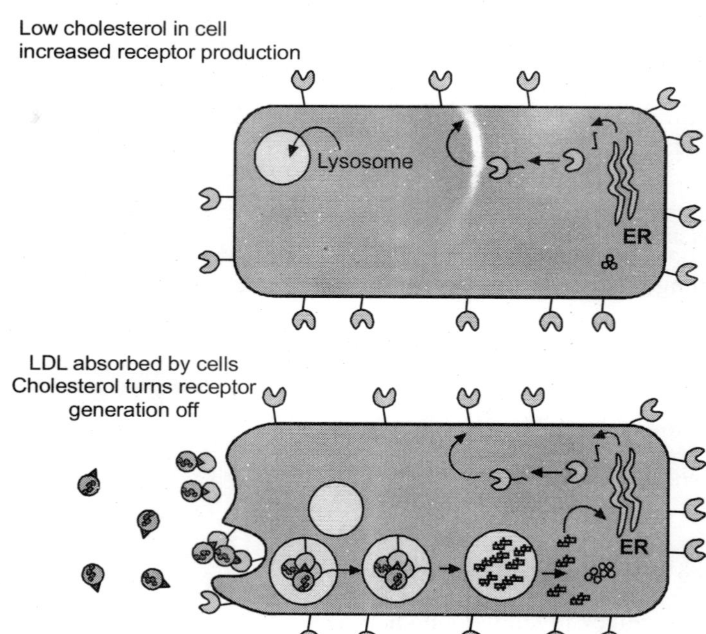

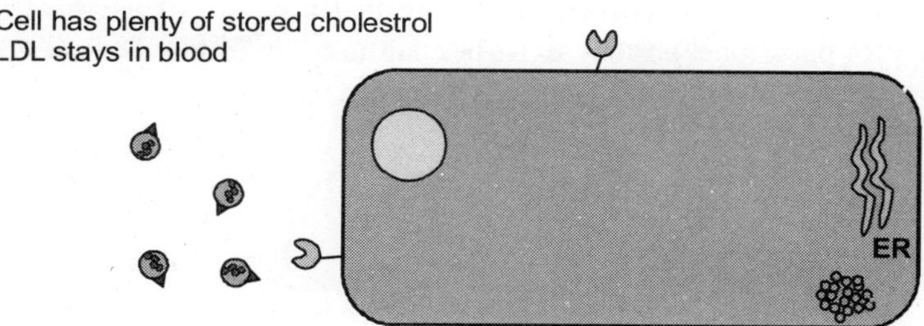

Arteriosclerosis

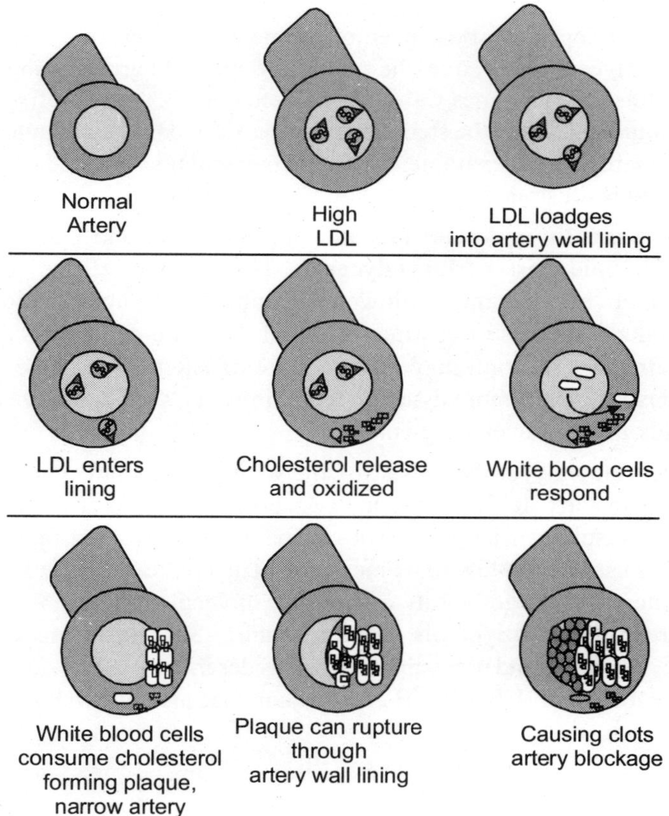

TRIACYLGLYCEROLS SIMPLE LIPIDS

Triacylglycerols: Nearly all the commercially important fats and oils of animal and plant origin consist almost exclusively of the simple lipid class triacylglycerols (termed "triglycerides" in the older literature). They consist of a glycerol moiety with each hydroxyl group esterified to a fatty acid. In nature, they are synthesised by enzyme systems, which determine that a centre of asymmetry is created about carbon-2 of the glycerol backbone, so they exist in enantiomeric forms, i.e. with different fatty acids in each position.

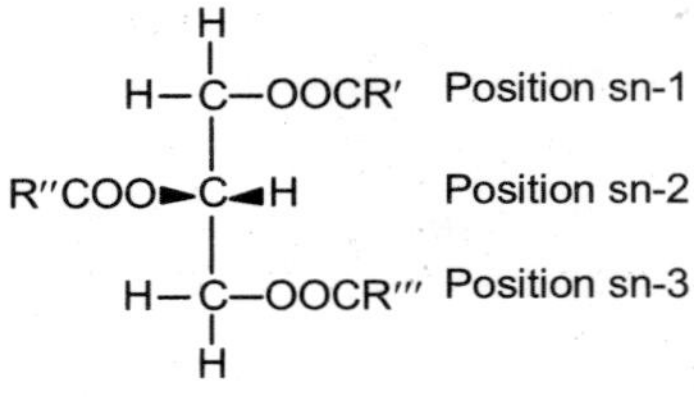

Position sn-1

Position sn-2

Position sn-3

Fischer projection of a triacyl-*sn*-glycerol

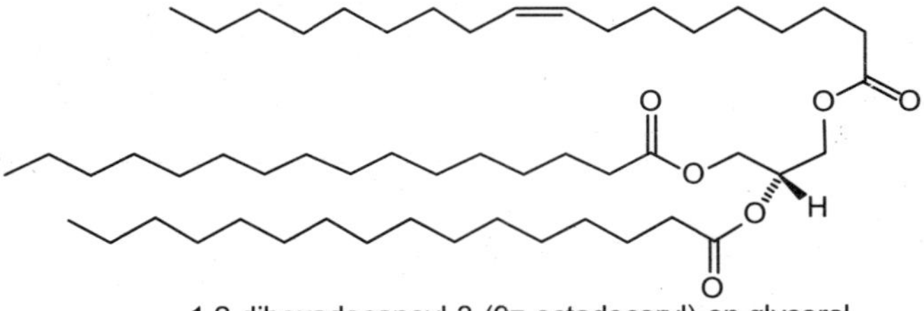

1,2-dihexadecanoyl-3-(9z-octadeceryl)-*sn*-glycerol

A stereospecific numbering system has been recommended to describe these forms. In a Fischer projection of a natural L-glycerol derivative, the secondary hydroxyl group is shown to the left of C-2; the carbon atom above this then becomes C-1 and that below is C-3. The prefix "*sn*" is placed before the stem name of the compound, when the stereochemistry is defined. Their primary biological function is to serve as a store of energy. As an example, the single molecular species 1,2-dihexadecanoyl-3-(9*Z*-octadecenoyl)-*sn*-glycerol is illustrated.

Nearly all the commercially important fats and oils of animal and plant origin consist almost exclusively of the simple lipid class - **triacylglycerols**. This includes all the vegetable oils, such as those from corn (maize), olive, palm, and sunflower, and animal fats, such as tallow, lard and butter, as well as commercial products such as margarines. Most of these are depots fats (from adipose tissue) or milk fats, where their main function may be as a store of energy, but some triacylglycerols (e.g. those of plasma or liver) may have a more dynamic role. Similarly, seed oils serve as a source of energy and structural fatty acids for the developing embryo.

In chemical terms, triacylglycerols consist of the trihydric alcohol glycerol esterified, almost invariably, with long-chain fatty acids. When the two primary hydroxyl groups are esterified with different fatty acids, the resulting triacylglycerol can be asymmetric and thus can display "optical activity", although this is usually too low to be measured. The conventional D/L or R/S systems could designate such enantiomers without ambiguity with simple molecules, but problems arise in application to the complex mixtures of triacylglycerols found in nature. Such problems can be avoided if the stereochemistry of triacylglycerols and other glycerolipids is described by the "stereospecific numbering" (*sn*) system as recommended by a IUPAC-IUB commission (see also our web page on Nomenclature).

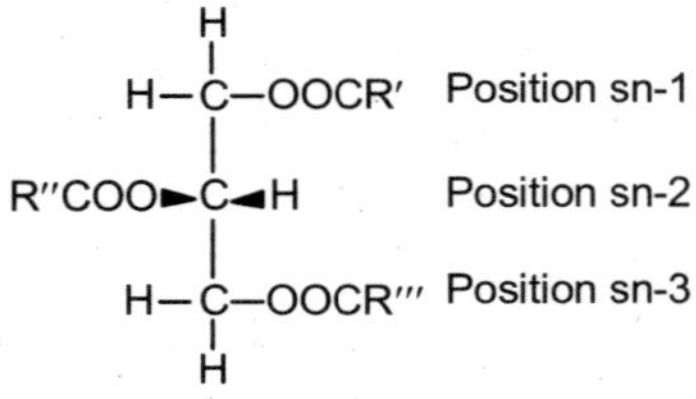

Position sn-1

Position sn-2

Position sn-3

Fischer projection of a triacyl-*sn*-glycerol

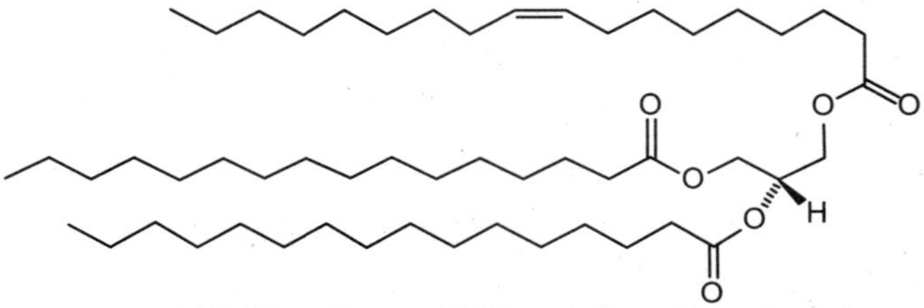

1,2-dihexadecanoyl-3-(9z-octadeceryl)-*sn*-glycerol

In a Fischer projection of a natural L-glycerol derivative as shown above, the secondary hydroxyl group is shown to the left of C-2; the carbon atom above this then becomes C-1 while that below becomes C-3, and the prefix *sn* is placed before the stem name of the compound. The term "triacyl-*sn*-glycerol" should then be used to designate the molecule rather than "triglyceride". The former is technically more accurate and is essential for the stereospecific numbering system, while the latter term may only be familiar now to older readers. When the detailed stereochemistry is not specified, the primary hydroxy groups are often termed the α- and α'-positions and the secondary, the β-position. As an example, the single molecular species 1,2-dihexadecanoyl-3-(9*Z*-octadecenoyl)-*sn*-glycerol is illustrated.

Differences in the distributions of fatty acids among the positions of the glycerol moiety in triacylglycerols from natural fats and oils were first demonstrated systematically by enzymatic hydrolysis procedures, specifically pancreatic lipase hydrolysis for the analysis of the fatty acids of position *sn*-2, before complex stereospecific hydrolysis procedures were developed that permitted the complete positional distributions of the fatty acids to be determined. Because of this historical development of the analytical procedures, there has been a tendency to assume that the composition of fatty acids esterified to the sole secondary hydroxyl group must have greater importance than those of the two primary positions. It is certainly true that the composition of position *sn*-2 is of great importance when triacylglycerols are consumed and digested by animals, since 2-monoacyl-*sn*-glycerols are then formed which can be absorbed by the intestines and utilised as such. On the other hand, the results of stereospecific analyses have shown that the compositions of all three positions in certain fats can be distinctive and can highlight important aspects of the biosynthetic processes.

The discussion that follows centres on the fatty acid compositions and the stereospecific distribution of fatty acids within triacyl-*sn*-glycerols with a few selected examples. Note that all data are presented as mol% not weight %. The second document here dealing with triacylglycerols (Triacylglycerols. Part 2) discusses the biochemistry of triacylglycerols and methods of analysis.

Triacylglycerols from Seed Oils

While seed and fruit oils within the plant kingdom as a whole can contain a vast range of different fatty acids, those that are most widely for food purposes tend to have relatively simple compositions in which C18 unsaturated fatty acids, especially oleate and linoleate are predominant. Thus, olive oils contains over 70% of oleic acid, while safflower oil can contain a similar proportion of linoleate, which is also the major component of maize (corn) oil and soybean oils. Linseed oil is used mainly for industrial purposes as it contains about 60% α-linolenic acid, which is not suitable for food purposes as it oxidizes too readily. Soybean oil, with 3 to 7% linolenic acid, is often hydrogenated before it is used as a food ingredient. Rapeseed oil is available in two forms, i.e. with high or low erucic acid contents. The former has many industrial uses, but is banned from foods because of fears about the safety of erucic acid for human consumption. Castor oil, with mainly industrial applications, contains up to 90% ricinoleic acid (see our web pages on hydroxy fatty acids). The so-called 'tropical oils', such as palm oil, contain higher amounts of saturated fatty acids than most other commercial oils. Palm kernel and coconut oils are rich sources of medium-chain fatty acids.

A considerable amount of data has been obtained by means of pancreatic lipase hydrolysis on the composition of position *sn*-2 of the triacyl-*sn*-glycerols of seed oils, and this has been enhanced in recent years by stereospecific analyses. The author has of necessity been selective in presenting results, and some representative analyses are listed in Table 2.1.

Table 2.1 : Positional distributions of fatty acids (mol %) in triacyl-*sn*-glycerols of seed oils

Oil	Position	Fatty acid					
		16:0	18:0	18:1	18:2	18:3	C_{20}-C_{24}
Peanut	TG	9	3	58	23		7
	1	14	5	59	19		4
	2	2	tr	59	39		1
	3	11	5	57	10		15
Rapeseed*	TG	3	2	26	17	10	43
	1	4	2	23	11		53
	2	1		37	36	6	6
	3	4	3	17	4	20	70
Soyabean	TG	9	4	24	54	8	
	1	14	6	23	48	9	
	2	1	tr	22	70	7	
	3	13	6	28	45	9	
Linseed	TG	6	4	16	17	57	
	1	10	6	15	16	53	
	2	2	1	16	21	60	
	3	6	4	17	13	59	
Maize (corn)	TG	11	2	29	57	1	
	1	1	3	28	50	1	
	2	18	tr	27	70	1	
	3	2	3	31	52	1	
Olive	TG	10	2	76	10	1	
	1	13	3	72	10	1	
	2	1		83	14	1	
	3	17	4	74	5		
Cacao butter	TG	24	35	36	3	tr	1
	1	34	50	12	1	1	1
	2	2	2	87	9		
	3	37	53	9	tr		2
Palm	TG	48	4	36	10		
	1	60	3	27	9		
	2	13	tr	68	18		
	3	72	8	14	3		

tr = trace (<0.5%). * High erucic acid rapeseed oil. TG = intact triacylglycerols

As was well known from studies involving hydrolysis with pancreatic lipase, position sn-2 of the triacylglycerols of seed oils is greatly enriched in polyunsaturated fatty acids (specifically linoleic and linolenic acids), while saturated fatty acids are concentrated in the primary positions, and monoenoic acids are relatively evenly distributed. There are exceptions to these rules and in cacao butter, for example, oleic acid is present largely in position sn-2. Minor differences only in the distributions of saturated and monoenoic fatty acids between positions sn-1 and sn-3 have been observed and too few samples have been analysed for definitive comment. Longer-chain fatty acids (C20-C24) are apparently concentrated in the primary positions with some small preference for position sn-3.

In those seed oils containing unusual fatty acids that have been subjected to stereospecific analysis, an allenic estolide was found entirely in position *sn*-3 in the triacyl-*sn*-glycerols of *Sapium sebiferum*, acetic acid appeared to be linked entirely to position *sn*-3 in *Euonymus verrucosus*, and much of the coriolic acid was in position *sn*-3 in *Monnina emarginata.*

In summary, seed oils containing the usual range of saturated and unsaturated fatty acids tend to have the polyunsaturated components in position *sn*-2, but relatively little difference between the primary positions; less-common fatty acids tend to be concentrated in position *sn*-3.

Triacylglycerols from Animal Fats—Adipose Tissue

Much of the triacylglycerols in animal tissues, including some of the commercially important fats such as lard or tallow, are contained within various adipose tissue sites, where they serve mainly as an energy store. The subcutaneous fats also help to insulate animals, while fat stores in fish and marine animals help to maintain buoyancy. Fatty acids are synthesised *de novo* in tissues (saturated and monoenoic acids), but the composition of the triacylglycerols also reflects the diet. In terrestrial animals, the composition tends to be quite simple, with C_{16} (mainly 16:0) and C_{18} fatty acids predominating. All the essential fatty acids, such as linoleic and α-linolenic acid, must come from the diet of course. Ruminant animals, such as the cow and sheep, have relatively saturated fats because the dietary unsaturated fatty acids are subjected to biohydrogenation in the rumen, a process that also generates ***trans*** fatty acids as by-products. These animals also tend to have relatively higher concentrations of odd- and branched-chain fatty acids from the rumen microflora. Non-ruminant herbivores, such as the horse, can have appreciable amounts of linolenic acid from grass in their adipose tissue, while that of marine mammals is characterized by high concentrations of long-chain mono- and polyenoic fatty acids because of their diet of fish.

Marked differences have been observed in the distributions of fatty acids among the three positions of the glycerol moiety in most species examined. There are also appreciable inter-species differences. A few representative results are listed in Table 2.2.

For most species, saturated fatty acids are found predominantly in position *sn*-1, although appreciable amounts of oleic acid are usually present also. Position *sn*-2 tends to contain mainly unsaturated fatty acids, especially linoleic acid, although some of the shorter-chain fatty acids also accumulate here in

Table 2.2 : Positional distributions of fatty acids (mol %) in triacyl-*sn*-glycerols of animal depot fats

Species	Position	Fatty acid						
		14:0	16:0	16:1	18:0	18:1	18:2	18:3
Human	TG	5	24	7	8	46	7	1
	1	4	42	3	15	27	6	1
	2	6	10	12	2	55	4	2
	3	4	19	6	6	57	11	1
Cattle	TG	5	27	6	17	33	5	1
	1	4	41	6	17	20	4	1
	2	9	17	6	9	41	5	1
	3	1	22	6	24	37	5	1
Sheep	TG	3	22	2	35	36	2	
	1	1	35	2	47	4	-	
	2	4	14	2	15	52	5	
	3	3	16	1	42	26	2	
Pig	TG	2	27	3	13	45	9	
	1	1	10	2	30	51	6	
	2	4	72	5	2	13	3	
	3	-	tr	2	7	70	18	
Rat	TG	2	23	5	6	35	26	1
	1	2	32	5	9	32	15	1
	2	1	10	4	1	37	45	1
	3	2	27	5	7	37	17	1
Rabbit	TG	3	28	9	3	29	20	4
	1	3	34	9	6	25	14	2
	2	6	25	12	1	26	23	5
	3	1	24	7	3	35	22	5
Chicken	TG	1	30	6	6	45	11	1
	1	1	47	7	8	31	5	1
	2	tr	13	5	6	55	19	1
	3	1	31	7	3	49	8	1

Results are listed for *cis*-18:1 isomers only; *trans*-18:1 was present in positions *sn*-1, *sn*-2 and *sn*-3 as 5, 2 and 6 %, respectively.

tr = trace (<0.5%). TG = intact triacylglycerols

some instances. There is some preference for the longer-chain fatty acids to be located in position *sn*-3. The main exception to these rules is in the pig and related species where it has been known for many years that palmitic acid can comprise more than 70% of the fatty acids in position *sn*-2. However, it is less well known though perhaps of equal interest that most of the stearic acid (75% of that in the tissue) is in position *sn*-1, while position *sn*-3 contains more than 70% of oleic acid. Although the absolute fatty acid compositions of adipose tissue at various sites in an animal can vary somewhat (subcutaneous fats tend to contain more unsaturated fatty acids than internal depot fats), the proportionate distributions of each fatty acid among the three positions does not vary significantly in any of these tissues in either in the pig or the sheep.

A variety of dietary factors can influence the fatty acid compositions and thence the structures of depot fat triacylglycerols, not least the composition of the diet. Few systematic studies of the relationship between triacylglycerol composition and structure have been performed, and in the example below the pig was the experimental animal. An example of the kind of result obtained is shown in Figure 2.19.

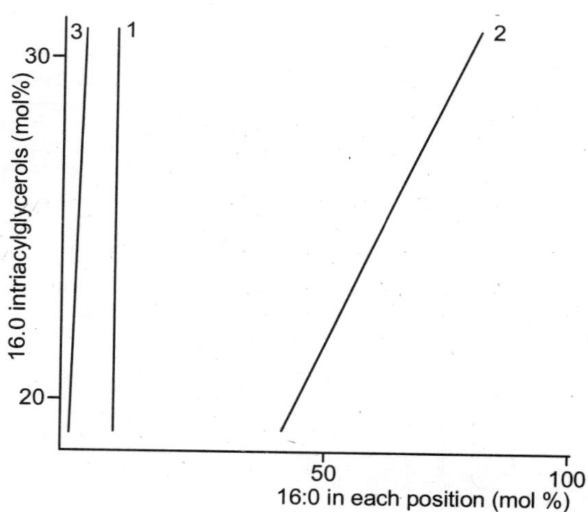

Fig. 2.19 : Variation of the composition of palmitic acid in position *sn*-1, *sn*-2 and *sn*-3 of the triacylglycerols of pig adipose tissue with changes in overall composition

As the proportion of palmitic acid in the triacylglycerols varied between 18 and 33%, the amount in position *sn*-1 remained constant at about 10%, that in position *sn*-3 increased linearly but relatively slowly from 2 to 4%, while that in position *sn*-2 was affected most and increased rapidly from less than 50% to more than 70%. Over the range of compositions studied, the amounts of all of the fatty acids varied in a characteristic manner that was linear in each of the three positions.

Triacylglycerols from Animal Fats - Milk Fat

Milk fats are the only animal fat designed by nature to serve as a food. The range of fatty acids is more extensive than in adipose tissue with often a higher proportion of short- and medium-chain fatty acids, which are not used as such for structural purposes but provide a rapid source of energy. Ruminants, such as the cow, have a range of saturated fatty acids from butyric upwards (indeed even acetate has been found in esterified form), and there are relatively low amounts only of polyunsaturated fatty acids because of biohydrogenation in the rumen. In other species, the compositions are less extreme. In human milk, for example, 12:0 and 14:0 fatty acids are more abundant than in other tissues, but the linoleic acid concentration varies between 10 and 20% depending on diet.

Severe technical problems were encountered in the stereospecific analysis of milk triacyl-*sn*-glycérols from ruminants, because of the presence of short-chain fatty acids, which give rise to difficulties in the isolation of the required partially hydrolysed intermediates. Overcame the problem by isolating fractions enriched in either long-chain or short-chain components by means of molecular distillation or thin-layer chromatography, and subjecting these separately to stereospecific analysis, recombining the results at the end of the procedure. These showed unequivocally that cows' milk is one of the most asymmetric of animal fats, containing all the butyric acid and most of the hexanoic acid in position *sn*-3. Some representative results for the cow and other species are listed in Table.

Table 2.3 : The composition of the fatty acids esterified to each position of the triacyl-*sn*-glycerols in the milk fats of various species

Species	Position	Fatty acid											
		4:0	6:0	8:0	10:0	12:0	14:0	16:0	16:1	18:0	18:1	18:2	18:3
Cow	TG	12	5	2	4	4	11	24	2	7	24	3	
	1	-	1	1	2	5	10	34	2	10	30	2	
	2	-	1	1	3	6	18	32	4	10	19	4	
	3	35	13	4	6	1	6	5	1	1	23	2	
Human	TG				tr	3	26	27	6	7	36	11	1
	1				tr	1	3	16	4	15	46	11	tr
	2				tr	2	7	58	5	3	13	7	1
	3				1	6	7	6	8	2	50	15	1
Rat	TG			6	19	14	12	21	2	3	13	10	1
	1			3	10	10	10	20	2	5	24	14	1
	2			6	20	16	18	29	2	1	3	5	1
	3			10	26	15	9	13	2	2	12	12	1
Pig	TG						4	32	9	5	39	10	1
	1						2	22	7	7	50	11	1
	2						7	58	11	1	15	8	1
	3						4	15	10	6	52	12	2

tr = trace (< 0.5 %). TG = intact triacylglycerols.

The pig and sheep have probably been most studied. With the latter in the liver triacylglycerols, for example, the stearic acid is distributed approximately equally between the three positions, and the oleic acid is present in high concentrations in positions *sn*-2 and *sn*-3, thus differing from the depot fats (Table 2.3 & 2.4). The distinctive feature of the sheep plasma triacylglycerols is a high concentration of palmitic acid in position *sn*-2, and in this they resemble those of the lymph, from which they are derived biosynthetically. In most other tissues of the sheep, the triacylglycerols resemble those of adipose tissue. However, in adrenal glands from which adhering adipose tissue had been carefully removed, triacylglycerols containing a high proportion of palmitic acid in position *sn*-2 were again found, and with long-chain polyunsaturated fatty acids (not listed) in positions *sn*-2 and *sn*-3. It is possible that triacylglycerols of this kind are more widespread than has been thought and that further research will reveal more examples.

Other distinctive triacylglycerols of animal origin are found in the tissues of the chicken. In the plasma triacylglycerols, for example, palmitic acid comprises over 70% of the fatty acids of position *sn*-1 with relatively small amounts in positions *sn*-2 and *sn*-3, while oleic acid comprised 60% of the fatty acids in position *sn*-1 and more than 70% of that in position *sn*-3. Virtually identical structures were found in the ovarian follicles and in the egg, suggesting a common biosynthetic origin. It is important to recognize that lipid structural analyses can rarely be used to prove the existence of particular biosynthetic pathways, but they can provide valuable pointers to the biochemist to potentially productive experimental approaches.

Triacylglycerols from Animal Fats - Other Tissues

The fatty acid compositions of triacyl-sn-glycerols animal tissues other than adipose tissue tend to resemble those of the latter, but the stereospecific distributions can differ. Adrenal tissues can contain longer-chain polyunsaturated fatty acids (10% or more). Such data are often of most value to the biochemist. The pig and sheep have probably been most studied. With the latter in the liver triacylglycerols, for example, the stearic acid is distributed approximately equally between the three positions, and the oleic acid is present in high concentrations in positions *sn*-2 and *sn*-3, thus differing from the depot fats (Table 2.4). The distinctive feature of the sheep plasma triacylglycerols is a high concentration of palmitic acid in position *sn*-2, and in this they resemble those of the lymph, from which they are derived biosynthetically. In most other tissues of the sheep, the triacylglycerols resemble those of adipose tissue. However, in adrenal glands from which adhering adipose tissue had been carefully removed, triacylglycerols containing a high proportion of palmitic acid in position *sn*-2 were again found, and with long-chain polyunsaturated fatty acids (not listed) in positions *sn*-2 and *sn*-3. It is possible that triacylglycerols of this kind are more widespread than has been thought and that further research will reveal more examples.

Other distinctive triacylglycerols of animal origin are found in the tissues of the chicken. In the plasma triacylglycerols, for example, palmitic acid comprises over 70% of the fatty acids of position *sn*-1 with relatively small amounts in positions *sn*-2 and *sn*-3, while oleic acid comprised 60% of the fatty acids in position *sn*-1 and more than 70% of that in position *sn*-3. Virtually identical structures were found in the ovarian follicles and in the egg, suggesting a common biosynthetic origin. It is

Table 2.4 : Positional distributions of fatty acids (mol %) in triacyl-*sn*-glycerols from animal tissues other than depot fats and milk

Source	Position	Fatty acid				
		16:0	16:1	18:0	18:1	18:2
Sheep liver[a]	TG	26	3	16	43	5
	1	47	2	17	21	1
	2	14	3	18	55	8
	3	16	3	13	54	5
Sheep plasma	TG	28	2	22	31	2
	1	34	2	33	21	2
	2	38	3	6	33	1
	3	11	2	28	40	2
Sheep adrenals	TG	23	2	16	32	3
	1	33	2	31	28	1
	2	41	4	4	26	3
	3	4	1	15	48	4
Chicken plasma	TG	29	5	5	50	11
	1	74	6	4	13	1
	2	4	3	2	60	31
	3	8	5	8	76	2
Chicken egg	TG	29	5	7	49	10
	1	71	5	4	17	2
	2	4	3	3	63	26
	3	12	6	14	67	1

A sheep tissues contain appreciable amounts of minor components not listed here (e.g. odd-and branched-chain, etc.).
TG = intact triacylglycerols.

important to recognize that lipid structural analyses can rarely be used to prove the existence of particular biosynthetic pathways, but they can provide valuable pointers to the biochemist to potentially productive experimental approaches.

Triacylglycerols from Fish Oils

The fatty acid compositions of triacylglycerols of fish oils reflect their diet and usually comprise high concentrations of long chain monoenoic and polyunsaturated fatty acids, especially those of the *omega*-3 biosynthetic family. The triacyl-*sn*-glycerols in depot fats from a number of fish and other marine animals have been subjected to stereospecific analysis, and some typical results are listed in Table 2.5.

Table 2.5 : Positional distributions of fatty acids (mol %) in triacyl-*sn*-glycerols of fish oils

Species	Position	Fatty acid										
		14:0	16:0	16:1	18:0	18:1	18:2	20:1	22:1	20:5	22:5	22:6
Herring	TG	7	12	9	1	11	2	17	23	9	2	5
	1	6	12	13	1	16	3	25	14	3	1	1
	2	10	17	10	1	10	3	6	5	18	3	13
	3	4	7	5	1	8	1	20	50	4	1	1
Mackerel	TG	6	14	7	2	17	2	11	16	9	2	9
	1	6	15	11	3	21	2	8	18	5	1	2
	2	10	21	6	1	9	1	5	5	12	3	20
	3	2	5	4	2	21	2	19	24	10	1	5
Skate	TG	2	13	8	2	23	1	13	9	7	3	18
	1	2	19	12	5	30	1	12	8	4	1	5
	2	3	15	7	1	9	1	8	5	6	7	37
	3	1	6	6	1	28	2	19	11	11	2	11
Cod	TG	6	13	13	3	20	2	12	6	12	2	9
	1	6	15	14	6	28	2	12	6	2	1	1
	2	8	16	12	1	9	2	7	5	12	3	20
	3	4	7	14	1	23	2	17	7	13	1	6

TG = intact triacylglycerols.

Myristic, palmitic and palmitoleic acids are preferentially esterified to positions *sn*-1 and *sn*-3. Oleic and longer-chain monoenoic fatty acids are located mainly in the primary positions, with a tendency for a higher proportion to be in position *sn*-3 as the chain-length increases. Polyunsaturated fatty acids are in greatest concentration in position *sn*-2 with substantial amounts also being found in position *sn*-3. There are significant differences in triacylglycerol structure between those of fish and of marine animals that consume fish.

Triacylglycerols of Bacteria and Yeast

Triacylglycerols are fairly widespread in eukaryotic organisms such as yeasts, moulds and fungi, but among the prokaryotes only species from the actinomycetes group, including some of the human mycobacterial pathogens, accumulate triacylglycerols to a significant extent. They occur as cytoplasmic inclusions or lipid droplets within the organisms. As in eukaryotes, triacylglycerols appear to function as a reserve of fatty acyl groups. Those from a few species only have been subjected to stereospecific analyses, and the fatty acid compositions of each are very different, so no general conclusions can be drawn. Three sets of data are listed in Table 2.6.

Table 2.6 : Positional distributions of fatty acids (mol %) in triacyl-*sn*-glycerols of bacteria and yeast

Species	Position	Fatty acid							
		14:0	16:0	16:1	17:0	18:0	18:1	18:2	Other
M. smegmatis	TG	3	24	10	1	10	29		22
	1	1	8	9	tr	7	60		10
	2	7	57	13	2	6	9		4
	3	1	7	7	tr	16	18		51
R. opacus	TG	4	26	10	13	3	22	-	22
	1	12	18	7	12	6	24	-	20
	2	11	59	tr	18	tr	tr	-	12
	3	-	-	21	7	5	42	-	34
L. lipoferus	TG	3	15	8		4	62	8	
	1	3	14	8		4	61	10	
	2	-	1	2		-	88	9	
	3	6	29	13		9	37	6	

tr = trace (<0.5%). TG = intact triacylglycerols. Other = branched, odd- or longer-chain.

In the triacyl-*sn*-glycerols of *Mycobacterium smegmatis*, oleic acid is the main component of position *sn*-1, palmitic acid is the main component of position *sn*-2, while C_{18} and longer-chain fatty acids are the principal constituents of position *sn*-3; indeed 90% of the 24:0 is in position *sn*-3. Although the fatty acid composition of *Rhodococcus opacus* is very different from this, the distribution of fatty acids is somewhat similar in that saturated and shorter-chain fatty acids are concentrated in position *sn*-2. In the triacylglycerols of the yeast *Lipomyces lipoferus*, oleic acid is the main fatty acid in positions *sn*-1 and *sn*-2, especially, while much of the palmitic, palmitoleic and stearic acids are found in position *sn*-3.

A further document on this site dealing with triacylglycerols (Triacylglycerols. Part 2) discusses the biochemistry and metabolism of triacylglycerols.

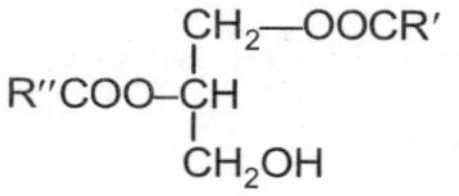

$$\begin{array}{l} \quad\quad\quad\;\; CH_2\text{—}OOCR' \\ \quad\quad\quad\;\;\; | \\ R''COO\text{–}CH \\ \quad\quad\quad\;\;\; | \\ \quad\quad\quad\;\; CH_2OH \end{array}$$

1,2-/2,3-diacylglycerol

Diacylglycerols (less accurately termed "diglycerides") and monoacylglycerols (monoglycerides) contain two moles and one mole of fatty acids per mole of glycerol, respectively, and exist in various isomeric forms. They are sometimes termed collectively "partial glycerides". Although they are rarely present at greater than trace levels in fresh animal and plant tissues, synthetic materials have importance in commerce. However, 1,2-diacyl-*sn*-glycerols are key intermediates in the biosynthesis of triacylglycerols and other lipids, and they are vital cellular messengers, generated on hydrolysis of phosphatidylinositol and related lipids by a specific phospholipase C Diacylglycerols as Components of Oils and Fats.

Diacylglycerols (or "diglycerides") are esters of the trihydric alcohol glycerol in which two of the hydroxyl groups are esterified with long-chain fatty acids. They can exist in three stereochemical forms (see our web document on Triacylglycerols for a discussion of nomenclature).

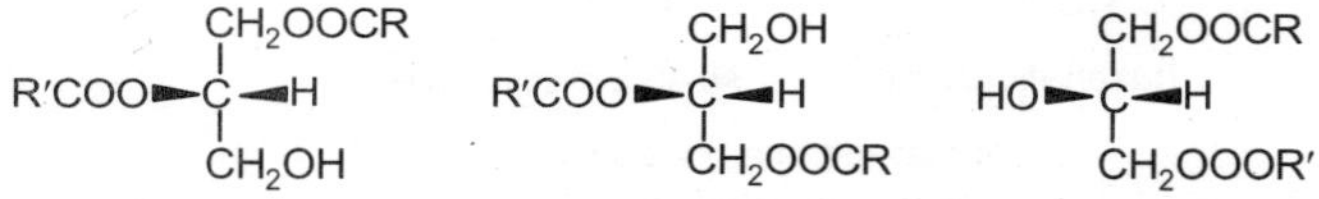

CH_2OOCR $R'COO$–C–H CH_2OH	CH_2OH $R'COO$–C–H CH_2OOCR	CH_2OOCR HO–C–H CH_2OOOR'
sn-1,2-diacylglycerol	*sn*-2,3-diacylglycerol	*sn*-1,3-diacylglycerol

A racemic mixture of *sn*-1,2- and 2,3-diacylglycerols are sometimes termed α,β'-diacylglycerols, while *sn*-1,3-diacylglycerols may be designated α,α'-diacylglycerols.

α, β-Diacylglycerols are formed as intermediates in the hydrolysis of triacylglycerols by pancreatic lipase and other hydrolytic enzymes in animal tissues, and they are generated in seed oils by the action of plant lipases. They are important technologically in commercial seed oils, as small amounts can have a profound influence on the physical properties.

Recently, edible oils consisting of 80% 1,3-diacylglycerols have been marketed in Japan as nutritional supplements. It is claimed that they are metabolized in a different way from triacylglycerols with beneficial nutritional effects. The 1,3-monoacylglycerols formed when they are digested are absorbed into tissues relatively poorly, apparently limiting the accumulation of fats in body tissues.

It should be noted that it is easy to generate diacylglycerols artefactually on storing and extracting tissues if inappropriate methods are used. Often, attempts are made to analyse 1,2-/2,3- and 1,3-diacylglycerols separately, but the data may not be meaningful as acyl migration occurs rapidly until an equilibrium mixture is formed that contains about 67% of the 1,3-isomer. Diacylglycerols will isomerize slowly on standing in inert solvents or in the dry state even at low temperatures.

Diacylglycerols can be recovered from tissues with minimal isomerization, if this is necessary, by extracting the tissues with non-alcoholic solvents such as diethyl ether or chloroform, taking care not to heat extracts at any stage. When pure positional isomers are required, it is necessary to chromatograph the partial glycerides on TLC plates coated with silica gel G impregnated with boric acid at a level of 10% of the adsorbent), using a solvent system of chloroform (alcohol-free)-acetone (96:4, v/v).

Routine determination of molecular species of diacylglycerols in oils and fats can be accomplished by various chromatographic methods of which high-temperature GC seems most appropriate, since information on the composition as well as the absolute amount is obtained in this way.

sn-1,2-Diacylglycerols in Tissues - Biological Importance

sn-1,2-Diacylglycerols tend to be minor components of most tissues in quantitative terms, but they are very important in animal tissues, as they function as second messengers in many cellular processes, modulating vital biochemical mechanisms. They arise by several mechanisms, For example, they are formed as intermediates both in the biosynthesis and catabolism of triacyl-***sn***-glycerols and in the biosynthesis of certain phospholipids. Thus, sn-1,2-diacylglycerol is a key intermediate in the formation of phosphatidylcholine, phosphatidylethanolamine and phosphatidylserine. In this instance, phosphatidate phosphatase is an important enzyme, which is present mainly in the endoplasmic reticulum of the cell, and converts phosphatidic acid to sn-1,2-diacylglycerol. The reverse reaction in which phosphatidic acid is produced by the action of a diacylglycerol kinase is also of great biological importance.

$$\begin{array}{c} CH_2OOCR \\ | \\ CHOOCR' \\ | \\ CH_2OPO_3H \end{array} \underset{2}{\overset{1}{\rightleftharpoons}} \begin{array}{c} CH_2OOCR \\ | \\ CHOOCR' \\ | \\ CH_2OH \end{array}$$

Phosphatidic acid *sn*-1, 2-diacylglycerol

1, phosphatidate phosphatase
2, diacylglycerol kinase

Most of the precursor phosphatidic acid is generated via the Kennedy pathway, but a second mechanism involves the action of a specific phospholipase D on phosphatidylcholine. The latter can also be a precursor for diacylglycerol production via the action of phospholipase C, or by an exchange reaction with ceramide in the biosynthesis of sphingomyelin (a related pathway may be important in some pathogenic fungi). These reactions occur more slowly but are of longer duration.

More significantly in relation to their signalling function, *sn*-1,2-diacylglycerols are formed along with the important signalling molecules, the water-soluble inositol phosphates, by the action of the enzyme phospholipase C (a family of eleven related enzymes in four sub-families exist) on phosphatidylinositol and the polyphosphoinositides and especially phosphatidylinositol-4,5-bisphosphate (see the web pages on these lipids for further information). These enzymes are activated by agonists at receptors on the cell surface, of which over one hundred types have been identified. The response is immediate if short-lived.

$$\begin{array}{c} CHOOCR \\ | \\ R'COO-CH \\ | \\ CH_2-O-P(=O)(O^-)(H^+)-O-\text{inositol}(OH)_5 \end{array} \xrightarrow{\text{Phospholipase C}} \begin{array}{c} CHOOCR \\ | \\ R'COO-CH \\ | \\ CH_2OH \end{array}$$

Phosphatidylinositol *sn*-1, 2-diacylglycerol

The fatty acid compositions of the diacylglycerols formed by these various routes then reflect the composition of the parent phospholipids. In particular, those derived from phosphatidylinositols are highly enriched in molecular species containing stearic in position *sn*-1 and arachidonic acid in position *sn*-2. There is evidence that the diacylglycerols must contain polyunsaturated fatty acids to fulfil their function as messengers optimally.

Diacylglycerols accumulate transiently in membranes, where they bind via strong hydrophobic interactions to specific proteins, and then cause changes in the physical properties of the bilayer. As their polar head group is small, they tend to form inverted micellar structures. In practice, this means that they introduce small areas of unstable negative curvature in membranes that facilitate membrane fission or fusion. By exposing small areas of the apolar regions of neighbouring lipids, they improve the hydrophobic interactions with proteins within these membrane regions thereby affecting their activities. The fusion of biological membranes is of great importance for the proper functioning of cells, and diacylglycerols in membranes are able to facilitate this process partly via their specific physical properties and partly through activation of certain proteins.

It appears that the formation of diacylglycerols may be initiated at the plasma membrane, but subsequently the reaction at internal membranes becomes more important. In particular, there is a significant production of diacylglycerols at the nucleus in response to stimuli. It is thought that diacylglycerols produced for purposes other than signalling are segregated spatially with the cell.

An especially important function of *sn*-1,2-diacylglycerols, and especially those derived from phosphatidylinositol, is that they affect vital processes in cell physiology by activating members of the protein kinase C family of enzymes, often acting in concert with the soluble phosphoinositides. The *sn*-1,2-configuration is essential for this activity. Diacylglycerols appear to function then by increasing the concentrations of calcium ions in the cell, which stimulate the translocation of the various iso-enzymes of protein kinase C to the inner face of the plasma membrane. By binding in a 1:1 ratio to the enzymes at a highly conserved cysteine-rich C1 domain, which consists of a sequence of 50 amino acids with a characteristic motif, they also increase their hydrophobicity and thus enhance their affinity for membrane lipids, especially phosphatidylserine, 4 to 10 molecules of which appear to be essential cofactors. This in turn increases the kinase activities of the enzymes by modifying their phosphorylation state. The protein kinase C enzymes are involved in both short- and long-term modifications of normal cellular physiology with more than a hundred substrates identified to date. Of particular importance is the finding that the tumour-promoting phorbol esters mimic the activity of diacylglycerols and activate the same enzymes. Diacylglycerols therefore have a key role in the pathophysiology of cancer and other disease states. In addition, the identification of non-kinase receptors of *sn*-1,2-diacylglycerols, many but not all of which have the conserved C1 domain, has revealed new and strategic functions in regulating cellular responses and in cytoskeletal remodelling.

Diacylglycerols also bind to protein kinase D, a cytosolic serine-threonine kinase that in turn binds to the trans-Golgi membrane network and regulates transport of proteins to the cell surface. Protein transport is blocked in the absence of diacylglycerols.

In addition, *sn*-1,2-diacylglycerols can serve as precursors for biosynthesis of phosphatidic acid via the action of diacylglycerol kinase (as illustrated above) and of 2-arachidonylglycerol, both of which also have second messenger functions. In particular in mammals, there is a family of at least ten diacylglycerol kinase isoenzymes, which are structurally related to the sphingosine kinase, each of which may have slightly different properties and functions, which may be segregated in distinct cellular organelles and activated by different means. Some are cytosolic, some are associated with membranes and some are located within the nucleus. In the brain different isoenzymes are expressed in different types of neuron, some of which have several isoenzymes. These enzymes have a negative effect on diacylglycerol signalling by reducing its concentration in cells, but they are believed to generate phosphatidic acid with specific signalling functions, rather than simply to serve as a precursor of other lipids.

In insects, although lipids are stored in the form of triacylglycerols in fat bodies, they are transported in hemolymph (the insect equivalent of plasma) in the form of *sn*-1,2-diacylglycerols bound to the lipoprotein lipophorin to tissues where they are required as a source of energy.

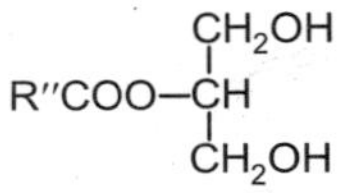

CH_2OH
R″COO—CH
CH_2OH
2-monoacylglycerol

2-Monoacyl-*sn*-glycerols are formed as intermediates or end-products of the enzymatic hydrolysis of triacylglycerols; these and other positional isomers are powerful surfactants. 2-Arachidonoylglycerol has important biological properties (as an endocannabinoid).

Acyl migration occurs rapidly in partial glycerides at room temperature, but especially on heating, in alcoholic solvents or in the presence of acid or base, so special procedures are required for their isolation or analysis if the stereochemistry is to be retained. Synthetic 1-/3-monoacylglycerols are important in commerce as surfactants.

Cholesterol

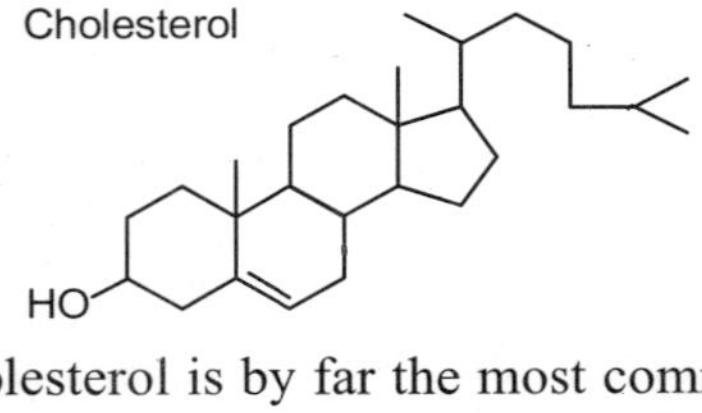

HO

Sterols and sterol esters: Cholesterol is by far the most common member of a group of steroids in *animal tissues*; it has a tetracyclic ring system with a double bond in one of the rings and one free hydroxyl group. It is found both in the free state, where it has an essential role in maintaining membrane fluidity, and in esterified form, i.e. as cholesterol esters. Other sterols are present in free and esterified form in animal tissues, but at trace levels only.

In *plants*, cholesterol is rarely present in other than small amounts, but such phytosterols as sitosterol, stigmasterol, avenasterol, campesterol and brassicasterol, and their fatty acid esters are usually found, and they perform a similar function.

Plant Sterols

Plants contain a rather different range of sterols from animals. Like cholesterol, to which they are related structurally and biosynthetically, plant sterols form a group of triterpenes with a tetracyclic

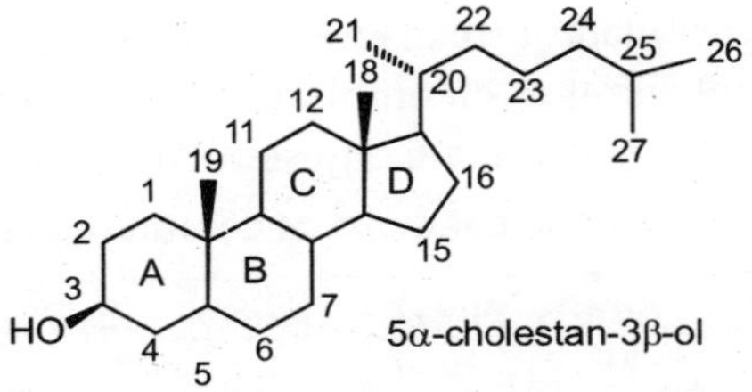

21 22 24 25 26 18 20 23 12 11 27 19 C D 16 1 2 A B 15 3 HO 4 5 6 7 5α-cholestan-3β-ol

cyclopenta[*a*]phenanthrene structure and a side chain at carbon 17. The four rings (A, B, C, D) have *trans* ring junctions, and the side chain and two methyl groups (C-18 and C-19) are at an angle to the rings above the plane with β stereochemistry (as for the hydroxyl group on C-3 also). The basic sterol from which other sterol structures are defined is 5α-cholesten-3β-ol.

The phytosterols (as opposed to zoosterols) include campesterol, â-sitosterol, stigmasterol and Δ5-avenasterol, some of which are illustrated below.

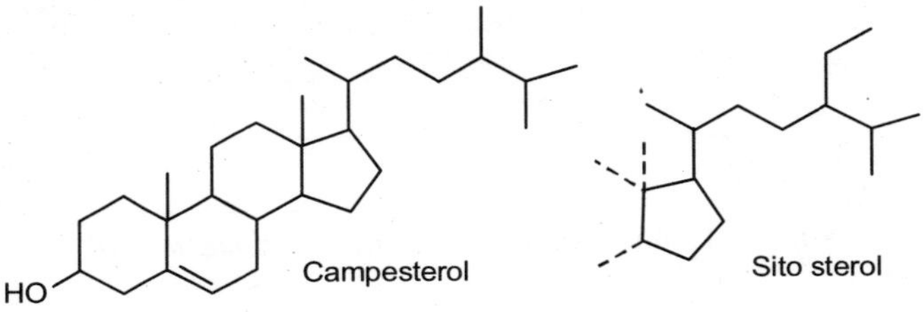

HO Campesterol Sito sterol

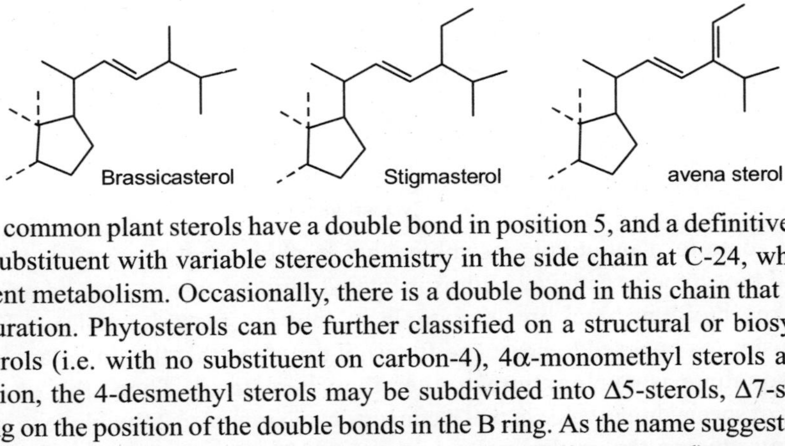

Brassicasterol Stigmasterol avena sterol

These more common plant sterols have a double bond in position 5, and a definitive feature – a one- or two-carbon substituent with variable stereochemistry in the side chain at C-24, which is preserved during subsequent metabolism. Occasionally, there is a double bond in this chain that can be of the *cis* or *trans* configuration. Phytosterols can be further classified on a structural or biosynthetic basis as 4-desmethyl sterols (i.e. with no substituent on carbon-4), 4α-monomethyl sterols and 4,4-dimethyl sterols. In addition, the 4-desmethyl sterols may be subdivided into Δ5-sterols, Δ7-sterols and Δ5,7-sterols depending on the position of the double bonds in the B ring. As the name suggests, brassicasterols are found mainly in the brassica family of plants. Phytostanols (fully saturated) are normally present at trace levels only in plants, but they are relatively abundant in cereal grains.

Many different sterols may be present in plant species (over 200 have been characterized), and the amounts and relative proportions of all are dependent on the plant species. As a rough generality, a typical plant sterol mixture would be 70% sitosterol, 20% stigmasterol and 5% campesterol. Table 1 contains data from some representative commercial seed oils. Cholesterol is usually a minor component only of plant sterols, but it is unwise to generalize too much as it can be the main sterol component of red algae and of some families of higher plants (Solanaceae, Liliaceae, Scrophylariaceae), for example. It can also be a significant constituent sterol of chloroplasts, shoots, pollen and leaf surface lipids in other plant families. Yeasts and fungi tend to contain ergosterol as their main sterol (Fig. 2.7).

Table 2.7 : Sterol composition in some seed oils of commercial importance (mg/Kg).

	Corn oil	cotton-seed	olive	palm	rape-seed	saff-lower	soy-bean	sun-flower
cholesterol	-	-	-	26	-	-	-	-
campesterol	2691	170	28	358	1530	452	720	313
stigmasterol	702	42	14	204	-	313	720	313
β-sitosterol	7722	3961	1310	1894	3549	1809	1908	2352
Δ5-avenasterol	468	85	29	51	122	35	108	156
Δ7-stigmastenol	117	-	58	25	306	696	108	588
Δ7-avenasterol	-	-	-	-	-	104	36	156
brassicasterol	-	-	-	-	612	-	-	-
other	-	-	-	-	-	69	-	39

Data from Gunstone, F.D., Harwood, J.L. and Padley, F.B. *The Lipid Handbook (Second Edition)* (Chapman & Hall, London) (1994).

HO

Cycloartenol

The biosynthetic route to plant sterols resembles that to cholesterol in many aspects in that it follows an isoprenoid biosynthetic pathway with isopentenyl pyrophosphate, derived primarily from mevalonate, as the key building block in the cytoplasm (but not plastids) at least. However, in photosynthetic organisms (as opposed to yeast and fungi), it differs in that the important intermediate in the route from squalene is **cycloartenol** rather than lanosterol. Then, the enzyme sterol methyltransferase 1 is of special importance in that it converts cycloartenol to 24-methylene cycloartenol, as the first step in introducing the methyl group onto C-24. There are more than thirty enzyme-catalysed steps in the overall process, each associated with membranes. The 4,4-dimethyl- and 4α-methylsterols are part of the biosynthetic pathway, but are only minor if ubiquitous sterol components of plants.

In addition, an alternative pathway for the biosynthesis of isopentenyl pyrophosphate and dimethylallyl pyrophosphate, the isoprene units, which does not use mevalonic acid as a precursor has been found in plant chloroplasts, algae, cyanobacteria, eubacteria, and some parasites, but not in animals. This route is variously termed the 'non-mevalonate', '1-deoxy-D-xylulose-5-phosphate' (DOXP) or better the

'2*C*-methyl-D-erythritol 4-phosphate (MEP)' pathway, since the last compound is presumed to be the first committed intermediate in sterol biosynthesis by this route. In the first step, pyruvate and glyceraldehyde phosphate are combined to form deoxyxylose phosphate, which is in turn converted to 2C-methyl-D-erythritol 4-phosphate. The pathway then proceeds via various erythritol intermediates until isopentenyl pyrophosphate and dimethylallyl pyrophosphate are formed, when sterol biosynthesis thereafter continues via squalene and cycloartenol. There is evidence that some of the isoprene units are exchanged between the cytoplasm and plastids.

Pyruvate + Glyceraldehyde phosphate → Deoxyxylose phosphate → 2-C-methyl-D-enthritol 4-phosphate → Various erythritol phosphate intermediates → 2-methylbutane-1,4-diol diphosphate → Isopentenyl diphosphate ⇌ Dimethylallyl diphosphate

Like cholesterol, plant sterols are amphiphilic and are vital constituents of all membranes, and especially of the plasma membrane, the mitochondrial outer membrane and the endoplasmic reticulum. The three-dimensional structure of the plant sterols is such that there are planar surfaces at both the top and the bottom of the molecules, which permit multiple hydrophobic interactions between the rigid sterol and the other components of membranes. Indeed, they must govern the physical properties of membranes to an appreciable extent. It is believed that sitosterol and 24-methylcholesterol are able to regulate membrane fluidity and permeability in plant membranes by restricting the mobility of fatty acyl chains in a similar manner to cholesterol in mammalian cells. They may be involved in how plant membranes adapt to changes in temperature. Stigmasterol has much less effect on lipid ordering and no effect on the permeability of membranes. In the plasma membrane, plant sterols associate with the glycosphingolipids such as glucosylceramide in raft-like sub-domains, analogous to those in animal cells.

Plant sterols can modulate the activity of membrane-bound enzymes. Thus, stigmasterol and cholesterol regulate the activity of the Na^+/K^+-ATPase in plant cells, probably in a manner analogous to that of cholesterol in animal cells. Stigmasterol may be required specifically for cell differentiation and proliferation. Certain sterols, such as campesterol in *Arabidopsis thaliana*, in minute amounts are precursors of oxidized steroids that act as growth hormones and are collectively named **brassinosteroids**, which have crucial importance for growth and development. Cholesterol is a precursor for the biosynthesis of some steroidal saponins and alkaloids, as well as of other steroids including the ecdysteroids (insect moulting hormones).

Substantial amounts of phytosterols are available as by-products of the refining of vegetable oils and of tall oil from the wood pulp industry. There is increasing interest in such commercial sources of plant sterols to be added as "nutraceuticals" to margarines and other foods, as it appears that they can inhibit the uptake of cholesterol from the diet and thereby reduce the levels of this in the plasma low-density lipoproteins. Hydrogenated phytosterols or *"stanols"* are also used for this purpose. In addition, dietary supplements of plant sterols have been reported to have anti-cancer effects. The actual absorption of dietary plant sterols and stanols in humans is low (0.02-3.5%) compared to cholesterol (35-70%). In some rare cases, increased levels of plant sterols in plasma serve as markers for an inherited lipid storage disease (phytosterolemia). There is evidence that while plant sterols can substitute for cholesterol in maintaining membrane function in mammalian cells, they can exert harmful effects by disrupting cholesterol homeostasis.

Phytosterols can be subjected to autoxidation in a similar manner to that of cholesterol in animals, resulting in ring products such as hydroxy-, keto-, epoxy- and triol-derivatives. In addition, enzymic reactions can oxidize the side chain. These oxy-phytosterols can enter the food chain and, although they are not efficiently absorbed, they have been detected in human plasma and have the potential to exert a variety of biological effects.

Sterol Esters

Sterol esters are present in plant tissues, but as minor components relative to the free sterols other than in waxes. Usually the sterol components of sterol esters are similar to the free sterols, although there may be relatively less of stigmasterol. The fatty acid components tend to resemble those of the other plant tissue lipids, but there can be significant differences on occasion. Sterol esters are presumed to serve as inert storage forms of sterols, as they are often enriched in the intermediates of sterol biosynthesis and can accumulate in lipid droplets within the cells. However, they have been found in membranes, especially in microsomes and mitochondrial preparations, although their function here is uncertain. They may also have a role in transport within cells and between tissues, as they have been found in the form of soluble lipoprotein complexes.

Biosynthesis of sterol esters in *Arabidopsis thaliana* is known to occur in the endoplasmic reticulum and involves transfer of a fatty acyl group to the sterol from position *sn*-2 of phosphatidylethanolamine. However, other enzymes may exist but have yet to be characterized. In yeasts, two sterol ester synthases have been identified that utilize CoA esters of fatty acids as substrate. Little appears to be known of the catabolism of sterol esters in plants, but specific sterol ester hydrolases have been characterized from yeasts.

Certain distinctive phytosterol esters occur in the aleurone cells of cereal grains, including *trans*-hydroxycinnamate, ferulate (4-hydroxy-3-methoxy cinnamate) and *p*-coumarate esters. Similarly, rice bran oil is a rich source of esters of ferulic acid and a mixture of sterols and triterpenols, termed 'γ-orizanol", and an example of one of these compounds is illustrated below. This is sold as a health food supplement, because of claimed beneficial effects, including cholesterol-lowering and antioxidant activities. It is also reputed to enhance muscle growth and sports performance. However, none of these effects have been confirmed by rigorous clinical testing.

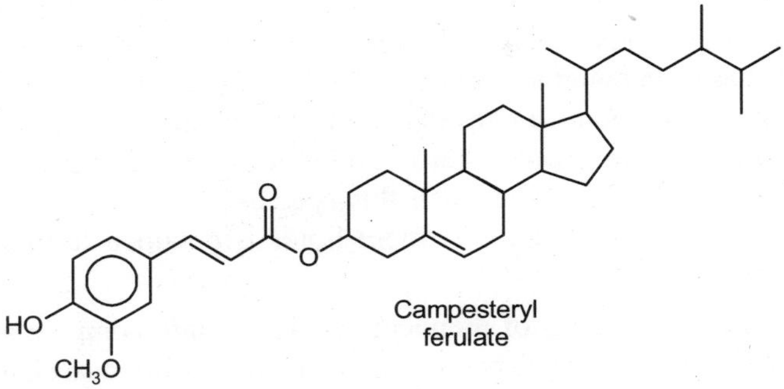

HO
CH_3O
Campesteryl ferulate

Sterol Glycosides

Leaf and other tissue in plants contain a range of **sterol glycosides** and **acyl sterol glycosides**, and typical examples (glucosides of β-sitosterol) are illustrated below.

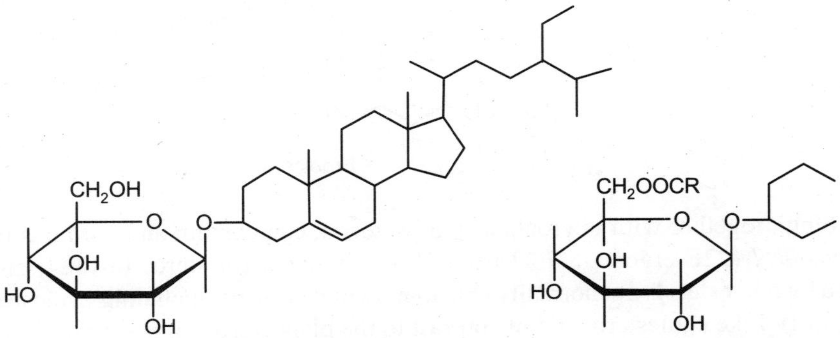

CH_2OH O O OH HO OH
Sterol glucoside

CH_2OOCR O O OH HO OH
Acyl sterol glucoside

Most of the common plant sterols occur in this form, and the carbohydrate moieties (glucose, xylose, arabinose) can vary also with the plant species. The carbohydrate moiety can sometimes be quite complex with up to five hexose units linked in a linear fashion. Similarly, the nature of the fatty acid component in the acyl sterol glycosides can vary as well as the hydroxyl group to which they are linked, although it is usually position 6. In potato tubers, for example, the 6'-palmitoyl-β-D-glucoside of β-sitosterol is the major species, while the corresponding linoleate derivative predominates in soybeans. Usually, the acyl sterol glycosides are present at concentrations that are two to tenfold greater than those of the non-acylated forms. Plant, animal, fungal and most bacterial steryl glycosides have a β-glycosidic linkage, but in a few bacterial species there is an α-linkage. Sterol glycosides and acyl sterol glycosides may be involved in the adaptation of plant membranes to low temperatures. Biosynthesis involves reaction of free sterols with uridine diphosphoglucose (UDP-glucose) in the presence of a

UDP-glucose:sterol glucosyltransferase in plant membranes, probably the plasma membrane. Sitosterol-β-D-glucoside in the plasma membrane is the primer molecule for cellulose synthesis in plants and it is required for the initiation of glucan polymerization. The sterol is eventually removed from the polymer by a specific cellulase enzyme.

Sterol glycosides have only rarely been reported from organisms other than plants, although cholesteryl glucoside has been found in a few animal tissues. The gram-negative bacterium *Helicobacter pylori*, one of the most common human pathogens, is an important exception in that it contains appreciable amounts of unique cholesterol glucosides, including cholesteryl-6-*O*-acyl-α-D-glucopyranoside, in its cell walls. Other than this species, cholesteryl glucosides are rather rare in bacteria. It is known that the cholesterol of these lipids is obtained from an animal host and is not synthesised by bacteria.

In addition, a number of species of monocotyledons contain complex steroidal **saponins**, which consist of a furostanol- or spirostanol-based aglycone and an oligosaccharide chain of two to five hexose or pentose moieties attached to the 3-hydroxyl group of the sterol. Further conjugates of this type include other triterpenes with saccharide units linked at various positions

Sterols in Yeasts and Fungi

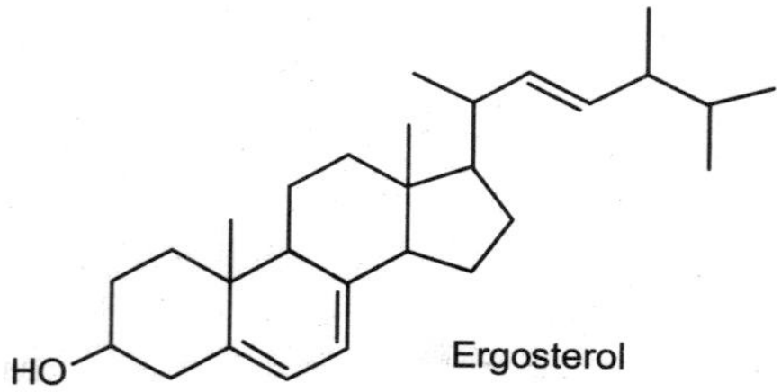

HO

Ergosterol

Yeasts and fungi, together with microalgae and protozoa, can contain an enormous range of different sterols. *Ergosterol* (22E-ergosta-5,7,22-trien-3β-ol) is the main sterol in fungi and yeasts, and is accompanied by other sterols not normally abundant in higher plants including zymosterol (5α-cholesta-8,24-dien-3β-ol). Like cholesterol and in contrast to the plant sterols, it is synthesised via lanosterol as the key intermediate. Some antifungal drugs are targeted against ergosterol biosynthesis. Under some conditions, especially those that retard growth, a high proportion of the sterols in yeasts can be in esterified form.

Many mutants defective in ergosterol biosynthesis have been isolated, and these have yielded a great deal of information on the features of the sterol molecule required for its structural role in membranes of yeast and fungi. Ergosterol stabilizes the liquid-ordered phase in the same manner as cholesterol, and also forms rafts with sphingolipids, whereas lanosterol does not. It is also evident that ergosterol has a multiplicity of functions in the regulation of yeast growth. Ergosterol esters are synthesised in yeast by enzymes (ARE1 and ARE2) related to ACAT-1 and ACAT-2 that perform this function in animals. Protozoans also synthesise many different sterols. For example, some species of *Acanthamoeba* and *Naegleria* produce both lanosterol and cycloartenol, but only latter is used for synthesis of other sterols, especially ergosterol. In some other protozoan species, sterol biosynthesis occurs via lanosterol.

Bacterial Sterols

It has long been recognised that some bacteria take up cholesterol and other sterols from host animals for use as membrane constituents. Indeed, an external source of sterols is required for growth in species of *Mycoplasma*. In addition, there have been a number of reports of biosynthesis of sterols by various bacterial species, although a high proportion of these appear now to have been discounted because of fungal contamination. In particular, the possibility of sterol biosynthesis in cyanobacteria has been controversial, and molecular biology studies have yet to detect the presence of the required enzyme squalene epoxide cyclase.

That said, there is good evidence that a few species of prokaryotes at least have the capacity to synthesise sterols *de novo*. Among the eubacteria certain methylotrophs (*Methylobacterium* and *Methylosphaera* species) produce mono- and dimethyl sterols, including lanosterol. Similarly, some soil bacteria produce 4-desmethylsterols. It has now been established from gene sequence studies that certain bacteria contain enzymes of the sterol biosynthesis pathway, but as these have no obvious evolutionary link it seems probable that they were acquired via lateral transfer from eukaryotes.

Analysis

In the analysis of animal and plant sterols, a sterol fraction is first isolated from lipid extracts by thin-layer or column chromatography, following hydrolysis if necessary. Individual components can then be determined by gas chromatography in the presence of an internal standard (e.g. epicoprostanol or betulin), often after conversion to trimethylsilyl ether derivatives to give sharper peaks. Mass spectrometry may be required for identification of individual components. Analysis of the minor oxysterols that may be found in plasma or foods is a rather specialized task, because they tend to be present at rather low levels and there is a danger of further oxidation or side reactions during the analytical process. Rigorous attention to detail is necessary for meaningful results.

Sterol esters are trans-methylated for GC analysis of the fatty acid components, although the reaction may again be much slower than with glycerolipids. Intact sterol esters are best analysed by reversed-phase HPLC. Analysis of sterol glycosides is a more specialized endeavour that can be more concerned with carbohydrate than with lipid chemistry. Hopanoids are related lipids produced by some bacterial species.

Waxes: In their most common form, wax esters consist of fatty acids esterified to long-chain alcohols with similar chain-lengths. The latter tend to be saturated or have one double bond only. Such compounds are found in animal, plant and microbial tissues and they have a variety of functions, such as acting as energy stores, waterproofing and lubrication.

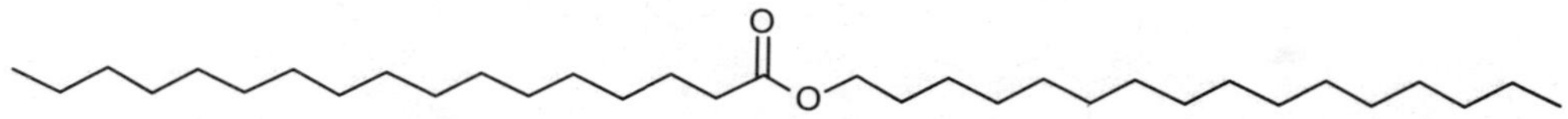

O
O

In some tissues, such as skin, avian preen glands or plant leaf surfaces, the wax components can be much more complicated in their structures and compositions. They can contain aliphatic diols, free alcohols, hydrocarbons (e.g. squalene), aldehydes and ketones.

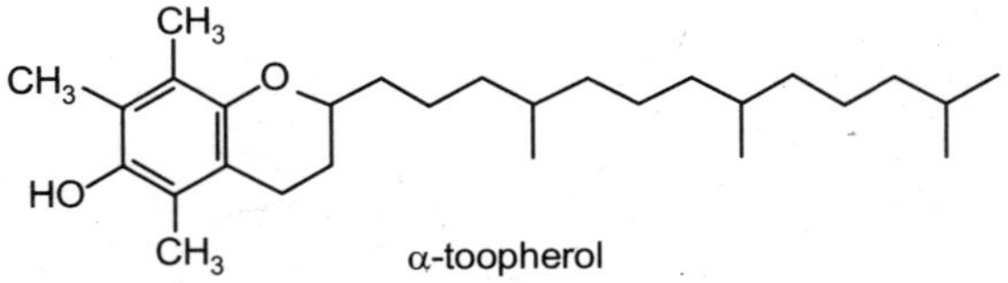

α-toopherol

Tocopherols are substituted benzopyranols (methyl tocols) that occur in vegetable oils. Different forms (α-, β-, γ- and δ-) are recognized according to the number or position of methyl groups on the aromatic ring. α-Tocopherol (Vitamin E) illustrated is an important natural antioxidant. Tocotrienols have similar ring structures but with three double bonds in the aliphatic chain. Tocopherols and Tocotrienols.

Structure, Composition, Biology and Analysis

Structure and Biosynthesis

Tocopherols constitute a series of related benzopyranols (or methyl tocols) that occur in plant tissues and vegetable oils and are powerful lipid-soluble antioxidants. In the tocopherols, the C_{16} side chain is saturated, and in the **tocotrienols** it contains three *trans* double bonds. Together, these two groups are termed the **tocochromanols.** In essence, the tocopherols have a 20-carbon phytyl tail (including the pyranol ring), and the tocotrienols a 20-carbon geranylgeranyl tail with double bonds at the 3', 7' and 11' positions, attached to the benzene ring. The side-chain methyl groups have *R,R,R* stereochemistry. The four main constituents of the two classes are termed-α (5,7,8-trimethyl), β(5,8-dimethyl), γ(7,8-dimethyl) and δ(8-methyl).

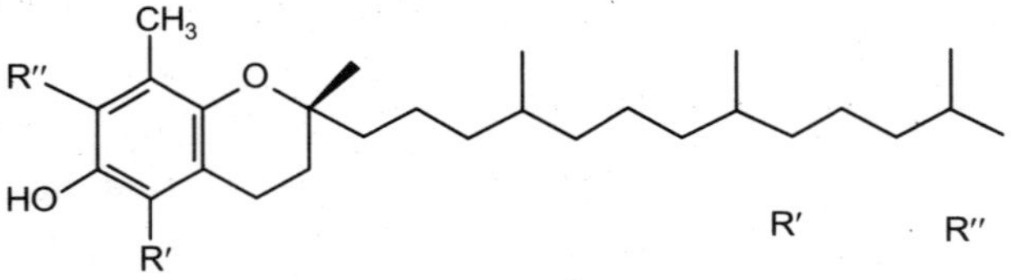

	R′	R″
α-tocophenol	—CH_3	—CH_3
β-tocophenol	—CH_3	—H
γ-tocophenol	—H	—CH_3
δ-tocophenol	—H	—H

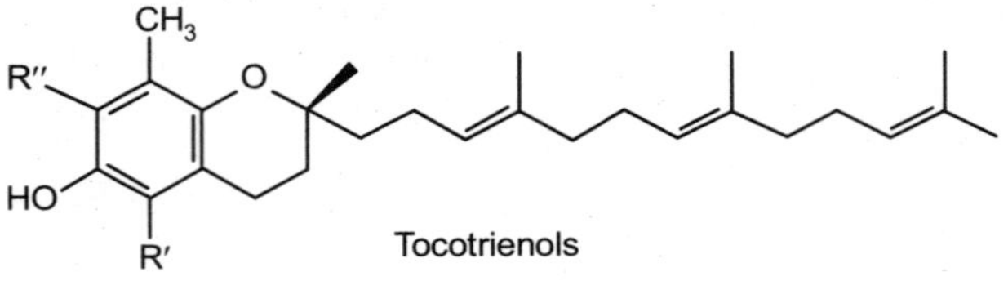

Tocotrienols

These compounds are only synthesised by plants and other oxygenic, photosynthetic organisms, but they are essential components of the diet of animals, and collectively they are termed 'vitamin E' (the individual tocopherols are 'vitamers'). In plants, there is a great range of tocochromanol contents and compositions, and photosynthetic plant tissues contain from 10 to 50 μg tocochromanols per g fresh weight. α-Tocopherol is often the main tocochromanol in leaves. Seed oils are a major source for the human diet and the compositions of tocopherols in some unrefined oils are listed in **Table 2.8.** Sunflower and olive oils are good sources of α-tocopherol and palm oil of the tocotrienols. In general, tocotrienols tend to be more abundant in seeds of monocots, such as wheat, rice and barley.

Table 2.8 : Tocopherol and tocotrienol contents (mg/Kg) in some seed oils

	α-T*	α-T	γ-T	δ-T	α-TT*	β-TT	γ-TT	δ-TT
palm	89	-	18	-	128	-	323	72
soybean	100	8	1021	421	-	-	-	-
maize	282	54	1034	54	49	8	161	6
sunflower	670	27	11	1	-	-	-	-
rapeseed	202	65	490	9	-	-	-	-

* Abbreviations: T, tocopherol; TT, tocotrienoi

The mechanism of biosynthesis of tocopherols is well understood, and involves coupling of phytyl diphosphate with homogentisic acid (2,5-dihydroxyphenylacetic acid), followed by cyclization and methylation reactions.

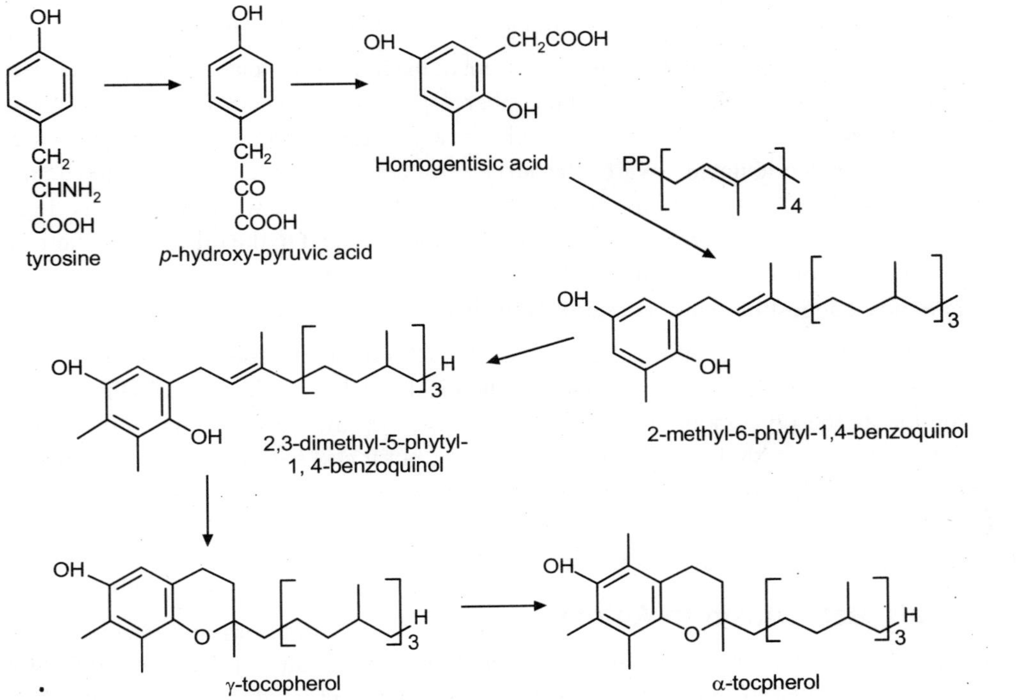

OH
CH_2
$CHNH_2$
COOH
tyrosine

OH
CH_2
CO
COOH
p-hydroxy-pyruvic acid

OH CH_2COOH OH
Homogentisic acid

PP [ ]4

OH OH [ ]3
2-methyl-6-phytyl-1,4-benzoquinol

OH OH [ ]3 H
2,3-dimethyl-5-phytyl-1, 4-benzoquinol

OH O [ ]3 H
γ-tocopherol

OH O [ ]3 H
α-tocpherol

The plant chloroplast is the site of biosynthesis, and the aromatic amino acid tyrosine can be considered the basic precursor. This is oxidized to *p*-hydroxypyruvic acid, which in the first committed step is converted to homogentisic acid by the enzyme *p*-hydroxyphenylpyruvate dioxygenase. Homogentisic acid is condensed with phytyl diphosphate in a reaction catalysed by a prenyl transferase to yield 2-methyl-6-phytyl-plastoquinol, which is first methylated to form 2,3-dimethyl-5-phytyl-1,4-benzoquinol and then converted by the enzyme tocopherol cyclase to γ-tocopherol. A further methylation reaction produces α-tocopherol, while modifications to the pathway produce β- and δ-tocopherols and plastoquinones. Tocotrienols result from a similar series of reactions but with geranylgeranyl diphosphate as substrate in the condensation step.

Fish contain an unusual tocopherol that has been termed marine-derived α-tocomonoenol (a related isomer has been found in palm oil). It is found together with α-tocopherol in a wide range of marine fish species and appears to be a more efficient scavenger of free radicals at low temperatures.

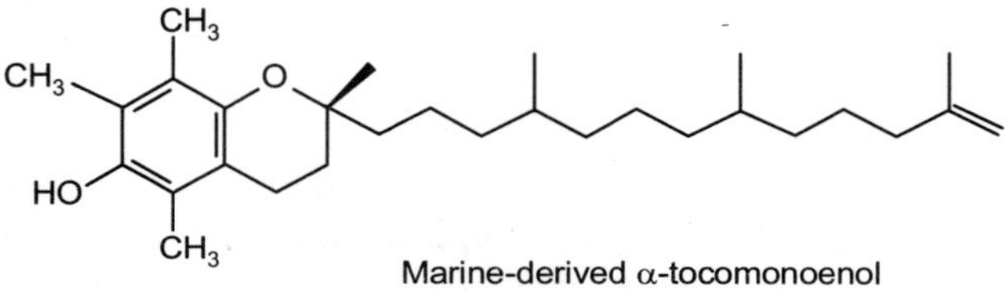

CH_3 CH_3 O HO CH_3

Marine-derived α-tocomonoenol

α-Tocopheryl phosphate has recently been detected at low levels in liver and adipose tissue, and it is possible that it may be a ubiquitous constituent of animal and plant tissues.

α-Tocopherol is a minor but ubiquitous component of the lipid constituents of animal cell membranes, with estimates ranging from one molecule of tocopherol to from 100 to 1000 molecules of phospholipid, depending on the membrane. The hydrophobic tail lies within the membrane, as might be expected, and the polar head group is orientated towards the surface but below the level of the phosphate moieties of the phospholipids. There may be some limited hydrogen bonding between the hydroxyl groups and phosphate depending on the degree of hydration of the membrane. On the other hand, there is a strong affinity of a-tocopherol for polyunsaturated fatty acids, where the chromanol unit may interact with the double bonds, suggesting that tocopherol is located deep within the membrane.

During the refining of vegetable oils, much of the natural tocopherols are lost or destroyed. Most commercial vitamin E is therefore prepared by chemical synthesis with trimethylhydroquinone and phytyl bromide as the precursors. The resulting product is a mixture of eight stereoisomers (from R', R'', R''' to S', S'', S''' methyl groups) of α-tocopherol, but it still has appreciable vitamin E activity. It is usually administered as the acetate derivative. Tocopherols are not usually regarded as effective antioxidants in the polyunsaturated seed oils of commerce, and at higher concentrations can even act as pro-oxidants, although the reasons for this are not understood.

Tocopherols Metabolism in Animals

In animals, all tocopherols are absorbed to a similar extent in the intestines and are transported to the liver in chylomicrons mainly, but α-tocopherol is preferentially utilized and re-exported. This process is mediated by a specific tocopherol-binding protein in the liver that has a marked affinity for α-tocopherol, transferring it to the plasma lipoproteins (mainly the LDL and HDL in humans) for transport to other tissues (together with lesser amounts of γ-tocopherol). The "α-tocopherol salvage pathway" results in a 20- to 30-fold enrichment of γ-tocopherol in plasma (average concentration 22-28 ìM) relative to the other tocopherols. Transfer of tocopherols from the lipoproteins to peripheral tissues is promoted by the enzyme lipoprotein lipase. Concentrations of tocopherols can vary appreciably amongst tissues, with most in adipose tissue and adrenals, less in kidney, heart and liver, and least in the erythrocytes. Dietary tocotrienols are also absorbed in the intestines but less efficiently.

The process of conservation of one specific tocopherol appears to determine the relative vitamin E activities of the tocopherols and tocotrienols *in vivo*, rather than their individual potencies as antioxidants as measured in model systems *in vitro*. Only α-tocopherol (including synthetic material) or natural mixtures containing this can be sold under the label 'Vitamin E'. However, the tocotrienols are more potent antioxidants, *in vitro* at least, while γ-tocopherol (which is relatively abundant in skin) has some specific biological properties that are distinct from those of α-tocopherol.

Most of the tocochromanols other than á-tocopherol, together with any excess of the latter, are metabolized. The unwanted surplus may be excreted in the urine in the form of the so-called 'Simon metabolites', α-tocopheronic acid and α-tocopheronolactone, after oxidative cleavage of much of the phytyl tail. However, these are normally in the form of conjugates as sulphate or glucuronidate esters.

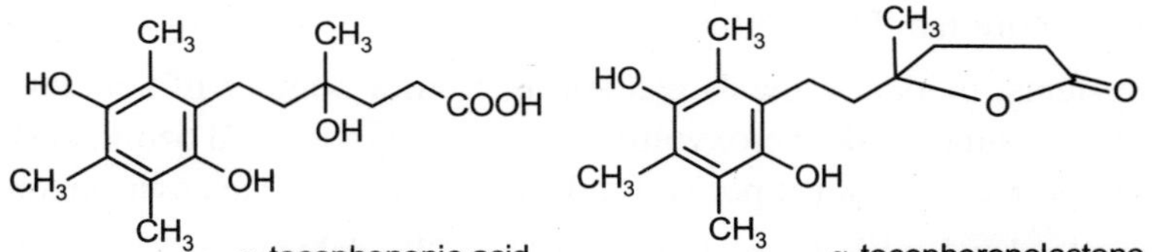

α-tocophenonic acid

α-tocopheronolactone

The first step in catabolism is ω-hydroxylation by cytochrome P450 (CYP4F2) at the 13' carbon to form a 13'-hydroxychromanol, followed by stepwise β-oxidation to cut off two or three carbon moieties from the phytyl chain in each cycle. Various carboxychromanol intermediates have been identified for all of the tocopherols together with sulphated forms of these in human cell cultures *in vitro*.

Tocopherols as Antioxidants

Although the syndrome associated with a lack of vitamin E in the diet of animals has been well known for decades, the mode of action and specific location of tocopherols in cell membranes are not clearly understood. Several theories have been proposed to explain their functions. Many argue that their primary task is to act as antioxidants to prevent free radical damage to unsaturated lipids or other membrane constituents and thence to tissues, while others now suggest that this may be secondary to more important biological functions (see page 186). That said there is no doubt that tocopherols are powerful antioxidants *in vitro* and surely have some such function *in vivo*. They are certainly extremely useful as antioxidants in non-biological systems, including foods, cosmetics, pharmaceutical preparations and so forth.

Because of their lipophilic character, tocopherols are located in the membranes or with storage lipids where that are immediately available to interact with lipid hydroperoxides. They react rapidly in a non-enzymic manner unlike many other cellular antioxidants, which are dependent on enzymes, to scavenging lipid peroxyl radicals, i.e. the chain-carrying species that propagate lipid peroxidation. In model systems *in vitro*, all the tocopherols ($\alpha > \gamma > \beta > \delta$) and tocotrienols are good antioxidants, with the tocotrienols being the most potent.

In general, the oxidation of lipids is known to proceed by a chain process mediated by a free radical, in which the lipid peroxyl radical serves as a chain carrier. In the initial step of chain propagation, a hydrogen atom is abstracted from the target lipid by the peroxyl radical as shown—

$$LOO- + LH \rightarrow LOOH + L+ \quad (1)$$

$$L^{+} + O_2 \rightarrow LOO^{-} \quad (2)$$

—where LH is a lipid, LOO is the lipid peroxyl radical and LOOH is the lipid hydroperoxide.

The main function of α-tocopherol is to scavenge the lipid peroxyl radical before it is able to react with the lipid substrate as—

$$LOO- + TOH \rightarrow LOOH + TOO^{-} \quad (3)$$

—where TOH is tocopherola nd TOO* is the tocopheroxyl radical

It thus prevents propagation of the chain reaction. The potency of an antioxidant is determined by the relative rates of reactions (1) and (2). Studies of the relative rates of chain propagation to chain inhibition by á-tocopherol in model systems have demonstrated that α-tocopherol is able to scavenge peroxyl radicals much more rapidly than the peroxyl radical can react with a lipid substrate.

In biological systems, oxidant radicals can spring from a number of sources, including singlet oxygen, alkoxyl radicals, superoxide, peroxynitrite, nitrogen dioxide and ozone. α-Tocopherol is most efficient at providing protection against peroxyl radicals in a membrane environment.

When a tocopheroxyl radical is formed, it is stabilized by delocalisation of the unpaired electron about the fully substituted chromanol ring system rendering it relatively unreactive. This also explains the high first order rate constant for hydrogen transfer from α-tocopherol to peroxyl radicals. Reaction of the tocopheroxyl radical with a lipid peroxyl radical, as illustrated, yields 8α-substituted tocopherones, which are readily hydrolysed to 8α-hydroxy tocopherones that rearrange spontaneously to form

To copheroxyl radical

8 α-alkyldioxytocopherone

tocopherol quinone

tocopherol hydroquinone

α-tocopherol quinones. In an alternative pathway, the tocopheroxyl radical reacts with the lipid peroxyl radical to form epoxy-8α-hydroperoxytocopherones, which hydrolyse and rearrange to epoxyquinones. Tocopherol dimers and trimers may also be formed as minor products.

In plant and animal tissues, tocopherols can be regenerated from the tocopheroxyl radicals in a redox cycle mediated by a number of endogenous antioxidants, including vitamins A and C and coenzyme Q, and this must greatly extend their biological potency. Vitamin C (ascorbate) may be especially important in aqueous systems, although it may also act at the surface of membranes.

Suggestions that dietary supplements of vitamin E may reduce the rate of oxidation of lipids in low-density lipoproteins and thence the incidence or severity of atherosclerosis now appear to be unfounded, although benefits in some conditions have been claimed. Indeed, there are suggestions that excessive vitamin E supplementation may even be harmful. A recent study has suggested that relatively high doses of natural α-tocopherol over a long period are required to demonstrate a significant reduction in the levels of F_2 isoprostanes in the urine, which are considered to be the most reliable marker for oxidative stress *in vivo*. This subject is highly contentious and I prefer to leave it to the clinical experts.

In plants, tocopherols are most abundant in the membranes of the chloroplasts, where they were long believed to be the most important antioxidants, limiting the damage from photosynthesis-derived reactive oxygen species during conditions of oxidative stress, including high-intensity light stress. However, recent studies seem to suggest that they are just one of a number of different components that are involved in photo-protection. Certainly, any tocochromanol peroxy radicals formed must be converted back to the original compounds by the concerted action of other plant antioxidants. On the other hand, there is no doubt that tocopherols are essential for the control of non-enzymatic lipid peroxidation during seed dormancy and germination of seedlings. In their absence, elevated levels of malondialdehyde and phytoprostanes are formed, and there can be inappropriate activation of plant defense responses.

Other Biological Functions of Tocopherols

With the discovery that the antioxidant effects of various tocopherols and tocotrienols have little relation to their vitamin E activities has come the realization that they have many other functions in tissues, most of which are specific to α-tocopherol. Indeed, it has even been suggested that these are so important that tocopherol may be protected from functioning as an antioxidant in tissues *in vivo* through a network of cellular antioxidant defences. Only when other antioxidants are exhausted are the tocopherols utilized. However, there is no experimental proof of this hypothesis, although it is certainly true that most other vitamins are essential cofactors for specific enzymes or transcription factors.

α-Tocopherol is believed to be a gene regulator, causing up-regulation of mRNA or protein synthesis that could be the result of effects on gene transcription, mRNA stability, protein translation, protein stability and post-translational events. β-Tocopherol has no such properties. Effects have also been observed on genes connected with tocopherol catabolism, lipid uptake, collagen synthesis, cellular adhesion, inflammation and cell signalling. Vitamin E modulates the activity of several enzymes involved in signal transduction, perhaps through influencing protein-membrane interactions. For example, α-tocopherol inhibits the enzyme protein kinase C, which in turn reduces the release of reactive oxygen species in various ways and has effects on gene expression. It may also have secondary roles in stabilizing the structure of membranes, in regulating haem biosynthesis, in modulating the immune response, and as a participant in electron transport chains. Some non-antioxidant effects of γ-tocopherol in tissues in relation to reactive nitrogen oxide species have been observed, but the specificity of these is not yet certain.

Tocotrienols have been shown to have neuroprotective effects, to inhibit cholesterol synthesis and to reduce the growth of breast cancer cells *in vitro*. These properties are largely distinct from those of the tocopherols.

The biological function α-tocopheryl phosphate is not known, but it has been suggested that that it may be a storage or transport form or it could be involved in cellular signalling. Synthetic phosphate derivatives of γ-tocopherol and α-tocopheryl succinate are known to have potent anti-cancer properties.

In plants, there is evidence that tocopherols also play a part in intracellular signalling in that they regulate the amounts of jasmonic acid in leaves and so influence plant development and stress responses. Thus, by controlling the degree of lipid peroxidation in chloroplasts, they limit the accumulation of lipid hydroperoxides required for synthesis of jasmonic acid, which in turn regulates the expression of genes that affect a number of stress conditions. In addition, tocopherols are required for the development of the cell walls in phloem transfer cells under cold conditions.

Analysis

Tocopherols can be analysed by gas chromatography, both with flame-ionization and mass spectrometric detection, but the methods that are usually recommended involve high-performance liquid chromatography with fluorescence detection. Related methods are used for ubiquinones and the isoprenoid alcohols.

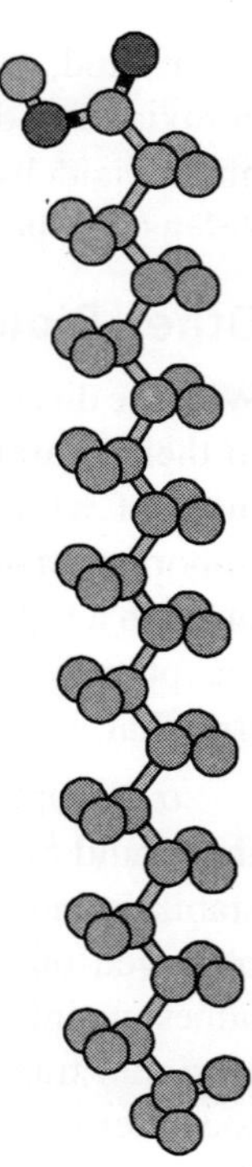

Free (unesterified) fatty acids are minor constituents of living tissues but are of biological importance as precursors of lipids as an energy source and as cellular messengers.

Free or unesterified fatty acids are ubiquitous if minor components of all living tissues. In animals, much of the dietary lipid is hydrolysed to free acids before it is absorbed and utilized for lipid synthesis. Intact lipids in tissues can be hydrolysed to free acids by a variety of lipolytic enzymes (*e.g.* lipoprotein lipase, hormone-sensitive lipase, phospholipase A), before being metabolized in various ways including oxidation, desaturation, elongation or re-esterification. As free acids can interact with a wide range of enzyme systems in both specific and non-specific ways, they must be rapidly sequestered in tissues by various means to ensure that their activities are closely regulated.

Monomeric fatty acids in the free state have very low solubilities in aqueous media. In serum, they are transported between tissues bound to the protein albumin, which has up to six strong binding sites and a large number of weak binding sites where non-polar interactions are possible between the fatty acid hydrocarbon chains and uncharged amino acid side chains. In this way, the concentration of a long-chain fatty acid in serum çan be increased by as much as 500 times above its normal maximum. However, the bound fatty acids can diffuse into the aqueous phase, where they are rapidly taken up into the outer leaflet of the plasma membrane by non-enzymatic mechanisms. It is then possible that fatty acids can then cross the membrane simply and rapidly by a biophysical process, i.e. by 'flip-flop'. On the other hand, there is also evidence that specific transporter proteins may be involved in part to activate by formation of acyl-coA prior to further esterification, but also to ensure vectorial transport so that specific fatty acids are directed towards particular purposes. Certainly within the cell, a family of fatty acid binding or transport proteins has essential functions in fatty acid trafficking pathways and in fatty acid activation. It appears that cells have several overlapping mechanisms that ensure sufficient uptake and directed intracellular movement of the fatty acids required for their physiological functions.

Unesterified fatty acids can act as second messengers required for the translation of external signals, as they can be produced rapidly as a consequence of the binding of specific agonists to plasma membrane receptors. In this way, they can substitute for the second messengers of the inositide pathways. Fatty acids are effective also in operating at specific intracellular locations reversibly to amplify or otherwise modify signals. For example, they influence the activities of protein kinases, phospholipases, G-proteins, adenylate and guanylate cyclases, and many other metabolic processes. Part of the action of fatty acids may occur indirectly via metabolism of arachidonic acid to eicosanoids. On the other hand, there is much evidence that fatty acids *per se* are messengers that mediate the responses of the cell to extracellular signals. Many of these reactions are specific to particular fatty acids. For example, polyunsaturated fatty acids, including docosahexaenoic and arachidonic acids, bind to the retinoid X receptor and induce activation. Some related processes appear to occur in plants.

In addition in animal tissues, long-chain polyunsaturated fatty acids are involved in regulating gene expression, mainly targeting genes that encode proteins with roles in fatty acid transport or metabolism. In this respect, (n-3) fatty acids are more potent than (n-6) fatty acids. Straight-chain saturated and monoenoic fatty acids do not appear to be involved in the process, but surprisingly poly-methyl-branched fatty acids, such as phytanic and pristanic, may have a function. In some circumstances, both the free acids *per se* and their coenzyme A esters may be involved.

The mechanisms by which modulation of gene transcription occurs are only partially resolved, and this is the subject of considerable research effort, especially with respect to the family of transcription factors, i.e. peroxisome proliferator-activated receptors (PPARs), in the nuclei of cells. However, it is believed that fatty acid-binding-proteins are intimately involved and in effect act as nutrient sensors. They bind long-chain fatty acids with high affinity in the cytoplasm and transport them to nuclei, which they enter via the nuclear pores, where they are able to form complexes with nuclear receptors enabling them to regulate receptor activation.

The effects can be highly specific, different fatty acids binding to or activating different types of PPAR, although the PPARá and hepatocyte nuclear factor 4α (HNF4α) are especially important. In particular, polyunsaturated fatty acids may exert beneficial effects by up-regulating the expression of genes encoding enzymes involved in oxidation of fatty acids, while at the same time down-regulating genes for enzymes involved in lipid synthesis. They also influence glucose metabolism. As a result, unesterified fatty acids may mitigate the undesirable symptoms of the metabolic syndrome and may even reduce the risk of heart disease. In contrast, abnormal PPAR activation can be a factor in the lipotoxicity observed with obesity, insulin resistance, type-2 diabetes and hyperlipidemia.

Similarly, in bacteria, it has been demonstrated that the bacterial fatty acid transport and trafficking system leads to fatty acid-responsive regulation of gene expression.

However, some of the mediator effects appear to be independent of PPARs and are characterized by involvement with cell surface receptors instead. Thus, multiple G-protein-coupled receptors for free fatty acids have been identified that function on the cell surface and have important roles in nutrition regulation. Some of these are activated by short-chain and others by medium- and long-chain free fatty acids. The former are expressed preferentially in pancreatic β-cells and mediate insulin secretion. In certain intestinal cells, there are believed to be such 'fatty acid sensing molecules', which recognise the presence of free long-chain acids in digesta. These stimulate the release of the hormone cholecystokinin, which in turn causes the gall-bladder to contract and release bile to aid digestion.

Free fatty acids have potent antimicrobial, antiviral and antifungal properties, and they exert such effects in some living systems, especially the skin and mucosa of the lung. As they are powerful detergents and will inhibit very many enzymes systems in a non-specific manner, it is not clear whether the biocidal properties are also non-specific. Unsaturated fatty acids seem to have the greatest effects, but this may be because they can insert more readily into membranes.

Analysis

Accurate measurement of free fatty acids concentrations in plasma and tissues can be a useful measure of metabolic status. Unfortunately, it is very easy to generate free acids artefactually by faulty storage or extraction. Lipases can continue to function slowly in some tissues, even at –20°C, and the process will accelerate if tissues are allowed to thaw prior to extraction. In one important series of experiments, it was shown that if animal tissues were pulverized and extracted at –70°C, very low levels only of free fatty acids were detected in comparison to more conventional procedures. The high concentrations reported occasionally in the literature are obviously impossible in living tissues and are due to inappropriate sample handling.

Following extraction of lipids from tissues by a suitable procedure, a free fatty acid fraction can be isolated by thin-layer chromatography or by solid-phase extraction chromatography on a bonded amine phase. This can be methylated and analysed by gas chromatography with an internal standard.

Glycerophospholipids

Phosphatidic acid or 1,2-diacyl-*sn*-glycerol-3-phosphate is found in trace amounts only in tissues under normal circumstances, but it has great metabolic importance as a biosynthetic precursor of all other glycerolipids. It is strongly acidic and is usually isolated as a mixed salt. One specific isomer is illustrated as an example.

Lysophosphatidic acid with one mole of fatty acid per mole of lipid (in position *sn*-1) is a marker for ovarian cancer, and is a key cellular messenger.

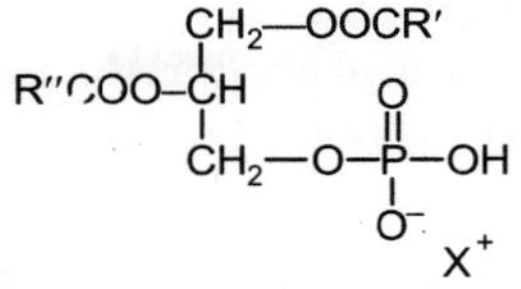

$$\begin{array}{l} CH_2—OOCR' \\ R''COO—CH \\ CH_2—O—P(=O)(—OH)—O^- \; X^+ \end{array}$$

Where, X = Na, K, H, Ca, etc.

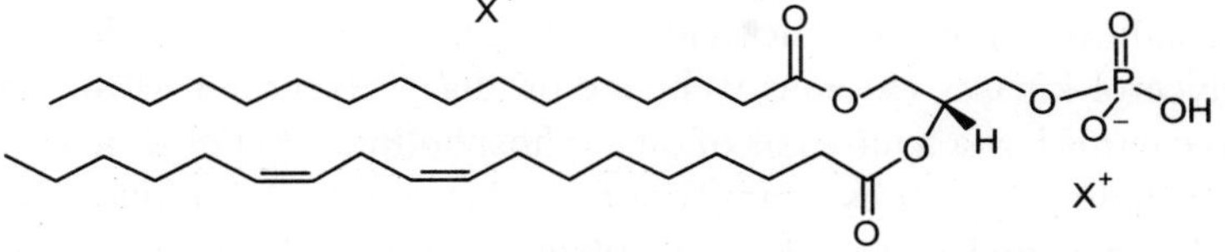

1-hexadecanoyl, 2-(9Z, 12Z-octadecadienoyl)-*sn*-glycero-3-phosphate
(phosphatidic acid)

Phosphatidic Acid—Occurrence and Biosynthesis

Phosphatidic acid is not an abundant lipid constituent of any living organism to my knowledge, but it is extremely important as an intermediate in the biosynthesis of triacylglycerols and phospholipids. Indeed, it is often over-estimated in tissues as it can arise by inadvertent enzymatic hydrolysis during inappropriate storage or extraction conditions during analysis. The molecule is acidic and bears a negative charge so requires a counter ion.

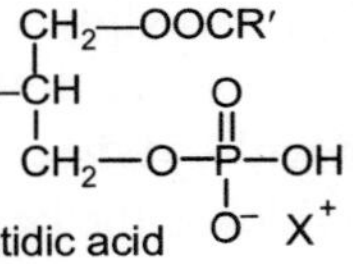

$$\begin{array}{l} CH_2—OOCR' \\ R''COO—CH \\ CH_2—O—P(=O)(—OH)—O^- \; X^+ \end{array}$$

Phosphatidic acid

Where, X = Na, K, Ca, etc.

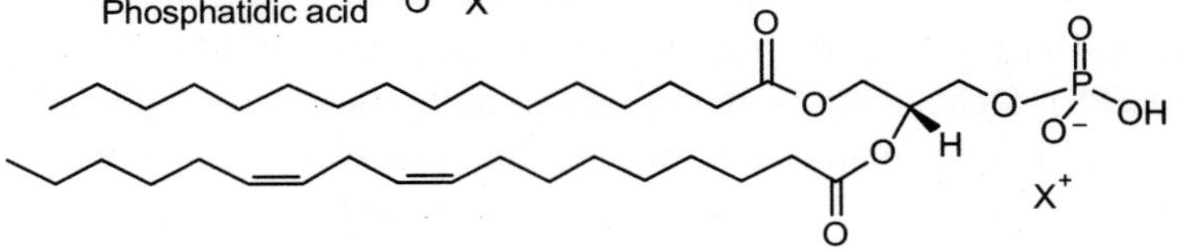

1-hexadecanoyl, 2-(9Z, 12Z-octadecadienoyl-*sn*-glycero-3-phosphate

The main biosynthetic route in plant and animal tissues involves sequential acylation of α-glycerophosphate, derived from catabolism of glucose, by acyl-coA derivatives of fatty acids as illustrated. Specific acyltransferases catalyse first the acylation of position *sn*-1 to form lysophosphatidic acid (1-acyl-*sn*-glycerol-3-phosphate) and then of position *sn*-2 to yield phosphatidic acid.

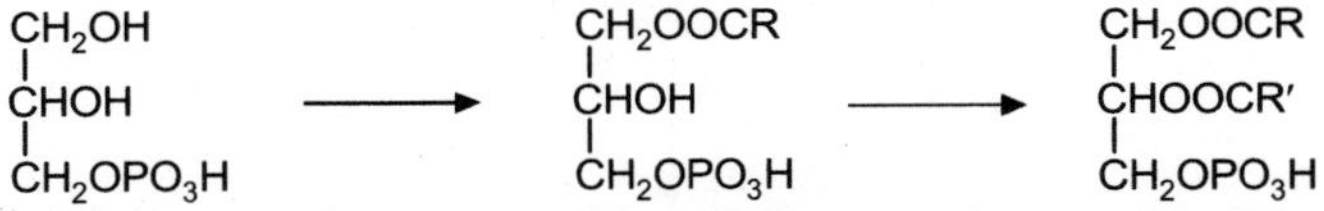

$$\begin{array}{l} CH_2OH \\ CHOH \\ CH_2OPO_3H \end{array} \longrightarrow \begin{array}{l} CH_2OOCR \\ CHOH \\ CH_2OPO_3H \end{array} \longrightarrow \begin{array}{l} CH_2OOCR \\ CHOOCR' \\ CH_2OPO_3H \end{array}$$

sn-glycerol-3-phosphate — 1-acyl-*sn*-glycerol-3-phosphate — Phosphatidic acid

In mammals, the glycerol-3-phosphate acyltransferase that catalyses the first step exists in four isoforms, two in mitochondria (designated GPAT1 and 2) and two in the endoplasmic reticulum (GPAT3 and 4). The activity in the endoplasmic reticulum predominates in adipose tissue, but the mitochondrial forms are responsible for half the activity in liver. All are membrane-bound enzymes, which are believed to span the membranes, but many questions remain regarding the regulation and function of the different isoforms. Similarly, at least two acyl-CoA:lysophosphatidic acid acyltransferases (LPAAT1 and LPAAT2) are known that catalyse the second step. Human LPAAT1 showed higher activity with 14:0-, 16:0- and 18:2-CoAs, while LPAAT2 prefers 20:4-CoA.

Under some conditions, phosphatidic acid can be generated from 1,2-diacyl-*sn*-glycerols by the action of diacylglycerol kinases (see our webpage on diacylglycerols). However, a more important route in quantitative terms is via hydrolysis of other phospholipids, but especially phosphatidylcholine, by the enzyme phospholipase D (or by a family or related enzymes of this kind). This enzyme is present in most animal cell types, and its activity is regulated by phosphatidylinositol-4,5-bisphosphate and protein kinase C.

$$\begin{array}{l} CH_2{-}OOCR' \\ R''COO{-}CH \\ CH_2{-}O{-}\overset{O}{\overset{\|}{P}}{-}O{-}CH_2CH_2\overset{+}{N}(CH_3)_3 \\ \qquad\quad O^- \end{array} \xrightarrow{\text{Phospholipase D}} \begin{array}{l} CH_2{-}OOCR' \\ R''COO{-}CH \\ CH_2{-}O{-}\overset{O}{\overset{\|}{P}}{-}OH \\ \qquad\quad O^- \;\; H^+ \end{array}$$

The subsequent steps in the utilization of phosphatidic acid in the biosynthesis of triacylglycerols and phospholipids are described in separate documents in this section of the website. In brief, hydrolysis of phosphatidic acid by the enzyme phosphatidate phosphatase is the source of ***sn***-1,2-diacylglycerols (DAG), which are the precursors for the biosynthesis of triacylglycerols (TAG), phosphatidylcholine (PC) and phosphatidylethanolamine (PE) via the so-called Kennedy pathway (also of monogalactosyldiacylglycerols in plants). Via reaction with cytidine triphosphate, phosphatidic acid is the precursor of cytidine diphosphate diacylglycerol, which is the key intermediate in the synthesis of phosphatidylglycerol (PG), and thence of cardiolipin, and of phosphatidylinositol (PI) and phosphatidylserine (PS). Depending on the organism and other factors, phosphatidylserine can be a precursor for phosphatidylethanolamine. Similarly, the latter can give rise to phosphatidylcholine by way of mono- and dimethylphosphatidylethanolamine intermediates.

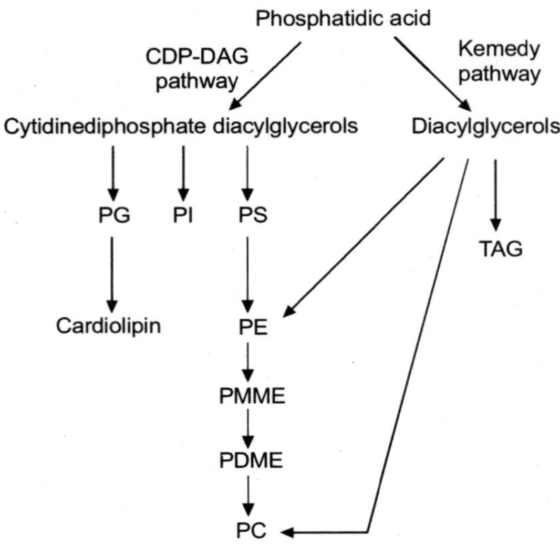

The fatty acid composition of phosphatidic acid can resemble that of the eventual products, though in many instances this can be much altered by re-modelling after synthesis via deacylation-reacylation reactions.

Phosphatidic Acid - Biological Functions

The phosphatidic acid generated by the action of phospholipase D and by diacylglycerol kinases may have signalling functions as a second messenger, although it is not certain whether all the activities suggested by studies *in vitro* operate *in vivo*. Nonetheless, phosphatidic acid has been implicated in many aspects of animal cell biochemistry and physiology, including cell proliferation and differentiation, cell transformation, tumor progression and survival signalling. It appears to regulate some membrane trafficking events, and it is involved in activation of the enzyme NADPH oxidase, which operates as part of the defense mechanism against infection and tissue damage during inflammation. It may have a role in promoting phospholipase A_2 activity, and it appears to function in vesicle formation and transport within the cell. By binding to targeted proteins, including protein kinases, protein phosphatases and G-proteins, it may increase or inhibit their activities. For example in yeast, phosphatidic acid on the endoplasmic reticulum binds directly to a specific transcriptional repressor to keep it inactive outside the nucleus; when the lipid precursor inositol is added, this phosphatidic acid is rapidly depleted, releasing the transcriptional factor so that it can be translocated to the nucleus where it is able to repression target genes. The overall effect is a mechanism to control phospholipid synthesis. In addition, the murine phosphatidylinositol 4-phosphate 5-kinase does not appear to function unless phosphatidic acid is bound.

In relation to signalling activities, it should be noted that phosphatidic acid can be metabolized to ***sn***-1,2-diacylglycerols or to lysophosphatidic acid (see next section), both of which have distinctive signalling functions in their own right. Conversely, both of these compounds can be in effect be de-activated by conversion back to phosphatidic acid.

In many cell types, vesicle trafficking, secretion and endocytosis may also require phosphatidic acid derived by the action of phospholipase D. For example, a specific form of this enzyme located on the outer surface of mitochondria acts upon cardiolipin to promote mitochondrial fusion.

Some of these effects may be explained simply by the physical properties of phosphatidic acid, which has a propensity to form a hexagonal II phase, especially in the presence of calcium ions. Thus, hydrolysis of phosphatidylcholine, a cylindrical, non-fusogenic lipid, converts it into cone-shaped, fusogenic phosphatidic acid, which promotes negative membrane curvature. It can effect membrane fusion in model systems, probably because of its ability to form non-bilayer phases. Also of relevance in this context is its overall negative charge, and it is not always clear whether some of the observed biological effects are specific to phosphatidic acid or simply to negatively charged phospholipids in general. However, it has been demonstrated that the positively charged lysine and arginine residues on proteins can bind with some specificity to phosphatidic acid through hydrogen bonding with the phosphate group, thus distinguishing it from other phospholipids.

Phosphatidic acid in plants: Phospholipase D activity and the phosphatidic acid produced may be even more significant in plants. They have long been recognized as of importance during germination and senescence, and they appear to have a role in response to stress damage and pathogen attack. A high content of phosphatidic acid induced by phospholipase D action during wounding or senescence brings about a loss of the membrane bilayer phase, as a consequence of the conical shape of this phospholipid in comparison to the cylindrical shape of structural phospholipids. As a result, cells lose their viability. The phosphatidic acid generated in this way is broken down further by phosphatases, acyl-hydrolases and lipoxygenases into fatty acids and other small molecules, which are subsequently absorbed and recycled. In addition, phosphatidic acid is important in the response to other forms of stress, including osmotic stress (salinity or drought), cold, and oxidation, although much remains to be learned of the mechanism by which it exerts its effects.

Phosphatidic acid is of considerable importance in cellular signalling in plants, for example in promoting pollen-tube growth, decreasing peroxide-induced cell death, and mediating the signalling processes that lead to responses to the plant hormone abscisic acid. Thus in the 'model' plant *Arabidopsis*, which contains twelve distinct members of the phospholipase D family, phosphatidic acid generated by the action of the enzymes interacts with a protein phosphatase to signal the closure of stomata promoted by abscisic acid; it interacts also with a further enzyme to mediate the inhibition of stomatal opening effected by abscisic acid. Together these reactions constitute a signalling pathway that regulates water loss from plants.

Lysophosphatidic Acid

Lysophosphatidic acid or 1-acyl-*sn*-glycerol-3-phosphate differs from phosphatidic acid in having only one mole of fatty acid per mole of lipid. As such, it is the simplest possible glycerophospholipid. Molecular species with both saturated and unsaturated fatty acid constituents occur in different tissues. Although it is present at very low levels only in animal tissues, it is extremely important biologically, influencing many biochemical processes. These activities seem to be shared by the 1-alkyl- and alkenyl-ether forms. It is also important as the biosynthetic precursor of phosphatidic acid.

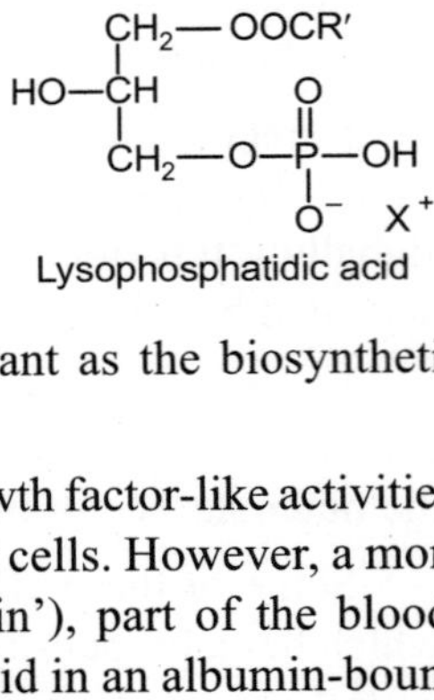

$CH_2—OOCR'$
HO—CH O
$CH_2—O—P—OH$
O^- X^+

Lysophosphatidic acid

In particular, lysophosphatidic acid is an intercellular lipid mediator with growth factor-like activities, and is rapidly produced and released from activated platelets to influence target cells. However, a more important source is the activity of a specific lysophospholipase D ('autotaxin'), part of the blood-clotting process, on lysophosphatidylcholine, which yields lysophosphatidic acid in an albumin-bound form. This is more abundant in serum (1-5 μM) than in plasma, where it accounts for much of the biological activity.

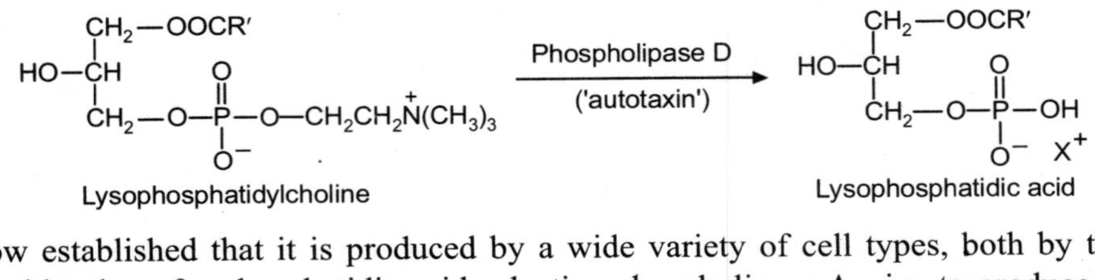

$CH_2—OOCR'$
HO—CH O
$CH_2—O—P—O—CH_2CH_2\overset{+}{N}(CH_3)_3$
O^-

Lysophosphatidylcholine

Phospholipase D ('autotaxin') →

$CH_2—OOCR'$
HO—CH O
$CH_2—O—P—OH$
O^- X^+

Lysophosphatidic acid

It is now established that it is produced by a wide variety of cell types, both by the action of autotoxin and by that of a phosphatidic acid-selective phospholipase A_1, i.e. to produce a *sn*-2-acyl-lysophosphatidic acid in this instance, and almost certainly also by the action of a lysophospholipase A_2. A surprising recent finding is that the activity of the phosphatidic acid-selective lysophospholipase A_1 is essential for normal hair growth in humans.

Most mammalian cells express receptors for lysophosphatidic acid, and lysophosphatidic acid may initiate signalling in the cells in which it is produced, as well as affecting neighbouring cells. In the last few years, the characterization of cloned lysophosphatidic acid receptors in combination with strategies of molecular genetics has allowed determination of both signalling and biological effects that are dependent on receptor mechanisms. At least seven G protein-coupled receptors that are specific for lysophosphatidic acid have now been identified. Experimental activation of these receptors has shown that a range of downstream signalling cascades mediate lysophosphatidic acid signalling. These include activation of protein kinases, adenyl cyclase and phospholipase C, release of arachidonic acid, and much more. There is evidence that lysophosphatidic acid is involved in cell survival in some circumstances, and in programmed cell death in others. In some instances, molecular species with specific fatty acid components may be involved.

There is particular interest in the activity of lysophosphatidic acid in various disease states, where intervention in its metabolism has the potential for beneficial health effects. For example, a finding that lysophosphatidic acid is markedly elevated in the plasma of ovarian cancer patients, compared to healthy controls may be especially significant. In particular, elevated plasma levels were found in patients in the first stage of ovarian cancer, suggesting that it may represent a useful marker for the early detection of the disease. Lysophosphatidic acid is believed to stimulate DNA synthesis and the proliferation of ovarian cancer cells, and it may induce cell migration. Therefore, it is a target of the pharmaceutical industry for cancer therapy.

In addition, lysophosphatidic acid generated by the action of a lysophospholipase D is believed to play an important role in reproductive biology. Under certain conditions, it can become athero- and thrombogenic and might aggravate cardiovascular disease. This may be especially important in cancer patients. Lysophosphatidic acid has also been found in saliva in significant amounts, and it has been suggested that it is involved in wound healing in the upper digestive organs such as the mouth, pharynx, and esophagus. It has similar effects when applied topically to skin wounds, probably by stimulating proliferation of new cells to seal the wound. There is also evidence that the lipid is involved in brain development and vascular remodelling.

Catabolic deactivation of lysophosphatidic acid is accomplished by dephosphorylation to monoacylglycerol by a family of three lipid phosphate phosphatases, which also dephosphorylate sphingosine-1-phosphate, phosphatidic acid and ceramide 1-phosphate in a non-specific manner. It can be converted back to phosphatidic acid by a membrane-bound *O*-acyltransferase (MBOAT2) specific for lysophosphatidic acid (and lysophosphatidylethanolamine) with a preference for oleoyl-CoA as substrate.

Other lysophospholipids and especially the sphingolipid analogue, sphingosine-1-phosphate, show a related range of activities.

Cyclic Phosphatidic Acid

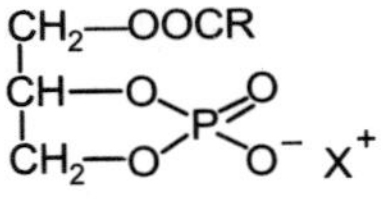

$$\begin{array}{l} CH_2\text{—OOCR} \\ |\\ CH\text{—O}\diagdown \\ | \qquad\qquad P(=O)O^-\ X^+ \\ CH_2\text{—O}\diagup \end{array}$$

Cyclic phosphatidic acid

Cyclic phosphatidic acid (sometimes termed 'cyclic lysophosphatidic acid') was isolated originally from a slime mould, but has now been detecte of cyclic phosphatidic acidd in a wide range of organisms including humans, especially in the brain but also bound to albumin in serum (at a concentration of 10^{-7}M, or a tenth that of lysophosphatidic acid). It has a cyclic phosphate at the *sn*-2 and *sn*-3 positions of the glycerol carbons, and this structure is absolutely necessary for its biological activity. It is most abundant in tissues subject to injury. In human serum, the main molecular species contains palmitic acid.

While cyclic phosphatidic acid may have some similar signalling functions to lysophosphatidic acid *per se* in that it binds to some of the same receptors, it also has some quite distinct activities in animal tissues. For example, cyclic phosphatidic acid is known to be a specific inhibitor of DNA polymerase-α. It has an appreciable effect on the inhibition of cancer cell invasion and metastasis, a finding that is currently attracting great pharmacological interest. In addition, it inhibits the platelet aggregation induced by lysophosphatidic acid.

Studies of the biosynthesis of cyclic phosphatidic acid in fetal bovine serum suggest that it is the product of an enzyme related to the human enzyme autotaxin, a serum lysophospholipase D that produces lysophosphatidic acid. This enzyme appears to produce cyclic phosphatidic acid in serum by an intramolecular transphosphatidylation reaction. However, it can also be formed artefactually by the addition of strong acid to serum.

Pyrophosphatidic Acid

Pyrophosphatidic acid or *sn*-1,2-diacylglycero-3-pyrophosphate is an unusual and little known phospholipid that was first identified as a minor component in yeasts, and is also know to be present in mushrooms and higher plants as a product of the enzyme phosphatidic acid kinase.

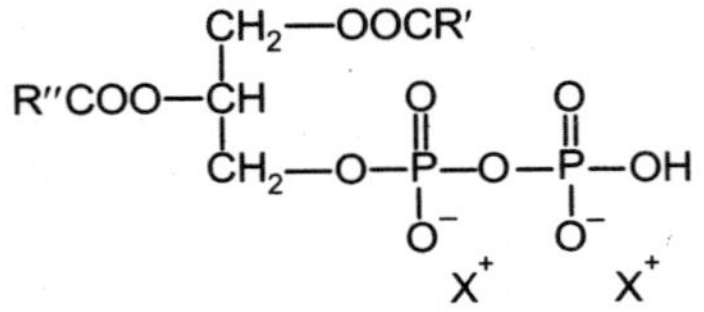

$$\begin{array}{l} \qquad\qquad\quad CH_2\text{—OOCR}' \\ R''COO\text{—}CH \\ \qquad\qquad\quad CH_2\text{—O—}\overset{O}{\overset{\|}{P}}\text{—O—}\overset{O}{\overset{\|}{P}}\text{—OH} \\ \qquad\qquad\qquad\qquad\ \ O^-\ X^+ \quad\ \ O^-\ X^+ \end{array}$$

diacyl-*sn*-glycerol-3-pyrophosphate

It is rapidly metabolized back to phosphatidic acid by a specific phosphatase and thence to diacylglycerols, and it may have a function in the phospholipase C and D signalling cascades in plants. Pyrophosphatidic acid is barely detectable in non-stimulated plant cells but its concentration increases very rapidly in response to stress situations, including osmotic stress and attack by pathogens. Such findings add to the belief that it is an important signalling molecule in plants under stress. In yeasts, it may have a role in the regulation of the synthesis and metabolism of phospholipids, especially phosphatidylserine.

Analysis

Phosphatidic acid and related lipids are not the easiest to analyse. On adsorption chromatography, retention times tend to be variable and may be dependent to some extent on the nature of the cations associated with the acidic lipids. However, two-dimensional TLC can give good results. Phosphatidic acid, lysobisphosphatidic acid and pyrophosphatidic acid are never easy to distinguish, and the best hope for success appears to lie with modern liquid chromatography-mass spectrometric methods.

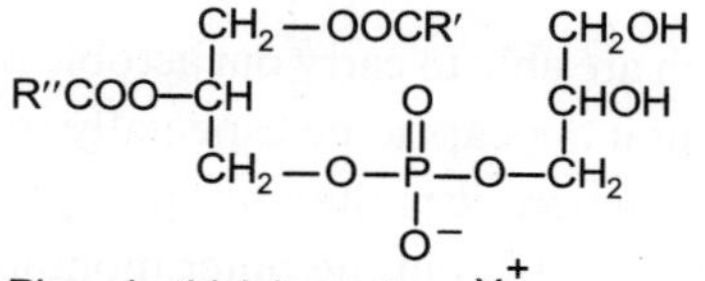

Phosphatidylglycerol X^+

Phosphatidylglycerol or 1,2-diacyl-*sn*-glycerol-3-phosphoryl-1′-*sn*-glycerol tends to be a trace constituent of most tissues, but it is often the main component of some bacterial membranes. It has important functions in lung surfactant, where its physical properties are significant, and in plant chloroplasts, where it appears to have an essential role in photosynthesis. Also, it is the biosynthetic precursor of cardiolipin. In some bacterial species, the 3′-hydroxyl of the phosphatidylglycerol moiety is linked to an amino acid (lysine, ornithine or alanine) to form an *O*-aminoacylphosphatidylglycerol or 'lipoamino acid'.

Phosphatidylglycerol in Bacteria and Plants

Phosphatidylglycerol is a ubiquitous lipid that can be the main component of some bacterial membranes, and it is found also in membranes of plants and animals where it appears to perform specific functions. The charge on the phosphate group means that it is an anionic lipid at neutral pH. The dihexadecanoyl species is illustrated as an example.

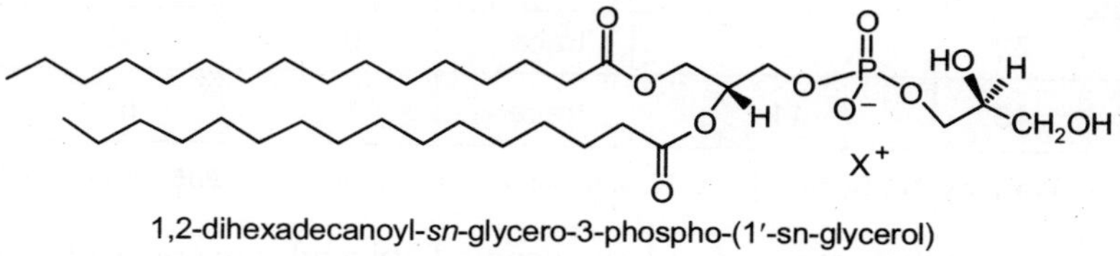

1,2-dihexadecanoyl-*sn*-glycero-3-phospho-(1′-sn-glycerol)

Phosphatidylglycerol is found in almost all bacterial types. For example, *Escherichia coli,* a widely studied organism, has up to 20% of phosphatidylglycerol in its membranes (phosphatidylethanolamine makes up much of the rest with a little cardiolipin). In many bacteria, the diacyl form of the lipid predominates, but in others the alkylacyl- and alkenylacyl forms are more abundant. Phosphatidylglycerol is synthesised only in mitochondria of non-photosynthetic eukaryotes, and it is used as the precursor for cardiolipin, which is located in the inner mitochondrial membrane and is required for proper functioning of the enzymes involved in oxidative phosphorylation.

There is conflicting evidence on as to whether *E. coli* has an absolute requirement for phosphatidylglycerol in its membranes. For example, studies with mutants deficient in phosphatidylglycerol have suggested that its absence results in defective DNA replication and a lack of a necessary modification to the main cellular lipoprotein or proteolipid, leading to membrane welding and eventually cell death. However, others have concluded from similar experiments that phosphatidylglycerol and cardiolipin are entirely dispensable and can be substituted by other anionic phospholipids such as phosphatidic acid. There is evidence that in some bacterial membranes, phosphatidylglycerol may be segregated into distinct domains, which differ in lipid and protein composition and degree of order from other regions.

In cyanobacteria and plants, which are able to carry out aerobic photosynthesis, phosphatidylglycerol is found in all cellular membranes, but it appears to be especially important in the thylakoid membrane, which surrounds the chloroplast, and where it is the only phospholipid comprising up to 10% of the total lipids with a high proportion (up to 70%) in the outer monolayer (much of the remaining lipid is glycosyldiacylglycerols). While sulfoquinovosyldiacylglycerol can substitute for phosphatidylglycerol to a certain extent, especially under conditions of phosphate deficiency and presumably to maintain a required level of anionic lipids, a minimum level of phosphatidylglycerol appears to be essential for photosynthesis and growth.

In the photosynthetic membranes of leaf tissue of higher plants, the phosphatidylglycerol is unique in that in contains a high proportion of *trans*-3-hexadecenoic acid, which is located exclusively in position *sn*-2 (**See below Table 2.9**). This fatty acid is not found in other lipids of the thylakoid membrane. The rate of its synthesis in leaves deprived of light is greatly reduced (with accumulation of the precursor palmitic acid).

Table 2.9 : Composition (mol %) of fatty acids in positions *sn*-1 and *sn*-2 of phosphatidylglycerol from leaves of *Arabidopsis thaliana*.

Position	Fatty acids					
	16:0	*trans*-3-16:1	18:0	18:1	18:2	18:3(n-3)
sn-1	22	-	trace	9	13	55
sn-2	43	41	trace	1	8	8

Data from: Browse, J., Warwick, N., Somerville, C.R. and Slack, C.R. *Biochem. J.*, **235**, 25-31 (1986).

It is interesting to note that saturated and monoenoic fatty acids are concentrated in position *sn*-2 and polyunsaturated in position *sn*-1, the opposite of that found for most animal phospholipids other than phosphatidylglycerol. This is because phosphatidylglycerol is synthesised in chloroplasts via the so-called "prokaryotic" pathway only. In some plant species, position *sn*-2 of the thylakoid phosphatidylglycerol is occupied exclusively with C_{16} fatty acids giving a rather distinctive molecular species distribution. While the role of phosphatidylglycerol in the photosynthetic apparatus of higher plants is still unclear, it is known that in cyanobacteria, it is essential for the oligomerization of photosystems I and II. Analysis of the crystal structure of the photosystem I of cyanobacteria has shown that it contains three molecules of phosphatidylglycerol and one of monogalactosylglycerol as integral components, while phosphatidylglycerol is one of 14 lipid molecules bound to the photosystem

II complex. This phospholipid also appears to be required for crystallization and polymerization of the light-harvesting complex II in pea chloroplasts, where it may be the 'glue' that binds the individual protein components. A report that *trans*-3-hexadecenoic acid in phosphatidylglycerol is essential for the latter process has been questioned.

Disaturated molecular species of phosphatidylglycerol in plants are believed to be an important factor in sensitivity to chilling, and experiments with genetic modifications to increase the degree of unsaturation of this lipid have produced plants with a greater resistance to cold. However, there are discrepancies between the results of different experimental approaches, and other factors are certainly involved.

In cyanobacteria in addition to its role in photosynthesis, phosphatidylglycerol is intimately involved in the processes of cell fission and division. Here its propensity to form non-bilayer structures in the presence of calcium ions may be important, aided by its ability to bind to specific proteins. There may be parallels with the division of plastids in higher plants.

A fully acylated phosphatidylglycerol, termed ***bis*-phosphatidic acid** or phosphatidyldiacylglycerol, and plasmalogen analogues have been found in bacteria. While this structure has on occasion been ascribed in error to other lipids in developing seeds or brain, it can indeed be formed in animal tissues. Two other unusual phosphatidylglycerol derivatives based on an archaeol backbone, i.e. phosphatidylglycerol sulfate and phosphatidylglycerol phosphate methyl ester, are unique constituents of the primitive organisms, the Haloarchaea. They are important constituents of bacteriorhodopsin, a retinal-containing integral membrane protein of the cytoplasmic membrane, which forms two-dimensional crystalline patches known as the purple membrane.

Phosphatidylglycerol in Animal Tissues

Phosphatidylglycerol is present at a level of 1-2% in most animal tissues, but it can be the second most abundant phospholipid in lung surfactant at up to 11% of the total (in a few species, it is replaced by another acidic lipid, phosphatidylinositol). It is well established that the concentration of phosphatidylglycerol increases during foetal development, coincident with the formation of stable lamellar phases, but its precise function is a matter of conjecture. For example, it may aid the spreading of dipalmitoyl-phosphatidylcholine, which is presumed to be the main functional component of lung surfactant. The fatty acid composition of lung tissue from several species is listed in Table 2.10.

In each, the content of saturated fatty acids is high while that of the polyunsaturated components is relatively low in comparison to phospholipids in other tissues. It has also been shown that lung phosphatidylglycerol in many animals contains a high proportion of disaturated molecular species, although this does not appear to be true of human lung surfactant. It seems that the acidic head-group is more important to surfactant function than the precise molecular species composition.

The lung aside, phosphatidylglycerol may be present in animal tissues merely as a precursor for diphosphatidylglycerol (cardiolipin). As an example of another tissue, the positional distribution of fatty

Table 2.10 : Fatty acid composition (weight % of the total) in lung phosphatidylglycerol from various species

Fatty acid	Species			
	pig	cow	rabbit	guinea pig
16:0	27	34	29	37
16:1	2	1	3	6
18:0	21	15	19	18
18:1	34	37	27	24
18:2	7	3	8	5
18:3	1	2	1	2
20:3	1	1	1	trace
20:4(n-6)	3	3	4	3
22:4(n-6)	1	1	3	2
22:5	1	1	3	1
22:6(n-3)	1	1	1	1

acids in rat liver phosphatidylglycerol is listed in **Table 2.11**. Like cardiolipin, there is a very high proportion of linoleate, much of which is concentrated in position *sn*-1.

Table 2.11 : Positional distribution of fatty acids in phosphatidylglycerol from rat liver

Position	Fatty acid					
	16:0	18:0	18:1	18:2	20:4	22:6
sn-1	7	3	3	81		
sn-2	3	1	34	50	2	1

Biosynthesis of Phosphatidylglycerol

In animal, plant and microbial tissues, phosphatidylglycerol is formed from phosphatidic acid by a sequence of enzymatic reactions that proceeds via the intermediate, cytidine diphosphate diacylglycerol (CDP-diacylglycerol), which is rarely detected as a normal component of tissues amounting to only 0.05% or so of the total phospholipids. Biosynthesis proceeds by condensation of phosphatidic acid and cytidine triphosphate with elimination of pyrophosphate via the action of an enzyme phosphatidate cytidyltransferase (or CDP-synthase). The same liponucleotide is an important intermediate in the biosynthesis of phosphatidylinositol, but rather different routes are taken to phosphatidylcholine and phosphatidylethanolamine.

CDP-diacylglycerol reacts with glycerol-3-phosphate via phosphatidylglycerophosphate synthase to form 3-*sn*-phosphatidyl-1′-*sn*-glycerol 3′-phosphoric acid, with release of cytidine monophosphate (CMP). Finally, phosphatidylglycerol is formed by the action of one of two phosphatases. As biosynthesis is via glycerol-3-phosphate, the second glycerol moiety is attached at position *sn*-1 to the phosphate group. However, there are other minor biosynthetic routes to phosphatidylglycerol, e.g. by phospholipase D-catalysed catabolism of diphosphatidylglycerol (cardiolipin) or by glycerolysis of other phospholipids (also catalysed by phospholipase D), which can change the stereochemistry in part (in effect, racemization).

Phosphatidic acid → Cytidine diphosphate diacylglycerol (CDP-DAG)

CDP-DAG + Glycerol-3-phosphate → 3-*sn*-phosphatidyl-1′-*sn*-glycerol 3′-phosphoric acid → Phosphatidylglycerol

In animal tissues, the eventual fatty acid composition is attained by the process of remodelling known as the Lands' cycle. The first step, is hydrolysis by a phospholipase A_2 to lysophosphatidylglycerol, followed by reacylation by means of an acyl-CoA:lysophosphatidylglycerol acyltransferase. The human form of the latter, designated LPGAT1, has been characterized and found to have a preference for 16:0-, 18:0-, and 18:1-CoAs as donors.

In cyanobacteria, a disaturated molecular species of phosphatidylglycerol is synthesised first, and the fatty acid in position *sn*-1 is subsequently desaturated by specific acyl-lipid desaturases. That in position *sn*-2 is not affected. In higher plants, phosphatidylglycerol is synthesised in three cellular compartments, plastids, endoplasmic reticulum and mitochondria. In the plastids, the selectivity of the acyltransferases is such that the initial molecular species formed contains oleic acid in position *sn*-1 and palmitic acid in position *sn*-2. Some of the palmitate in position *sn*-2 is desaturated to the *trans*-3 isomer, while the oleate in position *sn*-1 is desaturated to 18:2 and 18:3 fatty acids. In the endoplasmic reticulum in contrast, the initial molecular species contain palmitic and oleic acids in position *sn*-1 and oleic acid in position *sn*-2. The oleate in both positions, but not the palmitate, is further desaturated by acyl-lipid desaturases until the final fatty acid compositions are attained. These details of the biosynthetic processes that occur in mitochondria have still to be determined.

Phosphatidylglycerol is the biosynthetic precursor of cardiolipin, lysobisphosphatidic acid and many glycophospholipids, as well as bacterial proteolipids, lipoteichoic acids and the complex lipoamino acids.

Acylphosphatidylglycerol

Acylphosphatidylglycerol or (1,2-diacyl-*sn*-glycero-3-phospho-(3'-acyl)-1'-*sn*-glycerol) was first isolated as a minor component of the phospholipids of the bacterium *Salmonella typhimurium*, and it has since been found in a number of prokaryotic species, including Escherichia coli. In particular, it is a characteristic component of the membranes of Corynebacteria and is especially abundant in those species that lack mycolic acids. *C. amycolatum*, for example, contained 20-29% of this lipid, with mainly C_{14} to C_{18} saturated and monoenoic fatty acid components; the fatty acid on the head group glycerol was mainly oleate. It has also been found in parasitic protozoa, such as *Trichomonas vaginalis* and *T. foetus*. The only report of its occurrence in plants is from oats (*Avena sativa*), which are also known to contain *N*-acylphosphatidylethanolamine in appreciable amounts.

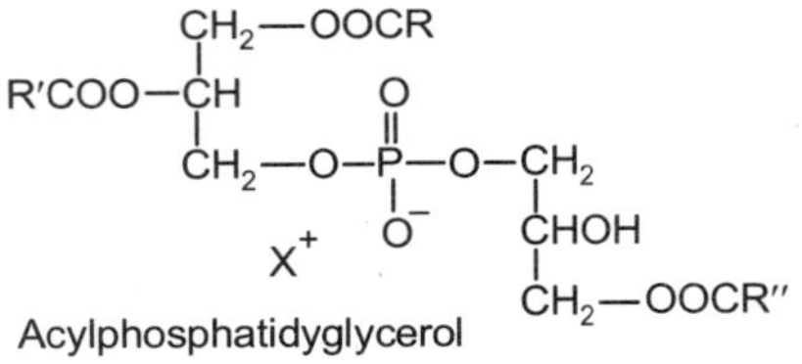

$$\begin{array}{l} CH_2{-}OOCR \\ R'COO{-}CH \quad\quad O \\ CH_2{-}O{-}P{-}O{-}CH_2 \\ \quad X^+ \quad\; O^- \quad CHOH \\ \quad\quad\quad\quad\quad\; CH_2{-}OOCR'' \end{array}$$

Acylphosphatidyglycerol

Acylphosphatidylglycerol is formed *in vitro* in experiments designed to study the biosynthesis of lysobisphosphatidic acid in animal cells, and in this instance the fatty acid on the glycerol head group is presumed to be in the *sn*-2' position. It is not clear whether it occurs naturally in animal tissues.

***Bis*-phosphatidic acid** or phosphatidyldiacylglycerol (fully acylated phosphatidylglycerol) can be produced as a minor component of animal cells by trans-phosphatidylation of phosphatidylcholine with diacylglycerol, catalysed by the enzyme phospholipase D, a possible mechanism for removing excess messenger diacylglycerol. In this instance, the stereochemistry of the glycerol is presumably different from that in normal phosphatidylglycerol, i.e. the phosphate will be attached to the sn-3/*sn*-3' positions. The *bis*-phosphatidic acid found in lysosomes is related to lysobisphosphatidic acid.

Complex Lipoamino Acids

In some Gram-positive bacterial species, the 3'-hydroxyl of the phosphatidylglycerol moiety may be esterified to an amino acid (lysine, ornithine or alanine) to form an *O*-aminoacylphosphatidylglycerol, or it can be linked to another fatty acid.

The former have been termed **lipoamino acids**, though it might be better to call them "complex lipoamino acids" to distinguish them from those consisting simply of a fatty acid linked to an amino acid.

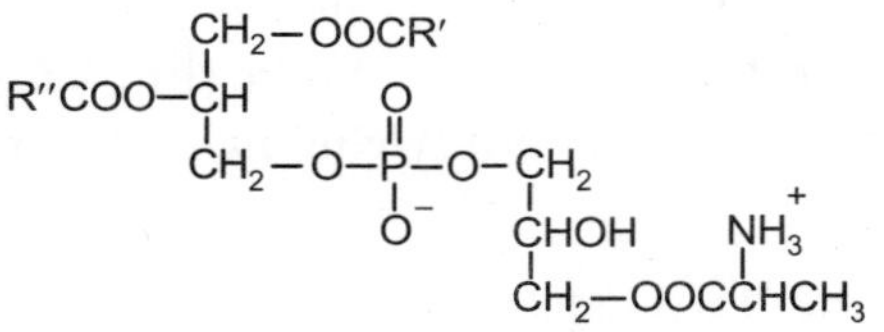

```
          CH2—OOCR'
           |
R''COO—CH           O
           |           ||
          CH2—O—P—O—CH2
                    |       |            +
                    O-    CHOH     NH3
                            |            |
                           CH2—OOCCHCH3
```

Alanylphosphatid ylglycerol

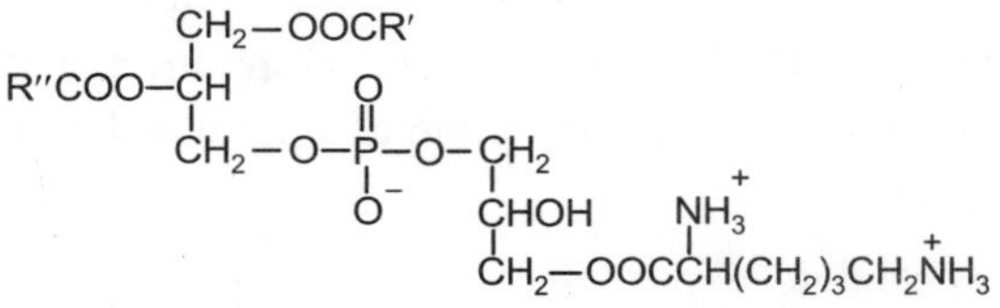

```
          CH2—OOCR'
           |
R''COO—CH           O
           |           ||
          CH2—O—P—O—CH2
                    |       |            +
                    O-    CHOH     NH3
                            |            |                 +
                           CH2—OOCCH(CH2)3CH2NH3
```

Lysylphosphatidylglycerol

There is evidence that the function of lysylphosphatidylglycerol and other lipoamino acids in the membranes of pathogenic bacteria is to protect them from antimicrobial cationic polypeptides produced by plants and animals. Membranes containing this lipid are much less permeable than those containing phosphatidylglycerol *per se*, especially when exposed to acidic conditions. The betaine lipids, together with phosphatidylserine, phosphatidylthreonine, and related lipids discussed elsewhere, could also be termed complex lipoamino acids.

Analysis

Phosphatidylglycerol is not the easiest phospholipid to analyse. It tends to elute close to phosphatidic acid in many chromatographic systems, but it can usually be resolved by two-dimensional thin-layer chromatography. Electrospray mass spectrometry under negative ionization conditions appears to be well suited to determination of molecular species composition. Similarly, modern mass spectrometric methods seem to be suited to the analysis of the complex lipoamino acids.

Cardiolipin (diphosphatidylglycerol or more precisely 1,3-*bis*(*sn*-3'-phosphatidyl)-*sn*-glycerol) is a unique phospholipid with in essence a dimeric structure, having four acyl groups and potentially carrying two negative charges (and is thus an acidic lipid). It is an important constituent of mitochondrial lipids especially, so heart muscle is a rich source. Amongst other functions, it plays a key role in modifying the activities of the enzymes concerned with oxidative phosphorylation.

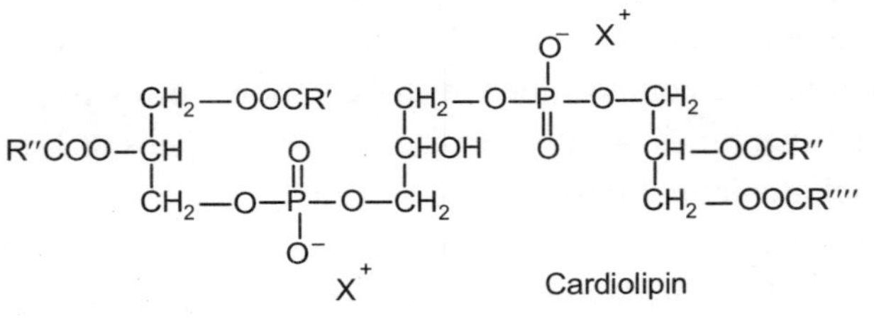

```
                                                     O-   X+
                                                      |
          CH2—OOCR'        CH2—O—P—O—CH2
           |                      |           ||       |
R''COO—CH            O        CHOH      O      CH—OOCR''
           |            ||         |                     |
          CH2—O—P—O—CH2                   CH2—OOCR''''
                     |
                     O-
                       X+
```

Cardiolipin

```
                 O
                 ||
          —O—P—O—
         |        |      |
HO —               O-       —OH
         |                   |
RCOO—                    —OOCR'
```

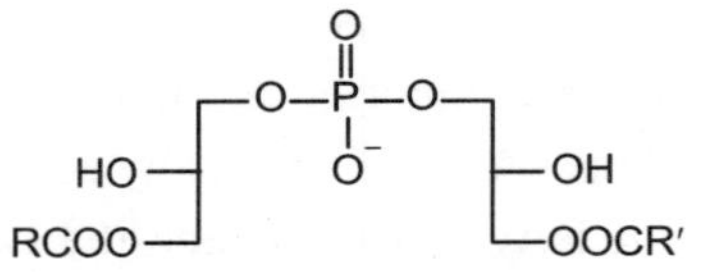

Lysobisphosphatidic acid

Phosphatidylcholine or 1,2-diacyl-*sn*-glycerol-3-phosphorylcholine (or "lecithin", although the term is now used more frequently for the mixed phospholipid by-products of seed oil refining) is usually the most abundant lipid in the membranes of animal tissues, and it is often a major lipid component of plant membranes, but only rarely of bacteria. With the other choline-containing phospholipid, sphingomyelin, it is a key structural component and constitutes much of the lipid in the external monolayer of the plasmamembrane of animal cells especially. **Phosphatidylcholine – Structure and Occurrence.**

Phosphatidylcholine (once given the trivial name 'lecithin') is usually the most abundant phospholipid in animal and plants, often amounting to almost 50% of the total, and as such it is obviously the key building block of membrane bilayers. In particular, it makes up a very high proportion of the outer leaflet of the plasma membrane. Phosphatidylcholine is also the principal phospholipid circulating in plasma, where it is an integral component of the lipoproteins, especially the HDL. On the other hand, it is less often found in bacterial membranes, perhaps 10% of species. It is a neutral or zwitterionic phospholipid over a pH range from strongly acid to strongly alkaline. In animal tissues, some of its membrane functions appear to be shared with the structurally related sphingolipid - sphingomyelin – although the latter has many unique properties of its own.

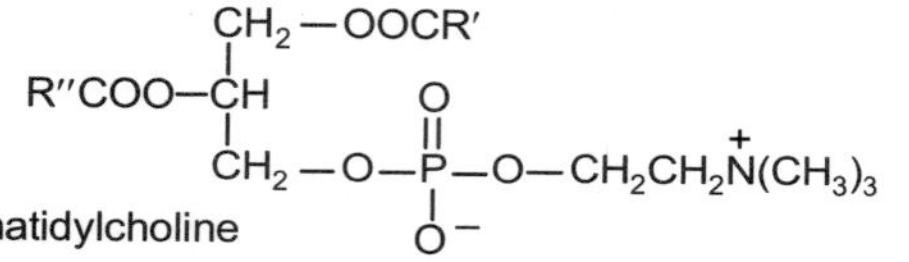

Phosphatidylcholine

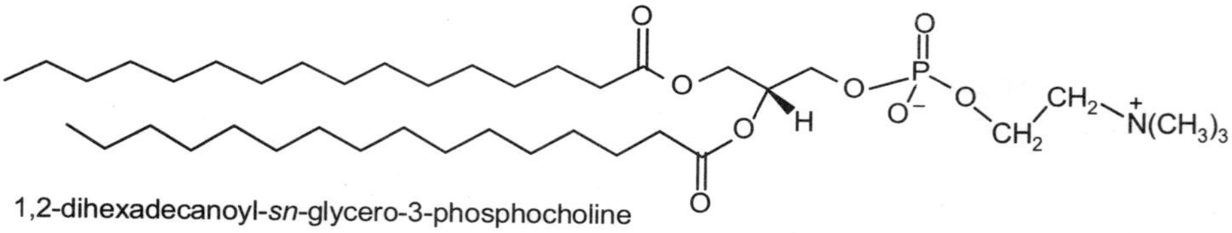

1,2-dihexadecanoyl-*sn*-glycero-3-phosphocholine

In animal tissues, phosphatidylcholine tends to exist in mainly in the diacyl form, but small proportions (in comparison to phosphatidylethanolamine and phosphatidylserine) of alkylacyl and alkenylacyl forms may also be present. Data for the compositions of these various forms from bovine heart muscle are listed in our web pages on Ether lipids. As a generalization, animal phosphatidylcholine tends to contain lower proportions of arachidonic and docosahexaenoic acids and more of the C_{18} unsaturated fatty acids than the other zwitterionic phospholipid, phosphatidylethanolamine. The saturated fatty acids are most abundant in position *sn*-1, while the polyunsaturated components are concentrated in position *sn*-2. Indeed, C_{20} and C_{22} polyenoic acids are exclusively in position *sn*-2. Dietary factors obviously influence fatty acid compositions, but in comparing animal species, it would be expected that the structure of the phosphatidylcholine in the same metabolically active tissue would be somewhat similar in terms of the relative distributions of fatty acids between the two positions. **Table 2.12** lists some representative data.

Table 2.12 : Positional distribution of fatty acids in the phosphatidylcholine of some animal tissues

Position	Fatty acid						
	16:0	16:1	18:0	18:1	18:2	20:4	22:6
Rat liver							
sn-1	23	1	65	7	1	trace	-
sn-2	6	1	4	13	23	39	7
Rat heart							
sn-1	30	2	47	9	11	-	-
sn-2	10	1	3	17	20	33	9
Rat lung							
sn-1	72	4	15	7	3	-	-
sn-2	54	7	2	12	11	10	1
Human plasma							
sn-1	59	2	24	7	4	trace	-
sn-2	3	1	1	26	32	18	5
Human erythrocytes							
sn-1	66	1	22	7	2	-	-
sn-2	5	1	1	35	30	16	4
Bovine brain (gray matter)							
sn-1	38	5	32	21	1	-	-
sn-2	33	4	trace	48	1	9	4
Chicken egg							
sn-1	61	1	27	9	1	-	-
sn-2	2	1	trace	52	33	7	4

There are some exceptions to the rule. The phosphatidylcholine in some organs contains relatively high proportions of disaturated molecular species. For example, it is well known that lung phosphatidylcholine in most if not all animal species studied to date contains a high proportion (50% or more) of dipalmitoylphosphatidylcholine. It appears that this is the main surface-active component, providing alveolar stability by decreasing the surface tension at the alveolar surface to a very low level. Also, the internal lipids of the animal cell nucleus (after the external membrane has been removed) contain a high proportion of disaturated phosphatidylcholine, amounting to 10% of the volume indeed. This is synthesised entirely within the nucleus, unlike phosphatidylinositol for example, and in contrast to other cellular lipids its composition cannot be changed by extreme dietary manipulation. It has been suggested that it may have a role in stabilizing or regulating the structure of the chromatin, as well as being a source of diacylglycerols with a signalling function.

The positional distributions of fatty acids in phosphatidylcholine in representative plants and yeast are listed in **Table 2.13**. In the leaves of the model plant *Arabidopsis thaliana*, saturated fatty acids are concentrated in position *sn*-1, but monoenoic fatty acids are distributed approximately equally between the two positions, and there is a preponderance of di- and triunsaturated fatty acids in position *sn*-2. The same is true for soybean 'lecithin'. The pattern is somewhat similar for the yeast *Lipomyces lipoferus*, except that much of the 16:1 is in position *sn*-1 in this instance.

Table 2.13 : Composition of fatty acids (mol %) in positions *sn*-1 and sn-2 in the phosphatidylcholine from plants and yeast

Position	Fatty acid					
	16:0	16:1	18:0	18:1	18:2	18:3
Arabidosis thaliana (leaves)						
sn-1	42		4	5	23	26
sn-2	1		trace	5	47	47
Soybean 'lecithin'						
sn-1	24		9	14	47	4
sn-2	5		1	13	75	6
Lipomyces lipoferus						
sn-1	24	18	trace	37	16	4
sn-2	4	5	trace	39	31	19

Phosphatidylcholine - Biosynthesis and Biological Function

There are several mechanisms for the biosynthesis of phosphatidylcholine in animals, plants and micro-organisms. Choline itself is not synthesised as such by animal cells and is an essential nutrient. It must be obtained from dietary sources or by degradation of existing choline-containing lipids, for example those produced by the second pathway described below. Once taken up into cells, choline is immediately phosphorylated by a choline kinase in the cytoplasm of the cell to phosphocholine, which is reacted with cytidine triphosphate (CTP) to form cytidine diphosphocholine. The membrane-bound enzyme CDP-choline:1,2-diacylglycerol cholinephosphotransferase in the endoplasmic reticulum catalyses the reaction of the last compound with *sn*-1,2-diacylglycerols to form phosphatidylcholine. This is the main pathway for the synthesis of phosphatidylcholine in animals and plants, and it is analogous to the biosynthesis of phosphatidylethanolamine.

$$\underset{\text{Choline}}{HOCH_2CH_2N(CH_3)_3} \xrightarrow{\text{ATP} \rightarrow \text{ADP}} \underset{\text{Phosphocholine}}{O{-}\overset{O}{\overset{\|}{P}}(O){-}OCH_2CH_2\overset{+}{N}(CH_3)_3} \xrightarrow{\text{CTP} \rightarrow \text{PPi}} \underset{\text{Cytidine diphosphocholine}}{\text{Cytidine}{-}O{-}\overset{-}{O}{-}P(O^-){-}O{-}\overset{O}{\overset{\|}{P}}(O^-){-}OCH_2CH_2\overset{+}{N}(CH_3)_3}$$

$$\text{CDP-choline} + \underset{sn\text{-1,2-diacylglycerol}}{\begin{array}{l} CH_2{-}OOCR' \\ R''COO{-}CH \\ CH_2OH \end{array}} \longrightarrow \underset{\text{phosphatidylcholine}}{\begin{array}{l} CH_2{-}OOCR' \\ R''COO{-}CH \\ CH_2{-}O{-}\overset{O}{\overset{\|}{P}}(O){-}O{-}CH_2CH_2\overset{+}{N}(CH_3)_3 \end{array}}$$

The discovery of the importance of this pathway depended a little on serendipity in that in experiments in the lab of Professor Eugene Kennedy, samples of adenosine triphosphate (ATP) contained some cytidine triphosphate (CTP) as an impurity. However, luck is of little value without receptive minds, and Kennedy and co-workers demonstrated that the impurity was an important metabolite that was essential for the formation of phosphatidylcholine.

The above reaction, together with the biosynthetic mechanism for phosphatidylethanolamine, is significantly different from that for phosphatidylglycerol, phosphatidylinositol and cardiolipin. Both make use of nucleotides, but with the latter, the nucleotide is covalently linked directly to the lipid intermediate, i.e. cytidine diphosphate diacylglycerol.

The source of the *sn*-1,2-diacylglycerol precursor, which is also a key intermediate in the formation of phosphatidylethanolamine and phosphatidylserine, and of triacylglycerols, is phosphatidic acid. In this instance, the important enzyme is phosphatidic acid phosphatase (or 'phosphatidate phosphatase' or 'lipid phosphate phosphatase'), which is present mainly in the endoplasmic reticulum of the cell.

$$\underset{\text{Phosphatidic acid}}{\begin{array}{l} CH_2OOCR \\ CHOOCR' \\ CH_2OPO_3H \end{array}} \longrightarrow \underset{sn\text{-1,2-diacylglycerol}}{\begin{array}{l} CH_2OOCR \\ CHOOCR' \\ CH_2OH \end{array}}$$

This enzyme is also important for the production of diacylglycerols as essential intermediates in the biosynthesis of triacylglycerols and of phosphatidylethanolamine. In mammals, much of the activity responsible for phospholipid biosynthesis resides in the endoplasmic reticulum (phosphatidic acid phosphatase-2 (or PAP2)), but in addition three related cytoplasmic proteins, termed lipin-1, lipin-2, and lipin-3 may also have a role.

The second pathway for biosynthesis of phosphatidylcholine involves sequential methylation of phosphatidylethanolamine, with *S*-adenosylmethionine as the source of methyl groups, with mono- and dimethyl-phosphatidylethanolamine as intermediates and catalysed by the enzyme phosphatidylethanolamine *N*-methyltransferase. A single enzyme (~20 Kda) catalyses all three reactions and is located mainly in the endoplasmic reticulum where it spans the membrane. This is a major pathway in the liver, but not in other animal tissues or in general in higher organisms. It may be the main route to phosphatidylcholine in those bacterial species that produce this lipid and in yeasts.

CH_2—OOCR′
R″COO—CH
CH_2—O—P(=O)(OH)—O—$CH_2CH_2NH_3$
Phosphatidylethanolamine

→

CH_2—OOCR′
R″COO—CH
CH_2—O—P(=O)(OH)—O—$CH_2CH_2NHCH_3$
Phosphatidylmonomethylethanolamine

↓

CH_2—OOCR′
R″COO—CH
CH_2—O—P(=O)(OH)—O—$CH_2CH_2NH(CH_3)_2$
Phosphatidyldimethylethanolamine

←

CH_2—OOCR′
R″COO—CH
CH_2—O—P(=O)(OH)—O—$CH_2CH_2\overset{+}{N}(CH_3)_3$
Phosphatidylcholne

This liver enzyme is especially important when choline is deficient in the diet. A by-product of the biosynthesis of phosphatidylcholine from phosphatidylethanolamine is the conversion of *S*-adenosylmethionine to *S*-adenosylhomocysteine, which is hydrolysed in the liver to adenosine and homocysteine. It is noteworthy that elevated plasma homocysteine is a risk factor for cardiovascular disease and myocardial infarction.

Phosphatidylcholine biosynthesis by both pathways in the liver is necessary for normal secretion of the plasma lipoproteins (VLDL and HDL), and it is relevant to a number of human physiological conditions.

In one bacterial species, a third pathway for phosphatidylcholine biosynthesis that is choline-dependent, has been found in which the lipid is formed in one step via condensation of choline directly with CDP-diacylglycerol. The yeast *Saccharomyces cerevisiae* is able to reacylate endogenously generated glycerophosphocholine with acyl-CoA in the microsomal membranes, first to lysophosphatidylcholine and then to phosphatidylcholine.

While phosphatidylcholine is a major lipid in yeasts, recent work suggests that it is not essential if suitable alternative growth substrates are available, unlike higher organisms where perturbation of phosphatidylcholine synthesis can lead to inhibition of growth or even cell death. Enhanced synthesis of phosphatidylcholine appears to occur in cancer cells and solid tumours, and this may prove to be a target for therapeutic agents.

Whatever the mechanism of biosynthesis in tissues, it is apparent that the fatty acid compositions and positional distributions on the glycerol moiety are determined post synthesis by extensive re-modelling involving hydrolysis (phospholipase A_2 mainly) and re-acylation, a process that is sometimes termed the **'Lands' Cycle'** after its discoverer W.E.M. (Bill) Lands. The re-acylation step is catalysed by a membrane-bound coenzyme A-dependent lysophosphatidylcholine acyltransferase (MBOAT5 also designated 'LPCAT3'), which has been located chiefly within the endoplasmic reticulum, though also in mitochondria and the plasma membrane, in organs such as the liver, adipose tissue and pancreas. This incorporates linoleoyl and arachidonoyl chains specifically into lysophosphatidylcholine (see below). Polyunsaturated fatty acids introduced by this route can then be transferred to 1-alkyl and 1-alkenyl phospholipids by CoA-independent transacylases. A second such enzyme LPCAT4 is known and has a clear preference for 18:1-CoA. Similarly, the highly saturated molecular species of phosphatidylcholine found in the nucleus are formed from species with a more conventional composition by remodelling, presumably by acyltransferases with somewhat different specificity. These and further related enzymes are involved in remodelling of all other phospholipids.

In plants, fatty acids esterified to phosphatidylcholine can serve as substrates for desaturases, and this means that the fatty acid composition changes also after the initial synthetic process. The process is further complicated in plants in that biosynthesis or partial synthesis (via lysophosphatidylcholine) occurs in different organelles, such as the endoplasmic reticulum, plastids and mitochondria, from different fatty acid pools or with differing specificities.

Because of the generally cylindrical shape of the molecule, phosphatidylcholine spontaneously organizes into bilayers, so it is ideally suited to serve as the bulk structural element of biological membranes. Such properties are essential to act as a balance to those lipids that do not form bilayers or that form specific micro-domains such as rafts.

In addition to its function as a membrane constituent, phosphatidylcholine may have a role in signalling via the generation of diacylglycerols, especially in the nucleus. Although the pool of the precursor is so great in many tissues that turnover is not easily measured, the presence of phospholipases C and D specific for phosphatidylcholine, which are activated by a number of agonists, suggests such a function especially in the cell nucleus. Diacylglycerols formed in this way would be much more saturated than those derived from phosphatidylinositol, and would not be expected to be as active. The plasmalogen form of phosphatidylcholine may also have a signalling function, as thrombin treatment of endothelial cells activates a selective hydrolysis (phospholipase A_2) of molecular species containing arachidonic acid in the *sn*-2 position, releasing this fatty acid for eicosanoid production. The diacyl form of phosphatidylcholine may have a related function in signal transduction in other tissues. In addition, it is known that the enzyme 3-hydroxybutyrate dehydrogenase requires to be bound to phosphatidylcholine before it can function optimally.

Phosphatidylcholine is the biosynthetic precursor of sphingomyelin and as such must have some influence on the many metabolic pathways that constitute the sphingomyelin cycle. It is also a precursor for phosphatidic acid, lysophosphatidylcholine and platelet-activating factor, each with important signalling functions, and of phosphatidylserine.

On catabolism of choline-containing lipids, much of the choline is re-used for phosphatidylcholine biosynthesis, often after being returned to the liver. Some is oxidized in the kidney and liver to betaine, which serves as a donor of methyl groups for S-adenosylmethionine production. A proportion is used in nervous tissues for production of acetylcholine, which is a neurotransmitter of importance to learning, memory and sleep. Some choline is lost through excretion of phosphatidylcholine in the bile.

3. Lysophosphatidylcholine

Lysophosphatidylcholine, with one mole of fatty acid per mole of lipid in position *sn*-1, is found in small amounts in most tissues. It is formed by hydrolysis of phosphatidylcholine by the enzyme phospholipase A_2, as part of the de-acylation/re-acylation cycle that controls its overall molecular species composition. It can also be formed inadvertently during extraction of lipids from tissues if the phospholipase is activated by careless handling. There is also a phospholipase A_1, which is able to cleave the *sn*-1 ester bond. As part of the Lands cycle, it can be converted back to phosphatidylcholine by specific acyl transferases.

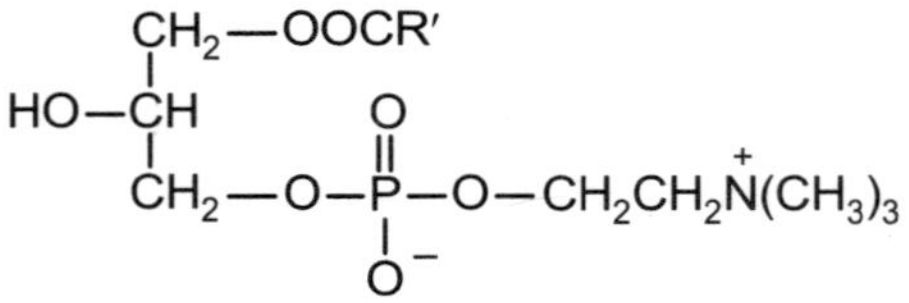

$$\begin{array}{l} \quad\;\; CH_2-OOCR' \\ HO-CH \qquad\quad O \\ \quad\;\; CH_2-O-\overset{\|}{P}-O-CH_2CH_2\overset{+}{N}(CH_3)_3 \\ \qquad\qquad\quad\;\; O^- \end{array}$$

Lysophosphatidylcholine

In plasma of animal species, appreciable amounts of lysophosphatidylcholine are formed by a specific enzyme system, lecithin:cholesterol acyltransferase (LCAT), which is secreted from the liver. The enzyme catalyses the transfer of the fatty acids of position *sn*-2 of phosphatidylcholine to the free cholesterol in plasma, with formation of cholesterol esters and of course of lysophosphatidylcholine. Identification of a highly specific phospholipase A_2 in peroxisomes that generates 2-arachidonoyl lysophosphatidylcholine suggests that this may be of relevance to eicosanoid generation and signalling.

Lysophosphatidylcholine has pro-inflammatory properties *in vitro* and it is known to be a pathological component of oxidized lipoproteins (LDL) in plasma and of atherosclerotic lesions. Recently, it has been found to have some functions in cell signalling, and specific receptors (coupled to G proteins) have been identified. It activates the specific phospholipase C that releases diacylglycerols and inositol triphosphate with resultant increases in intracellular Ca^{2+} and activation of protein kinase C. It also activates the mitogen-activated protein kinase in certain cell types. Stearoyl lysophosphatidylcholine has been shown to be protective against lethal sepsis in experimental animals by various mechanisms, including stimulation of neutrophils to eliminate invading pathogens through a peroxide-dependent reaction.

Platelet-Activating Factor

Platelet-activating factor (PAF) or 1-alkyl-2-acetyl-*sn*-glycero-3-phosphocholine is an ether analogue of phosphatidylcholine that is biologically active.

Analysis

Analysis of phosphatidylcholine presents no particular problems. It is readily isolated by thin-layer or high-performance liquid chromatography methods. Determination of the dipalmitoyl species in lung surfactant is a more demanding task, but specific methods have been published. Phospholipase A_2 from snake venom is used in methods to determine the position of fatty acids on the glycerol moiety. Modern mass spectrometry methodology has greatly simplified the task of molecular species analysis.

Lysophosphatidylcholine, which contains only one fatty acid moiety in each molecule, generally in position *sn*-1, is sometimes present as a minor component of tissues. It is a powerful surfactant and

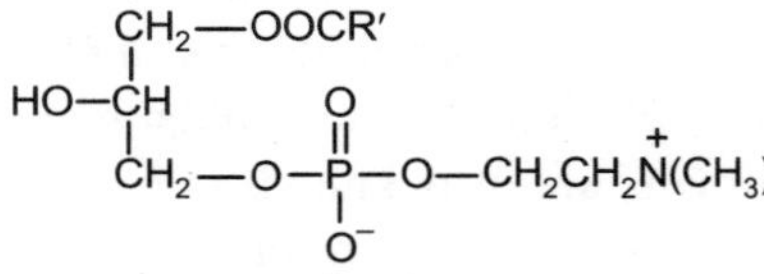

$$\begin{array}{l} CH_2—OOCR' \\ HO—CH \quad\quad O \\ \quad\quad | \quad\quad\quad \| \quad\quad\quad\quad\quad\quad + \\ CH_2—O—P—O—CH_2CH_2N(CH_3)_3 \\ \quad\quad\quad\quad\quad | \\ \quad\quad\quad\quad\quad O^- \end{array}$$

Lysophosphatidylcholine

is more soluble in water than most other lipids.

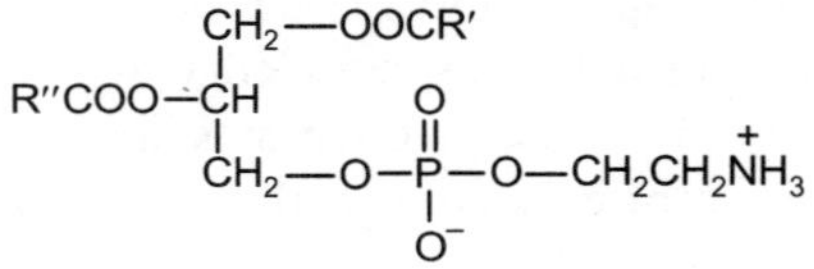

$$\begin{array}{l} CH_2—OOCR' \\ R''COO—CH \quad\quad O \\ \quad\quad | \quad\quad\quad \| \quad\quad\quad\quad\quad\quad + \\ CH_2—O—P—O—CH_2CH_2NH_3 \\ \quad\quad\quad\quad\quad | \\ \quad\quad\quad\quad\quad O^- \end{array}$$

Phosphatidylethanolamine

Phosphatidylethanolamine (once given the trivial name "cephalin") is usually the second most abundant phospholipid class in animal and plant tissues, and can be the major lipid class in microorganisms. As part of an important cellular process, the amine group can be methylated enzymically to yield first phosphatidyl-*N*-monomethylethanolamine and then phosphatidyl-*N,N*-dimethylethanolamine, but these never accumulate in significant amounts; the eventual product is phosphatidylcholine.

***N*-Acyl-phosphatidylethanolamine** is a minor component of some plant tissues, especially cereals, and it is occasionally found in animal tissues, where it is the precursor of some biologically active amides. Lysophosphatidylethanolamine contains only one mole of fatty acid per mole of lipid.

Phosphatidylserine is a weakly acidic lipid that is present in most tissues of animals and plants and is also found in microorganisms. It is located entirely on the inner monolayer surface of the plasma membrane and other cellular membranes. Phosphatidylserine is an essential cofactor for the activation of protein kinase C, and it is involved in many other biological processes, including blood coagulation and apoptosis (programmed cell death).

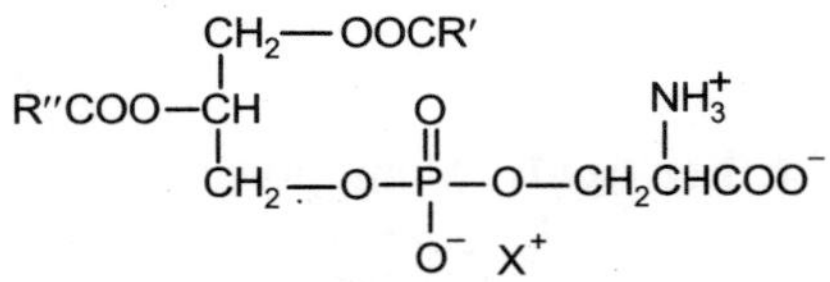

$$\begin{array}{l} CH_2—OOCR' \\ R''COO—CH \quad\quad O \quad\quad\quad\quad NH_3^+ \\ \quad\quad | \quad\quad\quad \| \quad\quad\quad\quad\quad\quad | \\ CH_2—O—P—O—CH_2CHCOO^- \\ \quad\quad\quad\quad\quad | \\ \quad\quad\quad\quad\quad O^- \; X^+ \end{array}$$

Phosphatidylserine

N-Acylphosphatidylserine has been detected in some animal tissues.

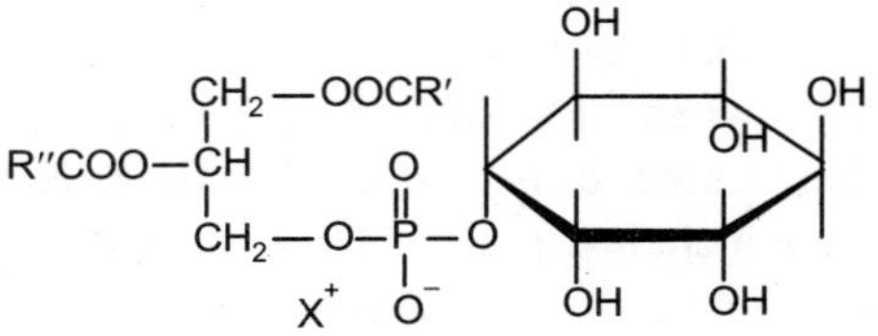

$$\begin{array}{l} CH_2—OOCR' \\ R''COO—CH \quad\quad O \\ \quad\quad | \quad\quad\quad \| \\ CH_2—O—P—O—(\text{inositol ring: OH, OH, OH, OH, OH}) \\ \quad\quad\quad\quad\quad | \\ \quad\quad\quad X^+ \; O^- \end{array}$$

Phosphatidylinositol

Phosphatidylinositol, containing the optically inactive form of inositol, *myo*-inositol, is a common constituent of animal, plant and microbial lipids. In animal tissues especially, it may be accompanied by small amounts of phosphatidylinositol 4-phosphate and phosphatidylinositol 4,5-bisphosphate ('polyphosphoinositides'). These compounds have a rapid rate of metabolism in animal cells, and are converted to metabolites such as diacylglycerols and inositol phosphates, which are important in regulating vital processes. For example, diacylglycerols regulate the activity of a group of enzymes known as protein kinase C, which in turn control many key cellular functions, including differentiation, proliferation, metabolism and apoptosis. In addition, phosphatidylinositol is the primary source of the arachidonate used for eicosanoid synthesis in animals, and it is known to be the anchor that can link a variety of proteins to the external leaflet of the plasma membrane via a glycosyl bridge (glycosylphosphatidylinositol(GPI)-anchored proteins).

Phosphonolipids are lipids with a phosphonic acid moiety esterified to glycerol, i.e. with a carbon-phosphorus bond that is not easily hydrolysed by chemical reagents. Phosphonylethanolamine, for example, is found mainly in marine invertebrates and in protozoa. A ceramide analogue is often found in the same organisms.

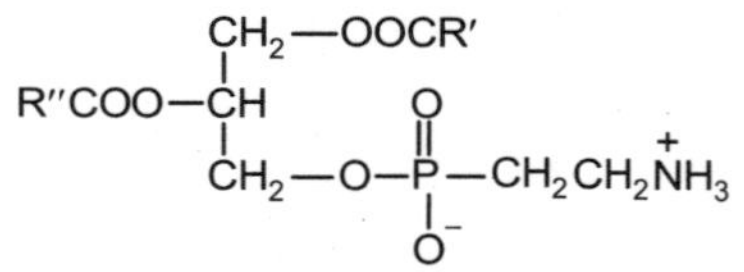

$CH_2—OOCR'$ / $R''COO—CH$ / $CH_2—O—P(=O)(O^-)—CH_2CH_2\overset{+}{N}H_3$

Phosphonylethanolamine

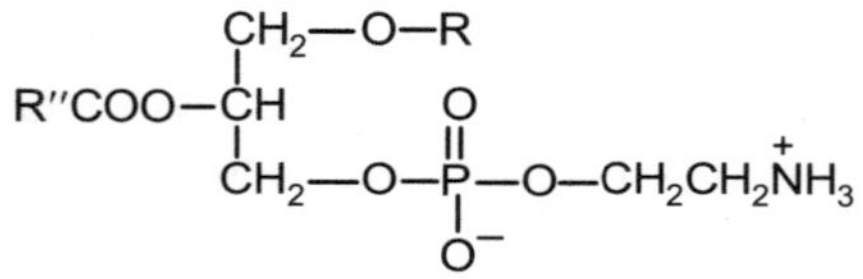

$CH_2—O—R$ / $R''COO—CH$ / $CH_2—O—P(=O)(O^-)—O—CH_2CH_2\overset{+}{N}H_3$

Plasmanylethanolamine

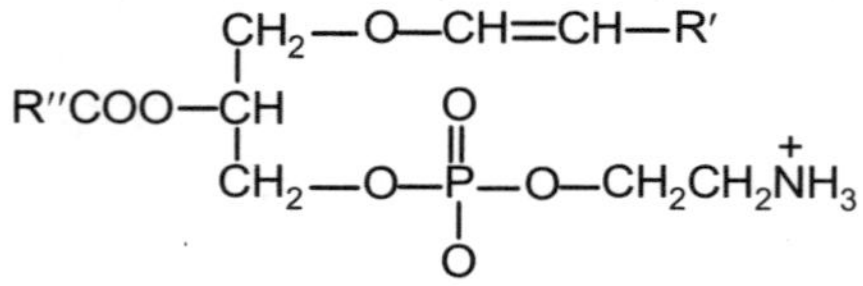

$CH_2—O—CH{=}CH—R'$ / $R''COO—CH$ / $CH_2—O—P(=O)(O)—O—CH_2CH_2\overset{+}{N}H_3$

Plasmanylethanolamine

Ether lipids: Many glycerolipids, but mainly phospholipids, and those of animal and microbial origin especially, contain aliphatic residues linked either by an ether bond or a vinyl ether bond to position 1 of *L*-glycerol. When a lipid contains a vinyl ether bond, the generic term "plasmalogen" is often used. They can be abundant in the phospholipids of animals and microorganisms, and especially in the phosphatidylethanolamine fraction. In this instance, it has been recommended that they should be termed "plasmanylethanolamine" and "plasmenylethanolamine", respectively.

On hydrolysis of glycerolipids containing an alkyl ether bond, 1-alkylglycerols are released that can be isolated for analysis. Similarly, when plasmalogens are hydrolysed under basic conditions, 1-alkenylglycerols are released. Aldehydes are formed on acidic hydrolysis. With both groups of compound, the aliphatic residues generally have a chain-length of 16 or 18, and they are saturated or may contain one additional double bond, that is remote from the ether linkage.

1-Alkyl-2,3-diacyl-*sn*-glycerols, analogues of triacylglycerols, tend to be present in trace amounts only in animal tissues, but can be major constituents of certain fish oils. Related compounds containing a 1-alk-1'-enyl moiety ('neutral plasmalogens') are occasionally present also.

'Platelet-activating factor' or 1-alkyl-2-acetyl-*sn*-glycerophosphorylcholine is an ether-containing phospholipid, that is presently being studied intensively because it can exert profound biological effects at minute concentrations. For example, it effects aggregation of platelets at concentrations as low as 10^{-11} M, and it induces a hypertensive response at very low levels. Also, it is a mediator of inflammation and has messenger functions.

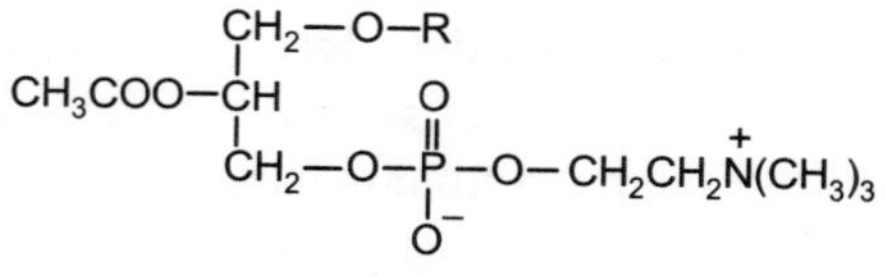

CH_2—O—R
CH_3COO—CH
CH_2—O—P(=O)(O^-)—O—$CH_2CH_2\overset{+}{N}(CH_3)_3$

Platelet-activating factor

Glycoglycerolipids

In plants, especially the photosynthetic tissues, a substantial proportion of the lipids consists of 1,2-diacyl-*sn*-glycerols joined by a glycosidic linkage at position *sn*-3 to a carbohydrate moiety. The main components are the **mono-** and **digalactosyldiacylglycerols**, but related compounds have been found with up to four galactose units, or in which one or more of these is replaced by glucose moieties. It is clear that these have an important role in photosynthesis, but many of the details have still to be worked out.

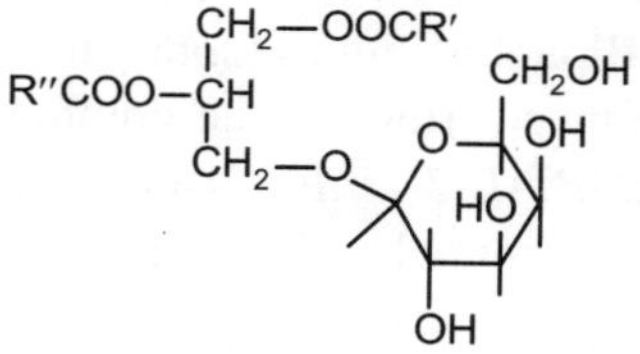

Monogalactosyldiacylglycerol

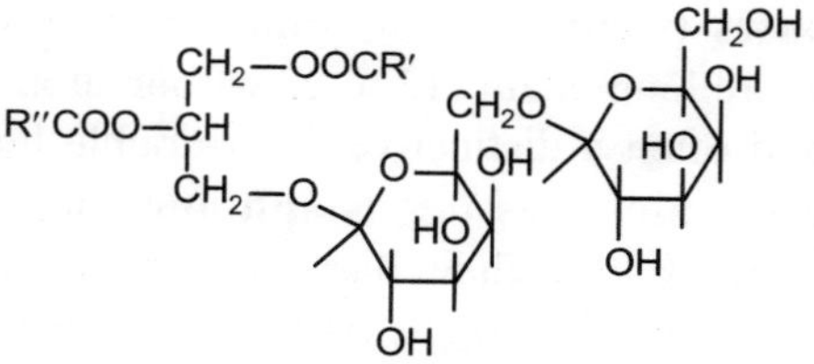

Digalactosyldiacylglycerol

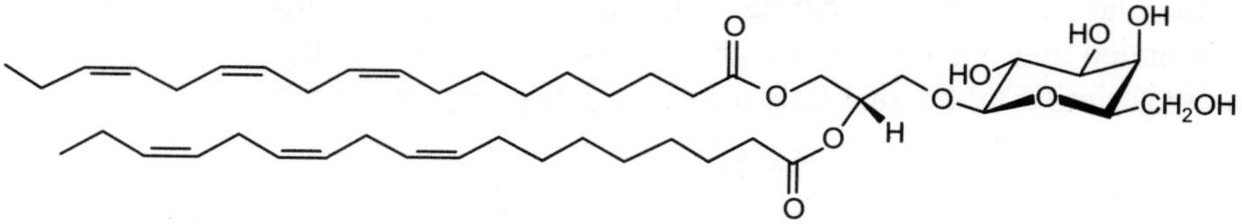

1,2-di-(9Z,12Z,15Z)-octadecatrienoyl-3-O-β-D-galactosyl-*sn*-glycerol

In addition, a 6-*O*-acyl-monogalactosyldiacylglycerol is occasionally a component of plant tissues. See our web pages dealing with plant galactolipids.

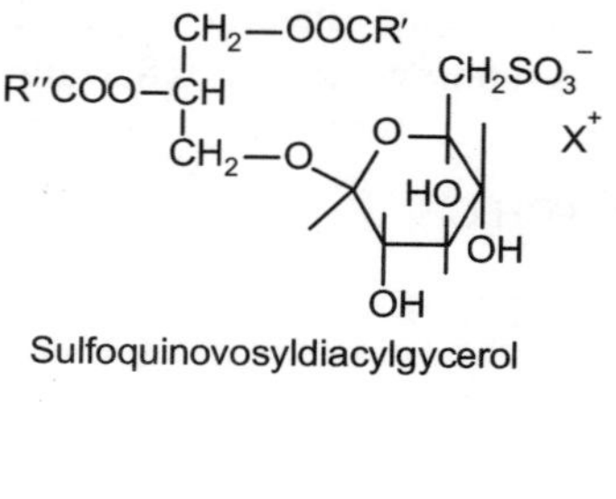

Sulfoquinovosyldiacylgycerol

$CH_3(CH_2)_{12}CH{=}CHCHCH$
trans $2HXH_22H_2OH$ sphingosine

A related unique plant glycolipid is **sulfoquinovosyldiacylglycerol** or the "plant sulfolipid". It contains a sulfonic acid residue linked by a carbon-sulfur bond to the 6-deoxyglucose moiety of a monoglycosyldiacylglycerol and is found exclusively in the chloroplasts.

Monogalactosyldiacylglycerols are not solely plant lipids as they have been found in small amounts in brain and nervous tissue in some animal species. A range of complex glyceroglycolipids have also been characterized from intestinal tract and lung tissue. They exist in both diacyl and alkyl acyl forms. Such compounds are destroyed by some of the methods used in the isolation of glycosphingolipids, so they may be more widespread than has been thought.

A complex glyco-glycero-sulfolipid, termed **seminolipid**, of which the main component is 1-*O*-hexadecyl-2-*O*-hexadecanoyl-3-*O*-(3'-sulfo-β-D-galactopyranosyl)-*sn*-glycerol, is the principal glycolipid in testis and sperm. See our web pages on animal glycosyldiacylglycerols.

A further range of highly complex glycolipids occur in bacteria and other micro-organisms. These include acylated sugars that do not contain glycerol.

Sphingomyelin and Glycosphingolipids

Sphingolipids consist of long-chain bases, linked by an amide bond to a fatty acid and via the terminal hydroxyl group to complex carbohydrate or phosphorus-containing moieties.

Long-chain bases (sphingoids or sphingoid bases) are the characteristic structural unit of sphingolipids. They are long-chain (12 to 22 carbon atoms) aliphatic amines, containing two or three hydroxyl groups, and often a distinctive *trans*-double bond in position 4. The commonest or most abundant of these in animal tissues is **sphingosine**, (2*S*,3*R*,4*E*-2-amino-4-octadecen-1,3-diol) (illustrated). More than 60 long-chain bases have been found in animals, plants and microorganisms, and many of these may occur in a single tissue, but almost always as part of a complex lipid as opposed to in the free form. The aliphatic chains can be saturated, monounsaturated and diunsaturated, with double bonds of either the *cis* or *trans* configuration, and they may sometimes have methyl substituents. In addition, saturated and monoenoic straight- and branched-chain trihydroxy bases are found. The aliphatic moiety is similar to that of a fatty acid in its physical properties.

$$CH_3(CH_2)_{13}.CHOH{>}CHOH.CHNH_2.CH_2OH$$

phytosphingosine

Phytosphingosine (2*S*,3*S*,4*R*-2-amino-octadecanetriol) is the most common long-chain base of plant origin.

For shorthand purposes, a nomenclature similar to that for fatty acids can be used, i.e. the chain-length and number of double bonds are denoted in the same manner with the prefix "d" or "t" to designate di- and trihydroxy bases respectively. Thus, sphingosine is d18:1 and phytosphingosine is t18:0.

$$\begin{array}{c} R.CHOH.CH.CH_2OH \\ | \\ NHOC,R' \end{array}$$

ceramide

Ceramides contain fatty acids linked by an amide bond to the amine group of a long-chain base. In general, they are present at low levels only in tissues, but they are key intermediates in the biosynthesis of the complex sphingolipids. In addition, they have important functions in cellular signalling, and especially in the regulation of apoptosis, and cell differentiation, transformation and proliferation.

Unusual ceramides have been located in the epidermis of the pig and humans; the fatty acids linked to the sphingoid base consist of C_{30} and C_{32} (ω-hydroxylated components, with predominantly the essential fatty acid, linoleic acid, esterified to the terminal hydroxyl group. They are believed to have a special role in preventing the loss of moisture through the skin.

Sphingomyelin is a sphingophospholipid and consists of a ceramide unit linked at position 1 to phosphorylcholine; it is found as a major component of the complex lipids of all animal tissues but not of plants or micro-organisms.

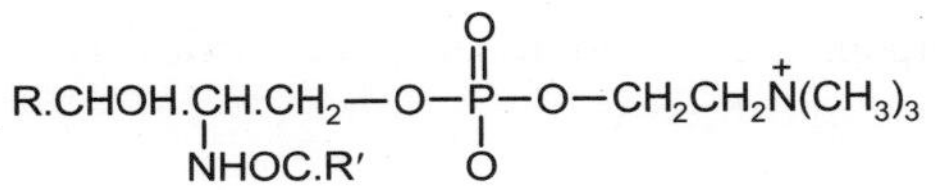

$$R.CHOH.\underset{NHOC.R'}{CH}.CH_2-O-\overset{O}{\underset{O}{P}}-O-CH_2CH_2\overset{+}{N}(CH_3)_3$$

Sphingomyelin

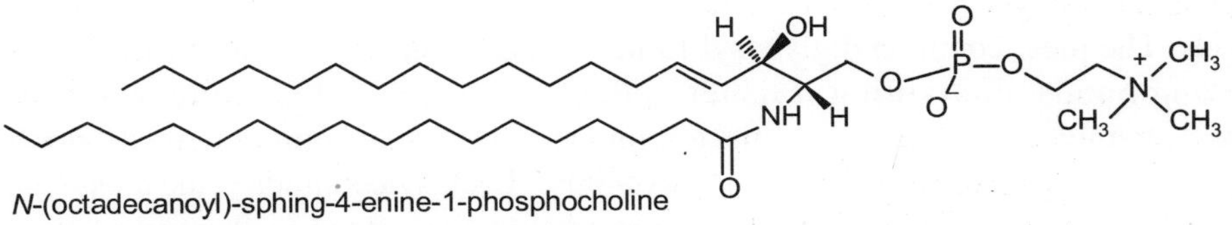

N-(octadecanoyl)-sphing-4-enine-1-phosphocholine

It resembles phosphatidylcholine in many of its physical properties, and can apparently substitute in part for this in membranes although it also has its own unique role. For example, it is a major constituent of the plasma membrane of cells, where it is concentrated together with sphingoglycolipids and cholesterol in tightly organized sub-domains termed 'rafts'. Sphingosine tends to be the most abundant long-chain base constituent, and it is usually accompanied by sphinganine and C_{20} homologues. Sphingomyelin is a precursor for a number of sphingolipid metabolites that have important functions in cellular signalling, including sphingosine-1-phosphate, as part of the 'sphingomyelin cycle'. A correct balance between the various metabolites is vital for good health. Niemann-Pick disease is a rare lipid storage disorder that results from of a deficiency in the enzyme responsible for the degradation of sphingomyelin.

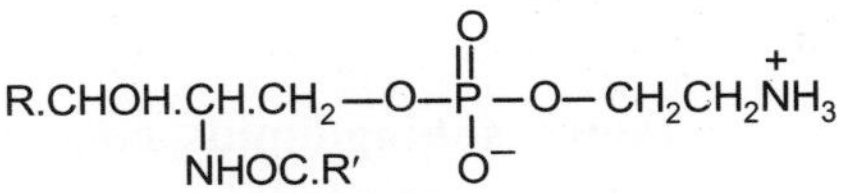

$$R.CHOH.\underset{NHOC.R'}{CH}.CH_2-O-\overset{O}{\underset{O^-}{P}}-O-CH_2CH_2\overset{+}{N}H_3$$

Ceramide phosphorylethanolamine

Ceramide phosphorylethanolamine is found in the lipids of insects and some fresh water invertebrates; the phosphonolipid analogue, ceramide 2-aminoethylphosphonic acid, has been detected in sea anemones and protozoa. Ceramide phosphorylinositol is also found in some organisms, and like phosphatidylinositol, it can be an anchor unit for oligosaccharide-linked proteins in membranes.

Neutral glycosylceramides: The most widespread glycosphingolipids are the **monoglycosylceramides** (or cerebrosides), and they consist of a basic ceramide unit linked by a glycosidic bond at carbon 1 of the long-chain base to glucose or galactose. They were first found in brain lipids, where the principal form is galactosylceramide, but they are now known to be ubiquitous constituents of animal tissues. Glucosylceramide is also found in animal tissues, and especially in skin, where it functions as part of the water permeability barrier. It is the biosynthetic precursor of lactosylceramide, and thence of the complex oligoglycolipids and gangliosides. In addition, glucosylceramide is found in plants, where the main long-chain base is phytosphingosine.

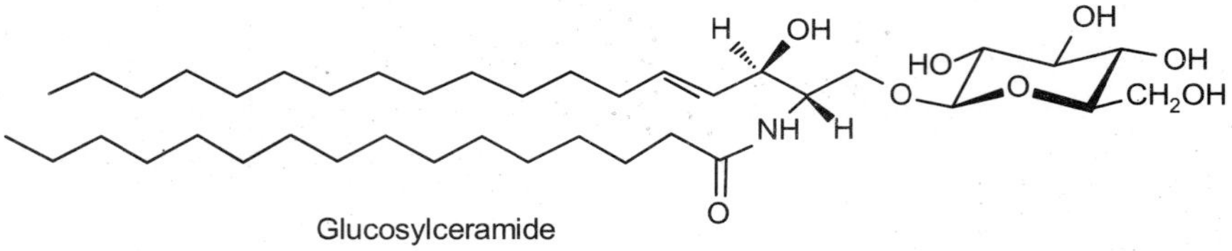

H OH HO OH O O CH_2OH NH H O

Glucosylceramide

O-acyl-glycosylceramides have been detected in small amounts in some tissues, as have cerebrosides with monosaccharides such as xylose, mannose and fucose.

Di-, tri- and tetraglycosylceramides (oligoglycosylceramides) are present also in most animal tissues at low levels. The most common diglycosyl form is **lactosylceramide**, and it can be accompanied by related compounds containing further galactose or galactosamine residues. Tri- and tetraglycosylceramides with a terminal galactosamine residue are sometimes termed "globosides", while glycolipids containing fucose are known as "fucolipids". Lactosylceramide is the biosynthetic precursor of most of these with further monosaccharide residues being added to the end of the carbohydrate chain (up to as many as twenty). They are an important element of the immune response system. For example some glycolipids are involved in the antigenicity of blood group determinants, while others bind to specific toxins or bacteria. As the complex glycosyl moiety is considered to be of primary importance in this respect, it has received most attention from investigators. However, certain of these lipids have been found on occasion to have distinctive long-chain base and fatty acid compositions, which enhance their biological activity. Some glycolipids accumulate in persons suffering from rare disease syndromes, characterized by deficiencies in specific enzyme systems related to glycolipid metabolism.

Sulphate esters of galactosylceramide and lactosylceramide (**sulfoglycosphingolipids** - often referred to as "sulfatides" or "lipid sulphates"), with the sulphate group linked to position 3 of the galactosyl moiety, are major components of brain lipids and they are found in trace amounts in other tissues.

Complex plant sphingolipids, **phytoglycosphingolipids**, containing glucosamine, glucuronic acid and mannose linked to the ceramide via phosphorylinositol, were isolated and characterized from seeds initially, but related compounds are also known to be present in other plant tissues and in fungi.

Gangliosides are highly complex oligoglycosylceramides, which contain one or more sialic acid

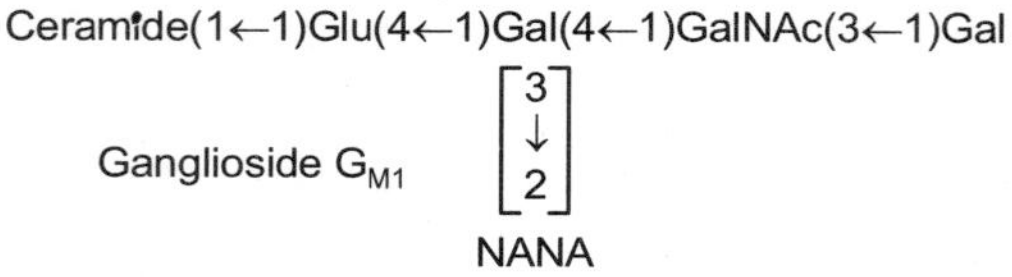

Ceramide(1←1)Glu(4←1)Gal(4←1)GalNAc(3←1)Gal

Ganglioside G_{M1} [3 ↓ 2]

NANA

groups (*N*-acyl, especially acetyl, derivatives of neuraminic acid, abbreviated to "NANA") in addition to glucose, galactose and galactosamine.

The polar and ionic nature of these lipids renders them soluble in water (contrary to some definitions of a lipid). They were first found in the ganglion cells of the central nervous system, hence the name, but are now known to be present in most animal tissues. The long-chain base and fatty acid components of gangliosides can vary markedly between tissues and species, and they are presumably related in some way to function. Gangliosides have been shown to control growth and differentiation of cells, and they have important roles in the immune defense systems. They act as receptors for a number of tissue metabolites and in this way may regulate cell signalling. Also, they bind specifically to various bacterial toxins, such as those from botulinum, tetanus and cholera. A number of unpleasant lipidoses have been identified involving storage of excessive amounts of gangliosides in tissues, the most important of which is Tay-Sachs disease.

Sphingosine-1-phosphate is one of the simplest sphingolipids structurally. It is present at low levels only in animal tissues, but it is a pivotal lipid in many cellular signalling pathways (together with ceramide and ceramide-1-phosphate). For example, within cells, sphingosine-1-phosphate promotes cellular division (mitosis), while in the blood it may play a critical role in platelet aggregation and thrombosis.

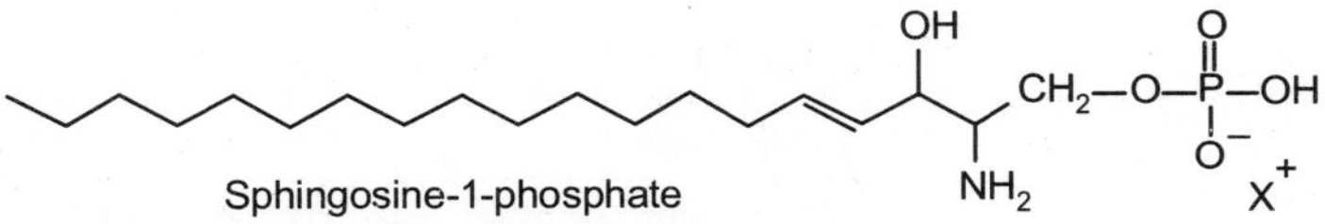

OH O CH_2—O—P—OH NH_2 O^- X^+

Sphingosine-1-phosphate

The fatty acids of sphingolipids: Although structures of fatty acids are discussed elsewhere, it is worth noting that the acyl groups of ceramides are very different from those in the glycerolipids. They tend to consist of long-chain (C_{16} up to C_{26} but occasionally longer) odd- and even-numbered saturated or monoenoic fatty acids and related 2-D-hydroxy fatty acids, both in plant and animal tissues. Linoleic acid may be present at low levels in sphingolipids from animal tissues, but polyunsaturated compounds are rarely found (although their presence is often reported in error).

Fatty Acids

The common fatty acids of plant tissues are C_{16} and C_{18} straight-chain compounds with zero to three double bonds of a *cis* (or *Z*) configuration. Such fatty acids are also abundant in animal tissues, together with other even numbered components with a somewhat wider range of chain-lengths and up to six *cis* double bonds separated by methylene groups (methylene-interrupted). The systematic and trivial names of those fatty acids encountered most often, together with their shorthand designations, are listed in table 2.14.

$$HOOC\,(CH_2)_{14}CH_3$$

palmitic acid

The most abundant saturated fatty acid in nature is **hexadecanoic** or palmitic acid. It can also be designated a "16:0" fatty acid, the first numerals denoting the number of carbon atoms in the aliphatic chain and the second, after the colon, denoting the number of double bonds. All the even-numbered saturated fatty acids from C_2 to C_{30} have been found innature, but only the C_{14} to C_{18} homologues are likely to be encountered in appreciable concentrations in glycerolipids, other than in a restricted range of commercial fats and oils.

Table 2.14 : The common fatty acids of animal and plant origin

Systematic name	Trivial name	Shorthand
Saturated fatty acids		
ethanoic	acetic	2:0
butanoic	butyric	4:0
hexanoic	caproic	6:0
octanoic	caprylic	8:0
decanoic	capric	10:0
dodecanoic	lauric	12:0
tetradecanoic	myristic	14:0
hexadecanoic	palmitic	16:0
octadecanoic	stearic	18:0
eicosanoic	arachidic	20:0
docosanoic	behenic	22:0
Monoenoic fatty acids		
cis-9-hexadecenoic	palmitoleic	16:1(n-7)
cis-6-octadecenoic	petroselinic	18:1(n-12)
cis-9-octadecenoic	oleic	18:1(n-9)
cis-11-octadecenoic	*cis-vaccenic*	18:1(n-7)
cis-13-docosenoic	erucic	22:1(n-9)
cis-15-tetracosenoic	nervonic	24:1(n-9)
Polyunsaturated fatty acids*		
9,12-octadecadienoic	linoleic	18:2(n-6)
6,9,12-octadecatrienoic	γ-linolenic	18:3(n-6)
9,12,15-octadecatrienoic	α-linolenic	18:3(n-3)
5,8,11,14-eicosatetraenoic	arachidonic	20:4(n-6)
5,8,11,14,17-eicosapentaenoic	EPA	20:5(n-3)
4,7,10,13,16,19-docosahexaenoic	DHA	22:6(n-3)

* all the double bonds are of the *cis* configuration

Oleic or ***cis*-9-octadecenoic** acid, the most abundant monoenoic fatty acid in nature, is designated as "18:1", or more precisely as "18:1(*n*-9)", to indicate that the last double bond is 9 carbon atoms from the terminal methyl group.

HOOC

Oleic acid

The latter form of the nomenclature is of special value to biochemists. Similarly, the most abundant *cis* monoenoic acids fall into the same range of chain-lengths, i.e. 16:1(*n*-7) and 18:1(*n*-9), though 20:1 and 22:1 are abundant in fish. Fatty acids with double bonds of the *trans* (or *E*) configuration are found occasionally in natural lipids, or are formed during food processing (hydrogenation) and so enter the food chain, but they tend to be minor components only of animal tissue lipids, other than of ruminants. Their suitability for human nutrition is currently a controversial subject.

The C_{18} polyunsaturated fatty acids, **linoleic** or *cis*-9,*cis*-12-octadecadienoic acid (18:2(*n*-6)) and **α-linolenic** or *cis*-9,*cis*-12,*cis*-15-octadecatrienoic acid (18:3(*n*-3)), are major components of most plant lipids, including many of the commercially important vegetable oils.

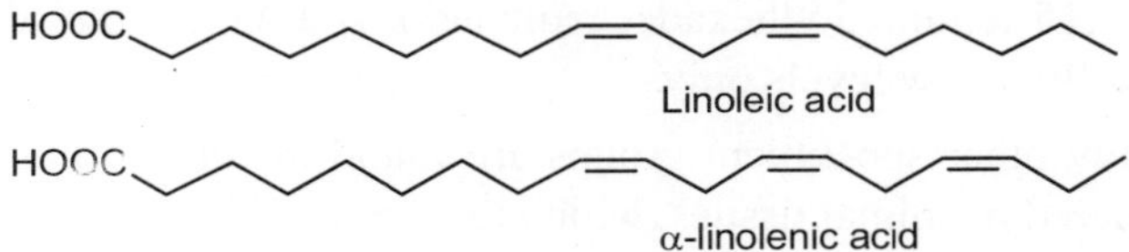

Linoleic acid

α-linolenic acid

They are essential fatty acids in that they cannot be synthesised in animal tissues. On the other hand, as linoleic acid is almost always present in foods, it tends to be relatively abundant in animal tissues. In turn, these fatty acids are the biosynthetic precursors in animal systems of C_{20} and C_{22} polyunsaturated fatty acids, with three to six double bonds, via sequential desaturation and chain-elongation steps (desaturases in animal tissues can only insert a double bond on the carboxyl side of an existing double bond). Those fatty acids derived from linoleic acid, especially arachidonic acid (20:4(*n*-6)), are important constituents of the membrane phospholipids in mammalian tissues, and are also the precursors of the prostaglandins and other eicosanoids. In fish, linolenic acid is the more important essential fatty acid, and polyunsaturated fatty acids of the (*n*-3) series, especially eicosapentaenoic acid (20:5(n-3) or EPA) and docosahexaenoic acid (22:6(n-3) or DHA), are found in greater abundance.

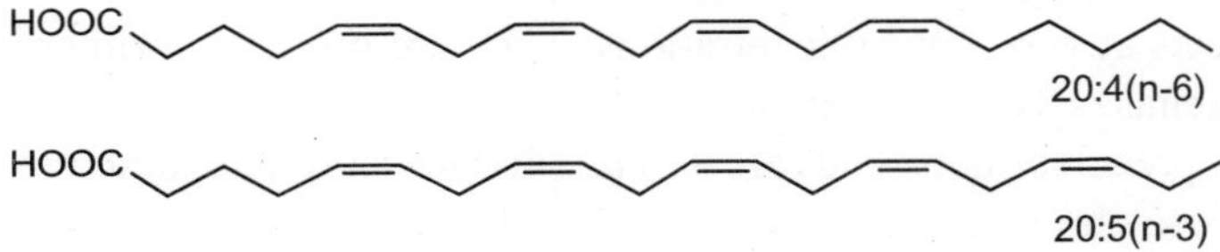

20:4(n-6)

20:5(n-3)

Many other fatty acids that are important for nutrition and health do of course exist in nature, and at present there is great interest in γ-linolenic acid (18:3(*n*-6)), the active constituent of evening primrose oil -

HOOC

γ-linolenic acid

- and in conjugated linoleic acid (mainly, 9-*cis*,11-*trans*-octadecadienoate) or ‘CLA’, a natural constituent of dairy products, that is claimed to have remarkable health-giving properties.

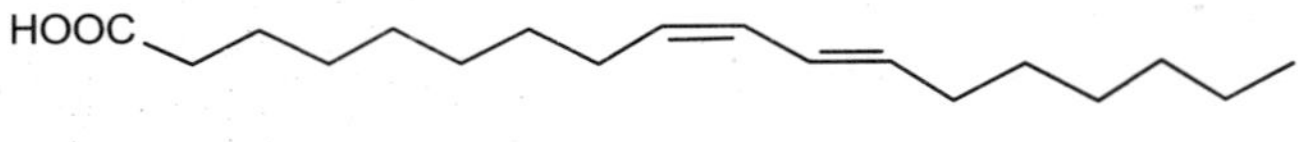

HOOC

9-*cis*,11-trans-octadecadienoic acid

Branched-chain fatty acids are synthesised by many microorganisms (most often with an *iso*- or an *anteiso*-methyl branch). and they are synthesised to a limited extent in higher organisms. They can also enter animal tissues via the diet, especially those of ruminants.

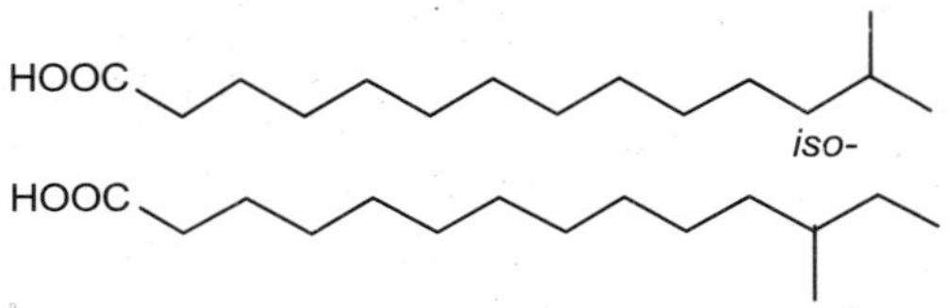

HOOC

iso-

HOOC

anteiso-

Phytanic acid, 3,7,11,15-tetramethylhexadecanoic acid, is a metabolite of phytol and is found in animal tissues, but generally at low levels only.

Fatty acids with many other substituent groups are found in certain plants and microorganisms, and they may be encountered in animal tissues, which they enter via the food chain. These substituents include acetylenic and conjugated double bonds, allenic groups, cyclopropane, cyclopropene, cyclopentene and furan rings, and hydroxy-, epoxy- and keto-groups. For example, 2-hydroxy fatty acids are synthesised in animal and plant tissues, and are often major constituents of the sphingolipids. 12-Hydroxy-octadec-9-enoic or ‘ricinoleic’ acid is the main constituent of castor oil.

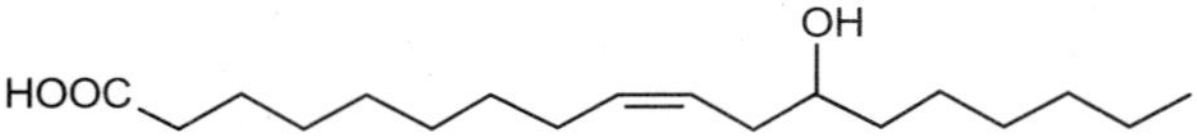

OH

HOOC

Ricinoleic acid

Eicosanoids and Related Lipids

The term eicosanoid is used to embrace biologically active lipid mediators (C_{20} fatty acids and their metabolites), including prostaglandins, thromboxanes, leukotrienes and other oxygenated derivatives, which exert their effects at very low concentrations. They are produced primarily by three classes of enzymes, cyclooxygenases (COX-1 and COX 2), lipoxygenases (LOX) and cytochrome P450 epoxygenase. The key precursor fatty acids are 8c,11c,14c-eicosatrienoic (dihomo-?-linolenic or 20:3(n-6)), 5c,8c,11c,14c-eicosatetraenoic (arachidonic or 20:4(n-6)) and 5c,8c,11c,14c,17c-eicosapentaenoic (20:5(n-3) or EPA) acids. More recently docosanoids (resolvins and protectins) derived from 4c,7c,10c,13c,16c,19c-docosahexaenoic acid (22:6(n 3) or DHA) have been described. Other eicosanoids are produced by non-enzymic means (isoprostanes).

Those derived from arachidonic acid appear to be of special importance and have been most studied. The prostaglandins and thromboxanes have cyclic structures, generated by cyclo-oxygenase enzymes, and are involved in the processes of inflammation. The hydroxy-eicosatetraenoic acids are generated by lipoxygenases, and of these the 5-lipoxygenase is especially important as it produces the first intermediate in the biosynthesis of leukotrienes. The resolvins and protectins have anti-inflammatory properties.

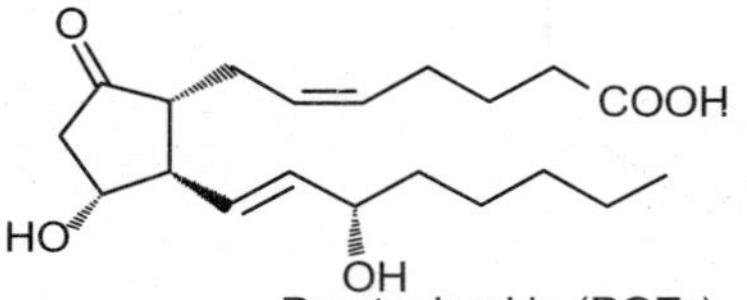

Prostagland in (PGE_2)

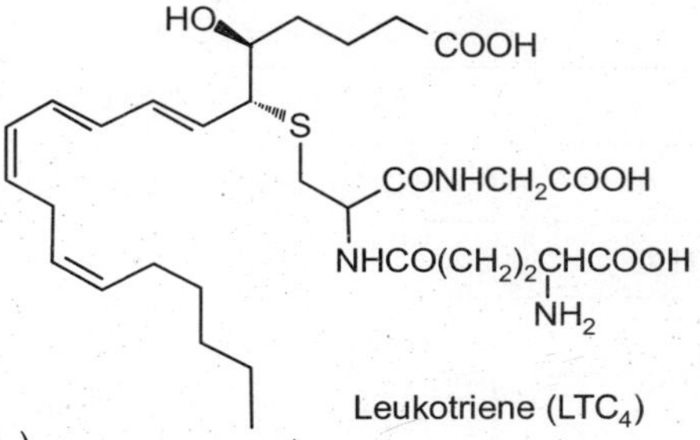

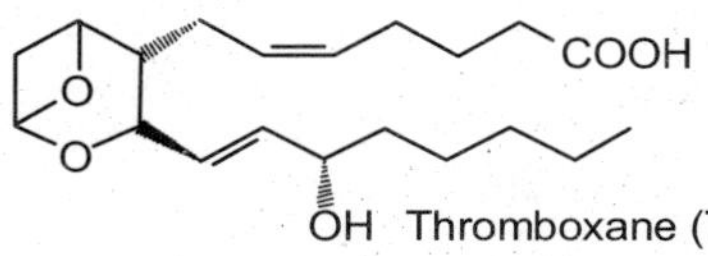

Leukotriene (LTC_4)

Thromboxane (TXA_2)

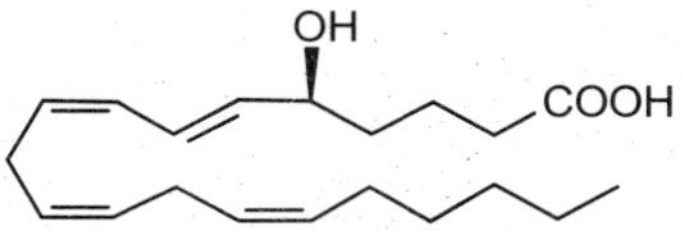

5-hydroxy-eico satetraenoic acid (5-HETE)

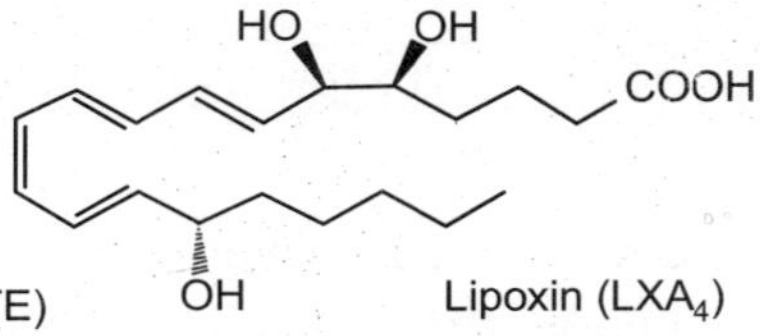

Lipoxin (LXA_4)

Plant products, such as the jasmonates and other oxylipins derived from 9c,12c,15c-octadecatrienoic (1-linolenic or 18:3(n-3)) are also generated by the action of lipoxygenases. They are involved in responses to physical damage by animals or insects, stress and attack by pathogens. There are obvious structural similarities between the jasmonates and prostanoids.

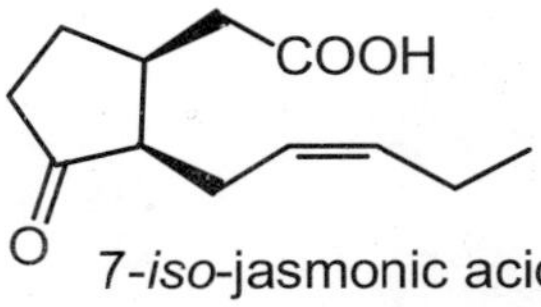

7-*iso*-jasmonic acid

NOMENCLATURE OF SATURATED AND UNSATURATED FATTY ACIDS

Fatty acids are merely carboxylic acids with long hydrocarbon chains. The hydrocarbon chain length may vary from 10-30 carbons (most usual is 12-18). The non-polar hydrocarbon alkane chain is an important counter balance to the polar acid functional group. In acids with only a few carbons, the acid functional group dominates and gives the whole molecule a polar character. However, in fatty acids, the non-polar hydrocarbon chain gives the molecule a non- polar character.

The most common fatty acids are listed. Note that there are two groups of fatty acids—saturated and unsaturated. Recall that the term **unsaturated** refers to the presence of one or more double bonds between carbons as in alkenes. A **saturated fatty acid** has all bonding positions between carbons occupied by hydrogens.

Table 2.15

Acid Name	Structure	Melt Point
SATURATED		
Lauric	$CH_3(CH_2)_{10}COOH$	+44
Palmitic	$CH_3(CH_2)_{14}COOH$	+63
Stearic	$CH_3(CH_2)_{16}COOH$	+70
UNSATURATED		
Oleic	$CH_3(CH_2)_7CH{=}CH(CH_2)_7COOH$	+16
Linoleic	$CH_3(CH_2)_4(CH{=}CHCH_2)_2(CH_2)_6COOH$	-5
Linolenic	$CH_3CH_2(CH{=}CHCH_2)_3(CH_2)_6COOH$	-11
Arachi-donic	$CH_3(CH_2)_4(CH{=}CHCH_2)_4(CH_2)_2COOH$	-50

The melting points for the saturated fatty acids follow the boiling point principle observed previously. Melting point principle: **as the molecular weight increases, the melting point increases.** This observed in the series lauric (C12), palmitic (C16), stearic (C18).

Room temperature is 25°C, Lauric acid which melts at 44° is still a solid, while arachidonic acid has long since melted at -50°, so it is a liquid at room temperature.

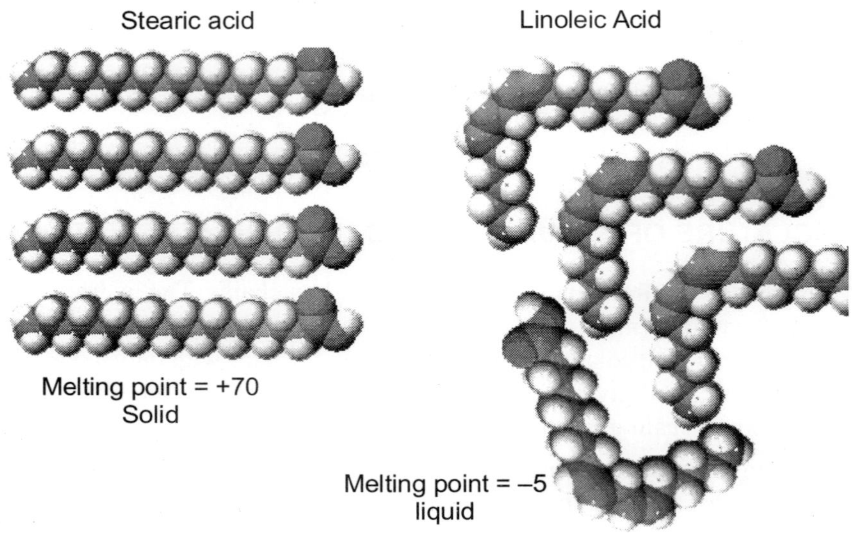

Fig. 2.20 : The Fatty Acids

-Melting Points of Saturated vs. Unsaturated Fatty Acids

The **unsaturated fatty acids have lower melting points than the saturated fatty acids**.

The reason for this phenomenon can be found by a careful consideration of molecular geometries. The tetrahedral bond angles on carbon results in a molecular geometry for saturated fatty acids that is relatively linear although with zigzags.

This molecular structure allows many fatty acid molecules to be rather closely "stacked" together. As a result, close intermolecular interactions result in relatively high melting points.

On the other hand, the introduction of one or more double bonds in the hydrocarbon chain in unsaturated fatty acids results in one or more "bends" in the molecule. The geometry of the double bond is almost always a cis configuration in natural fatty acids. These molecules do not "stack" very well. The intermolecular interactions are much weaker than saturated molecules. As a result, the melting points are much lower for unsaturated fatty acids.

Saturated Fatty Acids

They have commonly straight chains and even carbon number (4-30). They have the general formula: $CH_3(CH_2)_nCOOH$.

They are named from from the saturated hydrocarbon with the same number of carbon atoms, the final -e is changed to **-oic**. For example, the fatty acid with 18 carbon atoms is correctly termed octadecanoic acid but it has also a trivial name (as several common fatty acids), i.e. stearic acid. This compound may be defined also 18:0.

Table 2.16 : A List of the Most Common Saturated Fatty Acids

Systematic name	Trivial name	Shorthand designation	Molecular wt.	Melting point (°C)
1	2	3	4	5
butanoic	butyric	4:0	88.1	-7.9
pentanoic	valeric	5:0		
hexanoic	caproic	6:0	116.1	-3.4
octanoic	caprylic	8:0	144.2	16.7
nonanoic	pelargonic	9:0	158.2	12.5
decanoic	capric	10:0	172.3	31.6
dodecanoic	lauric	12:0	200.3	44.2
tetradecanoic	myristic	14:0	228.4	53.9

...(Contd.)

1	2	3	4	5
hexadecanoic	palmitic	16:0	256.4	63.1
heptadecanoic	margaric (daturic)	17:0	270.4	61.3
octadecanoic	stearic	18:0	284.4	69.6
eicosanoic	arachidic	20:0	312.5	75.3
docosanoic	behenic	22:0	340.5	79.9
tetracosanoic	lignoceric	24:0	368.6	84.2
hexacosanoic	cerotic	26:0	396.7	88
heptacosanoic	carboceric	27:0	410.7	
octacosanoic	montanic	28:0	424.8	
triacontanoic	melissic	30:0	452.9	
dotriacontanoic	lacceroic	32:0	481	
tritriacontanoic	ceromelissic (psyllic)	33:0	495	
tetratriacontanoic	geddic	34:0	509.1	
pentatriacontanoic	ceroplastic	35:0	523.1	

Solution Properties

Normal fatty acids exhibit appreciable solubility in water compared to the corresponding hydrocarbons due to the presence of the polar carboxyl group. The first members of the saturated fatty acid series are miscible with water in all proportions at room temperature.

The solubility behaviour of the fatty acids in organic solvents is of considerable theoretical and industrial importance. Solubility data for the most common saturated fatty acids are given in the table below (in grams per liter at 20°C).

Table 2.17 : Solubility in water at 20°C (in grams acid per liter)

Carbon number	Solubility
2	infinite
4	infinite
6	9.7
8	0.7
10	0.15
12	0.055
14	0.02
16	0.007
18	0.003

On the basis of solubility data, it can be concluded that the normal saturated fatty acids are generally more soluble in chloroform and less soluble in acetonitrile than in any of the organic solvents investigated.

Up to 6 (or 4) carbon atoms, organic acids are considered "short-chain organic acids", they have substantial solubility in water. Furthermore, they do not behave physiologically like other fatty acids since they are more rapidly digested and absorbed in the intestinal tract and have unique properties in regulating sodium and water absorption through the mucosal epithelium.

Table 2.18

Carbon number	Chloroform	Benzene	Cyclohexane	Acetone	Ethanol 95%	Acetic acid	Methanol	Acetonitrile
10	3260	3980	3420	4070	4400	5670	5100	660
12	830	936	680	605	912	818	1200	76
14	325	292	215	159	189	102	173	18
16	151	73	65	53.8	49.3	21.4	37	4
18	60	24.6	24	15.4	11.3	1.2	1	<1

Biochemically, they are more closely related to carbohydrates than to fats.

From 8 (or 6) to 10 (or 12) carbon atoms, fatty acids are said to have a medium chain. Physiological studies have shown that ingestion of triglycerides containing these medium-chain fatty acids may result, as for short-chain fatty acids, in increased energy expenditure via faster satiety. Thus, they facilitate weight control when included in the diet as a replacement for long-chain triglycerides Fatty acids which have 14 (or 12) and more carbon atoms are considered as long-chain fatty acids. Fatty acids with 4 to 12 carbon atoms are found mainly in milk fats (mainly butyric acid in cow and decanoic acid in sheep) but those with 10 and 12 carbon atoms are found also in certain seed oils such as coconut and other kernel fats of the palm family.

- **Butyric acid** (4:0) is the lowest member of the acetic acid series found in natural fats. It occurs (2 to 4%) as a component of milk fats. It gives a rancid odor to butter when triglycerides are hydrolyzed and is present in fermentation products of carbohydrates. This fatty acid has peculiar physiological properties in causing growth arrest and apoptosis in various cell types (*Urbano A et al., Leukemia 1998, 12, 930*). It was tested in the therapy of solid tumors or leukemia (*Kasukabe T et al., Br J Cancer 1997, 75, 850*).
- **Valeric acid** (5:0) has been identified in petroleum distillates and in oxidation products of oils and fats and fermentation of carbohydrates. It has a putrid odor.
- **Caproic acid** (6:0) occurs in milk fats to the extent of about 2%. It was first isolated from butter in 1816 by Chevreul. It has a characteristic odor of goats, hence its name (from the Latin *caper*, goat). Caproic acid is present as glucose ester in leaf trichomes of *Datura metel*.
- **Caprylic acid** (8:0) is widely distributed in animal and vegetable fats but rarely exceeding 8% of the total fatty acids, except in the seed oils of two Lythraceae, *Cuphea hookerina* and *C. painteri*, which contain about 70% caprylic acid (*Miller RW et al., JAOCS 1964, 41, 279*). It occurs to an extent of 1 to 4% in milk fats, and 6 to 8% in coconut and palm oils.
- **Pelargonic acid** (9:0) is the first example of the occurrence of an odd-numbered carbon fatty acid in natural products. It occurs in secretion of sebaceous glands and in essential oil of *Pelargonium roseum* from which it derives its name. It is also a primary product of oxidative fission of oleic acid.

- **Capric acid** (10:0) occurs as a minor component in the same fats that contain caprylic acid but also in the head oil of the sperm whale, and in wool and hair fats. It is a major constituent of elm seed oil (over 60% in *Ulmus americana* and over 70% in *Zelkova serrata)* but is absent in other Ulmaceae (*Apanthe, Morus*).
- **Lauric acid** (12:0) is one of the three most widely distributed saturated fatty acids found in nature. It occurs extensively in *Lauraceae* seeds (*Laurus nobilis*) where it was discovered (*Marsson T Ann 1842*). It is dominant in cinnamon oil (80-90%), coconut oil (40-60% as trilaurin) and is found also in *Cuphea* species (*Umbelliferae*) whose production was initiated in Germany. The recent uses of lauric acid are in the manufacture of flavourings, cocoa butter, margarine, alkyd resins, soaps, shampoos and other surface active agents, including special lubricants. Lauric acid as monoglyceride is known to the pharmaceutical industry for its good antimicrobial properties. It may play a role in combating lipid-coated RNA and DNA viruses. The major sources of lauric acid for human food are palm kernel, coconut and palm.
- **Myristic acid** (14:0) is present in major amounts in seeds of the family *Myristicaceae* (nutmeg oil - or oil of mace - from *Myristica fragrans* contains about 60-70% of trimyristin) where it was first discovered (*Playfair L Ann 1841*). Nutmeg is found in Moluccas and spice islands of Indonesia. Coconut and palm kernel are also convenient sources of 14:0 (trimyristine) which may be isolated in a pure form by distillation. It is also present in milk fats (8-12%) and in the head oil of the sperm whale (15%).
- **Palmitic acid** (16:0) is the commonest saturated fatty acids in plant and animal lipids.

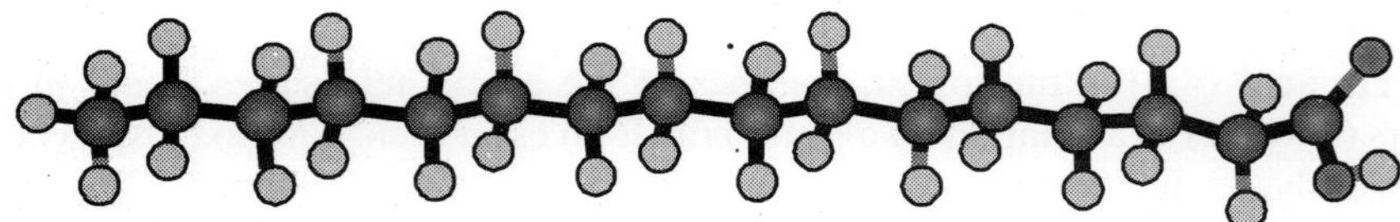

Fig. 2.21

It was purified first by **Chevreul** in his researches on butter and tallow, but was first surely characterized by Fremy E (*Ann 1840.*), who prepared it in pure form from palm oil, from which he named it. Despite its wide distribution, it is generally not present in fats in very large proportions. It usually forms less than 5% of the total fatty acids, sometimes as much as 10% in common vegetal oils (peanut, soybean, corn, coconut) and in marine-animal oils. Lard, tallow, cocoa butter palm oil contain 25 to 40% of this component.

- **Stearic acid** (18:0) was described by **Chevreul** (1823) in the course of his researches on fats. It is the highest molecular weight saturated fatty acid occurring abundantly in fats and oils. It occurs in small quantities in seed and marine oils. Milk fats (5-15%), lard (10%), tallow (15-30%), cocoa and shea butters ((30-35%) are the richest sources of stearic acid. It is the principal constituent of hydrogenated fats and oils (about 90%).

The longer chains are less frequent, they can be found in uncommon seed oils (C20-24 in *Leguminoseae* and *Sapindaceae*), in palm oil, in waxes (C24-30) and in some sphingolipids (C20-24).

Long-chain saturated fatty acids (from C24 to C28) are produced by microalgae and it was estimated that diatoms contribute from 30 to 80% of these components in sandy sediments. These long-chain fatty acids derive from higher plant waxes and are more abundant in deep than in surface sediments.

- **Arachidic acid** (20:0) occurs in appreciable quantities in groundnut (*Arachis hypogea)* oil (3%) where it was discovered in 1854 by Gössmann A (*Ann Chemie 1854, 89, 1*). Larger amounts are found in seeds of *Sapindaceae* (up to 20%). It is also found in the depot fat of some animals and in milk fats.
- **Behenic acid** (22:0) was first reported as a constituent of ben (behen) oil (seeds of *Moringa oleifera*). Except for the seed oils of the *Cruciferae* (between 0.5 and 3.4%), this fatty chain does not occur in the principal oils. Large amounts are found in hydrogenated animal and vegetal oils (8-57%).
- **Lignoceric acid** (24:0) is present at trace levels in plant oils except in groundnut oil (about 1%) and notably in a Leguminous seed oil (*Adenanthera pavonina*) where it may amount to about 25%.It is the principal fatty acid present in carnauba wax (30% of the normal fatty acids). A major source is rice-wax bran (about 40%). Without double bonds or other functional groups, these fatty acids are nearly chemically inert and thus can be subjected to drastic chemical conditions (temperature, oxidation).

Saturated fatty acids were shown to be the major constituents of **adipocere**, the white and soap-like decomposition product which forms due to the post-mortem conversion of body adipose tissue. Immediately following death, triglycerides are hydrolyzed into free fatty acids and glycerol. The free fatty acids (mainly myristic, palmitic, and stearic acids) present in characteristic relations are formed by hydrogenation of triglyceride components under suitable environmental conditions. During that conversion process a number of byproducts may be formed, such as hydroxy or keto fatty acid derivatives. The occurrence of salts of these saturated fatty acids has been suggested as resulting from reaction with the surrounding mineral environment. Saturated fatty acids with straight chain have been found in a number of sediments ranging in age from Precambrian to Recent. In most sediments, fatty acids with even-carbon chain are more abundant than those with odd-carbon chain. All fatty acids from C8 to C28 have been found in sediments. Experiments suggest that normal paraffins in petroleum may be produced from normal fatty acids of longer chain lengths by decarboxylation or other chemical reactions.

Branched-chain Fatty Acids

Branched chain fatty acids are found most frequently with an unsubstituted carbon chain (some branched polyunsaturated fatty acids are found in sponges) but may have one or several branched methyl groups.

Mono or multibranched chain fatty acids

Branched chain fatty acids (mono- branched) may have also a **methoxy** or a **hydroxy** substitution, they are found in exotic animals or bacteria.

Branched methoxy fatty acids and Branched hydroxy fatty acids (Mycolic acids)

Mono- and multibranched fatty acids as for hydrocarbons, they have usually either an iso-structure (methyl group at the penultimate carbon atom) or a anteiso-structure (methyl group on the third carbon from the end). Examples: 14-methyl pentadecanoic acid (isopalmitic) is of the iso-series and 13-methyl pentadecanoic acid is the corresponding anteiso-acid.

Monomethyl branched fatty acids are found in vegetals, animal, and microbial lipids but in small concentrations. In animals, some classic examples of these compounds include the 2- and 4-monomethylated fatty acids from the uropygial gland of ducks In vegetals, 14-methyl-16:0 has been identified in *Ginkgo biloba* and was found to be characteristic of pine seed oil (up to 1%). This fatty acid was found exclusively in *Pinaceae* (genera *Pinus, Abies, Cedrus, Picea*). The unsaturated 11-methyloctadec-12-enoic and 12-methyloctadec-10-enoic acids were identified in the seed oil of *Byrsocarpus coccineus* (Connaraceae). A *trans*-monounsaturated branched-chain fatty acid, 8-methyl-*trans*-6-nonenoic acid, is characteristic of specific compounds, the capsaicinoids, found in fruits of the genus Capsicum (Solanaceae). These plants (bell pepper, chili pepper) are among the oldest cultivated plants, their pungent fruits being used as spices for over 6000 years. Capsaicinoids are synthesized by an enzymatic condensation of vanillylamine and a medium chain branched acid. More than 20 compounds, different only in the fatty acid structures, have been described.The sponges contain also large quantities of C14 up to C30 fatty acids with branch as well as odd-chains. As an example, a new structure (20-methyl-26:0) has been elucidated in the sponge *Verongia aerophoba* from the Canary islands while some monomethyl polyunsaturated fatty acids were described in different marine sponges (24-methyl-5,9-pentacosadienoic acid or 2-methoxy-13-methyl-6-tetradecenoic acid) . Similarly, two new 2-methyl branched monoenoic very long chain fatty acids (2-methyl-24:1 n-7 and 2-methyl-26:1 n-9) were described in a marine sponge *Halichondria panicea* (*Imbs AB et al., Chem Phys Lipids 2004, 129, 173*).It must be noticed that the similarity between the composition of the midchain branching pattern of fatty acids in some sponges and in bacteria suggests the presence of bacteria in these sponges. *Calcarea*, the constant and prominent occurrence of iso- and anteiso-fatty acids (>40% of the total fatty acids in most species) suggests an origin of these compounds from the sponge cells rather than from bacterial lipids. Another indication for a sponge cell origin of these compounds is the major presence (7-15 % of the total) of anteiso-nonadecanoic acid (16-methyloctadecanoic acid), an exotic compound that has not yet been reported as a major fatty acid in bacteria. It has been found that *Caenorhabditis elegans* is able to synthesized iso-C15 and iso-C17 and that these branched-chain fatty acids are essential for the animal growth and development.

These results suggest that these fatty acids may play a potentially important role in other eukaryotes.

A new unsaturated methyl-branched fatty acid, 9-methyl-16:1(n-6) and the uncommon 11-methyl-18:1(n-6) were found in the lipid extract of a new strain of bacterium *Vibrio alginolyticus* associated with the alga *Cladophora coelothrix*. Another novel methyl-branched fatty acid, 10-methyl-18:1(n-9), was found in the lipid extract of the marine fungus *Microsphaeropis olivacea*.

Branched methyl-substituted fatty acids of bacterial origin are commonly found in lake or marine sediments, decreasing rapidly with depth. Long-chain monomethyl-branched anteiso acids were also identified in settlings particles and surface sediments from freshwater lakes where they may be useful molecular markers for lake acidity (*Fukushima*. 10-methyl octadecanoic acid (tuberculostearic acid) is present in the phosphatidylinositol moiety of lipoarabinomannans found mainly in *Mycobacterium tuberculosis* but also in other bacteria in the genus *Mycobacterium*. The detection of this fatty acid in cerebrospinal fluid was proposed as a possibility for rapid and specific diagnosis of tuberculous meningitis. The presence of this fatty acid in sputum lipids was successively utilized for the diagnosis of tuberculous pneumonia.

10-methyl nonadecanoic acid (**phytomonic acid**) is also found in *Mycobacterium*.

In animals, 17-methyl-6-octadecenoic and 17-methyl-7-octadecenoic acids were identified in the Australian mollusk *Siphonaria denticulata*.

An unusual complex and polyunsaturated fatty acid substituted with one hydroxyl and one aldehyde group has been described as a new polyene pigment, laetiporic acid, in the wood-rotting basidiomycete *Laetiporus sulphureus* (*Weber RW et al., Tetrahedron lett 2004, 45, 1075*). This orange pigment, with an UV-visible spectrum similar to that of carotenoids, bears an unprecedented decaene skeleton as part of its chromophore.

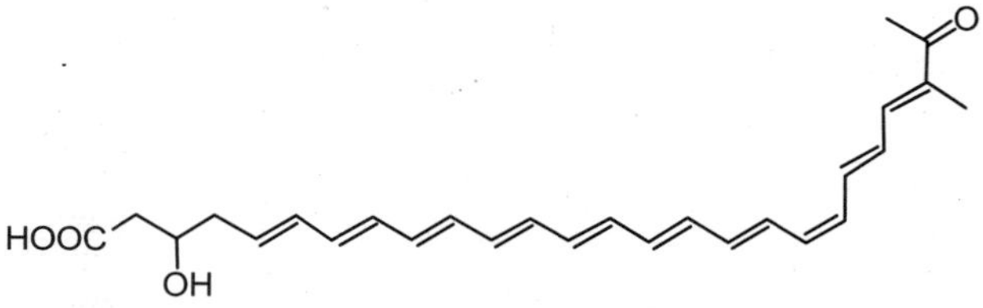

HOOC ... OH

Multimethyl branched acids are found mainly in bacteria.

Several dimethylated fatty acids (14 or 16 carbon atoms) with the first methyl substituent at carbon 2 or 4 have been isolated from a halophilic *Bacillus* species. Di- and tri-methylated fatty acids with 15 to 18 carbon atoms have been isolated from environmental subsurface sediments. These fatty acids may be of value for the knowledge of biomass and of the metabolic status of the viable microbial community. Multimethyl branched acids are abundant in cell wall lipids of *Mycobacteria*, each methyl group being on even carbon atoms (2,4,6,8...from the methyl end). Thus, forming waxes and glycolipids (mycosides), several multibranched fatty acids are commonly found - **mycoceranic** (2,4,6-trimethyloctacosanoic) and **mycolipenic** (2,4,6-trimethyl-trans-2-tetracosenoic) acids- **mycocerosic acid** (2,4,6,8-tetramethyl C32 fatty acid)

- **phthioceranic acids** which are hepta or octamethyl fatty acids, some of them being also hydroxylated (hydroxyphthioceranic acid).

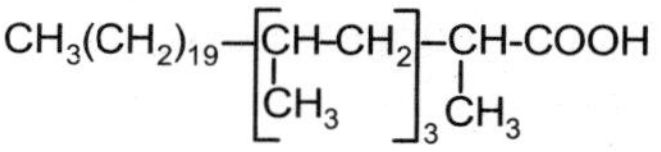

$$CH_3(CH_2)_{19}-[CH(CH_3)-CH_2]_3-CH(CH_3)-COOH$$

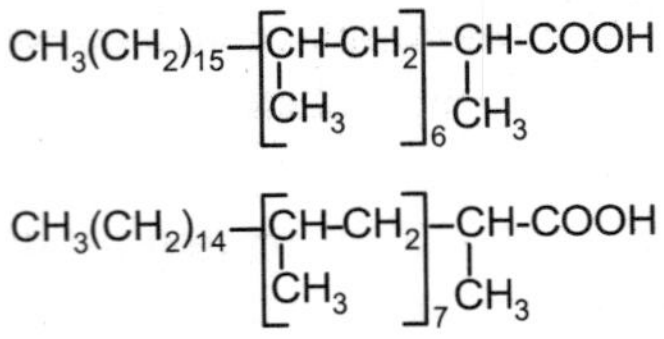

$$CH_3(CH_2)_{15}-[CH(CH_3)-CH_2]_6-CH(CH_3)-COOH$$

$$CH_3(CH_2)_{14}-[CH(CH_3)-CH_2]_7-CH(CH_3)-COOH$$

Branched acids but with shorter chains are also found in the depot fats of ruminant animals, in sebum and animal waxes (wool-wax). Ruminants produced a huge variety of C10-C18 acids with one to four branched methyl groups, they were detected in lamb adipose tissue. Multimethyl branched acids are also dominant in the uropygial waxes of the birds preen gland. Curiously, they were also identified in the lipid depot in the head of marine animals, location involved in their echo-locating abilities.

Unusual branched fatty acids have been isolated as minor components from the glycolipids (GL) fraction of freshwater sponges. It is possible that these neo acids could be of cyanobacterial origin. An example of one of them with the longest carbon chain is shown below.

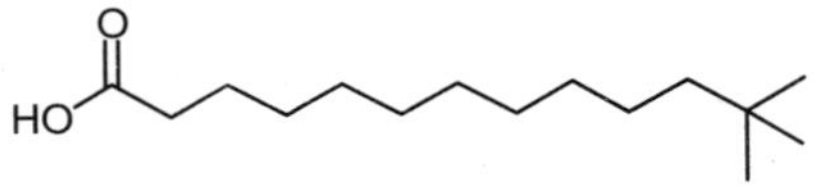

O

HO

Others have isoprenoid structures, thus coming from the diterpene phytol derived from chlorophyll and not by the *de novo* pathway. Among them, two are found in marine organisms, in geological sediments but, one of them (**phytanic acid** or 3,7,11,15-tetramethyl hexadecanoic acid) is present in human diet or in animal tissues where it may be derived from chlorophyll in plant extracts.

Two very unusual phytyl esters were obtained from the extract of the hornwort *Megaceros flagellaris (Bryophyte, Anthocerothae)*. The fatty acid moiety comprises 3,7,11,15-tetramethyl-16:1 or 3,7,11,15-tetramethyl-16:0, which is esterified to the corresponding tetramethyl unsaturated (16:1) alcohol (Phytanic acid derives from the corresponding alcohol, phytol, and is oxidized into **pristanic acid**.

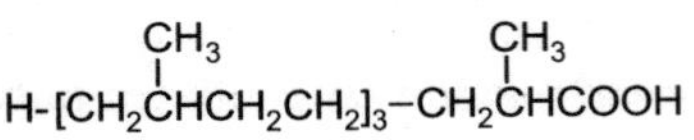

$$H-[CH_2CH(CH_3)CH_2CH_2]_3-CH_2CH(CH_3)COOH$$

Pristanic acid

$$H-[CH_2CH(CH_3)CH_2CH_2]_3-CH_2CH(CH_3)CH_2COOH$$

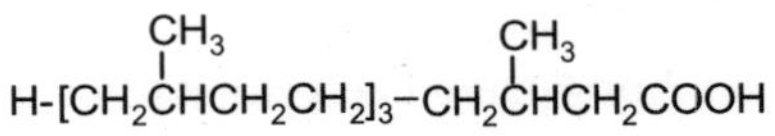

Phytanic acid

Pristanic acid was first isolated from butter (*Hansen RP et al., Biochemical Journal 1964, 93, 225*). The name of the substance is derived from pristane (2,6,10,14-tetramethylpentadecane), the corresponding hydrocarbon which was isolated from shark (*pristis* in Latin). It is also found in the lipids from many sources such as sponges, crustacea, milk fats, animal depot fat but also in petroleum samples.

Phytanic acid characterizes a precise human pathology, the Refsum's syndrome. This inherited neurological disorder is characterized by a accumulation of phytanic acid, normal metabolite of phytol, in blood and tissues. The disorder was later related to deficiency in the a-oxidation pathway in the liver. Both phytanic acid and pristanic acid have been shown to activate the peroxisome proliferator-activated receptor* (PPAR) in a concentration-dependent manner Freshwater sponges contain polymethyl branched fatty acids such as 4,8,12-trimethyltridecanoic, phytanic and pristanic acids. These acids may have chemotaxonomical significance for both marine and freshwater sponges. The isoprenoic 4,8,12-trimethyltridecanoic was found to be always present in the marine calcareous sponges (Calcarea) but in minor amounts, this acid being presumed to be derived from phytol, a degradation product of chlorophyll. Phytanic acid present in immature (recent) sediments is thought to derive from phytol, a supposition backed by stereochemical studies. Few sources, including sponges, contain branched polyunsaturated fatty acids. As an example, freshwater *Demospongia* (*Spongillidae*) were shown to contain di-, tri-, and tetramethyl substituted dienoic, tetraenoic, and hexaenoic fatty acids.

Isoprenoid Fatty Acids

Some isoprene derivatives (sesquiterpenes) synthesized by invertebrates from farnesoic acid have important endocrinological functions (juvenile hormones) such as molting, reproduction and metamorphosis. Addition of isoprene acids to proteins (prenylation), discovered in 1984, is an important post-translational modification of proteins which has been recognized as a key physiological process for facilitating cellular protein-protein interactions and membrane-associated protein trafficking. Protein prenylation occurs by the covalent addition of two types of isoprenoids, farnesyl pyrophosphate (a 15-carbon sesquiterpene) or geranylgeranyl pyrophosphate (a 20-carbon diterpene), to cysteine residues at or near the terminal carboxyl group . The largest family of prenylated proteins are the intracellular GTP-binding proteins that transduce extracellular signals into intracellular changes via downstream effectors Some isoprenoid fatty acids with conjugated double bonds are known. In this group, the most interesting is **retinoic acid** which derives from retinol but is synthetized ultimately from b-carotene (provitamin A). It has important functions in cell regulation.

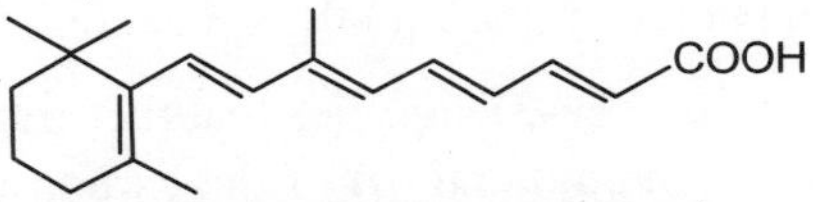

COOH

β-retinoic acid (*all*-trans isomer)

Abscissic acid, a methylated derivative of retinoic acid, plays a variety of roles in plant physiology.It is ubiquitous in plants, including algae.

Branched Methoxy Fatty Acids

Iso or anteiso a-methoxylated C15-C16 fatty acids were reported in phospholipids of a Caribean sponge.

Thus, 2-methoxy-13-methyltetradecanoic acid, 2-methoxy-14-methylpentadecanoic acid, and 2-methoxy-13-methyl pentadecanoic acid were identified by gas-liquid chromatography and mass spectrometry.

It has been suggested that these compounds have originated from a novel bacteria in symbiosis with the sponge.

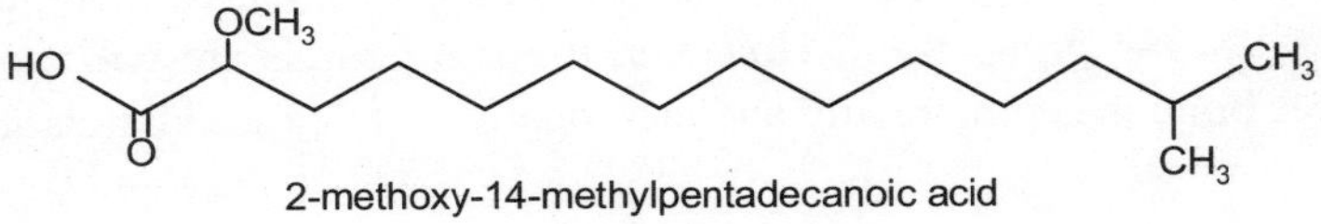

OCH_3 HO CH_3 O CH_3

2-methoxy-14-methylpentadecanoic acid

ESSENTIAL FATTY ACIDS
ω-3 Fatty Acids or Omega-3 Fatty Acids)

***n*–3 fatty acids** (popularly referred to as **ω–3 fatty acids** or **omega-3 fatty acids**) are a family of unsaturated fatty acids that have in common a carbon–carbon double bond in the *n*–3 position; that is, the third bond from the methyl end of the fatty acid.Important nutritionally essential *n*–3 fatty acids are: α-linolenic acid (ALA), eicosapentaenoic acid (EPA), and docosahexaenoic acid (DHA). The human body cannot synthesize *n*–3 fatty acids *de novo*, but it can form 20- and 22-carbon unsaturated *n*–3 fatty acids from the eighteen-carbon *n*–3 fatty acid, α-linolenic acid. These conversions occur competitively with *n*–6 fatty acids, which are essential closely related chemical analogues that are derived from linoleic acid. Both the *n*–3 α-linolenic acid and *n*–6 linoleic acid are essential nutrients which must be obtained from food. Synthesis of the longer *n*–3 fatty acids from linolenic acid within the body is competitively slowed by the *n*–6 analogues. Thus accumulation of long-chain *n*–3 fatty acids in tissues is more effective when they are obtained directly from food or when competing amounts of *n*–6 analogs do not greatly exceed the amounts of *n*–3.

Chemical structure of alpha-linolenic acid (ALA), an essential *n*–3 fatty acid, (18:3Δ9c,12c,15c, which means a chain of 18 carbons with 3 double bonds on carbons numbered 9, 12 and 15). Although chemists count from the carbonyl carbon (blue numbering), physiologists count from the *n* (ω) carbon (red numbering). Note that from the *n* end (diagram right), the first double bond appears as the third carbon-carbon bond (line segment), hence the name "*n*–3". This is explained by the fact that the *n* end is almost never changed during physiologic transformations in the human body, as it is more stable energetically, and other carbohydrates compounds can be synthetized from the other carbonyl end, for example in glycrids, or from double bonds in the middle of the chain.

The term ***n*–3** (also called **ω–3** or **omega-3**) signifies that the first double bond exists as the **third** carbon-carbon bond from the terminal methyl end (*n*) of the carbon chain.

n–3 fatty acids which are important in human nutrition are: α-linolenic acid (18:3, *n*–3; ALA), eicosapentaenoic acid (20:5, *n*–3; EPA), and docosahexaenoic acid (22:6, *n*–3; DHA). These three polyunsaturates have either 3, 5 or 6 double bonds in a carbon chain of 18, 20 or 22 carbon atoms, respectively. All double bonds are in the *cis*-configuration, i.e. the two hydrogen atoms are on the same side of the double bond.

Most naturally-produced fatty acids (created or transformed in animalia or plant cells with an even number of carbon in chains) are in *cis*-configuration where they are more easily transformable. The *trans*-configuration results in much more stable chains that is very difficult to further break or transform, forming longer chains that aggregate in tissues and lacking the necessary hydrophilic properties. This *trans*-configuration can be the result of the transformation in alkaline solutions, or of the action of some bacterias that are shortening the carbonic chains. Natural transforms in vegetal or animal cells more rarely affect the last *n*-3 group itself. However, *n*-3 compounds are still more fragile than *n*′6 because the last double bond is geometrically and electrically more exposed, notably in the natural *cis* configuration.

List of *n*-3 fatty acids

This table lists several different names for the most common *n*–3 fatty acids found in nature.

Table 2.19

Common name	Lipid name	Chemical name
	16:3 (n–3)	all-cis-7,10,13-hexadecatrienoic acid
α-Linolenic acid (ALA)	18:3 (n–3)	all-cis-9,12,15-octadecatrienoic acid
Stearidonic acid (STD)	18:4 (n–3)	all-cis-6,9,12,15-octadecatetraenoic acid
Eicosatrienoic acid (ETE)	20:3 (n–3)	all-cis-11,14,17-eicosatrienoic acid
Eicosatetraenoic acid (ETA)	20:4 (n–3)	all-cis-8,11,14,17-eicosatetraenoic acid
Eicosapentaenoic acid (EPA)	20:5 (n–3)	all-cis-5,8,11,14,17-eicosapentaenoic acid
Docosapentaenoic acid (DPA), Clupanodonic acid	22:5 (n–3)	all-cis-7,10,13,16,19-docosapentaenoic acid
Docosahexaenoic acid (DHA)	22:6 (n–3)	all-cis-4,7,10,13,16,19-docosahexaenoic acid
Tetracosapentaenoic acid	24:5 (n–3)	all-cis-9,12,15,18,21-docosahexaenoic acid
Tetracosahexaenoic acid (Nisinic acid)	24:6 (n–3)	all-cis-6,9,12,15,18,21-tetracosenoic acid

Biological Significances

The biological effects of the n–3 are largely mediated by their interactions with the n-6 fatty acids; see Essential fatty acid interactions for detail.

A 1992 article by biochemist William E.M. Lands provides an overview of the research into *n*-3-fatty acids, and is the basis of this section.

The 'essential' fatty acids were given their name when researchers found that they were essential to normal growth in young children and animals. (Note that the modern definition of 'essential' is more strict.) A small amount of *n*–3 in the diet (~1% of total calories) enabled normal growth, and increasing the amount had little to no additional effect on growth.

Likewise, researchers found that *n*–6 fatty acids (such as γ-linolenic acid and arachidonic acid) play a similar role in normal growth. However, they also found that *n*–6 was "better" at supporting dermal integrity, renal function, and parturition. These preliminary findings led researchers to concentrate their studies on *n*–6, and it was only in recent decades that *n*–3 has become of interest.

In 1963 it was discovered that the *n*–6 arachidonic acid was converted by the body into pro-inflammatory agents called prostaglandins. By 1979 more of what are now known as eicosanoids were discovered: thromboxanes, prostacyclins and the leukotrienes. The eicosanoids, which have important biological functions, typically have a short active lifetime in the body, starting with synthesis from fatty acids and ending with metabolism by enzymes. However, if the rate of synthesis exceeds the rate of metabolism, the excess eicosanoids may have deleterious effects. Researchers found that *n*–3 is also converted into eicosanoids, but at a much slower rate. Eicosanoids made from *n*–3 fats often have opposing functions to those made from *n*–6 fats (i.e., anti-inflammatory rather than inflammatory). If both *n*–3 and *n*–6 are present, they will "compete" to be transformed, so the ratio of *n*–3 : *n*–6 directly affects the type of eicosanoids that are produced.

This competition was recognized as important when it was found that thromboxane is a factor in the clumping of platelets, which leads to thrombosis. The leukotrienes were similarly found to be important in immune and inflammatory-system response, and therefore relevant to arthritis, lupus, and asthma. These discoveries led to greater interest in finding ways to control the synthesis of *n*–6 eicosanoids. The simplest way would be by consuming more *n*–3 and fewer *n*–6 fatty acids.

Health benefits On September 8, 2004, the U.S. Food and Drug Administration gave "qualified health claim" status to eicosapentaenoic acid (EPA) and docosahexaenoic acid (DHA) *n*–3 fatty acids, stating that "supportive but not conclusive research shows that consumption of EPA and DHA [*n*–3] fatty acids may reduce the risk of coronary heart disease.

People with certain circulatory problems, such as varicose veins, benefit from fish oil. Fish oil stimulates blood circulation, increases the breakdown of fibrin, a compound involved in clot and scar formation, and additionally has been shown to reduce blood pressure. There is strong scientific evidence that *n*–3 fatty acids significantly reduce blood triglyceride levels and regular intake reduces the risk of secondary and primary heart attack.

Some benefits have been reported in conditions such as rheumatoid arthritis and cardiac arrhythmias.

There is a promising preliminary evidence that *n*-3 fatty acids supplementation might be helpful in cases of depression and anxiety. Studies report highly significant improvement from *n*-3 fatty acids supplementation alone and in conjunction with medication.

Some research suggests that fish oil intake may reduce the risk of ischemic and thrombotic stroke. However, very large amounts may actually increase the risk of hemorrhagic stroke (see below). Lower amounts are not related to this risk, 3 grams of total EPA/DHA daily are considered safe with no increased risk of bleeding involved and many studies used substantially higher doses without major side effects (for example: 4.4 grams EPA/2.2 grams DHA in 2003 study).

Several studies report possible anti-cancer effects of *n*–3 fatty acids (particularly breast, colon and prostate cancer). No clear conclusion can be drawn at this time, however.

A 2006 report in the Journal of the American Medical Association concluded that their review of literature covering cohorts from many countries with a wide variety of demographic characteristics demonstrating a link between *n*–3 fatty acids and cancer prevention gave mixed results. This is similar to the findings of a review by the British Medical Journal of studies up to February 2002 that failed to find clear effects of long and shorter chain *n*–3 fats on total mortality, combined cardiovascular events and cancer.

In 1999, the GISSI-Prevenzione Investigators reported in the Lancet, the results of major clinical study in 11,324 patients with a recent myocardial infarction. Treatment 1 gram per day of *n*–3 fatty acids reduced the occurrence of death, cardiovascular death and sudden cardiac death by 20%, 30% and 45% respectively. These beneficial effects were seen already from three months onwards.

In April 2006, a team led by Lee Hooper at the University of East Anglia in Norwich, UK, published a review of almost 100 separate studies into *n*-3 fatty acids, found in abundance in oily fish. It concluded that they do not have a significant protective effect against cardiovascular disease. This meta-analysis was controversial and stands in stark contrast with two different reviews also performed in 2006 by the American Journal of Clinical Nutrition and a second JAMA review that both indicated decreases in total mortality and cardiovascular incidents (i.e. myocardial infarctions) associated with the regular consumption of fish and fish oil supplements. In addition *n*–3 has shown to aid in other mental disorders such as aggression and ADHD (Attention-deficit hyperactivity disorder).

Several studies published in 2007 have been more positive. In the March 2007 edition of the journal *Atherosclerosis*, Japanese men with unhealthy blood sugar levels were randomly assigned to receive 1800 mg daily of eicosapentaenoic acid (EPA—an *n*–3 essential fatty acid from fish oil) with the other half being a control group. The thickness of the carotid arteries and certain measures of blood flow were measured before and after supplementation. This went on for approximately two years. A total of 60 patients (30 in the EPA group and 30 in the control group) completed the study. Those given the EPA had a statistically significant decrease in the thickness of the carotid arteries along with improvement in blood flow. The authors indicated that this was the first demonstration that administration of purified EPA improves the thickness of carotid arteries along with improving blood flow in patients with unhealthy blood sugar levels.

In another study published in the American Journal of Health System Pharmacy March 2007, patients with high triglycerides and poor coronary artery health were given 4 grams a day of a combination of EPA and DHA along with some monounsaturated fatty acids. Those patients with very unhealthy triglyceride levels (above 500 mg/dl) reduced their triglycerides on average 45% and their VLDL cholesterol by more than 50%. VLDL is a bad type of cholesterol and elevated triglycerides can also be deleterious for cardiovascular health.

There was another study published on the benefits of EPA in *The Lancet* in March 2007. This study involved over 18,000 patients with unhealthy cholesterol levels. The patients were randomly assigned to receive either 1,800 mg a day of EPA with a statin drug or a statin drug alone. The trial went on for a total of five years. It was found at the end of the study those patients in the EPA group had superior cardiovascular function. Non-fatal coronary events were also significantly reduced in the EPA group. The authors concluded that EPA is a promising treatment for prevention of major coronary events, especially non-fatal coronary events.

Another study regarding fish oil was published in the *Journal of Nutrition* in April 2007. Sixty four healthy Danish infants from nine to twelve months of age received either cow's milk or infant formula alone or with fish oil. It was found that those infants supplemented with fish oil had improvement in immune function maturation with no apparent reduction in immune activation.

There was yet another study on *n*-3 fatty acids published in the April 2007 *Journal of Neuroscience*. A group of mice were genetically modified to develop accumulation of amyloid and tau proteins in the brain similar to that seen in people with poor memory. The mice were divided into four groups with one group receiving a typical American diet (with high ratio of *n*–6 to *n*–3 fatty acids being 10 to 1). The other three groups were given food with a balanced 1 to 1 *n*–6 to *n*–3 ratio and two additional groups supplemented with DHA plus long chain *n*–6 fatty acids. After three months of feeding, all the DHA supplemented groups were noted to have a lower accumulation of beta amyloid and tau protein. Some research suggests that these abnormal proteins may contribute to the development of memory loss in later years.

There is also a study published regarding *n*e3 supplementation in children with learning and behavioural problems. This study was published in the April 2007 edition of the *Journal of the Developmental and Behavioral Pediatrics* , where 132 children, between the ages of seven to twelve years old, with poor learning, participated in a randomized, placebo-controlled, double-blinded interventional trial. A total of 104 children completed the trial. For the first fifteen weeks of this study, the children were given polyunsaturated fatty acids (*n*e3 and *n*e6, 3000 mg a day), polyunsaturated fatty acids plus multi-vitamins and minerals or placebo. After fifteen weeks, all groups crossed over to the polyunsaturated fatty acids (PUFA) plus vitamins and mineral supplement. Parents were asked to rate their children's condition after fifteen and thirty weeks. After thirty weeks, parental ratings of behaviour improved significantly in nine out of fourteen scales. The lead author of the study, Dr. Sinn, indicated the present study is the largest PUFA trial to date with children falling in the poor learning and focus range. The results support those of other studies that have found improvement in poor developmental health with essential fatty acid supplementation.

Research in 2005 and 2006 has suggested that the in-vitro anti-inflammatory activity of *n*e3 acids translates into clinical benefits. Cohorts of neck pain patients and of rheumatoid arthritis sufferers have demonstrated benefits comparable to those receiving standard NSAIDs. Those who follow a Mediterranean-style diet tend to have less heart disease, higher HDL ("good") cholesterol levels and higher proportions of *n*–3 in tissue highly unsaturated fatty acids. Similar to those who follow a Mediterranean diet, Arctic-dwelling Inuit - who consume high amounts of *n*e3 fatty acids from fatty fish - also tend to have higher proportions of *n*–3, increased HDL cholesterol and decreased triglycerides (fatty material that circulates in the blood) and less heart disease. Eating walnuts (the ratio of *n*–3 to *n*–6 is circa 1:4 respectively) was reported to lower total cholesterol by 4% relative to controls when people also ate 27% less cholesterol.

A study examining whether omega-3 exerts neuroprotective action in Parkinson's disease found that it did, using an experimental model, exhibit a protective effect (much like it did for Alzheimer's disease as well). The scientists exposed mice to either a control or a high omega-3 diet from two to twelve months of age and then treated them with a neurotoxin commonly used as an experimental model for Parkinson's. The scientists found that high doses of omega-3 given to the experimental group completely prevented the neurotoxin-induced decrease of dopamine that ordinarily occurs. Since Parkinson's is a disease caused by disruption of the dopamine system, this protective effect exhibited could show promise for future research in the prevention of Parkinson's disease.

A study carried out involving 465 women showed serum levels of eicosapentaenoic acid is inversely related to the levels of anti-oxidized-LDL antibodies. Oxidative modification of LDL is thought to play an important role in the development of atherosclerosis.

In 2008 a German study suggested that Omega-3 fatty acids have a positive effect on atopic dermatitis.

Health Risks

In a letter published October 31, 2000, the United States Food and Drug Administration Center for Food Safety and Applied Nutrition, Office of Nutritional Products, Labeling, and Dietary Supplements noted that known or suspected risks of EPA and DHA *n*-3 fatty acids may include the possibility of:

- Increased bleeding if overused (normally over 3 grams per day) by a patient who is also taking aspirin or warfarin. However, this is disputed.
- Hemorrhagic stroke (only in case of very large doses).
- Oxidation of *n*–3 fatty acids forming biologically active oxidation products.
- Reduced glycemic control among diabetics.
- Suppression of immune and inflammation responses, and consequently, decreased resistance to infections and increased susceptibility to opportunistic bacteria.
- An increase in concentration of LDL cholesterol in some individuals.

Subsequent advices from the FDA and national counterparts have permitted health claims associated with heart health.

Cardiac Risk

Persons with congestive heart failure, chronic recurrent angina or evidence that their heart is receiving insufficient blood flow are advised to talk to their doctor before taking *n*-3 fatty acids. There have been concerns if such persons take *n*-3 fatty acids or eating foods that contain them in substantial amounts In a recent large study, *n*-3 fatty acids on top of standard heart failure therapy produced a small but statistically significant benefit in terms of mortality and hospitalization.

In congestive heart failure, cells that are only barely receiving enough blood flow become electrically hyperexcitable. This, in turn, can lead to increased risk of irregular heartbeats, which, in turn, can cause sudden cardiac death. *n*-3 fatty acids seem to stabilize the rhythm of the heart by effectively preventing these hyperexcitable cells from functioning, thereby reducing the likelihood of irregular heartbeats and sudden cardiac death. For most people, this is obviously beneficial and would account for most of the large reduction in the likelihood of sudden cardiac death. Nevertheless, for people with congestive heart failure, the heart is barely pumping blood well enough to keep them alive. In these patients, *n*-3 fatty acids may eliminate enough of these few pumping cells that the heart would no longer be able to pump sufficient blood to live, causing an increased risk of cardiac death.

Dietary sources and Daily valuesAs macronutrients, fats are not assigned recommended daily allowances. Macronutrients have AI (Acceptable Intake) and AMDR (Acceptable Macronutrient Distribution Range) instead of RDAs. The AI for *n*-3 is 1.6 grams/day for men and 1.1 grams/day for women while the AMDR is 0.6% to 1.2% of total energy.

"A growing body of literature suggests that higher intakes of α-linolenic acid (ALA), eicosapentaenoic acid (EPA), and docosahexaenoic acid (DHA) may afford some degree of protection against coronary heart disease. Because the physiological potency of EPA and DHA is much greater than that for α-linolenic acid, it is not possible to estimate one AMDR for all *n*-3 fatty acids. Approximately 10 percent of the AMDR can be consumed as EPA and/or DHA." There was insufficient evidence as of 2005 to set a UL (upper tolerable limit) for *n*-3 fatty acids.

A perceived risk of fish oil *n*-3 supplementation has been heavy metal poisoning by the body's accumulation of traces of heavy metals, in particular mercury, lead, nickel, arsenic and cadmium as well as other contaminants (PCBs, furans, dioxins), which potentially might be found especially in less-refined fish oil supplements. An independent test in 2006 of 44 fish oils on the US market found that all of the products passed safety standards for potential contaminants. The FDA recommends that total dietary intake of *n*–3 fatty acids from fish not exceed 3 grams per day, of which no more than 2 grams per day are from nutritional supplements.

Historically, the Council for Responsible Nutrition (CRN) and the World Health Organization (WHO) have published acceptable standards regarding contaminants in fish oil. The most stringent current standard is the International Fish Oils Standard (IFOS). Fish oils that typically make this highest grade are those that are molecularly distilled under vacuum, and have virtually no measurable level of contaminants (measured parts per billion and parts per trillion).

n-3 supplementation in food has been a significant recent trend in food fortification, with global food companies launching *n*-3 fortified bread, mayonnaise, pizza, yogurt, orange juice, children's pasta, milk, eggs, confections and infant formula.

Fish

The most widely available source of EPA and DHA is cold water oily fish such as salmon, herring, mackerel, anchovies and sardines. Oils from these fish have a profile of around seven times as much n–3 as n–6. Other oily fish such as tuna also contain n–3 in somewhat lesser amounts. Consumers of oily fish should be aware of the potential presence of heavy metals and fat-soluble pollutants like PCBs and dioxins which may accumulate up the food chain. Some supplement manufacturers remove heavy metals and other contaminants from the oil through various means, such as molecular distillation (see above), which increases purity, potency and safety. In addition, some companies use distillation processes to enhance the DHA to EPA ratio of the fish oil.

Even some forms of fish oil may not be optimally digestible. Of four studies that compare bioavailability of the triglyceride form of fish oil vs. the ester form, two have concluded that the natural triglyceride form is bettei, and the other two studies did not find a significant difference. No studies have shown the ester form to be superior although it is cheaper to manufacture.

Although fish is a dietary source of n–3 fatty acids, fish do not synthesize them; they obtain them from the algae in their diet.

Flax seeds produce linseed oil, which has a very high *n*-3 content

Six times richer than most fish oils in *n*-3, Flax (aka linseed) (Linum usitatissimum) and its oil are perhaps the most widely available botanical source of *n*-3. Flaxseed oil consists of approximately 55% ALA (alpha-linolenic acid). Flax, like chia, contains approximately three times as much *n*-3 as *n*-6.

15 grams of flaxseed oil provides ca. 8 grams of ALA, which is converted in the body to EPA and then DHA at an efficiency of 2-15% and 2-5%, respectively.

Botanical sources of *n*-3 fatty acids **(Fig. 2.20)**

Table 2.20 : *n*-3 content as the percentage of ALA in the seed oil

Common Name	Alternative name	Linnaean Name	% *n*-3
Chia	chia sage	Salvia hispanica	64
Kiwifruit	Chinese gooseberry	Actinidia chinensis	62
Perilla	shiso	Perilla frutescens	58
Flax	linseed	Linum usitatissimum	55
Lingonberry	Cowberry	Vaccinium vitis-idaea	49
Camelina	Gold-of-pleasure	Camelina sativa	36
Purslane	Portulaca	Portulaca oleracea	35
Black Raspberry		Rubus occidentalis	33

Table 2.21 : *n*-3 content as the percentage of ALA in the whole food.

Common name	Linnaean name	% *n*–3
Flaxseed	Linum usitatissimum	18.1
Butternuts	Juglans cinerea	8.7
Walnuts	Juglans regia	6.3
Pecan nuts	Carya illinoinensis	0.6
Hazel nuts	Corylus avellana	0.1

Eggs

Eggs produced by chickens fed a diet of greens and insects produce higher levels of *n*-3 fatty acids (mostly ALA) than chickens fed corn or soybeans. In addition to feeding chickens insects and greens, fish oils may be added to their diet to increase the amount of fatty acid concentrations in eggs.

Meat

The *n*-6 to *n*-3 ratio of grass-fed beef is about 2:1, making it a more useful source of *n*-3 than grain-fed beef, which usually has a ratio of 4:1.

Commercially available lamb is almost always grass-fed, and subsequently higher in *n*-3 than other common meat sources.

The omega-3 content of chicken meat may be enhanced by increasing the animals' dietary intake of grains such as flax, chia, and canola.

Other Sources

Milk and cheese from grass-fed cows may also be good sources of *n*-3. One UK study showed that half a pint of milk provides 10% of the recommended daily intake (RDI) of ALA, while a piece of organic cheese the size of a matchbox may provide up to 88%".

Krill, which are small, shrimp-like zooplankton, also contain the *n*-33 fatty acids EPA and DHA. One advantage of extracting *n*-33 fatty acids from krill, as opposed to sources higher in the food chain, is that krill contain fewer heavy metals and PCBs harmful to humans. However, in comparison to higher animals, they also contain fewer *n*-33 fatty acids per gram.

The microalgae Crypthecodinium cohnii and Schizochytrium are rich sources of DHA (22:6 *n*-33) and can be produced commercially in bioreactors. Oil from brown algae (kelp) is a source of EPA. Walnuts are one of few nuts that contain appreciable *n*-33 fat, with approximately a 1:4 ratio of *n*-33 to *n*-36. Acai palm fruit also contains *n*-33 fatty acids.

Omega 3 is also found in softgels in pharmacies and nowadays it is also found in combination with omega 6, omega 9 and shark liver oil.

Some vegetables, too, contain a noteworthy amount of *n*-3, including strawberries and broccoli.

The n-36 to n-33 ratio

Clinical studies indicate that the ingested ratio of *n*-36 to *n*-33 (especially Linoleic vs Alpha Linolenic) fatty acids is important to maintaining cardiovascular health.

Both *n*-33 and *n*-36 fatty acids are essential, i.e. humans must consume them in the diet. *n*–33 and *n*-36 compete for the same metabolic enzymes, thus the *n*-36:*n*-33 ratio will significantly influence the ratio of the ensuing eicosanoids hormones, e.g. prostaglandins, leukotrienes, thromboxanes etc., and will alter the body's metabolic functionGenerally, grass-fed animals accumulate more *n*-33 than do grain-fed animals which accumulate relatively more *n*-36. Metabolites of *n*-36 are significantly more inflammatory (esp. arachidonic acid) than those of *n*-33. This necessitates that *n*-33 and *n*-36 be consumed in a balanced proportion; healthy ratios of *n*-36:n-33 range from 1:1 to 4:1. Studies suggest that the evolutionary human diet, rich in game animals, seafood and other sources of *n*-33, may have provided such a ratio.

Typical Western diets provide ratios of between 10:1 and 30:1 - i.e., dramatically skewed toward *n*-36. Here are the ratios of *n*-36 to *n*-33 fatty acids in some common oils: canola 2:1, soybean 7:1, olive 3-13:1, sunflower (no *n*-33), flax 1:3, cottonseed (almost no *n*-33), peanut (no *n*-33), grapeseed oil (almost no *n*–33) and corn oil 46 to 1 ratio of *n*-36 to *n*-33. It should be noted that olive, peanut and canola oils consist of approximately 80% monounsaturated fatty acids, (i.e. neither *n*-36 nor *n*-33) meaning that they contain relatively small amounts of *n*-33 and *n*-36 fatty acids. Consequently, the *n*-36 to *n*-33 ratios for these oils (i.e. olive, canola and peanut oils) are not as significant as they are for corn, soybean and sunflower oils.

Conversion Efficiency of ALA to EPA and DHA

It has been reported that conversion of ALA to EPA and further to DHA in humans is limited, but varies with individuals. Women have higher ALA conversion efficiency than men, probably due to the lower rate of utilization of dietary ALA for beta-oxidation. This suggests that biological engineering of ALA conversion efficiency is possible. In the online book of The Benefits of Omega-3 Fatty Acids found in Seal Oil, as Opposed to Fish and Flaxseed Oils, Dr. Ho listed the several factors that inhibit the ALA conversion, which again indicate that the efficiency of ALA conversion could be adjusted by altering one's dietary habits, such as rebalancing the ratio of *n*-33 and *n*-36 fatty acid intake, restraining direct alcohol consumptions, and supplementing vitamins and minerals. However, Goyens et al. argues that it is the absolute amount of ALA, rather than the ratio of *n*-33 and *n*-36 fatty acids, which affects the conversion.

PROSTAGLANDINS : BIOSYNTHESIS AND FUNCTIONS

Prostaglandin

The name *prostaglandin* derives from the prostate gland. When prostaglandin was first isolated from seminal fluid in 1935 by the Swedish physiologist Ulf von Euler, and independently by M.W. Goldblatt, it was believed to be part of the prostatic secretions (in actuality prostaglandins are produced by the seminal vesicles); it was later shown that many other tissues secrete prostaglandins for various functions.

In 1971, it was determined that aspirin-like drugs could inhibit the synthesis of prostaglandins. The biochemists Sune K. Bergström, Bengt I. Samuelsson and John R. Vane jointly received the 1982 Nobel Prize in Physiology or Medicine for their research on prostaglandins.

A **prostaglandin** is any member of a group of lipid compounds that are derived enzymatically from fatty acids and have important functions in the animal body. Every prostaglandin contains 20 carbon atoms, including a 5-carbon ring. They are mediators and have a variety of strong physiological effects; although they are technically hormones, they are rarely classified as such.

The prostaglandins together with the thromboxanes and prostacyclins form the prostanoid class of fatty acid derivatives; the prostanoid class is a subclass of eicosanoids.

Biosynthesis

Prostaglandins are found in virtually all tissues and organs. They are produced by all nucleated cells except lymphocytes. They are autocrine and paracrine lipid mediators that act upon platelets, endothelium, uterine and mast cells, among others. They are synthesized in the cell from the essential fatty acids (EFAs).

An intermediate is created by phospholipase-A_2, then passed into one of either the cyclooxygenase pathway or the lipoxygenase pathway to form either prostaglandin and thromboxane or leukotriene. The cyclooxygenase pathway produces thromboxane, prostacyclin and prostaglandin D, E and F. The lipoxygenase pathway is active in leukocytes and in macrophages and synthesizes leukotrienes.

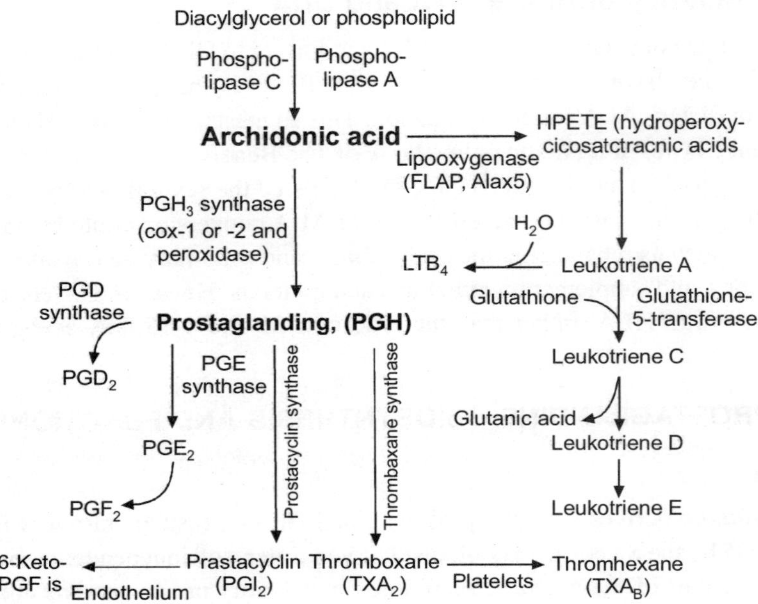

Fig. 2.22 : Biosynthesis of eicosanoids.

Name	EFA Type	Series
Gamma-linolenic acid (GLA) via DGLA	ω-6	series-1
Arachidonic acid (AA)	ω-6	series-2
Eicosapentaenoic acid (EPA)	ω-3	series-3

Release of Prostaglandins from the Cell

Prostaglandins were originally believed to leave the cells via passive diffusion because of their high lipophilicity. The discovery of the prostaglandin transporter (PGT, SLCO2A1), which mediates the cellular uptake of prostaglandin, demonstrated that diffusion cannot explain the penetration of prostaglandin through the cellular membrane. The release of prostaglandin has now also been shown to be mediated by a specific transporter, namely the multidrug resistance protein 4 (MRP4, ABCC4), a member of the ATP-binding cassette transporter superfamily. Whether MRP4 is the only transporter releasing prostaglandins from the cells is still unclear.

Cyclooxygenases

Prostaglandins are produced following the sequential oxidation of AA, DGLA or EPA by cyclooxygenases (COX-1 and COX-2) and terminal prostaglandin synthases. The classic dogma is as follows:

- COX-1 is responsible for the baseline levels of prostaglandins.
- COX-2 produces prostaglandins through stimulation.

However, while COX-1 and COX-2 are both located in the blood vessels, stomach and the kidneys, prostaglandin levels are increased by COX-2 in scenarios of inflammation. A third form of COX, termed COX-3 is thought to exist in the brain and may be associated with relief of Headaches when on NSAID therapy.

Prostaglandin E synthase

Prostaglandin E_2 (PGE_2) is generated from the action of prostaglandin E synthases on prostaglandin H_2 (PGH_2). Several prostaglandin E synthases have been identified. To date, microsomal prostaglandin E synthase-1 emerges as a key enzyme in the formation of PGE_2.

Other Terminal Prostaglandin Synthases

Terminal prostaglandin synthases have been identified that are responsible for the formation of other prostaglandins. For example, hematopoietic and lipocalin prostaglandin D synthases (hPGDS and lPGDS) are responsible for the formation of PGD_2 from PGH_2. Similarly, prostacyclin (PGI_2) synthase (PGIS) converts PGH_2 into PGI_2. A thromboxane synthase (TxAS) has also been identified. Prostaglandin F synthase (PGFS) catalyzes the formation of 9α,11β-$PGF_{2\alpha,\beta}$ from PGD_2 and $PGF_{2\alpha}$ from PGH_2 in the presence of NADPH. This enzyme has recently been crystallyzed in complex with PGD_2 and bimatoprost (a synthetic analogue of $PGF_{2\alpha}$).

Function

There are currently nine known prostaglandin receptors on various cell types. Prostaglandins ligate a subfamily of cell surface seven-transmembrane receptors, G-protein-coupled receptors. These receptors are termed DP1-2, EP1-4, FP, IP, and TP, corresponding to the receptor that ligates the corresponding prostaglandin (e.g., DP1-2 receptors bind to PGD2).

These varied receptors mean that Prostaglandins thus act on a variety of.cells, and have a wide variety of actions:

- cause constriction or dilation in vascular smooth muscle cells
- cause aggregation or disaggregation of platelets
- sensitize spinal neurons to pain
- decrease intraocular pressure
- regulate inflammatory mediation
- regulate calcium movement
- control hormone regulation
- control cell growth

Prostaglandins are potent but have a short half-life before being inactivated and excreted. Therefore, they exert only a paracrine (locally active) or autocrine (acting on the same cell from which it is synthesized) function.

Types

Following is a comparison of the prostaglandin types Prostaglandin I_2 (PGI_2), Prostaglandin E_2 (PGE_2) and Prostaglandin $F_{2\alpha}$ ($PGF_{2\alpha}$).

Type	Receptor	Function
PGI_2	IP	• vasodilation
		• inhibit platelet aggregation
		• bronchodilatation
	EP_1	• bronchoconstriction
		• GI tract smooth muscle contraction
	EP_2	• bronchodilatation
		• GI tract smooth muscle relaxation
		• vasodilatation
PGE_2		• ↓ gastric acid secretion
		• ↑ gastric mucus secretion
	EP_3	• uterus contraction (when pregnant)
		• GI tract smooth muscle contraction
		• lipolysis inhibition
		• ↑ autonomic neurotransmitters
	Unspecified	• hyperalgesia
		• pyrogenic
$PGF_{2\alpha}$	FP	• uterus contraction
		• bronchoconstriction

Role in Pharmacology

Inhibition

Examples of prostaglandin antagonists are:

- NSAIDs (inhibit cyclooxygenase)
- Corticosteroids (inhibit phospholipase A2 production)
- COX-2 selective inhibitors or coxibs

However, both NSAIDs and Coxibs can raise the risk of myocardial infarction.

Clinical Uses

Synthetic prostaglandins are used:

- To induce childbirth (parturition) or abortion (PGE_2 or PGF_2, with or without mifepristone, a progesterone antagonist);

- To prevent closure of patent ductus arteriosus in newborns with particular cyanotic heart defects (PGE1)
- To prevent and treat peptic ulcers (PGE)
- As a vasodilator in severe Raynaud's phenomenon or ischemia of a limb
- In pulmonary hypertension
- In treatment of glaucoma (as in bimatoprost ophthalmic solution, a synthetic prostamide analog with ocular hypotensive activity)
- To treat erectile dysfunction or in penile rehabilitation following surgery (PGE1 as alprostadil).
- To treat egg binding in small birds.

LEUCOTRIENS AND THROMBOXANES

The name *leukotriene*, introduced by Swedish biochemist Bengt Samuelsson in 1979, comes from the words *leukocyte* and *triene* (indicating the compound's three conjugated double bonds). What would be later named leukotriene C, "slow reaction smooth muscle-stimulating substance" (SRS) was originally described between 1938 and 1940 by Feldberg and Kellaway. The researchers isolated SRS from lung tissue after a prolonged period following exposure to snake venom and histamine.

Leukotrienes are commercially available to the research community. Leukotrienes are naturally produced eicosanoid lipid mediators, which may be responsible for the effects of an inflammatory response. Leukotrienes use both autocrine signalling and paracrine signalling to regulate the body's response. Leukotrienes are produced in the body from arachidonic acid by the enzyme 5-lipoxygenase. Their production by the body is part of a complex response that usually includes the production of histamine.

Biochemistry and Synthesis

Leukotrienes are synthesized in the cell from arachidonic acid by 5-lipoxygenase. The catalytic mechanism involves the insertion of an oxygen moiety at a specific position in the arachidonic acid backbone.The lipoxygenase pathway is active in leukocytes, including mast cells, eosinophils, neutrophils, monocytes and basophils. When such cells are activated, arachidonic acid is liberated from cell membrane phospholipids by phospholipase A2, and donated by the 5-lipoxygenase activating protein (FLAP) to 5-lipoxygenase.

5-lipoxygenase (5-LO) uses FLAP to convert arachidonic acid into 5-hydroperoxyeicosatetraenoic acid (5-HPETE), which spontaneously reduces to 5-hydroxyeicosatetraenoic acid (5-HETE). The enzyme 5-LO acts again on 5-HETE to convert it into leukotriene A_4 (LTA_4), an unstable epoxide.

In cells equipped with LTA_4 hydrolase, such as neutrophils and monocytes, LTA_4 is converted to the dihydroxy acid leukotriene LTB_4, which is a powerful chemoattractant for neutrophils acting at BLT_1 and BLT_2 receptors on the plasma membrane of these cells.

In cells that express LTC_4 synthase, such as mast cells and eosinophils, LTA_4 is conjugated with the tripeptide glutathione to form the first of the cysteinyl-leukotrienes, LTC_4. Outside the cell, LTC_4

can be converted by ubiquitous enzymes to form successively LTD_4 and LTE_4, which retain biological activity.

The cysteinyl-leukotrienes act at their cell-surface receptors CysLT1 and CysLT2 on target cells to contract bronchial and vascular smooth muscle, to increase permeability of small blood vessels, to enhance secretion of mucus in the airway and gut, and to recruit leukocytes to sites of inflammation.

Both LTB4 and the cysteinyl-leukotrienes (LTC_4, LTD_4, LTE_4) are partly degraded in local tissues, and ultimately become inactive metabolites in the liver.

Function

Leukotrienes act principally on a subfamily of G protein coupled receptors. They may also act upon peroxisome proliferator-activated receptors. Leukotrienes are involved in asthmatic and allergic reactions and act to sustain inflammatory reactions; several leukotriene receptor antagonists (e.g. montelukast and zafirlukast) are used to treat asthma. Recent research points to a role of 5-lipoxygenase in cardiovascular and neuropsychiatric illnesses.

Leukotrienes are very important agents in the inflammatory response. Some such as LTB4 have a chemotactic effect on migrating neutrophils, and as such help to bring the necessary cells to the tissue. Leukotrienes also have a powerful effect in bronchoconstriction, they also increase vascular permeability.

Leukotrienes in Asthma

Leukotrienes assist in the pathophysiology of asthma, causing or potentiating the following symptoms:

- airflow obstruction
- increased secretion of mucus
- mucosal accumulation
- bronchoconstriction
- infiltration of inflammatory cells in the airway wall

Role of Cysteinyl Leukotrienes

Cysteinyl leukotriene receptors **CysLT1** and **CysLT2** are present on mast cells, eosinophil and endothelial cells. During cysteinyl leukotriene interaction, they can stimulate proinflammatory activities such as endothelial cell adherence and chemokine production by mast cells. As well as mediating inflammation, they induce asthma and other inflammatory disorders, thereby reducing the airflow to the alveoli. In excess, the cysteinyl leukotrienes can induce anaphylactic shock.

Zileuton blocks 5-lipoxygenase inhibiting the synthetic pathway of leukotriene metabolism. Zileuton affects the LTB4 pathway, montelukast doesn't.

Thromboxane is a member of the family of lipids known as eicosanoids. The two major thromboxanes are thromboxane A2 and thromboxane B2.

Thromboxane is named for its role in clot formation (thrombosis).

Production

Enzymes and substrates associated with thromoboxane and prostacyclin synthesis. It is produced in platelets by thromboxane-A synthase from the endoperoxides produced by the cyclooxygenase (COX) enzyme from arachidonic acid.

Mechanism

Thromboxane acts by binding to any of the thromboxane receptors, G-protein coupled receptors coupled to the G protein G_q

Functions

Thromboxane is a vasoconstrictor and a potent hypertensive agent, and it facilitates platelet aggregation.

It is in homeostatic balance in the circulatory system with prostacyclin, a related compound. The mechanism of secretion of thromboxanes from platelets is still unclear.

Role of A2 in Platelet Aggregation

Thromboxane A_2 (TXA_2), produced by activated platelets, has prothrombotic properties, stimulating activation of new platelets as well as increasing platelet aggregation.

Platelet aggregation is achieved by mediating expression of the glycoprotein complex GP IIb/IIIa in the cell membrane of platelets. Circulating fibrinogen binds these receptors on adjacent platelets, further strengthening the clot.

Pathology

It is believed that the vasoconstriction caused by thromboxanes plays a role in Prinzmetal's angina.

Suppression

The widely used drug aspirin acts by inhibiting the ability of the COX enzyme to synthesize the precursors of thromboxane within platelets. Low-dose, long-term aspirin use irreversibly blocks the formation of thromboxane A_2 in platelets, producing an inhibitory effect on platelet aggregation. This anticoagulant property makes aspirin useful for reducing the incidence of heart attacks. 40 mg of aspirin a day is able to inhibit a large proportion of maximum thromboxane A_2 release provoked acutely, with the prostaglandin I_2 synthesis being little affected; however, higher doses of aspirin are required to attain further inhibition. One side effect of this is that people who regularly take aspirin will suffer from excessive bleeding whenever the skin is perforated.

CLASSIFICATION AND STRUCTURE OF PHOSPHOLIPIDS A SURFACTANT GLYCOLIPIDS

These are carbohydrate-attached lipids. Their role is to provide energy and also serve as markers for cellular recognition.

They occur where a carbohydrate chain is associated with phospholipids on the exoplasmic surface of the cell membrane. The carbohydrates are found on the outer surface of all eukaryotic cell membranes.

They extend from the phospholipid bilayer into the aqueous environment outside the cell where it acts as a recognition site for specific chemicals as well as helping to maintain the stability of the membrane and attaching cells to one another to form tissues.

Types of Glycolipids

The following is an incomplete listing of glycolipid types.

- Glyceroglycolipids
  - Galactolipids
  - Sulpholipids (SQDG)
- Glycosphingolipids
  - Cerebrosides
    - Galactocerebrosides
    - Glucocerebrosides
      - * Glucobicaranateoets
  - Gangliosides (the most complex animal glycolipids; contain negatively charged oligosacchrides with one or more sialic acid residues; more than 40 different ganglidosides have been identified; they are most adundant in cells)
  - Globosides
  - Sulfatides
  - Glycophosphosphingolipids (complex glycophospholipids from fungi, including yeasts, and in plants, where they were originally called "phytoglycolipidds" by dCdarter, et al., may comprise as complicated a ddof compounds as the negatively charged gandgliosides in animals. The head group of a glycolipid is composed of sugars.

glycolipids are membrane components composed of lipids that are covalently bonded to monosaccharides or polysaccharides. One type of glycolipid found in human red blood cells is involved in the ABO blood type antigens. The Fig. 2.23 shows the relationship between the ABO blood type, the RBC glycolipid and antibodies to the glycolipids in the blood plasma.

Glycolipids are present ubiquitously in cell surface membranes. The most common glycolipids in mammals are a class of sphingolipids, glycosphingolipids (GSLs). Major GSLs including gangliosides are derived from glucosylceramide (GlcCer) that is synthesized by a specific glycosyltransferase, ceramide glucosyltransferase (GlcT-1, UGCG). Galactosylceramide (GalCer) formed by ceramide galactosyltransferase (CGT) is also present as a dominant glycolipid in myelin sheath. The class of glyceroglycolipids is found in particular tissues: galactosylalkylacylglyceride (seminolipid) in the testis, and phosphatidyglucose in human cord blood cells. Glucosylated cholesterol is found in mammalian cells.

blood type	glycolipid	antibodies in blood
O	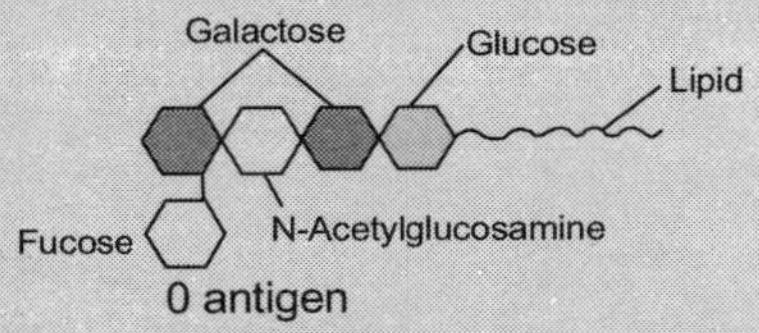 Galactose Glucose Lipid Fucose N-Acetylglucosamine 0 antigen RBC of O type blood have this glycolipids in their RBCs plasma membranes.	**anti A and anti B** Individuals with type O blood have antibodies against the type A and type B antigens
A	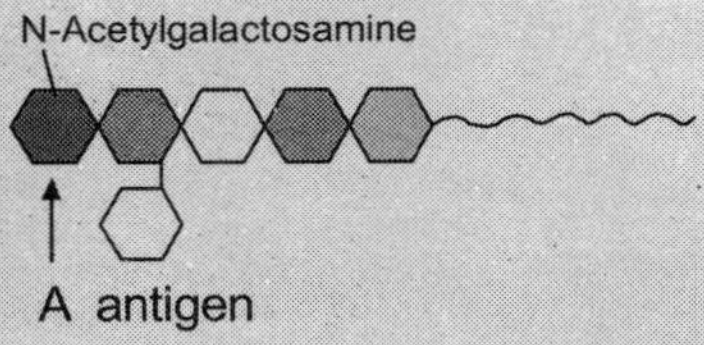 N-Acetylgalactosamine A antigen Individuals with type A blood have this type of glycolipid in their RBCs plasma membranes. The A type glycolipid has the same carbohydrate composition as does the type O antigen with the addition of a additional N-Acetylgalactosamine which is the A type antigenic determinant.	**anti B** individuals of type A blood have antibodies against the type B determinant
B	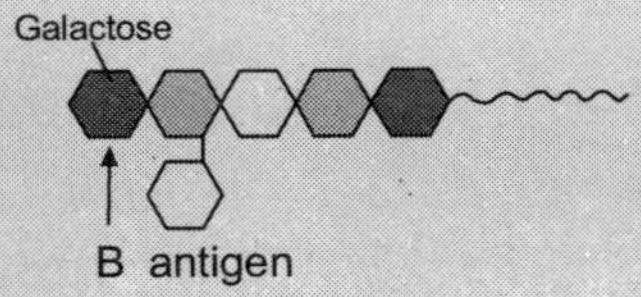 Galactose B antigen Individuals with type B blood have the above kind of glycolipid in their RBC plasma membrane. The B type glycolipid has the same carbohydrate composition as does the type O antigen with the addition of an additional galactose which is the B type antigenic determinant.	**anti A** Individuals of type B blood have antibodies against the type A antigenic determinant
AB	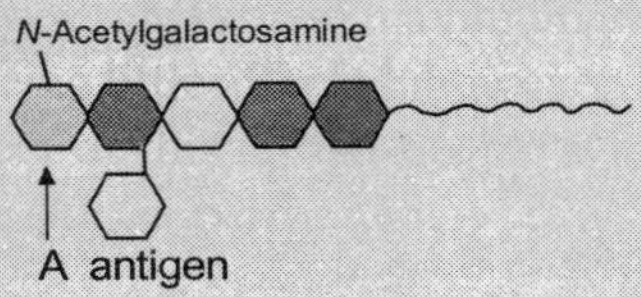 *N*-Acetylgalactosamine A antigen 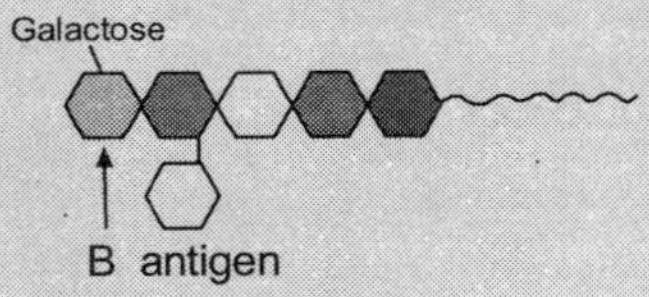 Galactose B antigen Individuals with type AB blood have both of the above glycolipids in their RBC plasma membranes	**neither A or B** Individuals with they AB blood have antibodies against neither the the A nor B antigens

Fig. 2.23 : Glycolipids in blood plasma

Table 2.22 : Roles of monoglycosylated lipids in mammals: almost all lipids are glycosylated

Glycolipid	Distribution	Synthetic enzyme	Role
GlcCer	ubiquitous	UGCG (human, mouse, rat)	precursor for GSLs biosynthesis, negative regulator for ceramide, axonal elongation of neuron, multidrug resistance
GalCer	myelin, kidney,	CGT (human, mouse, rat)	insulative function of myelin
ÉøGalCer	unknown	unknown	ligand for NKT cell
FucCer	colon cancer, lung cancer	unknown	unknown
Cholesteryl glucoside	human skin fibroblast	unknown	heat shock response
Galactosyldiacyl glycerol	sperm (seminolipid), intestine, the nervous system	CGT (the same as GalCer synthase)	sperm development
phosphatidylglucoside	cord blood cell, HL60	unknown	unknown

NKT: natural Killer T

Phosphatidylinositol is not included.

Table 2.23 : Gene targeted mice: glycolipid functions at the whole animal level

Gene disrupted	Phenotype Biochemistry	General
GlcCer synthase (UGCG)	loss of all GSLs except GalCer	embryonic lethality at E7.5, enhancemant of apoptosis in ectodermal layer
GalCer synthase (CGT)	loss of GalCer and appearance of GlcCer in myelin, loss of seminolipid in sperm	ataxia, loss of insulative function of myelin, sterility in male
Sulfatide synthase (CST)	loss of sulfatide in myelin and of sulfated seminolipid in sperm	similarity to CGT KO mouse
GM2/GD2 (GalNAc-T) synthase	loss of higher gangliosides (only GM3, GD3 and GT3)	normal development, defect in nerve regeneration
GD3 synthase	loss of higher gangliosides (only monosialo gangliosides)	normal development defect in nerve regeneration
GM2/GD3 double KO	loss of gangliosides except GM3	normal development, lethality to sound stress

UDCG:UDP-Glc Ceramide Glucosyltransferase

CGT: Ceramide Galactosyltransferase

CST: Cerebroside Sulfotransferase

Recent cell biological studies show that GSLs in cell membranes are preferentially distributed into lipid domains, so-called rafts (Fig. 2.24,). Raft-like lipid microdomains can be isolated biochemically as a low density Triton X-100 insoluble fraction. The fraction isolated contains sphingolipids (GSLs and sphingomyelin) and cholesterol. Importantly, the glycolipid-enriched membrane domain also includes proteins related to signal transduction such as src-family kinases. The lipid domains are suggested to play roles in cell-cell adhesion and receptor-mediated signal transduction. A glycolipid-enriched lipid micodomain is also involved in target for host pathogens (Vibrio cholera, O157, HIY) and their toxin bindings (Fig. 2.24).

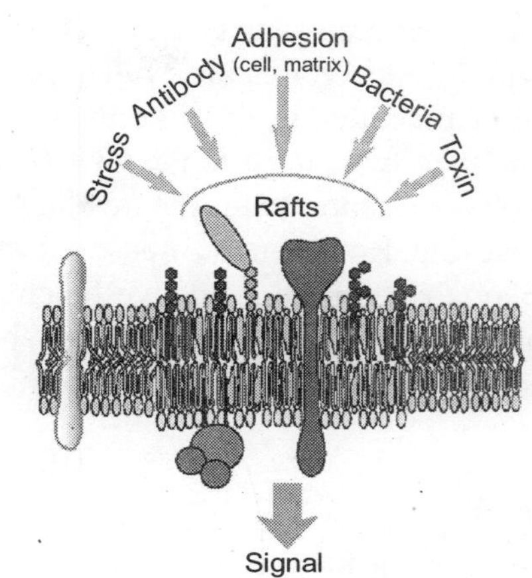

Fig. 2.24 : Showing the Diagramatic representation of Biological action of GSL

A cell line deficient in an entire group of GSLs was established from B16 mouse melanoma. This mutant cell line is defective in a GlcCer synthase activity. *In vitro* studies on glycolipid function using the mutant cells show that the glycolipid biosynthesis is not essential for cell survival and proliferation, at least in the melanoma cells. Sphingomyelin as the sole sphingolipid on the plasma membrane can act as a substitute. Similarly, GSLs are not a critical component for membrane domain formation. These results from *in vitro* experiments indicate that GSLs apparently do not have house-keeping functions when studied at the single cell level. Thus, GSL research is now focusing on multicellular systems or whole animals. Up to now, however, there is no case report of human genetic disease associated with glycolipid synthetic enzyme. Therefore, the physiological functions of GSLs remain unclear in humans.

To solve the riddle of glycolipid functions, a method to eliminate a gene of glycolipid glycosyltransferase in mice has been developed using a homologous recombination system and several knockout mice have been generated. Surprisingly, mutant mice deficient in complex gangliosides are born normally, showing no apparent abnormalities. Even a double knockout mouse of GD3 synthase and GM2/GD2 synthase, expressing only GM3 ganglioside, results in normal development. Studies with these mutant mice prove that complex gangliosides are not involved in cellular differentiation but rather in the maintenance (homeostasis) and regeneration of nervous system tissues (see "Glyco Word" by Furukawa). Very recently, Kawai et al. have shown that the double mutant is highly sensitive to sound stimulus and displays a sudden death phenotype. In 1999, a mutant mouse for GlcCer synthase (UGCG) was generated. Since the mutant mouse is deficient in all GSLs except for GalCer, it gives us valuable information on the general functions of GSLs. The mice die at around embryonic day 7.5, indicating clear evidence of the essential role of GSL biosynthesis during multicellular development. The mutant mice exhibit an elevated apoptosis in the ectoderm. However, it is difficult to explain clearly why ablation of the UGCG gene causes the enhancement of apoptotic cell death because UGCG has at least two distinct functions one of which is a negative regulatory factor for intracellular ceramide content through ceramide glycosylation. The other is a positive regulatory role of the glycolipid, GlcCer in cell growth. Moreover, it is not evident if there is any relationship between the embryonic lethality and glycolipid functions in raft-like membrane microdomains. GalCer knockout mouse was first generated in 1996. In contrast to the UGCG mutant, the CGT knockout mouse is born normally and exhibits no obvious abnormalities in myelin structure, although GalCer and sulfatide, both enriched in myelin, are lacking. However, the mouse shows severe generalized tremor and mild ataxia, proving that the GSLs play a role in the insulative function of myelin. The same mousé shows that CGT is also essential for sperm development. As shown above, glycolipid research is now regarded as a potent and promising research target of the post-genomic era.

DERIVED LIPIDS

Cholesterol

Cholesterol is a lipid like, waxy alcohol found in the cell membranes and transported in the blood plasma of all animals. It is an essential component of mammalian cell membranes where it is required to establish proper membrane permeability and fluidity. Cholesterol is the principal sterol synthesized by animals while smaller quantities are synthesized in other eukaryotes such as plants and fungi. In contrast, cholesterol is almost completely absent among prokaryotes such as bacteria. Cholesterol is classified as a sterol (a contraction of steroid and alcohol).

While minimum levels of cholesterol are essential for life, excess levels in the circulation are associated with atherosclerosis. Cholesterol is synthesized by virtually all the body cells but significant quantities can also be absorbed from the diet.

The name cholesterol originates from the Greek *chole-* (bile) and *stereos* (solid), and the chemical suffix *-ol* for an alcohol, as researchers first identified cholesterol in solid form in gallstones by François Poulletier de la Salle in 1769. However, it is only in 1815 that chemist Eugène Chevreul named the compound "cholesterine".

Physiology and Function Cholesterol is required to build and maintain cell membranes; it regulates membrane fluidity over a wide range of temperatures. The liver produces about 1 gram of cholesterol per day through the bile, which contains cholesterol. The hydroxyl group on cholesterol interacts with the polar head groups of the membrane phospho- and sphingolipids, while the bulky steroid and the hydrocarbon chain is embedded in the membrane, alongside the nonpolar fatty acid chains of the other lipids. Some research indicates that cholesterol may act as an antioxidant. Cholesterol also aids in the manufacture of bile (which is stored in the gallbladder and helps digest fats), and is also important for the metabolism of fat soluble vitamins, including vitamins A, D, E and K. It is the major precursor for the synthesis of vitamin D and of the various steroid hormones (which include cortisol and aldosterone in the adrenal glands, and the sex hormones progesterone, the various estrogens, testosterone, and derivatives). It is the base structure for steroids. It also covers nerves in the form of myelin. This helps to conduct nerve impulses.

Recently, cholesterol has also been implicated in cell signaling processes, where it has been suggested that it assists in the formation of lipid rafts in the plasma membrane. It also reduces the permeability of the plasma membrane to hydrogen ions (protons) and sodium ions.

Cholesterol is essential for the structure and function of invaginated caveolae and clathrin-coated pits, including the caveolae-dependent endocytosis and clathrin-dependent endocytosis. The role of cholesterol in caveolae-dependent and clathrin-dependent endocytosis can be investigated by using methyl beta cyclodextrin (MβCD) to remove cholesterol from the plasma membrane.

Synthesis and absorption Most of the cholesterol in the body is synthesized by the body and some has dietary origin. Cholesterol is more abundant in tissues which either synthesize more or have more abundant densely-packed membranes, for example, the liver, spinal cord and brain. It plays a central role in many biochemical processes, such as the composition of cell membranes and the synthesis of steroid hormones.

Cholesterol is required in the membrane of mammalian cells for normal cellular function, and is either synthesized in the endoplasmic reticulum, or derived from the diet, in which case it is delivered by the bloodstream in low-density lipoproteins. These are taken into the cell by LDL receptor-mediated endocytosis in clathrin-coated pits, and then hydrolysed in lysosomes.

Cholesterol is primarily synthesized from acetyl CoA through the HMG-CoA reductase pathway in many cells and tissues. About 20-25% of total daily production (~1 g/day) occurs in the liver; other sites of higher synthesis rates include the intestines, adrenal glands and reproductive organs. For a person of about 150 pounds (68 kg), typical total body content is about 35 g, typical daily internal production is about 1 g and typical daily dietary intake is 200-300 mg in the United States and societies with similar dietary patterns. Of the cholesterol input to the intestines via bile production, 92-97% is reabsorbed in the intestines and recycled via enterohepatic circulation.

Konrad Bloch and Feodor Lynen shared the Nobel Prize in Physiology or Medicine in 1964 for their discoveries concerning the mechanism and regulation of the cholesterol and fatty acid metabolism.

Plasma transport Since cholesterol is insoluble in blood, it is transported in the circulatory system within lipoproteins, complex spherical particles which have an exterior composed mainly of water-soluble proteins; fats and cholesterol are carried internally. There is a large range of lipoproteins within blood, generally called, from larger to smaller size: chylomicrons, very low density lipoprotein (VLDL), intermediate density lipoprotein (IDL), low density lipoprotein (LDL) and high density lipoprotein (HDL). The cholesterol within all the various lipoproteins is identical.

Cholesterol is minimally soluble in water; it cannot dissolve and travel in the water-based bloodstream. Instead, it is transported in the bloodstream by lipoproteins-protein "molecular-suitcases" that are water-dispersible and carry cholesterol and triglycerides internally. The apolipoproteins forming the surface of the given lipoprotein particle determine from what cells cholesterol will be removed and to where it will be supplied.

Cholesterol is transported towards peripheral tissues by the lipoproteins chylomicrons, very low density lipoproteins (VLDL) and low-density lipoproteins (LDL). Large numbers of small dense LDL (sdLDL) particles are strongly associated with the presence of atheromatous disease within the arteries. For this reason, LDL is referred to as “bad cholesterol”.

On the other hand, high-density lipoprotein (HDL) particles transport cholesterol back to the liver for excretion in a process known as reverse cholesterol transport (RCT). Having large numbers of large HDL particles correlates with better health outcomes. In contrast, having small numbers of large HDL particles is independently associated with atheromatous disease progression within the arteries.

Homeostasis Biosynthesis of cholesterol is directly regulated by the cholesterol levels present, though the homeostatic mechanisms involved are only partly understood. A higher intake from food leads to a net decrease in endogenous production, while lower intake from food has the opposite effect. The main regulatory mechanism is the sensing of intracellular cholesterol in the endoplasmic reticulum by the protein SREBP (Sterol Regulatory Element Binding Protein 1 and 2). In the presence of cholesterol, SREBP is bound to two other proteins: SCAP (SREBP-cleavage activating protein) and Insig1. When cholesterol levels fall, Insig-1 dissociates from the SREBP-SCAP complex, allowing the complex to migrate to the Golgi apparatus, where SREBP is cleaved by S1P and S2P (site-1 and -2 protease), two enzymes that are activated by SCAP when cholesterol levels are low. The cleaved SREBP then migrates to the nucleus and acts as a transcription factor to bind to the SRE (sterol regulatory element) of a number of genes to stimulate their transcription. Among the genes transcribed are the LDL receptor and HMG-CoA reductase. The former scavenges circulating LDL from the bloodstream, whereas HMG-CoA reductase leads to an increase of endogenous production of cholesterol.

A large part of this signaling pathway was clarified by Dr. Michael S. Brown and Dr. Joseph L. Goldstein in the 1970s. In 1985, they received the Nobel Prize in Physiology or Medicine for their work. Their subsequent work shows how the SREBP pathway regulates expression of many genes that control lipid formation and metabolism and body fuel allocation.

Metabolism and excretion: Cholesterol is oxidized by the liver into a variety of bile acids. These in turn are conjugated with glycine, taurine, glucuronic acid, or sulfate. A mixture of conjugated and non-conjugated bile acids along with cholesterol itself is excreted from the liver into the bile. Approximately 95% of the bile acids are reabsorbed from the intestines and the remainder lost in the feces. The excretion and reabsorption of bile acids forms the basis of the enterohepatic circulation which is essential for the digestion and absorption of dietary fats. Under certain circumstances, when more concentrated, as in the gallbladder, cholesterol crystallises and is the major constituent of most gallstones, although lecithin and bilirubin gallstones also occur less frequently.

Dietary sources: Animal fats are complex mixtures of triglycerides, with lesser amounts of phospholipids and cholesterol. Consequently all foods containing animal fat contain cholesterol to varying extents. Cholesterol is not present in plant based food sources unless it has been added during the foods preparation. However plant products such as flax seeds and peanuts contain healthy cholesterol-like compounds called phytosterols, which are suggested to help lower serum cholesterol levels. Major dietary sources of cholesterol include cheese, egg yolks, beef, pork, poultry, and shrimp. Human breast milk also contains significant quantities of cholesterol.

Dietary cholesterol plays a smaller role in blood cholesterol levels in comparison to fat intake. A number of measures can be taken to reduce blood cholesterol levels through changes in lifestyle, one of which is a change in diet. Trans and saturated fats are significant contributors to elevated cholesterol levels in the blood stream. Avoiding animal products may decrease the cholesterol levels in the body not through dietary cholesterol reduction alone, but primarily through a reduced saturated fat intake. Those wishing to reduce their cholesterol through a change in diet should aim to consume less than 7% of their daily calories from saturated fat and less than 200mg of cholesterol per day.

Clinical Significance (Hypercholesterolemia)

According to the lipid hypothesis, abnormally high cholesterol levels (hypercholesterolemia), or, more correctly, higher concentrations of LDL and lower concentrations of functional HDL are strongly associated with cardiovascular disease because these promote atheroma development in arteries (atherosclerosis). This disease process leads to myocardial infarction (heart attack), stroke and peripheral vascular disease. Since higher blood LDL, especially higher LDL particle concentrations and smaller LDL particle size, contribute to this process more than the cholesterol content of the LDL particles, LDL particles are often termed "bad cholesterol" because they have been linked to atheroma formation. On the other hand, high concentrations of functional HDL, which can remove cholesterol from cells and atheroma, offer protection and are sometimes referred to colloquially as "good cholesterol". These balances are mostly genetically determined but can be changed by body build, medications, food choices and other factors.

Conditions with elevated concentrations of oxidized LDL particles, especially "small dense LDL" (sdLDL) particles, are associated with atheroma formation in the walls of arteries, a condition known as *atherosclerosis*, which is the principal cause of coronary heart disease and other forms of cardiovascular disease. In contrast, HDL particles (especially large HDL) have been identified as a mechanism by which cholesterol and inflammatory mediators can be removed from atheroma. Increased concentrations of HDL correlate with lower rates of atheroma progressions and even regression. A 2007 study pooling data on almost 900,000 subjects in 61 cohorts demonstrated that blood total cholesterol levels have an exponential effect on cardiovascular and total mortality, with the association more pronounced in younger subjects. Still, because cardiovascular disease is relatively rare in the younger population, the impact of high cholesterol on health is still larger in older people.

Elevated levels of the lipoprotein fractions, LDL, IDL and VLDL are regarded as atherogenic (prone to cause atherosclerosis). Levels of these fractions, rather than the total cholesterol level, correlate with the extent and progress of atherosclerosis. Conversely, the total cholesterol can be within normal limits, yet be made up primarily of small LDL and small HDL particles, under which conditions atheroma growth rates would still be high. In contrast, however, if LDL particle number is low (mostly large particles) and a large percentage of the HDL particles are large, then atheroma growth rates are usually low, even negative, for any given total cholesterol concentration Recently, a post-hoc analysis of the IDEAL and the EPIC prospective studies found an association between high levels of HDL cholesterol (adjusted for apolipoprotein A-I and apolipoprotein B) and increased risk of cardiovascular disease, casting doubt on the cardioprotective role of "good cholesterol".

Multiple human trials utilizing HMG-CoA reductase inhibitors, known as statins, have repeatedly confirmed that changing lipoprotein transport patterns from unhealthy to healthier patterns significantly lowers cardiovascular disease event rates, even for people with cholesterol values currently considered low for adults. As a result, people with a history of cardiovascular disease may derive benefit from statins irrespective of their cholesterol levels, and in men without cardiovascular disease there is benefit from lowering abnormally high cholesterol levels ("primary prevention"). Primary prevention in women is practiced only by extension of the findings in studies on men, since in women, none of the large statin trials has shown a reduction in overall mortality or in cardiovascular end points.

The 1987 report of National Cholesterol Education Program, Adult Treatment Panels suggest the total blood cholesterol level should be: < 200 mg/dL normal blood cholesterol, 200-239 mg/dL borderline-high, > 240 mg/dL high cholesterol. The American Heart Association provides a similar set of guidelines for total (fasting) blood cholesterol levels and risk for heart disease:

Level mg/dL	Level mmol/L	Interpretation
< 200	< 5.0	Desirable level corresponding to lower risk for heart disease
200-240	5.2-6.2	Borderline high risk
> 240	> 6.2	High risk

However, as today's testing methods determine LDL ("bad") and HDL ("good") cholesterol separately, this simplistic view has become somewhat outdated. The desirable LDL level is considered to be less than 100 mg/dL (2.6 mmol/L), although a newer target of < 70 mg/dL can be considered in higher risk individuals based on some of the above-mentioned trials. A ratio of total cholesterol to HDL-another useful measure-of far less than 5:1 is thought to be healthier. Of note, typical LDL values for children before fatty streaks begin to develop is 35 mg/dL.

Total cholesterol is defined as the sum of HDL, LDL, and VLDL. Ordinarily just the total, HDL, and triglycerides are actually measured. The VLDL is estimated as one-fifth of the triglycerides. It is important to fast for at least eight hours before the blood test because the triglyceride level varies significantly with food intake.

Given the well-recognized role of cholesterol in cardiovascular disease, it is surprising that some studies have shown an inverse correlation between cholesterol levels and mortality in subjects over 50 years of age-an 11% increase overall and 14% increase in CVD mortality per 1 mg/dL per year drop in cholesterol levels. In the Framingham Heart Study, the researchers attributed this phenomenon to the fact that people with severe chronic diseases or cancer tend to have below-normal cholesterol levels. This explanation is not supported by the Vorarlberg Health Monitoring and Promotion Programme, in which men of all ages and women over 50 with very low cholesterol were increasingly likely to die of cancer, liver diseases, and mental diseases. This result indicates that the low cholesterol effect occurs even among younger respondents, contradicting the previous assessment among cohorts of older people that this is a proxy or marker for frailty occurring with age.

A small group of scientists, united in The International Network of Cholesterol Skeptics, continues to question the link between cholesterol and atherosclerosis. However, the vast majority of doctors and medical scientists accepts the link as fact.

Hypocholesterolemia Abnormally low levels of cholesterol are termed hypocholesterolemia. Research into the causes of this state is relatively limited, but some studies suggest a link with depression, cancer and cerebral hemorrhage. Generally, the low cholesterol levels seem to be a consequence of an underlying illness, rather than a cause.

STEROID HORMONES

The steroid hormones are all derived from cholesterol. Moreover, with the exception of vitamin D, they all contain the same cyclopentanophenanthrene ring and atomic numbering system as cholesterol. The conversion of C_{27} cholesterol to the 18-, 19-, and 21-carbon steroid hormones (designated by the nomenclature C with a subscript number indicating the number of carbon atoms, e.g. C19 for androstanes) involves the rate-limiting, irreversible cleavage of a 6-carbon residue from cholesterol, producing pregnenolone (C_{21}) plus isocaproaldehyde. Common names of the steroid hormones are widely recognized, but systematic nomenclature is gaining acceptance and familiarity with both nomenclatures is increasingly important. Steroids with 21 carbon atoms are known systematically as pregnanes, whereas those containing 19 and 18 carbon atoms are known as androstanes and estranes, respectively. The important mammalian steroid hormones are shown below along with the structure of the precursor, pregneolone. Retinoic acid and vitamin D are not derived from pregnenolone, but from vitamin A and cholesterol respectively.

Pregnenolone: produced directly from cholesterol, the precusor molecule for all C18, C19 and C21 steroids.

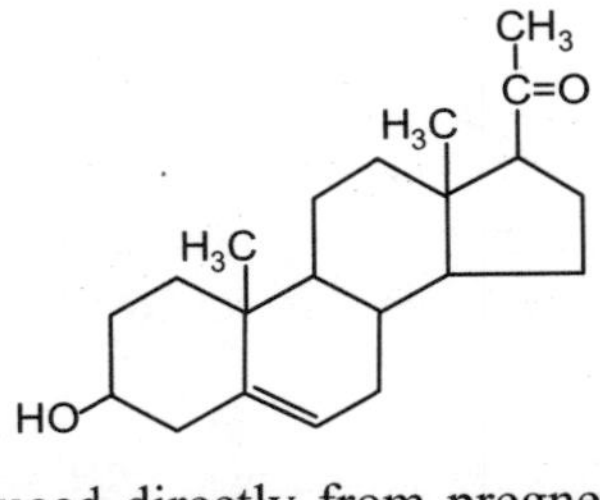

CH_3
C=O
H_3C
H_3C
HO

Progesterone: a progestin, produced directly from pregnenolone and secreted from the corpus luteum, responsible for changes associated with luteral phase of the menstrual cycle, differentiation factor for mammary glands.

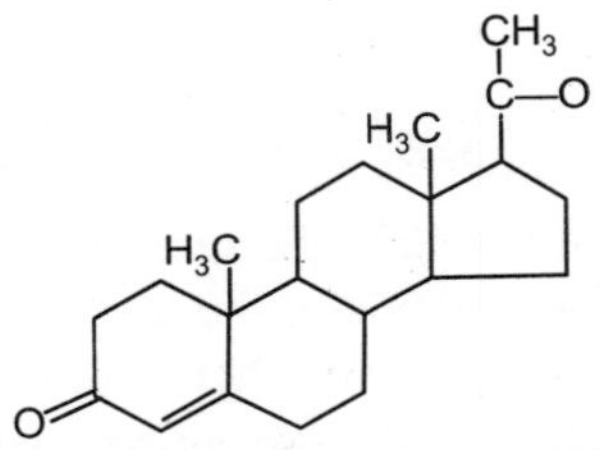

CH_3
C—O
H_3C
H_3C
O

Aldosterone: the principal mineralocorticoid, produced from progesterone in the zona glomerulosa of adrenal cortex, raises blood pressure and fluid volume, increases Na^+ uptake.

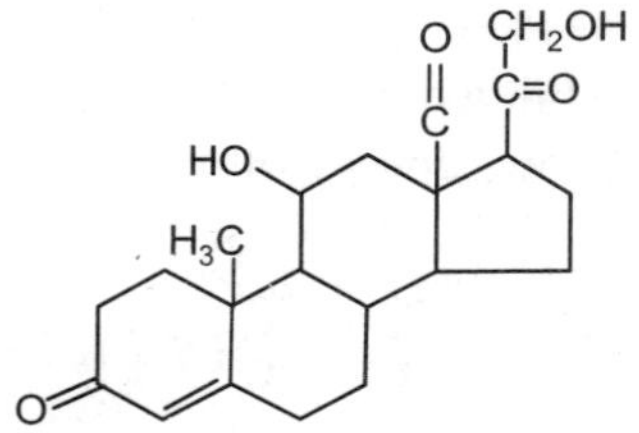

Testosterone: an androgen, male sex hormone synthesized in the testes, responsible for secondary male sex characteristics, produced from progesterone.

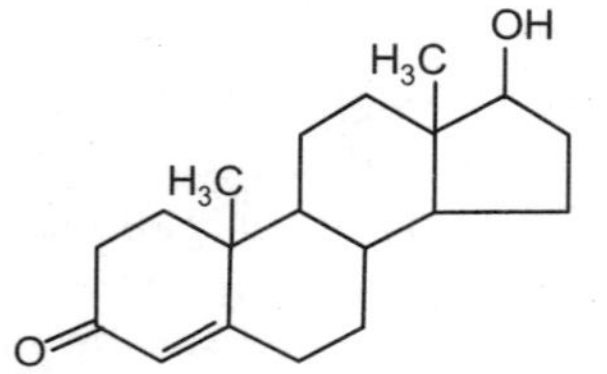

Estradiol: an estrogen, principal female sex hormone, produced in the ovary, responsible for secondary female sex characteristics.

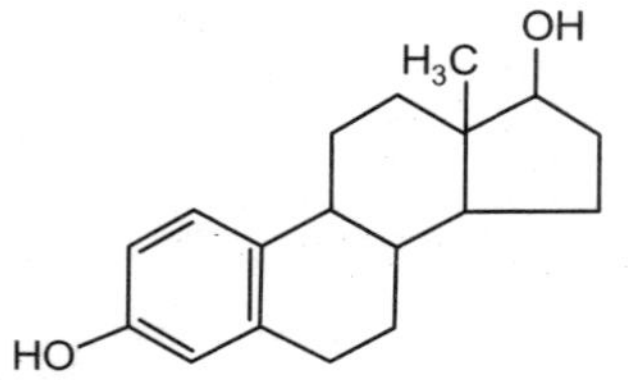

Cortisol: dominant glucocorticoid in humans, synthesized from progesterone in the *zona fasciculata* of the adrenal cortex, involved in stress adaptation, elevates blood pressure and Na^+ uptake, numerous effects on the immune system.

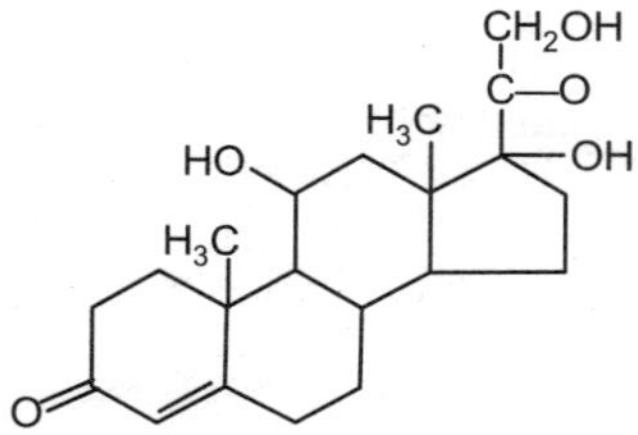

All the steroid hormones exert their action by passing through the plasma membrane and binding to intracellular receptors. The mechanism of action of the thyroid hormones is similar; they interact with intracellular receptors. Both the steroid and thyroid hormone-receptor complexes exert their action by binding to specific nucleotide sequences in the DNA of responsive genes. These DNA sequences are identified as hormone response elements, HREs. The interaction of steroid-receptor complexes with DNA leads to altered rates of transcription of the associated genes.

Steroid Hormone Biosynthesis Reactions

The particular steroid hormone class synthesized by a given cell type depends upon its complement of peptide hormone receptors, its response to peptide hormone stimulation and its genetically expressed complement of enzymes. The following indicates which peptide hormone is responsible for stimulating the synthesis of which steroid hormone:

Luteinizing Hormone (LH): progesterone and testosterone

Adrenocorticotropic hormone (ACTH): cortisol

Follicle Stimulating Hormone (FSH): estradiol

Angiotensin II/III: aldosterone

The first reaction in converting cholesterol to C_{18}, C_{19} and C_{21} steroids involves the cleavage of a 6-carbon group from cholesterol and is the principal committing, regulated, and rate-limiting step in steroid biosynthesis. The enzyme system that catalyzes the cleavage reaction is known as P450-linked side chain cleaving enzyme (P450ssc)or desmolase, and is found in the mitochondria of steroid-producing cells, but not in significant quantities in other cells.

Mitochondrial desmolase is a complex enzyme system consisting of cytochrome P450, and adrenadoxin (a P450 reductant). The activity of each of these components is increased by 2 principal cAMP- and PKA-dependent processes. First, cAMP stimulates PKA, leading to the phosphorylation of a cholesteryl-ester esterase and generating increased concentrations of cholesterol, the substrate for desmolase. Second, long-term regulation is effected at the level the gene for desmolase. This gene contains a cAMP regulatory element (CRE) that binds cAMP and increases the level of desmolase RNA transcription, thereby leading to increased levels of the enzyme. Finally, cholesterol is a negative feedback regulator of HMG CoA reductase activity (see regulation of cholesterol synthesis). Thus, when cytosolic cholesterol is depleted, *de novo* cholesterol synthesis is stimulated by freeing HMG CoA reductase of its feedback constraints. Subsequent to desmolase activity, pregnenolone moves to the cytosol, where further processing depends on the cell (tissue) under consideration.

The various hydroxylases involved in the synthesis of the steroid hormones have a nomenclature that indicates the site of hydroxylation (e.g. 17α-hydroxylase introduces a hydroxyl group to carbon 17). These hydroxylase enzymes are members of the cytochrome P450 class of enzymes and as such also have a nomenclature indicative of the site of hydroxylation in addition to being identified as P450 class enzymes (e.g. the 17α-hydroxylase is also identified as P450c17). The officially preferred nomenclature for the cytochrome P450 class of enzymes is to use the prefix **CYP**. Thus, 17α-hydroxylase should be identified as CYP17A1. There are currently 57 identified CYP genes in the human genome.

Steroids of the Adrenal Cortex

The adrenal cortex is responsible for production of 3 major classes of steroid hormones: glucocorticoids, which regulate carbohydrate metabolism; mineralocorticoids, which regulate the body levels of sodium and potassium; and androgens, whose actions are similar to that of steroids produced by the male gonads. Adrenal insufficiency is known as Addison disease, and in the absence of steroid hormone replacement therapy can rapidly cause death (in 1 - 2 weeks).

The adrenal cortex is composed of 3 main tissue regions: zona glomerulosa, zona fasciculata, and zona reticularis. Although the pathway to pregnenolone synthesis is the same in all zones of the cortex, the zones are histologically and enzymatically distinct, with the exact steroid hormone product dependent on the enzymes present in the cells of each zone. Many of the enzymes of adrenal steroid hormone synthesis are of the class called cytochrome P450 enzymes. These enzymes all have a common nomenclature and a standardized nomenclature. The standardized nomenclature for the P450 class of enzymes is to use the abbreviation CYP. For example the P450ssc enzyme (also called 20,22 desmolase or cholesterol desmolase) is identified as CYP11A1. In order for cholesterol to be converted to pregnenolone in the adrenal cortex it must be transported into the mitochondria where CYP11A1 resides. This transport process is mediated by steroidogenic acute regulatory protein (StAR). This transport process is the rate-limiting step in steroidogenesis *(See Fig. 2.25)*.

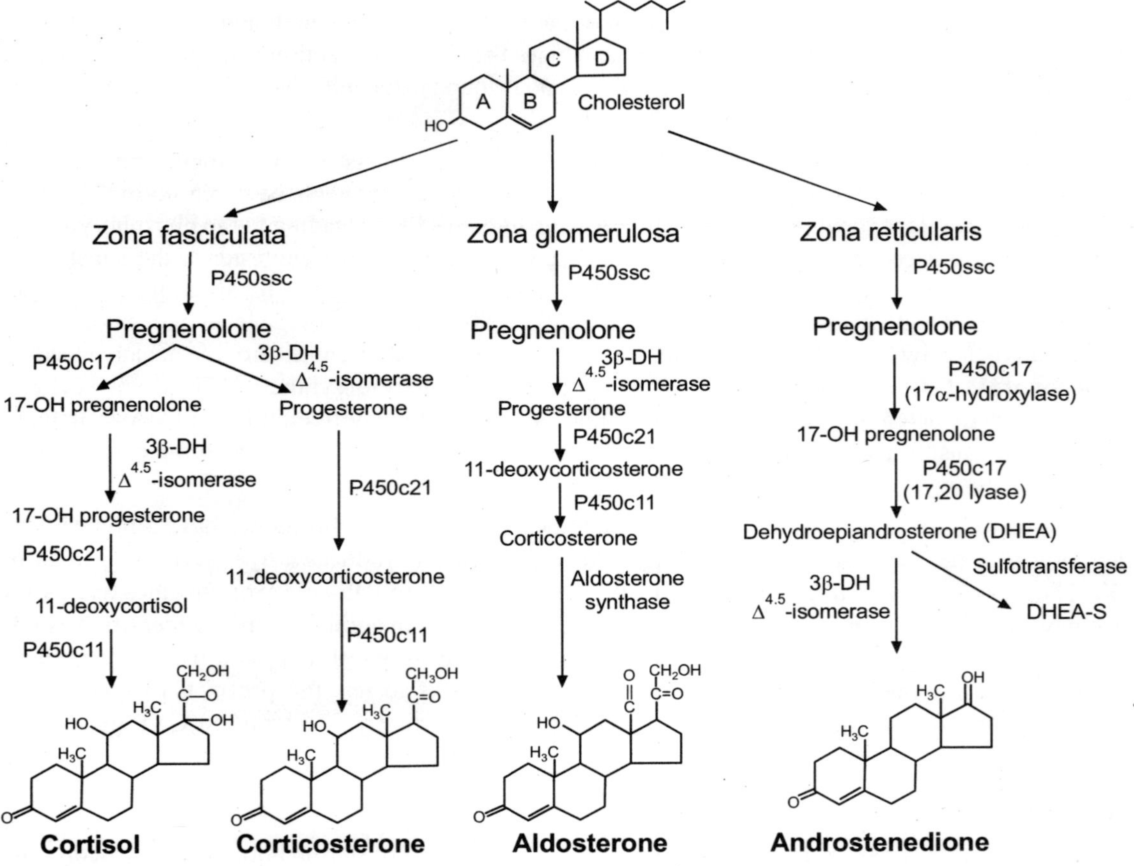

Fig. 2.25 : Steroid genesis

Synthesis of the various adrenal steroid hormones from cholesterol. Only the terminal hormone structures are included. 3β-DH and $\Delta^{4,5}$-isomerase are the two activities of 3β-hydroxysteroid dehydrogenase type 1 (gene symbol HSD3B2), P450c11 is 11β-hydroxylase (CYP11B1), P450c17 is CYP17A1. CYP17A1 is a single microsomal enzyme that has two steroid biosynthetic activities: 17α-hydroxylase which converts pregnenolone to 17-hydroxypregnenolone (17-OH pregnenolone) and 17,20-lyase which converts 17-OH pregnenolone to DHEA. P450c21 is 21-hydroxylase (CYP21A2, also identified as CYP21 or CYP21B). Aldosterone synthase is also known as 18á-hydroxylase (CYP11B2). The gene symbol for sulfotransferase is SULT2A1.

Conversion of prenenolone to progesterone requires the two enzyme activities of HSD3B2: the 3â-hydroxysteroid dehydrogenase and $\Delta^{4,5}$-isomerase activities. Zona glomerulosa cells lack the P450c17 that converts pregnenolone and progesterone to their C_{17} hydroxylated analogs. Thus, the pathways to the glucocorticoids (deoxycortisol and cortisol) and the androgens [dehydroepiandosterone (DHEA) and androstenedione] are blocked in these cells. Zona glomerulosa cells are unique in the adrenal cortex in containing the enzyme responsible for converting corticosterone to aldosterone, the principal and most potent mineralocorticoid. This enzyme is P450c18 (or 18α-hydroxylase, CYP11B2), also called aldosterone synthase. The result is that the zona glomerulosa is mainly responsible for the conversion of cholesterol to the weak mineralocorticoid, corticosterone and the principal mineralocorticoid, aldosterone.

Cells of the zona fasciculata and zona reticularis lack aldosterone synthase (P450c18) that converts corticosterone to aldosterone, and thus these tissues produce only the weak mineralocorticoid corticosterone. However, both these zones do contain the P450c17 missing in zona glomerulosa and thus produce the major glucocorticoid, cortisol. Zona fasciculata and zona reticularis cells also contain P450c17, whose 17,20-lyase activity is responsible for producing the androgens, dehydroepiandosterone (DHEA) and androstenedione. Thus, fasciculata and reticularis cells can make corticosteroids and the adrenal androgens, but not aldosterone.

As noted earlier, P450ssc is a mitochondrial activity. Its product, pregnenolone, moves to the cytosol, where it is converted either to androgens or to 11-deoxycortisol and 11-deoxycorticosterone by enzymes of the endoplasmic reticulum. The latter 2 compounds then re-enter the mitochondrion, where the enzymes are located for tissue-specific conversion to glucocorticoids or mineralocorticoids, respectively.

Regulation of Adrenal Steroid Synthesis

Adrenocorticotropic hormone (ACTH), of the hypothalamus, regulates the hormone production of the zona fasciculata and zona reticularis. ACTH receptors in the plasma membrane of the cells of these tissues activate adenylate cyclase with production of the second messenger, cAMP. The effect of ACTH on the production of cortisol is particularly important, with the result that a classic feedback loop is prominent in regulating the circulating levels of corticotropin releasing hormone (CRH), ACTH, and cortisol.

Mineralocorticoid secretion from the zona glomerulosa is stimulated by an entirely different mechanism. Angiotensins II and III, derived from the action of the kidney protease renin on liver-derived angiotensinogen, stimulate zona glomerulosa cells by binding a plasma membrane receptor coupled to phospholipase C. Thus, angiotensin II and III binding to their receptor leads to the activation of PKC and elevated intracellular Ca^{2+} levels. These events lead to increased P450ssc activity and increased production of aldosterone. In the kidney, aldosterone regulates sodium retention by stimulating gene expression of mRNA for the Na^+/K^+-ATPase responsible for the reaccumulation of sodium from the urine.

The interplay between renin from the kidney and plasma angiotensinogen is important in regulating plasma aldosterone levels, sodium and potassium levels, and ultimately blood pressure. Among the drugs most widely employed used to lower blood pressure are the angiotensin converting enzyme (ACE) inhibitors. These compounds are potent competitive inhibitors of the enzyme that converts angiotensin I to the physiologically active angiotensins II and III. This feedback loop is closed by potassium, which is a potent stimulator of aldosterone secretion. Changes in plasma potassium of as little as 0.1 millimolar can cause wide fluctuations (+/- 50%) in plasma levels of aldosterone. Potassium increases aldosterone secretion by depolarizing the plasma membrane of zona glomerulosa cells and opening a voltage-gated calcium channel, with a resultant increase in cytoplasmic calcium and the stimulation of calcium-dependent processes.

Although fasciculata and reticularis cells each have the capability of synthesizing androgens and glucocorticoids, the main pathway normally followed is that leading to glucocorticoid production. However, when genetic defects occur in the 3 enzyme complexes leading to glucocorticoid production, large amounts of the most important androgen, dehydroepiandrosterone (DHEA), arė produced. These lead to hirsutism and other masculinizing changes in secondary sex characteristics.

Functions of the Adrenal Steroid Hormones

Glucocorticoids: The glucocorticoids are a class of hormones so called because they are primarily responsible for modulating the metabolism of carbohydrates. Cortisol is the most important naturally occurring glucocorticoid. As indicated in the Figure above, cortisol is synthesized in the zona fasciculata of the adrenal cortex. When released to the circulation, cortisol is almost entirely bound to protein. A small portion is bound to albumin with more than 70% being bound by a specific glycosylated α-globulin called transcortin or corticosteroid-binding globulin (CBG). Between 5% and 10% of circulating cortisol is free and biologically active. Glucocorticoid function is exerted following cellular uptake and interaction with intracellular receptors as discussed below. Cortisol inhibits uptake and utilization of glucose resulting in elevations in blood glucose levels. The effect of cortisol on blood glucose levels is further enhanced through the increased breakdown of skeletal muscle protein and adipose tissue triglycerides which provides enegry and substrates for gluconeogenesis. Glucocorticoids also increase the synthesis of gluconeogenic enzymes. The increased rate of protein metabolism leads to increased urinary nitrogen excretion and the induction of urea cycle enzymes.

In addition to the metabolic effects of the glucocorticoids, these hormones are immunosuppressive and anti-inflammatory. Hence, the use of related drugs such as prednisone, in the acute treatment of inflammatory disorders. The anti-inflammatory activity of the glucocorticoids is exerted, in part, through inhibition of phospholipase A_2 (PLA_2) activity with a consequent reduction in the release of arachidonic acid from membrane phospholipids. Arachidonic acid serves as the precursor for the synthesis of various eicosanoids. Glucocorticoids also inhibit vitamin D-mediated intestinal calcium uptake, retard the rate of wound healing, and interfere with the rate of linear growth.

Mineralocorticoids: The major circulating mineralocorticoid is aldosterone. Deoxycorticosterone (DOC) exhibits some mineralocorticoid action but only about 3% of that of aldosterone. As the name of this class of hormones implies, the mineralocorticoids control the excretion of electrolytes. This occurs primarily through actions on the kidneys but also in the colon and sweat glands. The principle effect of aldosterone is to enhance sodium reabsorption in the cortical collecting duct of the kidneys. However, the action of aldosterone is exerted on sweat glands, stomach, and salivary glands to the same effect, i.e. sodium reabsorption. This action is accompanied by the retention of chloride and water resulting in the expansion of extracellular volume. Aldosterone also enhances the excretion of potassium and hydrogen ions from the medullary collecting duct of the kidneys.

Androgens: The androgens, androstenedione and DHEA, circulate bound primarily to sex hormone-binding globulin (SHBG). Although some of the circulating androgen is metabolized in the liver, the majority of interconversion occurs in the gonads, skin, and adipose tissue. DHEA is rapidly converted to the sulfated form, DHEA-S, in the liver and adrenal cortex. The primary biologically active metabolites of the androgens are testosterone and dihydrotestosterone which function by binding intracellular receptors, thereby effecting changes in gene expression and thereby, resulting in the manifestation of the secondary sex characteristics.

Clinical Significance of Adrenal SteroidogenesisDefective synthesis of the steroid hormones produced by the adrenal cortex can have profound effects on human development and homeostasis. In 1855 Thomas Addison identified the significance of the "suprarenal capsules" when he reported on the case of a patient who presented with chronic adrenal insufficiency resulting from progressive lesions of the adrenal glands caused by tubersclerosis. Addison disease thus represents a disorder characterized by adrenal insufficiency. In addition to diseases that result from the total absence of adrenocortical function, there are syndromes that result from hypersecretion of adrenocortical hormones. In 1932 Harvey Cushing reported on several cases of adrenocortical hyperplasia that were the result of basophilic adenomas of the anterior pituitary. As with Addison disease, disorders that manifest with adrenocortical hyperplasia are referred to as Cushing syndrome. Despite the characterizations of adrenal insufficiency and adrenal hyperplasia, there remained uncertainty about the relationship between adrenocortical hyperfunction and virilism. (premature development of male secondary sex characteristics). In 1942 this confusion was resolved by Fuller Albright when he delineated the differences between children with Cushing syndrome and those with adrenogenital syndromes which are more commonly referred to as congenital adrenal hyperplasias (CAH).The CAH are a group of inherited disorders that result from loss-of-function mutations in one of several genes involved in adrenal steroid hormone synthesis. In the virilizing forms of CAH the mutations result in impairment of cortisol production and the consequent accumulation of steroid intermediates proximal to the defective enzyme. All forms of CAH are inherited in an autosomal recessive manner. There are two common and at least three rare forms of CAH that result in virilization. The common forms are caused by defects in either CYP21A2 (21-hydroxylase, also identified as just CYP21 or CYP21B) or CYP11B1 (11β-hydroxylase). The majority of CAH cases (90-95%) are the result of defects in CYP21A2 with a frequency of between 1 in 5,000 and 1 in 15,000. Three rare forms of virilizing CAH result from either defects in 3β-hydroxysteroid dehydrogenase (HSD3B2), placental aromatase or P450-oxidoreductase (POR). An additional CAH is caused by mutations that affect either the 17α-hydroxylase, 17,20-lyase or both activities encoded in the CYP17A1 gene. In individuals harboring CYP17A1 mutations that result in severe loss of enzyme activity there is absent sex steroid hormone production accompanied by hypertension resulting from mineralocorticoid excess.

Gonadal Steroid Hormones

Although many steroids are produced by the testes and the ovaries, the two most important are testosterone and estradiol. These compounds are under tight biosynthetic control, with short and long negative feedback loops that regulate the secretion of follicle stimulating hormone (FSH) and luteinizing hormone (LH) by the pituitary and gonadotropin releasing hormone (GnRH) by the hypothalamus. Low levels of circulating sex hormone reduce feedback inhibition on GnRH synthesis (the long loop), leading to elevated FSH and LH. The latter peptide hormones bind to gonadal tissue and stimulate P450ssc

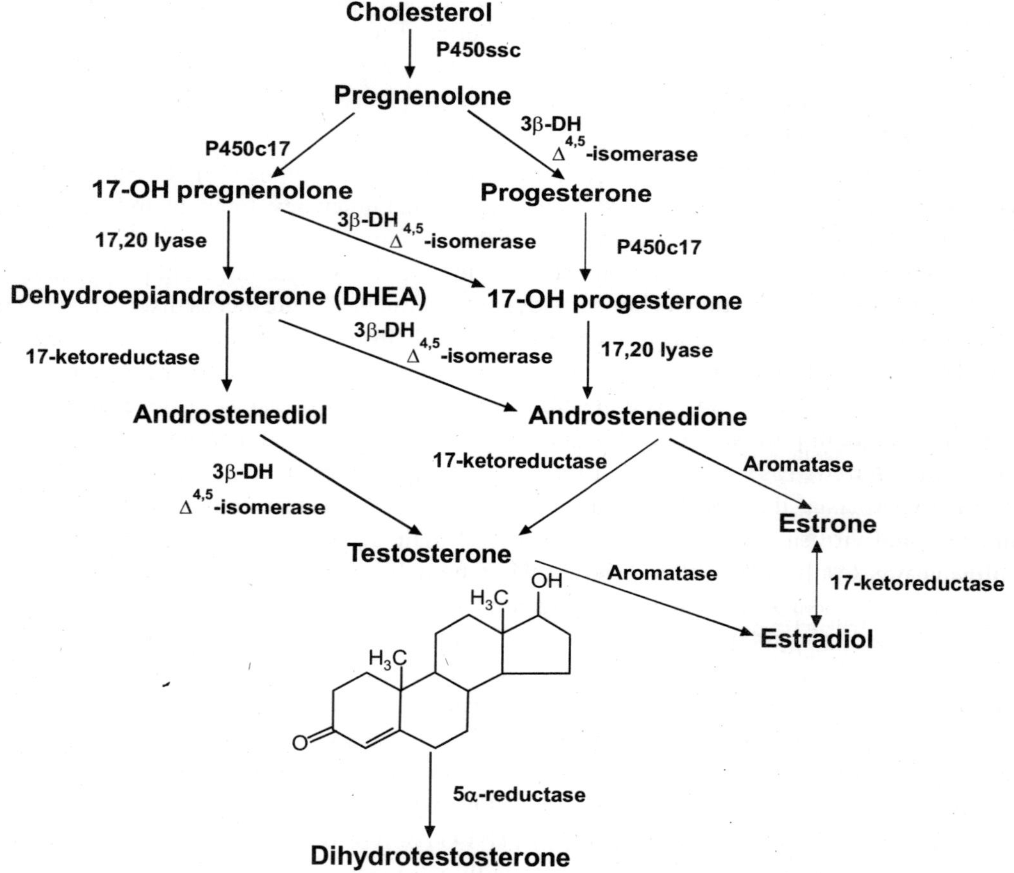

Fig. 2.26 : Synthesis of the male sex hormones in Leydig cells of the testis. P450SSC, 3β-DH, and P450c17 are the same enzymes as those needed for adrenal steroid hormone synthesis.

activity, resulting in sex hormone production via cAMP and PKA mediated pathways. The roles of cAMP and PKA in gonadal tissue are the same as that described for glucocorticoid production in the adrenals, but in this case adenylate cyclase activation is coupled to the binding of LH to plasma membrane receptors.The biosynthetic pathway to sex hormones in male and female gonadal tissue includes the production of the androgens, androstenedione and dehydroepiandrosterone. Testes and ovaries contain an additional enzyme, a 17β-hydroxysteroid dehydrogenase, that enables androgens to be converted to testosterone.In males, LH binds to Leydig cells, stimulating production of the principal Leydig cell hormone, testosterone. Testosterone is secreted to the plasma and also carried to Sertoli cells by androgen binding protein (ABP). In Sertoli cells the Δ4 double bond of testosterone is reduced, producing dihydrotestosterone. Testosterone and dihydrotestosterone are carried in the plasma, and delivered to target tissue, by a specific gonadal-steroid binding globulin (GBG). In a number of target tissues, testosterone can be converted to dihydrotestosterone (DHT). DHT is the most potent of the male steroid hormones, with an activity that is 10 times that of testosterone. Because of its relatively lower potency, testosterone is sometimes considered to be a prohormone.

17,20-lyase is the same activity of CYP17A1 described above for adrenal hormone synthesis. Aromatase (also called estrogen synthetase) is CYP19A1. 17-ketoreductase is also called 17β-hydroxysteroid dehydrogenase type 3 (gene symbol HSD17B3). The full name for 5α-reductase is 5α-reductase type 2 (gene symbol SRD5A2). Place your mouse over structure names to see chemical structures.

Testosterone is also produced by Sertoli cells but in these cells it is regulated by FSH, again acting through a cAMP- and PKA-regulatory pathway. In addition, FSH stimulates Sertoli cells to secrete androgen-binding protein (ABP), which transports testosterone and DHT from Leydig cells to sites of spermatogenesis. There, testosterone acts to stimulate protein synthesis and sperm development.

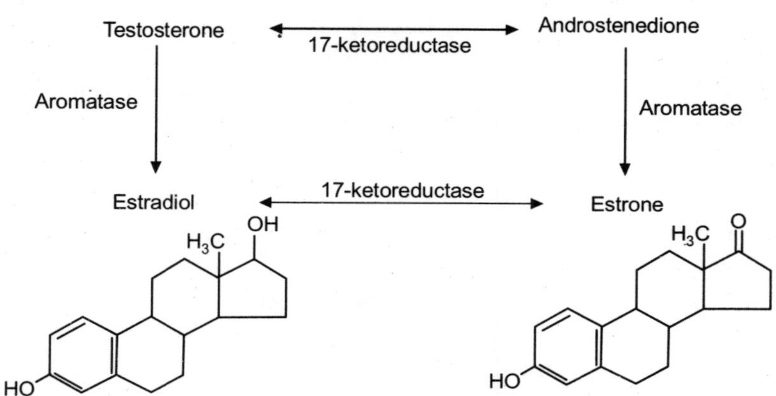

Fig. 2.27 : Synthesis of the major female sex hormones in the ovary

In females, LH binds to thecal cells of the ovary, where it stimulates the synthesis of androstenedione and testosterone by the usual cAMP- and PKA-regulated pathway. An additional enzyme complex known as aromatase is responsible for the final conversion of the latter 2 molecules into the estrogens. Aromatase is a complex endoplasmic reticulum enzyme found in the ovary and in numerous other tissues in both males and females. Its action involves hydroxylations and dehydrations that culminate in aromatization of the A ring of the androgens.

Synthesis of testosterone and androstenedione from cholesterol occurs by the same pathways as indicated for synthesis of the male sex hormones. Aromatase (also called estrogen synthetase) is CYP19A1.

Aromatase activity is also found in granulosa cells, but in these cells the activity is stimulated by FSH. Normally, thecal cell androgens produced in response to LH diffuse to granulosa cells, where granulosa cell aromatase converts these androgens to estrogens. As granulosa cells mature they develop competent large numbers of LH receptors in the plasma membrane and become increasingly responsive to LH, increasing the quantity of estrogen produced from these cells. Granulosa cell estrogens are largely, if not all, secreted into follicular fluid. Thecal cell estrogens are secreted largely into the circulation, where they are delivered to target tissue by the same globulin (GBG) used to transport testosterone.

Steroid Hormone Receptors

The receptors to which steroid hormones bind are ligand-activated proteins that regulate transcription of selected genes. Unlike peptide hormone receptors, that span the plasma membrane and bind ligand outside the cell, steroid hormone receptors are found in the cytosol and the nucleus. The steroid hormone receptors belong to the steroid and thyroid hormone receptor super-family of proteins, that includes receptors for steroid hormones, thyroid hormones, vitamin D and vitamin A (retinoic acid).

When these receptors bind ligand they undergo a conformational change that renders them activated to recognize and bind to specific nucleotide sequences. These specific nucleotide sequences in the DNA are referred to as hormone-response elements (HREs). When ligand-receptor complexes interact with DNA they alter the transcriptional level (responses can be either activating or repressing) of the associated gene. Thus, the steroid-thyroid family of receptors all have three distinct domains: a ligand-binding domain, a DNA-binding domain and a transcriptional regulatory domain. Although there is the commonly observed effect of altered transcriptional activity in response to hormone-receptor interaction, there are family member-specific effects with ligand-receptor interaction. Binding of thyroid hormone to its receptor results in release of the receptor from DNA. Several receptors are induced to interact with other transcriptional mediators in response to ligand binding. Binding of glucocorticoid leads to translocation of the ligand-receptor complex from the cytosol to the nucleus.

The receptors for the retinoids (vitamin A and its derivatives) are identified as RARs (for retinoic acid, RA receptors) and exist in at least three subtypes, RARα, RARβ and RARγ. In addition, there is another family of nuclear receptors termed the retinoid X receptors (RXRs) that represents a second class of retinoid-responsive transcription factors. The RXRs have been shown to enhance the DNA-binding activity of RARs and the thyroid hormone receptors (TRs). There are also three distinct RXRs (α, β and γ). The major difference between the RARs and RXRs is that the former exhibit highest affinity for all-*trans*-retinoic acid (all-*trans*-RA) and the latter for 9-*cis*-RA.

Additional super-family members are the peroxisome proliferator-activated receptors (PPARs). These receptors were originally discovered as proteins activated by agents that stimulate proliferation of peroxisomes in rat liver. An intracellular lipid-binding protein identified as aP2 is expressed exclusively in differentiated adipocytes. An adipocyte-specific enhancer of the aP2 gene is a target for peroxisome proliferators, fatty acids and 9-*cis*-RA. Subsequent to these observations it was found that there is an adipocyte-specific PPAR family identified as PPARγ. The PPARγ proteins form heterodimers with RXRs to activate adipocyte-specific enhancers such as the one in the aP2 gene.

Recent evidence has demonstrated a role for PPARγ proteins in the etiology of type 2 diabetes. A relatively new class of drugs used to increase the sensitivity of the body to insulin are the thiazolidinedione drugs. These compounds bind to and alter the function of PPARγ. Mutations in the gene for PPARγ have been correlated with insulin resistance. It is still not completely clear how impaired PPARγ signaling can affect the sensitivity of the body to insulin or indeed if the observed mutations are a direct or indirect cause of the symptoms of insulin resistance.

BILE ACIDS

Bile acids (also known as bile salts) are steroid acids found predominantly in the bile of mammals. In humans, taurocholic acid and glycocholic acid (derivatives of cholic acid) represent approximately eighty percent of all bile acids. The two major bile acids are cholic acid, and chenodeoxycholic acid. They, their glycine and taurine conjugates, and their 7-alpha-dehydroxylated derivatives (deoxycholic acid and lithocholic acid) are all found in human intestinal bile. An increase in bile flow is exhibited with an increased secretion of bile acids. Bile acid's main function is to facilitate the formation of micelles, which promotes dietary fat processing.

Production and Distribution

Bile acids are produced in the liver by the oxidation of cholesterol. They are conjugated with taurine or the amino acid glycine, or with a sulfate or a glucuronide, and are then stored in the gallbladder. Bile acid also serves the purpose of breaking down fats. Upon eating a meal containing fat, the contents of the gallbladder are secreted into the intestine. In humans, the rate limiting step is the addition of a hydroxyl group on position 7 of the steroid nucleus by the enzyme cholesterol 7 alpha-hydroxylase. Bile acids serve multiple functions, which include: eliminating cholesterol from the body; driving the flow of bile to eliminate catabolites from the liver; emulsifying lipids and fat soluble vitamins in the intestine; and aiding in the reduction of the bacteria flora found in the small intestine and biliary tract.

The term *bile acid* refers to the unconjugated form. *Bile salt* refers to the conjugated form. Bile salts are more efficient at emulsifying fats because at intestinal pH, bile salts are more electrically charged than bile acids.

Synthesis of bile acids is a major consumer of cholesterol in most species (other than humans). The body produces about 800 mg of cholesterol per day and about half of that is used for bile acid synthesis. In total about 20-30 grams of bile acids are secreted into the intestine daily; about 90% of excreted bile acids are reabsorbed (by active transport in the ileum) and recycled. This is referred to as the enterohepatic circulation. Bile is also used to break down fat goblets into tiny droplets. Bile from slaughtered animals can be mixed to use as soap.

Types

Bile salts constitute a large family of molecules, composed of a steroid structure with four rings, a five or eight carbon side-chain terminating in a carboxylic acid, and the presence and orientation of different numbers of hydroxyl groups. The four rings are labelled from left tó right (as commonly drawn) A, B, C, and D, with the D-ring being smaller by one carbon than the other three. The hydroxyl groups have a choice of being in 2 positions, either up (or out) termed α (often drawn by convention as a solid line), or down, termed α (seen as a dashed line in drawings). All bile acids have a hydroxyl group on position 3, which was derived from the parent molecule, cholesterol. In cholesterol, the 4 steroid rings are flat and the position of the 3-hydroxyl is α.

In many species, the initial step in the formation of a bile acid is the addition of a 7-α hydroxyl group. Subsequently, in the conversion from cholesterol to a bile acid, the junction between the first two steroid rings (A and B) is altered, making the molecule bent, and in this process, the 3-hydroxyl is converted to the α orientation. Thus, the default simplest bile acid (of 24 carbons) has two hydroxyl groups at positions 3-α and 7-α. The chemical name for this compound is 3-α,7-α-dihydroxy-5-β-cholan-24-oic acid, or as it is commonly known, chenodeoxycholic acid. This bile acid was first isolated from the domestic goose, from which the "cheno" portion of the name was derived.

Another bile acid, cholic acid (with 3 hydroxyl groups) had already been described, so the discovery of chenodeoxcholic acid (with 2 hydroxyl groups) made the new bile acid a "deoxycholic acid" in that it had one less hydroxyl group than cholic acid. The 5-β portion of the name denotes the orientation of the junction between rings A and B of the steroid nucleus (in this case, they are bent). The term "cholan" denotes a particular steroid structure of 24 carbons, and the "24-oic acid" indicates that the carboxylic acid is found at position 24, which happens to be at the end of the side-chain. Chenodeoxycholic acid is made by many species, and is quite a functional bile acid. Its chief drawback lies in the ability of intestinal bacteria to remove the 7-α hydroxyl group, a process termed dehydroxylation. The resulting bile acid has only a 3-α hydroxyl group and is termed lithocholic acid (litho = stone). It is poorly water-soluble and rather toxic to cells. Bile acids formed by synthesis in the liver are termed "primary" bile acids, and those made by bacteria are termed "secondary" bile acids. As a result, chenodeoxycholic acid is a primary bile acid, and lithocholic acid is a secondary bile acid.

To avoid the problems associated with the production of lithocholic acid, most species add a third hydroxyl group to chenodeoxycholic acid. In this manner, the subsequent removal of the 7-α hydroxyl group by intestinal bacteria will result in a less toxic, still functional dihydroxy bile acid. Over the course of vertebrate evolution, a number of positions have been chosen for placement of the third hydroxyl group. Initially, the 16-α position was favoured, particularly in birds. Later, this position was superseded by a large number of species selecting position 12-α. Primates (including humans) utilize 12-α for their third hydroxyl group position. The resulting primary bile acid in humans is 3-α,7-α,12-α-trihydroxy-5-β-cholan-24-oic acid, or as it is commonly called, cholic acid.

In the intestine, cholic acid is dehydroxylated to form the dihydroxy bile acid deoxycholic acid. In many vertebrate orders still subject to speciation, new species are discarding 12-α hydroxylation in favour of a hydroxy group on position 23 of the side-chain. It should be noted that vertebrate families and species exist that have experimented with and utilize just about every position imaginable on the steroid nucleus and side-chain.

The principal bile acids are:

I. Cholic acid

II. Chenodeoxycholic acid

III. Glycocholic acid

IV. Taurocholic acid

In humans, the most important bile acids are cholic acid, deoxycholic acid, and chenodeoxycholic acid, Prior to secretion by the liver, they are conjugated with either the amino acid glycine or taurine. Conjugation increases their water solubility, preventing passive re-absorption once secreted into the small intestine. As a result, the concentration of bile acids in the small intestine can stay high enough to form micelles and solubilize lipids. The term "critical micellar concentration" refers to both an intrinsic property of the bile acid itself and amount of bile acid necessary to function in the spontaneous and dynamic formation of micelles.

Regulation

As surfactants or detergents, bile acids are potentially toxic to cells and their levels are tightly regulated. They function directly as signaling molecules in the liver and the intestines by activating a nuclear hormone receptor known as FXR also known by its gene name *NR1H4*. This results in inhibition of bile acid synthesis in the liver when bile acid levels are too high. Emerging evidence associates FXR activation with alterations in triglyceride metabolism, glucose metabolism and liver growth.

Clinical Significance

Since bile acids are made from endogenous cholesterol, the enterohepatic circulation of bile acids may be disrupted as a way to lower cholesterol. This is the mechanism of action behind bile acid sequestrants. Bile acid sequestrants bind bile acids in the gut, preventing their reabsorption. In so doing, more endogenous cholesterol is shunted into the production of bile acids, thereby lowering cholesterol levels. The sequestered bile acids are then excreted in the feces.

Cholic acid is a bile acid, a white crystalline substance insoluble in water (soluble in alcohol and acetic acid), with a melting point of 200-201 °C. Salts of cholic acid are called **cholates**. Cholic acid, along with chenodeoxycholic acid, is one of two major bile acids produced by the liver where it is synthesized from cholesterol. Of the two major bile acids, cholate derivatives represent approximately eighty percent of all bile acids. These derivatives are made from cholyl-CoA which forms a conjugate with either glycine, or taurine, yielding glycocholic and taurocholic acid respectively.

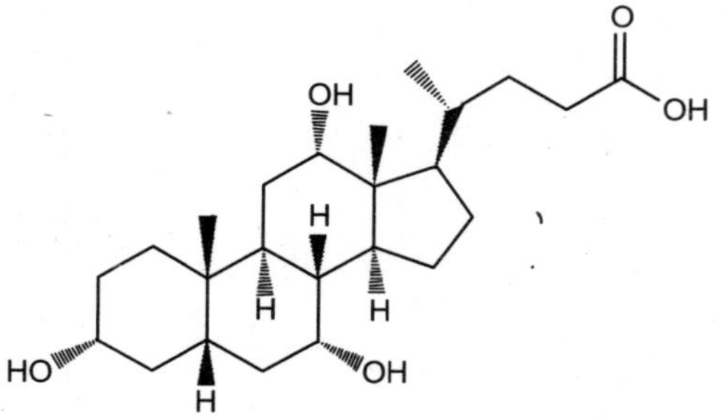

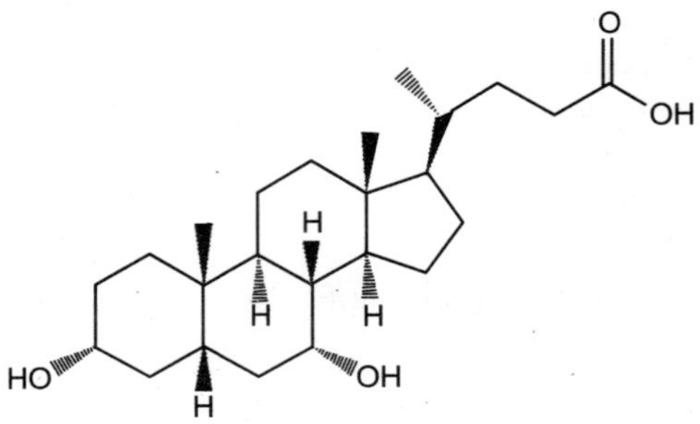

Cholic acid and chenodeoxycholic acid are the most important human bile acids. Some other mammals synthesize predominantly deoxycholic acid.

Chenodeoxycholic acid (also known as **chenodesoxycholic acid**) is a bile acid. It occurs as a white crystalline substance insoluble in water but soluble in alcohol and acetic acid, with melting point at 165-167 °C. Salts of this carboxylic acid are called **chenodeoxycholates**. Chenodeoxycholic acid is one of the 4 main organic acids produced by the liver.

Chenodeoxycholic acid is synthesized in the liver from cholesterol.

This compound, when altered by bacteria in the colon, will result in conversion to its secondary bile acid known as lithocholic acid. Both of these bile acids, in addition to the others, can be conjugated to taurine or glycine. Conjugation, a function carried out by the liver will result in a lowered pKa and therefore, the compounds will remain ionized. These ionized compounds will stay in the gastrointestinal tract until reaching the ileum where they will be reabsorbed. The purpose of this conjugation is to keep the bile acids in the tract until the end to facilitate lipid digestion all the way to the ileum.

In cases where bacteria overgrow in the small intestine, often due to a blind loop in the intestine retaining chyme in one place, the bacteria will de-conjugate the bile acids and therefore impede fat digestion and absorption. This can lead to steatorrhea.

Chenodeoxycholic acid and cholic acid are the most important human bile acids. Some other mammals synthesize predominantly deoxycholic acid.

Potential Applications

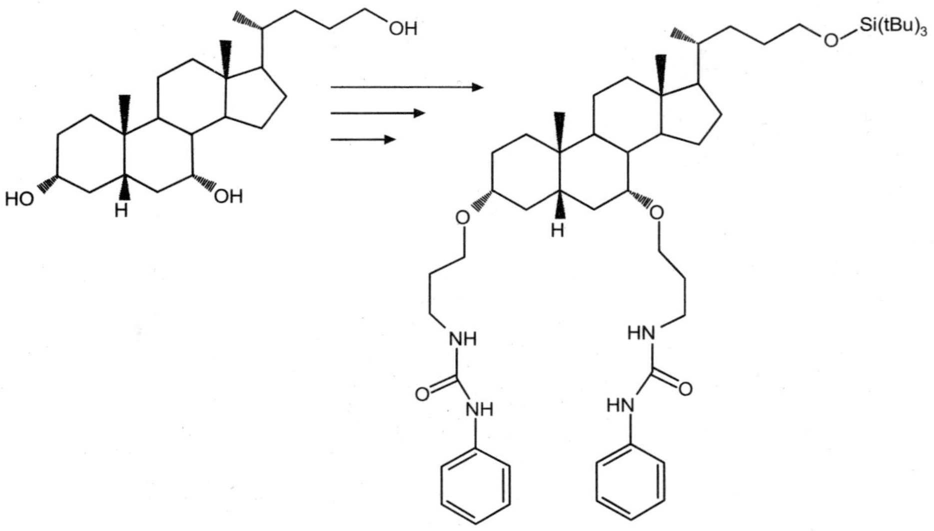

The Australian biotechnology company Giaconda has developed a treatment for Hepatitis C infection that combines chenodeoxycholic acid with bezafibrate.

In supramolecular chemistry molecular tweezers based on a **chenodeoxycholic acid** scaffold is a urea receptor which can contain anions in its binding pocket in order of affinity: $H_2PO_4^-$ (dihydrogen phosphate) > Cl^- > Br^- > I^- reflecting their basicities (tetrabutylammonium counter ion).

Glycocholic acid, or **cholylglycine**, is a crystalline bile acid involved in the emulsification of fats. It occurs as a sodium salt in the bile of mammals. It is a conjugate of cholic acid with glycine. Its anion is called **glycocholate**.

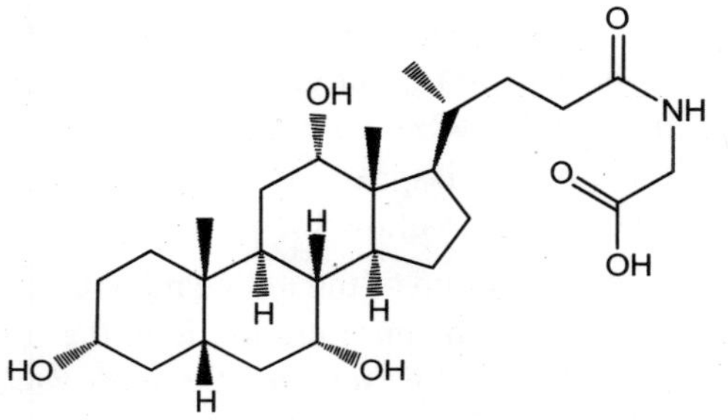

Taurocholic acid, known also as **cholaic acid**, **cholyltaurine**, or **acidum cholatauricum**, is a deliquescent yellowish crystalline bile acid involved in the emulsification of fats. It occurs as a sodium salt in the bile of mammals. It is a conjugate of cholic acid with taurine. In medical use, it is administered as a cholagogue and choleretic.Hydrolysis of taurocholic acid yields taurine, a nonessential amino acid.For commercial use, taurocholic acid is manufactured from cattle bile, a byproduct of the meat-processing industry.

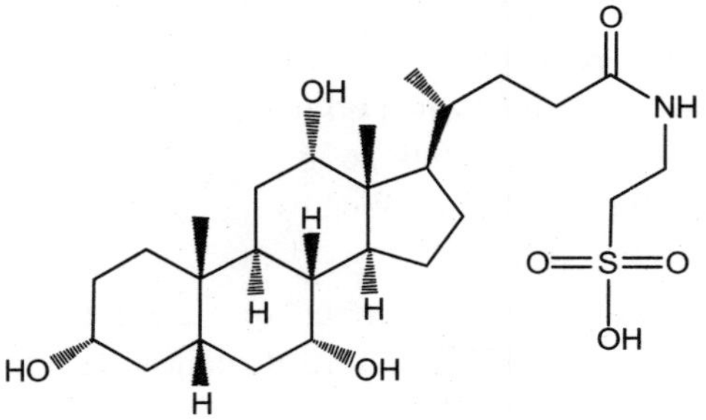

CHARACTERISATION OF LIPIDS

Saponification Number, Iodine Number, Acid Number, Acetyl Number, Polensky Number, and R.M. Number

Saponification value (or "saponification number", also referred to as "sap" in short) represents the number of milligrams of potassium hydroxide or sodium hydroxide required to saponify 1g of fat under the conditions specified. It is a measure of the average molecular weight (or chain length) of all the fatty acids present. As most of the mass of a fat/triester is in the 3 fatty acids, it allows for comparison of the average fatty acid chain length.

If more moles of base are required to saponify N grams of fat then there are more moles of the fat the chain lengths are relatively small, given the following relation:

Number of moles = mass of oil/relative atomic mass

Handmade soapmakers who aim for bar soap use NaOH sap values which are derived from the saponification value calculated by laboratories (KOH sap value). To convert KOH values to NaOH values, divide the KOH values by the ratio of the molecular weights of KOH and NaOH (1.403).

Standard methods for analysis are for example: ASTM D 94 (for petroleum) and DIN 51559.

The calculated saponification value is not applicable to fats and oils containing high amounts of unsaponifiable material, free fatty acids (>0.1%), or mono- and diacylglycerols (>0.1%).

The **iodine value** (or "iodine adsorption value" or "iodine number" or "iodine index") in chemistry is the mass of iodine in grams that is consumed by 100 grams of a chemical substance. An iodine solution is yellow/brown in colour and any chemical group in the substance that reacts with iodine will make the color disappear at a precise concentration. The amount of iodine solution thus required to keep the solution yellow/brown is a measure of the amount of iodine sensitive reactive groups.

One application of the iodine number is the determination of the amount of unsaturation contained in fatty acids. This unsaturation is in the form of double bonds which react with iodine compounds. The higher the iodine number, the more unsaturated fatty acid bonds are present in a fat. In a typical procedure the acid is treated with an excess of the **Hanus solution** which is a solution of **iodobromine** (BrI). Unreacted iodobromine is reacted with potassium iodide which converts it to iodine. The iodine concentration is then determined by titration with sodium thiosulfate.

Standard methods for analysis are for example ASTM D5768-02(2006) and DIN 53241.

For a simple analysis, 0.2 grams of the fat is mixed with 20cm^3 Wij's solution and 10cm^3 1,1,1-trichloroethane. It is then left in the dark for 30mins. Next, 15cm^3 of 10% potassium iodide solution and 10cm^3 of deionized water is added. This is then titrated against 0.1M sodium thiosulphate solution. 1cm^3 of 0.1M sodium thiosulfate solution = 0.01269g of iodine. The difference between a control titration and the titration with the fat present multiplied by this factor gives the mass of iodine absorbed by the oil.

In chemistry, **acid value** (or "neutralization number" or "acid number" or "acidity") is the mass of potassium hydroxide (KOH) in milligrams that is required to neutralize one gram of chemical substance. The acid number is a measure of the amount of carboxylic acid groups in a chemical compound, such as a fatty acid, or in a mixture of compounds. In a typical procedure, a known amount of sample dissolved in organic solvent is titrated with a solution of potassium hydroxide with known concentration and with phenolphthalein as a colour indicator.

The acid number is used to quantify the amount of acid present, for example in a sample of biodiesel. It is the quantity of base, expressed in milligrams of potassium hydroxide, which is required to neutralize the acidic constituents in 1 g of sample.

$$AN = \left(V_{eq} - b_{eq}\right) N \frac{56.1}{W_{oil}}$$

V_{eq} is the amount of titrant (ml) consumed by the crude oil sample and 1ml spiking solution at the equivalent point, b_{eq} is the amount of titrant (ml) consumed by 1 ml spiking solution at the equivalent

For the determination of the iodine number see appendix of this book also consult book A chemical Analyser's Guide by the same author.

point, and 56.1 is the molecular weight of KOH.

The molarity concentration of titrant (N) is calculated as such:

$$N = \frac{1000W_{KHP}}{204.23V_{eq}}$$

In which W_{KHP} is the amount (g) of KHP in 50 ml of KHP standard solution, V_{eq} is the amount of titrant (ml) consumed by 50 ml KHP standard solution at the equivalent point, and 204.23 is the molecular weight of KHP.

There are standard methods for determining the acid number, such as ASTM D 974 and DIN 51558 (for mineral oils, biodiesel), or specifically for Biodiesel using the European Standard EN 14104 and ASTM D664 are both widely utilised worldwide. Acid number (mg KOH/g oil) for biodiesel should to be lower than 0.50 mgKOH/g in both EN 14214 and ASTM D6751 standard fuels. This is since the FFA produced may corrode automotive parts and these limits protect vehicle engines and fuel tanks.

As oil-fats rancidify, triglycerides are converted into fatty acids and glycerol, causing an increase in acid number. A similar observation is observed with Biodiesel aging through analogus oxidation processes and when subjected to prolonged high temperatures (ester thermolysis) or through exposure to acids or bases (acid/base ester hydrolysis).

Acetyl Number Number of milligrams of potassium hydroxide required to neutralize the acetic acid set free from 1 gram of acetylated compound when the latter is subjected to hydrolysis.

The **Polenske value** (also known as the **Polenske number**) is a value determined when examining fat. The Polenske value is an indicator of how much volatile fatty acid can be extracted from fat through saponification. It is equal to the number of mililiters of 0.1 normal hydroxide solution necessary for the neutralization of the water-insoluble volatile fatty acids distilled and filtered from 5 grams of a given saponified fat. (The hydroxide solution used in such a titration is typically made from sodium hydroxide, potassium hydroxide, or barium hydroxide.)

The value is named for the chemist who developed it, Eduard Polenske.

The Reichert value and Kirschner value are related numbers based on similar tests.

PROTEINS

The term protein has derived from the Greek word "protos" meaning primary and no doubt well chosen. It represents an enormous group of complex nitrogenous compounds, mebers of which occur in all animal and plant protoplasm.

There is greater diversity in the chemical composition of proteins than in that of any other group of biologically important compounds.

The Protein content of tissues varies widely, asn shown in the table below:

Source	Protein content in percentage
Fresh mammalian muscle (striated† and unstriated)	18 – 20%
Blood Plasma	6.5 – 7.5%
Brain	8%
Egg Yolk	15%
Egg White	12%
Cow Milk	3.3%
Cheese	14 – 49%
Lettuce	1.2%
Asperegus	1.8%
Cabbage	1.6%
Potatoes	2%
Beets	1.6%
Cereal Grains	10 – 15%
Navy Beens	22%
Soya Beens	**37%**
Pea Nuts	26%
Almonts	21%
Fruits	0.4 – 1.5%
Fish	25%
Meat	21%
Beef	28%

Plant readilysynthesize proteins from simple substances such as nitrogen compounds, Animals, however cannot do this therefore are directly or indirectly dependant on plants for their protein supply. Althouigh animals utilize food carbohydrate and lipids to a small extent as structural components of protoplasm; their chief utilization is as sources of energy for operating the anatomical mechanism. On the otyehr hand the main function of food proteins is to provide the major organic structuires of the protoplasmic machine itself, although excess is utilized as asource of energy.

Much of the regulation and integration of physiological process uin the body is also accomplished by hormones which are secreated by endochine glands. Several proteins, such as insulin of the pancreas and the hormones of the anterior pituitary glands.

The chemical process in volved in the digestion of the food and also in the utilization or metabiolism of foods in animal tissue are in general catalized and directed by substances known "enzymes". The chemical processes of plants are similarly under the control of enzymes. Manyenzymes have been isolated purified and analyzed, without exception, they have all been found to be protein in character.

Certain deseases are caused by some critical agents commonly known as " virus" they are formation of complicated protein structures.

It should be noted that all the proteins no matter what has the basic composition C, O, H, N and generally S and sometimes P. Elements I, Fe, Cu, Zn are also occationaly present. The approximate average elementary composition of protein is as follows.

The average composition chart has been shown in the table below:

Element	Percent
Carbon	50
Hydrogen	7
Oxygen	23
Nitrogen	16
Sulphur	0-3
Phosphorous	0-3

PROPERTIES AND FUNCTIONS OF PROTEINS

Proteins cannot be classified purely on chemical basis at the present time because of deficiencies of knowledge of their composition and structure, and they must be classified on the basis of both physical properties of most of the values for classification purposes are those of solubility and heat coagulability. The classification given below is based upon the recommendation of a joint comette of the Americal Society of Biological Chemists.

According to this system main groups are :

1. Simple Proteins
2. Conjugated Proteins
3. Derived Proteins

And each of thease main groups are subdivided into classes.

Simple Protein

Albumin (Latin: albus, white) refers generally to any protein with water solubility, which is moderately soluble in concentrated salt solutions, and experiences heat coagulation (protein denaturation). Substances containing albumin, such as egg white, are called *albuminoids*. It has a molecular weight of about 65,000 and consists of 584 amino acids and contains no carbohydrate.

Types

Serum albumin

The most well-known type of albumin is the serum albumin in the blood.

Serum albumin is the most abundant blood plasma protein and is produced in the liver and forms a large proportion of all plasma protein. The human version is human serum albumin, and it normally constitutes about 70% of human plasma protein.

Serum albumins are important in regulating blood volume by maintaining the oncotic pressure (also known as colloid osmotic pressure) of the blood compartment. They also serve as carriers for molecules of low water solubility, including lipid soluble hormones, bile salts, unconjugated bilirubin, free fatty acids (apoprotein), calcium, iron (transferrin), and some drugs like warfarin, phenobutazone, clofibrate & phenytoin. Competition between drugs for albumin binding sites may cause drug interaction by increasing the free fraction of one of the drugs, thereby affecting potency.

Specific types include:

- human serum albumin
- bovine serum albumin (cattle serum albumin) or BSA, often used in medical and molecular biology labs.

Low albumin (hypoalbuminaemia) may be caused by liver disease, nephrotic syndrome, burns, protein-losing enteropathy, malabsorption, malnutrition, late pregnancy, artefact, genetic variations and malignancy.

High albumin is almost always caused by dehydration. In some cases of retinol (Vitamin A) deficiency the albumin level can become raised to High-normal values(ex: 4.9 g/dL). This is because retinol causes cells to swell with water (this is also the reason too much Vitamin A is toxic). In lab experiments it has been shown that All-trans retinoic acid down regulates human albumin production

Normal range of human serum albumin in adults (> 3 y.o.) is 3.5 to 5 g/dL. For children less than three years of age, the normal range is broader, 2.5-5.5 g/dL.

Other types

Other types include the storage protein ovalbumin in egg white, and different storage albumins in the seeds of some plants.

- Note that the protein 'albumin' is spelled with an 'i', while "albumen" with an 'e', is the white of an egg which contains (among other things) several dozen types of albumin (with an 'i'), mostly ovalbumin.

Albumin Measurement

Plasma albumin is a component of the Liver Function Tests (LFTs) but may be ordered separately. Albumin can be measured in Serum (yellow-top tube), plain tube with no additives red-top tube) or heparin plasma (green-top tube). The reference interval is 36 - 52 g/L. (note upper limit increased from 47 g/L on the 15th June 2007) One of the method used is Bromocresol green on a Roche Modular or Olympus AU2700 analyser.

α 1-Antitrypsin or **α_1-antitrypsin** (A1AT) is a glycoprotein and generally known as **serum trypsin inhibitor**. The correct name, however, is α-1 proteinase inhibitor (A1PI) because it is a serine protease inhibitor (serpin), inhibiting a wide variety of proteases. It protects tissues from enzymes of inflammatory cells, especially elastase, and has a reference range in blood of 1.5 - 3.5 gram/liter, but the concentration can rise manyfold upon acute inflammation. In its absence, elastase is free to break down elastin, which contributes to the elasticity of the lungs, resulting in respiratory complications such as emphysema, or COPD (chronic obstructive pulmonary disease) in adults and cirrhosis in adults or children.

Function

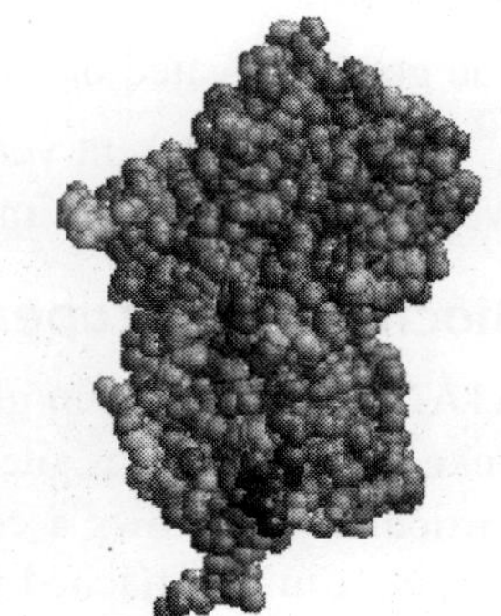

A1AT is a 52-kDa serpin (serine protease inhibitor), and, in medicine, it is considered the most prominent serpin; the terms *α1-antitrypsin* and *protease inhibitor* (P_i) are often used interchangeably.

Most serpins inactivate enzymes by binding to them covalently, requiring very high levels to perform their function. In the acute phase reaction, a further elevation is required to "limit" the damage caused by activated neutrophil granulocytes and their enzyme elastase, which breaks down the connective tissue fiber elastin.

Like all serine protease inhibitors, A1AT has a characteristic secondary structure of β sheets and alpha helices. Mutations in these areas can lead to non-functional proteins that can polymerise and accumulate in the liver (infantile hepatic cirrhosis).

Role in Disease

Disorders of the enzyme include alpha 1-antitrypsin deficiency, a hereditary disorder in which lack of alpha 1-antitrypsin leads to a chronic uninhibited tissue breakdown. This causes the subsequent degradation especially of lung tissue and to the manifestation of pulmonary emphysema.[3] Evidence has shown that cigarette smoke can lead to oxidation of methionine 358 of α_1-antitrypsin, a residue essential for binding elastase; this is the mechanism by which cigarette smoking (or second-hand smoke) can lead to emphysema. Because A1AT is created in the liver, certain mutations in the DNA code for the enzyme can cause misfolding and impaired secretion of the enzyme, which can lead to liver cirrhosis.

A remarkable form of P_i, (termed $P_{iPittsburgh}$) functions as an antithrombin (a related serpin), due to a mutation (Met358Arg). One patient with this abnormality has been described; he died of a lethal bleeding diathesis.

Nomenclature

The protein was called "antitrypsin" because of its ability to covalently bind and irreversibly inactivate the enzyme trypsin in vitro. Trypsin, a type of peptidase, is a digestive enzyme active in the duodenum and elsewhere.

The term *alpha-1* refers to the enzyme's behaviour on protein electrophoresis. On electrophoresis, the protein component of the blood is separated by electric current. There are several *clusters*, the first being albumin, the second being the α, the third β and the fourth γ immunoglobulins. The non-albumin proteins are referred to as globulins.

The *alpha* region can be further divided into two sub-regions, termed '1' and '2'. α1-antitrypsin is the main enzyme of the alpha-globulin 1 region.

Another name used is *α-1 proteinase inhibitor* (α_1-PI).

Genetics

The gene is located on the long arm of the fourteenth chromosome (14q32.1).

Over 80 different versions of α_1-antitrypsin have been described in various populations. North-Western Europeans are most at risk for carrying a mutant form of A1AT.

Biochemical Properties

A1AT is a single-chain glycoprotein consisting of 394 amino acids in the mature form. The three N-linked glycosylations sites are mainly equipped with so-called diantennary N-glycans. However, one particular site shows a considerable amount of heterogeneity since tri- and even tetraantennary N-glycans can be attached to the Asparagine 107 (ExPASy amino acid nomenclature). These glycans carry different amounts of negatively-charged sialic acids, this causes the heterogeneity observed on normal A1AT when analysed by isoelectric focussing. In addition, the fucosylated triantennary N-glycans were shown to have the fucose as part of a so-called Sialyl Lewis x epitope, which could confer this protein particular protein-cell recognition properties. The single cysteine residue of A1AT in position 256 (ExPASy nomenclature) is found to be covalently linked to a free single cysteine by a disulfide bridge.

Analysis

As protein electrophoresis is imprecise, A1AT is analysed by electrofocusing (isoelectric focusing analysis), where the protein is passed along a pH gradient.

Normal A1AT is termed *M*, as it is neutral and does not run very far. Other variants are less functional, and are termed A-L and N-Z, dependent on whether they run proximal or distal to the M band. The presence of deviant bands on electrofocusing can signify the presence of alpha 1-antitrypsin deficiency.

As every person has two copies of the A1AT gene, a heterozygote with two different copies of the gene may have two different bands showing on electrofocusing, although heterozygote with one null mutant that abolishes expression of the gene will only show one band.

In blood test results, the electrofocusing results are notated as in P_iMM, where P_i stands for protease inhibitor and "MM" is the banding pattern of that patient.

α1-antitrypsin levels in the blood depend on the genotype. Some mutant forms fail to fold properly and are, thus, targeted for destruction in the proteasome, whereas others have a tendency to polymerise, being retained in the endoplasmic reticulum. The serum levels of some of the common genotypes are:

- PiMM: 100% (normal)
- PiMS: 80% of normal serum level of A1AT

- PiSS: 60% of normal serum level of A1AT
- PiMZ: 60% of normal serum level of A1AT
- PiSZ: 40% of normal serum level of A1AT
- PiZZ: 10-15% (severe α1-antitrypsin deficiency)
- PiZ is caused by a glutamate to lysine mutation at position 342
- PiS is caused by a glutamate to valine mutation at position 264

Other rarer forms have been described; in all there are over 80 variants.

Therapeutic Use

Recombinant α1-antitrypsin is not yet commercially available, but is under investigation as a therapy for α1-antitrypsin deficiency. Therapeutic concentrates are prepared from the blood plasma of blood donors.

Globulin is one of the two types of serum proteins, the other being albumin. This generic term encompasses a heterogeneous series of families of proteins, with larger molecules and less soluble in pure water than albumin, which migrate less than albumin during serum electrophoresis. The normal range in blood is 2 to 3.5 g/dl.

It is sometimes used synonymously with globular protein. However, albumin is also a globular protein, but not a globulin. All other serum globular proteins are globulins.

Protein electrophoresis is used to categorize globulins into the following four categories:

- α1 globulins
- α2 globulins
- β globulins
- Gamma globulins (one group of γ-globulins are immunoglobulins, that function as antibodies)

αGlobulins are a group of globular proteins in plasma, which are highly mobile in alkaline or electrically charged solutions. They inhibit certain blood protease and inhibitor activity.

α1 globulins

α_1-antitrypsin

Alpha 2 globulins

- Haptoglobin
- α_2-macroglobulin
- Ceruloplasmin
- Thyroxine-binding globulin

Haptoglobin (abbreviated as **Hp**) is a protein in the blood plasma that binds free hemoglobin released from erythrocytes with high affinity and thereby inhibits its oxidative activity. The haptoglobin-hemoglobin complex will then be removed by the reticuloendothelial system (mostly the spleen). In clinical settings, the haptoglobin assay is used to screen for and monitor intravascular hemolytic anemia.

Clinical significance

Since the reticuloendothelial system will remove the haptoglobin-hemoglobin complex from the body, haptoglobin levels will be decreased in hemolytic anemias. In the process of binding hemoglobin, haptoglobin sequesters the iron within hemoglobin, preventing iron-utilizing bacteria from benefitting from hemolysis. It is theorized that because of this, haptoglobin has evolved into an acute phase protein.

Test order protocol

Haptoglobin is ordered whenever a patient exhibits symptoms of anemia, such as pallor, fatigue, shortness of breath along with physical signs of hemolysis, such as jaundice or dark-coloured urine. The test is also commonly ordered as a hemolytic anemia battery which also includes a reticulocyte count and a peripheral blood smear. It can also be ordered along with a Direct Antiglobulin Test when a patient is suspected of having a transfusion reaction or symptoms of autoimmune hemolytic anemia. Finally, it may be ordered in conjunction with a bilirubin.

Results interpretation

A decrease in haptoglobin can support a diagnosis of hemolytic anemia, especially when correlated with a decreased RBC count, Hemoglobin, and Hematocrit, and also an increased reticulocyte count.

If the reticulocyte count is increased, but the haptoglobin level is normal, this may indicate that cellular destruction is occurring in the spleen and liver, which may indicate a drug induced hemolysis, or a red cell dysplasia. The spleen and liver recognize an error in the red cells (either Drug coating the red cell membrane, or a dysfunctional red cell membrane), and destroy the cell. This type of destruction does not release hemoglobin into the peripheral blood, so the haptoglobin cannot bind to it. Thus, the haptoglobin will stay normal.

If there are symptoms of anemia but both the reticulocyte count and the haptoglobin level are normal, the anemia is most likely not due to hemolysis, but instead some other error in cellular production, such as aplastic anemia

Haptoglobin levels which are decreased but do not accompany signs of anemia may indicate liver damage, as the liver is not producing enough haptoglobin to begin with.

As haptoglobin is indeed an acute phase protein, any inflammatory process (infection, extreme stress, burns, major crush injury, allergy, etc) may increase the levels of plasma haptoglobin.

Structure

Haptoglobin is produced mostly by hepatocytes but also by other tissues: e.g. skin, lung, and kidney. According to Trayburn and Woods (2004) several studies have shown that the Haptoglobin gene is expressed in murine and human adipose tissue. Haptoglobin, in its simplest form, consists of two α- and two β-chains, connected by disulfide bridges. The chains originate from a common precursor protein which is proteolytically cleaved during protein synthesis.

Hp exists in two allelic forms in the human population, so called *Hp1* and *Hp2*; the latter one having arisen due to the partial duplication of *Hp1* gene. Three phenotypes of Hp, therefore are found in humans: Hp1-1, Hp2-1, and Hp2-2. Hp of different phenotypes have been shown to bind hemoglobin with different affinities, with Hp2-2 being the weakest binder.

alpha-2-Macroglobulin, also known as **α_2-macroglobulin** and abbreviated as **α2M** and **A2M**, is a large plasma protein found in the blood. It is produced by the liver, and is a major component of the α-2 band in protein electrophoresis.

Structure

Alpha-2-macroglobulin is composed of four identical subunits bound together by -S-S- bonds.

Function

Alpha-2-macroglobulin is able to inactivate an enormous variety of proteinases (including serine-, cysteine, aspartic- and metalloproteinases).

Alpha-2-macroglobulin has in its structure a 35 aminoacid "bait" region. Proteinases binding and cleaving the bait region become bound to α2M. The proteinase-α2M complex is recognised by macrophage receptors and cleared from the system.

It functions as an inhibitor of coagulation by inhibiting thrombin.

It functions as an inhibitor of fibrinolysis by inhibiting plasmin and kallikrein

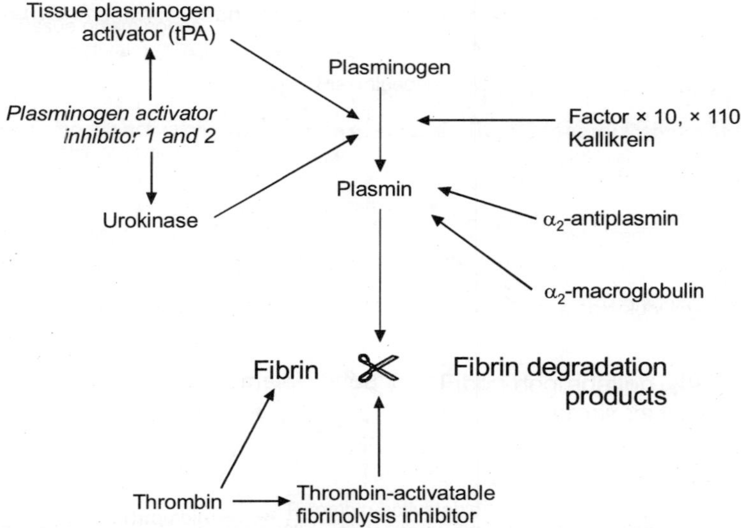

Fig. 2.28 : Fibrinolysis (simplified). Blue arrows denote stimulation, and red arrows inhibition

Disease

α-2-macroglobulin levels are increased in nephrotic syndrome, a condition wherein the kidneys start to leak out some of the smaller blood proteins. Because of its size, α2-macroglobulin is retained in the bloodstream. Increased production of all proteins means á2-macroglobulin concentration increases. This increase has little adverse effect on the health, but is used as a diagnostic clue. Longstanding chronic renal failure can lead to amyloid by *alpha-2-macroglobulin* (*see main article: amyloid*).

A common variant (29.5%) (polymorphism) of α2-macroglobulin leads to increased risk of Alzheimer's disease, although the mechanism is unknown.

α-2-macroglobulin binds to and removes the active forms of the gelatinase (MMP-2 and MMP-9) from the circulation via scavenger receptors on the phagocytes.

Ceruloplasmin (or **caeruloplasmin**) is officially known as **ferroxidase** or **iron(II):oxygen oxidoreductase**. It is the major copper-carrying protein in the blood, and in addition plays a role in iron metabolism. It was first described in 1948.

Function: It is an enzyme (EC 1.16.3.1) synthesized in the liver containing 6 atoms of copper in its structure. Ceruloplasmin carries 90% of the copper in our plasma. The other 10% is carried by albumin, albumin may be confused at times to have a greater importance as a copper carrier because it binds copper less tightly than ceruloplasmin. Ceruloplasmin exhibits a copper-dependent oxidase activity, which is associated with possible oxidation of Fe^{2+} (ferrous iron) into Fe^{3+} (ferric iron), therefore assisting in its transport in the plasma in association with transferrin, which can only carry iron in the ferric state. The molecular weight of human ceruloplasmin is reported as 151kDa.

Pathology

Like any other plasma protein, levels drop in patients with hepatic disease due to reduced synthesizing capabilities.

- Mechanisms of low ceruplasmin levels:
  - Gene expression genetically low: aceruloplasminemia
  - Copper levels are low in general:
  - Malnutrition/trace metal deficiency in the food source
  - Copper does not cross the intestinal barrier due to ATP7A deficiency in Menkes disease
  - Delivery of copper into the lumen of the ER-Golgi network is absent in hepatocyte due to absent ATP7B in Wilson's disease.

Copper availability doesn't affect the translation of the nascent protein. However, the apoenzyme without copper is unstable. Apoceruloplasmin is largely degraded intracellularly in the hepatocyte and the small amount that is released has a short circulation half life of 5 hours as compared to the 5.5 days for the holo-ceruloplasmin.

Mutations in the ceruloplasmin gene can lead to the rare genetic human disease aceruloplasminemia, characterized by iron overload in the brain, liver, pancreas, and retina.

Interpretation

Reference ranges for blood tests, comparing blood content of ceruloplasmin (shown in light blue) with other constituents.

Decreased levels

Lower-than-normal ceruloplasmin levels may indicate:

- Menkes disease (Menke's kinky hair syndrome) (very rare)
- Wilson's disease (a rare copper storage disease)
- Overdose of Vitamin C
- Copper deficiency
- Aceruloplasminemia

Elevated levels

Greater-than-normal ceruloplasmin levels may indicate:

- pregnancy
- lymphoma
- acute and chronic inflammation (it is an acute-phase reactant)
- rheumatoid arthritis
- Alzheimer's Disease

Thyroxine-binding globulin (TBG) binds Thyroid hormone in circulation. It is one of three proteins (along with transthyretin and albumin) responsible for carrying the thyroid hormones thyroxine (T4) and 3,5,3-triiodothyronine (T3) in the bloodstream. Of these three proteins, TBG has the highest affinity for T4 and T3, but is present in the lowest concentration. Despite its low concentration, TBG carries the majority of T4 in serum. Due to the very low serum concentration of T4 & T3, TBG is rarely more than 25% saturated with its ligand. Unlike transthyretin and albumin, TBG has a single binding site for T4/T3. TBG is synthesized primarily in the liver as a 54 kDa protein. Genomically, TBG is a serpin, although it has no inhibitory function like many other members of this class of proteins.

Role in Diagnosis

TBG tests are sometimes used in finding the reason for elevated or diminished levels of thyroid hormone. This is done by measuring resin binding to labelled thyroid hormone, which only happens when the labeled thyroid hormone is free.

The patient's serum is mixed with the labelled thyroid hormone; then, the resin is added to the whole mixture, to measure the amount of free labelled thyroid hormone. So, for instance, if the patient is truly hypothyroid, and TBG levels are normal, then there are many sites open for binding on the TBG, since the total thyroid hormone level is low. Therefore, when the labelled hormone is added, it will mostly bind to the TBG, leaving little of it left for binding to the resin. Conversely, if the patient is truly hyperthyroid, and TBG levels are normal, the patient's endogenous hormone will saturate the TBG binding sites more, leaving less room for the labelled hormone, which allows greater binding to the resin.

In the situations described above, TBG testing is not very useful. However, total thyroid hormone levels point to hypothyroidism or hyperthyroidism, without the expected symptoms, the utility of TBG testing becomes more evident, since TBG production can be modified by other factors such as estrogen, corticosteroids, or liver failure. If, for example, the TBG level is high, which can occur when estrogen levels are high, the TBG will bind more thyroid hormone, decreasing the free hormone available in the blood, which leads to stimulation of TSH, and the production of more thyroid hormone. In this case, the total thyroid hormone level will be high. However, when labelled hormone is added, since TBG is so high, it will bind to the TBG, leaving little free labelled hormone for uptake into the resin. Conversely, in the presence of corticosteroids, which lower TBG levels, the total thyroid hormone (bound and free) in the blood will be low. However, when the labeled hormone is added, since so little TBG is available in the blood, only a small portion of it will bind, leaving plenty available for uptake by the resin.

Beta globulins are a group of globular proteins in plasma that are more mobile in alkaline or electrically charged solutions than gamma globulins, but less mobile than α globulins.

Examples of β globulins include:

- α-2 microglobulin
- plasminogen
- angiostatins
- properdin
- sex hormone binding globulin
- transferrin

β_2 microglobulin also known as **B2M** is a component of MHC class I molecules, which are present on almost all cells of the body (red blood cells are a notable exception).[1][2]

β_2 microglobulin lies lateral to the α_3 chain on the cell surface. Unlike α_3, β_2 has no transmembrane region. Directly above β_2 (i.e. away from the cell) lies the α_1 chain, which itself is lateral to the α_2.

β_2 microglobulin associates not only with the alpha chain of MHC class I molecules **See Fig. 2.29**, but also with class I-like molecules such as CD1 and Qa.

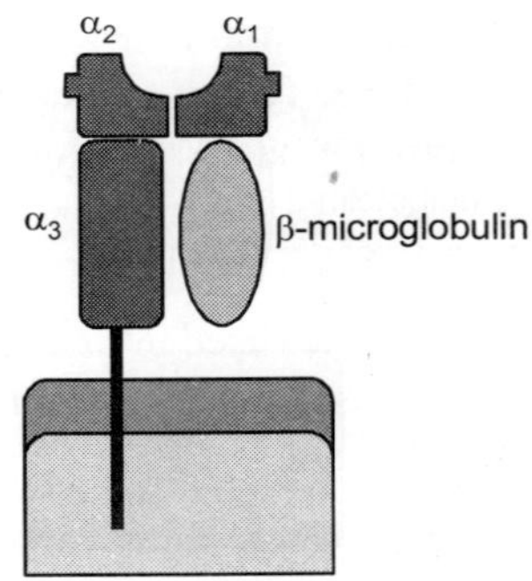

Fig. 2.29 : Schematic representation of MHC class I

Clinical significance

In patients on long-term hemodialysis, it can aggregate into amyloid fibers that deposit in joint spaces, a disease known as dialysis-related amyloidosis.

Mice models deficient for the **β_2 microglobulin** gene have been engineered. These mice demonstrate that **β_2 microglobulin** is necessary for cell surface expression of MHC class I and stability of the peptide binding groove. In fact, in the absence of **β_2 microglobulin**, very limited amounts of MHC class I (classical and non-classical) molecules can be detected on the surface. In the absence of MHC class I, CD8 T cells cannot develop. (CD8 T cells are a subset of T cells involved in the development of acquired immunity.)

Levels of β-2 microglobulin can be elevated in multiple myeloma and lymphoma.

Plasmin also known as is an important enzyme (EC 3.4.21.7) present in blood that degrades many blood plasma proteins, most notable, fibrin clots. The degradation of fibrin is termed fibrinolysis. In humans, the plasmin protein is encoded by the *PLG* gene.

Function

Plasminogen (PLG) is a circulating zymogen that is converted to the active enzyme plasmin by cleavage of the peptide bond between Arg-560 and Val-561, which is mediated by urokinase and tissue plasminogen activator. The main function of plasmin is to dissolve fibrin blood clots. Plasmin, like trypsin, belongs to the family of serine proteases.

It is a serine protease that is released as **plasminogen** from the liver into the circulation and activated by tissue plasminogen activator (tPA), urokinase plasminogen activator (uPA), thrombin, fibrin, and factor XII (Hageman factor). It is inactivated by alpha 2-antiplasmin, a serine protease inhibitor (serpin).

Apart from fibrinolysis, plasmin proteolyses proteins in various other systems: It activates collagenases, some mediators of the complement system and weakens the wall of the Graafian follicle (leading to ovulation). It cleaves fibrin, fibronectin, thrombospondin, laminin, and von Willebrand factor.

Apart from fibrinolysis, plasminogen is shown to play important role in wound healing, liver repair as well as the mantainance of liver homeostasis.

Angiostatin is a naturally occurring protein found in several animal species, including humans. It is an endogenous angiogenesis inhibitor (i.e., it blocks the growth of new blood vessels), and it is currently undergoing clinical trials for its use in anticancer therapy.

Structure

Angiostatin is a 38 kDa fragment of a larger protein, plasmin (itself a fragment of plasminogen) enclosing three to five contiguous Kringle modules. Each module contains two small β sheets and three disulfide bonds.

Generation

Angiostatin is produced, for example, by autoproteolytic cleavage of plasminogen, involving extracellular disulfide bond reduction by phosphoglycerate kinase. Furthermore angiostatin can be cleaved from plasminogen by different metalloproteinases (MMPs), elastase, prostata-specific antigen (PSA), 13 KD serine protease, or 24KD endopeptidase.

Biological activity

Angiostatin is known to bind many proteins, especially to angiomotin and endothelial cell surface ATP synthase but also integrins, annexin II, C-met receptor, NG2-proteoglycans, tissue-type plasminogen activator, chondroitin sulfate proteoglycans, and CD26. Additionally, smaller fragments of angiostatin may bind several other proteins. There is still considerable uncertainty on its mechanism of action, but it seems to involve inhibition of endothelial cell migration, proliferation and induction of apoptosis. It has been proposed that angiostatin activity is related, among other things, to the coupling of its mechanical and redox properties.

Properdin or factor P is a globulin protein found in the blood serum of higher animals. In the complement system, an innate-immunity series of proenzymes dissolved in the circulation, it is also called "Factor P".

Function

It is known that it participates in some specific immune responses. It plays a part in tissue inflammation as well as the engulfing of pathogens by phagocytes. In addition it is known to help to neutralize some viruses.

As a component of the alternative pathway for complement activation (otherwise known as the 'properdin pathway'), it complexes with another protein, C3b, to stabilize the alternative C3 convertase (C3bBb) that then cleaves more C3.

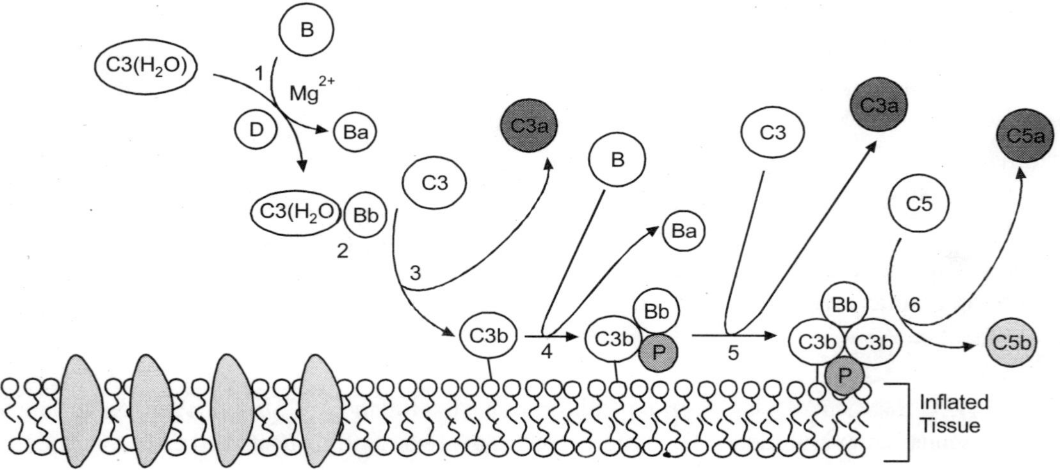

Fig. 2.30 : Alternative pathway. Properdin is the 'P' in the blue circles.

The alternative pathway is not dependent on antibodies. This branch of the complement system is activated by IgA immune complexes and bacterial endotoxins, polysaccharides, and cell walls, and results in producing anaphylatoxins, opsonins, chemotactic factors, and the membrane attack complex, all of which help fight pathogens.

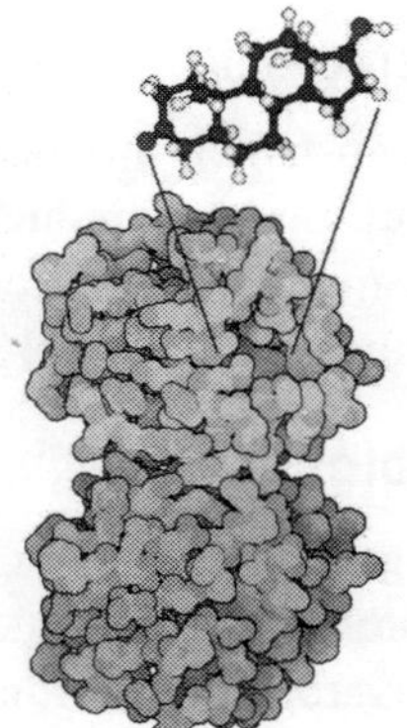

Fig. 2.31

Sex hormone-binding globulin (**SHBG**) is a glycoprotein that binds to sex hormones, specifically testosterone and estradiol. Other steroid hormones such as progesterone, cortisol, and other corticosteroids are bound by transcortin.

Transport of sex hormones

Testosterone and estradiol circulate in the bloodstream, bound mostly to SHBG and to some degree bound to serum albumin. Only a small fraction is unbound, or "free," and thus biologically active and able to enter a cell and activate its receptor. The SHBG inhibits the function of these hormones. Thus bioavailability of sex hormones is influenced by the level of SHBG.

SHBG production

SHBG is produced by the liver cells and is released into the bloodstream. Other sites that produce SHBG are the brain, uterus, and placenta and vagina. In addition SHBG is produced by the testes; testes-produced SHBG is also called androgen-binding protein. The gene for SHBG is located on chromosome 17.

Control

SHBG levels appear to be controlled by a delicate balance of enhancing and inhibiting factors. Its level is decreased by high levels of insulin and insulin-like growth factor 1 (IGF-1). Also, high androgen levels decrease SHBG, while high estrogen and thyroxine levels increase it.

However, recent evidence suggests that it is the liver's production of fats that reduces SHBG levels, not any direct effect of insulin and specific genetic mechanisms have been found that do this.

Conditions with high or low levels

Conditions with low SHBG include polycystic ovary syndrome, diabetes, and hypothyroidism. Conditions with high SHBG include pregnancy, hyperthyroidism, and anorexia nervosa. There has recently been research to link high SHBG levels with breast and testicular cancer as well.

Measurement of sex hormones

When determining levels of circulating estradiol or testosterone, either a total measurement could be done that includes the 'free' and the bound fractions, or only the 'free' hormone could be measured. A free androgen index expresses the ratio of testosterone to the sex hormone binding globulin and can be used to summarise the activity of free testosterone.

The total testosterone is likely the most accurate measurement of testosterone levels and should always be measured at 8 O'clock in the morning. Sex hormone binding globulin can be measured separate from the total fraction of testosterone.

Transferrin is a blood plasma protein for iron ion delivery which in humans is encoded by the *TF* gene. Transferrin is a glycoprotein, which binds iron very tightly but reversibly. Although iron bound to transferrin is less than 0.1% (4 mg) of the total body iron, dynamically it is the most important iron pool, with the highest rate of turnover (25 mg/24 h). Transferrin has a molecular weight of around 80 kiloDaltons and contains 2 specific high affinity Fe(III) binding sites. The affinity of transferrin for Fe(III) is extremely high (10^{23} M^{-1} at pH 7.4) but decreases progressively with decreasing pH below neutrality.

When not bound to iron, it is known as 'apo-transferrin' (see also apoprotein).

Transport mechanism

When a transferrin protein loaded with iron encounters a transferrin receptor on the surface of a cell (importantly, to erythroid precursors in the bone marrow), it binds to it and is consequently transported into the cell in a vesicle. The pH of the vesicle is reduced by hydrogen ion pumps (H^+ ATPases), causing transferrin to release its iron ions. The receptor (with its ligand, transferrin, bound) is then transported through the endocytic cycle back to the cell surface, ready for another round of iron uptake. Each transferrin molecule has the ability to carry two iron ions in the ferric form (Fe^{3+}).

The gene coding for transferrin in humans is located in chromosome band 3q21. Research on king snakes by Dessauer and Zwiefel in 1981 revealed that the inheritance of transferrin is a codominant trait.

Medical professionals may check serum transferrin level in iron deficiency, hemochromatosis and other iron overload disorders.

Immune system

Transferrin is also associated with the innate immune system. Transferrin is found in the mucosa and binds iron, thus creating an environment low in free iron, where few bacteria are able to survive. The levels of transferrin decreases in inflammation, seeming contradictory to its function

A decrease in the amount of transferrin would result in hemosiderin in the liver.

Other effects

The metal binding properties of transferrin have a great influence on the biochemistry of plutonium in humans. Transferrin has a bacteriocidal effect on bacteria, in that it makes Fe^{3+} unavailable to the bacteria.

Pathology

A deficiency is associated with atransferrinemia.

Gamma globulins, or γ- Ig's, are a class of proteins in the blood, identified by their position after serum protein electrophoresis. The most significant gamma globulins are immunoglobulins.

Injections

γ-globulin injections are usually given in an attempt to temporarily boost a patient's immunity against disease. Injections are most commonly used on patients who have been exposed to hepatitis A or measles, or to make a kidney donor and recipient compatible regardless of blood type of tissue match. Injections are also used to boost immunity in patients who cannot produce gamma globulins naturally because of an immune deficiency, such as X-linked agammaglobulinemia and hyper IgM syndrome. Such injections are less common in modern medical practice than they were previously, and injections of gamma globulin previously recommended for travellers have largely been replaced by the use of hepatitis A vaccine.

γ-globulin infusions are also used to treat immunological diseases, such as idiopathic thrombocytopenia purpura (ITP), a disease in which the platelets are being attacked by antibodies, leading to seriously low platelet counts. Gamma globulin apparently causes the spleen to ignore the antibody-tagged platelets, thus allowing them to survive and function.

Gamma globulin injections also provide substantial benefit to many suffering from Chronic Fatigue Syndrome, also known as Chronic Fatigue and Immune Dysfunction Syndrome; Myalgic Encephalitis; Chronic Epsteinn-Barr; Chronic Mono.

Another theory on how gamma globulin administration works in autoimmune disease is by overloading the mechanisms which degrade gamma globulins. Overloading the degradation mechanism causes the harmful gamma globulins to have a much shorter halflife in sera.

Pathology

An excess is known as hypergammaglobulinemia.

A disease of gamma globulins is called a "gammopathy" (for example, in monoclonal gammopathy of undetermined significance.)

Disease Treatments

• Kawasaki disease

Kidney Transplant: Intravenous Gamma globulin was FDA approved in 2004 to reduce antibodies in a patient with kidney failure to allow that person to accept a kidney from a donor who has a different blood type, (ABO incompatible) or is an unacceptable tissue match. Dr. Stanley Jordan at Cedars-Sinai Medical Center in Los Angeles pioneered this treatment.

Glutelins are soluble in dilute acids or bases, detergents, chaotropic or reducing agents. They are generally prolamin-like proteins in certain grass seeds. Glutenin is the most common glutelin as it is found in wheat and is responsible from some of the refined baking properties in bread wheat. The glutelins of barley and rye have also been identified. Typically there are HMW and LMW glutelins in these species, they crosslink with themselves and other proteins during baking via disulfide bonds. A HMW glutelin (glutenin) of the grass tribe *Triticeae* can be sensitizing agents for coeliac disease in individuals who possess the HLA-DQ8 class II antigen receptor gene. (Not yet characterized to the epitope level)

Prolamins are a group of plant storage proteins having a high proline content and found in the seeds of cereal grains: wheat (gliadin), barley (hordein), rye (secalin), corn (zein) and as a minor protein, avenin in oats. They are characterised by a high glutamine and proline content and are generally soluble only in strong alcohol solutions. Some prolamins, notably gliadin, and similar proteins found in the tribe Triticeae (see Triticeae glutens) may induce coeliac disease in genetically predisposed individuals

Scleroproteins are one of the two main classes of protein tertiary structure (the other being globular proteins).

They are also called fibrous proteins.

Characteristics

They form long protein filaments, rod- or wire-like shapes. They are usually inert structural or storage proteins. They are generally water-insoluble and are found as an aggregate due to hydrophobic R groups that stick out of the molecule. The amino acid sequences they are made from often have limited residues with repeats. These can form unusual secondary structures, e.g. collagen triple helix. The structures often contain 'cross-links' between chains, for example cys-cys disulfide bonds between keratin chains.

Globular proteins tend to denature more easily than fibrous proteins.

Functions

They usually play a role which is protective or supportive.

They are usually used to construct connective tissues, tendons, bone matrix and muscle fiber.

Attempts at artificial synthesis have been made.

Examples

Examples of include keratins, collagens and elastins.

Another example is fibroin.

Conjugated Proteins

A **nucleoprotein** is any protein which is structurally associated with nucleic acid (either DNA or RNA). The prototypical example is any of the histone class of proteins, which are identifiable on strands of chromatin. Telomerase, a RNP (RNA/protein complex), and Protamines are also nucleoproteins.

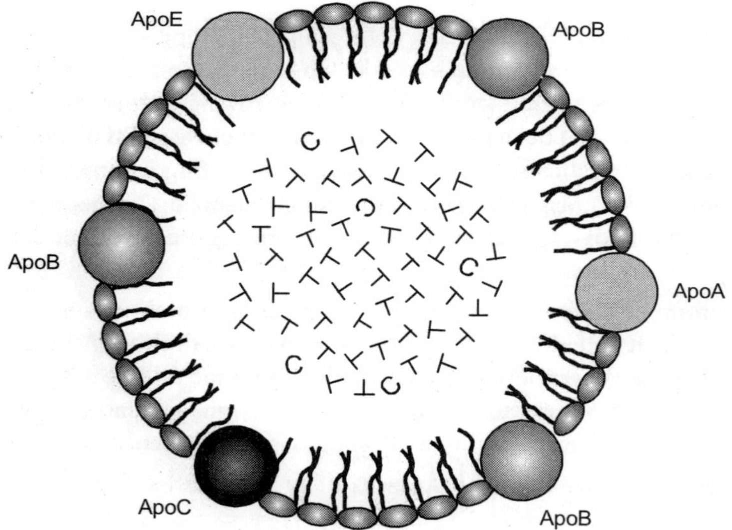

Fig. 2.32 : Chylomicron structure

Phosphoproteins are a group of proteins which are chemically bonded to a substance containing phosphoric acid (see phosphorylation for more). The category of organic molecules that includes Fc receptors, Ulks, Calcineurins, K chips, and urocortins.

A **lipoprotein** is a biochemical assembly that contains both proteins and lipids. The lipids or their derivatives may be covalently or non-covalently bound to the proteins. Many enzymes, transporters, structural proteins, antigens, adhesins and toxins are lipoproteins. Examples include the high density and low density lipoproteins which enable fats to be carried in the blood stream, the transmembrane proteins of the mitochondrion and the chloroplast, and bacterial lipoproteins.

Function

The function of lipoprotein particles is to transport lipids (fats) around the body in the aqueous blood, in which they would not normally dissolve.

All cells use and rely on fats and, for all animal cells, cholesterol as building blocks to create the multiple membranes which cells use to both control internal water content, internal water soluble elements and to organize their internal structure and protein enzymatic systems.

The protein particles have hydrophilic groups aimed outward so as to attract water molecules; this makes them soluble in the salt water based blood pool. Triglyceride-fats and cholesterol are carried internally, shielded from the water by the protein particle.

The interaction of the proteins forming the surface of the particles with (a) enzymes in the blood, (b) with each other and (c) with specific proteins on the surfaces of cells, determine whether triglycerides and cholesterol will be added to or removed from the lipoprotein transport particles.

Regarding atheroma development and progression vs. regression, the key issue has always been cholesterol transport patterns, not cholesterol concentration itself.

Transmembrane Lipoproteins

The lipids are often an essential part of the complex, even if they seem to have no catalytic activity themselves. To isolate transmembrane lipoproteins from their associated membranes, detergents are often needed.

Classification

By Density

General categories of lipoproteins, listed in order from larger and less dense (more fat than protein) to smaller and denser (more protein, less fat):

- Chylomicrons - carry triacylglycerol (fat) from the intestines to the liver, skeletal muscle, and to adipose tissue.
- Very low density lipoproteins (VLDL) - carry (newly synthesised) triacylglycerol from the liver to adipose tissue.
- Intermediate density lipoproteins (IDL) - are intermediate between VLDL and LDL. They are not usually detectable in the blood.
- Low density lipoproteins (LDL) - carry cholesterol from the liver to cells of the body. Sometimes referred to as the "bad cholesterol" lipoprotein.

- High density lipoproteins (HDL) - collects cholesterol from the body's tissues, and brings it back to the liver. Sometimes referred to as the "good cholesterol" lipoprotein.

Density (g/mL)	Class	Diameter (nm)	% protein	% choles-terol	% phospho-lipid	% triacyl-glycerol
>1.063	HDL	5-15	33	30	29	8
1.019-1.063	LDL	18-28	25	50	21	4
1.006-1.019	IDL	25-50	18	29	22	31
0.95-1.006	VLDL	30-80	10	22	18	50
<0.95	Chylomicrons	100-1000	<2	8	7	84

Metabolism

The handling of lipoproteins in the body is referred to as lipoprotein metabolism. It is divided into two pathways, exogenous and endogenous, depending in large part on whether the lipoproteins in question are composed chiefly of dietary (exogenous) lipids or whether they originated in the liver (endogenous).

Exogenous Pathway

Epithelial cells lining the small intestine readily absorb lipids from the diet. These lipids, including triglycerides, phospholipids, and cholesterol, are assembled with apolipoprotein B-48 into chylomicrons. These nascent chylomicrons are secreted from the intestinal epithelial cells into the lymphatic circulation in a process that depends heavily on apolipoprotein B-48. As they circulate through the lymphatic vessels, nascent chylomicrons bypass the liver circulation and are drained via the thoracic duct into the bloodstream. In the bloodstream, HDL particles donate apolipoprotein C-II and apolipoprotein E to the nascent chylomicron; the chylomicron is now considered mature. Via apolipoprotein C-II, mature chylomicrons activate lipoprotein lipase (LPL), an enzyme on endothelial cells lining the blood vessels. LPL catalyzes a hydrolysis reaction that ultimately releases glycerol and fatty acids from the chylomicrons. Glycerol and fatty acids can be absorbed in peripheral tissues, especially adipose and muscle, for energy and storage.

The hydrolyzed chylomicrons are now considered chylomicron remnants. The chylomicron remnants continue circulating until they interact via apolipoprotein E with chylomicron remnant receptors, found chiefly in the liver. This interaction causes the endocytosis of the chylomicron remnants, which are subsequently hydrolyzed within lysosomes. Lysosomal hydrolysis releases glycerol and fatty acids into the cell, which can be used for energy or stored for later use.

Endogenous Pathway

The liver is another important source of lipoproteins, principally VLDL. Triacylglycerol and cholesterol are assembled with apolipoprotein B-100 to form VLDL particles. Nascent VLDL particles are released into the bloodstream via a process that depends upon apolipoprotein B-100.

As in chylomicron metabolism, the apolipoprotein C-II and apolipoprotein E of VLDL particles are acquired from HDL particles. Once loaded with apolipoproteins C-II and E, the nascent VLDL particle is considered mature. Again like chylomicrons, VLDL particles circulate and encounter LPL expressed on endothelial cells. Apolipoprotein C-II activates LPL, causing hydrolysis of the VLDL particle and the release of glycerol and fatty acids. These products can be absorbed from the blood by peripheral tissues, principally adipose and muscle. The hydrolyzed VLDL particles are now called VLDL remnants or intermediate density lipoproteins (IDLs). VLDL remnants can circulate and, via an interaction between apolipoprotein E and the remnant receptor, be absorbed by the liver, or they can be further hydrolyzed by hepatic lipase.

Hydrolysis by hepatic lipase releases glycerol and fatty acids, leaving behind IDL remnants, called low density lipoproteins (LDL), which contain relatively high cholesterol content. LDL circulates and is absorbed by the liver and peripheral cells. Binding of LDL to its target tissue occurs through an interaction between the LDL receptor and apolipoprotein B-100 or E on the LDL particle. Absorption occurs through endocytosis, and the internalized LDL particles are hydrolyzed within lysosomes, releasing lipids, chiefly cholesterol.

Lipoprotein(a) (also called Lp(α)) is a lipoprotein subclass. Studies have identified Lp(α) as a putative risk factor for atherosclerotic diseases such as coronary heart disease and stroke.

Lipoprotein(α) was dicovered in 1963 by Kåre Berg and the human gene encoding this protein was cloned in 1987.

Structure

Lipoprotein(α) [Lp(α)] consists of an LDL-like particle and the specific apolipoprotein(a) [apo(a)], which is covalently bound to the apoB of the LDL like particle. Lp [apo(a)] plasma concentrations are highly heritable and mainly controlled by the apolipoprotein(a) gene [LPA] located on chromosome 6q26-27. Apo(a) proteins vary in size due to a size polymorphism [KIV-2 VNTR], which is caused by a variable number of so called kringle IV repeats in the LPA gene. This size variation at the gene level is expressed on the protein level as well, resulting in apo(a) proteins with 10 to > 50 kringle IV repeats (each of the variable kringle IV consists of 114 amino acids). These variable apo(a) sizes are known as "apo(a) isoforms". There is a general inverse correlation between the size of the apo(a)isoform and the Lp(a) plasma concentration which is caused by a variable rate of degradation before the apo(a) protein has matured for Lp(a) assembly. Apo(a) is expressed by the liver cells (hepatocytes), and the assembly of apo(a) and LDL particles seems to take place at the outer hepatocyte surface. The half-life of Lp(a) in the circulation is about 3 to 4 days.

Catabolism and Clearance

The mechanism and sites of Lp(a) catabolism are largely unknown. Uptake via the LDL receptor is not a major pathway of Lp(a) metabolism. The kidney has been identified as playing a role in Lp(a) clearance from plasma.

Populations

Lp(a) concentrations vary over one thousandfold between individuals, from < 0.2 to > 200 mg/dL. This range of concentrations is observed in all populations studied so far. The mean and median concentrations between different world populations show distinct particularities, the main being the two- to threefold higher Lp(a) plasma concentration of populations of African descent compared to Asian, Oceanic, or European populations. The general inverse correlation between apo(a) isoform size and Lp(a)plasma concentration is observed in all populations, however, mean Lp(a) associated with certain apo(a) isoforms varies between populations.

Function

The physiological function of Lp(a)/apo(a) is still unknown. A function within the coagulation system seems plausible, given the aspect of the high homology between apo(a) and plasminogen. In fact, the LPA gene derives from a duplication of the plasminogen gene.

Other functions have been related to recruitment of inflammatory cells through interaction with Mac-1 integrin, angiogenesis, and wound healing.

However, individuals without Lp(a) or with very low Lp(a) levels seem to be healthy. Thus plasma Lp(a) is certainly not vital, at least under normal environmental conditions. Since apo(a)/Lp(a) derived rather recently in mammalian evolution - only old world monkeys and humans have been shown to harbour Lp(a) - its function might not be vital but just evolutionary advantageous under certain environmental conditions, e.g. in case of exposure to certain infectious diseases.

Pathology

Lipoprotein's structure is similar to plasminogen and tPA (tissue plasminogen activator) and it competes with plasminogen for its binding site, leading to reduced fibrinolysis. Also because Lp(a) stimulates secretion of PAI-1 it leads to thrombogenesis. In addition, because of LDL cholesterol content, Lp-a contributes to atherosclerosis.

Lipoprotein(a) and Disease

High Lp(a) in blood is a risk factor for coronary heart disease (CHD), cerebrovascular disease (CVD), atherosclerosis, thrombosis, and stroke. Lp-a concentrations may be affected by disease states, but are only slightly affected by diet, exercise, and other environmental factors. Commonly prescribed lipid-reducing drugs have little or no effect on Lp(a) concentration. Niacin (nicotinic acid) and aspirin are two relatively safe, easily available and inexpensive drugs known to significantly reduce the levels of Lp(a) in some individuals with high Lp(a); they should be used under the supervision of a qualified physician.

High Lp(a) predicts risk of early atherosclerosis similar to high LDL, but in advanced atherosclerosis, Lp(a) is an independent risk factor not dependent on LDL. Lp(a) then indicates a coagulant risk of plaque thrombosis. Apo(a) contains domains that are very similar to plasminogen (PLG). Lp(a) accumulates in the vessel wall and inhibits binding of PLG to the cell surface, reducing plasmin generation which increases clotting. This inhibition of PLG by Lp(a) also promotes proliferation of smooth muscle cells. These unique features of Lp(a) suggest Lp(a) causes generation of clots and atherosclerosis.

Metalloprotein is a generic term for a protein that contains a metal ion cofactor. Metalloproteins have many different functions in cells, such as enzymes, transport and storage proteins, and signal transduction proteins. Indeed, about one quarter to one third of all proteins require metals to carry out their functions. The metal ion is usually coordinated by nitrogen, oxygen or sulfur atoms belonging to amino acids in the polypeptide chain and/or a macrocyclic ligand incorporated into the protein. The presence of the metal ion allows metalloenzymes to perform functions such as redox reactions that cannot easily be performed by the limited set of functional groups found in amino acids.

Storage and Transport Metalloproteins

Oxygen Carriers

Hemoglobin, which is the principal oxygen carrier in humans has four sub-units in which the iron(II) ion is coordinated by the planar, macrocyclic ligand protoporhyrin IX (PIX) and the imidazole nitrogen atom of a histidine residue. The sixth coordination site contains a water molecule or a dioxygen molecule. myoglobin has only one such unit. The active site is located in an hydrophobic pocket. This is important as, without it, the iron(II) would be irreveribly oxidised to iron(III). The equilibrium constant for the formation of HbO_2 is such that oxygen is taken up or released depending on the partial pressure of oxygen in the lungs or in muscle. In hemoglobin the four sub-units show a cooperativity effect which allows for easy oxygen transfer from hemoglobin to myoglobin.

In both hemoglobin and myoglobin it is sometimes incorrectly stated that the oxygenated species contains iron(III). It is now known that the diamagnetic nature of these species is due to the fact that the iron(II) is in the low-spin state. In oxyhemoglobin the iron atom is located in the plane of the porphyrin ring, but in the paramagnetic deoxyhemoglobin the iron atom lies above the plane of the ring. The change in spin state is a cooperative effect of higher crystal field splitting and smaller ionic radius of Fe^{2+} in the oxy- moiety.

Hemerythrin is another iron-containing oxygen carrier. The oxygen binding site is a binuclear iron center. The iron atoms are coordinated to the protein through the carboxylate side chains of a glutamate and aspartate and five histidine residues. The uptake of O_2 by hemerythrin is accompanied by two-electron oxidation of the reduced binuclear center to produce bound peroxide (OOH-). The mechanism of oxygen uptake and release have been worked out in detail.

Hemocyanins carry oxygen in the blood of most molluscs, and some arthropods such as the horseshoe crab. They are second only to hemoglobin in biological popularity of use in oxygen transport. On oxygenation the two copper(I) atoms at the active site are oxidised to copper(II) and the dioxygen molecules is reduced to peroxide, O_2^{2-}

Cytochromes

Oxidation and reduction are reactions are not common in organic chemistry as few organic molecules can act as oxidizing or reducing agents. Iron(II), on the other hand, can easily be oxidized to iron(III). This functionality is used in cytochromes which function as electron-tranfer vectors. The iron atom in most cytochromes is contained in a heme group. The differences between those cytochromes lies in the different side-chains. For instance Cytochrome a has a heme a prosthetic group and cytochrome b has a heme b prosthetic group. These differences result in different Fe^{2+}/Fe^{3+} redox potentials such that various cytochromes are involved in the mitochondrial electron transport chain.

Cytochrome P450 enzymes perform the function of inserting an oxygen atom into a C—H bond, an oxidation reaction.

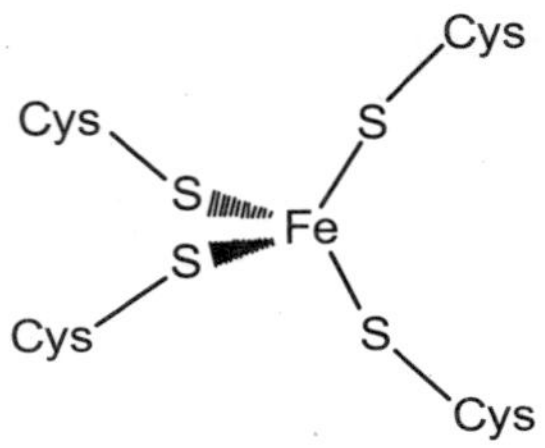

rubredoxin active site.

Rubredoxin

Rubredoxin is an electron-carrier found in sulfur-metabolizing bacteria and archaea. The active site contains an iron ion which is coordinated by the sulphur atoms of four cysteine residues forming an almost regular tetrahedron. Rubredoxins perform one-electron transfer processes. The oxidation state of the iron atom changes between the +2 and +3 states. In both oxidation states the metal is high spin, which helps to minimize structural changes.

Iron Storage and Transfer

Iron is stored as iron(III) in ferritin. The exact nature of the binding site has not yet been determined. The iron appears to be present as an hydrolysis product such as FeO(OH). Iron is transported by transferrin whose binding site structure is also as yet unknown.

The human body has no mechanism for iron excretion. This can lead to iron-overload problems in patients treated with blood transfusions, as, for instance, with β-thallasemia.

Ceruloplasmin

Ceruloplasmin is the major copper-carrying protein in the blood. Ceruloplasmin exhibits oxidase activity, which is associated with possible oxidation of Fe^{2+} (ferrous iron) into Fe^{3+} (ferric iron), therefore assisting in its transport in the plasma in association with transferrin, which can only carry iron in the ferric state.

Metalloenzymes

Metalloenzymes all have one feature in common, namely, that the metal ion is bound to the protein with one coordination site free. As with all enzymes, the shape of the active site is crucial. The metal ion is usually located in a pocket whose shape fits the substrate. The metal ion catalyzes reactions which are difficult to achieve in organic chemistry.

Carbonic Anhydrase *(See Fig. 2.33)*

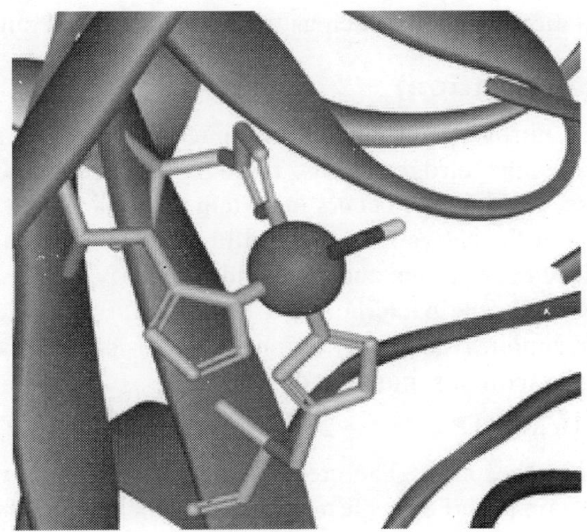

Fig. 2.33 : Active site of carbonic anhydrase. The three coordinating histidine residues are shown in pink, hydroxide in red and white, and the zinc in purple.

$$CO_2 + H_2O \rightleftharpoons H_2CO_3$$

This reaction is very slow in the absence of a catalyst, but quite fast in the presence of the hydroxide ion

$$CO_2 + OH^- \rightleftharpoons HCO_3^-$$

A reaction similar to this is almost instantaneaous with carbonic anhydrase. The structure of the active site in carbonic anhydrases is well-known from a number of crystal structures. It consists of a zinc ion coordinated by three imidazole nitrogen atoms from three histidine units. The fourth coordination site is occupied by a water molecule. The coordination sphere of the zinc ion is approximately tetrahedral. The positively charged zinc ion polarizes the coordinated water molecule and nucleophilic attack by the negatively charged hydoxide portion on carbon dioxide (carbonic anhydride) proceeds rapidly. The catalytic cycle produces the bicarbonate ion and the hydrogen ion as the equilibrium

$$H_2CO_3 \rightleftharpoons HCO_3^- + H^+$$

favours dissociation of carbonic acid at biological pH values.

Vitamin B12-dependent Enzymes

Vitamin B12 catalyzes the tranfer of methyl ($-CH_3$) groups between two molecules, which involves the breaking of C-C bonds, a process that is energetically expensive in organic reactions. The metal ion lowers the activation energy for the process by forming a transient Co-CH_3 bond. The structure of the coenzyme was famously determined by Dorothy Hodgkin and co-workers, for which she received a Nobel prize. It constists of a cobalt(II) ion coordinated by four nitrogen atons of a corrin rings and a fifth nirogen atom from an imidazole group. In the resting state there is a Co—C α bond with the 5' carbon atom of adenosine.[13] This is a naturally occurring organometallic compound, which explains its function in trans-methylation reactions, such as the reaction carried out by methionine synthase.

Nitrogenase (nitrogen fixation)

The fixation of atmospheric nitrogen is a very energy-intensive process, as it involves breaking the very stable triple bond between the nitrogen atoms. The enzyme nitrogenase is one of the few enzymes that can catalyze the process. The enzyme occurs in certain bacteria. There are three components to its action: a molybdenum atom at the active site, Iron-sulfur clusters which are involved in transporting the electrons needed to reduce the nitrogen and an abundant energy source. The energy is provided by a symbiotic relationship between the bacteria and a host plant, often a legume. The relationship is symbiotic because the plant supplies the energy by photosynthesis and benefits by obtaining the fixed nitrogen. The reaction may be written symbollically as:

$$N_2 + 16\ MgATP + 8e^- \rightarrow 2NH_3 + 16\ MgADP + 16\ P_i + H_2$$

where P_i stands for inorganic phosphate. The precise structure of the active site has been difficult to determine. It appears to contain a $MoFe_7S_8$ cluster which is able to bind the dinitrogen molecule and, presumably, enable the reduction process to begin. The electrons are transported by the associated "P" cluster, which contains two cubical Fe_4S_4 clusters joined by sulphur bridges.

SuperOxide Dismutase

The superoxide ion, O_2^- is generated in biological systems by reduction of molecular oxygen. It has an unpaired electron, so it behaves as a free radical. It is a powerful oxidising agent. These properties render the superoxide ion very toxic and are deployed to advantage by phagocytes to kill invading micro organisms. Otherwise, the superoxide ion must be destroyed before it does unwanted damage in a cell. The superoxide dismutase enzymes perform this function very efficiently.

The formal oxidation state of the oxygen atoms is ½. In solutions at neutral pH, the superoxide ion disproportionates to molecular oxygen and hydrogen peroxide.

$$2\ O_2^- + 2\ H^+ \rightarrow O_2 + H_2O_2$$

In biology this type of reaction is call a dismutation reaction. It involves both oxidation and reduction of superoxide ions. The superoxide dismutase group of enzymes, abbreviated as SOD, increase the rate of reaction to near the diffusion limited rate. The key to the action of these enzymes is a metal ion with variable oxidation state which can act as either an oxidizing agent or as a reducing agent.

$$\text{Oxidation: } M^{(n+1)+} + O_2^- \rightarrow M^{n+} + O_2$$

$$\text{Reduction: } M^{n+} + O_2^- + 2H^+ \rightarrow M^{(n+1)+} + H_2O_2.$$

In human SOD the active metal is copper, as Cu^{2+} or Cu^+, coordinated tetrahedrally by four histidine residues. This enzyme also contains zinc ions. Other isozymes may contain iron, manganese or nickel. Ni-SOD is particularly interesting as it involves nickel(III), an unusual oxidation state for this element. The active site Ni geometry cycles from square planar Ni(II), with thiolate (Cys2 and Cys6) and backbone nitrogen (His1 and Cys2) ligands, to square pyramidal Ni(III) with an added axial His1 side chain ligand.

Chlorophyll

Chlorophyll plays a crucial role in photosynthesis. In contains a magnesium enclosed in a chlorin ring. However, the magnesium ion is not directly involved in the photosynthetic function and can be replaced by other divalent ions with a little loss of activity. Rather, the photon is absorbed by the chlorin ring, whose electronic structure is well-adapted for its purpose.

Initially the absorption of a photon causes an electron to be excited into a singlet state of the Q band. This is the reason why chlorophyll is green. The excited state undergoes an intersystem crossing from the singlet state to a triplet state in which there are two electrons with parallel spin. This species is, effectively, a free radical, and is very reactive and allows an electron to be transferred to acceptors which are adjacent to the chlorophyll in the chloroplast. In the process chlorophyll is oxydised. Later in the phosynthetic cycle it reduced by accepting an electron and the final oxidation product is molecular oxygen whose atoms come from water molecules.

Signal-transduction Metalloproteins

Calmodulin

Calmodulin is an example of a signal-transduction protein. It is a small protein which contains four EF-hand motifs, each of which can bind a Ca^{2+} ion.

In an EF-hand loop the calcium ion is coordinated in a pentagonal bipyramidal configuration. Six Glutamic acid and Aspartic acid residues involved in the binding are in positions 1, 3, 5, 7, 9 of the polypeptide chain. At position 12 there is a glutamate or aspartate ligand which behaves as a (bidentate ligand), providing two oxygen atoms. The ninth residue in the loop is necessarily glycine due to the conformational requirements of the backbone. The coordination sphere of the calcium ion contains only carboxylate oxygen atoms and no nitrogen atoms. This is consistent with the hard nature of the calcium ion.

The protein has two approximately symmetrical domains, separated by a flexible 'hinge' region. Binding of calcium causes a conformational change to occur in the protein. Calmodulin participates in an intracellular signalling system by acting as a diffusible second messenger to the initial stimuli.

Transcription Factors

Many transcription factors contain a structure known as a zinc finger, this is a structural module where a region of protein folds around a zinc ion. The zinc does not directly contact the DNA that these proteins bind to, instead the cofactor is essential for the stability of the tightly-folded protein chain. In these proteins the zinc ion is usually coordinated by pairs of cysteine and histidine side chains.

Other metalloenzymes

Ion	Examples of enzymes containing this ion
Magnesium	Glucose 6-phosphatase Hexokinase DNA polymerase
Vanadium	vanabins
Manganese	arginase
Iron	Catalase Hydrogenase IRE-BP Aconitase
Nickel Hydrogenase	Urease
Copper Plastocyanin Laccase	Cytochrome oxidase
Zinc	Alcohol dehydrogenase carboxypeptidase Aminopeptidase
Molybdenum	Nitrate reductase
Selenium	Glutathione peroxidase
various	Metallothionein Phosphatase

Derived Proteins

This class of proteins as the name depicts includes those substances formed from simple and conjugated proteins. It is the least well defined of the protein groups. Derived proteins are subdivided into primary derived proteins and secondary derived proteins.

Primary Derived proteins: These proteins derivative are formed by process which cause only slight changes in the protein molecule and its properties. There is a little or no division of peptide bonds. The primary derived proteins are synonymous with denaturated proteins.

Proteans: The proteans are insoluble products formed by the incident action of water, very dilute acids and are particularly formed from globulins in being insoluble in dilute salt solutions. In general they have the physical characteristic of the naturally occurring glutelins.

Examples: Myosan from myosin; edystan from edestin and fibrin from fibrinogen

Metaproteins The Metaptroteins are formed by further action of acids and alkaliess upon proteins. They are generally soluble in very dilute acids and alkalies but sinsoluble in neutral solvents. Examples include acid and alkali metaproteins such as acid alkali albuminates.

Coagulated proteins The coagulated proteins are insoluble products formed by the action of heat or alcohol upon natural protein. Similar substances may also be formed by action of ultra violet light X-Ray, very high pressure, and mechanical shaking upon protein solutions at the iso – electric pH Examples are cooked egg albumin, cooked meal and other proteins and alcohols precipitated proteins.

Secondary Derived Protein These are also called secondary derived proteins, they are formed in the progressive hydrolytic division of the peptide union of protein molecules. They represent a gereat complexity of molecules of different sizes and amino acid composition. They are roughly grouped into proteoses, peptones and peptides, according to relative average molecular complexity. Each group is composed of many different substances.

Thrombin (activated **Factor II [IIa]**) is a coagulation protein that has many effects in the coagulation cascade. It is a serine protease (EC 3.4.21.5) that converts soluble fibrinogen into insoluble strands of fibrin, as well as catalyzing many other coagulation-related reactions.

Genetics

The prothrombin gene is located on the eleventh chromosome (11p11-q12). The molecular weight of prothrombin is approximately 72000 $gmol^{-1}$; in contrast, the molecular weight of thrombin is 36000 $gmol^{-1}$. Once activated, the catalytic domain of prothrombin is released from prothrombin fragment 1.2. There are an estimated 30 people in the world that have been diagnosed with the congenital form of Factor II deficiency (Degen, 1995), which should not be confused with a mutation of prothrombin. The prothrombin gene mutation is called **Factor II mutation**. Factor II mutation is congenital. The Factor II mutated gene is not usually accompanied by other factor mutations (i.e. the most common is Factor V Leiden). The gene may be inherited **heterozygous**, or much more rarely, **homozygous**, and is not related to gender or blood type. Homozygous mutations increase the risk of thrombosis more than heterozygous mutations, but the relative increased risk is well documented. Other potential risks for thrombosis, such as **oral contraceptives** may be additive. The previously reported relationship of inflammatory bowel disease (i.e. Crohn's disease or Ulcerative Colitis) and prothrombin mutation or Factor V Leiden mutation have been contradicted by research.

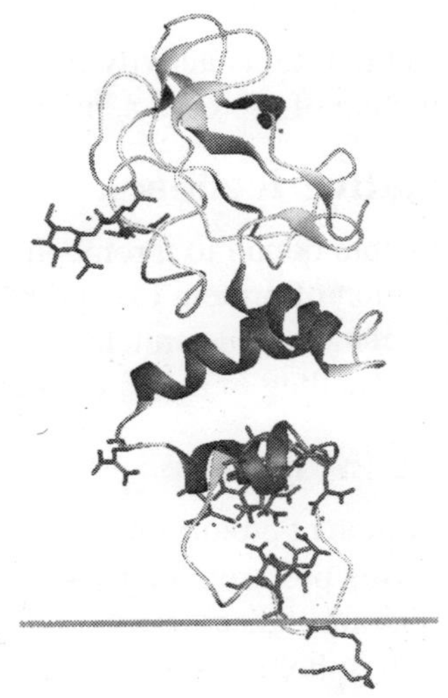

Achoring of bovine prothrombin to the membrane through its Gla domain

Physiology and Generation

Thrombin is produced by the enzymatic cleavage of two sites on prothrombin by activated Factor X (Xa). The activity of factor Xa is greatly enhanced by binding to activated Factor V (Va), termed the prothrombinase complex. Prothrombin is produced in the liver and is post-translationally modified in a vitamin K-dependent reaction that converts ten glutamic acids on prothrombin to gamma-carboxyglutamic acid (Gla). In the presence of calcium, the Gla residues promote the binding of thrombin to phospholipid bilayers (see the picture). Deficiency of vitamin K or administration of the anticoagulant warfarin inhibits the production of gamma-carboxyglutamic acid residues, slowing the activation of the coagulation cascade.

In human beings the level of prothrombin in the blood stream increases after birth and typically peaks on the 8th day after which the prothrombin level lowers to normal levels.

Action

Thrombin converts fibrinogen to an active form that assembles into fibrin. Thrombin also activates factor XI, factor V, and factor VIII. This positive feedback accelerates the production of thrombin.

Factor XIII is also activated by thrombin. Factor XIIIa is a transglutaminase that catalyzes the formation of covalent bonds between lysine and glutamine residues in fibrin. The covalent bonds increase the stability of the fibrin clot.

Platelets

In addition to its activity in the coagulation cascades, thrombin also promotes platelet activation, via activation of protease-activated receptors on the platelet.

Negative feedback

Thrombin bound to thrombomodulin activates protein C, an inhibitor of the coagulation cascade. The activation of protein C is greatly enhanced following the binding of thrombin to thrombomodulin, an integral membrane protein expressed by endothelial cells. Activated protein C inactivates factors Va and VIIIa. Binding of activated protein C to protein S leads to a modest increase in its activity.

Role in disease

Activation of prothrombin is crucial in physiological and pathological coagulation. Various rare diseases involving prothrombin have been described (e.g., hypoprothrombinemia). Anti-thrombin antibodies in autoimmune disease may be a factor in the formation of the lupus anticoagulant also known as (antiphospholipid syndrome). Hyperprothrombinemia can be caused by a mutation at 20210a.

Thrombin, a potent vasoconstrictor and mitogen, is implicated as a major factor in vasospasm following subarachnoid hemorrhage. Blood from a ruptured cerebral aneurysm clots around a cerebral artery, releasing thrombin. This can induce an acute and prolonged narrowing of the blood vessel, potentially resulting in cerebral ischemia and infarction (stroke).

Biotechnology

Due to its high proteolytic specificity, thrombin is a valuable biochemical tool. The thrombin cleavage site (Leu-Val-Pro-Arg-Gly-Ser) is commonly included in linker regions of recombinant fusion protein constructs. Following purification of the fusion protein, thrombin can be used to selectively cleave between the Arginine and Glycine residues of the cleavage site, effectively removing the purification tag from the protein of interest with a high degree of specificity.

Pharmacology

Prothrombin complex concentrate and fresh frozen plasma are prothrombin-rich coagulation factor preparations that can be used to correct deficiencies (usually due to medication) of prothrombin. Indications include intractable bleeding due to warfarin.

Manipulation of prothrombin is central to the mode of action of most anticoagulants. Warfarin and related drugs inhibit vitamin K-dependent carboxylation of several coagulation factors, including prothrombin. Heparin increases the affinity of antithrombin to thrombin (as well as factor Xa). The direct thrombin inhibitors, a newer class of medication, directly inhibit thrombin by binding to its active

Conventions

One convention is that those peptide chains that are short enough to be made synthetically from the constituent amino acids are called peptides rather than proteins. However, with the advent of better synthetic techniques, peptides as long as hundreds of amino acids can be made, including full proteins like ubiquitin. Native chemical ligation has given access to even longer proteins, so this convention seems to be outdated.

Another convention places an informal dividing line at approximately 50 amino acids in length (some people claim shorter lengths). However, this definition is somewhat arbitrary. Long peptides, such as the amyloid beta peptide linked to Alzheimer's disease, can be considered proteins; and small proteins, such as insulin, can be considered peptides.

Peptide Classes

Here are the major classes of peptides, according to how they are produced:

Ribosomal peptides

Are synthesized by translation of mRNA. They are often subjected to proteolysis to generate the mature form. These function, typically in higher organisms, as hormones and signalling molecules. Some organisms produce peptides as antibiotics, such as microcins. Since they are translated, the amino acid residues involved are restricted to those utilized by the ribosome. However, these peptides frequently have posttranslational modifications, such as phosphorylation, hydroxylation, sulfonation, palmitylation, glycosylation and disulfide formation. In general, they are linear, although lariat structures have been observed. More exotic manipulations do occur, such as racemization of L-amino acids to D-amino acids in platypus venom.

Nonribosomal Peptides

These peptides are assembled by enzymes that are specific to each peptide, rather than by the ribosome. The most common non-ribosomal peptide is glutathione, which is a component of the antioxidant defenses of most aerobic organisms.[4] Other nonribosomal peptides are most common in unicellular organisms, plants, and fungi and are synthesized by modular enzyme complexes called *nonribosomal peptide synthetases*. These complexes are often laid out in a similar fashion, and they can contain many different modules to perform a diverse set of chemical manipulations on the developing product. These peptides are often cyclic and can have highly-complex cyclic structures, although linear nonribosomal peptides are also common. Since the system is closely related to the machinery for building fatty acids and polyketides, hybrid compounds are often found. Oxazoles, thiazoles often indicate that the compound was synthesized in this fashion.

Peptones

Are derived from animal milk or meat digested by proteolytic digestion. In addition to containing small peptides, the resulting spray-dried material includes fats, metals, salts, vitamins and many other biological compounds. Peptone is used in nutrient media for growing bacteria and fungi.

Peptide Fragments

Refer to fragments of proteins that are used to identify or quantify the source protein. Often these are the products of enzymatic degradation performed in the laboratory on a controlled sample, but can also be forensic or paleontological samples which have been degraded by natural effects.

Peptides in Molecular Biology

Peptides have received prominence in molecular biology in recent times for several reasons. The first and most important is that peptides allow the creation of *peptide antibodies* in animals without the need to purify the protein of interest. This involves synthesizing antigenic peptides of sections of the protein of interest. These will then be used to make antibodies in a rabbit or mouse against the protein.

Another reason is that peptides have become instrumental in mass spectrometry, allowing the identification of proteins of interest based on peptide masses and sequence. In this case the peptides are most often generated by in-gel digestion after electrophoretic separation of the proteins.

Peptides have recently been used in the study of protein structure and function. For example, synthetic peptides can be used as probes to see where protein-peptide interactions occur.

Inhibitory peptides are also used in clinical research to examine the effects of peptides on the inhibition of cancer proteins and other diseases.

Well-known Peptide Families in Humans

The peptide families in this section are all ribosomal peptides, usually with hormonal activity. All of these peptides are synthesized by cells as longer 'propeptides' or 'proproteins' and truncated prior to exiting the cell. They are released into the bloodstream where they perform their signalling functions.

The Tachykinin peptides

- Substance P
- Kassinin
- Neurokinin A
- Eledoisin
- Neurokinin B

Vasoactive intestinal peptides

- VIP (*__V__asoactive __I__ntestinal __P__eptide*; PHM27)
- PACAP *(__P__ituitary __A__denylate __C__yclase __A__ctivating __P__eptide*)
- Peptide PHI 27 (*__P__eptide __H__istidine __I__soleucine 27*)
- GHRH 1-24 *(__G__rowth __H__ormone __R__eleasing __H__ormone 1-24)*
- Glucagon
- Secretin

Pancreatic polypeptide-related peptides

- NPY
- PYY *(__P__eptide __YY__)*
- APP *(__A__vian __P__ancreatic __P__olypeptide)*
- PPY (*__P__ancreatic __P__ol__Y__peptide*)

Opioid peptides

- Proopiomelanocortin (POMC) peptides
- Enkephalin pentapeptides
- Prodynorphin peptides

Calcitonin peptides

- Calcitonin
- Amylin
- AGG01

Other peptides

- B-type Natriuretic Peptide (BNP) - produced in myocardium & useful in medical diagnosis

Notes on terminology

- A *polypeptide* is a single linear chain of amino acids.

- A *protein* is one or more polypeptides more than about 50 amino acids long.
- An *oligopeptide* or (simply) a *peptide* is a polypeptide less than 30-50 amino acids long.
- A *dipeptide* has two amino acids.
- A *tripeptide* has three amino acids.
- A *tetrapeptide* has four amino acids.
- A *pentapeptide* has five amino acids.
- An *octapeptide* has eight amino acids (e.g., angiotensin II).
- A *nonapeptide* has nine amino acids (e.g., oxytocin).
- A *decapeptide* has ten amino acids (e.g., gonadotropin-releasing hormone & angotensin I).
- A *neuropeptide* is a peptide that is active in association with neural tissue.
- A *peptide hormone* is a peptide that acts as a hormone.

STRUCTURE OF PROTEIN

Protein Structure

The primary structure of a segment of a polypeptide chain or of a protein is the amino-acid sequence of the polypeptide chain(s), without regard to spatial arrangement (apart from configuration at the alpha-carbon atom). This definition does not include the positions of disulphide bonds, and is, therefore, not identical with "covalent structure" (IUPAC-IUB, 1970). The commonly occurring amino acids are of 20 different kinds which contain the same dipolar ion group $H_3N^+.CH.COO^-$. They all have in common a central carbon atom to which are attached a hydrogen atom, an amino group (NH_2) and a carboxyl group (COOH). The central carbon atom is called the C_α-atom and is a chiral centre. All amino acids found in proteins encoded by the genome have the L-configuration at this chiral centre.

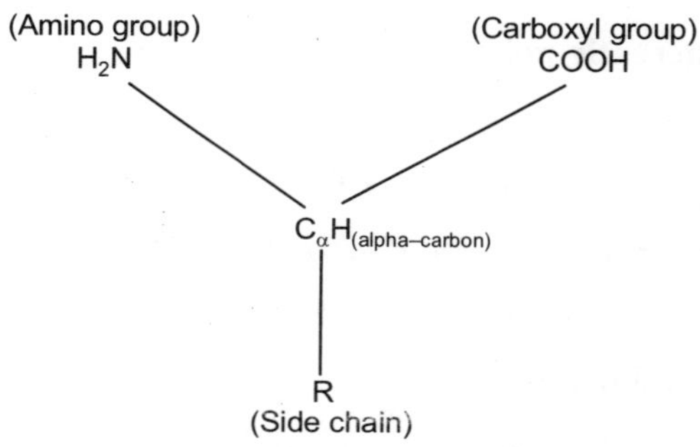

This configuration can be remembered as the **CORN** law. Imagine looking along the H-C_α bond with the H atom (Fig. 2.34).

When read clockwise, the groups attached to the C_α spell the word CORN (Richardson, 1981). There are 20 side chains found in proteins encoded by the genetic machinery of the cell. The side chains confer important properties on a protein such as the ability to bind ligands and catalyse biochemical reactions. They also direct the folding of the nascent polypeptide and

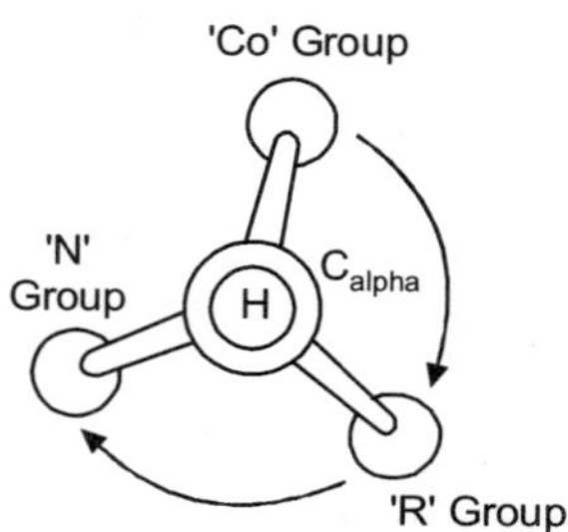

Fig. 2.34 : Showing the C_{alpha} bond

stabilise its final conformation. In molecular graphics, atoms can be represented in different ways. For expedience, molecules are often displayed only as lines or vectors between the atoms bonded together covalently. An elegant representation is the ball-and-stick type in which atoms are drawn as coloured spheres and their bonds as rod-like connections. Another useful display is the space-filling representation in which a surface is drawn around the atoms to indicate their van der Waals radii. This surface can be drawn as a series of dots or as a solid entity (Lesk, 1991). Amino acids in proteins(or polypeptides) are joined together by peptide bonds. The sequence of R-groups along the chain is called the **primary structure**.

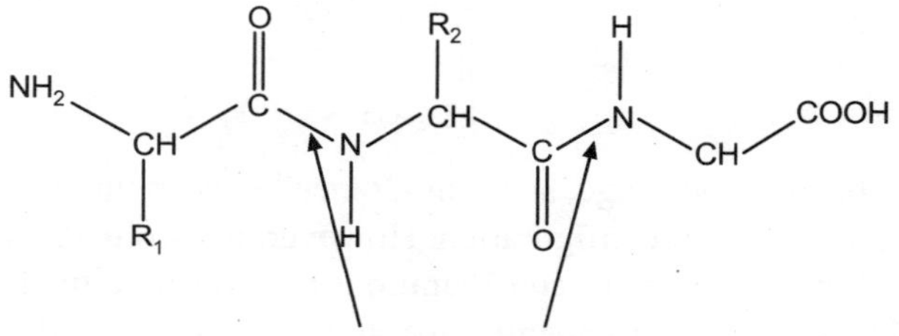

NH_2 CH R_1 C O N H CH R_2 C O N H CH COOH

Peptide bonds

The Peptide bond

Pauling *et al.* (1951) analysed the geometry and dimensions of the peptide bonds in the crystal structures of molecules containing either one or a few peptide bonds. Their results are summarised in the diagram below where the consensus bond lengths are shown in Angstrom units. Bond angles in degrees are also shown for the peptide N and C atoms. It should be noted that the C-N bond length of the peptide is 10% shorter than that found in usual C-N amine bonds (Schulz and Schirmer, 1990; Creighton, 1993).

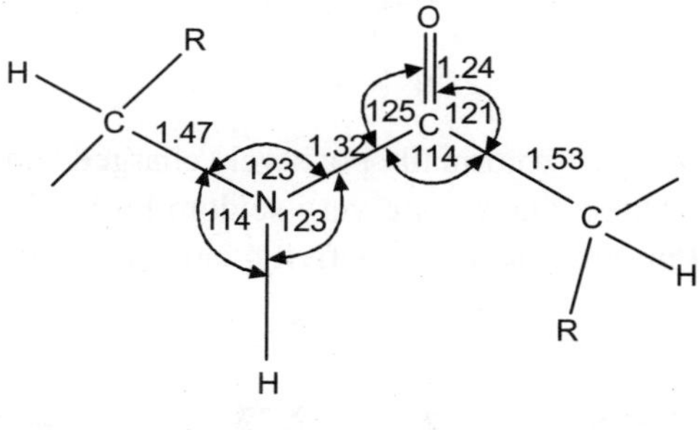

H R C 1.47 123 N 114 123 H 1.32 125 C 121 O 1.24 114 1.53 C R H

This is because the peptide bond has some double bond character (40%) due to resonance which occurs with amides. The two canonical structures are:

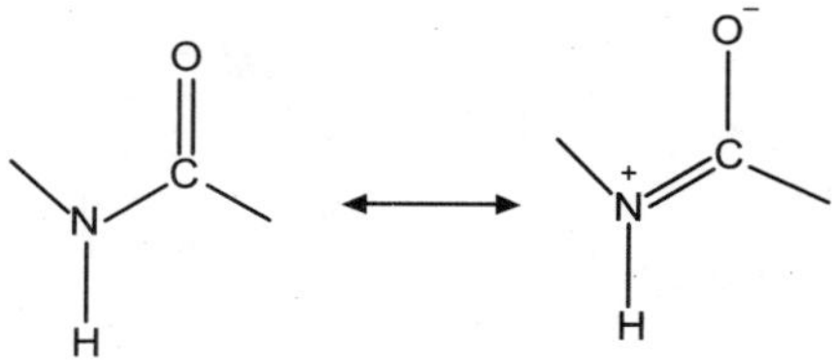

N H C O ⟷ N^+ H C O^-

As a consequence of this resonance all peptide bonds in protein structures are found to be almost planar, i.e. atoms $C_\alpha(i)$, C(i), O(i), N(i+1) and $C_\alpha(i+1)$ are approximately co-planar. This rigidity of the peptide bond reduces the degrees of freedom of the polypeptide during folding. The peptide bond nearly always has the *trans* configuration since it is more favourable than *cis*, which is sometimes found to occur with proline (Pro) residues (Schulz and Schirmer, 1990).

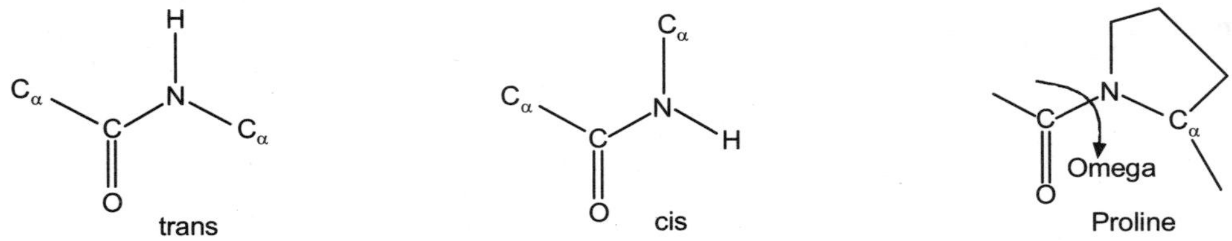

trans　cis　Proline

As can be seen from the previous page, steric hindrance between the functional groups attached to the C_α atoms will be greater in the *cis* configuration. However, for proline residues, the cyclic nature of the side chain means that both *cis* and *trans* configurations have more equivalent energies. Thus proline is found in the *cis* configuration more frequently than the other amino acids.

Properties of amino acids

The sequence and properties of side chains determine all that is unique about a particular protein, including its biological function and its specific three-dimensional structure.

Histidine (His) is the only side chain that titrates near physiological pH, making it especially useful for enzymatic reactions.

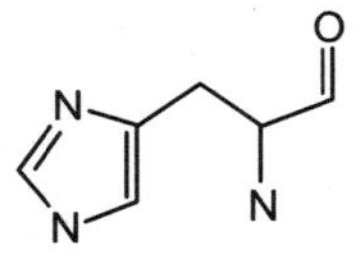

Lysine (Lys) and arginine (Arg) are normally positively charged and aspartate (Asp) and glutamate (Glu) are negatively charged. These charges are very seldom buried in protein interiors except when they are serving some special purpose, as in the activity and activation of chymotrypsin.

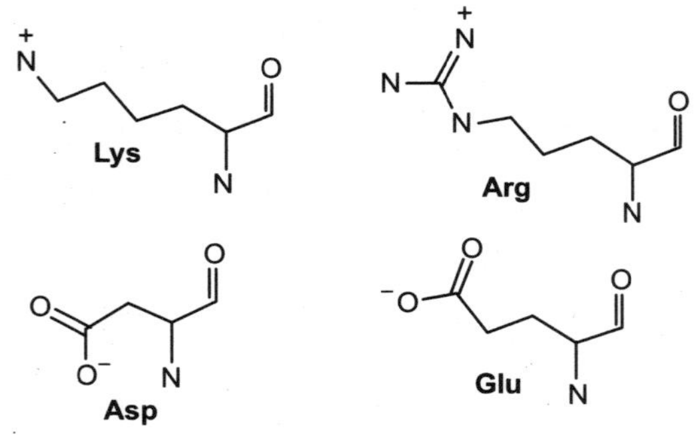

Lys　Arg

Asp　Glu

Asparagine (Asn) and glutamine (Gln) have interesting hydrogen-bonding properties, since they resemble the backbone peptides. The hydrophobic residues provide a very strong driving force for folding, through the indirect effect of their ceasing to disrupt the water structure once they are buried (Kauzmann, 1959). They also, however, affect the structure in a highly specific manner because their varied sizes and shapes fit together in very efficient packing (Lee and Richards, 1971).

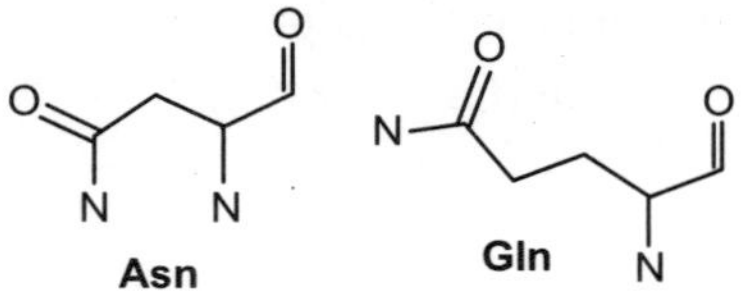

Asn Gln

Proline has stronger stereochemical constraints than any other residue, with only one instead of two variable backbone angles, and it lacks the normal backbone NH for hydrogen bonding. It is both disruptive to regular secondary structure and also good at forming turns in the polypeptide chain, so that in spite of its hydrophobicity it is usually found at the edge of the protein (Richardson, 1981).

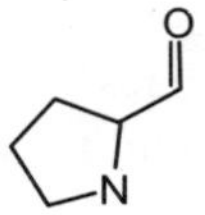

Glycine (Gly) has three different unique capabilities. As the smallest side group (only a hydrogen), it is often found where main chains approach each other very closely. In addition Gly can assume conformations normally restricted by close contacts of the β-carbon and finally it is more flexible than other residues, thus contributing to parts of the backbone that need to move or hinge (Richardson, 1981).

Serine (Ser) and threonine (Thr) carry aliphatic hydroxyl groups capable of forming hydrogen bonds with suitable donor or acceptor groups, such as the imino nitrogen or the carbonyl oxygen of the main polypeptide chain. Serine reacts with phosphate by an ester bond, forms part of the catalytic site of many hydrolytic enzymes (Dickerson and Geis, 1969) and contributes to the lining of ion channels. Serine, threonine, and asparagine are also the binding sites of carbohydrates that are attached to the surface of many proteins. Carbohydrates bound to serine and threonine form O-glycosidic bonds and those linked to asparagine form N-glycosidic bonds (Perutz, 1992).

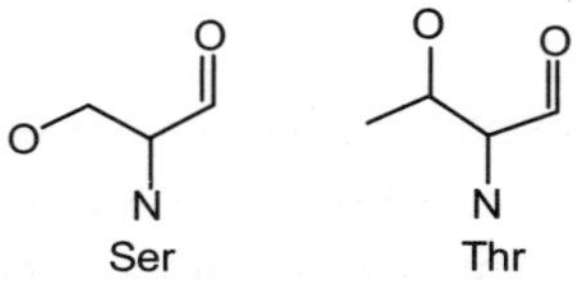

Ser Thr

Cysteine (Cys) carries the highly reactive sulphydryl group. This does not ionise at physiological pH nor form hydrogen bonds of significant strength, but two cysteines placed some distance apart along a polypeptide chain, or forming part of different chains, can be joined by oxidation to form the disulphide bridge of cystine which plays an important part in stabilizing protein structures. Disulphide bonds increase the conformational stability mainly by constraining the unfolded conformations of the protein and thereby decreasing their conformational entropy (Pace, 1990). Cysteines also bind zinc, copper, and iron ions. The sulphur atom in methionine is unreactive and generally serves no function other than imposing a special configuration on the aliphatic sidechain, but in cytochrome c it forms the link between the protein and the heme iron (Olson, 1992).

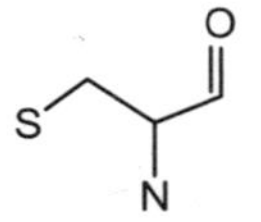

S O N

Protein Structure Determination

In terms of the accuracy of protein structure determinations, all of the bond lengths are invariant. Bond angles are also essentially invariant, except perhaps for *T*, the backbone N-C_α-C angle. The α-carbon is tetrahedral, which would give 110°, but there are indications from accurately refined protein structures that can sometimes stretch to larger values in order to accommodate other strains in the structure. The dihedral angle ω at the peptide is very close to 180° (producing a trans, planar peptide with the neighbouring α-carbons and the N, H, C, and O between them all lying in one plane). The remaining dihedral angles are the source of essentially all the interesting variability in protein conformation. The backbone dihedral angles are φ and ψ in sequence order on either side of the α-carbon, so that φ is the dihedral angle around the N-C_α bond and around the Cα-C bond. The side chain dihedral angles are x_1, x_2, etc. An extremely useful device for studying protein conformation is the Ramachandran plot (Ramachandran *et al.*, 1963) which plots φ and ψ. The values of φ and ψ that are possible are constrained geometrically due to steric clashes between non-neighboring atoms. The permitted values of and are indicated on a two-dimensional map of the - plane.

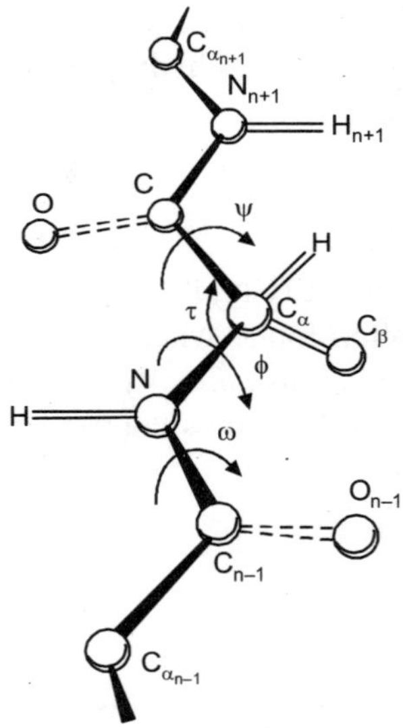

$C_{\alpha n+1}$ N_{n+1} H_{n+1} C O ψ H τ C_α C_β N φ H ω O_{n-1} C_{n-1} $C_{\alpha n-1}$

Fig. 2.35 : The Alpha helix

Secondary Structure

The secondary structure of a segment of polypeptide chain is the local spatial arrangement of its main-chain atoms without regard to the conformation of its side chains or to its relationship with other segments (IUPAC-IUB, 1970). There are three common secondary structures in proteins, namely **alpha helices, beta sheets** and **turns**. That which cannot be classified as one of the standard three classes is usually grouped into a category called “other” or “random coil”. This last designation is unfortunate as no portion of protein three dimensional structure is truly random and it is not a coil either. Regular secondary structure conformations in segments of a polypeptide chain occur when all the ϕ bond angles in that polypeptide segment are equal to each other, and all the ψ bond angles are equal. The rotational angles for ϕ and ψ bonds for common regular secondary structures are shown in the table below.

Parameters of regular secondary structures. n is the number of residues per helical turn, p is the helical pitch, and A is the atoms in H-bonded loop

Structure	ϕ	ψ	n	p(Å)	A	H-bond (CO, HN)
Right-handed α-helix[3.6_{13}helix]	-57	-47	3.6	5.4	13	i,i+2
3_{10}-helix	-74	-4	3.0	6.0	10	i,i+3
pi-helix	-57	-70	4.4	5.0	16	i,i+4
Parallel beta strand	-119	113	2.0	6.4		
Antiparallel beta strand	-139	135	2.0	6.8		

The α-helix and β-structure conformations for polypeptide chains are generally the most thermodynamically stable of the regular secondary structures. However, particular amino acid sequences of a primary structure in a protein may support regular conformations of the polypeptide chain other than α-helical or β-structure. Thus, whereas α-helical or β-structure are found most commonly, the actual conformation is dependent on the particular physical properties generated by the sequence present in the polypeptide chain and the solution conditions in which the protein is dissolved. In addition, in most proteins there are significant regions of unordered structure in which the ϕ and ψ angles are not equal. A large proportion of (85%) of helices are distorted in some way i.e. radius of curvature greater than 90Å and deviation of axis from straight line is equal to or greater than 0.25Å. These may be due to a number of reasons:

- CO groups form hydrogen bonds with NH groups 3 residues along the chain forming a 3_{10} helix. A substantial amount of all 3_{10} helices occur at the ends of α-helices. They are called 3_{10} because there are 3 residues per turn and 10 atoms enclosed in a ring formed by each hydrogen bond. Dipoles are not aligned as in a normal right-handed alpha-helix.
- Packing of buried helices against other secondary structural elements in the core of a protein can lead to distortions since the side chains are on the surface of helices.
- Proline residues induce distortions of around 20deg in the direction of a helix. This is because proline cannot form a regular alpha-helix due to steric hindrance arising from its cyclic sidechain which blocks the main chain NH group. Proline causes 2 hydrogen bonds in the helix to be broken. Helices containing proline are usually long because shorter helices would be destabilized.
- Exposed helices are often bent away from the solvent. This is because the exposed C=O groups tend to point towards solvent to maximize their hydrogen bonding capacity i.e tend to form hydrogen-bonds to sòlvent as well as N-H groups. This gives rise to a bend in the helix axis.

- The π helix is an extremely rare secondary structural element in proteins. Like the 3_{10} helix, one turn of π helix is sometimes found at the ends of regular alpha helices. The infrequency of this particular form of secondary structure stems from the following properties: (*i*) the π and ψ angles of the pure π helix (-57.1, -69.7) lie at the very edge of an allowed minimum energy region of the Ramachandran map, (*ii*) the π helix requires that the angle tau (N-C_{alpha}-C') be larger (114.9) than the standard tetrahedral angle of 109.5degrees, (*iii*) the large radius of the π helix means the polypeptide backbone is no longer in van der Waals contact across the helical axis forming an axial hole too small for solvent water to fill, and (*iv*) side chains are more staggered than the ideal 3_{10} helix but not as well as the α-helix (Low and Grenville-Wells, 1953; Schulz and Schirmer, 1990).

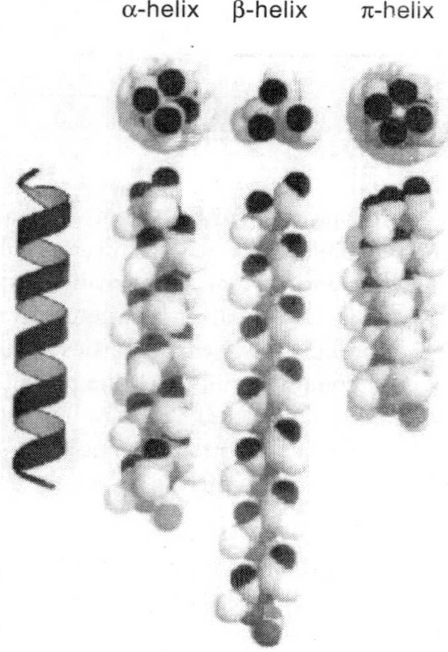

Fig. 2.36 : The Alpha helix Beta Sheets and π helix

Besides the α-helix, β-sheets are another major structural element in globular proteins containing 20-28% of all residues (Kabsch and Sander, 1983; Creighton, 1993). The basic unit of a β-sheet is a beta strand (which can be thought of as a helix with *n*=2 residues/turn) with approximate backbone dihedral angles $\phi = -120$ and $\psi = +120$ producing a translation of 3.2 to 3.4 Å / residue for residues in antiparallel and parallel strands, respectively. The β strand is then like the α-helix, a repeating secondary structure. However, since there are no intra-segment hydrogen bonds and van der Waals interactions between atoms of neighbouring residues are not significant due to the extended nature of the chain, this extended conformation is only stable as part of a β-sheet where contributions from hydrogen bonds and van der Waals interactions between aligned strands exert a stabilizing influence. The β-sheet is sometimes called the β pleated sheet since sequential neighbouring C_α atoms are alternately above and below the plane of the sheet giving a pleated appearance. β-sheets are found in two forms designated as "Antiparallel" or "Parallel" based on the relative directions of two interacting beta strand (as shown below).

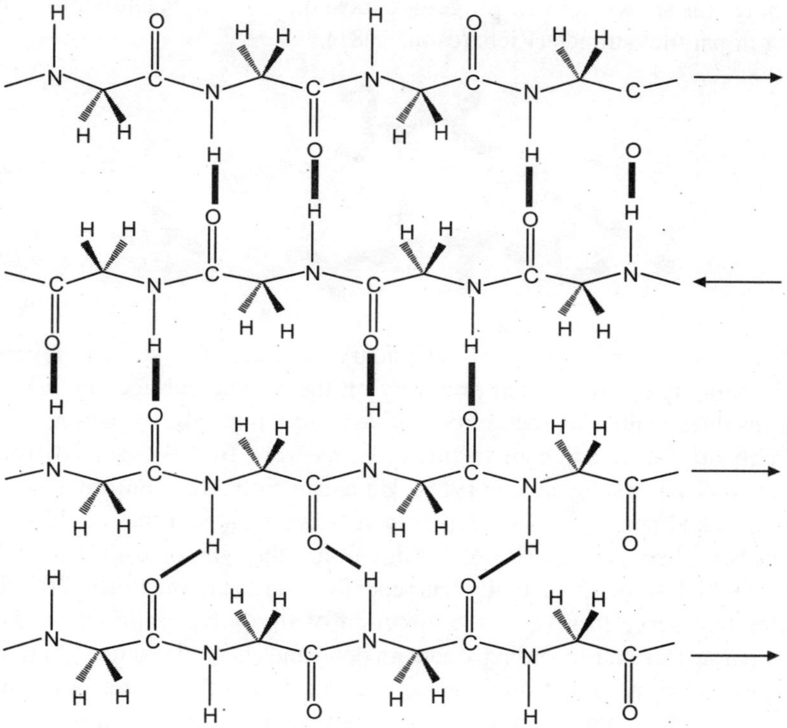

Hydrogen bond patterns in β-sheets. A four-stranded β-sheet is drawn schematically which contains three **antiparallel** and one **parallel** strand. Hydrogen bonds are indicated with red lines (antiparallel strands) and blue lines (parallel strands) connecting the hydrogen and receptor oxygen.

Like α-helices, β-sheets have the potential for amphiphilicity with one face polar and the other apolar. However, unlike alpha-helices which are composed of residues from a continuous polypeptide segment (i.e., hydrogen bonds between CO of residue I and NH of residue I+3), β-sheets are formed from strands that are very often from distant portions of the polypeptide sequence. Hydrogen bonds in β-sheets are on average 0.1 Å shorter than those found in α-helices (Baker and Hubbard, 1984).

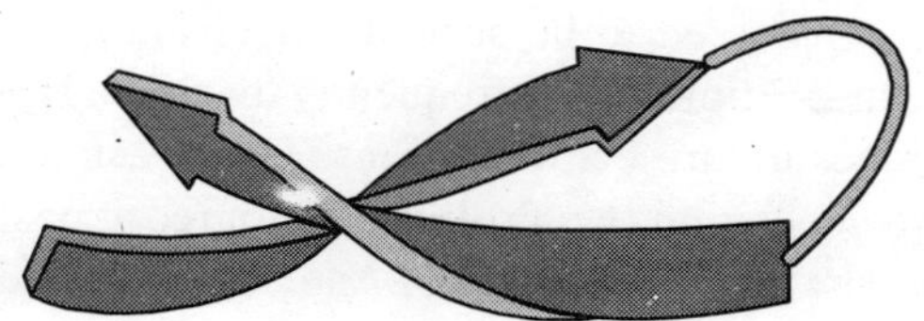

The classical β-sheets originally proposed are planar but most sheets observed in globular proteins are twisted (0 to 30 degrees per residue).

Antiparallel β-sheets are more often twisted than parallel sheets. Another irregularity found in antiparallel β-sheets is the hydrogen-bonding of two residues from one strand with one residue from another called a β bulge (as shown above). Bulges are most often found in antiparallel sheets with ~5% of bulges occurring in parallel strands (Richardson, 1981).

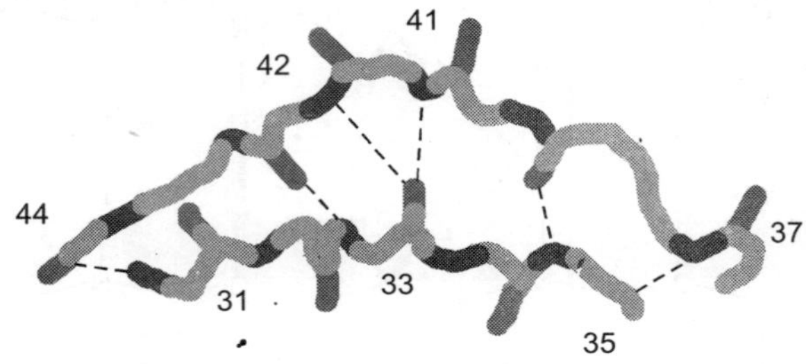

Turns are the third of the three "classical" secondary structures that serve to reverse the direction of the polypeptide chain. They are located primarily on the protein surface and accordingly contain polar and charged residues. Antibody recognition, phosphorylation, glycosylation, and hydroxylation sites are found frequently at or adjacent to turns. Turns were first recognised from a theoretical conformational analysis by Venkatachalam (1968). He considered what conformations were available to a system of three linked peptide units (or four successive residues) that could be stabilised by a backbone hydrogen bond between the CO of residue n and the NH of residue n+3. He found three general types, one of which (type III) actually has repeating Turns are the third of the three "classical" secondary structures that serve to reverse the direction of the polypeptide chain. They are located primarily on the protein surface and accordingly contain polar and charged residues. Antibody recognition, phosphorylation, glycosylation, and hydroxylation sites are found frequently at or adjacent to turns. Turns were first recognised from a theoretical conformational analysis by Venkatachalam (1968). He considered what conformations were available to a system of three linked peptide units (or four successive residues) that could be stabilised by a backbone hydrogen bond between the CO of residue n and the NH of residue n+3. He found three general types, one of which (type III) actually has repeating ϕ, ψ values of -60deg, -30deg and is identical with the 3_{10}-helix. The three types each contain a hydrogen bond between the carbonyl oxygen of residue i and the amide nitrogen of i+3. These three types of turns are designated I, II, and III. Many have speculated on the role of this type of secondary structure in globular proteins. Turns may be viewed as a weak link in the polypeptide chain, allowing the other secondary structures (helix and sheet) to determine the conformational outcome. In contrast (based on the recent experimental finding of "turn-like" structures in short peptides in aqueous solutions (Dyson et al., 1988), turns are considered to be structure-nucleating segments, formed early in the folding process. Type I turns occur 2-3 times more frequently than type II. There are position dependent amino acid preferences for residues in turn conformations. Type I can tolerate all residues in position i to i+3 with the exception of Pro at position i+2. Proline is favoured at position i+1 and Gly is favoured at i+3 in type I and type II turns. The polar sidechains of Asn, Asp, Ser, and Cys often populate position i where they can hydrogen bond to the backbone NH of residue i+2.

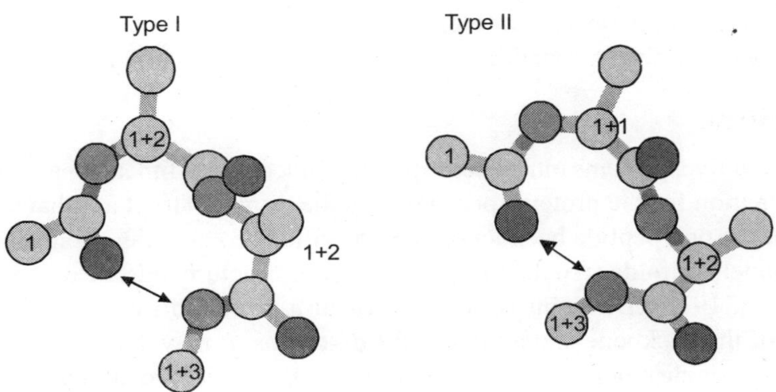

Supersecondary Structure

Rao and Rossmann (1973) observed that there were structural components comprising a few β-helices or β-strands which were frequently repeated within structures, called 'supersecondary structures' (being intermediate to secondary and tertiary structure) and suggested that these structures might be due to convergence. A variety of recurring structures were subsequently recognised such as the 'Greek key' (Richardson, 1977). Some of these structural motifs can be associated with a particular function whilst others have no specific biological function alone but are part of larger structural and functional assemblies. The first comprehensive attempt to classify proteins on the basis of structural comparisons was due to Levitt and Chothia (1976) who used four major classifications:

- proteins containing mostly α-helix;
- proteins containing mostly β-sheet;
- proteins that contain α-helices and β-strands in an irregular sequence, and
- α/β proteins with alternate segments of α-helices and β-strands.

The simplest motif with a specific function consists of two α-helices joined by a loop region. Two such motifs are (*i*) a motif specific for DNA binding and (*ii*) a motif specific for calcium binding and is present in parvalbumin, calmodulin, troponin-C, and other proteins that bind calcium and thereby regulate cellular activities.

Tertiary Structure

The tertiary structure of a protein molecule, or of a subunit of a protein molecule, is the arrangement of all its atoms in space, without regard to its relationship with neighbouring molecules or subunits (IUPAC-IUB, 1970). As an example of a protein's tertiary structure, the structure of lysozyme is shown below.

Quaternary Structure

The quaternary structure of a protein molecule is the arrangement of its subunits in space and the ensemble of its intersubunit contacts and interactions, without regard to the internal geometry of the subunits (IUPAC-IUB, 1970). The subunits in a quaternary structure must be in noncovalent association.

Haemoglobin contains four polypeptide chains ($\alpha_2\beta_2$) held together noncovalently in a specific conformation as required for its function.

Protein Stability

To be biologically active, proteins must adopt specific folded three-dimensional, tertiary structures. Yet the genetic information for the protein specifies only the primary structure, that is the linear sequence of amino acids in the polypeptide backbone. Many purified proteins can spontaneously refold *in vitro* after being completely unfolded, so the three-dimensional structure must be determined by the primary structure (Anfinsen, 1973). Different conformations of a protein differ only in the angle of rotation about the bonds of the backbone and amino acid side-chains. It may, therefore, appear surprising that a protein folds into a single unique conformation from all the possible rotational conformations available around single bonds in the primary structure of a protein. The question arises as to when a protein folds up to its native conformation, does this structure represent a local or a global energy minimum? When a protein folds it samples a number of conformations. Does the structure which results from the folding depend on its stability or on the energy barriers encountered by the polypeptide? The polypeptide whilst folding may become trapped in the local energy well and cannot fold to the global energy minimum (kinetic hypothesis of protein folding, Wetlaufer, 1973; Wetlaufer and Ristow, 1973).

The native structure of the protein may correspond to a metastable state with a very long lifetime. If proteins are only metastable, their structures must be grossly different from the most stable one. The polypeptide may adopt a structure corresponding to the global minimum. This means the final structure does not depend on the size of the energy barriers (thermodynamic hypothesis of protein folding, Epstein et al., 1963). Anfinson's work on ribonuclease provided some evidence for the thermodynamic hypothesis (Haber and Anfinsen, 1962). The initial stages of folding were considered to be nearly random. However, if the rest of the folding pathway was a random search, it would not be feasible for any protein to try out all of its conformations on a practical time scale. For example, if each residue of a 100 residue polypeptide had only three conformations, the total number of conformations would be $3^{100} = 5 \times 10^{47}$. Since conformational changes occur on the timescale of 10^{-13} seconds, the time required by the 100 residue protein to search all conformations would be $5 \times 10^{47} \times 10^{-13} \approx 10^{37}$ years. Nevertheless, proteins are observed to fold in $10^{-1} - 10^{3}$ seconds both *in-vivo* and *in-vitro* (Creighton, 1993). The conclusion, therefore, was that proteins do not fold by sampling all possible conformations randomly until the one with the lowest free energy is encountered.

This led to the idea of a biased random search resulting in faster folding since a proportion of the conformations would be sterically disallowed. A framework model proposed (Baldwin, 1989; Kim and Baldwin, 1990) that the folded structure was formed by packing together pre-existing individual elements of secondary structure which had significant stability in the unfolded protein. Another mechanism postulated the unfolded polypeptide chain to undergo rapid hydrophobic collapse under refolding conditions (Dill, 1985), perhaps to something approximating the molten globule state. Simply constraining the polypeptide chain to be compact might greatly increase the probability of the final folded conformation being encountered (Gregoret and Cohen, 1991). It became evident through some experiments that the equilibrium of unfolding of proteins does not always follow a simple two-state model in which only the native and fully unfolded states are significantly populated (Wong and Tanford, 1973). An intermediate compact structure known as the molten globule which is different from the native structure and whose formation is determined mainly by non-specific interactions of amino acid residues with their environment was presented. Specific interactions could direct the folding pathway by stabilizing folded conformations. The best studied example is bovine pancreatic trypsin inhibitor (Creighton, 1978). For this protein it was shown that formation of a disulphide bridge stabilizes secondary structure elements, and the protein refolds by a specific pathway of disulphide bond formation and rearrangement (Directed folding model). Since noncovalent forces act on the primary structure to cause a protein to fold into a unique conformational structure and then stabilize the native structure against denaturation processes, it is of importance to understand the properties of these forces.

Non-covalent Forces

Non-covalent forces are weak forces of bonding strength of 1-7kcal mol^{-1} (4-29kJ mol^{-1}) as compared to the strength of covalent bonds which have a bonding strength of at least 50 kcal mol^{-1}. The non-covalent bonding forces are just higher than the average kinetic energy of molecules at 37°C (0.6 kcal mol^{-1}). Apart from their involvement in the stabilization of molecules, they contribute to the ability of molecules to undergo changes in conformation and interact with each other. Since a major part of this project (Friedli, 1996) involves a study of interactions, an explanation of the forces involved is crucial for an understanding of the mechanisms at the molecular level. The major forces and interactions are:

- Ion - Ion forces, for example those involving; NaCl or $MgSO_4$;
- Ion - Dipole forces, for example those between NaCl and H_2O;
- van der Waals interactions including;
  1. Dipole - Dipole, such as the hydrogen bond;
  2. Dipole - Induced Dipole, for example: $H_2O + I_2$ and
  3. Dispersion forces including Induced Dipole - Induced Dipole, for example between two aliphatic hydrogens.
- Electrostatic interaction or salt bridge between charged residues and
- Hydrophobic forces between nonpolar residues.

Ion - dipole attractions depend on:

- the distance between the ion and the dipole thus the closer the ion and dipole are, the stronger the attraction;
- the charge on the ion - the greater the ion charge, the stronger the attraction, and
- the magnitude of the dipole - the greater the magnitude of the dipole, the stronger the attraction.

Water is an excellent example of a polar molecule, a molecule with positive and negative electrical poles. Hydrogen bonding is a special form of dipole - dipole attraction. When H is attached to a very electronegative atom X, the interaction between other molecules and the H - X bond dipole is significantly greater than expected for ordinary dipole - dipole attraction. This interaction is called hydrogen bonding because it occurs only when H is part of one or both of the interacting dipoles. The electronegativities of N (3.0), O (3.5), and F (4.0) are among the highest of all of the elements, while that of H (2.1) is considerably less (Bohinski, 1987).

A polar water molecule can induce a dipole in non-polar O_2. For the dipole to be induced depends on the atom's or molecule's polarizability. As the molecular mass of a molecule increases, either there is an increase in the number of valence electrons or the valence electrons are less tightly held. Therefore, the ease of polarization of the valence electron cloud generally increases with mass. Since a dipole is more readily induced as the polarizability increases, the strengths of dipole - induced dipole interactions generally increase with mass. Also, since the solubility of substances such as CO_2 or O_2 depends on the strength of the dipole - induced dipole interaction, the solubility of non-polar substances in polar solvents generally increases with mass (Kotz and Purcell, 1991).

The weakest of all intermolecular forces is between two induced dipoles. Such forces are often called London forces or dispersion forces. When atoms or molecules approach each other, each experiences the electric field provided by the other. This electric field distorts the charge distribution (Webster, 1990). The attraction of a molecule to its own distorted charge distribution creates an attractive force between the molecules even when they are a long distance apart. This force acts to bind the approaching atoms or molecules. Fritz London (1930) gave an approximate result for the attractive energy resulting from the interaction between two induced dipoles. The interaction energy, E^{London}/J, which is attractive for all inter-atomic, or inter-molecular distances, R/m, is expressed by the formula:

$$E^{london} = -3/2\{I_A I_B l(I_A + I_B)\}\left[(\alpha_A \alpha_B)/(4\pi\varepsilon_0)^2 R^6\right]$$

where I_A/J and I_B/J are the first ionization energies for A and B, and $(\alpha_A / 4\pi\varepsilon_0)/m^3$ and $(a_B / 4\pi\varepsilon_0)/m^3$ are their polarizability volumes. E^{London} is called the London dispersion energy. The London energy will be larger in magnitude as R decreases but the key feature is that it creates an attraction even at large R values. For this reason it is usually classified as a long-range interaction. The London dispersion force, however, is only one component of van der Waals forces. A second component, the John Lennard-Jones potential (LJ potential), U(R)/J, is defined by:

$$U(R) = 4\varepsilon\left[(\sigma / R)^{12} - (\sigma / R)^{6}\right]$$

The parameter ε represents the binding energy of a van der Waals molecule at its equilibrium geometry. The parameter σ is the value of R, the inter-atomic or inter-molecular distance, when U(R) = 0, other than at R = ∞. From equation (2), the term R^{-6}, approximates the London dispersion energy which is attractive and this is counteracted by a repulsive term having an R^{-12} dependence. An alternative expression which approximates the energy, U(R)/kJmol^{-1}, of van der Waals molecule is :

$$U(R) = b \exp(-aR) - d/R^6$$

The van der Waals force is of great importance in biopolymer structure. This force has an attractive term dependent on the 6th. power of the distance between two interacting atoms and a repulsive term dependent on the 12th. power of the distance between them. The attractive component is due to the induction of complementary partial charges or dipoles in the electron density of adjacent atoms when the electron orbitals of the two atoms approach to a close distance whereas the repulsive component of the van der Waals force predominates at closer distances, when the electro orbitals of the adjacent atoms begin to overlap. This type of repulsion is commonly called steric hindrance. The distance of maximum favourable interaction between two atoms is known as the van der Waals contact distance, which is equal to the sum of the van der Waals radii for the two atoms. While a van der Waals - London dispersion interaction between any two atoms in a protein is usually less than 1 kcal mol^{-1}, the total number of these weak interactions in a protein molecule is in the thousands. Thus the sum of the attractive and repulsive van der Waals - London dispersion forces are extremely important to protein folding and stability. The van der Waals contact distances of 2.8-4.1 Å are longer than hydrogen-bond distances of 2.6–3.1 Å, and at least twice as long as normal covalent bond distances of 1.0–1.6 Å between C, H, N, and O atoms. Although the latter bonds are shorter than the van der Waals contact distance, a repulsive van der Waals force must be overcome in forming hydrogen bonds and covalent bonds between atoms.

The fundamental law of electrostatics namely Coulomb's law, expresses the inverse square law of force between two electric charges q_1 q_2 separated by a distance R in a vacuum in the form:

$$F = q1\ q2/(4\pi\varepsilon_0 R^2)$$

where ε_0 is the permittivity of free space (vacuum). If q_1 and q_2 have the same sign, the force is a repulsion; if they are of opposite sign, the force is an attraction. In the presence of a material medium surrounding both charges, the force is reduced by a factor $\varepsilon_x = \varepsilon/\varepsilon_0$, the relative permittivity (or dielectric constant) of the medium. The work done in bringing two charges together from infinite separation to a distance R in a medium of permittivity ε is, therefore, given by:

$$\Delta W = \int_{\infty}^{R} F dR = q1q2/(4\pi\varepsilon R) = E$$

and measures the electrical free energy of the system relative to that at infinite separation. Electrostatic interactions between charged groups are of importance to particular protein structures, in the binding of charged ligands and substrates to proteins, interaction between basic and acidic proteins, and repulsion between charges of the same sign as between SWP and sodium alginate (Friedli *et al.*, 1995). The strength of the electrostatic force (E) is directly dependent on the charge (q) for each ion, and is inversely dependent on the dielectric constant (ε) of the solvent and the distance between the charges (R). Water has a high dielectric constant ($\varepsilon = 80$), and ionic charge interactions in water are relatively weak in comparison to electrostatic interactions in the interior of a protein, where the dielectric constant ($\varepsilon = 2$-40) is approximately a factor of 1 : 40 to 1 : 2 that of water. Consequently, the strength of an electrostatic interaction in the interior of a protein, where the dielectric constant is low, may be of significant energy. However, most charged groups of proteins are on the surface of the protein where they do not strongly interact with other charged groups from the protein or other biopolymers due to the high dielectric constant of the water solvent, but are stabilized by hydrogen bonding and polar interactions to the water. Electrostatic interactions in water are less than those in other solvents because of water's high dielectric constant, which results from the tendency of the large dipoles of water molecules to align with any electric field. The dielectric constant of pure water at 25°C is 78.5 and it decreases at higher temperatures because thermal motion overcomes the orienting effects of the water dipoles. This effect of temperature explains some of the findings in. When small diffusible ions such as Na^+ and Cl^- are present in water, the apparent dielectric constant of the solution increases because the ions tend to concentrate in the vicinity of charges of the opposite sign. Since the present project considered biopolymer interactions, most of the experiments were performed with water in order to avoid interference from salts.

The strong inclination of water molecules to form hydrogen bonds with each other influences their interactions with non-polar molecules that are incapable of forming hydrogen bonds (e.g., alkanes, hydrocarbons inert atoms etc.). When water molecules come in contact with such a molecule they are faced with an apparent dilemma that whichever way the water molecules face, it would appear that one or more of the four charges per molecule (ST2 model of water, named after Stillinger and Rahman, 1974) will have to point towards the inert solute molecule and thus be lost to hydrogen bond formation. Clearly the best configuration would have the least number of tetrahedral charges pointing towards the unaccommodating species so that the other charges can point towards the water phase and, therefore, participate in hydrogen bond attachments much as before. There are many options to salvage lost hydrogen bonds. If the non-polar solute molecule is not too large, it is possible for water molecules to pack around it without giving up any of their hydrogen-bonding sites, thus forming clathrate 'cages' around a dissolved non-polar solute molecule. Such structures are not rigid but labile, and their hydrogen bonds are not stronger than in pure water, but the water molecules forming these cages are more ordered than in the bulk liquid.

It is also clear that the sizes and shapes of non-polar solute molecules are fairly critical in determining the water structure adopted around them. This is often referred to as hydrophobic solvation or hydrophobic hydration. At present there is no simple theory of such solute-solvent interactions. However, both theoretical and experimental studies indicate that the re-orientation, or re-structuring, of water around non-polar solutes or surfaces is entropically very unfavourable, since it disrupts the existing water structure and imposes a new and more ordered structure on the surrounding water molecules. This immiscibility of inert substances with water, and the mainly entropic nature of this incompatibility is known as the hydrophobic effect (Kauzmann, 1959; Tanford, 1980), and such substances, e.g., hydrocarbons and fluorocarbons, are known as hydrophobic substances. Similarly, hydrophobic surfaces are not 'wetted' by water but when water comes into contact with such surfaces it rolls up into small lenses and subtends a large contact angle on them.

Closely related to the hydrophobic effect is the hydrophobic interaction, which describes the unusually strong attraction between hydrophobic molecules and surfaces in water which are often stronger than their attraction in free space. For example, the van der Waals interaction energy between two contacting methane molecules in free space is -2.5×10^{-21} J, while in water it is -14×10^{-21} J. Because of its strength it was originally believed that some sort of 'hydrophobic bond' was responsible for this interaction. But at present it is known that there is no bond associated with this mainly entropic phenomenon, which arises primarily from the rearrangements of hydrogen bond configurations in the overlapping solvation zones as two hydrophobic species come together, and which is also much of a longer range than any typical bond. To date there have been very few direct measurements of the hydrophobic interaction between dissolved non-polar molecules, mainly because they are so insoluble. Tucker *et al.* (1981) reported values of -8.4 and -11.3 kJ mol^{-1} for the free energies of dimerization of benzene-benzene and cyclohexane-cyclohexane, respectively, and Ben Naim et al. (1973) deduced a value of about -8.5kJ mol^{-1} for two methane molecules. There is as yet no satisfactory theory of the hydrophobic interaction, though a number of promising theoretical approaches have been proposed (Dashevsky and Sarkisov, 1974; Pratt and Chandler, 1977; Marcelja *et al.*, 1977; Pangali *et al.*, 1979; Nicholson and Parsonage, 1982). Israelachvili and Pashley (1982) measured the hydrophobic force law between two macroscopic curved surfaces in water and found that in the range 0-10nm, the force decayed exponentially with distance with a decay length of about 1nm. Based on these findings Israelachvili and Pashley proposed that for small solute molecules, the hydrophobic free energy of dimerization increased in proportion with their diameter s according to:

ΔG (hydrophobic pair potential) $\approx -20\ \sigma$ kJ mol^{-1}

where σ is in nanometres.

The hydrophobic interaction plays a central role in;

- many surface phenomena;
- molecular self-assembly;
- micelle formation;
- biological membrane structure;
- determining the conformations of proteins, and
- in protein-protein interaction and gelation.

While there is no phenomenon actually known as the hydrophilic effect or the hydrophilic interaction, such effects can be recognised in the propensity of certain molecules and groups to be water soluble and to repel each other strongly in water, in contrast to the strong attraction exhibited by hydrophobic groups. Hydrophilic (i.e., water-loving) groups prefer to be in contact with water rather than with each other, and they are often hygroscopic (taking up water from vapour). Strongly hydrated ions and zwitterions are hydrophilic; but some uncharged and even non-polar molecules can be hydrophilic if they have the right geometry and if they contain electronegative atoms capable of associating with the hydrogen bond network in water, for example, the O atoms in alcohols and polyethylene oxide and the N atoms in amines. It is also important to note that a polar group is not necessarily hydrophilic and that a non-polar group is also not always hydrophobic. While hydrophobic molecules tend to increase the ordering of water molecules around them, hydrophilic molecules are believed to have a disordering effect. Certain highly polar molecules are so effective in altering or disrupting the local water structure that when dissolved in water they can have a drastic effect on other solute molecules. For example, when urea, $(NH_2)_2C{=}O$, is dissolved in water it can cause proteins to unfold. Such non-ionic but highly potent compounds are commonly referred to as chaotropic agents, a term that was coined to convey the idea that their disruption of the local water structure leads to chaos. It appears, therefore, that the hydrophilic and hydrophobic interactions, unlike electrostatic and dispersion interactions are inter-dependent and therefore not additive. Indeed, one would not expect them to be independent of each other, since both ultimately rely on the structure of the water hydrogen bonds adopted around dissolved groups. Because some of the systems studied in the present project were dispersions instead of solutions it is appropriate that a discussion on their interactions should involve long-range forces like the electric double-layer forces, van der Waals forces and steric polymer interactions.

As mentioned earlier, the van der Waals force between similar particles in a medium is always attractive, so that if only van der Waals forces were operating, we might expect all dissolved particles to stick together (coagulate) immediately and precipitate out of solution as a mass of solid material. This normally does not happen, because particles suspended in water and any liquid of high dielectric constant are usually charged and can be prevented from coalescing by repulsive electrostatic forces. Other repulsive forces that can prevent coalescence are solvation and steric forces. The charging of a surface can come about by the ionisation or dissociation of surface groups (e.g., the dissociation of protons from surface carboxylic groups [$-COOH \rightarrow -COO^- + H^+$], which leaves behind a negatively charged surface). The surface charge is balanced by an equal but oppositely charged region of counterions, some of which are bound, usually transiently, to the surface within the so-called Stern or Helmholtz layer, while others form an atmosphere of ions in rapid thermal motion close to the surface, known as the diffuse electric double layer

The electric potential in the solution falls off exponentially with distance from the surface.

$$\psi = \psi_0 \exp(-kx)$$

where ψ_0 is the potential at the potential determining surface and ψ the potential at a distance x from the surface in the electrolyte solution. The quantity k is called the Debye-Hückel parameter. The quantity 1/k is referred to as the thickness of the double layer (Sennett and Olivier, 1965; Shaw, 1986; Ross and Morrison, 1988; Everett, 1989; Hunter, 1993; Atkins, 1994). As two charged surfaces come together, their double layers overlap and as a rough approximation, the electrical potentials arising from the two surfaces are additive. This implies an increase in the electrical contribution to the free energy of the system. A dispersion represents a state of higher free energy than that corresponding to the material in bulk. Passage to a state of lower free energy will, therefore, tend to occur spontaneously unless there is a substantial energy barrier preventing the elimination of the dispersed state (Everett, 1989). In the presence of such a barrier the system will be metastable and may remain in that state for a long time. On the other hand, if conditions are adjusted so that the energy barrier becomes negligibly small, or disappears altogether, then the dispersion becomes unstable and aggregate. The energy necessary to carry the system over the energy barrier comes from Brownian motion of the particles, which results from the random bombardment of the surface of the particles by molecules of the medium. Instability will ensue if the ratio of the energy barrier height to kT is reduced. This may arise in various ways. In principle, if the absolute height remained constant, then instability could be induced by an increase in temperature. The barrier height is also influenced by concentration and ionic strength (Hunter, 1993; Israelachvili, 1995). The double layer repulsion depends on the ionic strength of the medium as follows:

- the curves may show a high repulsive barrier at low ionic strength;
- a so-called secondary minimum at intermediate ionic strength, and
- a negligibly small barrier, or none at all, at higher ionic strengths.

The secondary minimum is of the order of a few kT, and that is where weakly bound aggregates (flocs) form. Although the aggregates formed at the secondary minimum are sufficiently stable and cannot be completely dissociated by Brownian motion, they may disintegrate under externally applied hydrodynamic forces, such as vigorous stirring. If the particles are able to overcome the primary maximum (energy barrier) and enter the primary minimum, then there is no turning back, the aggregates become irreversible. These are some of the situations encountered in the project as the proteins interact with each other or with other proteins.

Covalent Forces

When two or more atoms come together to form a molecule, as when two hydrogen atoms and one oxygen atom combine to form a water molecule, the forces that tightly bind the atoms together within the molecule are called covalent forces, and the inter-atomic bonds formed are called covalent bonds. In a covalent bond electrons are shared between two or more atoms so that the discrete nature of the atoms is lost. Depending on the position an atom (or element) occupies in the periodic table, it can participate in a certain number of covalent bonds with other atoms. This number or stoichiometry is known as the atomic valency. A further characteristic of covalent bonds is their directionality, that is, they are directed or oriented at well-defined angles relative to each other. Covalent forces are short range, that is, they operate over very short distances of the order of inter-atomic separations (0.1 - 0.2 nm) and tend to decrease in strength with increasing bond length. There are two types of covalent bonds in proteins, the peptide bond and the disulphide bond.

Disulphide bonds occur between the sulphurs of two cysteine side chains. They predominantly occur in extracellular proteins and are part of the primary structure. Inside the cell the sulphydryl is maintained in a reduced state by glutathione, but extracellularly, in the presence of oxygen, thiols are unstable relative to S-S bridges (Fahey et al., 1977). Although the disulphides are part of the primary structure, it has been shown that some native S-S bridges are only formed once the secondary and even tertiary structure of the protein has been achieved (Creighton, 1978). In extracellular proteins of known sequence, which contain disulphides, there is rarely more than one free -SH group. Ovalbumin, with one S-S and four cysteines, was the only extracellular exception found (Nisbet et al., 1981). Sulphydryls are very reactive in an extracellular environment and readily oxidise to form disulphides. Consequently if a cysteine is external this may lead to disastrous polymerisation or make folding more complex. When the disulphide conformation of all the proteins in the Protein Data Bank (PDB) were examined in detail, it was found that they could be grouped into two major categories (Richardson, 1981):

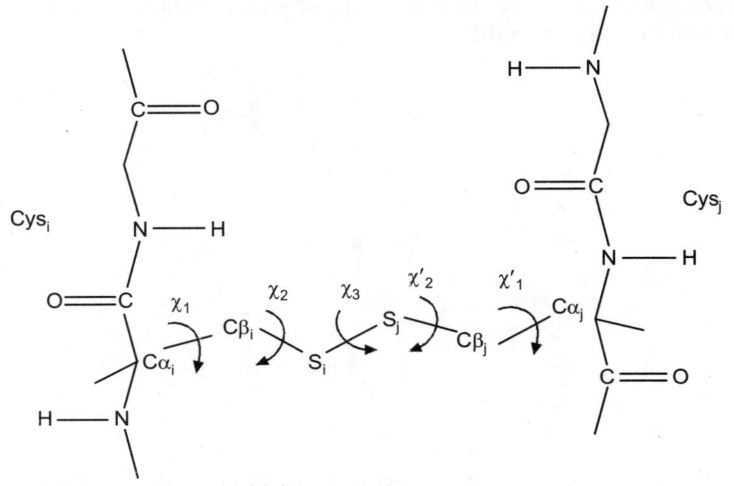

- The left-handed spiral conformation, with

$$x_1 = -60°, x_2 = -90°, x_3 = -90°, x_2^1 = -90°, x_1^1 = -60° \text{ and}$$

- The right-handed hook conformation, with

$$x_1 = -60°,\ x_2 = +120°, x_3 = +90°, x_2^1 = -50°, x_1^1 = -60°$$

A detailed analysis was done on the number of residues between half-cystines (Thornton, 1981) and it was found that the most frequent separation is 10 to 14 residues. The shortest connection found was two residues and connections longer than 150 residues are rare. A study was carried out to find out how disulphides fit into the three-dimensional structure of a protein (Sternberg and Thornton, 1976). The disulphides were divided into two groups: local disulphides, with half-cystines separated by less than 45 residues and non-local disulphides with a separation greater than 45 residues. Thirty-three out of the 34 local disulphides investigated had one of the structures shown below. The most common being the coil-coil (cc), followed by c-bb-c. No helix-helix (aa) local disulphides were observed, although (aa) disulphides were found between non-local half-cystines.

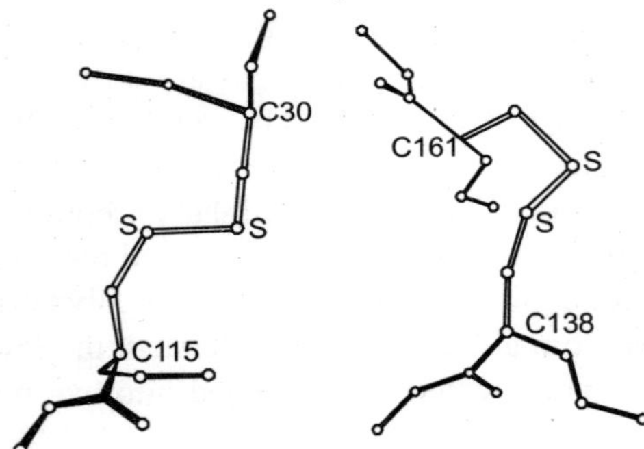

Fig. 2.37 : A left-handed spiral disulphide from hen egg white lysozyme (left) and a right-handed hook disulphide from carboxypeptidase A (right).

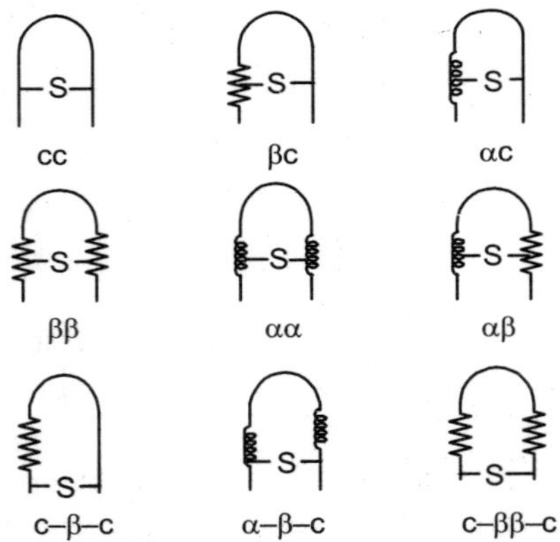

Local disulphide "single loop" topologies. c=coil, beta=strand, alpha=helix. (Thornton, 1981)

In summary,

- local disulphides are preferred where possible;
- disulphides between the amino and carboxyl termini are also favourable, even though they are non-local;
- secondary structure (alpha-helix and beta-sheet) often prevents formation of local disulphides
- disulphides are rarely found between cystines separated by more than 150 residues or between cystines in different domains;
- Half-cystines are predominantly in coiled regions in contrast to cysteines, which are preferentially helical, and
- local disulphides tend to be right handed whilst non-local disulphides are left-handed and adopt the left-handed spiral structure.

Disulphide bond formation was found to be important in the present study. Exposure of cysteines in BSA and beta-lactoglobulin through heating led to aggregation and gel. Glutenin polymerises through inter-disulphide bond formation. At room temperature, soluble wheat protein (SWP) was made to gel through reduction of its disulphide bonds. As the disulphide bond in proteins are reduced, the protein unfolds or denatures.

Protein Denaturation

Protein denaturation has been defined in several ways, for example as a change in solubility (Mirsky, 1941) or by simultaneous changes in chemical, physical and biological properties (Neurath et al., 1944; Langmuir, 1938) under some standard reference set of conditions (Timasheff and Gibbs, 1957). These changes in physical, and to a lesser extent chemical properties are manifestations of configurational changes taking place in the polypeptide chains. The denaturation process presumably involves an unfolding or at least an alteration in the nature of the folded structure (Foster & Samsa, 1951). Most denaturation changes consist of changes in secondary bonds: ion-dipole, hydrogen and Van der Waals, and in the rotational positions about single bonds which are controlled by the secondary bond structure (Lumry and Eyring, 1954). The term denaturation denotes the response of the native protein to heat, acid, alkali, and a variety of other chemical and physical agents which cause marked changes in the protein structure. Rice et al. (1958) suggested denaturation to mean a class of reactions which lead to changes in the structure of the macromolecule with no change in molecular weight. Timasheff and Gibbs (1957), pointed out that the approaches used to define the concept of denaturation can be classified into two types:

- molecular, in terms of actual structural changes taking place on the molecule, and
- operational, in terms of changes in measurable properties.

The operational approach to denaturation has the advantage of being purely phenomenological, but it cannot lead to a precise definition, since the concept of property, even restricted to measurable property, is itself without finite boundary. On the other hand, the molecular point of view, although still involves a few assumptions despite the important advances in investigations of protein structure, will allow us to reach a definition which is very convenient, owing to its general validity. Whatever it is, changes can be quantitatively described by comparison with the native state. This seems simple, although in fact difficulties arise in attempting to obtain an accurate quantitative description of the native state since it is not a state that is static and fixed. Moreover, physico-chemical measurements are ordinarily not made in situ, but with extracted proteins, and it cannot be excluded that extractions and purification treatments do, sometimes, modify, to a certain extent, the structure of the protein molecules. Nevertheless, in practice, the state of the molecules prepared by the mildest extraction method is conventionally assumed to be the native state. Following these preliminary remarks, denaturation may be defined in very general terms as any modification of the secondary, tertiary or quaternary structure of the protein molecule, excluding any breaking of covalent bond (Joly, 1965). The degree to which the three-dimensional structure of a protein may differ from the native state may vary from a change in a single noncovalent bond or side-chain orientation to the case where almost no atom exist in the same spatial relationship to others except for the constraints of the primary structure.

The special importance attached to an understanding of the denaturation process is due to the fact that denaturation is usually a prerequisite for gelation. Since this project investigationed the elucidation of the gelation process it was considered that the control of denaturation could lead to a specific gelation properties. Evidence suggests that the various individual non-covalent bonds do not act independently but that there is a cooperative action of particular groups of 'bonds' or contacts in stabilizing various segments of the structure or even the total conformation (Tanford, 1968). Thus conformational transitions are found to pass through a few intermediate states or take place by an all-or-none type of mechanism between two states with no intermediates occurring in substantial concentration (Creighton, 1994).

The denaturation process can be achieved by any one of the following methods: increasing temperature, changing pH, using denaturants (i.e. urea, guanidine hydrochloride, beta-mercaptoethanol, dithiothreitol), inorganic salts (i.e lithium bromide, potassium thiocyanate, sodium iodide), organic solvents and (i.e. formamide, dimethylformamide, dichloro- and trichloroacetic acids and their salts), detergents (i.e. sodium dodecyl sulphate), high pressure and ultrasonic homogenisation.

The temperatures at which various proteins unfold vary enormously. Most proteins unfold at elevated temperatures, and some unfold at very low temperatures. Many proteins unfold at temperatures only a few degrees higher than those at which they function. Others are stable to much higher temperatures such as the gluten proteins. The driving force for denaturation is the increase in entropy that accompanies the transition of a single conformation into an ensemble of random ones. With increasing temperature the contribution of this entropy increases and becomes more significant, and at some temperature it overcomes the energy effect (the protein is heat denatured). It is interesting to consider possible intermediate structures. The early unlocking of the tertiary structure deletes a large number of the bonds holding the structure together but increases the randomness only insignificantly. The later stages of denaturation lead to larger increases in entropy. Thus, the intermediate states are relatively unstable, and heat denaturation is often an all-or-none phenomenon. The unfolding of the protein exposes the buried non-polar amino acid residues. Their intermolecular clustering leads to aggregation of the denatured protein. Consequently, heat denaturation is essentially irreversible.

In chemical denaturation the secondary bonds holding the protein segments together are disrupted by some chemicals capable of forming equally strong or stronger bonds with the groups holding the conformation together. For disrupting the hydrogen bonds, urea or guanidine hydrochloride are used. At high concentrations of these substances (e.g, 8M urea or 5M guanidine hydrochloride) many proteins adopt a highly unfolded conformation in solution. Proteins of multiple subunits are likely to be separated into their constituent polypeptide chains. Other proteins aggregate upon denaturation in urea or guanidine hydrochloride which is frequently due to the formation of disulphide bridges between sulphydryl groups made accessible by the unfolding of the polypeptide chains. Such reactions may be inhibited by the addition of iodoacetate (Friedli, PhD thesis, chapter 3 and 5). Under these conditions, the denatured molecules remain in solution and may revert into native molecules if the denaturing agent is slowly dialysed away. Powerful detergents like SDS disrupt both hydrophobic and hydrogen bonds and effectively solvate the denatured molecule. beta-Mercaptoethanol and dithiothreitol (DTT) disrupt disulphide bonds and can be used in conjunction with urea or SDS to fully solubilise protein molecules.

Protein-Protein Interactions

Protein-protein interactions occur widely. These can either involve specific binding or non-specific interactions. Numerous examples of specific binding can be observed in biological systems:

- where proteins are directed to the correct compartments of cells by binding to other proteins;
- protein messengers bind to protein receptors on the outer surface of cell membranes to send signals between cells;
- proteins form structural connections between cells;
- some inhibitors of enzymes are proteins;
- proteins are modified and degraded by enzymes;
- interaction between different protein subunits are the basis of allosteric changes in multimers, and
- protein-protein interactions underlie very large-scale movements in organisms, such as muscle contraction.

However only a small selection of molecules are involved in specific binding (Zuckerkandl,1975) which is a property of individual proteins. The only category of proteins which has specialised in creating binding sites for practically all kinds of molecules are the immunoglobulins. In addition the substrates of serine proteases are proteins, therefore, the complex formation between these enzymes and their substrates can provide useful information on protein-protein interactions.

Protein-protein interactions are generally favoured under conditions which reduce the net charge on the molecules, i.e. pH values near the isoelectric point. High ionic strength tends to reduce electrostatic repulsion between proteins due to the shielding of ionizable groups by mobile ions. Protein-protein association involves the specific complementary recognition of two macromolecules to form a stable assembly (Jones and Thornton, 1995). Fundamental to the stabilization of protein association is the hydrophobic interaction (Chothia and Janin, 1975). The term hydrophobic interaction is used to describe the gain in free energy which occurs when non-polar residues of proteins associate in an aqueous environment (Kauzmann, 1959). The process of folding and protein-protein aggregation reduces the surface area in contact with water. When the protein-solvent interaction is attractive, the protein can reduce its total energy by surrounding itself with solvent molecules, conversely, when the interaction is repulsive, the solvent is excluded (Tanaka, 1981). The aggregation of protein subunits buries the hydrophobic residues of the proteins, and hence minimizes the number of thermodynamically unfavourable solute-solvent interactions as found when SWP is hydrated in distilled water at room temperature.

Most of the proteins in food systems are denatured to varying degrees depending upon the type of processing used. The functionality required dictates the type and concentrations of ingredients and their environment. Protein-protein or protein-polysaccharide interactions under these circumstances are not specific, like in biological systems, but rather, depend mainly on physico-chemical forces. Proteins spontaneously aggregate when hydrated, therefore, molecular interactions are best studied in dilute systems. As their concentrations increase, their behaviour is better explained with colloid chemistry. Physical functions associated with proteins in a food system typically includes hydration and water binding which affect viscosity and gelation; modification of surface and interfacial activity which control emulsification and foaming ability and chemical reactivity leading to altered states of cohesion/ adhesion and a potential for texturization (Fligner and Mangino, 1991).

A particular functional property may be a manifestation of a specific component of the food protein used or due to interactions involving the biopolymers in the system. Schoen, (1977) pointed out that one major objective of protein functionality research, as it relates to foods, is to understand how proteins interact with each other and with other components in mixed systems. Protein interactions under certain circumstances lead to gelation.

Gelation

Gels may be defined by their ability to immobilize liquid, macromolecular structure, textural or rheological properties (Kinsella, 1976). The Collins English Dictionary defines a gel as a semi-rigid jelly-like colloid in which liquid is dispersed in a solid (Latin, gelare, meaning 'to freeze'). Bungenberg de Jong (1949), defined a gel as a system of solid character in which the colloidal particles somehow constitute a coherent structure. Hermans (1949), defined a gel with three propositions : (a) they are coherent colloid systems of at least two components; (b) they exhibit mechanical properties characteristic of a solid; (c) both the dispersed component and the dispersion medium extend themselves continuously throughout the whole system. Flory (1974), classified gels, based on their structure into four types as follows :

- Well-ordered lamellar structures, including gel mesophases;
- Covalent polymeric network ; completely disordered;
- Polymer networks formed through physical aggregation;
- predominantly disordered, but with regions of local order, and
- Particulate, disordered structures.

An example of the first type is a phospholipid with well-ordered lamellar structure as in biological membranes. Gels of the second type are completely disordered with structural units covalently linked to one another. The units are either bi-functional or poly-functional with a network which pervades the entire system giving it an infinite molecular weight for example elastin and condensation polymers. Gels of the third type are usually primary macromolecules bound together via crystalline domains or multi-stranded helices where the network is caused by physical aggregation of chains, previously disordered, but with regions of local order. The orientation of the ordered regions are mutually uncorrelated and their locations may likewise be independent. Chains between the ordered domains are random. Whereas, the network in the second type of gels are covalently linked making them irreversible, the strands of the third type gels are linked by non-covalent forces and can under certain conditions be reversible. An example of this type of gel is gelatin, where triple helices like those in native collagen are formed and these appear to be further aggregated to form small crystalline domains at higher concentrations (Flory and Garrett, 1958). At low concentrations the triple helices appear to be individually dispersed without aggregation as crystallites (Flory and Garrett, 1958; Peniche-Covas et al., 1974; Eagland et al., 1974). Only a small fraction of the gelatin may be involved in the ordered domains. Gels formed by aggregation of proteins are of the fourth type, and they usually occur under conditions of partial denaturation. The proteins may be fibrillar or globular. Although specific interactions as in clotting of fibrin and antibody-antigen interactions may be responsible for the formation of gel-like aggregates, the aggregation of other proteins to a gel may be non-specific (Goldberg, 1952; Flory, 1974).

Biopolymer gels differ from synthetic polymer gels in a number of ways:

- They contain large amounts of solvent (usually water) than their synthetic counter-part
- The point covalent cross-links (e.g, vulcanised rubber) in synthetic polymer networks are replaced in biopolymer networks by a combination of many physical interactions which occur between sizeable regions of the biopolymer (Clark *et al.*, 1990).
- Elasticity of synthetic polymer networks is entropic in origin. In these networks, strands between cross-links can assume numerous conformations due to Brownian motion. Deformations of the network lowers the entropy of the strands and this is responsible for the resistance against permanent deformation (Walstra *et al.*, 1991). The elastic properties of such networks can be described by the Theory of Rubber Elasticity (Treolar, 1975), in which the shear modulus, G is expressed as: ($G = N_e R_g T / V_{mol}$) where N_e is the number of Elastically Active Network Chains (EANC) (Dobson and Gordon, 1965) per monomer and V_{mol} is the molar volume of a monomer. The elastic properties of biopolymer networks are caused by enthalpic rather than entropic changes.

The mechanical spectrum (frequency-dependence of the dynamic moduli, G' and G") for a gel shows the storage modulus (G') to be greater than the loss modulus (G"), and both moduli are independent of frequency (i.e. rate of deformation). whereas, the dynamic viscosity decreases steeply with increasing frequency, and the slope of log h* vs log w approaches -1 when G' and G" are constant.

And on a shorter timescale the behaviour of non-covalently cross-linked gels approximates to that expected for a permanent network (Morris, 1983).

Tanaka (1981), defined a gel as a form of matter intermediate between a solid and a liquid which consists of polymers, or long chain molecules, cross-linked to create a tangled network immersed in a liquid medium. The liquid prevents the polymer network from collapsing into a compact mass, whilst the network prevents the liquid from flowing away. Tanaka identified three forces which interact to either expand or shrink polymer networks. The forces are; rubber elasticity, the polymer-polymer affinity and the hydrogen-ion pressure. The sum of these three forces was called the osmotic pressure of the gel, because it determines whether the gel tends to take up fluid or to expel it. The rubber elasticity is the elasticity of the individual polymer strands (i.e. the resistance the strands offer to either stretching or compression). The polymer-polymer affinity is the interaction between the polymer strands and the solvent. Such interactions can be either attractive or repulsive, depending on the electrical properties of the molecules. The hydrogen-ion pressure, is associated with the ionization of the polymer network, which releases an abundance of positively charged hydrogen ions (H^+) into the gel fluid. The hydrogen-ion give rise to pressure in the gel. A strand can be represented by a chain of rigid, jointed segments, each of which is in constant thermal motion. If the chain is stretched almost taut, the random movements of the segments give rise to a tension that pulls the ends of the chain inwards. If the chain is compressed into a compact ball, the force is directed outward. At an intermediate length of the chain the average force is zero. The rubber elasticity is proportional to the absolute temperature because thermal agitation is the ultimate root of the force. The polymer-polymer affinity decides if two polymers will aggregate and thereby exclude the solvent from the space between them or if there is a greater force of attraction between the polymer and the solvent than a polymer and another polymer.

A morphological classification of gels was proposed by Russo (1987) as :

- Fishnet gels; where cross-links, whether reversible or covalent, provide the 'strong points' of the structure and are separated by flexible strands which provide the elasticity.
- Lattice gels; where the division of the structure into cross-links and strands is inappropriate, i.e: the mechanical distinction between cross-links and strands is obscure, but nonetheless, a space-filling structure exits.
- Transient gels; a gel in which the structure is not permanent or due to high concentrations cause the polymers to overlap leading to temporary 'cross-links' due to chain entanglement.

Under this system of clasification, gelatin gels would be considered fishnet gels, globular protein network gels as lattice gels and yogurt as a transient gel. Aqueous solutions of biopolymers can often be converted to gels by thermal/or chemical means. Clark (1986) classified thermal set gels into heat setting and cold setting. An example of a heat set gel is the aggregation of thermally unfolded globular proteins (Barbu and Joly, 1953; Clark *et al.*, 1981a; Clark *et al.*, 1981b; Tombs, 1970;). The cross-linking of gelatin, a fibrous protein and various types of polysaccharide via disorder-to-order transition induced by cooling, is typical of the cold-setting process (Eagland *et al.*, 1974; Peniche-Covas *et al.*, 1974; Morris *et al.*, 1980; Rees and Welsh, 1977).

Burchard and Ross-Murphy (1988), proposed a phenomenological definition, stating that : all gels possess at least one property which can stand as the operational definition of a gel. They possess a plateau in the real part of the complex modulus extending over an appreciable window of frequencies - i.e. they are, or can be coaxed under appropriate conditions to be, viscoelastic solids.

Almdal *et al.*, (1993) also suggested a phenomenological definition of a gel based on two criteria :

- A gel is a soft, solid or solid-like material of two or more components, one of which is a liquid present in substantial quantity.
- Solid-like gels are characterised by the absence of an equilibrium modulus, by a storage modulus,

G', which exhibits a pronounced plateau extending to times at least of the order of seconds, and a loss modulus, G", which is considerably smaller than the storage modulus in the plateau region.

Morris (1985), characterised modes of gelation based on two different classes of materials, i.e. structural or storage. Storage materials are those which, even after extraction and purification, possess highly ordered structures (i.e. globular proteins). To gel aqueous solutions or dispersions it is usually necessary to first heat the samples in order to partially disrupt the ordered structure. Cooling leads to gelation which involves the formation of new intermolecular network. The original ordered structure cannot be recovered and such heat-set mechanisms are generally irreversible. Structural polymers are normally extracted from plant or animal cell walls (i.e. gelatin, alginate, pectin and carrageenans). Gelation occurs through a disorder-order transition which is reversible.

Most of the theories discussed so far are related to single component gels. The present project was carried out on mixed binary systems, therefore, a brief discussion on two component gels proposed by Morris (1985) would be appropriate. In a two component gel, the polymer that forms the network structure was labelled as 'active' and the other which is merely contained within the network structure as 'non-active'. A gel is called type-I if it is made up of both active and non-active polymers, and type II, if both polymers are active. Complex formation between the active and non-active polymers in type I gels is unlikely. The non-active polymers will tend to concentrate the active polymers promoting intra- and inter-molecular interactions between them. The network structure in type II gels will depend on the relative values of the two active components and their degree of phase separation prior to gelation. The network formed were classified as:

- Interpenetrating networks, consisting of two independent network structures which interact only through mutual entanglements.
- Phase separated networks formed from the incompatability of the active polymers resulting in composites with upper and lower bound moduli.
- Coupled networks which form through inter-molecular interactions between the active polymers.

If conditions necessary for the gelation of the two active polymers differ, then different network structures could be produced by manipulating those conditions, for example gelling one polymer on heating and the other on cooling (Kasapis, 1994).

Gelation of Globular Proteins

Several factors affect gelation, such as protein type, protein concentration, temperature, ionic strength, type of ion and pH (Mulvihill and Kinsella, 1987). Probably the two most important factors in gelation are the protein concentration and heating temperature. If either, or both the temperature (Dunkerley and Hayes, 1980) and protein concentration (Ross-Murphy, 1991) are too low, gelation will not occur. Once these factors are above their critical values, gel strength increases and gelation time decreases with increasing temperature (Schmidt and Illingworth, 1978; Dunkerley and Hayes, 1980; Dunkerley and Zadow, 1984) and concentration (Ross-Murphy, 1991; Plock *et al.*, 1992). Ferry (1948) suggested a two-step mechanism of gelation which involves: (1) an initiation step involving unfolding or dissociation of the protein molecules, followed by (2) an aggregation step in which association or aggregation reactions occur, resulting in gel formation under appropriate conditions. For the formation of a highly ordered gel, it is essential that the aggregation step proceeds at a slower rate than the unfolding step (Hermansson, 1978, 1979). Richardson and Ross-Murphy (1981) concluded from their gelation experiments, using 9% BSA solutions at pH 6.5 and at various temperatures, that the unfolding of BSA is rate-determining at temperatures below 57°C and that aggregation is rate-determining above that temperature. This finding is consistent with the views of Ferry (1948) that unfolding precedes aggregation and provides the driving force for protein-protein interaction.

There are a number of ways in which two or more proteins can interact with each other which will affect the properties of the gel formed. After mixing, proteins may be qualitatively considered incompatible, semicompatible, or compatible, depending on whether two immiscible phases are formed, partial mixing takes place at the molecular level, or a single thermodynamically stable phase is formed (Manson and Sperling, 1976). Composite or multicomponent gels are produced from mixtures of two or more gelling agents, or a single gelling and nongelling components. A second protein capable of gel formation may act as a non-gelling component if it is present in the mixture at a concentration below its critical concentration for gel formation (Ziegler and Foegeding, 1990). The second protein component may behave like a filler, interspersed throughout the primary gel network. The gel may be single phase with the filler remaining soluble (Figure 2.37A), or two phase, where thermodynamic incompatibility causes phase separation to occur, with the filler existing as dispersed particles of liquid or as a secondary gel network (Figure 2.37). The single phase system was labelled by Tolstoguzov (1986) as type-I and the

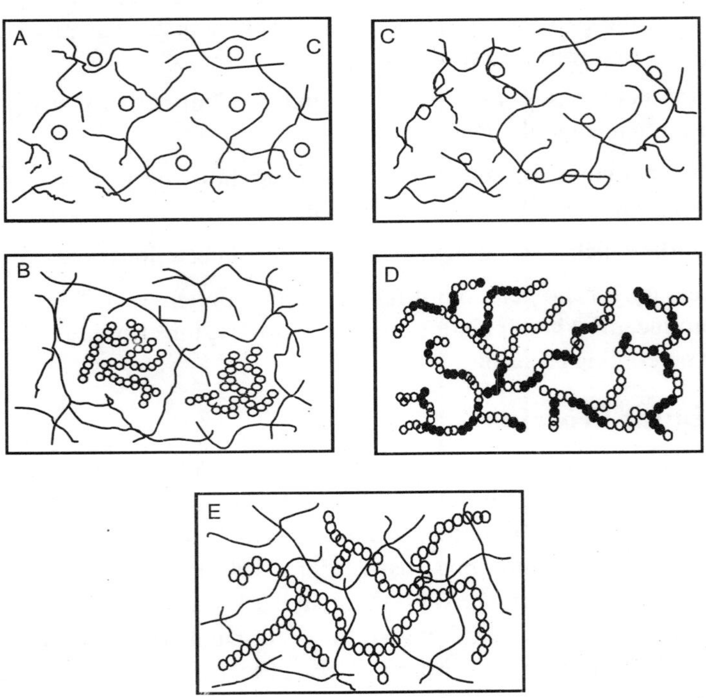

Fig. 2.38 : Types of mixed gels

phase separated system as type-II filled gel. In (Fig. 2.37C), the non-gelling component associates with the primary network in a random fashion via nonspecific interactions which may reduce the flexibility of the primary network chains and add rigidity to the gel. The two protein components may co-polymerise to form a single, heterogenous network (Figure 2.37D). An example of this type is the polymerisation which occurs in BSA-ovalbumin gels (Clark *et al.*, 1982). (Fig. 2.37E is an example of the interpenetrating polymer network where both components form separate continuous network throughout the system.

Protein-Polysaccharide Interactions

Protein-polysaccharide interactions in food systems often plays a role in determining the functional properties of these systems (Stainsby, 1980). Understanding the mechanisms involved in the interactions between proteins and polysaccharides and the way in which these interactions are affected during processing is important when these components are added into foods to improve their functional properties (Ledward, 1979; Stainsby, 1980). Although there is evidence to indicate that the major forces responsible for these interactions are electrostatic in nature (Imeson et al., 1977), other interactions such as hydrogen, hydrophobic or covalent bonds may also be significant in the stabilization of the interaction.

Proteins and polysaccharide combinations have been used to stabilize emulsions (Dickinson and Euston, 1991; Dickinson and Galazka, 1991) through electroststic interactions and when covalently linked (Kato et al., 1990; Dickinson and Galazka, 1991; Dickinson and Galazka, 1992). The formation of soluble protein-polysaccharide complex in the pH range that would lead to protein precipitation has been utilised in the preparation of fruit flavoured milk beverages in which fruity flavour is best expressed at pH 4.5-5.0. CMC has been found to be particularly effective in keeping milk proteins in solution. In addition to holding milk proteins in solution, excess polysaccharide may resolubilize a precipitated complex at low pH. This behaviour is explained as a result of two reactions (Hidalgo and Hansen, 1969): a primary ionic reaction leading to the formation of insoluble complex, followed by a 'peptization' reaction, involving redistribution of protein molecules on the polysaccharide, giving rise to increased hydration and thus solubilization. The protein-polysaccharide interaction also inhibits precipitation of some water soluble proteins following denaturation

PHYSICAL AND CHEMICAL PROPERTIES OF PROTEINS

Taste: Pure proteins are generally tasteless, though the predominant taste of protein hydrolysates (proteoses, peptones, peptides, amino acids) is' bitter.

Odor: Pure proteins are odorless. When heated, they turn brown and char and give off the odor of burning feathers or hair.

Molecular weights of proteins: As previously indicated, protein molecules are exceedingly complex in structure and are very large. Because of the size of their molecules, proteins belong to the so-called colloidal state of matter.

The most important method for the determination of protein molecular weights is based upon the ultracentrifuge. The student should review that discussion at this time.

The minimal molecular weight of a protein may be calculated from the content of some characteristic component, if this is known. For example, horse hemoglobin contains 0.335 per cent iron. At least one iron atom must be present per molecule of hemoglobin. Since an atom of iron represents 55.85 units of weight and this is 0.335 per cent of the total molecular weight of hemoglobin (if only one Fe is present), the minimal molecular weight of hemoglobin is given by the expression:

0.335%: 100%: : 55.85 : mol. wt.

mol. wt. = 16,671

Actually the molecular weight of hemoglobin is four times the minimal molecular weight as calculated on the assumption of one iron atom per molecule, which means that there are four iron atoms in each hemoglobin molecule, the true molecular weight is $4 \times 16{,}671 = 66{,}684$.

The amino acid content of a protein may also be used for calculating minimal molecular weights. The methods of analysis for tyrosine, cystine, and tryptophan give quite accurate results, and determination of these acids in proteins may be used for the calculation of minimal molecular weights. Minimal molecular weights are of value in checking on actual molecular weights determined by the ultracentrifuge or other physical methods. In some instances, by calculating the minimal molecular weight for several constituents, it is possible to calculate the actual molecular weight.

The molecular weights of proteins also may be obtained from osmotic pressure measurements and light scattering.

Table below gives the molecular weights of a number of proteins, most of which were obtained by the ultra centrifugal method.

Table 2.24 : Some Physical Properties of Proteins

Protein	S_{20}*	D_{20} (cm^2/sec)@	f/f_0#	V_{20} ml/g$	M_2%	IpH approximate
1	2	3	4	5	6	7
Actomyosin	12.0	0.30		(0.75)	3,900,000	
Adrenocorticotropic hormone (sheep)	2.1	10.5		(0.75)	20,000	
Albumin (serum, horse)	4.46	6.1	1.27	0.748	70,000	
Albumin (serum, human)	4.6	6.1	1.28	0.733	69,000	4.7
Aldolase	7.3	4.63	1.31	0.74	147,000	
β-Amylase (sweet potato)	8.9	5.77		(0.749)	152,000	
Carbonic anhydrase	2.8	9.0		0.749	30,000	5.3
Carboxyhemoglobin (cow)	4.6			0.749		
Catalase	11.3	4.1	1.25	0.73	250,000	
Chymotrypsinogen	2.54	9.5	1.19	0.72	23,200	9.5
Cytochrome c (horse heart)	1.9	10.1	1.29	0.707	15,600	10.65

...(Contd.)

1	2	3	4	5	6	7
Diphtheria toxin	4.6	6.0	1.22	0.736	74,000	
Enolase	5.59	8.08	1.01	0.735	63,700	
Fibrinogen (cow)	7.9	2.02	2.34	0.706	330,000	
Fumarase	8.51	4.05		(0.75)	204,000	
γ-Globulin (man)	7.1	3.84		(0.745)	176,000	6.4
Growth hormone (pituitary)	3.60	7.15	1.31	0.76	49,000	6.85
Hemoglobin (man)	4.46	6.9	1.16	(0.749)	63,000	6.7
Hexokinase	3.1	2.9	2.37	0.740	96,600	
Insulin (monomer)	1.6	15.0		0.749	12,000	
Insulin (tetramer)	3.68	7.45	1.18	47,800	5.3	
Lactablbumin (cow)	1.9	10.6		0.751	17,400	5.12
β-Lactoglobulin	3.12	7.3	1.26	0.7514	41,500	5.19
Lysozyme (chicken)	2.11	10.2-11.2	1.14-1.21	0.722	17,200	11.0
Myoglobin	2.04	11.3	1.11	0.741	16,900	7.0
Myosin	7.2	0.87	4.0	0.74	829,000-880,000	6.2-6.6
Ovalbumin	3.55	7.76	1.16	0.749	44,000	4.6
Pepsin	3.3	9.0	1.08	0.750	35,500	<1.1
Peroxidase	3.48	7.05	1.36	0.699	39,800	
Phosphorylase (rabbit muscle)	13.7	3.2-3.8		(0.74)	340,000-400,000	5.8
Pyruvic oxidase	40.4	0.91	1.51	(0.74)	4,000,000	
Urease	18.6	3.46	1.19	0.73	480,000	5.0-5.1
Virus (rabbit papilloma)	280.0	0.51	1.65	0.756	47,100,000	
Virus (tobacco mosaic, ordinary)	174.0	0.3	2.9	0.727	59,000,000	
Zein	1.9	4.0	2.7	0.776	50,000	

*S_{20} = sedimentation constant in Svedberg units reduced to water at 20° C.
@D_{20} = diffusion constant in units of 1×10^{-7} reduced to water at 20°C.
#f/f_0 = frictional ratio; the ratio of determined molar frictional constant, *f*, to the calculated molar frictional constant, f_0, for nonsolvated spherical molecules of the same mass.
$$V_{20}$ = partial specific volume of the protein at 20°C. The partial specific volume of a protein represents the volume increment when 1 g of dried protein is added to a very large volume of water (solvent). Values in brackets in the table are assumed values based upon determinations on closely related proteins. One gram of protein in solution generally occupies a volume of 0.70-0.75 ml, representing a protein density of 1.33-1.43 in water solution.
%M_8 = mulecular weight by sedimenttion velocity and diffusion measurements.

A number of the naturally occurring proteins designated by a definite name really represent complex mixtures of different. proteins with molecules of widely different sizes. The crude material isolated from milk and known as "casein" is shown by the ultracentrifuge to be a complex mixture of molecules varying in molecular weight from 75,000 to 375,000. If the casein is exhaustively purified according to Hammarsten's method, a fraction of molecular weight 375,000 is obtained. Yon Hippel and Waugh found that soluble casein at pH 12 and 0° C exists as monomeric units, *a*- and /3-casein monomers, with molecular weights in the range of 15,000 to 25,000. As the temperature is raised and the pH decreased to 7, these monomers polymerize into larger aggregates. The molecules of gelatin vary in mass from 10,000 to 100,000. Myoglobin of muscle apparently is composed of 17,000, 34,000, and 68,000 molecular weight fractions. The globulin fraction of blood serum is complex.

The molecular weights of proteins cannot be determined with a high degree of accuracy in most cases, and the values reported are more or less approximate.

Proteins as ampholytes. Isoelectric pH values of proteins. The properties of proteins as electrolytes are determined by the ionizable groups present. Since each open peptide chain contains only one free a-amino group and one free a-carboxyl group, those groups contribute relatively little to these properties. However, a number of the constituent amino acids contain ionizable groups not involved in peptide bond formation. These groups include the famino group of lysine, the guanidine group of arginine, the imidazole group of histidine, the, a-carboxyl of aspartic acid, the 'Y-carboxyl of glutamic acid, the phenolic hydroxyl of tyrosine, and the sulfhydryl group of cysteine. The extra carboxyl groups of aspartic and glutamic acids occur both free and combined with ammonia as the amide group (asparagine and glutamine). The acidic properties of phosphoproteins such as casein are partly due to the phosphoric acid groups present the amino acids in acid solutions exist as cations

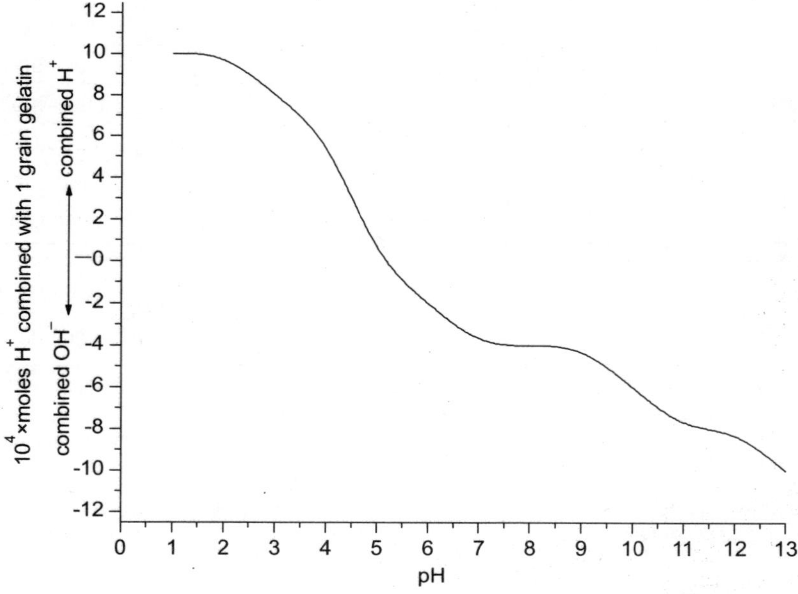

Fig. 2.39 : Dissociation curve of a standard gelatin preparation in hydrochloric acid or sodium hydroxide solution, as obtained from pH measurements at 300 C. The curve was drawn to fit the experimental points. (From Hitchcock, D. I.: *J.* Gen. *Physiol.,* 15, 125, 1932)

which dissociate to give H+ ions in stages to which acid dissociation constants K 1, K 2, K 3, etc., may be assigned. The zwitterion form of the amino acid is an intermediate stage in the process. Proteins in a similar manner exist as complex cations in acid solution and when titrated with alkali show successive overlapping stages of H^+ ion dissociation with the formation of zwitterions, and finally protein anions. Although the processes of protein dissociation represent many ionizable groups, numbers of which may be functioning simultaneously, the general process may be represented as follows, in which the numbers of charges and H+ ions involved are not indicated:

$$\underset{\text{cation}}{\text{protein}^+} \rightleftharpoons H^+ + \underset{\text{Zwitterion}}{\text{protein}} \rightleftharpoons H^+ + \underset{\text{Anion}}{\text{protein}}$$

Because protein molecules contain numbers of groups that interact reversi-bly with H+ ion~ over a wide pH range, their solutions have excellent buffer action, and much of the buffer capacity of body tissues is due to the protein buffers.

The dissociation curves, or titration curves, of proteins corresponding to the above equilibria with H+ ions may be determined, and an example is given in Figure 2.39. These curves extend over a wide range of pH and do not show the sharp breaks characteristic of the curves of monobasic weak acids. This is due to the large number of groups which ionize successively and simultaneously throughout most of the pH range.The nature of the ionizing groups in proteins may be determined from the titration curves under various conditions, and the heats of ionization, *AH,* of the amino acids and simple peptides.

The numbers of different ionizable groups in various proteins have been estimated. For example, in the case of ,B-lactoglobulin with a molecular weight of 40,000, there are 4 imidazolium groups (histidine), 27 e-ammonium groups (lysine), 7 guanidinium groups (arginine), 30, B-:carboxyl groups (aspartic acid), 59 'Y-carboxyl groups (glutamic acid), and 30 amide groups (asparagine and glutamine). Since the amide groups are formed from some of the carboxyls of aspartic and glutamic acids, the total of the free carboxyl groups is 30 + 59 – 30 = 59.

The isoelectric pH of a protein is the pH at which the protein does not migrate in an electric field. At this pH the protein exists as the zwitterion form in which the total positive charges equal the total negative, and the net charge is zero, though the total charge may be high. The isoionic point of a protein is the pH at which the number of H+ ions dissociated from the protein is equal to the number taken up from the solution by the protein. The isoionic and isoelectric pH values are the same when the protein does not combine with ions other than the H+ ion. In general, in the presence of salts the anions and cations of the salt are likely to associate to somewhat different degrees with the protein charges and change appreciably the migration in an electric field and the isoelectric pH; consequently, accurate isoelectric pH values can be given for proteins only under specified conditions of salt concentration and ionic strength.

The isoelectric pH of a protein is of great importance in relation to the physical and chemical properties of the protein. In general, the properties of a protein are minimal at the isoelectric pH. At this pH the net charge on the protein is least and the electrical conductivity is least. Also, at this pH the osmotic pressure, swelling capacity, viscosity, and solubility are minimum. Proteins exist as cations on the acid side and as anions on the alkaline side of the isoelectric pH.

A number of methods are used for the determination of the isoelectric pH values of proteins. These methods are generally based upon determination of minima in properties, such as the pH at which electrical migration is least or the pH of minimum solubility.

The migration of particles in a solution between electrodes is referred to as "electrophoresis." Tiselius has developed a very useful electrophoretic apparatus for studying the migration of proteins and other colloids. The solution to be examined is well buffered and is placed in a special square U-tube provided with compartments and with electrodes attached to the ends. Upon passage of a. current, the charged protein particles migrate toward the electrode of opposite charge. If the buffer solution is alkaline to the isoelectric pH of the protein, the particles migrate toward the positive electrode, while they migrate toward the negative electrode when the buffer solution is acid to the isoelectric pH of the protein. Migration of the particles ceases or is minimal when the pH of the buffer solution is the same as the isoelectric pH of the protein. The moving boundaries of the migrating protein particles are observed by photographing them through a special optical system. If the protein particles under observation possess the same isoelectric pH and are of the same size and shape, they migrate at the same speed and a sharp boundary is obtained. On the other hand, if a mixture of proteins in which the components have different isoelectric pH values, molecular sizes, etc., is observed in the apparatus, each molecular species will migrate at a different rate and become concentrated in different compartments of the cell, from which they can be removed. The Tiselius apparatus is of much importance in the separation and purification of proteins, and in the determination of isoelectric pH values.

The isoelectric pH values of a number of proteins are given in Table 2.30. Since the IpH value varies with the ionic strength of the solution, the values given are only approximate for different ionic strengths. The values of the table are largely for an ionic strength of 0.1 and range all the way from around 1 for pepsin, with many acidic groups, to 11 for lysozyme, with many basic groups. The protamin salmin, containing much basic arginine, has an IpH of around 12. Table 2.31 gives the distribution of a number of groups in several proteins which determine the isoelectric pH values and various chemical properties.

Proteins act as buffers on both sides of the isoelectric pH. The isoelectric pH values of most of the body proteins are below the pH of blood and tissue fluids and cells in which the proteins are found. This means that the body proteins are generally held at a pH alkaline to their isoelectric points and exist to a considerable extent as negative ions or salts with the cations Na^+, K^+, Ca^{++}, and Mg^{++}. In other words, the body proteins exist largely as buffers composed of protein- /H . protein. It is of interest that, in general, the imidazolium groups of histidine and the a-ammonium groups are responsible for the buffer action of proteins at physiologic pH. It will be seen from Table 2.30 that these are the only groups in proteins dissociating appreciably at physiologic pH.

Crystallization of proteins. Many of the proteins have been obtained in crystalline condition. Vegetable proteins in general are easy to crystallize. Among the animal proteins some hemoglobins crystallize very readily, while proteins such as serum albumin and ovalbumin are difficult to crystallize.

Many of the enzyme proteins have been crystallized. Among these are urease, pepsin, trypsin, and catalase.

Table 2.25 : Distribution of Groups in Some Proteins
Values are numbers of groups per 1000,000 grams of protein

Protein	Indole	Hydroxyl	Phenoxyl	Amide	-SH	Csteine/2 cystine–S–	Cations	Free anions	Total ionic
Aldolase	11.3	132.4	29.3	64.7	0	9.3	129.0	86.0	215.0
Casein (whole)	5.9	101.1	34.8	114.3	–	2.8	99.7	91.2	190.9
β-Casein	3.2	107.8	17.7	114.3	0	0	84.1	80.5	164.6
Chymotrypsinogen	27.3	193.9	14.9	109.4	10.8	27.5	76.7	23.0	99.7
Collagen	0	158.3	5.5	47.0	0	0	91.7	77.2	168.9
Edestin	7.2	92.4	23.5	126.8	4.2	7.2	128.6	104.2	232.8
Fibrinogen (human)	16.2	118.5	30.4	106.4	3.3	51.7	124.9	90.8	215.7
Gliadin	2.9	64.4	17.7	321.0	—	21	31.95	0	31.95
Insulin	0	67.3	69.0	99.5	0	104	68.4	78.5	146.9
β-Lactoglobulin	9.4	82.1	20.4	76.5	9.1	19.1	104.3	141.5	245.8
Myoglobin (horse)	11.5	71.4	13.3	47.1	0	0	173.6	126.5	300.1
Myosin (rabbit)	3.9	84.1	18.8	85.7	0	11.7	139.2	131.5	270.7
Salmin	0	87.0	0	0	0	0	490	0	490
Serum albumin (human)	0.98	70.6	26.0	63.0	5.8	46.7	142.7	120.2	262.9

The crystallization of proteins often may be expedited by the addition of a salt such as ammonium sulfate or sodium chloride and adjustment toward the isoelectric pH. The addition of definite amounts of alcohol or acetone is occasionally advantageous. The added substances and adjustment to the isoelectric pH decrease the solubility of the protein. The protein is also least dissociated at the isoelectric pH and may crystallize best in this condition. However, some proteins crystallize best in the form of protein salt.

The relative ease of crystallization of proteins as compared with polysaccharides is due to the high polarity *or* the protein molecules, giving rise to strong directional fields of force which orient the molecules and promote crystal formation.

Viscosity of protein solutions: The viscosity of protein solutions varies widely with the kind of protein and concentration. The viscosity of a protein in solution is closely related to molecular shape, long molecules giving higher viscosity than those more nearly globular in form. Viscosity measurements are of value in calculating the dimensions (axial lengths) of protein molecules. For a given shape large protein molecules give higher viscosities than small ones. Viscosity is least at the isoelectric pH, which means that protein ions orient in solution to increase frictional resistance to flow more than do neutral molecules.

Hydration of proteins. While proteins, as well as many other substances, combine with and hold a certain amount of water through the action of poorly defined adsorptive forces of attraction, the peptide chains constituting protein molecules contain many polar groups (-NH_2, -COOH, -OH, -CO:-,

-NH-, etc.) which tend to combine with water and become hydrated. The nitrogen and oxygen containing groups of proteins contain unshared electron pairs, and these unite with the hydrogen of water to form more or less loose complexes of the type:

```
H           H
N: H:O: H   N: H:O: H   O:H:O: H    C
H                       H           O:H:O: H
```

Since the water molecules combined with protein polar groups also possess unshared electron pairs, they may combine with more water to form aggregates attached to the polar groups such as:

```
H
N: H:O: H
H     H
     :O+:H:O:H
      H
```

Ionized acidic groups of proteins possess more attraction for water than the non-ionized groups because of the attraction of the negative charge the hydrogen of water:

```
  O:
C
  O:H:O: H
```

Also, the positively charged ions of proteins coordinate readily with elec-1S of the oxygen of water:

```
  H     H
H:N+:H:O+:
  H     H
```

accordingly, proteins hydrate to a greater extent when in the form of ions protein salts) than they do as undissociated molecules (in the isoelectric Idition) .

The water complexes with proteins are dissociable:

```
H    H         H
N:H:O+:  ⇌   N  +  H:O:H
H    H         H
```

When electrolytes, sugars, alcohols, and other substances which form complexes with water are added to protein solutions, there is competition for the titer, and the degree of hydration of the protein is decreased.

In general, the combination of proteins with water is, within limits, dependent upon the concentration of the protein solution, the pH of the solution in relation to the isoelectric pH of the protein, the presence of other dissolved substances which combine with water or form complexes with the protein, and the temperature of the solution.

That proteins do combine with water has been proved by observations showing changes in light absorption, dielectric values, protein density, etc., f-proteins treated with varying quantities of water.

Since the water complexes of proteins are dissociable, it is very difficult to determine the degree of hydration with accuracy, since the method used nay markedly affect the dissociation equilibrium.

Some dozen different procedures have been applied to the problem, among which determination of the freezing-point depression and vapor-pressure effects of crystalloids added; 0 protein solutions may be mentioned. In applying these methods, the assumption is made that the water not bound by protein constitutes the solvent for the crystalloid added, and the lowering of the freezing point or vapor pressure beyond that calculated if all water present acted as solvent is taken as an index of bound water. Another method, the dilatometric method, is based upon the volume changes of protein-water systems when cooled through a wide range of temperature as compared with an equal weight of water alone. In each case the volume progressively decreases, but the effect is greater in the protein-water mixture. The difference is utilized in calculating bound water. The values for bound water obtained by different methods vary over such a wide range that the validity of such determinations is extremely doubtful. For example, results on 1 per cent gelatin solutions by different methods indicate from 1 to 7 g of water to be bound per gram of gelatin.

Despite the fact that the combination of proteins with water represents reversible equilibria, some of the water is held with great tenacity, and the removal of the last traces of water from proteins by ordinary drying procedures is very difficult and often incomplete.

The swelling of proteins when placed in aqueous solutions is definitely associated with hydration of the molecules, though a number of additional processes may be involved. The swelling of protein gels is discussed in the chapter dealing with the Donnan membrane equilibrium and is intimately related to osmotic effects within the gel induced by the Donnan membrane phenomenon.

The combination of living tissues with water is largely determined by hydration of the tissue proteins. A certain degree of tissue hydration is essential for normal function. Severe dehydration of tissue proteins such as that caused by water deprivation leads to serious derangements of function.

Solubility and salting out of proteins: The solubility of a protein depends upon the proportions and distribution of polar hydrophilic groups and nonpolar hydrophobic groups in the molecule and the resulting protein dipole moment. The ionic polar groups of protein molecules interact electrostatically both within the same molecule and with surrounding molecules, tending to form aggregates and opposing solubility. This interaction between protein-charged groups is decreased in pure water with a high dielectric constant; that is, the degree of interaction is inversely proportional to the dielectric constant of the solvent. The polar water molecules interact with the polar groups of proteins, tending to increase solubility. In the case of albumins, pseudoglobulins, and other water-soluble proteins the protein-water interactions promoting solubility predominate over the protein-protein interactions opposing solubility, and the proteins are water-soluble.

The addition of an organic solvent, such as acetone or alcohol, to a solution of protein in water decreases the dielectric constant of the solvent and also displaces some of the water molecules associated with the protein and decreases the concentration of water present in the solution. These effects tend to decrease the solubility of the protein, and the addition of such solvents is often utilized in the precipitation of proteins from solution. This is generally done at low temperatures to avoid protein denaturation.

When small amounts of salt are added to a protein dissolved in pure water, the activity coefficient of the protein is decreased and its solubility is increased. Likewise, proteins, such as euglobulins, which are insoluble in pure water become soluble in the presence of small amounts of salts. This phenomenon, often called "salting in," is due to the forces of attraction between the protein ions and the salt ions. At low salt concentrations the increase in the logarithm of protein solubility, *S*, is proportional to the ionic strength of the solvent.

At high concentrations of very soluble saits such as ammonium sulfate, sodium sulfate, and phosphate buffers, proteins are "salted out" of solution. This salting out is due to a decrease in the activity of water, which diminishes the solubilizing interactions between water and the polar protein groups. The solubility, *S*, of many proteins at high salt concentrations decreases logarithmically as the salt concentration increases:

$$\log S = \beta - K_s \omega$$

where S = the protein solubility in the salt solution, ω= the ionic strength of the salt solution, β = the solubility of the protein in pure water (generally hypothetical and obtained by extrapolation of the solubility curve to $\omega = 0$), and $K.$ = the salting-out constant. The values of $K.$ vary with protein and salt. For example, the $K.$ values for horse hemoglobin and $MgSO_4$, $(NH_4)_2S0_4$, and Na_2SO_4 are 0.33,0.71, and 0.76, respectively; while the $K.$ values for fibrinogen and NaCl and $(NH_4)_2SO_4$are 1.07 and 1.46.

The relation of solubility of several proteins to ionic strength is shown in Fig. 2.39 taken from Cohn and Edsall. It will be observed from this figure that salting out may be used elrectivelyin the separation of proteins from solution.

In general, precipitation of a protein from solution is most complete (its solubility is least) at or near its isoelectric pH, as would be expected, because the protein is least soluble at this pH. However, there are some instances in which the addition of much salt causes the precipitation of a protein salt which separates best at a pH other than the isoelectric pH. For example, Sorenson found that the solubility of horse carboxyhemoglobin shows minimum solubilities in concentrated ammonium sulfate at pH 6.6.

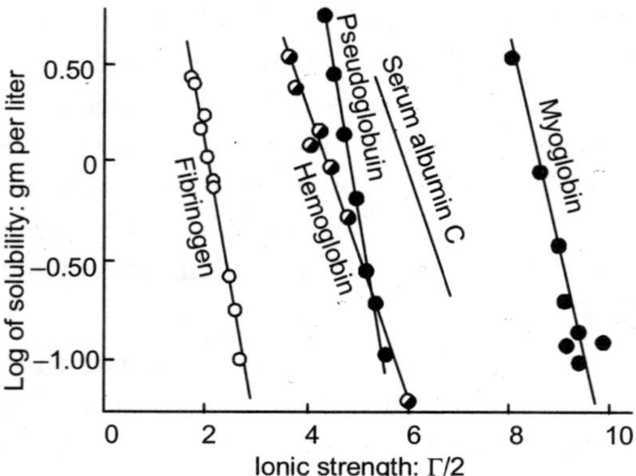

Fig. 2.40 : The Solubility of Proteins in ammonium sulphate

(nearthe isoelectric pH) and also at pH 5.4. The precipitate at pH 5.4 was combined with much sulfate and considered to be carboxyhemoglobin sulphate.

The effect of temperature upon the solubility of proteins in strong salt solutions is highly variable. Horse carboxyhemoglobin is about ten times more soluble at 0° than at 25° C in strong phosphate solution. Vegetable seed globulins are more soluble in salt solutions at higher than at lower temperatures. The solubility of egg albumin in salt solutions shows little variation with temperature.

The precipitation of proteins by salting out is one of the most used and valuable methods for the isolation and purification of proteins.

Precipitation of proteins with positive and negative ions: Combination with dyes. Proteins may be precipitated from solution by a variety of positive and negative ions. Such precipitations are of importance in the isolation of proteins, in the deproteinization of blood and other biologic fluids and extracts for analysis, and in the preparation of useful protein derivatives.

The positive ions most commonly used for protein precipitation are those of heavy metals, such as Zn^{++}, Cd^{++}, Hg^{++}, Fe^{+++}, Cu^{++}, and Pb^{++}. These ions precipitate proteins from solutions alkaline to the protein isoelectric pH, because at this pH the protein is dissociated as protein-which combines with the positive metal ion to give an insoluble precipitate of metal proteinate. The metal ions may be removed from the metal proteinate by acidification (protein-converted to protein+) or by precipitation of the metal by hydrogen sulfide or other agent. Metal protein precipitates are often dissolved by the addition of strong alkali. It is evident from the above facts that the pH of the protein solution (within a certain range) is of primary importance for the most complete precipitation of proteins by heavy metal ions.

The positive ions of organic bases combine with proteins on the alkaline side of the protein isoelectric pH, and some of these combinations are insoluble. Proteins of the protamin class, which are very basic in character, readily form insoluble compounds with various ordinary proteins. A notable example is the combination of protamin with insulin to form protamin insulinate (protamin-insulin), which is widely used in the treatment of diabetes mellitus. The isoelectric pH of insulin is about 5.3, while the isoelectric pH of protamin is high on the alkaline side (9.7-12.4). When insulin and protamin are mixed in solution at pH 7, the insulin is alkaline to its isoelectric pH and exists as insulin-, while the protamin is acid to its isoelectric pH and is present as protamin+. Accordingly, the positive and negative ions combine to form protamin insulinate. When this material is injected for the treatment of diabetes, it is gradually decomposed to yield insulin slowly to the body, thereby more nearly simulating pancreatic secretion of insulin than when insulin is injected alone.

Negative ions combine with proteins when in the form of protein+ (pH of the protein solution is acid to the protein isoelectric pH) to form protein salts. Several of these salts are insoluble and provide valuable methods for the precipitation of proteins from solution. Among the more common protein precipitants involving negative ion precipitation are tungstic acid, phosphotungstic acid, trichloracetic acid, picric acid, tannic acid, ferrocyanic acid, and sulfosalicylic acid. When the above agents are added to protein solutions at the proper pH, precipitates of protein tungstate, phosphotungstate, picrate, tannate, etc., are formed. These precipitates are generally dissolved, and the protein salts are decomposed upon the addition of alkali (forms protein-). Tungstic acid, phosphotungstic acid, trichloracetic acid, and picric acid are commonly used for the preparation of protein-free ffitrates of blood and other biologic materials preliminary to analysis. Hides are converted to leather by treatment with tannic acid, which reacts with skin proteins to form insoluble protein tannates.

Rawlins and Schmidt have shown that acid dyes (dissociate as acids → dye$^-$), such as Biebrich scarlet and naphthylamine brown, combine in definite proportions with proteins acid to the isoelectric pH (protein+), the amount of dye combining being a function of pH and protein+ concentration. Similarly, basic dyes (base → dye$^+$), such as methylene blue, induline scarlet, and safraniney, combine with proteins on the alkaline side of the isoelectric pH (protein). Since the proteindye combinations are ften highly insoluble, it is readily possible to determine the amount of dye combined with a given weight of protein. The above workers were able to lot dye-protein titration curves and to calculate the equivalent combining eights of proteins from their results.

The combination of proteins with dyes is of much importance in the stain-19 of tissues and microorganisms for microscopic examination. The bacricidal action of various dyes used in the treatment of infections such as Driflavin is probably related to combination of the dye with cellular proteins of the infecting organism rendering them incapable of normal function.

Precipitation of proteins by specific antibody proteins: The injection of a foreign protein into an animal generally causes the appearance of so-called specific immune bodies or antibodies against the protein injected. 'or example, if a rabbit is given a series of injections with egg albumin at frequent intervals, its blood serum will develop the property of forming a precipitate with very minute amounts of egg albumin but not with other proteins. The injection of egg albumin causes the production of antibody proteins in the rabbit that specifically precipitate the egg albumin (antigen). The antibody proteins are present in the γ-globulin fraction of blood serum and represent normal serum proteins which have been altered as a result of the presence of the injected protein (antigen).

The precipitation of proteins by their specific immune sera antibody proteins (precipitin reaction) affords a highly sensitive and specific method for the detection of proteins.

Reaction of proteins with nitrous acid: Since proteins generally contain some free amino groups, they react with nitrous acid to liberate nitrogen, with replacement of the amino group by a hydroxyl group, as is the ase in the reaction of amino acids with nitrous acid. Treatment of proteins with nitrous acid accordingly destroys the free amino groups, and produces deaminized proteins.

Acylation of proteins: Acid groups R-CO- may be introduced into he free hydroxyl and amino groups of proteins by treatment with acid chloides and anhydrides. Acetylated, 'benzoylated, and benzenesulfonated proteins may be readily prepared. Ketene, $CH_2=C=O$, is a valuable reagent or the acetylation of proteins. It reacts with amino and hydroxyl groups as follows:

$$R - NH_2 + CH_2 = C = O \rightarrow R - NH - CO\,CH_3$$

$$R - OH + CH_2 = C = O \rightarrow R - O - CO - CH_2$$

By controlling the conditions it is possible to acetylate the amino groups of proteins with ketene without acetylating the hydroxyl groups.

The acylation of proteins is of importance in studies of the influence of 'fee amino and hydroxyl groups upon the physiologic properties of the protein. For example, acetylation of only the amino groups of insulin reduces ts physiologic activity by about 20%, while acetylation of both amino and tyrosine hydroxyl groups abolishes its activity, indicating that free phenolic hydroxyl groups are much more important to the function of insulin than are free amino groups.

Slphufhydryl-disulphide exchange reactions of peptides and proteins. Much evidence has accumulated that peptides and proteins containing intramolecular disulfide linkages form aggregates linked by intermolecular disulphide bonds, –S–S– These aggregates are formed in a chain reaction initiated by a substance containing a free sulphuhydryl group, -SH, such as glutathione or another protein or peptide. The process may be illustrated diagrammatically as follows:

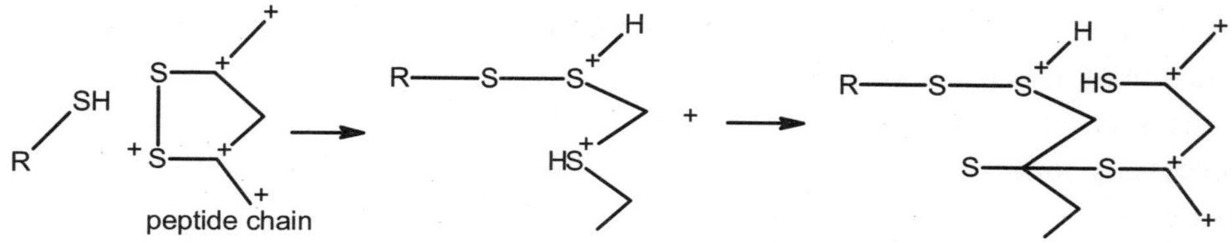

SH R + S S peptide chain → R—S—S H HS + → R—S—S H HS S S

R-SH reacts with the peptide chain A, containing an intramolecular -&-S- group, to form B, containing a free –SH group. B then reacts with another A to produce C, containing two A groups (also R-&-S-), and since C contains a free –SH group, it in turn may react with A to form an aggregate with three A groups, etc. The denaturation of proteins appears often to involve such aggregations. The process appears to be involved in the formation of cross –S–S– linkages when fibrinogen is aggregated into fibrin, and there is some evidence to indicate its operation in the formation of cross linkages in the mitotic apparatus of cell division.

Cleavage of –S–S– groups of proteins. Since the –SH groups formed by the reduction of –S–S– groups in proteins are unstable and may reoxidize to –S–S– compounds, other methods which give more stable cleavage products are desirable when proteins are to be split into simpler peptides for studies on amino acid sequence and other properties.

As indicated previously, Sanger split the –S–S– linkages of insulin to yield chain peptides by oxidation with performic acid, which breaks the -&-S- bond to form cysteic acid groups. If we designate a protein con-taining an -&-S- group by R-S-&- R', the reaction may be expressed as follows:

$$R\text{- }S\text{- }S\text{- }R' + 6H - CO_3H \uparrow R\text{- }SO_3H + R'\text{- }SO_3H + 6H{-}COO^- + 4H^+$$

where $R{-}SO_3H$ and $R'{-}SO_3H$ = peptides with a terminal cysteic acid group. Oxidation by performic acid has the disadvantage especially of disrupting tryptophan groups in the protein.

Swan has found that the –S–S– group in proteins is split by sulfite in the presence of cupric ions to form two thiosulfate groups (S-sulfocysteic acid groups). The reaction appears to take place in stages in which Curt is reduced to Cu^{++}

$$3R - S - S - R + 3SO_3^- \rightarrow 3R - S - SO_3 + 3RS^-$$

$$3RS^- + 2CU^{++} + SO_3^- \rightarrow R - S - SO_3^- + 2R - S - Cu$$

$$2RS - Cu + 4Cu^{++} + 2SO_3^- \rightarrow 2R - S - SO_3^- + 2Cu^{++}$$

over-all reaction is:

$$3R - S - S - R + 6SO_3^- + 6Cu^{++} \rightarrow 6R - S - SO_3^- + 6Cu^{++}$$

method of –S–S– cleavage has been applied to trypsinogen and chymotrypsinogen by and to wool keratin.

Cystine may exist in proteins as a component of a single peptide chain or two peptide chains linked by the –S–S– linkage. The –S–S– linkagcs of proteins may be reduced to two –SH groups by treatment with reducing agents such as sodium cyanide, cysteine, thioycolic acid, and glutathione. Reduction of structures such as A forms a olecule of free cysteine and leaves a cysteine molecule in the peptide chain. Reduction of a structure such as B breaks the cross –S–S– linkage between the peptide chains, separating them, and leaving cysteine molecules resent in each chain. The presence of certain –S–S– groups is necessary for the full activity of insulin. Their reduction to -SH groups results in the loss of about 50 per cent of the insulin activity.

Oxidation of –SH groups of proteins. 80me proteins contain free -SH groups which may be readily oxidized by agents such as ferricyanide, cystine, tetrathionate, porphyrindin, and Iodine. These oxidations apparently consist in the oxidation of two -8H groups to an –S–S– group presumably such an oxidation may involve two -8H groups in the same peptide chain, or in different peptide chains which would cause linkage of the chains and formation of a different and more complex protein structure.

The above oxidizing agents may be used for the quantitative estimation, of free –SH groups in proteins. The –S–S– linkages of proteins may be determined by reduction of the protein with thioglycolic acid, which reduces –S–S– to –SH groups which are then estimated. The –S–S– groups may be obtained from the difference in free –8H groups in the protein before and after reduction.

The determination of free –8H groups in proteins is of importance studying their relation to –properties of the protein (insulin, activity, for example) and other agencies. In general, the denaturation of a protein causes an increase in the number of free –SH groups.

Halogenation of proteins: When proteins are treated with alkaline iodine solution (NaOI), iodine is incorporated into the protein molecule. Hydrolysis of iodized protein yields diiodotyrosine, indicating that iodine substitutes into the tyrosine ring. More drastic methods of treatment lead to halogenation of the indole and imidazole groups of tryptophan and histidine. When proteins are treated with alkaline bromine solution (NaOBr), the cystine groups are oxidatively destroyed.

Iodine of iodides taken by mouth is utilized by the body in forming the iodized protein thyroglobulin, the precursor cf thyroxine, the hormone of the thyroid gland. When thyroglobulin is hydrolyzed, it yields the iodine-containing compounds diiodotyrosine and thyroxin. Harington and Rivers found thyroxin in the hydrolysis products of casein which had been treated with iodine. Practically all iodized proteins have some thyroid activity (increase metabolic rate).

Reaction of proteins with formaldehyde: Formaldehyde combines with proteins to change their properties markedly. Egg albumin becomes incoagulable by heat when treated with formaldehyde. Formaldehyde converts casein, soybean proteins, and other proteins into inert hard horny masses which are produced commercially as plastics (artificial ivories) in large quantities. Formaldehyde reacts with the proteins of tissues and hardens them when they are preserved in its solutions.

The chemical reactions involved in the action of formaldehyde upon proteins are obscure. Undoubtedly one of the reactions is combination of formaldehyde with various amino groups in a manner sipllar to its reaction with the amino groups of amino acids.

Color reactions of proteins and amino acids: Because of their peptide structure and the presence of different amino acid groups in their molecules, proteins react with a variety of agents to form colored products. Several of these color reactions of proteins are of importance in the qualitative detection and quantitative estimation of proteins and of their constituent amino acids.

(a) The ninhydrin reaction. The reaction of ninhydrin with amino acids which should be consulted for the mechanism of reaction. Ninhydrin reacts with free a-amino acids and with proteins, proteoses, pep tones, and peptides to give a blue color. The test is both the most general and one of the most delicate reactions known for the qualitative detection of proteins and their hydrolytic products. It may be used to show the presence of amino acids in urine and in deproteinized body fluids. Apparently all amino acids of protein hydrolysis give the ninhydrin reaction. The colors given by different amino acids vary in shade and depth. The ninhydrin reaction is of value in detecting the end point of protein hydrolysis, at which time the color value is constant.

When urea is heated to about 180°C it decomposes to form biuret.

Biuret Test: The Biuret Test is a general test for proteins. When a protein reacts with copper(II) sulphate (blue), the positive test is the formation of a violet colored complex.

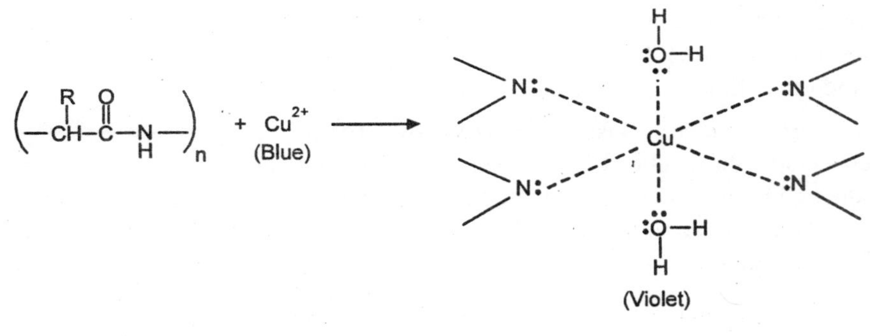

$$\left(-\overset{R}{\overset{|}{CH}}-\overset{O}{\overset{\|}{C}}-\underset{H}{N}-\right)_n + \underset{\text{(Blue)}}{Cu^{2+}} \longrightarrow \text{Cu complex (4 :N, 2 } H_2O\text{)}$$

(Violet)

In this reaction an alkaline solution of biuret is treated with very dilute copper sulphate a violet colour is obtained. This reaction is given by substances containing two $-CONH_2$ groups joined either directly or through a carbon or nitrogen atom, Compounds containing $-CH_2NH_2$ $-C(OH)NH_2$ and $-CSNH_2$ in place of $-CONH_2$ groups give test. Peptide structures are found in proteins and their derivatives which contain

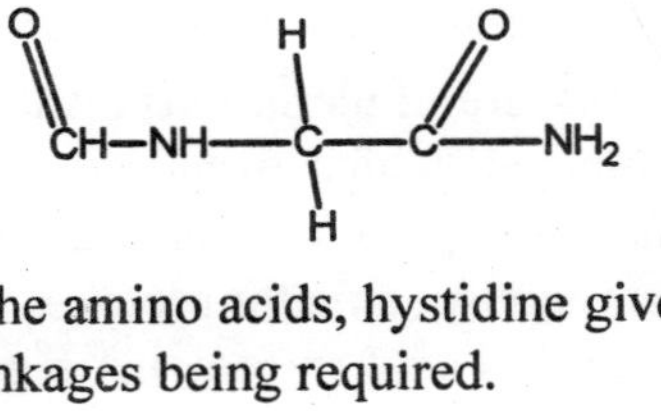

$$\overset{O}{\overset{\|}{C}}H-NH-\overset{H}{\underset{H}{C}}-\overset{O}{\overset{\|}{C}}-NH_2$$

linkage also give the biuret test. Of the amino acids, hystidine gives a positive test. Dipeptides do not give this test, two or more peptide linkages being required.

When protein solutions are made strongly alkaline with sodium or potassium hydroxide and very little copper sulphate solution is added, apurplish or pinkish violet colour is obtained, this colour depends upon the complexity of the protein.

The biuret test is apparently used to the coordination of cupric ions with the unshared electrons pair of peptide nitrogen and trhe oxygen of water to form a coloured co-ordination complex, which may be represented in the diagram below:

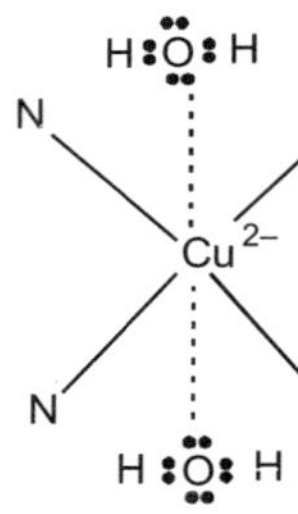

The presence of magnesium sulphate in the solution to be tested interferes with the biuret reaction because of the precipitation of magenisum hydroxide, Large amounts of ammonium salits also interfere with the test, but this may be minimized by the addition of a large excess of alkali. The biuret reaction is extensively used as a delicate test for presence of proteins in biologic materials. It has been converted into an excellent method for the quatitative determination of proteins in blood serum and other fluids.

The buiret reaction is extensively used as a delicate test for the presence of proteins in biological material. It has been converted into an excellent method for the quantitative dermination of proteins in blood serum and other fluids.

The Xanthotropic Reaction: Also kwon as yellow protein reaction, which includes the addition of nitric acid in concentrated condition produces an initial white precipitate that turns yellow on heating; the colour turns orange, when the solution is made alkaline. Insoluble protein are turned yellow or orange on the surface.

> The yellow stains upon the skin caused by nitric acid are the result of the xanthotrcpic reaction

$$HO-C_6H_4-CH_2-CH(NH_2)-COOH + HNO_3 \longrightarrow HO-C_6H_3(NO_2)-CH_2-CH(NH_2)-COOH$$

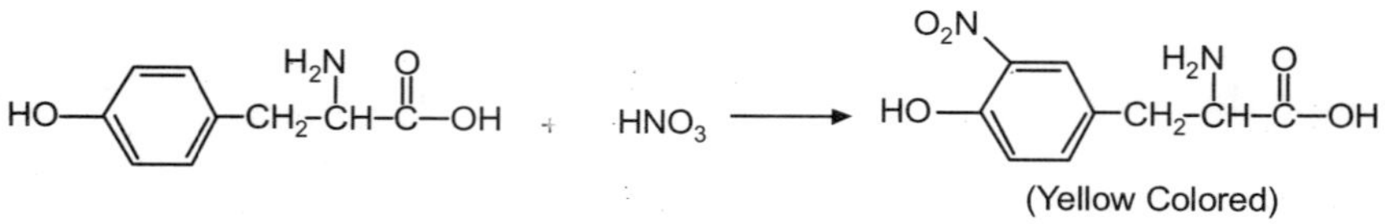

(Yellow Colored)

The xanthotropic reaction is due to nitration of phenyl rings present in tyrosine, phenyl amine and truptophan to give yellow nitro-substitution products which proteins give the xanthotripic reaction.

Millon's reagent is a composite mixture of mercurous nitriles and nitric acid. When it is added to protein solution, a white ppt is produced, which turns red upon heating. Secondary rpteoses and peptones give only red solution. Insoluble proteins turn red when they are suspended in water and heated with the reagent. The reaction is not specific for proteins, since it is given by phenols in general. Since tyrosine is the only phenolic amino acid in proteins a positive test indicates the presence of tyrosine. Solutions which are alkaline must be neutralized with acid (not HCl) before testing otherwise the mecury of the reagent is precipitated. Chloride interfere with the reaction probably because the nitric acid forms free chloride from them which destroyes the coloured compound. The reaction is dependant upon the formation of a coloured mercury compound with the hydrooxyphenyl group.

The Hopkin's Cole or glyoxalic reaction for tryptophan. The reagent, which may be prepared by reduction of oxalic acid with sodium amalgam or magnesium powder.

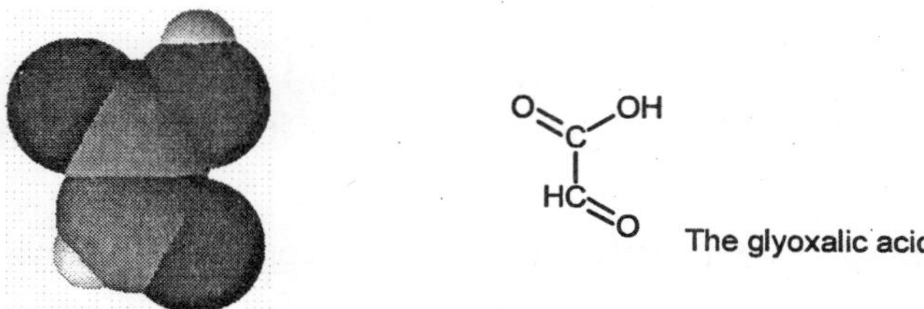

The glyoxalic acid

When a solution of tryptophan or protein which contains tryptophan is mixed with the reagent and the mixture is layered over concentrated sulphuric acid, a violet ring appears at the junction of the two liquids, the ring is the characteristics of indole ring, and other compounds than tryptophan hive the test. The nature of the coloured products fromed in the test is unknown. Chlorates nitrates and excess chlorides prevent the reaction A number of aldehydes other than glyoxalic acid give similar colour reaction with truptophan. Formaldehyde used similarly gives violet colour appears.

The Nitroprusside Reaction Proteins which contains free –SN groups (cystene) give a reddish colour with sodium nitroprusside, $Na_2Fe(CN)_5NO.2H_2O$ in ammonical solution. Many proteins which give negatibve test react positive after heat coagulation or denaturization by other means, indicating the liberation of free – SH groups. The cystene – S – S groups in protein may be reduced to –SH groups by reducing agents such as NaCN, after which they give the nitroprusside reaction.

Follin's Reaction Follin's amino acids reagent in sodium 1 – 2 napthoquinone – 4 sulphonate. This reagent gives deep red colour with proteins in alkaline medium.

The Sullivan Reaction This reaction involves a colourimetric method for the quantitative determination of cysteine and cystine. (After reduction of cysteine in protein hydrolysates. The method is based upon the red colour produced when cysteine is treated with sodiuk 1 – 2 naptholquinone – 4 – sulphonate in alkaline solution in the presence of sodium dithionate $Na_2S_2O_4$. The method has been widely used in the quantitative determination of proteins.

Sullivan's procedure provides a valuable colour test for cysteine, but experience of his quantitative methods in this laboratory has not been satisfactory.Thus it was found: (1) The depth of red colour is variable and far from proportional to the amount of cysteine or cystine present. Rimington [1929] had already noted this poor proportionality and advised having the unknown within 15 % of the standard in estimation. (2) Even when compared with a standard of nearly the same colour, the results for the unknown are often irregular. (3) The presence of relatively considerable amounts of other amino-acids diminishes the amount of red colour due to cysteine or cystine, and the yellow colour due to the other amino-acids interferes when a pure cysteine or cystine standard is used for comparison. The diminution in red colour was found to be attributable in part to the buffering effects of the amino-acids present upon the pH of the solution'.Two directions in which the accuracy of the method might be improved suggested themselves: (1) to maintain the same PE in standard and assay, and (2) to swamp both standard and assay with a considerable amount of aminoacid. It was found possible to affect both objects by adding glycine with appropriate amounts of alkali, and by this means, two of the most serious drawbacks to Sullivan's procedure (unreliability and lack of proportionality) were overcome. It was found that the reaction between the cystine and cyanide was the double decomposition:

RSSR + NaCN :- RSNa + RSCN

discovered by Pulewka and Winzer [1928]. Under conditions that may be varied considerably, this reaction proceeds so nearly to completion that any deviation from it cannot be detected colorimetrically.

Sakaguchi Reaction Is a terst for arginine either free or combined in proteins. The test consists in treating the sample solution with α-naphthol and sodium hypochlorite, with slow development of a internse red colour. Various substances which contain the guanidine group give the reaction, but since arginione is the only amino acid in proteins which is the only amino acid in proteins which contains this group the test is specific for arginine when applied to protein free agrinine is stated to give a positive test in concentration of 0.0004 mg per ml. The reaction is modified and adopted to the quantitative determination for arginine.

Unoxidised sulphur test This test isto prove the presence of sulphur containing amino acids. The protein or amino acid is boiled with atrong alkali to split out sulphur in the form of N_2 detected by the addition of lead acetate, which causes the formation of brown or black lead sulphide.

GLOBULAR AND FIBROUS PROTEIN

Globular proteins, or **spheroproteins** are one of the two main protein classes, comprising 'globe'-like proteins that are more or less soluble in aqueous solutions (where they form colloidal solutions). This main characteristic helps distinguishing them from fibrous proteins (the other class), which are practically insoluble.

The term globin can refer more specifically to proteins including the globin fold.

Globular Structure and Solubility

The term globular protein is quite old (dating probably from the 19th century) and is now somewhat archaic given the hundreds of thousands of proteins and more elegant and descriptive structural motif vocabulary. The globular nature of these proteins can be determined without the means of modern techniques, but only by using ultracentrifuges or dynamic light scattering techniques.

The spherical structure is induced by the protein's tertiary structure. The molecule's apolar (hydrophobic) amino acids are bounded towards the molecule's interior whereas polar (hydrophilic) amino acids are bound outwards, allowing dipole-dipole interactions with the solvent, which explains the molecule's solubility.

A Wide Range of Roles in the Organism

Unlike fibrous proteins which only play a structural function, globular proteins can act as:

- Enzymes, by catalyzing organic reactions taking place in the organism in mild conditions and with a great specificity. Different esterases fulfill this role.
- Messengers, by transmitting messages to regulate biological processes. This function is done by hormones, i.e. insulin etc.
- Transporters of other molecules through membranes
- Stocks of amino acids.
- Regulatory roles are also performed by globular proteins rather than fibrous proteins.

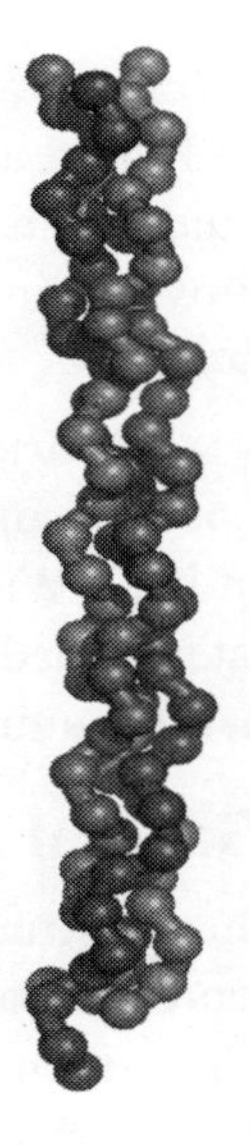

Members

Among the most known globular proteins is Hemoglobin, a member of the globin protein family. Other globular proteins are the immunoglobulins (IgA, IgD, IgE, IgG and IgM), and alpha, beta and gamma globulins. See protein electrophoresis for more information on the different globulins. Nearly all enzymes with major metabolic functions are globular in shape, as well as many signal transduction proteins.

Fibrous proteins, also called **scleroproteins**, are long filamentous protein molecules

Fibrous proteins are only found in animals.

Fibrous proteins form 'rod' or 'wire' -like shapes and are usually inert structural or storage proteins. They are generally water-insoluble. Fibrous proteins are usually used to construct connective tissues, tendons, bone matrix and muscle fiber.

Examples of fibrous proteins include keratins, collagens and elastins.

AMINO ACIDS

In chemistry, an amino acid is a molecule containing both amine and carboxyl functional groups. These molecules are particularly important in biochemistry, where this term refers to α-amino acids with the general formula $H_2NCHRCOOH$, where R is an organic substituent. In the alpha α acids, the amino and carboxylate groups are attached to the same carbon, which is called the α-carbon. The various α amino acids differ in which side chain (R group) is attached to their alpha carbon. They can vary in size from just a hydrogen atom in glycine through a methyl group in alanine to a large heterocyclic group in tryptophan.

Amino acids are critical to life, and have a variety of roles in metabolism. One particularly important function is as the building blocks of proteins, which are linear chains of amino acids. Amino acids are also important in many other biological molecules, such as forming parts of coenzymes, as in S-adenosylmethionine, or as precursors for the biosynthesis of molecules such as heme. Due to this central role in biochemistry, amino acids are very important in nutrition.

The amino acids are commonly used in food technology and industry. For example, monosodium glutamate is a common flavor enhancer that gives foods the taste called *umami.* Beyond the amino acids that are found in all forms of life, amino acids are also used in industry, with the production of biodegradable plastics, drugs and chiral catalysts being particularly important applications.

Overview α-amino acids are the building blocks of proteins. Amino acids combine in a condensation reaction, that is, through dehydration synthesis, that releases water and the new "amino acid residue" that is held together by a peptide bond. Proteins are defined by their unique sequence of amino acid residues; this sequence is the primary structure of the protein. Just as the letters of the alphabet can be combined to form an almost endless variety of words, amino acids can be linked in varying sequences to form a vast variety of proteins. Twenty standard amino acids are used by cells in protein biosynthesis, and these are specified by the general genetic code. These 20 amino acids are biosynthesized from other molecules, but organisms differ in which ones they can synthesize and which ones must be provided in their diet. The ones that cannot be synthesized by an organism are called *essential amino acids.*

History: The first few amino acids were discovered in the early 1800s. In 1806, French chemist, Louis-Nicolas Vauquelin, isolated a compound in asparagus that proved to be the amino acid, asparagine. In 1812, William Hyde Wollaston found a substance in urine that he identified as a cystic oxide, and was later named cystine. And in 1820, another French chemist, Henri Braconnot, discovered the first two natural amino acids, glycine and leucine

General Structure

In the structure shown at the top of the page, R represents a side chain specific to each amino acid. The carbon atom next to the carbonyl group is called the α-carbon and amino acids with a side chain bonded to this carbon are referred to as *α amino acids*. These are the most common form found in nature. In the alpha amino acids, the α-carbon is a chiral carbon atom (with the exception of glycine). In amino acids that have a carbon chain attached to the α-carbon, as in lysine on the right, the carbons are labelled in order as α, β, γ, δ, and so on. In some amino acids, the amine group is attached to the β or γ-carbon, and these are therefore referred to as *beta* or *gamma amino acids*.

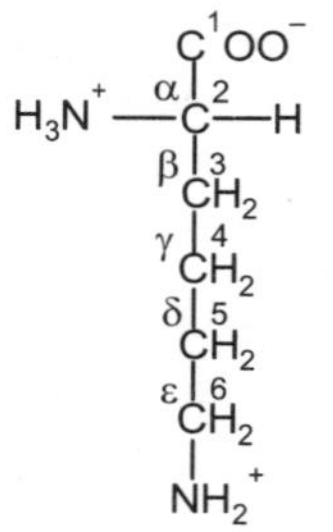

$$
\begin{array}{c}
\overset{1}{C}OO^- \\
H_3N^+ - \overset{\alpha\ 2}{C} - H \\
\beta\ |\ 3 \\
CH_2 \\
\gamma\ |\ 4 \\
CH_2 \\
\delta\ |\ 5 \\
CH_2 \\
\varepsilon\ |\ 6 \\
CH_2 \\
| \\
NH_2^+
\end{array}
$$

Fig. 2.41 : Lysine with the carbon atoms in the side chain labeled

Amino acids are usually classified by the properties of their side chain into four groups. The side chain can make them behave like a weak acid, a weak base, a hydrophile if they are polar, and hydrophobe if they are nonpolar. The chemical structures of the 20 standard amino acids, along with their chemical properties, are catalogued in the list of standard amino acids.

The phrase "branched-chain amino acids" or BCAA is sometimes used to refer to the amino acids having aliphatic side chains that are non-linear; these are leucine, isoleucine, and valine. Proline is the only proteinogenic amino acid whose side group links to the α-amino group and, thus, is also the only proteinogenic amino acid containing a secondary amine at this position. Proline has sometimes been termed an imino acid, but this is not correct in the current nomenclature.

Isomerism

Of the standard α-amino acids, all but glycine can exist in either of two optical isomers. While L-amino acids represent the vast majority of amino acids found in proteins, D-amino acids are found in some proteins produced by exotic sea-dwelling organisms, such as cone snails. They are also abundant components of the peptidoglycan cell walls of bacteria. and D-serine may act as a neurotransmitter in the brain. The L and D convention for amino acid configuration refers not to the optical activity of the amino acid itself, but rather to the optical activity of the isomer of glyceraldehyde from which that amino acid can theoretically be synthesized (D-glyceraldehyde is dextrorotary; L-glyceraldehyde is levorotary). Alternatively, the *(S)* and *(R)* designators are used to indicate the absolute stereochemistry. Almost all of the amino acids in proteins are *(S)* at the α carbon, with cysteine being *(R)* and glycine nonchiral. Cysteine is unusual since it has a sulfur atom at the first position in its side-chain, which has a larger atomic mass than the groups attached to the α-carbon in the other standard amino acids, thus the *(R)* instead of *(S)*.

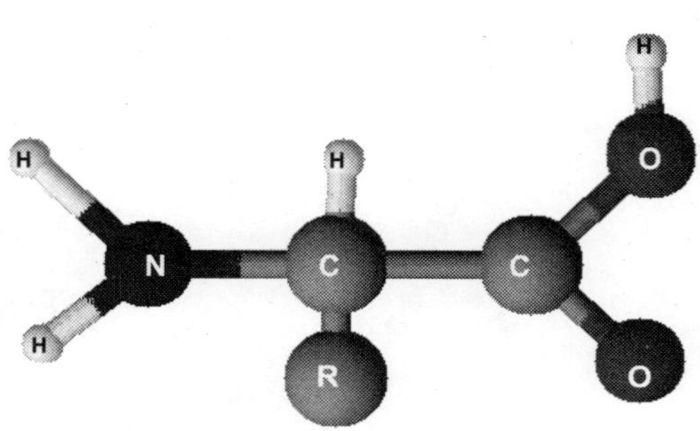

Fig. 2.41A. The general structure of Amino acid

An amino acid in its (1) unionized and (2) zwitterionic forms

Zwitterions

As amino acids have both an amine and a carboxylic acid functional group and are therefore both acid and base at the same time. At a certain compound-specific pH known as the isoelectric point, the number of protonated ammonium groups with a positive charge and deprotonated carboxylate groups with a negative charge are equal, resulting in a net neutral charge These ions are known as a *zwitterion,* which comes from the German word *Zwitter* meaning "hybrid". Amino acids are zwitterions in solid phase and in polar solutions such as water and depending on the pH, but not in the gas phase. Zwitterions have minimal solubility at their isolectric point and amino acids are often isolated by precipitation from water after adjusting the pH to their isolectric point.

Occurrence and Functions in Biochemistry

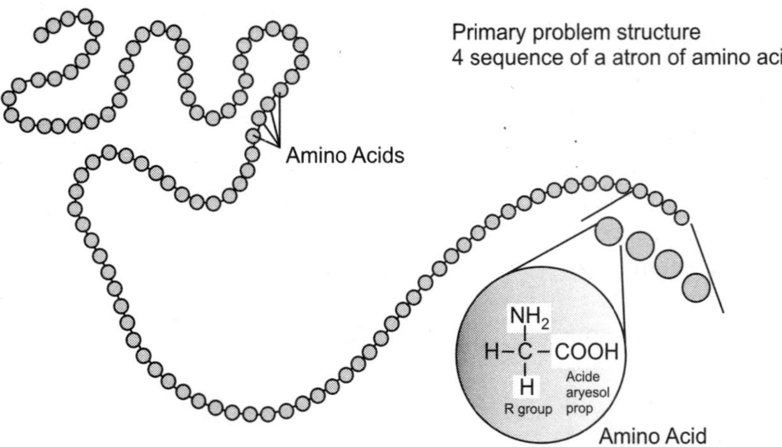

Fig. 2.42 : A polypeptide is a chain of amino acids

Standard Amino Acids

Amino acids are the basic structural building units of proteins. They form short polymer chains called peptides or longer chains called either polypeptides or proteins. These polymers are linear and unbranched. The process of making proteins is called *translation* and involves the step-by-step addition of amino acids to a growing protein chain by a ribozyme that is called a ribosome.[13] The order in which the amino acids are added is read through the genetic code from an mRNA template, which is a RNA copy of one of the organism's genes. Twenty amino acids are encoded by the standard genetic code and are called proteinogenic or *standard amino acids.*

The amino acid selenocysteine

Non-standard Amino Acids

Aside from the twenty standard amino acids, there are a vast number of "non-standard" amino acids. Two of these can be specified by the genetic code, but are rather rare in proteins. Selenocysteine is incorporated into some proteins at a UGA codon, which is normally a stop codon. Pyrrolysine is used by some methanogenic archaea in enzymes that they use to produce methane. It is coded for with the codon UAG. Other non-standard amino acids found in proteins are formed by post-translational modification, which is modification after translation in protein synthesis. These modifications are often essential for the function or regulation of a protein; for example, the carboxylation of glutamate allows for better binding of calcium cations, and the hydroxylation of proline is critical for maintaining connective tissues. Such modifications can also determine the localization of the protein, e.g., the addition of long hydrophobic groups can cause a protein to bind to a phospholipid membrane.

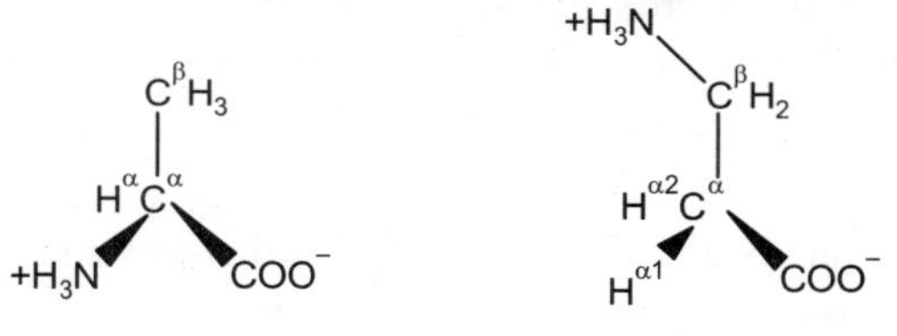

L-α-alanine β-alanine

Fig. 2.43 : α-alanine and its β-alanine isomer

Examples of nonstandard amino acids that are not found in proteins include lanthionine, 2-aminoisobutyric acid, dehydroalanine and the neurotransmitter gamma-aminobutyric acid. Nonstandard amino acids often occur as intermediates in the metabolic pathways for standard amino acids—for example ornithine and citrulline occur in the urea cycle, part of amino acid catabolism (see below). A rare exception to the dominance of α-amino acids in biology is the β-amino acid beta alanine (3-aminopropanoic acid), which is used in plants and microorganisms in the synthesis of pantothenic acid (vitamin B5), a component of coenzyme A.

*In human nutrition*When taken up into the body in the diet, the 20 standard amino acids are either used to synthesize proteins and other biomolecules or oxidized to urea and carbon dioxide as a source of energy. The oxidation pathway starts with the removal of the amino group by a transaminase, the amino group is then fed into the urea cycle. The other product of transamidation is a keto acid that enters the citric acid cycle. Glucogenic amino acids can also be converted into glucose, through gluconeogenesis.

Of the 20 standard amino acids, 8 are called essential amino acids because the human body cannot synthesize them from other compounds at the level needed for normal growth, so they must be obtained from food. However, the situation is quite complicated since cysteine, taurine, tyrosine, histidine and arginine are semiessential amino acids in children, because the metabolic pathways that synthesize these amino acids are not fully developed. The amounts required also depend on the age and health of the individual, so it is hard to make general statements about the dietary requirement for some amino acids.

Essential	Nonessential
Isoleucine	Alanine
Leucine	Asparagine
Lysine	Aspartate
Methionine	Cysteine*
Phenylalanine	Glutamate
Threonine	Glutamine*
Tryptophan	Glycine*
Valine	Proline*
	Serine*
	Tyrosine*
	Arginine*
	Histidine*

(*) Essential only in certain cases.

Several common mnemonics have evolved for remembering the ten amino acids often described as essential. PVT TIM HALL ("Private Tim Hall") uses the first letter of each of these amino acids. Another mnemonic that frequently occurs in student practice materials beneath "AH TV TILL Past Midnight", is "These ten valuable amino acids have long preserved life in man".

Non-protein Functions

In humans, non-protein amino acids also have important roles as metabolic intermediates, such as in the biosynthesis of the neurotransmitter gamma-aminobutyric acid. Many amino acids are used to synthesize other molecules, for example:

- Tryptophan is a precursor of the neurotransmitter serotonin.
- Glycine is a precursor of porphyrins such as heme.
- Arginine is a precursor of nitric oxide.
- Ornithine and S-adenosylmethionine are precursors of polyamines.
- Aspartate, glycine and glutamine are precursors of nucleotides.

However, not all of the functions of other abundant non-standard amino acids are known, for example taurine is a major amino acid in muscle and brain tissues, but although many functions have been proposed, its precise role in the body has not been determined.

Some non-standard amino acids are used as defenses against herbivores in plants. For example canavanine is an analogue of arginine that is found in many legumes, and in particularly large amounts in *Canavalia gladiata* (sword bean). This amino acid protects the plants from predators such as insects and can cause illness in people if some types of legumes are eaten without processing. The non-protein amino acid mimosine is found in other species of legume, particularly *Leucaena leucocephala*. This compound is an analogue of tyrosine and can poison animals that graze on these plants.

Uses in Technology

Amino acids are used for a variety of applications in industry. The major use for these compounds is as an additive to animal feed, since many of the bulk components of these products, such as soybeans, either have low levels or lack some of the essential amino acids: lysine, methionine, threonine, and tryptophan are most important in the production of these feeds. The food industry is also a major consumer of amino acids, particularly glutamic acid, which is used as a flavour enhancer, and Aspartame (aspartyl-phenylalanine-1-methyl ester) as a low-calorie artificial sweetener. The remaining production of amino acids is used in the synthesis of drugs and cosmetics.

Amino acid derivative	Pharmaceutical application
5-HTP (5-hydroxytryptophan)	Experimental treatment for depression.
L-DOPA (L-dihydroxyphenylalanine)	Treatment for Parkinsonism.
Eflornithine	Drug that inhibits ornithine decarboxylase and is used in the treatment of sleeping sickness.

Chiral Catalysts

In the chemical industry, amino acids are important as low-cost feedstocks in chiral synthesis. Here, these compounds are used in chiral pool synthesis as enantiomerically-pure building blocks that can be assembled into the desired chiral product. Alternatively, amino acids can be used to create chiral catalysts, such as by incorporating a ruthenium atom into proline to produce a catalyst that can carry out asymmetric hydrogenation reactions.

Biodegradable Plastics

Amino acids are under development as components of a range of biodegradable polymers. These materials have applications as environmentally-friendly packaging and in medicine in drug delivery and the construction of prosthetic implants. These polymers include polypeptides, polyamides, polyesters, polysulfides and polyurethanes with amino acids either forming part of their main chains, or bonded as side chains to modify the physical properties and reactivities of the polymers. An interesting example of such materials is polyaspartate, a water-soluble biodegradable polymer that may have applications in disposable diapers and agriculture. Due to its solubility and ability to chelate metal ions, polyaspartate is also being used as a biodegradeable anti-scaling agent and a corrosion inhibitor. In addition, the aromatic amino acid tyrosine is being developed as a possible replacement for toxic phenols such as bisphenol A in the manufacture of polycarbonates.

Reactions: As amino acids have both a primary amine group and a primary carboxyl group, these chemicals can undergo most of the reactions associated with these functional groups. These include nucleophilic addition, amide bond formation and imine formation for the amine group and esterification, amide bond formation and decarboxylation for the carboxylic acid group. The multiple side chains of amino acids can also undergo chemical reactions. The types of these reactions are determined by the groups on these side chains and are discussed in the articles dealing with each specific type of amino acid.

$$R\text{-}CHO \xrightarrow[NH_4Cl]{KCN} R\text{-}CH(NH_2)\text{-}C{\equiv}N \xrightarrow{H^+} R\text{-}CH(NH_2)\text{-}COOH$$

The Strecker Amino Acid Synthesis

Chemical synthesis Several methods exist to synthesize amino acids. One of the oldest uses the Hell-Volhard-Zelinsky halogenation to introduce a bromine atom on the α-carbon. Nucleophilic substitution of the bromine with ammonia then yields the amino acid. Alternatively, the Strecker amino acid synthesis involves the treatment of an aldehyde with potassium cyanide and ammonia, this produces an α-amino nitrile as an intermediate. Hydrolysis of the nitrile in acid then yields a α-amino acid. Using ammonia or ammonium salts in this reaction gives unsubstituted amino acids, while substituting primary and secondary amines will yield substituted amino acids. Likewise, using ketones, instead of aldehydes, gives α,α-disubstituted amino acids. The classical synthesis gives racemic mixtures of α-amino acids as products, but several alternative procedures using asymmetric auxiliaries or asymmetric catalysts have been developed.

Peptide Bond Formation

As both the amine and carboxylic acid groups of amino acids can react to form amide bonds, one amino acid molecule can react with another and become joined through an amide linkage. This polymerization of amino acids is what creates proteins. This condensation reaction yields the newly formed peptide bond and a molecule of water. In cells, this reaction does not occur directly; instead the amino acid is first activated by attachment to a transfer RNA molecule through an ester bond. This aminoacyl-tRNA is produced in an ATP-dependent reaction carried out by an aminoacyl tRNA synthetase. This aminoacyl-tRNA is then a substrate for the ribosome, which catalyzes the attack of the amino group of the elongating protein chain on the ester bond. As a result of this mechanism, all proteins made by ribosomes are synthesized starting at their N-terminus and moving towards their C-terminus.

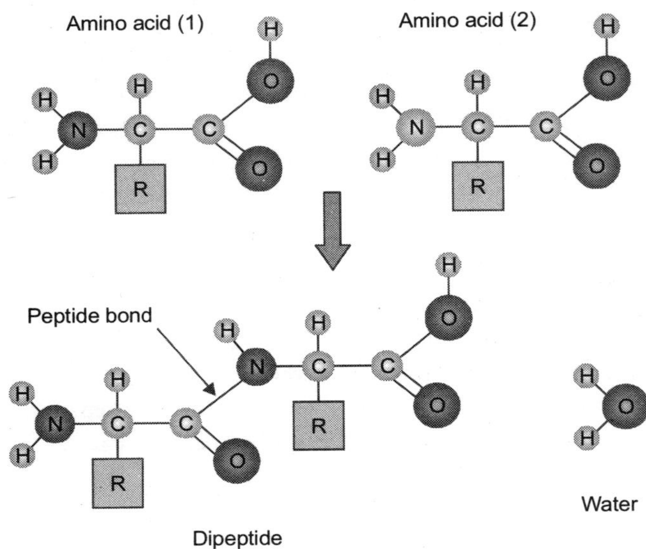

Fig. 2.44 : The condensation of two amino acids to form a peptide bond

However, not all peptide bonds are formed in this way. In a few cases, peptides are synthesized by specific enzymes. For example, the tripeptide glutathione is an essential part of the defenses of cells against oxidative stress. This peptide is synthesized in two steps from free amino acids. In the first step gamma-glutamylcysteine synthetase condenses cysteine and glutamic acid through a peptide bond formed between the side-chain carboxyl of the glutamate (the gamma carbon of this side chain) and the amino group of the cysteine. This dipeptide is then condensed with glycine by glutathione synthetase to form glutathione.

In chemistry, peptides are synthesized by a variety of reactions. One of the most used in solid-phase peptide synthesis, which uses the aromatic oxime derivatives of amino acids as activated units. These are added in sequence onto the growing peptide chain, which is attached to a solid resin support. The ability to easily synthesize vast numbers of different peptides by varying the types and order of amino acids (using combinatorial chemistry) has made peptide synthesis particularly important in creating libraries of peptides for use in drug discovery through high-throughput screening.

Biosynthesis and catabolism In plants, nitrogen is first assimilated into organic compounds in the form of glutamate, formed from alpha-ketoglutarate and ammonia in the mitochondrion. In order to form other amino acids, the plant uses transaminases to move the amino group to another alpha-keto carboxylic acid. For example, aspartate aminotransferase converts glutamate and oxaloacetate to alpha-ketoglutarate and aspartate. Other organisms use transaminases for amino acid synthesis too. Transaminases are also involved in breaking down amino acids. Degrading an amino acid often involves moving its amino group to alpha-ketoglutarate, forming glutamate. In many vertebrates, the amino group is then removed through the urea cycle and is excreted in the form of urea. However, amino acid degradation can produce uric acid or ammonia instead. For example, serine dehydratase converts serine to pyruvate and ammonia.

Nonstandard amino acids are usually formed through modifications to standard amino acids. For example, homocysteine is formed through the transsulfuration pathway or by the demethylation of methionine via the intermediate metabolite S-adenosyl methionine, while hydroxyproline is made by a posttranslational modification of proline.

Microorganisms and plants can synthesize many uncommon amino acids. For example, some microbes make 2-aminoisobutyric acid and lanthionine, which is a sulfide-bridged alanine dimer. Both these amino acids are found in peptidic lantibiotics such as alamethicin. While in plants, 1-aminocyclopropane-1-carboxylic acid is a small disubstituted cyclic amino acid that is a key intermediate in the production of the plant hormone ethylene.

Hydrophilic and Hydrophobic Amino Acids

Depending on the polarity of the side chain, amino acids vary in their hydrophilic or hydrophobic character. These properties are important in protein structure and protein-protein interactions. The importance of the physical properties of the side chains comes from the influence this has on the amino acid residues' interactions with other structures, both within a single protein and between proteins. The distribution of hydrophilic and hydrophobic amino acids determines the tertiary structure of the protein, and their physical location on the outside structure of the proteins influences their quaternary structure.

For example, soluble proteins have surfaces rich with polar amino acids like serine and threonine, while integral membrane proteins tend to have outer ring of hydrophobic amino acids that anchors them into the lipid bilayer, and proteins anchored to the membrane have a hydrophobic end that locks into the membrane. Similarly, proteins that have to bind to positively-charged molecules have surfaces rich with negatively charged amino acids like glutamate and aspartate, while proteins binding to negatively-charged molecules have surfaces rich with positively charged chains like lysine and arginine. Recently a new scale of hydrophobicity based on the free energy of hydrophobic association has been proposed.

Hydrophilic and hydrophobic interactions of the proteins do not have to rely only on the sidechains of amino acids themselves. By various posttranslational modifications other chains can be attached to the proteins, forming hydrophobic lipoproteins, or hydrophilic glycoproteins.

Table of standard amino acid abbreviations and side chain properties

List of standard amino acids

Amino Acid	3-Letter	1-Letter	Side polarity	chain Side chain charge (pH 7)	Hydropathy index
Alanine	Ala	A	nonpolar	neutral	1.8
Arginine	Arg	R	polar	positive	-4.5
Asparagine	Asn	N	polar	neutral	-3.5
Aspartic acid	Asp	D	polar	negative	-3.5
Cysteine	Cys	C	nonpolar	neutral	2.5
Glutamic acid	Glu	E	polar	negative	-3.5
Glutamine	Gln	Q	polar	neutral	-3.5
Glycine	Gly	G	nonpolar	neutral	-0.4
Histidine	His	H	polar	positive	-3.2
Isoleucine	Ile	I	nonpolar	neutral	4.5
Leucine	Leu	L	nonpolar	neutral	3.8
Lysine	Lys	K	polar	positive	-3.9
Methionine	Met	M	nonpolar	neutral	1.9
Phenylalanine	Phe	F	nonpolar	neutral	2.8
Proline	Pro	P	nonpolar	neutral	-1.6
Serine	Ser	S	polar	neutral	-0.8
Threonine	Thr	T	polar	neutral	-0.7
Tryptophan	Trp	W	nonpolar	neutral	-0.9
Tyrosine	Tyr	Y	polar	neutral	-1.3
Valine	Val	V	nonpolar	neutral	4.2

In addition to the specific amino acid codes, placeholders were used historically in cases where chemical or crystallographic analysis of a peptide or protein could not conclusively determine the identity of a residue.

Ambiguous Amino Acids	3-Letter	1-Letter
Asparagine or aspartic acid	Asx	B
Glutamine or glutamic acid	Glx	Z
Leucine or Isoleucine	Xle	J
Unspecified or unknown amino acid	Xaa	X

Unk is sometimes used instead of Xaa, but is less standard.

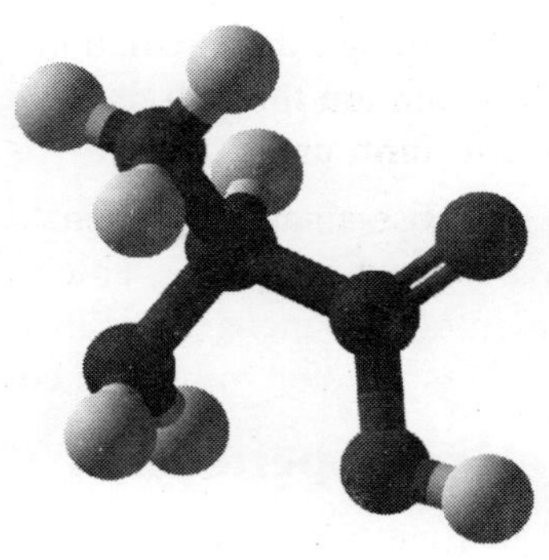

Alanine (abbreviated as Ala or A) is an α-amino acid with the chemical formula $CH_3CH(NH_2)COOH$. The L-isomer is one of the 20 proteinogenic amino acids, i.e. the building blocks of proteins. Its codons are GCU, GCC, GCA, and GCG. It is classified as a nonpolar amino acid. L-alanine is second only to leucine, accounting for 7.8% of the primary structure in a sample of 1,150 proteins. D-alanine occurs in bacterial cell walls and in some peptide antibiotics.

Structure: The α-carbon atom of alanine is bound with a methyl group ($-CH_3$), making it one of the simplest α-amino acids with respect to molecular structure and also resulting in alanine being classified as an aliphatic amino acid. The methyl group of alanine is non-reactive and is thus almost never directly involved in protein function.

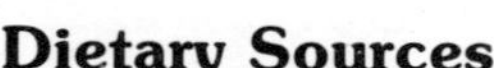

Dietary Sources

Alanine is a nonessential amino acid, meaning it can be manufactured by the human body, and does not need to be obtained directly through the diet. Alanine is found in a wide variety of foods, but is particularly concentrated in meats.

Good sources of alanine include:

- Animal sources: meat, seafood, caseinate, dairy products, eggs, fish, gelatin, lactalbumin
- Vegetarian sources: beans, nuts, seeds, soy, whey, brewer's yeast, brown rice bran, corn, legumes, whole grains.

Biosynthesis: Alanine can be manufactured in the body from pyruvate and branched chain amino acids such as valine, leucine, and isoleucine.

Alanine is most commonly produced by reductive amination of pyruvate. Because transamination reactions are readily reversible and pyruvate pervasive, alanine can be easily formed and thus has close links to metabolic pathways such as glycolysis, gluconeogenesis, and the citric acid cycle. It also arises together with lactate and generates glucose from protein via the alanine cycle.

Chemical Synthesis

Racemic alanine can be prepared via the condensation of acetaldehyde with ammonium chloride in the presence of potassium cyanide by the Strecker reaction.

Physiological function As a carrier of ammonia and of the carbon skeleton of pyruvate in alanine cycle. Alanine plays a key role in glucose-alanine cycle between tissues and liver. In muscle and other tissues that degrade amino acids for fuel, amino groups are collected in the form of glutamate by transamination. Glutamate can then transfer its amino group through the action of alanine aminotransferase to pyruvate, a product of muscle glycolysis, forming alanine and α-ketoglutarate. The alanine formed is passed into the blood and transported to the liver. A reverse of the alanine aminotransferase reaction takes place in liver. Pyruvate regenerated forms glucose through gluconeogenesis, which returns to muscle through the circulation system. Glutamate in the liver enters mitochondria and degrades into ammonium ion through the action of glutamate dehydrogenase, which in turn participate in the urea cycle to form urea.

Glucose-alanine cycle enables pyruvate and glutamate to be removed from muscle and find their ways to liver. Glucose is able to be regenerated from pyruvate and returned to muscle. The energetic burden of gluconeogenesis is thus imposed on the liver instead of the muscle. All available ATP in muscle is devoted to muscle contraction.

Link to Hypertension

An international study led by Imperial College London found a correlation between high levels of alanine and higher blood pressure, energy intake, cholesterol levels, and body mass index.

Chemical Properties

Free Radical Stability

The deamination of an alanine molecule produces a stable alkyl free radical, $CH_3C^{\bullet}HCOO^-$. Deamination can be induced in solid or aqueous alanine by radiation.

This property of alanine is used in dosimetric measurements in radiotherapy. When normal alanine is irradiated, the radiation causes certain alanine molecules to become free radicals, and, as these radicals are stable, the free radical content can later be measured in order to find out how much radiation the alanine was exposed to. In this way, one can be assured that complex radiotherapy treatment plans will deliver the intended pattern of radiation dose.

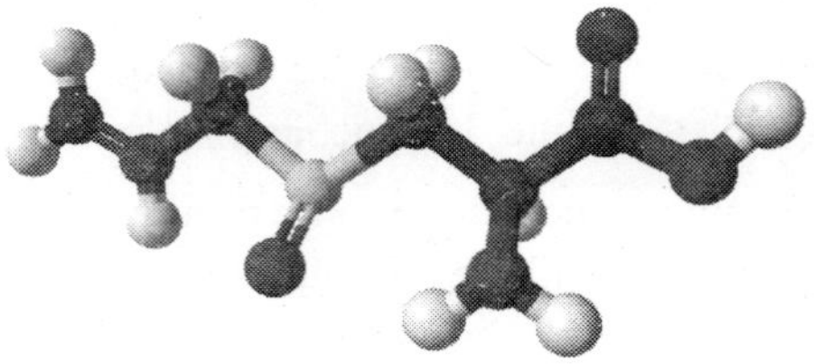

Alliin is a sulfoxide that is a natural constituent of fresh garlic. It is a derivative of the amino acid cysteine. When fresh garlic is chopped or crushed, the enzyme alliinase converts alliin into allicin which is primarily responsible for the aroma of fresh garlic. Garlic has been used since antiquity as a therapeutic remedy for oxygen toxicity, and when this was investigated, garlic did indeed show strong antioxidant and hydroxyl radical scavenging properties, possibly owing to the alliin contained within.

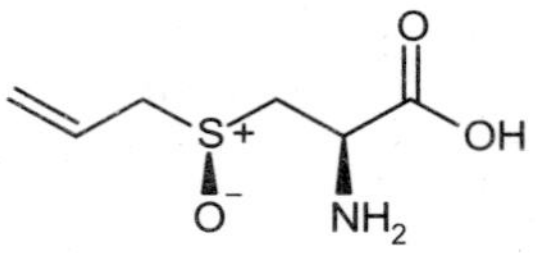

When the effect of alliin is observed on blood cells in vitro, a noted increase in the engulfing capacity of phagocyting cells is seen.

Arginine (abbreviated as **Arg** or **R**) is an α-amino acid. The L-form is one of the 20 most common natural amino acids. Its codons are CGU, CGC, CGA, CGG, AGA, and AGG. In mammals, arginine is classified as a semiessential or conditionally essential amino acid, depending on the developmental stage and health status of the individual. Infants are unable to meet their requirements and thus arginine is nutritionally essential for infants. Arginine was first isolated from a lupin seedling extract in 1886 by the Swiss chemist Ernst Schultze.

Structure

Arginine consists of a 4-carbon aliphatic straight chain, the distal end of which is capped by a complex guanidinium group. With a pK_a of 12.48, the guanidinium group is positively charged in neutral, acidic and even most basic environments, and thus imparts basic chemical properties to arginine. Because of the conjugation between the double bond and the nitrogen lone pairs, the positive charge is de-localized, enabling the formation of multiple H-bonds.

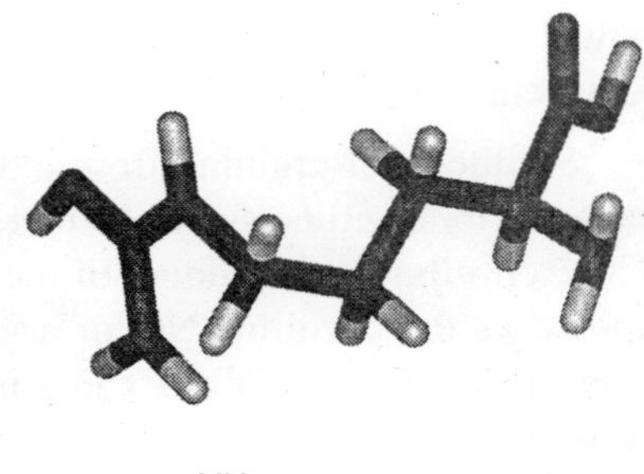

Sources

Dietary Sources

Arginine is a conditionally essential amino acid, meaning most of the time it can be manufactured by the human body, and does not need to be obtained directly through the diet. The biosynthetic pathway however does not produce sufficient arginine, and some must still be consumed through diet. Individuals who have poor nutrition or certain physical conditions may be advised to increase their intake of foods containing arginine. Arginine is found in a wide variety of foods, including:

- **Animal sources**: dairy products (e.g., cottage cheese, ricotta, milk, yogurt, whey protein drinks), beef, pork (e.g., bacon, ham), poultry (e.g., chicken and turkey light meat), wild game (e.g., pheasant, quail), seafood (e.g., halibut, lobster, salmon, shrimp, snails, tuna in water)
- **Vegetable sources**: wheat germ and flour, buckwheat, granola, oatmeal, nuts (coconut, pecans, cashews, walnuts, almonds, Brazil nuts, hazelnuts, pinenuts, peanuts), seeds (pumpkin, sesame, sunflower), chick peas, cooked soybeans.

Biosynthesis

Arginine is synthesized from citrulline by the sequential action of the cytosolic enzymes argininosuccinate synthetase (ASS) and argininosuccinate lyase (ASL). This is energetically costly, as the synthesis of each molecule of argininosuccinate requires hydrolysis of adenosine triphosphate (ATP) to adenosine monophosphate (AMP); *i.e.*, two ATP equivalents.

Citrulline can be derived from multiple sources:

- from arginine via nitric oxide synthase (NOS)
- from ornithine via catabolism of proline or glutamine/glutamate
- from asymmetric dimethylarginine (ADMA) via DDAH

The pathways linking arginine, glutamine, and proline are bidirectional. Thus, the net utilization or production of these amino acids is highly dependent on cell type and developmental stage.

On a whole-body basis, synthesis of arginine occurs principally via the intestinal–renal axis, wherein epithelial cells of the small intestine, which produce citrulline primarily from glutamine and glutamate, collaborate with the proximal tubule cells of the kidney, which extract citrulline from the circulation and convert it to arginine, which is returned to the circulation. Consequently, impairment of small bowel or renal function can reduce endogenous arginine synthesis, thereby increasing the dietary requirement.

Synthesis of arginine from citrulline also occurs at a low level in many other cells, and cellular capacity for arginine synthesis can be markedly increased under circumstances that also induce iNOS. Thus, citrulline, a coproduct of the NOS-catalyzed reaction, can be recycled to arginine in a pathway known as the citrulline-NO or arginine-citrulline pathway. This is demonstrated by the fact that in many cell types, citrulline can substitute for arginine to some degree in supporting NO synthesis. However, recycling is not quantitative because citrulline accumulates along with nitrate and nitrite, the stable end-products of NO, in NO-producing cells.

Function

Arginine plays an important role in cell division, the healing of wounds, removing ammonia from the body, immune function, and the release of hormones. Arginine, taken in combination with proanthocyanidins or yohimbine, has also been used as a treatment for erectile dysfunction.

The benefits and functions attributed to oral ingestion of L-arginine include:

- Precursor for the synthesis of nitric oxide (NO).
- Stimulation of the release of growth hormone.
- Improves immune function.
- Reduces healing time of injuries (particularly bone).
- Quickens repair time of damaged tissue.
- Reduces risk of heart disease.
- Increases muscle mass.
- Reduces adipose tissue body fat.
- Helps improve insulin sensitivity.
- Helps decrease blood pressure.
- Alleviates male infertility, improving sperm production and motility.
- Increases circulation throughout the body, including the sex organs.

In proteins

The geometry, charge distribution and ability to form multiple H-bonds make arginine ideal for binding negatively charged groups. For this reason arginine prefers to be on the outside of the proteins where it can interact with the polar environment. Incorporated in proteins, arginine can also be converted to citrulline by PAD enzymes. In addition, arginine can be methylated by protein methyltransferases.

As a precursor

Arginine is the immediate precursor of NO, urea, ornithine and agmatine; is necessary for the synthesis of creatine; and can also be used for the synthesis of polyamines (mainly through ornithine and to a lesser degree through agmatine), citrulline, and glutamate. For being a precursor of NO, (relaxes blood vessels), arginine is used in many conditions where vasodilation is required. The presence of asymmetric dimethylarginine (ADMA), a close relative, inhibits the nitric oxide reaction; therefore, ADMA is considered a marker for vascular disease, just as L-arginine is considered a sign of a healthy endothelium.

Treatment of Herpes Simplex Virus

A low ratio of arginine to lysine may be of benefit in the treatment of herpes simplex virus. For more information, refer to Herpes - Treatment.

Possible Increases in Risk of Death from Heart Disease

A clinical trial found that patients taking an L-arginine supplement following a heart attack didn't improve in their vascular tone or their hearts' ability to pump. In fact, more patients died who were taking L-arginine than placebo and the study was stopped early with the recommendation the supplement not be used by heart attack patients. The supplement still is widely marketed.

Lung Inflammation and Asthma

The Mayo Clinic web page on L-arginine reports that inhalation of L-arginine can increase lung inflammation and worsen asthma. No primary source for this observation is cited, making it difficult to judge the validity and clinical significance of this effect.

Growth Hormone

Arginine increases the production of growth hormone. Reports of its effects on male muscular development are not clearly proven.

Asparagine (abbreviated as **Asn** or **N**; **Asx** or **B** represent either asparagine or aspartic acid) is one of the 20 most common natural amino acids on Earth. It has carboxamide as the side chain's functional group. It is not an essential amino acid. Its codons are AAU and AAC.

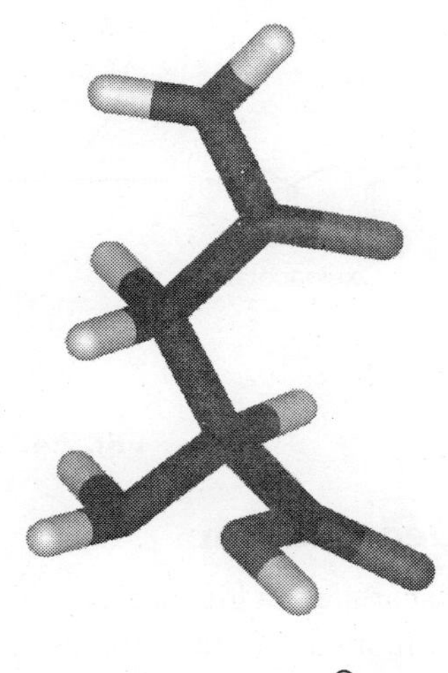

A reaction between asparagine and reducing sugars or reactive carbonyls produces acrylamide (acrylic amide) in food when heated to sufficient temperature. These products occur in baked goods such as french fries, potato chips, and roasted coffee.

History Asparagine was first isolated in 1806 from asparagus juice, in which it is abundant—hence its name—becoming the first amino acid to be isolated. The characteristic smell observed in the urine of individuals after their consumption of asparagus is attributed to various metabolic byproducts of asparagine.

Structural function in proteins Since the asparagine side chain can form hydrogen bond interactions with the peptide backbone, asparagine residues are often found near the beginning and the end of alpha-helices, and in turn motifs in beta sheets. Its role can be thought as "capping" the hydrogen bond interactions which would otherwise be satisfied by the polypeptide backbone. Glutamines, with an extra methylene group, have more conformational entropy and thus are less useful in this regard.

Asparagine also provides key sites for N-linked glycosylation, modification of the protein chain with the addition of carbohydrate chains.

Dietary Sources: Asparagine is not an essential amino acid, which means that it can be synthesized from central metabolic pathway intermediates in humans and is not required in the diet. Asparagine is found in:

- **Animal sources**: dairy, whey, beef, poultry, eggs, fish, lactalbumin, seafood.
- **Vegetarian sources**: asparagus, potatoes, legumes, nuts, seeds, soy, whole grains.

Biosynthesis The precursor to asparagine is oxaloacetate. Oxaloacetate is converted to aspartate using a transaminase enzyme. The enzyme transfers the amino group from glutamate to oxaloacetate producing α-ketoglutarate and aspartate. The enzyme asparagine synthetase produces asparagine, AMP, glutamate, and pyrophosphate from aspartate, glutamine, and ATP. In the asparagine synthetase reaction, ATP is used to activate aspartate, forming β-aspartyl-AMP. Glutamine donates an ammonium group which reacts with β-aspartyl-AMP to form asparagine and free AMP.

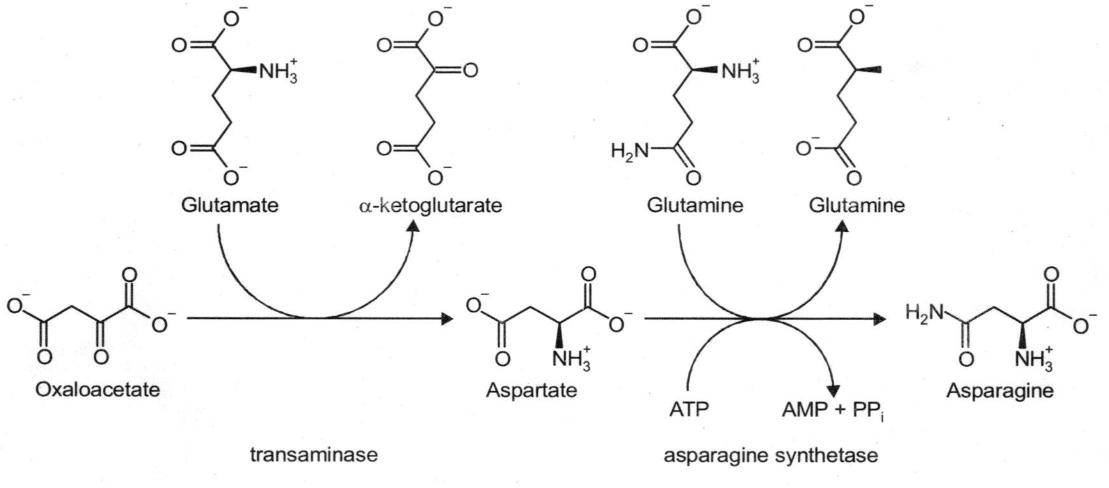

Glutamate α-ketoglutarate Glutamine Glutamine

Oxaloacetate Aspartate Asparagine

ATP AMP + PP_i

transaminase asparagine synthetase

Fig. 2.45 : The biosynthesis of asparagine from oxaloacetate

Degradation

Aspartate is a glucogenic amino acid. L-asparaginase hydrolyzes the amide group to form aspartate and ammonium. A transaminase converts the aspartate to oxaloacetate which can then be metabolized in the citric acid cycle or gluconeogenesis.

Aspartic acid (abbreviated as **Asp** or **D**; **Asx** or **B** represent either aspartic acid or asparagine) is an α-amino acid with the chemical formula $HO_2CCH(NH_2)CH_2CO_2H$. The carboxylate anion of aspartic acid is known as **aspartate**. The L-isomer of aspartate is one of the 20 proteinogenic amino acids, i.e., the building blocks of proteins. Its codons are GAU and GAC.

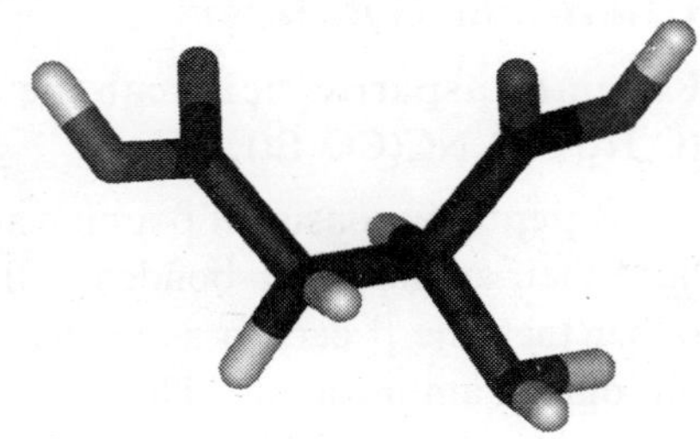

Aspartic acid is, together with glutamic acid, classified as an acidic amino acid with a pKa of 4.0. Aspartic acid is pervasive in biosynthesis. As with all amino acids, the location of acid protons depends on the pH of the solution and the crystallization conditions.

Role in Biosynthesis of Amino Acids

Aspartic acid is non-essential in mammals, being produced from oxaloacetate by transamination. In plants and microorganisms, aspartic acid is the precursor to several amino acids, including four that are essential: methionine, threonine, isoleucine, and lysine. The conversion of aspartic acid to these other amino acids begins with reduction of aspartic acid to its "semialdehyde," $HO_2CCH(NH_2)CH_2CHO$. Asparagine is derived from aspartic acid via transamidation:

$HO_2CCH(NH_2)CH_2CO_2H + GC(O)NH_2\ HO_2CCH(NH_2)CH_2CONH_2 + GC(O)OH$
(where $GC(O)NH_2$ and $GC(O)OH$ are glutamine and glutamic acid, respectively)

Other Biochemical Roles

Aspartate is also a metabolite in the urea cycle and participates in gluconeogenesis. It carries reducing equivalents in the malate-aspartate shuttle, which utilizes the ready interconversion of aspartate and oxaloacetate, which is the oxidized (dehydrogenated) derivative of malic acid. Aspartic acid donates one nitrogen atom in the biosynthesis of inositol, the precursor to the purine bases.

Neurotransmitter

Aspartate (the conjugate base of aspartic acid) stimulates NMDA receptors, though not as strongly as the amino acid neurotransmitter glutamate does. It serves as an excitatory neurotransmitter in the brain and is an excitotoxin.

As a neurotransmitter, aspartic acid may provide resistance to fatigue, and, thus, leads to endurance, although the evidence to support this idea is not strong.

Dietary Sources

Aspartic acid is not an essential amino acid, which means that it can be synthesized from central metabolic pathway intermediates in humans.

Aspartic acid is found in:

- **Animal sources**: luncheon meats, sausage meat, wild game.
- **Vegetable sources**: sprouting seeds, oat flakes, avocado, asparagus.

Chemical Synthesis

Racemic aspartic acid can be synthesized from diethyl sodium phthalimidomalonate, $(C_6H_4(CO)_2NC(CO_2Et)_2)$.

β-peptides consist of β amino acids, which have their amino group bonded to the β carbon rather than the β carbon as in the 20 standard biological amino acids. The only commonly naturally occurring β amino acid is β-alanine; although it is used as a component of larger bioactive molecules, β-peptides in general do not appear in nature. For this reason β-peptide-based antibiotics are being explored as ways of evading antibiotic resistance. Pioneering studies in this field were published in 1996 by the group of Dieter Seebach and that of Gellman

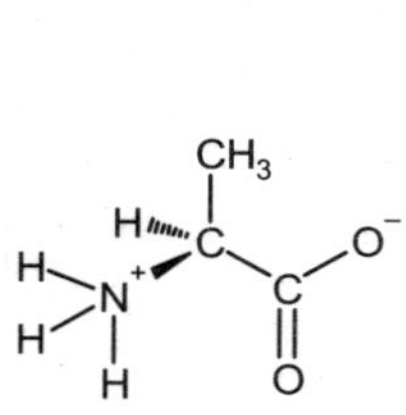

L-α-alanine

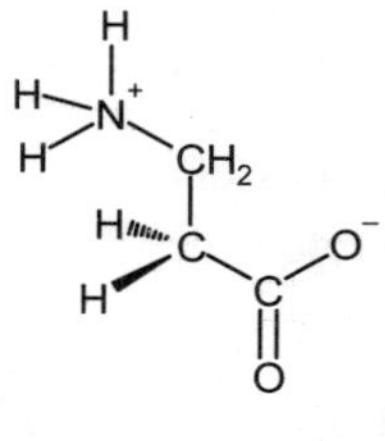

β-alanine

Chemical Structure and Synthesis

In α amino acids (molecule at left), both the carboxylic acid group (red) and the amino group (blue) are bonded to the same carbon, termed the α carbon (C^{α}) because it is one atom away from the carboxylate group. In β amino acids, the amino group is bonded to the β carbon (C^{β}), which is found in most of the 20 standard amino acids. Only glycine lacks a β carbon, which means that there is no β-glycine molecule.

The chemical synthesis of β amino acids can be challenging, especially given the diversity of functional groups bonded to the β carbon and the necessity of maintaining chirality. In the alanine molecule shown, the β carbon is achiral; however, larger amino acids have a chiral C^{β} atom. A number of synthesis mechanisms have been introduced to efficiently form β amino acids and their derivatives notably those based on the Arndt-Eistert synthesis.Two main types of β-peptides exist: those with the

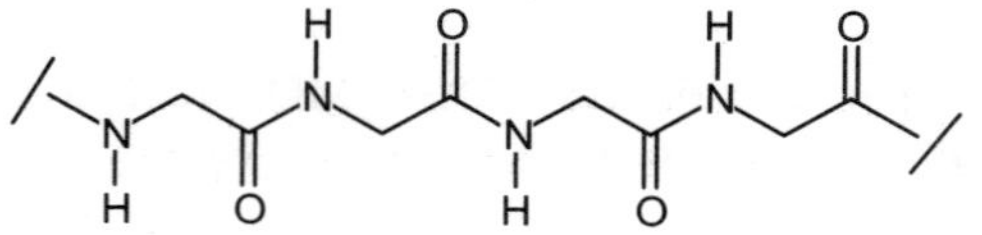

α-peptide

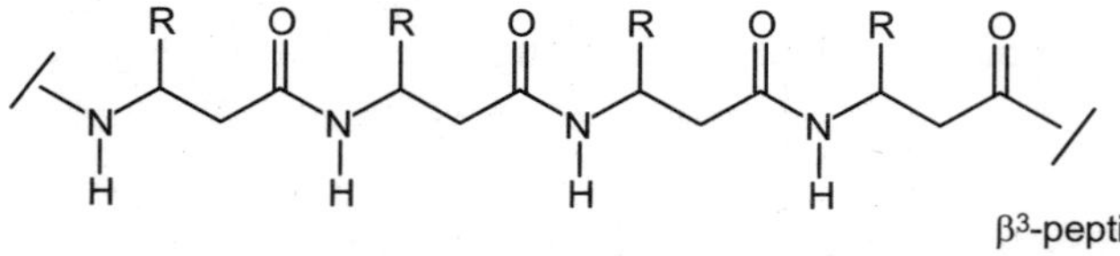

β^3-peptide

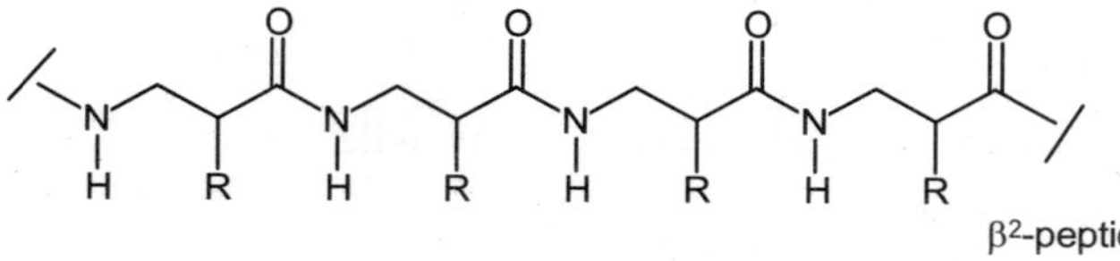

β^2-peptide

organic residue (R) next to the amine are called β^3-peptides and those with position next to the carbonyl group are called β^2-peptides.

Secondary Structure

Because the backbones of β-peptides are longer than those of peptides that consist of α-amino acids, β-peptides form different secondary structures. The alkyl substituents at both the and β positions in a â amino acid favour a gauche conformation about the bond between the α-carbon and β-carbon. This also affects the thermodynamic stability of the structure.

Many types of helix structures consisting of β-peptides have been reported. These conformation types are distinguished by the number of atoms in the hydrogen-bonded ring that is formed in solution; 8-helix, 10-helix, 12-helix, 14-helix, and 10/12-helix have been reported. Generally speaking, β-peptides form a more stable helix than α-peptides

The β-peptide *zwit-1F* with a fully described quaternary structure is called a **β-protein** because it has many characteristics of an actual protein. Eight 12-helix units self-assemble in water to a superstructure with a hydrophobic inner core.

Clinical Potential

β-peptides are stable against proteolytic degradation in vitro and in vivo, an important advantage over natural peptides in the preparation of peptide-based drugs. β-peptides have been used to mimic natural peptide-based antibiotics such as magainins, which are extremely powerful but difficult to use as drugs because they are degraded by proteolytic enzymes in the body.

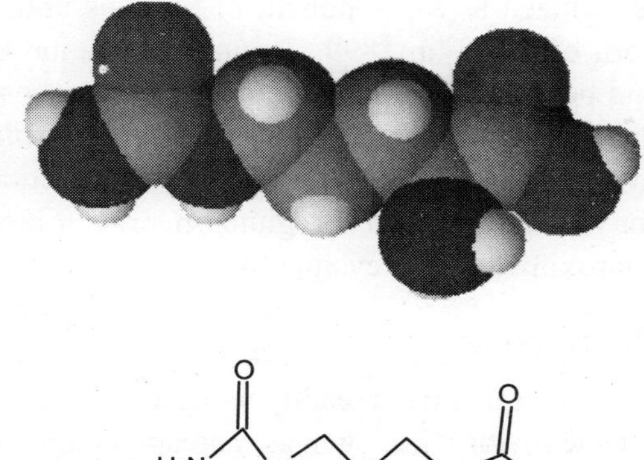

Citrulline is an α-amino acid. Its name is derived from *citrullus*, the Latin word for watermelon, from which it was first isolated in 1930. It has the idealized formula $H_2NC(O)NH(CH_2)_3CH(NH_2)CO_2H$. It is a key intermediate in the urea cycle, the pathway by which mammals excrete ammonia.

Biosynthesis

Citrulline is made from ornithine and carbamoyl phosphate in one of the central reactions in the urea cycle. It is also produced from arginine as a by-product of the reaction catalyzed by NOS family (NOS; EC 1.14.13.39). Arginine is first oxidized into N-hydroxyl-arginine, which is then further oxidized to citrulline concomitant with release of nitric oxide.

Function

Although citrulline is not coded for by DNA directly, several proteins are known to contain citrulline as a result of a posttranslational modification. These citrulline residues are generated by a family of enzymes called peptidylarginine deiminases (PADs), which convert arginine into citrulline in a process called citrullination or deimination. Proteins that normally contain citrulline residues include myelin basic protein (MBP), filaggrin, and several histone proteins, whereas other proteins, such as fibrin and vimentin are susceptible to citrullination during cell death and tissue inflammation.

Patients with rheumatoid arthritis often have detectable antibodies against proteins containing citrulline. Although the origin of this immune response is not known, detection of antibodies reactive with citrulline (anti-citrullinated protein antibodies) containing proteins or peptides is now becoming an important help in the diagnosis of rheumatoid arthritis.

In recent studies, citrulline has been found to relax blood vessels.

Sources

Citrulline in the form of citrulline malate is sold as a performance-enhancing athletic dietary supplement which is said to reduce muscle fatigue.

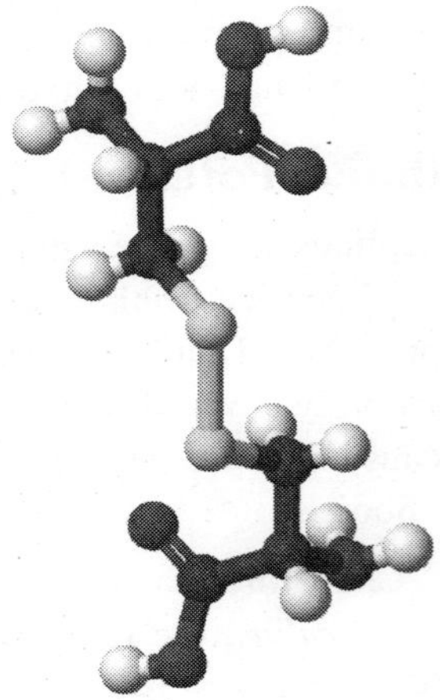

The rind of watermelon (*Citrullus lanatus*) is a good natural source of citrulline.

Cystine is the amino acid dimer formed when a pair of cysteine molecules are joined by a disulfide bond. It is described by the formula $(SCH_2CH(NH_2)CO_2H)_2$. It is a colourless solid, and melts at 247-249 °C. It was discovered in 1810 by William Hyde Wollaston but was not recognized as a component of proteins until it was isolated from the horn of a cow in 1899. Through formation of disulfide bonds within and between protein molecules, cystine is a significant determinant of the tertiary structure of most proteins. Disulfide bonding, along with hydrogen bonding and hydrophobic interactions is partially responsible for the formation of the gluten matrix in bread. Human hair contains approximately 5% cystine by mass.

Properties

The disulfide link is readily reduced to give the corresponding thiol, cysteine. This reaction is typically effected with thiols such as mercaptoethanol or dithiothreitol.

$$(SCH_2CH(NH_2)CO_2H)_2 + 2\,RSH \rightarrow 2\,HSCH_2CH(NH_2)CO_2H + RSSR$$

Nutritional Sources

Supplemental N-acetyl cysteine is claimed to be a source of cystine, but the dose of this supplement is limited by side effects. One of the richest nutritional sources of cystine in the diet is undenatured whey proteins from milk The disulfide-bonded cystine is not digested or significantly hydrolized by the stomach, but is transported by the blood stream to the tissues of the body. Here, within the cells of the body, the weak disulfide bond is cleaved to give cysteine, from which glutathione can be synthesized.

In Animal Feed

Disulfide bonds can be broken at temperatures above about 150 °C, especially at low moisture levels (below about 20%).

Side Effects

Nutritional sources of cystine are virtually free of the toxic side effects associated with the single molecule of cysteine, N-acetyl cysteine. The greatest dietary source of cystine is bio-active, unpasteurized or low-heat pasteurized undenatured whey proteins.

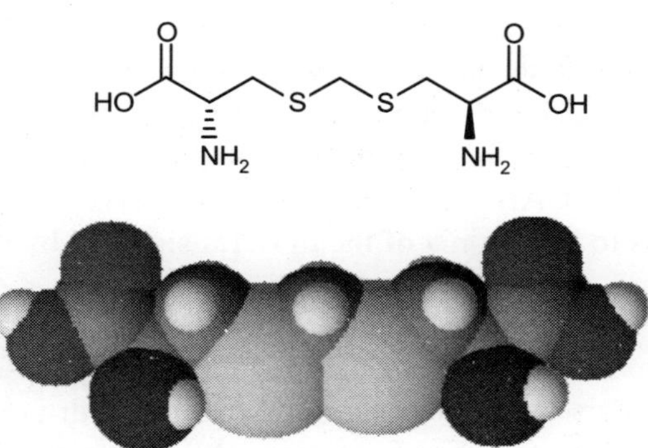

Djenkolic acid (or sometimes **jengkolic acid**) is a sulfur-containing non-protein amino acid naturally found in djenkol beans of the South-East Asian legumes jengkol (*Archidendron jiringa*). This compound consists of two cysteine radicals connected by a methylene group between the sulfurs, or it is 2-amino-3-[(2-amino-3-hydroxy-3-oxopropyl) sulfanylmethyl sulfanyl] propanoic acid. It is toxic to humans, especially causing nephrotoxicity.

Toxicity

The toxicity of djenkolic acid in humans arises from its poor solubility under acidic conditions after consumption of the jenkol bean. The amino acid precipitates into crystals which cause mechanical irritation of the renal tubules and urinary tract, resulting in symptoms such as abdominal discomfort, loin pains, severe colic, nausea, vomiting, dysurea, gross hematuria, and oliguria, occurring 2 to 6 hours after the beans were ingested. Urine analysis of patients reveals erythrocytes, epithelial cells, protein, and the needle-like crystals of djenkolic acid. Urolithiasis can also happen, with djenkolic acid as the nucleus. In young children it has also been reported to produce painful swelling of the genitalia.

Treatment for this toxicity requires hydration to increase urine flow and alkalinization of urine by sodium bicarbonate. Furthermore, this poisoning can be prevented when consuming djenkol beans by boiling them beforehand, since djenkolic acid is removed from the beans.

Discovery and Synthesis

Djenkolic acid was first isolated by Van Veen and Hyman from the urine of the natives of Java who had eaten the djenkol bean and were suffering from poisoning. They then succeeded in isolating the djenkolic acid crystals from the djenkol beans treated with $Ba(OH)_2$ at 30°C for a prolonged period of time.

Djenkolic acid was later reported to be found as much as 20 grams in every kilogram of dry djenkol beans, and it has also been reported to be found to a lesser extent in the seeds of other leguminous plants such as *Leucaena esculenta* (2.2 g/kg) and *Pithecolobium ondulatum* (2.8 g/kg).

Du Vigneaud and Patterson managed to synthesize djenkolic acid by the condensation of methylene chloride with 2 moles of L-cysteine in liquid ammonia and to show that their synthetic compound was identical with the naturally occurring djenkolic acid. Later on, Armstrong and du Vigneaud prepared djenkolic acid by the direct combination of 1 mole of formaldehyde with 2 moles of L-cysteine in a strongly acid solution.

Gamma-aminobutyric acid (GABA) is the chief inhibitory neurotransmitter in the mammalian central nervous system. It plays an important role in regulating neuronal excitability throughout the nervous system. In humans, GABA is also directly responsible for the regulation of muscle tone. In insect species GABA acts only on excitatory nerve receptors.

While technically an amino acid, GABA is rarely referred to as such in the scientific or medical communities, because the term "amino acid," used without a qualifier, refers to the alpha amino acids, which GABA is not, nor is it incorporated into proteins. In spastic diplegia in humans, GABA absorption by some nerves becomes damaged, which leads to hypertonia of the muscles signaled by those nerves.

Function as Neurotransmitter

In vertebrates, GABA acts at inhibitory synapses in the brain by binding to specific transmembrane receptors in the plasma membrane of both pre- and postsynaptic neuronal processes. This binding causes the opening of ion channels to allow the flow of either negatively charged chloride ions into the cell or positively charged potassium ions out of the cell. This action results in a negative change in the transmembrane potential, usually causing hyperpolarization. Three general classes of GABA receptor are known: $GABA_A$ and $GABA_C$ ionotropic receptors, which are ion channels themselves, and $GABA_B$ metabotropic receptors, which are G protein-coupled receptors that open ion channels via intermediaries (G proteins).

Neurons that produce GABA as their output are called GABAergic neurons, and have chiefly inhibitory action at receptors in the adult vertebrate. Medium Spiny Cells are a typical example of inhibitory CNS GABAergic cells. In contrast, GABA exhibits excitatory actions in insects, mediating muscle activation at synapses between nerves and muscle cells, and also the stimulation of certain glands. Whether GABA is excitatory or inhibitory depends on the direction (into or out of the cell) and magnitude of the ionic currents controlled by the $GABA_A$ receptor. When net positive ionic current is directed into the cell, GABA is excitatory, when the net positive current is directed out of the cell, GABA is inhibitory. A developmental switch in the molecular machinery controlling the polarity of this current is responsible for the changes in the functional role of GABA between the neonatal and adult stages. That is to say, GABA's role changes from excitatory to inhibitory as the brain develops into adulthood.

Development

In hippocampus and neocortex of the mammalian brain, GABA has primarily excitatory effects early in development, and is in fact the major excitatory neurotransmitter in many regions of the brain before the maturation of glutamate synapses

In the developmental stages preceding the formation of synaptic contacts, GABA is synthesized by neurons and acts both as an autocrine (acting on the same cell) paracrine (acting on nearby cells) signalling mediator.

GABA regulates the proliferation of neural progenitor cells the migration and differentiation the elongation of neurites and the formation of synapses. GABA also regulates the growth of embryonic and neural stem cells. GABA can inûuence the development of neural progenitor cells via brain-derived neurotrophic factor (BDNF) expression. GABA activates the $GABA_A$ receptor, causing cell cycle arrest in the S-phase, limiting growth.

Structure and Conformation

GABA is found mostly as a zwitterion, that is, with the carboxyl group deprotonated and the amino group protonated. Its conformation depends on its environment. In the gas phase, a highly folded conformation is strongly favoured because of the electrostatic attraction between the two functional groups. The stabilization is about 50 kcal/mol, according to quantum chemistry calculations. In the solid state, a more extended conformation is found, with a trans conformation at the amino end and a gauche conformation at the carboxyl end. This is due to the packing interactions with the neighboring molecules. In solution, five different conformations, some folded and some extended are found as a result of solvation effects. The conformational flexibility of GABA is important for its biological function, as it has been found to bind to different receptors with different conformations. Many GABA analogues with pharmaceutical applications have more rigid structures in order to control the binding better.

History

Gamma-aminobutyric acid was first synthesized in 1883, and was first known only as a plant and microbe metabolic product. In 1950, however, GABA was discovered to be an integral part of the mammalian central nervous system.

Synthesis

Organisms synthesize GABA from glutamate using the enzyme L-glutamic acid decarboxylase and pyridoxal phosphate (which is the active form of B6) as a cofactor. This process converts the principal excitatory neurotransmitter (glutamate) into the principal inhibitory one (GABA).

Pharmacology

Drugs that act as agonists of GABA receptors (known as GABA analogues or *GABAergic* drugs) or increase the available amount of GABA typically have relaxing, anti-anxiety and anti-convulsive effects. Many of the substances below are known to cause anterograde amnesia and retrograde amnesia.

It has been suggested that orally administered GABA increases the amount of Human Growth Hormone, but this is questionable since it is unknown whether GABA can pass the blood-brain barrier. However, orally administered GABA does have effects outside of the central nervous system (e.g. decreased muscle tone).

Drugs that affect GABA receptors:

- alcohol (ethanol)
- avermectins—doramectin, selamectin, ivermectin
- barbiturates
- bicucullines - GABA antagonist
- benzodiazepines
- baclofen
- baicalin and baicalein from skullcap scutellaria lateriflora
- carbamazepines
- cyclopyrrolone derivatives such as zopiclone
- fluoroquinolones
- gabazine (SR-95531)
- gamma-Hydroxybutyric acid (GHB)
- gamma-amino-beta-hydroxybutyric acid
- imidazopyridine derivatives such as zolpidem
- kavalactones
- meprobamate
- muscimol
- manganese
- modafinil
- phenytoin
- picamilon
- picrotoxin
- progabide
- propofol
- phenibut
- pyrazolopyrimidine derivatives such as zaleplon
- thujone—GABA antagonist
- Valerian extract
- pregabalin (tradename Lyrica) - GABA analogue

Drugs that affect GABA in other ways:

- tiagabine—potentiates by inhibiting uptake into neurons and glia
- vigabatrin—potentiates by inhibiting GABA-T, preventing GABA breakdown
- valproate—potentiates by inhibiting GABA-T

- tetanospasmin—primary toxin of tetanus bacteria, blocks release of GABA
- hyperforin—inhibits the reuptake of GABA
- thujone-A main ingredient in absinthe
- theanine - an amino acid found in green tea and the basidiomycete mushroom, increases GABA production

Glutamic acid (abbreviated as **Glu** or **E**) is one of the 20 proteinogenic amino acids and its codons are GAA and GAG. It is a non-essential amino acid. The carboxylate anions and salts of glutamic acid are known as **glutamates**.

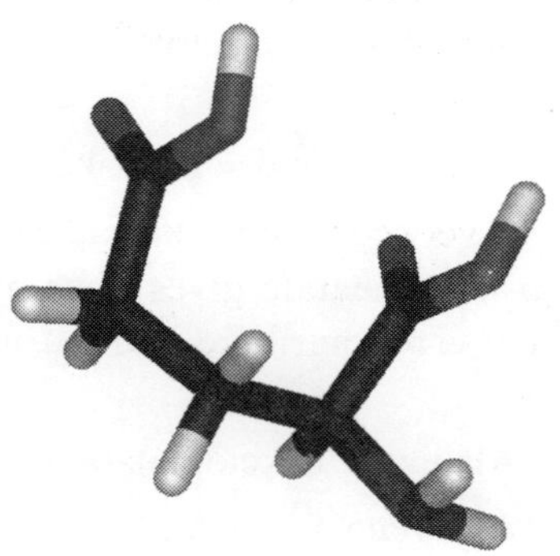

Chemistry

The side chain carboxylic acid functional group has pK_a of 4.1 and exists in its negatively charged deprotonated carboxylate form at physiological pH.

History

This compound was discovered in 1908 by the professor Kikunae Ikeda, who worked in the Imperial University of Tokyo. He loved seaweed which is used like spice in traditional Japanese food. He tried to find the root of this flavour. He discovered that the origin of this taste is glutamic acid. He isolated crystals of glutamic acid using a Kombu soup (one hundred grams of Kombu has nearly one gram of glutamic acid).

Moreover, he discovered that glutamate gave unique flavour to other foods. He called it "Umami" (meaning "yumminess" in Japanese). This distinctive flavor has brought glutamate the title of "the elusive fifth taste" to join the more traditional flavors, sweet, salty, sour, and bitter.

Biosynthesis

Reactants	Products	Enzymes
Glutamine + H_2O	→ Glu + NH_3	GLS, GLS2
NAcGlu + H_2O	→ Glu + Acetate	(unknown)
α-ketoglutarate + NADPH + NH_4^+	→ Glu + $NADP^+$ + H_2O	GLUD1, GLUD2
α-ketoglutarate + α-amino acid	→ Glu + α-oxo acid	transaminase
1-pyrroline-5-carboxylate + NAD^+ + H_2O	→ Glu + NADH	ALDH4A1
N-formimino-L-glutamate + FH_4	→ Glu + 5-formimino-FH_4	FTCD

Function and Uses

Metabolism

Glutamate is a key molecule in cellular metabolism. In humans, dietary proteins are broken down by digestion into amino acids, which serves as metabolic fuel for other functional roles in the body. A key process in amino acid degradation is transamination, in which the amino group of an amino acid is transferred to an α-ketoacid, typically catalysed by a transaminase. The reaction can be generalised as such:

R_1-amino acid + R_2-α-ketoacid $\rightleftharpoons$ R_1-α-ketoacid + R_2-amino acid

A very common α-ketoacid is α-ketoglutarate, an intermediate in the citric acid cycle. Transamination of α-ketoglutarate gives glutamate. The resulting α-ketoacid product is often a useful one as well, which can contribute as fuel or as a substrate for further metabolism processes. Examples are as follows:

Alanine + α-ketoglutarate $\rightleftharpoons$ pyruvate + glutamate

Aspartate + α-ketoglutarate $\rightleftharpoons$ oxaloacetate + glutamate

Both pyruvate and oxaloacetate are key components of cellular metabolism, contributing as substrates or intermediates in fundamental processes such as glycolysis, gluconeogenesis and also the citric acid cycle.

Glutamate also plays an important role in the body's disposal of excess or waste nitrogen. Glutamate undergoes deamination, an oxidative reaction catalysed by glutamate dehydrogenase, as follows:

glutamate + water + $NADP^+$ $\rightarrow$ α-ketoglutarate + NADPH + ammonia + H^+

Ammonia (as ammonium) is then excreted predominantly as urea, synthesised in the liver. Transamination can thus be linked to deamination, effectively allowing nitrogen from the amine groups of amino acids to be removed, via glutamate as an intermediate, and finally excreted from the body in the form of urea.

Neurotransmitter

Glutamate is the most abundant excitatory neurotransmitter in the mammalian nervous system. At chemical synapses, glutamate is stored in vesicles. Nerve impulses trigger release of glutamate from the pre-synaptic cell. In the opposing post-synaptic cell, glutamate receptors, such as the NMDA receptor, bind glutamate and are activated. Because of its role in synaptic plasticity, it is believed that glutamic acid is involved in cognitive functions like learning and memory in the brain.

Glutamate transporters are found in neuronal and glial membranes. They rapidly remove glutamate from the extracellular space. In brain injury or disease, they can work in reverse and excess glutamate can accumulate outside cells. This process causes calcium ions to enter cells via NMDA receptor channels, leading to neuronal damage and eventual cell death, and is called excitotoxicity.

The mechanisms of cell death include:

- Damage to mitochondria from excessively high intracellular Ca^{2+}

- Glu/Ca^{2+}-mediated promotion of transcription factors for pro-apoptotic genes, or downregulation of transcription factors for anti-apoptotic genes.

Excitotoxicity due to glutamate occurs as part of the ischemic cascade and is associated with stroke and diseases like amyotrophic lateral sclerosis, lathyrism, autism, some forms of mental retardation and Alzheimer's disease.

Glutamic acid has been implicated in epileptic seizures. Microinjection of glutamic acid into neurons produces spontaneous depolarisations around one second apart, and this firing pattern is similar to what is known as paroxysmal depolarizing shift in epileptic attacks. This change in the resting membrane potential at seizure foci could cause spontaneous opening of voltage-activated calcium channels, leading to glutamic acid release and further depolarization.

Experimental techniques to detect glutamate in intact cells include using a genetically-engineered nanosensor. The sensor is a fusion of a glutamate-binding protein and two fluorescent proteins. When glutamate binds, the fluorescence of the sensor under ultraviolet light changes by resonance between the two fluorophores. Introduction of the nanosensor into cells enables optical detection of the glutamate concentration. Synthetic analogs of glutamic acid that can be activated by ultraviolet light and two-photon excitation microscopy have also been described. This method of rapidly uncaging by photostimulation is useful for mapping the connections between neurons, and understanding synapse function.

Brain Nonsynaptic Glutamatergic Signaling Circuits

Extracellular glutamate in Drosophila brains has been found to regulate postsynaptic glutamate receptor clustering, via a process involving receptor desensitization. A gene expressed in glial cells actively transports glutamate into the extracellular space, while in the nucleus accumbens stimulating group II metabotropic glutamate receptors, this gene was found to reduce extracellular glutamate levels. This raises the possibility that this extracellular glutamate plays an "endocrine-like" role as part of a larger homeostatic system.

GABA Precursor

Glutamic acid also serves as the precursor for the synthesis of the inhibitory GABA in GABA-ergic neurons. This reaction is catalyzed by glutamic acid decarboxylase (GAD), which is most abundant in the cerebellum and pancreas.

Stiff-man syndrome is a neurologic disorder caused by anti-GAD antibodies, leading to a decrease in GABA synthesis and therefore, impaired motor function such as muscle stiffness and spasm. Since the pancreas is also abundant for the enzyme GAD, a direct immunological destruction occurs in the pancreas and the patients will have diabetes mellitus.

Flavor Enhancer

Free glutamic acid is present in a wide variety of foods, including cheese and soy sauce and is responsible for one of the five basic tastes of the human sense of taste (umami). Glutamic acid is often used as a food additive and flavour enhancer in the form of its sodium salt, monosodium glutamate (MSG).

Nutrient

All meats, poultry, fish, eggs, dairy products, as well as kombu are excellent sources of glutamic acid. Some protein-rich plant foods also serve as sources. Ninety-five percent of the dietary glutamate is metabolized by intestinal cells in a first pass.

Plant Growth

Auxigro is a plant growth preparation that contains 30% glutamic acid.

Production

China-based Fufeng Group Limited is the largest producer of glutamic acid in the world, with capacity increasing to 300,000 tons at the end of 2006 from 180,000 tons during 2006, putting them at 25%–30% of the Chinese market. Meihua is the second largest Chinese producer. Together, the top five producers have roughly 50% share in China. Chinese demand is roughly 1.1 million tons per year, while global demand, including China, is 1.7 million tons per year.

Pharmacology

The drug phencyclidine (more commonly known as PCP) antagonizes glutamic acid non-competitively at the NMDA receptor. For the same reasons, sub-anaesthetic doses of Ketamine have strong dissociative and hallucinogenic effects. Glutamate does not easily pass the blood brain barrier, but instead this transport is mediated by a high affinity transport system. It can also be converted into glutamine.

Role in Sickle-cell Disease

A point mutation (valine in place of glutamic acid at position 6) in the β-globin chain of hemoglobin forms HbS. This variant of hemoglobin can cause sickle-cell anemia, where the abnormal hemoglobin are prone to polymerization when deoxygenated, thus distorting the erythrocyte which are removed by the spleen or cause microvascular obstruction (ischemic crises). This trait and disease is common in areas with a high prevalence of *Plasmodium falciparum* (one of three Plasmodium species that causes malaria).

Glycine (abbreviated as **Gly** or **G**) is the organic compound with the formula NH_2CH_2COOH. It is the smallest of the 20 amino acids commonly found in proteins, coded by codons GGU, GGC, GGA and GGG. Glycine is unique among the proteinogenic amino acids in that it is not chiral. Most proteins contain only small quantities of glycine. A notable exception is collagen, which contains about 35% glycine. In its solid, i.e., crystallized, form, glycine is a free-flowing, sweet-tasting crystalline material.

Production

Glycine is manufactured industrially, either by treating chloroacetic acid with ammonia:

$$ClCH_2COOH + NH_3 \rightarrow H_2NCH_2COOH + HCl$$

or via the Strecker amino acid synthesis. There are two producers of glycine in the United States. Chattem Chemicals, Inc., purchased by Sun Pharmaceutical, Mumbai, India and GEO Specialty

Chemicals, Inc., who purchased the glycine and naphthalene sulfonate production facilities of Dow/ Hampshire Chemical Corp.

Biosynthesis

Glycine is not essential to the human diet, since it is biosynthesized in the body from the amino acid serine, which is in turn derived from 3-phosphoglycerate. In most organisms, the enzyme Serine hydroxymethyltransferase catalyses this transformation by removing one carbon atom; pyridoxal phosphate is also necessary:

$$\text{Serine} + \text{tetrahydrofolate} \rightarrow \text{Glycine} + N^5,N^{10}\text{-Methylene tetrahydrofolate} + H_2O$$

In the liver of vertebrates, glycine synthesis is catalyzed by glycine synthase (also called glycine cleavage enzyme). This conversion is readily reversible:

$$CO_2 + NH_4^+ + N^5,N^{10}\text{-Methylene tetrahydrofolate} + \text{NADH} + H^+ \rightarrow \text{Glycine} + \text{tetrahydrofolate} + \text{NAD}^+$$

Degradation

Glycine is degraded via three pathways. The predominant pathway in animals involves the catalysis of glycine cleavage enzyme, the same enzyme also involved in the biosynthesis of glycine.

The degradation pathway is the reverse of this synthetic pathway:

$$\text{Glycine} + \text{tetrahydrofolate} + \text{NAD}^+ \rightarrow CO_2 + NH_4^+ + N^5, N^{10}\text{-Methylene tetrahydrofolate} + \text{NADH} + H^+$$

In the second pathway, glycine is degraded in two steps. The first step is the reverse of glycine biosynthesis from serine with serine hydroxymethyl transferase. Serine is then converted to pyruvate by serine dehydratase.

In the third pathway of glycine degradation, glycine is converted to glyoxylate by D-amino acid oxidase. Glycoxylate is then oxidized by hepatic lactate dehydrogenase to oxalate in an NAD^+-dependent reaction.

Physiological Function

As a Biosynthetic Intermediate

Glycine is a building block to numerous natural products. In higher eukaryotes, D-Aminolevulinic acid, the key precursor to porphyrins, is biosynthesized from glycine and succinyl-CoA. Glycine provides the central C_2N subunit of all purines.

As a Neurotransmitter

Glycine is an inhibitory neurotransmitter in the central nervous system, especially in the spinal cord, brainstem, and retina. When glycine receptors are activated, chloride enters the neuron via ionotropic receptors, causing an Inhibitory postsynaptic potential (IPSP). Strychnine is a strong antagonist at ionotropic glycine receptors, whereas bicuculline is a weak one. Glycine is a required co-agonist along with glutamate for NMDA receptors. In contrast to the inhibitory role of glycine in the spinal cord, this behaviour is facilitated at the (NMDA) glutaminergic receptors which are excitatory. The LD_{50} of glycine is 7930 mg/kg in rats (oral), and it usually causes death by hyperexcitability.

As a Potential Antipsychotic

Dr. Daniel Javitt a clinical researcher had studied people who were addicted to PCP ("angel dust") and Ketamine ("special K"), (Javitt, DC, Negative Schizophrenic Symptomatology and the Phencyclydine (PCP) Model of Schizophrenia, Hillside Journal of Psychiatry 1987 9:12-35). Their brains had been damaged by the use of this drug. In studies, it was found that their glutamate receptors had been damaged. Since use of PCP and ketamine creates psychosis similar to schizophrenia, it was hypothesized that giving glycine to people with schizophrenia would potentially reduce their psychotic symptoms. In a controlled study people with schizophrenia who were given glycine had their symptoms reduced in a measurable sense, primarily in the area of negative and cognitive symptoms when used as an adjunct to current antipsychotics. There have been some psychiatrists who have used it out of study as a primary antipsychotic with benefits on positive as well as negative and cognitive symptoms. Glycine's primary drawback is its required use in powdered format. However, as an NMDA receptor modulator, it is part of a class of antipsychotics in study that do not cause tardive dyskinesia or diabetes, the current long term side effects of dopaminergic antipsychotics as well as not creating extrapyramidal side effects (movement disorders), weight gain or sedation. These medications along with other new classes of medications in study may eventually replace the current antipsychotics which, from Thorazine to Abilify, have all been based on the dopamine hypothesis and in depleting the levels of dopamine create tardive dykinesia and other Parkinsonian movement disorders and potentially tardive psychosis which is still in study.

Glycine, is part of a promising new class of treatment for schizophrenia that may promote a full recovery without debilitating physical side effects. Research continues on glycine and other NMDA receptors modulates, which are currently in Phase II FDA Study. Although other NMDA receptor modulates in study have shown promise as primary antipsychotics, in the specific controlled studies glycine is used as an adjunct antipsychotic and although available should only be taken under the supervision of a psychiatrist.

Industrial Uses

Glycine is used as a sweetener/taste enhancer, buffering agent in antiperspirants based on zirconium/aluminum chlorohydrate (ZACH) salts complexed with glycine (aminoethanoic acid), which buffers ZACH salts without hindering performance, reabsorbable amino acid, chemical intermediate, metal complexing agent, and dietary supplement as well as in certain pharmaceuticals.

Presence in the Interstellar Medium

In 1994 a team of astronomers at the University of Illinois, led by Lewis Snyder, claimed that they had found the glycine molecule in space. It turned out that, with further analysis, this claim could not be confirmed. Nine years later, in 2003, Yi-Jehng Kuan from National Taiwan Normal University and Steve Charnley claimed that they detected interstellar glycine toward three sources in the interstellar medium.[13] They claimed to have identified 27 spectral lines of glycine utilizing a radio telescope. According to computer simulations and lab-based experiments, glycine was probably formed when ices containing simple organic molecules were exposed to ultraviolet light.

In October 2004, Snyder and collaborators reinvestigated the glycine claim in Kuan *et al.* (2003). In a rigorous attempt to confirm the detection, Snyder showed that glycine was not detected in any of the three claimed sources.

In 2008, the glycine like molecule amino acetonitrile was discovered in the Large Molecule Heimat, a giant gas cloud near the galactic center in the constellation Sagittarius by the Max Planck Institute for Radio Astronomy

Should the glycine claim be substantiated, the finding would not prove that life exists outside the Earth, but certainly makes that possibility more plausible by showing that amino acids can be formed in the interstellar medium.

Histidine is one of the 20 standard amino acids present in proteins. In the nutritional sense, in humans, histidine is considered an essential amino acid, but mostly only in children. Its codons are CAU and CAC. Histidine was first isolated by German physician Albrecht Kossel in 1896.

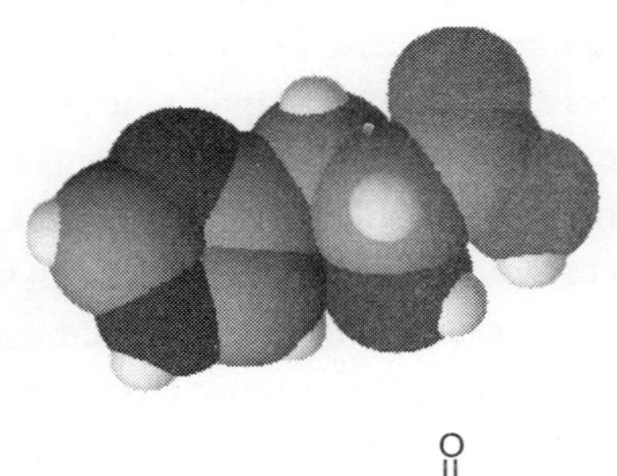

Chemical Properties

The imidazole side chains and the relatively neutral pKa of histidine (ca 6.0) mean that relatively small shifts in cellular pH will change its charge. For this reason, this amino acid side chain finds its way into considerable use as a coordinating ligand in metalloproteins, and also as a catalytic site in certain enzymes. The imidazole side chain has two nitrogens with different properties: One is bound to hydrogen and donates its lone pair to the aromatic ring and as such is slightly acidic, whereas the other one donates only one electron to the ring so it has a free lone pair and is basic. These properties are exploited in different ways in proteins. In catalytic triads, the basic nitrogen of histidine is used to abstract a proton from serine, threonine or cysteine to activate it as a nucleophile. In a histidine proton shuttle, histidine is used to quickly shuttle protons, it can do this by abstracting a proton with its basic nitrogen to make a positively-charged intermediate and then use another molecule, a buffer, to extract the proton from its acidic nitrogen. In carbonic anhydrases, a histidine proton shuttle is utilized to rapidly shuttle protons away from a zinc-bound water molecule to quickly regenerate the active form of the enzyme.

Histidine can be aromatic. When it is deprotonated, and uncharged, it is not aromatic. It no longer obeys Huckel's rule because 8 electrons are in the ring system (an extra two from the deprotonated nitrogen). Histidine obeys Huckel's rule when it is protonated, so then is aromatic. It can form stacking interactions, but is complicated by the positive charge. It doesn't absorb at 280nm in either state, but does in the lower UV range more than some amino acids.

Metabolism

The amino acid is a precursor for histamine and carnosine biosynthesis.

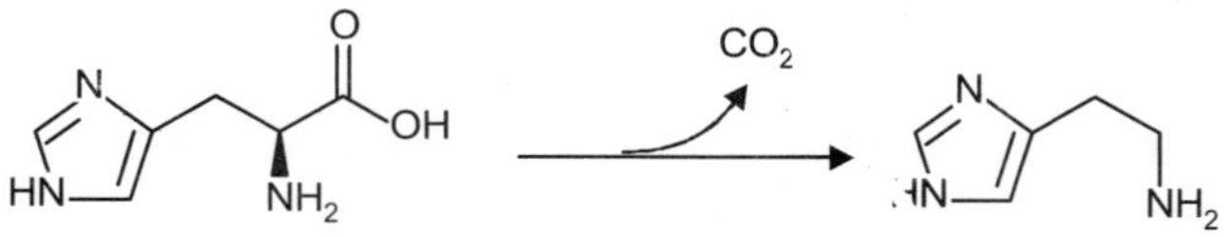

CO_2

Conversion of **histidine** to histamine by histidine decarboxylase

The enzyme histidine ammonia-lyase converts histidine into ammonia and urocanic acid. A deficiency in this enzyme is present in the rare metabolic disorder histidinemia.

Supplementation

Supplementation of Histidine has been shown to cause rapid zinc excretion in rats with an excretion rate 3 to 6 times normal. Some people take histidine in an attempt to raise their histamine levels. Histamine is released under conditions of higher osmolality (dehydration/salt). Raising histamine may be more effective by adapting the body to a higher osmolality by eating sodium without greatly increasing water intake and taking supplements that help the body adapt to those conditions (zinc/taurine/B1/B12). Histidine supplementation likely tricks the body into believing it is producing less histamine and assumes it is well or overhydrated and is better off without the nutrients used to maintain cellular hydration in a hyperosmotic state. Histidine supplementation however is inadvisable because one of the hallmarks of today's chronic diseases (diabetes, Alzheimer's, Parkinson's) is zinc depletion.

The histidine bound heme group of succinate dehydrogenase, an electron carrier in the mitochondrial electron transfer chain. The large semi-transparent sphere indicates the location of the iron ion. From PDB 1YQ3.

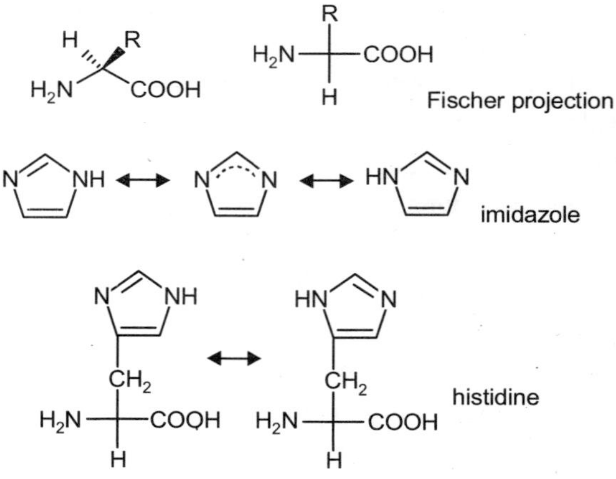

Fischer projection

imidazole

histidine

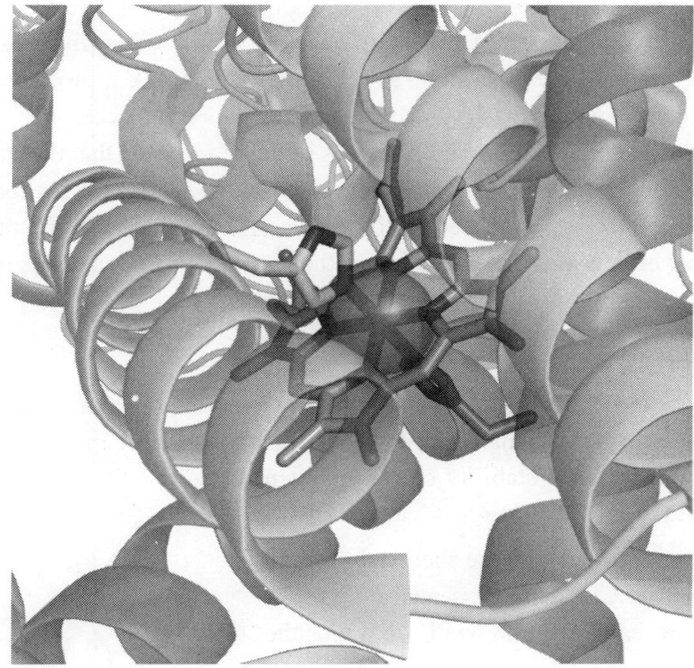

Fig. 2.46 : The histidine bound heme group of succinate dehydrogenase, an electron carrier in the mitochondrial electron transfer chain. The large semi-transparent sphere indicates the location of the iron ion. From PDB 1YQ3

4-Hydroxyproline, or hydroxyproline ($C_5H_9O_3N$), is an uncommon amino acid, abbreviated as **HYP.**

Structure

Hydroxyproline differs from proline by the presence of a hydroxyl (OH) group attached to the Cγ-atom.

Other hydroxyprolines also exist in nature, notably 2,3-cis-3,4-trans-3,4-dihydroxyproline which occurs in diatom cell walls and is postulated to have a role in silica deposition. Hydroxyproline is also found in the walls of oomycetes, fungus-like protists related to diatoms.

Production and Function

Hydroxyproline is produced by hydroxylation of the amino acid proline by the enzyme prolyl hydroxylase following protein synthesis (as a post-translational modification). The enzyme catalysed reaction takes place in the lumen of the endoplasmic reticulum.

Hydroxyproline is a major component of the protein collagen. Hydroxyproline and proline play key roles for collagen stability. They permit the sharp twisting of the collagen helix. In the canonical collagen Xaa-Yaa-Gly triad (where Xaa and Yaa are any amino acid), a proline occupying the Yaa position is hydroxylated to give a Xaa-Hyp-Gly sequence. This modification of the proline residue increases the stability of the collagen triple helix. It was initially proposed that the stabilization was due to water molecules forming a hydrogen bonding network linking the prolyl hydroxyl groups and the main-chain carbonyl groups. It was subsequently shown that the increase in stability is primarily through stereoelectronic effects and that hydration of the hydroxyproline residues provides little or no additional stability.

Clinical Significance

Proline hydroxylation requires ascorbic acid. The most obvious, first effects (gum and hair problems) of absence of ascorbic acid in humans come from the resulting defect in hydroxylation of proline residues of collagen, with reduced stability of the collagen molecule causing scurvy.

Isoleucine is an α-amino acid with the chemical formula $HO_2CCH(NH_2)CH(CH_3)CH_2CH_3$. It is an essential amino acid, which means that humans cannot synthesize it, so it must be part of our diet. Its codons are AUU, AUC and AUA.

With a hydrocarbon side chain, isoleucine is classified as a hydrophobic amino acid. Together with threonine, isoleucine is one of two common amino acids that have a chiral side chain. Four stereoisomers of isoleucine are possible, including two possible diastereomers of L-isoleucine. However, isoleucine present in nature exists in one enantiomeric form, (2*S*,3*S*)-2-amino-3-methylpentanoic acid.

CH_3 O H_3C OH NH_2

Biosynthesis

As an essential amino acid, isoleucine is not synthesized in animals, hence it must be ingested, usually as a component of proteins. In plants and microorganisms, it is synthesized via several steps, starting from pyruvic acid and alpha-ketoglutarate.

Enzymes involved in this biosynthesis include:

1. Acetolactate synthase (also known as acetohydroxy acid synthase).
2. Acetohydroxy acid isomeroreductase.
3. Dihydroxyacid dehydratase.
4. Valine aminotransferase.

Isomers of Isoleucine

Forms of Isoleucine							
Common name	isoleucine	D-isoleucine	L-isoleucine	DL-isoleucine	allo-D-isoleucine	allo-L-isoleucine	allo-DL-isoleucine
Synonyms:		(R)-Isoleucine	L(+)-Isoleucine	(R*,R*)-isoleucine		alloisoleucine	
PubChem:	CID 791	CID 94206	CID 6306	CID 76551			
EINECS number:	207-139-8	206-269-2	200-798-2		216-143-9	216-142-3	221-464-2
CAS number:	443-79-8	319-78-8	73-32-5		1509-35-9	1509-34-8	3107-04-8

Synthesis

Isoleucine can be synthesized in a multistep procedure starting from 2-bromobutane and diethylmalonate. Synthetic isoleucine was originally reported in 1905.

Leucine (abbreviated as **Leu** or **L**) is an α-amino acid with the chemical formula $HO_2CCH(NH_2)CH_2CH(CH_3)_2$. It is an essential amino acid, which means that humans cannot synthesise it. Its codons are UUA, UUG, CUU, CUC, CUA, and CUG. With a hydrocarbon side chain, leucine is classified as a hydrophobic amino acid. It has an isobutyl R group. Leucine is a major component of the sub units in ferritin, astacin and other 'buffer' proteins.

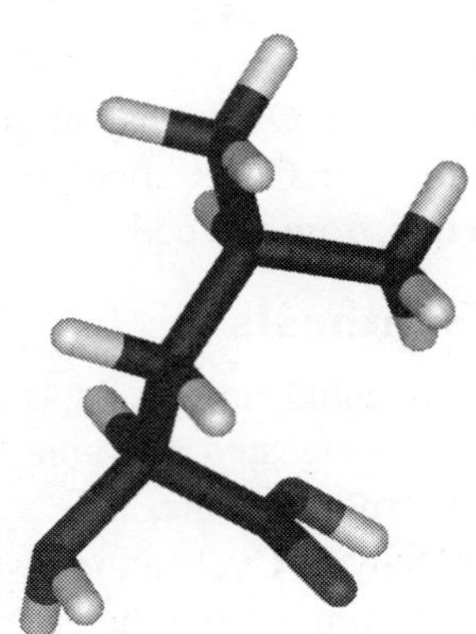

Biosynthesis

As an essential amino acid, leucine is not synthesized in animals, hence it must be ingested, usually as a component of proteins. It is synthesized in plants and microorganisms via several steps starting from pyruvic acid. The initial part of the pathway also leads to valine. The intermediate α-ketovalerate is converted to α-isopropylmalate and then β-isopropylmalate, which is dehydrogenated to α-ketoisocaproate, which in the final step undergoes reductive amination.

Enzymes involved in a typical leucine biosynthesis include:

- Acetolactate synthase,
- Acetohydroxy acid isomeroreductase,
- Dihydroxyacid dehydratase,
- α-Isopropylmalate synthase,
- α-Isopropylmalate isomerase,
- Leucine aminotransferase.

Biological Uses

As a dietary supplement, leucine has been found to slow the degradation of muscle tissue by increasing the synthesis of muscle proteins. Leucine is utilized in the liver, adipose tissue, and muscle tissue. In adipose and muscle tissue, leucine is used in the formation of sterols, and the combined usage of leucine in these two tissues is seven times greater than its use in the liver.

Chemical Properties

Racemic leucine had been subjected to circularly polarized synchrotron radiation in order to better understand the origin of biomolecular asymmetry. An enantiomeric enhancement of 2.6 % had been induced, indicating a photochemical origin of biomolecules' homochirality.

Food Additive

As a food additive L-Leucine has E number **E641** and is classified as a flavour enhancer

Lysine (abbreviated as **Lys** or **K**) is an α-amino acid with the chemical formula $HO_2CCH(NH_2)(CH_2)_4NH_2$. This amino acid is an essential amino acid, which means that humans cannot synthesize it. Its codons are AAA and AAG.

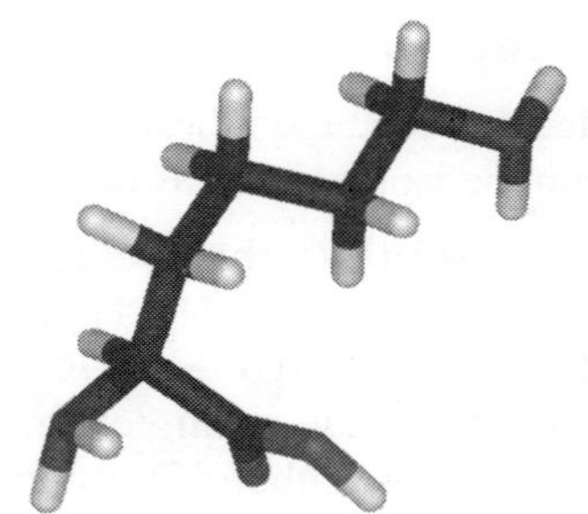

Lysine is a base, as are arginine and histidine. The ε-amino group often participates in hydrogen bonding and as a general base in catalysis. Common posttranslational modifications include methylation of the ε-amino group, giving methyl-, dimethyl-, and trimethyllysine. The latter occurs in calmodulin. Other posttranslational modifications at lysine residues include acetylation and ubiquitination. Collagen contains hydroxylysine which is derived from lysine by lysyl hydroxylase. *O*-Glycosylation of lysine residues in the endoplasmic reticulum or Golgi apparatus is used to mark certain proteins for secretion from the cell.

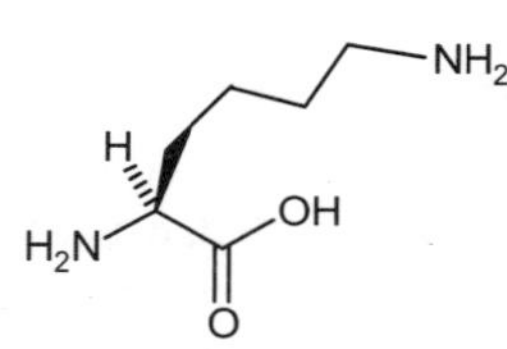

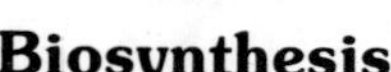

Biosynthesis

As an essential amino acid, lysine is not synthesized in animals, hence it must be ingested as lysine or lysine-containing proteins. In plants and microorganisms, it is synthesized from aspartic acid, which is first converted to β-aspartyl-semialdehyde. Cyclization gives dihydropicolinate, which is reduced to Δ^1-piperidine-2,6-dicarboxylate. Ring-opening of this heterocycle gives a series of derivatives of pimelic acid, ultimately affording lysine.

Enzymes involved in this biosynthesis include:

1. Aspartokinase
2. β-aspartate semialdehyde dehydrogenase
3. Dihydropicolinate synthase
4. Δ^1-piperidine-2,6-dicarboxylate dehydrogenase
5. *N*-succinyl-2-amino-6ketopimelate synthase
6. Succinyl diaminopimelate aminotransferase
7. Succinyl diaminopimelate desuccinylase

8. Diaminopimelate epimerase
9. Diaminopimelate decarboxylase

Metabolism

Lysine is metabolised in mammals to give acetyl-CoA, via an initial transamination with α-ketoglutarate. The bacterial degradation of lysine yields cadaverine by decarboxylation.

Synthesis

Synthetic, racemic lysine has long been known. A practical synthesis starts from caprolactam.

Dietary Sources

The human nutritional requirement is 1–1.5 g daily. It is the limiting amino acid (the essential amino acid found in the smallest quantity in the particular foodstuff) in all cereal grains, but is plentiful in all pulses (legumes). Plants that contain significant amounts of lysine include:

- Buffalo Gourd (10,130–33,000 ppm) in seed
- Berro, Watercress (1,340–26,800 ppm) in herb.
- Soybean (24,290–26,560 ppm) in seed.
- Carob, Locust Bean, St.John's-Bread (26,320 ppm) in seed;
- Common Bean (Black Bean, Dwarf Bean, Field Bean, Flageolet Bean, French Bean, Garden Bean, Green Bean, Haricot, Haricot Bean, Haricot Vert, Kidney Bean, Navy Bean, Pop Bean, Popping Bean, Snap Bean, String Bean, Wax Bean) (2,390–25,700 ppm) in sprout seedling;
- Ben Nut, Benzolive Tree, Jacinto (Sp.), Moringa (aka Drumstick Tree, Horseradish Tree, Ben Oil Tree), West Indian Ben (5,370–25,165 ppm) in shoot.
- Lentil (7,120–23,735 ppm) in sprout seedling.
- Asparagus Pea, Winged Bean (aka Goa Bean) (21,360–23,304 ppm) in seed.
- Fat Hen (3,540–22,550 ppm) in seed.
- Lentil (19,570–22,035 ppm) in seed.
- White Lupin (19,330–21,585 ppm) in seed.
- Black Caraway, Black Cumin, Fennel-Flower, Nutmeg-Flower, Roman Coriander (16,200–20,700 ppm) in seed.
- Spinach (1,740–20,664 ppm).
- Amaranth, Quinoa
- Kasha

Good sources of lysine are foods rich in protein including meat (specifically red meat, lamb, pork, and poultry), cheese (particularly Parmesan), certain fish (such as cod and sardines), and eggs.

Properties

L-Lysine is a necessary building block for all protein in the body. L-Lysine plays a major role in calcium absorption; building muscle protein; recovering from surgery or sports injuries; and the body's production of hormones, enzymes, and antibodies.

Modifications

Lysine can be modified through acetylation, methylation, ubiquitination, sumoylation, neddylation, biotinylation and carboxylation which tends to modify the function of the protein of which the modified lysine residue(s) are a part.

Clinical Significance

It has been suggested that lysine may be beneficial for those with herpes simplex infections. However, more research is needed to fully substantiate this claim. For more information, refer to Herpes simplex - Lysine.

There are Lysine conjugates that show promise in the treatment of cancer, by causing cancerous cells to destroy themselves when the drug is combined with the use of phototherapy, while leaving non-cancerous cells unharmed.

In Popular Culture

The 1993 film *Jurassic Park*, which is based on the 1990 Michael Crichton novel *Jurassic Park*, features dinosaurs that were genetically altered so they could not produce lysine. This was supposed to prevent the cloned dinosaurs from leaving the park, forcing them to depend on lysine supplements provided by the park's veterinary staff. Most vertebrates cannot produce lysine by default (it is an essential amino acid).

Linus Pauling believed people with existing atherosclerotic plaque can benefit from vitamin c and the amino acid lysine. Pauling filmed a video lecture in which he recommended that heart patients take between 2,000 and 6,000 mg of lysine daily with their vitamin C (more if serum Lp(a) is elevated).

In the sitcom The Big Bang Theory, the episode The Friendship Algorythm (air date 1/19/2009), it is revealed that Sheldon's favourite amino acid is Lysine. Although Raj had thought it was l-glutamine.

Methionine (abbreviated as **Met** or **M**) is an α-amino acid with the chemical formula $HO_2CCH(NH_2)CH_2CH_2SCH_3$. This Essential amino acid is classified as nonpolar. Together with cysteine, methionine is one of two sulfur-containing proteinogenic amino acids. Its derivative S-adenosyl methionine (SAM) serves as a methyl donor. Methionine is an intermediate in the biosynthesis of cysteine, carnitine, taurine, lecithin, phosphatidylcholine, and other phospholipids. Improper

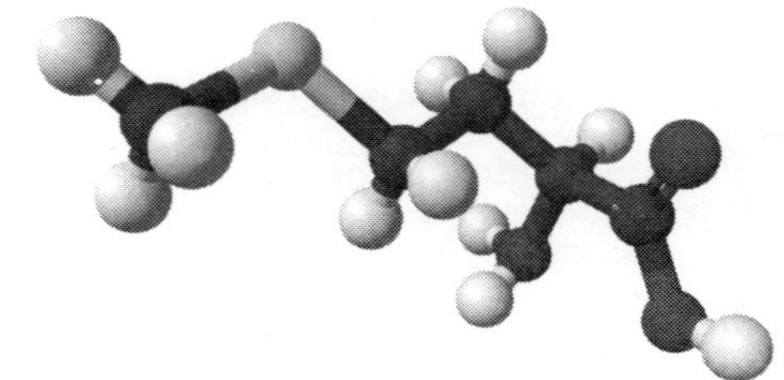

conversion of methionine can lead to atherosclerosis. Methionine is one of only two amino acids encoded by a single codon (AUG) in the standard genetic code (tryptophan, encoded by UGG, is the other). The codon AUG is also the "Start" message for a ribosome that signals the initiation of protein translation from mRNA. As a consequence, methionine is incorporated into the N-terminal position of all proteins in eukaryotes and archaea during translation, although it is usually removed by post-translational modification.

Biosynthesis

Aspartate → (1; ATP → ADP) → Aspartyl-β–phosphate → (2; NADPH+H^+ → $NADP^+$+P_i) → Aspartyl-β–phosphate

β-aspartate-semialdehyde → (3; NADPH+H^+ → $NADP^+$) → homoserine → (4; Succinyl-CoA → CoA-SH) → O-Succinylhomoserine

O-Succinylhomoserine → (5; Cyseine → Succinate) → Cystathionine → (6; → Pyruvate + NH_4^+) → Homocysteine

Homocysteine → (7; N^5-methyl THF → THF) → Methionine

As an essential amino acid, methionine is not synthesized in humans, hence we must ingest methionine or methionine-containing proteins. In plants and microorganisms, methionine is synthesized via a pathway that uses both aspartic acid and cysteine. First, aspartic acid is converted via β-aspartyl-semialdehyde into homoserine, introducing the pair of contiguous methylene groups. Homoserine converts to *O*-succinyl homoserine, which then reacts with cysteine to produce cystathionine, which is cleaved to yield homocysteine. Subsequent methylation of the thiol group by folates affords methionine. Both cystathionine-γ-synthase and cystathionine-β-lyase require Pyridoxyl-5'-phosphate as a cofactor, whereas homocysteine methyltransferase requires Vitamin B12 as a cofactor.

Enzymes involved in methionine biosynthesis:

1. aspartokinase
2. β-aspartate semialdehyde dehydrogenase
3. homoserine dehydrogenase
4. homoserine acyltransferase
5. cystathionine-γ-synthase
6. cystathionine-β-lyase
7. methionine synthase (in mammals, this step is performed by homocysteine methyltrans ferase)

Other Biochemical Pathways

Although mammals cannot synthesize methionine, they can still utilize it in a variety of biochemical pathways:

Methionine is converted to S-adenosylmethionine (SAM) by: (1) methionine adenosyltransferase. SAM serves as a methyl-donor in many; (2) methyltransferase reactions and is converted to S-adenosylhomocysteine (SAH); (3) adenosylhomocysteinase converts SAH to homocysteine.

There are two fates of homocysteine:

- Methionine can be regenerated from homocysteine via methionine synthase: It can also be remethylated using glycine betaine (NNN-trimethyl glycine) to methionine via the enzyme Betaine-homocysteine methyltransferase (E.C.2.1.1.5, BHMT). BHMT makes up to 1.5% of all the soluble protein of the liver, and recent evidence suggests that it may have a greater influence on methionine and homocysteine homeostasis than methionine synthase.
- Homocysteine can be converted to cysteine: Cystathionine-β-synthase (a PLP-dependent enzyme) combines homocysteine and serine to produce cystathionine. Instead of degrading cystathionine via cystathionine-β-lyase, as in the biosynthetic pathway, cystathionine is broken down to cysteine and α-ketobutyrate via cystathionine-γ-lyase. α-ketoacid dehydrogenase converts α-ketobutyrate to propionyl-CoA, which is metabolized to succinyl-CoA in a three-step process.

PROPIONYL-COA, PATHWAY

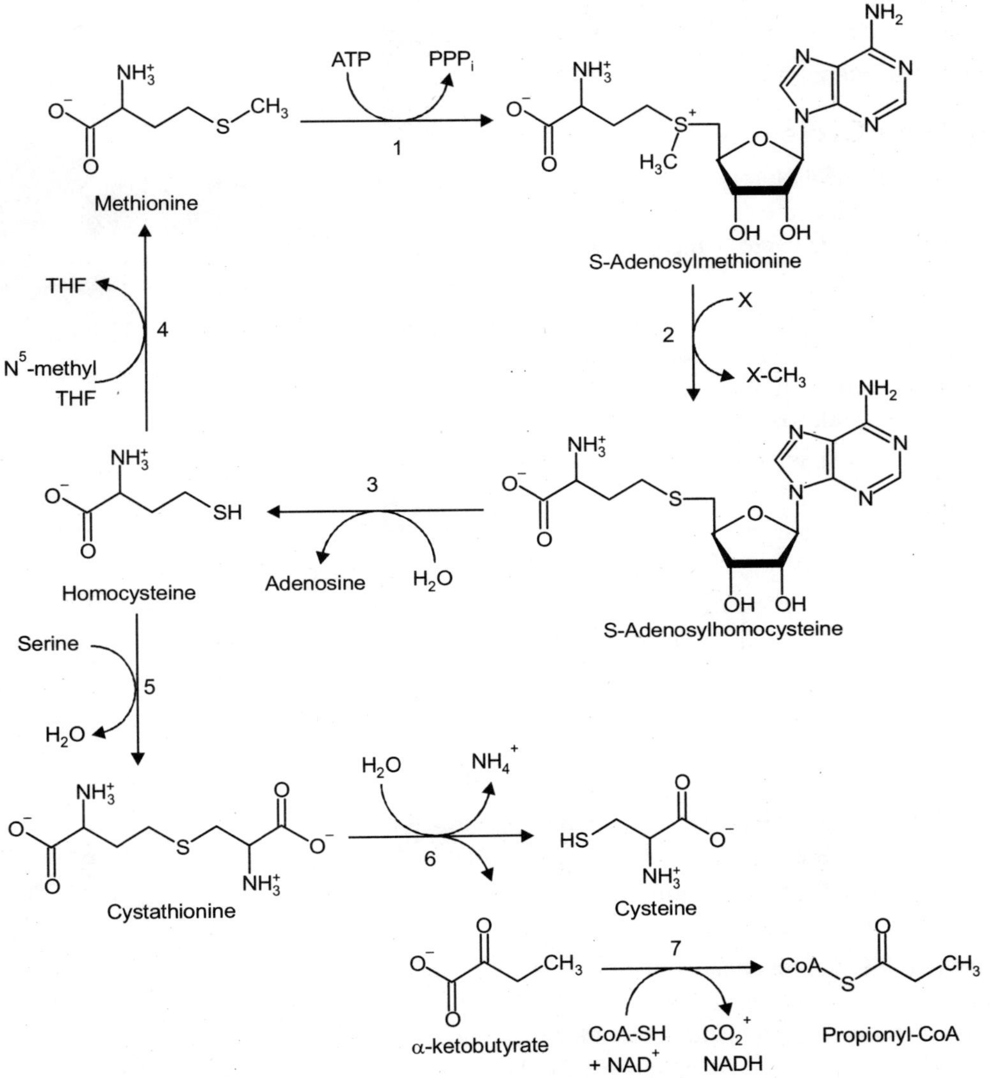

Synthesis

Racemic methionine can be synthesized from diethyl sodium phthalimidomalonate by alkylation with chloroethylmethylsulfide ($ClCH_2CH_2SCH_3$) followed by hydrolysis and decarboxylation.

Dietary Aspects

High levels of methionine can be found in sesame seeds, Brazil nuts, fish, meats, and some other plant seeds. Most fruits and vegetables contain very little of it; however, some have significant amounts, such as spinach, potatoes, and boiled corn. Most legumes, though high in protein, are also low in methionine. DL-methionine is sometimes added as an ingredient to pet foods. Methionine, cysteine, and soy protein heated in a small amount of water creates a meat-like aroma.

Ornithine is an amino acid which plays a role in the urea cycle.

Role in Urea Cycle

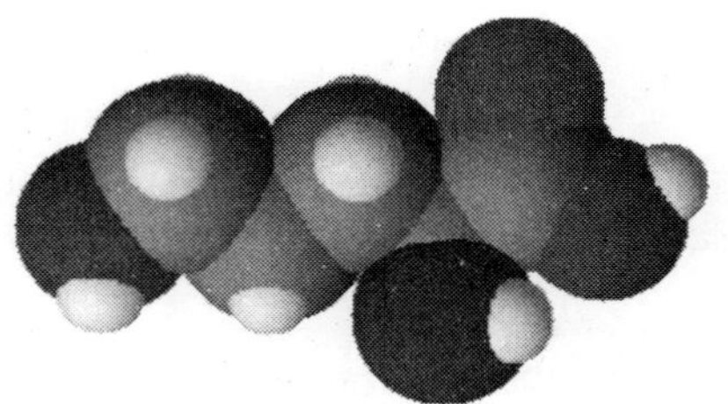

L-Ornithine is one of the products of the action of the enzyme arginase on L-arginine, creating urea. Therefore, ornithine is a central part of the urea cycle, which allows for the disposal of excess nitrogen.

Ornithine is not an amino acid coded for by DNA, and, in that sense, is not involved in protein synthesis. However, in mammalian non-hepatic tissues, the main use of the urea cycle is in arginine biosynthesis, so as an intermediate in metabolic processes, ornithine is quite important.

Lactamization

Ornithine lactamization

It is believed to not be a part of genetic code because polypeptides containing unprotected ornithines undergo spontaneous lactamization. This proved to be a problem when ornithine was artificially incorporated in 21st amino acid systems.

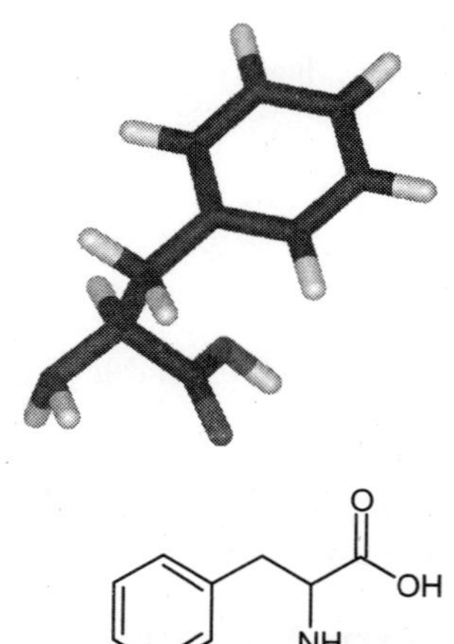

Phenylalanine (abbreviated as Phe or F) is an α-amino acid with the formula $HO_2CCH(NH_2)CH_2C_6H_5$, which is found naturally in the breast milk of mammals and manufactured for food and drink products and are also sold as nutritional supplements for their reputed analgesic and antidepressant effects. Phenylalanine is structurally closely related to dopamine, epinepherine (adrenaline) and tyrosine.

This essential amino acid is classified as nonpolar because of the hydrophobic nature of the benzyl side chain. The codons for L-phenylalanine are UUU and UUC. It is a white, powdery solid. L-Phenylalanine (LPA) is an electrically-neutral amino acid, one of the twenty common amino acids used to biochemically form proteins, coded for by DNA.

The most important consideration for the layman is to know that Phenylalanine has a laxative effect, it also gives you gas for extended periods. It is used in chewing gum such as Wrigleys Extra Fusion - a warning is displayed on the packet in very small text.

Other Biological Roles

L-phenylalanine can also be converted into L-tyrosine, another one of the DNA-encoded amino acids. L-tyrosine in turn is converted into L-DOPA, which is further converted into dopamine, norepinephrine (noradrenaline), and epinephrine (adrenaline). The latter three are known as the catecholamines.

Phenylalanine uses the same active transport channel as tryptophan to cross the blood-brain barrier, and, in large quantities, interferes with the production of serotonin.

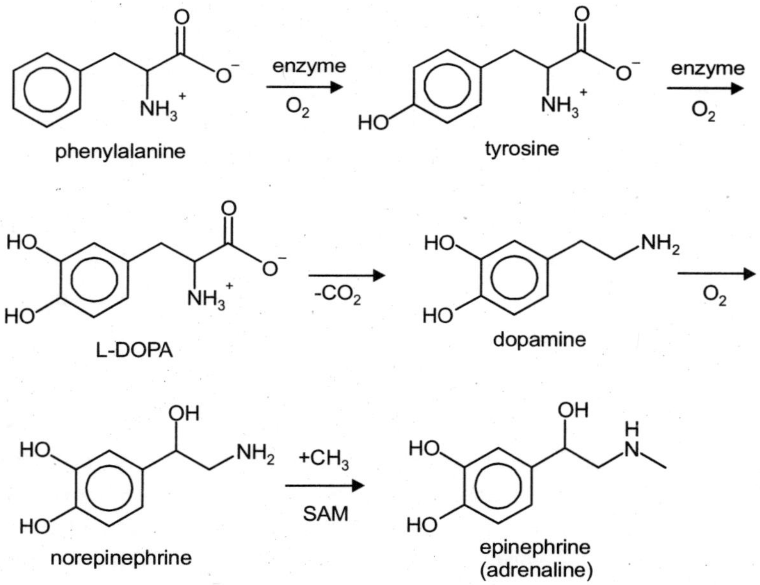

Lignan is derived from phenylalanine and from tyrosine. Phenylalanine is converted to cinnamic acid by the enzyme phenylalanine ammonia lyase.

Phenylketonuria

The genetic disorder phenylketonuria (PKU) is the inability to metabolize phenylalanine. Individuals with this disorder are known as "phenylketonurics" and must abstain from consumption of too much phenylalanine. This dietary restriction also applies to pregnant women with hyperphenylalanine (high levels of phenylalanine in blood) because they do not properly metabolize the amino acid phenylalanine. Persons suffering from PKU must monitor their intake of protein to control the buildup of phenylalanine as their bodies convert protein into its component amino acids.

A non food source of phenylalanine is the artificial sweetener aspartame. This compound, sold under the trade names "Equal" and "NutraSweet", is metabolized by the body into several chemical byproducts including phenylalanine. The breakdown problems phenylketonurics have with protein and the attendant build up of phenylalanine in the body also occurs with the ingestion of aspartame, although to a lesser degree. Accordingly, all products in Australia, the U.S. and Canada that contain aspartame must be labelled:

"Phenylketonurics: Contains phenylalanine." In the UK, foods containing aspartame must carry ingredients panels that refer to the presence of "aspartame or E951" and they must be labelled with a warning "Contains a source of phenylalanine." These warnings are specifically placed to aid individuals who suffer from PKU so that they can avoid such foods.

In Non-Humans

Geneticists have recently sequenced the genome of macaques. Their investigations have found "some instances where the normal form of the macaque protein looks like the diseased human protein" including markers for PKU.

D- and DL-phenylalanine

The unnatural stereoisomer D-phenylalanine (DPA) can be produced by conventional organic synthesis, either as a single enantiomer or as a component of the racemic mixture. It does not participate in protein biosynthesis although it is found in proteins in small amounts - particularly aged proteins and food proteins that have been processed. The biological functions of D-amino acids remain unclear although some, such as D-phenylalanine, may have pharmacological activity.

DL-Phenylalanine (DLPA) is marketed as a nutritional supplement for its supposed analgesic and antidepressant activities. The reputed analgesic activity of DL-phenylalanine may be explained by the possible blockage by D-phenylalanine of enkephalin degradation by the enzyme carboxypeptidase A. The mechanism of DL-phenylalanine's supposed antidepressant activity may be accounted for by the precursor role of L-phenylalanine in the synthesis of the neurotransmitters, norepinephrine and dopamine. Elevated brain levels of norepinephrine and dopamine are thought to have an antidepressant effect. Following ingestion, D-Phenylalanine is absorbed from the small intestine and transported to the liver via the portal circulation. A small amount of D-phenylalanine appears to be converted to L-phenylalanine. D-Phenylalanine is distributed to the various tissues of the body via the systemic circulation. It appears to cross the blood-brain barrier less efficiently than L-phenylalanine, and so a small amount of an ingested dose of D-phenylalanine is not absorbed but excreted in the urine.

History

The genetic codon for phenylalanine was first discovered by J. Heinrich Matthaei and Marshall W. Nirenberg in 1961. They showed that by using *m*-RNA to insert multiple uracil repeats into the bacterium *E. coli*, the bacterium produced a new protein consisting solely of repeated phenylalanine amino acids. This discovery lead to the determination of the relationship between RNA and amino acids, which was foundational to the understanding of the Genetic Code.

Phenylalanine is contained in many foodstuffs and byproducts. Many products are advertised as containing this ingredient including the soft drink 'lilt', 'Diet 7-Up', 'Tango', '5-Hour Energy', 'Caffeine Free Diet Pepsi', 'Stride' Gum, 'Red Bull Sugar free', 'Fisherman's Friend', 'Diet Coca-Cola' and others.

Proline (abbreviated as **Pro** or **P**) is an α-amino acid, one of the twenty DNA-encoded amino acids. Its codons are CCU, CCC, CCA, and CCG. It is not an essential amino acid, which means that humans can synthesize it. It is unique among the 20 protein-forming amino acids because the α-amino group is secondary.

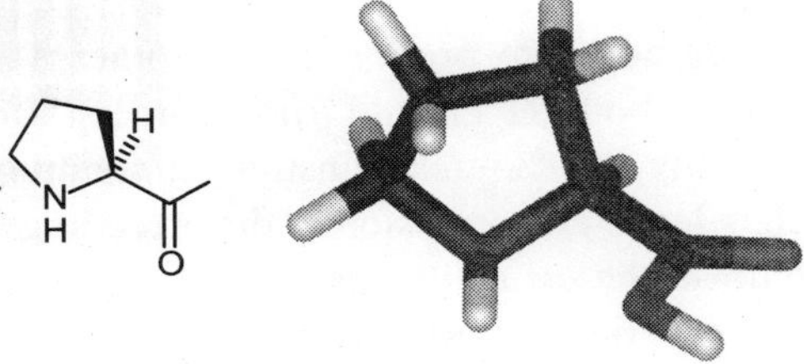

History

Hermann Emil Fischer discovered Proline between 1899 and 1908.

Biosynthesis

Proline is biosynthetically derived from the amino acid L-glutamate and its immediate precursor is the imino acid (*S*)-1-pyrroline-5-carboxylate (P5C). Enzymes involved in a typical biosynthesis include:

1. Glutamate kinase (ATP-dependent).
2. Glutamate dehydrogenase (requires NADH or NADPH).
3. Pyrroline-5-carboxylate reductase (requires NADH or NADPH).

Structural Properties

The distinctive cyclic structure of proline's side chain locks its φ backbone dihedral angle at approximately -75°, giving proline an exceptional conformational rigidity compared to other amino acids. Hence, proline loses less conformational entropy upon folding, which may account for its higher prevalence in the proteins of thermophilic organisms. Proline acts as a structural disruptor in the middle of regular secondary structure elements such as alpha helices and beta sheets; however, proline is commonly found as the first residue of an alpha helix and also in the edge strands of beta sneets. Proline is also commonly found in turns, which may account for the curious fact that proline is usually solvent-exposed, despite having a completely aliphatic side chain. Because proline lacks a hydrogen on the amide group, it cannot act as a hydrogen bond donor, only as a hydrogen bond acceptor.

Multiple prolines and/or hydroxyprolines in a row can create a polyproline helix, the predominant secondary structure in collagen. The hydroxylation of proline by prolylhydroxylase (or other additions of electron-withdrawing substituents such as fluorine) increases the conformational stability of collagen significantly. Hence, the hydroxylation of proline is a critical biochemical process for maintaining the connective tissue of higher organisms. Severe diseases such as scurvy can result from defects in this hydroxylation, e.g., mutations in the enzyme prolylhydroxylase or lack of the necessary ascorbate (vitamin C) cofactor.

Sequences of proline and 2-aminoisobutyric acid (Aib) also form a helical turn structure.

In 2006, scientists at ASU discovered that solutions of TiO_2 illuminated with ultraviolet radiation can serve as an extremely cost-effective and accurate protein cleavage catalyst. The TiO_2 catalyst preferentially and rapidly cleaves protein at sites where proline is present, while taking much longer to degrade the protein from its endpoints.

Cis-trans Isomerization

Peptide bonds to proline, and to other *N*-substituted amino acids (such as sarcosine), are able to populate both the *cis* and *trans* isomers. Most peptide bonds overwhelmingly adopt the *trans* isomer (typically 99.9% under unstrained conditions), chiefly because the amide hydrogen (*trans* isomer) offers less steric repulsion to the preceding C^{α} atom than does the following C^{α} atom (*cis* isomer). By contrast, the *cis* and *trans* isomers of the X-Pro peptide bond (where X represents any amino acid) both experience steric clashes with the neighboring substitution and are nearly equal energetically. Hence, the fraction of X-Pro peptide bonds in the *cis* isomer under unstrained conditions ranges from 10-40%; the fraction depends slightly on the preceding amino acid, with aromatic residues favouring the *cis* isomer slightly.

From a kinetic standpoint, *Cis-trans* proline isomerization is a very slow process that can impede the progress of protein folding by trapping one or more proline molecules crucial for folding in the non-native isomer, especially when the native protein requires the *cis* isomer. This is because proline residues are exclusively synthesized in the ribosome as the *trans* isomer form. All organisms possess prolyl isomerase enzymes to catalyze this isomerization, and some bacteria have specialized prolyl isomerases associated with the ribosome. However, not all prolines are essential for folding, and protein folding may proceed at a normal rate despite having non-native conformers of many X-Pro peptide bonds.

Uses

Proline and its derivatives are often used as asymmetric catalysts in organic reactions. The CBS reduction and proline catalysed aldol condensation are prominent examples.

L-Proline is an osmoprotectant and therefore is used in many pharmaceutical, biotechnological applications.

Specialities

Proline is one of the two amino acids that do not follow along with the typical Ramachandran plot, along with glycine. Due to the ring formation connected to the Beta-carbon, the ψ and φ angles about the peptide bond have less allowable degrees of rotation.

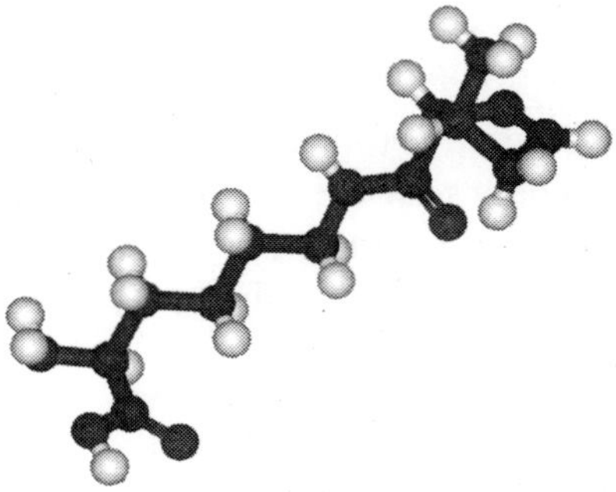

Pyrrolysine is a naturally occurring, genetically coded amino acid used by some methanogenic archaea in enzymes that are part of their methane-producing metabolism.

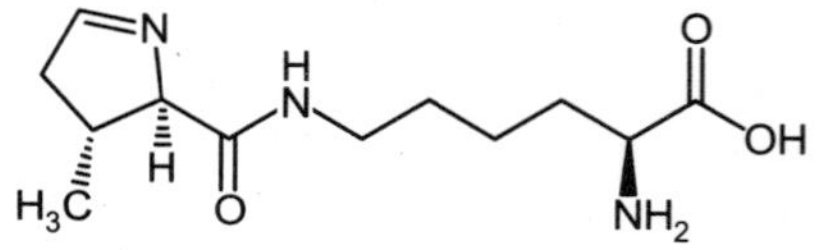

This lysine derivative is encoded by the UAG codon (normally the 'amber' stop codon), possibly modified by the presence of a specific downstream sequence, named PYLIS, which forms a stem-loop in the mRNA, forcing the incorporation of pyrrolysine instead of terminating translation. It is also of interest to note that UAG appears to be used much less often than other stop codons and whenever it is found in an open reading frame it is always followed by one or more of the other two stop codons shortly after.

Near a methyltransferase gene cluster of *Methanosarcina barkeri* is the *pylT* gene, which encodes an unusual transfer RNA (tRNA) with a CUA anticodon. The adjacent *pylS* gene encodes a class II aminoacyl-tRNA synthetase that charges the *pylT*-derived tRNA with pyrrolysine. The operon containing *pylT* and *pylS* are also found in the genomes of other sequenced members of the *Methanosarcinaceae* family. Homologs of *pylS* and *pylT* are found in a Gram-positive bacterium, *Desulfitobacterium hafniense*, although the function of these putative genes in this organism is unknown. It was initially shown that *pylT* encoded tRNA (CUA) can be charged with lysine by PylS. Recently, it has been shown that the tRNA(CUA) can be charged with lysine *in vitro* by the concerted action of the *M. barkeri* Class I and Class II Lysyl-tRNA synthetases. Charging a tRNA(CUA) with lysine was originally hypothesized to be the first step in translating UAG amber codons as pyrrolysine in certain methanogens. The current model based on *in vitro* and *in vivo* data favours direct charging of pyrrolysine on to the tRNA(CUA) by the protein product of the *pylS* gene. This makes pyrrolysine the 22nd genetically encoded natural amino acid. The mechanism of encoding makes it the 21st natural directly encoded amino acid.

The joint nomenclature committee of the IUPAC/IUBMB has officially recommended the three-letter symbol **Pyl** and the one-letter symbol **'O'** for pyrrolysine.

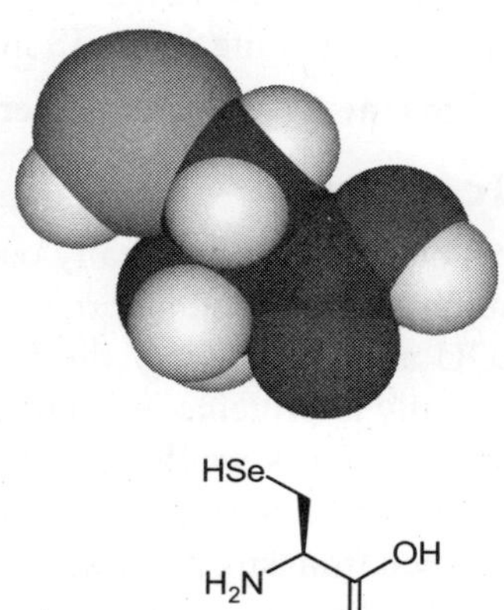

Selenocysteine is an amino acid that is present in several enzymes (for example glutathione peroxidases, tetraiodothyronine 5' deiodinases, thioredoxin reductases, formate dehydrogenases, glycine reductases and some hydrogenases).

Nomenclature:

The joint nomenclature committee of the IUPAC/IUBMB has officially recommended the three-letter symbol **Sec** and the one-letter symbol **U** for selenocysteine.

Structure

Selenocysteine has a structure similar to cysteine, but with an atom of selenium taking the place of the usual sulfur. Proteins that contain one or more selenocysteine residues are called selenoproteins.

Biology

Unlike other amino acids present in biological proteins, Selenocysteine is not coded for directly in the genetic code. Instead, it is encoded in a special way by a UGA codon, which is normally a stop codon. The UGA codon is made to encode selenocysteine by the presence of a SECIS element (SElenoCysteine Insertion Sequence) in the mRNA. The SECIS element is defined by characteristic nucleotide sequences and secondary structure base-pairing patterns. In bacteria, the SECIS element is located immediately following the UGA codon within the reading frame for the selenoprotein. In archaea and in eukaryotes, the SECIS element is in the 3' untranslated region (3' UTR) of the mRNA, and can direct multiple UGA codons to encode selenocysteine residues. When cells are grown in the absence of selenium, translation of selenoproteins terminates at the UGA codon, resulting in a truncated, nonfunctional enzyme.

Like the other amino acids used by cells, selenocysteine has a specialized tRNA. The primary and secondary structure of selenocysteine tRNA, tRNA(Sec), differ from those of standard tRNAs in several respects, most notably in having an 8-base (bacteria) or 10-base (eukaryotes) pair acceptor stem, a long variable region arm, and substitutions at several well-conserved base positions. The selenocysteine tRNAs are initially charged with serine by seryl-tRNA ligase, but the resulting Ser-tRNA(Sec) is not used for translation because it is not recognised by the normal translation factor (EF-Tu in bacteria, EF1-alpha in eukaryotes). Rather, the tRNA-bound seryl residue is converted to a selenothiomise-residue by the pyridoxal phosphate-containing enzyme selenocysteine synthase. Finally, the resulting Sec-tRNA(Sec) is specifically bound to an alternative translational elongation factor (SelB or mSelB) which delivers it in a targeted manner to the ribosomes translating mRNAs for selenoproteins. The specificity of this delivery mechanism is brought about by the presence of an extra protein domain (in bacterial SelB) or an extra subunit (SBP-5 for eukaryotic mSelB) which bind to the corresponding RNA secondary structures formed by the SecIS elements in selenoprotein mRNAs. The SecIS elements of bacterial selenoproteins (as far as analysed) are located within the coding sequences immediately following the UGA codons for selenocysteine, those of Eukarya and Archaea are located in the 3' UTR of the respective mRNAs. In addition, at least one case has been described for an archaeal selenoprotein mRNA containing its SecIS in the 5' UTR.

Serine (abbreviated as **Ser** or **S**) is an organic compound with the formula $HO_2CCH(NH_2)CH_2OH$.

Occurrence

It is one of the 20 naturally occurring proteinogenic amino acids. Its codons are UCU, UCC, UCA, UCG, AGU and AGC. Only the L-stereoisomer appears naturally in proteins. It is not essential to the human diet, since it is synthesized in the body from other metabolites, including glycine. Serine was first obtained from silk protein, a particularly rich source, in 1865. Its name is derived from the Latin for silk, *sericum.* Serine's structure was established in 1902. By virtue of the hydroxyl group, serine is classified as *a polar amino acid.*

Biosynthesis

The biosynthesis of serine starts with the oxidation of 3-phosphoglycerate to 3-phosphohydroxypyruvate and NADH. Reductive amination of this ketone followed by hydrolysis gives serine. Serine hydroxymethyltransferase catalyzes the reversible, simultaneous conversions of L-serine to glycine (retro-aldol cleavage) and 5,6,7,8-tetrahydrofolate to 5,10-methylenetetrahydrofolate (hydrolysis).

Chemical Synthesis

Racemic serine can be prepared from methyl acrylate via several steps. It is also naturally produced when UV light illuminates simple ices such as a combination of water, methanol, hydrogen cyanide, and ammonia, suggesting that it may be easily produced in cold regions of space.

Metabolic Function

Serine is important in metabolism in that it participates in the biosynthesis of purines and pyrimidines. It is also the precursor to several amino acids, including glycine, cysteine, and, in bacteria, tryptophan. It is also the precursor to numerous of other metabolites, including sphingolipids. Serine is also a precursor to folate, which is the principal donor of one carbon fragments in biosynthesis.

Structural Role

Serine plays an important role in the catalytic function of many enzymes. It has been shown to occur in the active sites of chymotrypsin, trypsin, and many other enzymes. The so-called nerve gases and many substances used in insecticides have been shown to act by combining with a residue of serine in the active site of acetylcholine esterase, inhibiting the enzyme completely. The unmetabolized acetylcholine cannot be recycled into the nerve for signaling. This results in depletion of acetylcholine at the neuromuscular junction, resulting in the inability to control muscles, which results in asphyxiation, and death.

As a constituent (residue) of proteins, its side chain can undergo O-linked glycosylation, which may be functionally related to diabetes. It is one of three amino acid residues that are commonly phosphorylated by kinases during cell signaling in eukaryotes. Phosphorylated serine residues are often referred to as phosphoserine. Serine proteases are a common type of protease.

Signaling

D-serine, synthesized by serine racemase from L-serine, serves as a neuronal signal by activating NMDA receptors in the brain.

Threonine (abbreviated as **Thr** or **T**) is an α-amino acid with the chemical formula $HO_2CCH(NH_2)CH(OH)CH_3$. Its codons are ACU, ACA, ACC, and ACG. This essential amino acid is classified as polar. Together with serine and tyrosine, threonine is one of three proteinogenic amino acids bearing an alcohol group.

The threonine residue is susceptible to numerous posttranslational modifications. The hydroxy side chain can undergo *O*-linked glycosylation. In addition, threonine residues undergo phosphorylation through the action of a threonine kinase. In its phosphorylated form, it can be referred to as phosphothreonine.

CH_3 O
HO OH
NH_2

Allo-threonine

With two chiral centers, threonine can exist in four possible stereoisomers, or two possible diastereomers of L-threonine. However, the name L-threonine is used for one single enantiomer, (2*S*,3*R*)-2-amino-3-hydroxybutanoic acid. The second diastereomer (2*S*,3*S*), which is rarely present in nature, is called L-*allo*-threonine.

Biosynthesis

As an essential amino acid, threonine is not synthesized in humans, hence we must ingest threonine in the form of threonine-containing proteins. In plants and microorganisms, threonine is synthesized from aspartic acid via α-aspartyl-semialdehyde and homoserine. Homoserine undergoes *O*-phosphorylation; this phosphate ester undergoes hydrolysis concomitant with relocation of the OH group.

Enzymes involved in a typical biosynthesis of threonine include:

1. aspartokinase.
2. α-aspartate semialdehyde dehydrogenase.
3. homoserine dehydrogenase.
4. homoserine kinase.
5. threonine synthase.

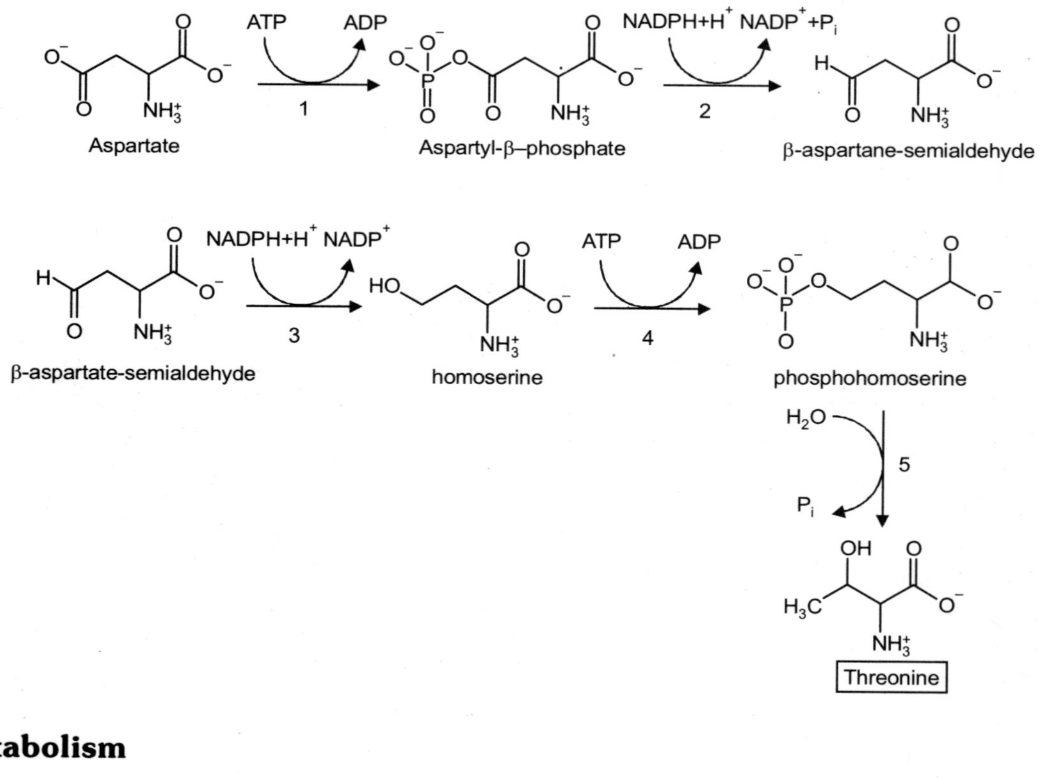

ATP ADP 1 NADPH+H$^+$ NADP$^+$+P$_i$ 2

NH_3^+ Aspartate — Aspartyl-β–phosphate — β-aspartane-semialdehyde

NADPH+H$^+$ NADP$^+$ 3 ATP ADP 4

β-aspartate-semialdehyde — homoserine — phosphohomoserine

H_2O 5 P_i

OH O H_3C NH_3^+

Threonine

Metabolism

Threonine is metabolized in two ways:

- It is converted to pyruvate via threonine dehydrogenase. An intermediate in this pathway can undergo thiolysis with CoA to produce Acetyl-CoA and glycine.

- In humans, it is converted to alpha-ketobutyrate in a less common pathway via the enzyme serine dehydratase, and thereby enters the pathway leading to succinyl-CoA.

Sources

Foods high in threonine include cottage cheese, poultry, fish, meat, lentils, and sesame seeds.

Racemic threonine can be prepared from crotonic acid by alpha-functionalization using mercury(II) acetate.

Tryptophan (abbreviated as **Trp** or **W**) is one of the 20 standard amino acids, as well as an essential amino acid in the human diet. It is encoded in the standard genetic code as the codon *UGG*. Only the L-stereoisomer of tryptophan is used in structural or enzyme proteins, but the D-stereoisomer is occasionally found in naturally produced peptides (for example, the marine venom peptide contryphan). The distinguishing structural characteristic of tryptophan is that it contains an indole functional group.

Isolation

The isolation of tryptophan was first reported by Sir Frederick Hopkins in 1901 through hydrolysis of casein. From 600 grams of crude casein one obtains 4-8 grams of tryptophan.

Biosynthesis and Industrial Production

Plants and microorganisms commonly synthesize tryptophan from shikimic acid or anthranilate. The latter condenses with phosphoribosylpyrophosphate (PRPP), generating pyrophosphate as a by-product. After ring opening of the ribose moiety and following reductive decarboxylation, indole-3-glycerinephosphate is produced, which in turn is transformed into indole. In the last step, tryptophan synthase catalyzes the formation of tryptophan from indole and the amino acid, serine.

The industrial production of tryptophan is also biosynthetic and is based on the fermentation of serine and indole using either wild-type or genetically modified *E. coli*. The conversion is catalyzed by the enzyme tryptophan synthase.

FUNCTION OF TRYPTOPHAN

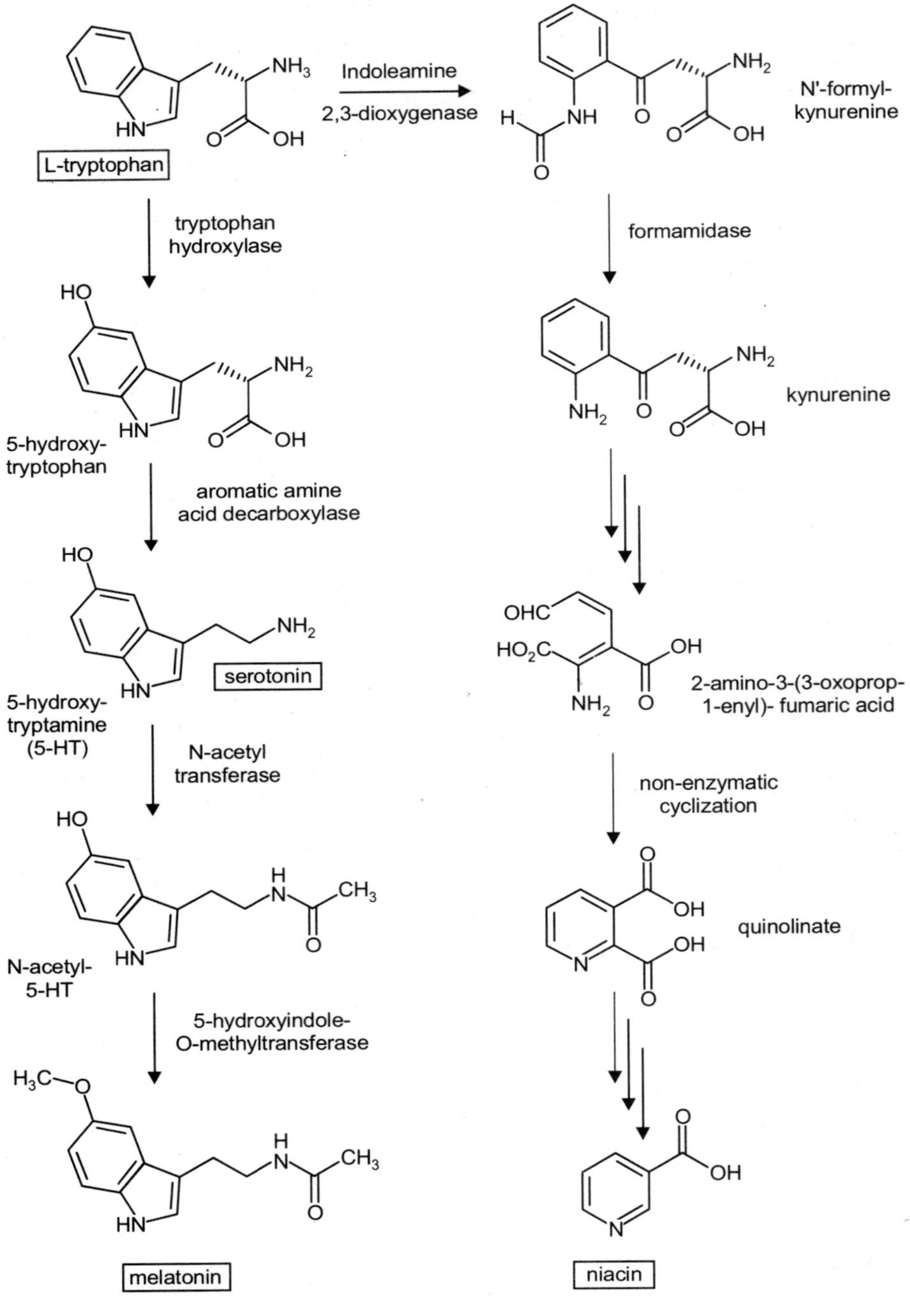

Metabolism of L-tryptophan into serotonin and melatonin (left) and niacin (right).

For many organisms (including humans), tryptophan is an essential amino acid. This means that it cannot be synthesized by the organism and therefore must be part of its diet. Amino acids, including tryptophan, act as building blocks in protein biosynthesis. In addition, tryptophan functions as a biochemical precursor for the following compounds (see also figure to the right):

- Serotonin (a neurotransmitter), synthesized via tryptophan hydroxylase. Serotonin, in turn, can be converted to melatonin (a neurohormone), via N-acetyltransferase and 5-hydroxyindole-O-methyltransferase activities.
- Niacin is synthesized from tryptophan via kynurenine and quinolinic acids as key biosynthetic intermediates.

The disorder Fructose Malabsorption causes improper absorption of tryptophan in the intestine, reduced levels of tryptophan in the blood and depression.

In bacteria that synthesize tryptophan, high cellular levels of this amino acid activate a repressor protein, which binds to the trp operon. Binding of this repressor to the tryptophan operon prevents transcription of downstream DNA that codes for the enzymes involved in the biosynthesis of tryptophan. So high levels of tryptophan prevent tryptophan synthesis through a negative feedback loop and, when the cell's tryptophan levels are reduced, transcription from the trp operon resumes. The genetic organisation of the trp operon thus permits tightly regulated and rapid responses to changes in the cell's internal and external tryptophan levels.

Dietary Sources

Tryptophan is a routine constituent of most protein-based foods or dietary proteins. It is particularly plentiful in chocolate, oats, bananas, durians, mangoes, dried dates, milk, yogurt, cottage cheese, red meat, eggs, fish, poultry, sesame, chickpeas, sunflower seeds, pumpkin seeds, spirulina, and peanuts. It is found in turkey at a level typical of poultry in general.

Table 2.26 : Tryptophan (Trp) Content of Various Foods

Food	Protein [g/100 g of food]	Tryptophan [g/100 g of food]	Tryptophan/Protein [%]
1	2	3	4
egg, white, dried	81.10	1.00	1.23
spirulina, dried	57.47	0.93	1.62
cod, atlantic, dried	62.82	0.70	1.11
soybeans, raw	36.49	0.59	1.62
cheese, Parmesan	37.90	0.56	1.47
caribou	29.77	0.46	1.55
sesame seed	17.00	0.37	2.17
cheese, cheddar	24.90	0.32	1.29
sunflower seed	17.20	0.30	1.74

...(Contd.)

1	2	3	4
pork, chop	19.27	0.25	1.27
turkey	21.89	0.24	1.11
chicken	20.85	0.24	1.14
beef	20.13	0.23	1.12
salmon	19.84	0.22	1.12
lamb, chop	18.33	0.21	1.17
perch, Atlantic	18.62	0.21	1.12
egg	12.58	0.17	1.33
wheat flour, white	10.33	0.13	1.23
milk	3.22	0.08	2.34
rice, white	7.13	0.08	1.16
potatoes, russet	2.14	0.02	0.84
banana	1.03	0.01	0.87

Use as a Dietary Supplement

For some time, tryptophan has been available in health food stores as a dietary supplement, although it is common in dietary protein. Many people found tryptophan to be a safe and reasonably effective sleep aid, probably due to its ability to increase brain levels of serotonin (a calming neurotransmitter when present in moderate levels) and/or melatonin (a sleep-inducing hormone secreted by the pineal gland in response to darkness or low light levels).

Clinical research has shown mixed results with respect to tryptophan's effectiveness as a sleep aid, especially in normal patients and for a growing variety of other conditions typically associated with low serotonin levels or activity in the brain such as premenstrual dysphoric disorder and seasonal affective disorder. In particular, tryptophan has been showing considerable promise as an antidepressant alone, and as an "augmenter" of antidepressant drugs. However, the reliability of these clinical trials has been questioned.

Metabolites

5-Hydroxytryptophan (5-HTP), a metabolite of tryptophan, has been suggested as a treatment for epilepsy and depression, although clinical trials are regarded inconclusive and lacking.

5-HTP readily crosses the blood-brain barrier and in addition is rapidly decarboxylated to serotonin (5-hydroxytryptamine or 5-HT) and therefore may be useful for the treatment of depression. However serotonin has a relatively short half-life since it is rapidly metabolized by monoamine oxidase, and therefore is likely to have limited efficacy. It is marketed in Europe for depression and other indications under the brand names Cincofarm and Tript-OH.

In the United States, 5-HTP does not require a prescription, as it is covered under the Dietary Supplement Act. However, since the quality of dietary supplements is not regulated by the U.S. Food and Drug Administration, the quality of dietary and nutritional supplements tends to vary, and there is no guarantee that the label accurately depicts what the bottle contains.

Turkey Meat and Drowsiness

One belief is that heavy consumption of turkey meat (as for example in a Thanksgiving or Christmas feast) results in drowsiness, which has been attributed to high levels of tryptophan contained in turkey. While turkey does contain high levels of tryptophan, the amount is comparable to that contained in most other meats. Furthermore, postprandial Thanksgiving sedation may have more to do with what is consumed along with the turkey, in particular carbohydrates and alcohol, rather than the turkey itself.

It has been demonstrated in both animal models and in humans that ingestion of a meal rich in carbohydrates triggers release of insulin. Insulin in turn stimulates the uptake of large neutral branched-chain amino acids (LNAA) but not tryptophan (trp) into muscle, increasing the ratio of trp to LNAA in the blood stream. The resulting increased ratio of tryptophan to large neutral amino acids in the blood reduces competition at the large neutral amino acid transporter resulting in the uptake of tryptophan across the blood-brain barrier into the central nervous system (CNS). Once inside the CNS, tryptophan is converted into serotonin in the raphe nuclei by the normal enzymatic pathway. The resultant serotonin is further metabolised into melatonin by the pineal gland. Hence, these data suggest that "feast-induced drowsiness," and in particular, the common post-Christmas and American post-Thanksgiving dinner drowsiness, may be the result of a heavy meal rich in carbohydrates which, via an indirect mechanism, increases the production of sleep-promoting melatonin in the brain.

Fluorescence

The fluorescence of a folded protein is a mixture of the fluorescence from individual aromatic residues. Most of the intrinsic fluorescence emissions of a folded protein are due to excitation of tryptophan residues, with some emissions due to tyrosine and phenylalanine; but be aware that di-sulfide bonds also have appreciable absorption in this wavelength range. Typically, tryptophan has a wavelength of maximum absorption of 280 nm and an emission peak that is solvatochromic, ranging from ca. 300 to 350 nm depending in the polarity of the local environment. Hence, protein fluorescence may be used as a diagnostic of the conformational state of a protein. Furthermore, tryptophan fluorescence is strongly influenced by the proximity of other residues (*i.e.*, nearby *protonated* groups such as Asp or Glu can cause quenching of Trp fluorescence). Also, energy transfer between tryptophan and the other fluorescent amino acids is possible, which would affect the analysis, especially in cases where the Förster acidic approach is taken. In addition, tryptophan is a relatively rare amino acid; many proteins contain only one or a few tryptophan residues. Therefore, tryptophan fluorescence can be a very sensitive measurement of the conformational state of individual tryptophan residues. The advantage compared to extrinsic probes is that the protein itself is not changed. The use of intrinsic fluorescence for the study of protein conformation is in practice limited to cases with few (or perhaps only one) tryptophan residues, since each experiences a different local environment, which gives rise to different emission spectra.

Tyrosine (abbreviated as **Tyr** or **Y**) or **4-hydroxyphenylalanine**, is one of the 20 amino acids that are used by cells to synthesize proteins. This is a non-essential amino acid and it is found in casein. Tyrosine has a non-polar/hydrophobic side group. In fact, the word "tyrosine" is from the Greek *tyros*, meaning *cheese*, as it was first discovered in 1846 by German chemist Justus von Liebig in the protein casein from cheese.

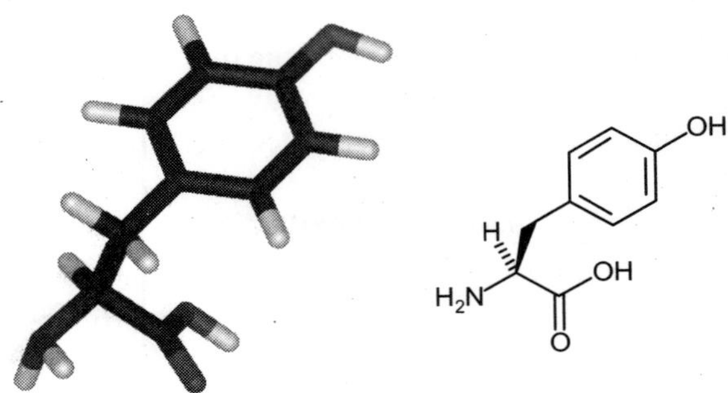

Functions

Aside from being a proteogenic amino acid, tyrosine has a special role by virtue of the phenol functionality. It occurs in proteins that are part of signal transduction processes. It functions as a receiver of phosphate groups that are transferred by way of protein kinases (so-called receptor tyrosine kinases). Phosphorylation of the hydroxyl group changes the activity of the target protein.

A tyrosine residue also plays an important role in photosynthesis. In chloroplasts (photosystem II), it acts as an electron donor in the reduction of oxidized chlorophyll. In this process, it undergoes deprotonation of its phenolic OH-group. This radical is subsequently reduced in the photosystem II by the four core manganese cluster.

Dietary Sources

Tyrosine, which can be synthesised in the body from phenylalanine, is also found in many high protein food products such as soy products, chicken, turkey, fish, peanuts, almonds, avocados, bananas, milk, cheese, yogurt, cottage cheese, lima beans, pumpkin seeds, and sesame seeds. Tyrosine can also be obtained through supplementation.

Biosynthesis

Plant biosynthesis of tyrosine from shikimic acid.

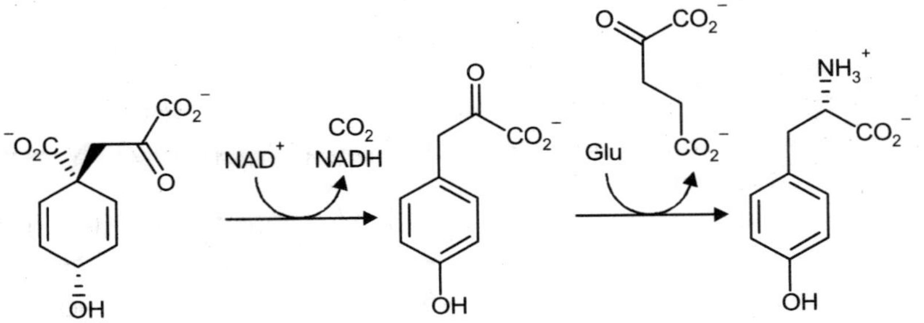

^-O_2C CO_2^- O OH

NAD^+ → CO_2 NADH

O CO_2^- OH

Glu → O CO_2^- CO_2^-

NH_3^+ CO_2^- OH

In plants and most microorganisms, **tyr** is produced via prephenate, an intermediate on the shikimate pathway. Prephenate is oxidatively decarboxylated with retention of the hydroxyl group to give *p*-hydroxyphenylpyruvate, which is transaminated using glutamate as the nitrogen source to give tyrosine and α-ketoglutarate.

Mammals synthesize tyrosine from the essential amino acid phenylalanine (**phe**), which is derived from food. The conversion of **phe** to **tyr** is catalyzed by the enzyme phenylalanine hydroxylase, a monooxygenase. This enzyme catalyzes the reaction causing the addition of an hydroxyl group to the end of the 6-carbon aromatic ring of phenylalanine, such that it becomes tyrosine.

Metabolism

Tyrosine → (Tyrosine hydroxylase) → Dihydroxy phenylalanine (L-DOPA) → (DOPA decarboxylase) → Dopamine → (Dopamine b-hydroxylase) → Norepinephrine → (Phenethanolamine N-methyltransferase) → Epinephrine

Catecholamine hormones produced from tyrosine metabolism.

Phosphorylation and Sulphation

Some of the tyrosine residues can be *tagged* with a phosphate group (phosphorylated) by protein kinases. (In its phosphorylated state, it is referred to as **phosphotyrosine**). Tyrosine phosphorylation is considered to be one of the key steps in signal transduction and regulation of enzymatic activity. Phosphotyrosine can be detected through specific antibodies. Tyrosine residues may also be modified by the addition of a sulfate group, a process known as tyrosine sulfation. Tyrosine sulfation is catalyzed by tyrosylprotein sulfotransferase (TPST). Like the phosphotyrosine antibodies mentioned above, antibodies have recently been described that specifically detect sulfotyrosine.

Precursor to Hormones

In the adrenal gland, tyrosine is converted to levodopa by the enzyme tyrosine hydroxylase (TH). TH is also the rate-limiting enzyme involved in the synthesis of the catecholamine hormones dopamine, norepinephrine (noradrenaline), and epinephrine.

The thyroid hormones triiodothyronine (T_3) and thyroxine (T_4) in the colloid of the thyroid also are derived from tyrosine.

Precursor to Alkaloids

In *Papaver somniferum*, the opium poppy, tyrosine is used to produce the alkaloid morphine.

Precursor to Pigments

Tyrosine is also the precursor to the pigment *melanin.*

Degradation

Tyrosine → (Tyrosine-transaminase) → p-Hydroxyphenylpyruvate → (p-Hydroxyphenylpyruvate-dioxygenase) → Homogentisate → (Homogentisate-dioxygenase) → 4-Maleylacetoacetate → (Maleylacetoacetate-cis-trans-isomerase) → 4-Fumarlacetoacetate → (4-Fumaryl-acetoacetate) → Acetoacetate + Fumarate

The decomposition of tyrosine to acetoacetate and fumarate. Two dioxygenases are necessary for the decomposition path. The end products can then enter into the citric acid cycle.

The decomposition of L-tyrosine (syn. *para*-hydroxyphenylalanine) begins with an α-ketoglutarate dependent transamination through the tyrosine transaminase to *para*-hydroxyphenylpyruvate. The positional description *para*, abbreviated *p*, mean that the hydroxyl group and side chain on the phenyl ring are across from each other.

The next oxidation step catalyzes by *p*-hydroxylphenylpyruvate-dioxygenase and splitting off CO_2 homogentisate (2,5-dihydroxyphenyl-1-acetate). In order to split the aromatic ring of homogentisate, a further dioxygenase, homogentistate-oxygenase is required. Thereby, through the incorporation of a further O_2 molecule, maleylacetoacetate is created.

Fumarylacetate is created maleylacetoacetate-*cis-trans*-isomerase through rotation of the carboxyl group created from the hydroxyl group via oxidation. This *cis-trans*-isomerase contains glutathione as a coenzyme. Fumarylacetoacetate is finally split via fumarylacetoacetate-hydrolase through the addition of a water molecule.

Thereby fumarate (also a metabolite of the citric acid cycle) and acetoacetate (3-ketobutyroate) are liberated. Acetoacetate is a ketone body, which is activated with succinyl-CoA, and thereafter it can be converted into acetyl-CoA which in turn can be oxidized by the citric acid cycle or be used for fatty acid synthesis.

Ortho- and Meta-tyrosine

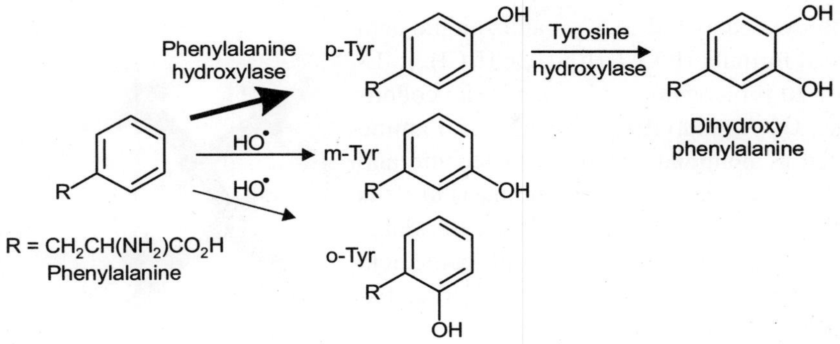

Enzymatic oxidation of tyrosine by phenylalanine hydroxylase (top) and non-enyzmatic oxidation by hydroxyl free radicals.

Three isomers of tyrosine are known. In addition to common amino acid L-tyrosine which is the para isomer (*para*-tyr, *p*-tyr or 4-hydroxyphenylalanine) there are two additional regioisomers, namely *meta*-tyrosine (*m*-tyr or 3-hydroxyphenylalanine or **L-m-tyrosine**) and *ortho*-tyrosine (*o*-tyr or 2-hydroxyphenylalanine) which occur in nature. The *m*-tyr and *o*-tyr isomers, which are rare, arise through non-enzymatic free-radical hydroxylation of phenylalanine under conditions of oxidative stress.

m-Tyrosine and analogues (rare in nature and therefore available synthetically) have shown application in Parkinson's Disease, Alzheimer's disease and arthritis.

Medical Use

Tyrosine is a precursor to neurotransmitters and increases plasma neurotransmitter levels (particularly dopamine and norepinephrine) but has little if any effect on mood. The effect on mood is more noticeable in humans subjected to stressful conditions.

A number of studies have found tyrosine to be useful during conditions of stress, cold, fatigue, prolonged work and sleep deprivation, with reductions in stress hormone levels, reductions in stress-induced weight loss seen in animal trials, improvements in cognitive and physical performance seen in human trials. Because tyrosine hydroxylase is the rate limiting enzyme, however, effects are less significant than those of L-dopa. Tyrosine does not seem to have any significant effect on mood, cognitive or physical performance in normal circumstances. A daily dosage supported in the literature is about 100 mg/kg for an adult. The usual dosage amounts to 500-1500 mg per day (dose suggested by most manufacturers; usually an equivalent to 1-3 capsules of pure tyrosine). It is not recommended to exceed 12000 mg (12 g) per day. In fact, too high doses result in reduced levels of dopamine. Tyrosine may decrease the absorption of other amino acids in high or chronic doses. It decreases absorption of L-dopa.

Drug Detoxification

Tyrosine appears to be a successful addition to conventional treatment for cocaine abuse and withdrawal.

It may be used in conjunction with other amino acids such as tryptophan. Some individuals using tyrosine have also reported successful withdrawal from caffeine and nicotine.

Valine (abbreviated as **Val** or **V**) is an α-amino acid with the chemical formula $HO_2CCH(NH_2)CH(CH_3)_2$. L-Valine is one of 20 proteogenic amino acids. Its codons are GUU, GUC, GUA, and GUG. This essential amino acid is classified as nonpolar. Along with leucine and isoleucine, valine is a branched-chain amino acid. It is named after the plant valerian. In sickle-cell disease, valine substitutes for the hydrophilic amino acid glutamic acid in hemoglobin. Because valine is hydrophobic, the hemoglobin does not fold correctly.

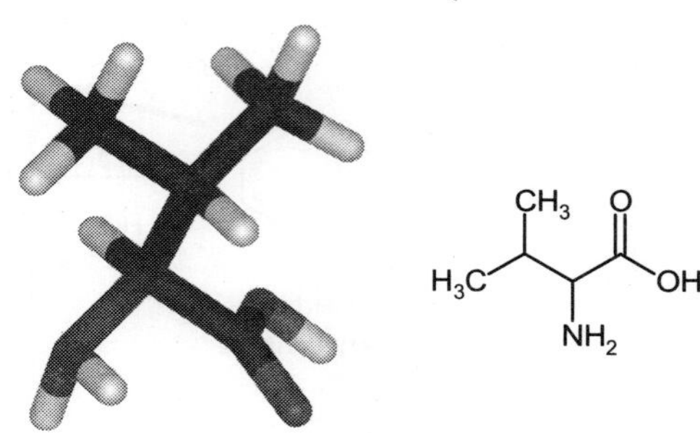

Biosynthesis

Valine is an essential amino acid, hence it must be ingested, usually as a component of proteins. It is synthesized in plants via several steps starting from pyruvic acid. The initial part of the pathway also leads to leucine. The intermediate α-ketovalerate undergoes reductive amination with glutamate.

Enzymes involved in this biosynthesis include:

1. Acetolactate synthase (also known as acetohydroxy acid synthase).
2. Acetohydroxy acid isomeroreductase.
3. Dihydroxyacid dehydratase.
4. Valine aminotransferase.

Synthesis

Racemic valine can be synthesized by bromination of isovaleric acid followed by amination of the α-bromo derivative

$$HO_2CCH_2CH(CH_3)_2 + Br_2 \rightarrow HO_2CCHBrCH(CH_3)_2 + HBr$$

$$HO_2CCHBrCH(CH_3)_2 + 2\,NH_3 \rightarrow HO_2CCH(NH_2)CH(CH_3)_2 + NH_4Br$$

Dietary Aspects

Nutritional sources of valine include cottage cheese, fish, poultry, peanuts, sesame seeds, and lentils.

A CHEMICAL REACTIONS OF AMINO ACID

The amino acids in general give the reactions characteristic of the carboxyl groups. Some of these reactions take place with more difficulty or under more restricted consitions than in the case of ordinary organic acids because of participation of carboxyl group of amino acids in Zwiterion structure.

A **zwitterion** (first part pronounced "tsvitter", from German "*Zwitter*"—"hybrid," "hermaphrodite") is a chemical compound that carries a total net charge of 0, thus electrically neutral but carries formal positive and negative charges on

Salt Formation and tritration Both the acidic and the basic groups of amino acids form salts. However owing to the Zwitterion formation the titration of amino acid in aqueous solutions is impossible because no sharp endpoint is obtained. The carboxyl group can be titrated with alkali in alcoholic or Acetone medium. In which the end point of the indicatior is shifted more towards the alkaline pH. The basic groups of amiono acids may be titrated with glacial acetic acid solution with standard sulphuric hydrobromic or perchloric acid using as indicator such as crytal violet.

The titration curves of amino acids in aqueous may be run with the glass of hydrogen electrodes. Theoretical curves may be calculated by the Henderson – Hasselbalch equation* from pK values.

These differences all point to internal salt formation by a proton transfer from the acidic carboxyl function to the basic amino group. The resulting ammonium carboxylate structure, commonly referred to as a **zwitterion**, is also supported by the spectroscopic characteristics of alanine.

$$CH_3CH(NH_2)CO_2H \rightleftharpoons CH_3CH(NH_3)^{(+)}CO_2^{(-)}$$

As expected from its ionic character, the alanine zwitterion is high melting, insoluble in nonpolar solvents and has the acid strength of a 1°-ammonium ion. To the right above is a Jmol display of an L-amino acid. Examples of a few specific amino acids may also be viewed in their favored neutral

---

Some chemists restrict this term to refer to compounds with non-adjacent positive and negative charges. This would exclude compounds such as *N*-oxides. Zwitterions are polar and are usually very water-soluble, but poorly soluble in most organic solvents.

Ampholytes are molecules that contain both acidic and basic groups (and are therefore amphoteric) and will exist mostly as zwitterions in a certain range of pH. The pH at which the average charge is zero is known as the molecule's isoelectric point.

Applications:

Ampholytes are used to establish a stable pH gradient for **use** in isoelectric focusing.

Typical examples of zwitterions are:

- Most amino acids at physiological pH are for the most part zwitterionic.
- Used as buffering agents in Good's buffers:
  - The amino-sulfonic acid based MES, MOPS, HEPES, PIPES or CAPS
  - The amino-carboxylic acid (amino acid) based glycine, its derivatives bicine and tricine, and alanine
- Used as detergents:
  - CHAPSO
- Natural products like the alkaloids psilocybin and lysergic acid.
- Betaines

Less common examples of zwitterions are:

- Quinonoid zwitterions.
- Drugs such as Fexofenadine (Allegra) and Cephaloridine.
- decaphenylferrocene $[(ç^5\text{-}C_5Ph_5)_2Fe]$ has been shown to have a zwitterionic linkage isomer $[(ç^5\text{-}C_5Ph_5)Fe^+(ç^6\text{-}C_6H_5C_5Ph_4^-)]$.

zwitterionic form. Note that in lysine the amine function farthest from the carboxyl group is more basic than the alpha-amine. Consequently, the positively charged ammonium moiety formed at the chain terminus is attracted to the negative carboxylate, resulting in a coiled conformation.

Decarboxylation: When amino acids are heated, preferably in the presence of barium hydroxide or diphenylamine, carbon dioxide is last and amine is formed. The Decarboxylation of histidine may ve taken as an example.

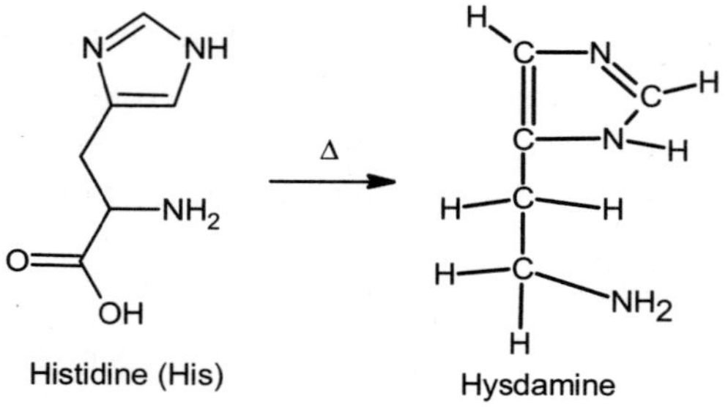

Histidine (His) —Δ→ Hysdamine

Decarboxylation of amino acids is caused also by dicarboxylase enzymes of bacteria. This occurs in the large intestine as relut of the action putrefactive organism. Some of the amines produced, such as histamine tryamine (from tyrosine) are pharmacologically active.

Amide formation: Amides can be prepared by treatment of the amino acid esters with alcoholc or anhydrous ammonia.

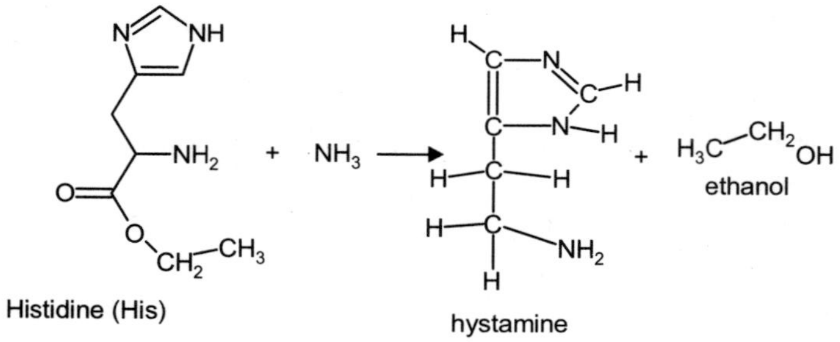

Histidine (His) + NH_3 → hystamine + $H_3C{-}CH_2{-}OH$ ethanol

Methylation of the Amino acids: The amino groups of the amino acids may be exhaustively methylated by treatment with methyl iodie or dimethyl sulphate in alkaline solution. The methylation takes place in stages but the final product is a betaine of the aminop acid. The reaction may be represented.

The basicity of the amino group is greatly increased by methylation to the quaternary nitrogen stage of the betaine structure, and the betaines represent internal salt or Zwitterion structure. The betaines forms well –crystallized double salts with the chlorides of platinum gold, and mercury which are useful in their separation and purification. Mixtures of amino acids have been separated by methylation to betained followed by formation and separation of the betaine heavy metal double salts. Various betains occur in plants.

+ $2CH_3I$ + 2NaOH ⟶ 2NaI + $2H_2O$

+ CH_3I ⟶ ⟶ NaI +

Reaction with Sanger's reagent: A very important reaction of the amino groups of amino aciod is with the 1 – fluro-2-4- dinitrobenzene, FDNB also called Sanger's reagent. The reagent condenses with free amino groups in the cold in mild alkaline solution (bicarbonate of sodium or potassium).

amino acid + 1-fluoro-2,4-dinitrobenzene Sanger's ⟶ 2-[(2,4-dinitrophenyl)amino]propanoic acid + HF hydrogen fluoride

Most of the DNP amino acid are coloured yellow and most are soluble in ether. The DNP bond is generally more stable in acid hydrolysis than peptide bonds. The importance of the reaction is largely due to the fact the terminal amino acid group in peptide or protein chainforms a DNP derivative which may be hydrolysis with acid to give the DNP amino acid. In this way it is possible to identify one of the terminal amino acid in a paptide or protein.

Reaction of Nitrous acid with amino acid: Amino group of the acid reacts with nitrous acid to form the corresponding hydroxyl acids with liberation of nitrogen gas.

amino acid + Nitrous acid ⟶ 2-hydroxypropanoic acid + N_2 + H_2O

Since each amino group gives one molecule of nitrogen, which can be accurately measured. Van Slyke has utilized the treaction in an excellent method for the estimation of free amino group in the amino acid, peptides and proteins. As commonly carried out, the amino acid, protein or peptide solution is treated with sodium nitrite and acetic acid, which react to form nitrous acid in a mild acidic medium. Under these conditions the α-amino group react quantitatively in these with three to four minutes while the amino group of the lysine react more slowly.

The Van Slyke's amino–nitrogen method may be used conveniently to follow the rate of hydrolysis of proteins. Since for each peptide linkage hydrolysed a α-amino group is liberated. The amino nitrogen thus progressively increases until hydrolysis of the protein is complete.

Carboxylic Acid Esterification: Amino acids undergo most of the chemical reactions characteristic of each function, assuming the pH is adjusted to an appropriate value. Esterification of the carboxylic acid is usually conducted under acidic conditions, as shown in the two equations written below. Under such conditions, amine functions are converted to their ammonium salts and carboxyic acids are not dissociated. The first equation is a typical Fischer esterification involving methanol. The initial product is a stable ammonium salt. The amino ester formed by neutralization of this salt is unstable, due to acylation of the amine by the ester function. The second reaction illustrates benzylation of the two carboxylic acid functions of aspartic acid, using p-toluenesulfonic acid as an acid catalyst. Once the carboxyl function is esterified, zwitterionic species are no longer possible and the product behaves like any 1°-amine.

HCl / CH_3OH ; CH_3OH / HCl ; $NaHCO_3$ or other base — unstable

$C_7H_7SO_3H$ / 2 $C_6H_5CH_2OH$

Amine Acylation: In order to convert the amine function of an amino acid into an amide, the pH of the solution must be raised to 10 or higher so that free amine nucleophiles are present in the reaction system. Carboxylic acids are all converted to carboxylate anions at such a high pH, and do not interfere with amine acylation reactions. The following two reactions are illustrative. In the first, an acid chloride serves as the acylating reagent. This is a good example of the superior nucleophilicity of nitrogen in acylation reactions, since water and hydroxide anion are also present as competing nucleophiles. A similar selectivity favouring amines was observed in the Hinsberg test. The second reaction employs an anhydride-like reagent for the acylation. This is a particularly useful procedure in peptide synthesis, thanks to the ease with which the t-butylcarbonyl (t-BOC) group can be removed at a later stage. Since amides are only weakly basic ($pK_a \sim -1$), the resulting amino acid derivatives do not display zwitterionic character, and may be converted to a variety of carboxylic acid derivatives.

$NaOH$ / C_6H_5COCl ; $H_3O^{\oplus}$

di-tert-butyl dicarbonate (an anhydride) ; $NaOH$ / THF & H_2O

Specific Oxidation: The mild oxidant iodine reacts selectively with certain amino acid side groups. These include the phenolic ring in tyrosine, and the heterocyclic rings in tryptophan and histidine, which all yield products of electrophilic iodination. In addition, the sulfur groups in cysteine and methionine are also oxidized by iodine. Quantitative measurent of iodine consumption has been used to determine the number of such residues in peptides. The basic functions in lysine and arginine are onium cations at pH less than 8, and are unreactive in that state. Cysteine is a thiol, and like most thiols it is oxidatively dimerized to a disulfide, which is sometimes listed as a distinct amino acid under the name cystine. Disulfide bonds of this kind are found in many peptides and proteins. For example, the two peptide chains that constitute insulin are held together by two disulfide links. Our hair consists of a fibrous protein called keratin, which contains an unusually large proportion of cysteine. In the manipulation called "permanent waving", disulfide bonds are first broken and then created after the hair has been reshaped. Treatment with dilute aqueous iodine oxidizes the methionine sulfur atom to a sulfoxide.

2 HS–CH₂–CH(NH₂)–COOH ⇌ [O] / [H] HOOC–CH(NH₂)–CH₂–S–S–CH₂–CH(NH₂)–COOH

Cysteine

Cysteine-Cystine Interconversion

ISOELECTRIC POINT

The titration curve for alanine, shown below, demonstrates this relationship. At a pH lower than 2, both the carboxylate and amine functions are protonated, so the alanine molecule has a net positive charge. At a pH greater than 10, the amine exists as a neutral base and the carboxyl as its conjugate base, so the alanine molecule has a net negative charge. At intermediate pH's the zwitterion concentration increases, and at a characteristic pH, called the **isoelectric point (pI)**, the negatively and positively charged

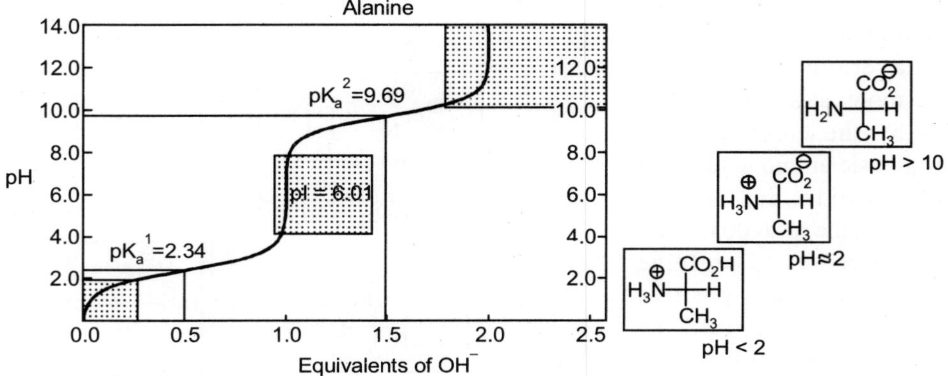

Fig. 2.47 : The Titration Curve of amino acid

molecular species are present in equal concentration. This behaviour is general for simple (difunctional) amino acids. Starting from a fully protonated state, the pK_a's of the acidic functions range from 1.8 to 2.4 for $-CO_2H$, and 8.8 to 9.7 for $-NH_3^{(+)}$. The isoelectric points range from 5.5 to 6.2. Titration curves show the neutralization of these acids by added base, and the change in pH during the titration.

The distribution of charged species in a sample can be shown experimentally by observing the movement of solute molecules in an electric field, using the technique of **electrophoresis**. For such experiments an ionic buffer solution is incorporated in a solid matrix layer, composed of paper or a crosslinked gelatin-like substance. A small amount of the amino acid, peptide or protein sample is placed near the center of the matrix strip and an electric potential is applied at the ends of the strip, as shown in the following diagram. The solid structure of the matrix retards the diffusion of the solute molecules, which will remain where they are inserted, unless acted upon by the electrostatic potential. In the example shown here, four different amino acids are examined simultaneously in a pH 6.00 buffered medium. To see the result of this experiment, click on the illustration. Note that the colours in the display are only a convenient reference, since these amino acids are colourless.

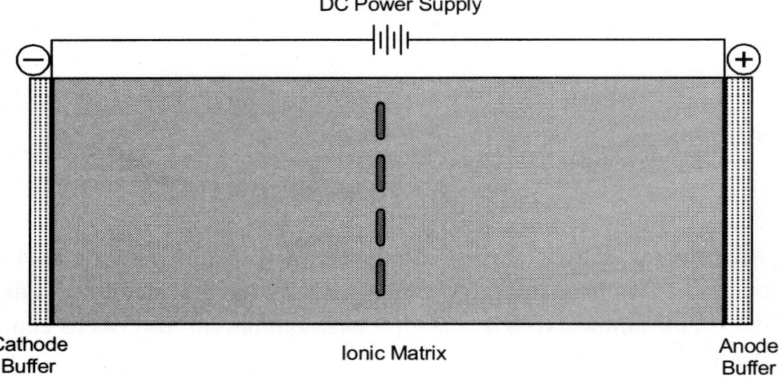

Fig. 2.48 : The Electrophoresis

At pH 6.00 alanine and isoleucine exist on average as neutral zwitterionic molecules, and are not influenced by the electric field. Arginine is a basic amino acid. Both base functions exist as "onium" conjugate acids in the pH 6.00 matrix. The solute molecules of arginine therefore carry an excess positive charge, and they move toward the cathode. The two carboxyl functions in aspartic acid are both ionized at pH 6.00, and the negatively charged

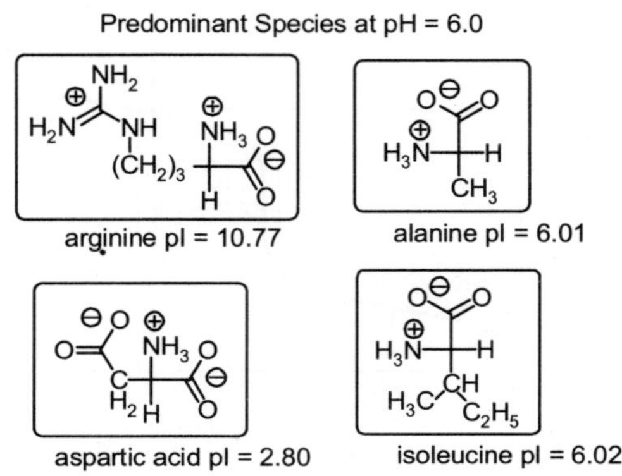

Fig. 2.49 : The Structures for all these species

solute molecules move toward the anode in the electric field. Structures for all these species are shown in the fig 2.49.

pK_a Values of Polyfunctional Amino Acids

Amino Acid	α-CO_2H pK_a^1	α-NH_3 pK_a^2	Side Chain pK_a^3	pI
Arginine	2.1	9.0	12.5	10.8
Aspartic Acid	2.1	9.8	3.9	3.0
Cysteine	1.7	10.4	8.3	5.0
Glutamic Acid	2.2	9.7	4.3	3.2
Histidine	1.8	9.2	6.0	7.6
Lysine	2.2	9.0	10.5	9.8
Tyrosine	2.2	9.1	10.1	5.7

It should be clear that the result of this experiment is critically dependent on the pH of the matrix buffer. If we were to repeat the electrophoresis of these compounds at a pH of 3.00, the aspartic acid would remain at its point of origin, and the other amino acids would move toward the cathode. Ignoring

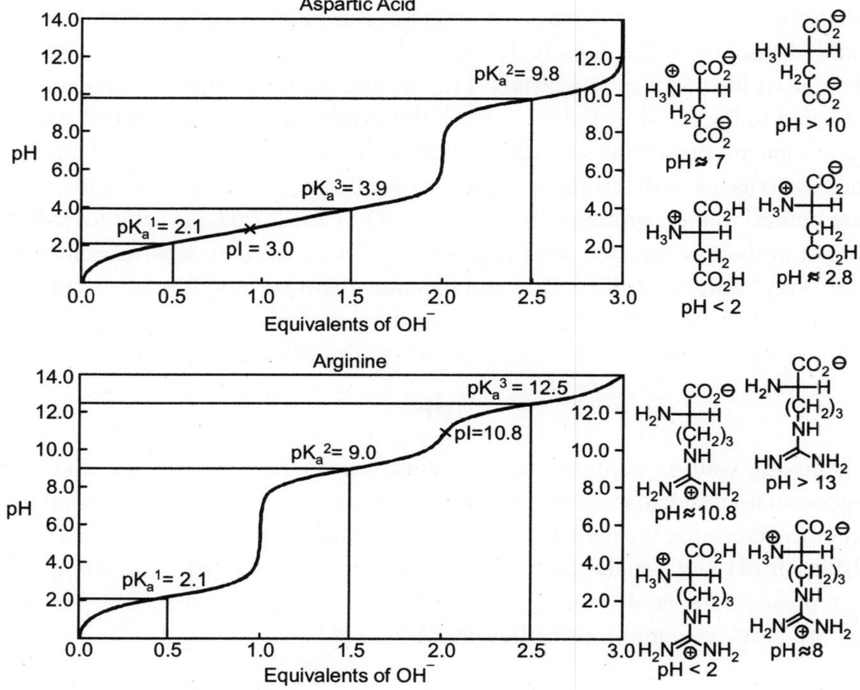

Fig. 2.50 : The pK values of Aspartic acid and Agrinine

differences in molecular size and shape, the arginine would move twice as fast as the alanine and isoleucine because its solute molecules on average would carry a double positive charge.

As noted earlier, the titration curves of simple amino acids display two inflection points, one due to the strongly acidic carboxyl group (pK_a^1 = 1.8 to 2.4), and the other for the less acidic ammonium function (pK_a^2 = 8.8 to 9.7). For the 2º-amino acid proline, pK_a^2 is 10.6, reflecting the greater basicity of 2º-amines.

Some amino acids have additional acidic or basic functions in their side chains. These compounds are listed in the table on the right. A third pK_a, representing the acidity or basicity of the extra function, is listed in the fourth column of the table. The pI's of these amino acids (last column) are often very different from those noted above for the simpler members. As expected, such compounds display three inflection points in their titration curves, illustrated by the titrations of arginine and aspartic acid shown below. For each of these compounds four possible charged species are possible, one of which has no overall charge. Formulas for these species are written to the right of the titration curves, together with the pH at which each is expected to predominate. The very high pH required to remove the last acidic proton from arginine reflects the exceptionally high basicity of the guanidine moiety at the end of the side chain.

The Isoelectric Point

As defined above, the isoelectric point, **pI**, is the pH of an aqueous solution of an amino acid (or peptide) at which the molecules on average have no net charge. In other words, the positively charged groups are exactly balanced by the negatively charged groups. For simple amino acids such as alanine, the pI is an average of the pK_a's of the carboxyl (2.34) and ammonium (9.69) groups. Thus, the pI for alanine is calculated to be: (2.34 + 9.69)/2 = 6.02, the experimentally determined value. If additional acidic or basic groups are present as side-chain functions, the pI is the average of the pK_a's of the two most similar acids. To assist in determining similarity we define two classes of acids. The first consists of acids that are neutral in their protonated form (e.g. CO_2H & SH). The second includes acids that are positively charged in their protonated state (e.g. $-NH_3^+$). In the case of aspartic acid, the similar acids are the alpha-carboxyl function (pK_a = 2.1) and the side-chain carboxyl function (pK_a = 3.9), so pI = (2.1 + 3.9)/2 = 3.0. For arginine, the similar acids are the guanidinium species on the side-chain (pK_a = 12.5) and the alpha-ammonium function (pK_a = 9.0), so the calculated pI = (12.5 + 9.0)/2 = 10.75.

pH

Since amino acids, as well as peptides and proteins, incorporate both acidic and basic functional groups, the predominant molecular species present in an aqueous solution will depend on the pH of the solution. In order to determine the nature of the molecular and ionic species that are present in aqueous solutions at different pH's, we make use of the **Henderson-Hasselbach Equation**, (written below). Here, the pK_a represents the acidity of a specific conjugate acid function (HA). When the pH of the solution equals pK_a, the concentrations of HA and $A^{(-)}$ must be equal (log 1 = 0).

$$\text{Henderson-Hasselbach Equation}: pK_a = pH + \log \frac{[HA]}{[A^-]}$$

ACID BASE PROPERTIES OF AMINO ACIDS

This sub-section looks at what happens to amino acids as you change the pH by adding either acids or alkalis to their solutions.For simplicity, the page only looks at amino acids which contain a single $-NH_2$ group and a single -COOH group.

Amino acids as zwitterions and Zwitterions in simple amino acid solutions:

An amino acid has both a basic amine group and an acidic carboxylic acid group.

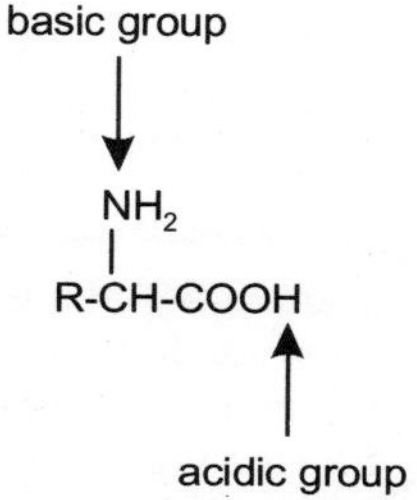

There is an internal transfer of a hydrogen ion from the -COOH group to the $-NH_2$ group to leave an ion with both a negative charge and a positive charge.

This is called a ***zwitterion***.

NH_3^+
R-CH-COO^-

a *zwitterion*

This is the form that amino acids exist in even in the solid state. If you dissolve the amino acid in water, a simple solution also contains this ion.

Adding an alkali to an amino acid solution: If you increase the pH of a solution of an amino acid by adding hydroxide ions, the hydrogen ion is removed from the $-NH_3^+$ group.

NH_3^+ + OH^- ⟶ NH_2 + H_2O
R-CH-COO^- R-CH-COO^-

Notice that this is now a negative ion
-no longer a z witterion

You could show that the amino acid now existed as a negative ion using ***electrophoresis***.

In its simplest form, electrophoresis can just consist of a piece of moistened filter paper on a microscope slide with a crocodile clip at each end attached to a battery. A drop of amino acid solution is placed in the centre of the paper.

Although the amino acid solution is colourless, its position after a time can be found by spraying it with a solution of ***ninhydrin***. If the paper is allowed to dry and then heated gently, the amino acid shows up as a coloured spot.

The amino acid would be found to travel towards the anode (the positive electrode).

Adding an acid to an amino acid solution: If you decrease the pH by adding an acid to a solution of an amino acid, the $-COO^-$ part of the zwitterion picks up a hydrogen ion.

$$R\text{-}CH(NH_3^+)\text{-}COO^- + H^+_{(aq)} \longrightarrow R\text{-}CH(NH_3^+)\text{-}COOH$$

Notice that this is now a negative ion
-no longer a z witterion

This time, during electrophoresis, the amino acid would move towards the cathode (the negative electrode).

You might have expected the isoelectric point to be at pH 7 - when the solution is neither acidic nor alkaline. In fact, the isoelectric point for many amino acids is about pH 6. Explaining why it isn't at pH 7 is quite long-winded.

Shifting the pH from one extreme to the other Suppose you start with the ion we've just produced under acidic conditions and slowly add alkali to it. That ion contains two acidic hydrogens - the one in the -COOH group and the one in the $-NH_3^+$ group. The more acidic of these is the one in the -COOH group, and so that is removed first - and you get back to the zwitterion.

$$R\text{-}CH(NH_3^+)\text{-}COOH + OH^- \longrightarrow R\text{-}CH(NH_3^+)\text{-}COO^- + H_2O$$

So when you have added just the right amount of alkali, the amino acid no longer has a net positive or negative charge. That means that it wouldn't move towards either the cathode or anode during electrophoresis. The pH at which this lack of movement during electrophoresis happens is known as the ***isoelectric point*** of the amino acid. This pH varies from amino acid to amino acid.

If you go on adding hydroxide ions, you will get the reaction we've already seen, in which a hydrogen ion is removed from the $-NH_3^+$ group.

$$R\text{-}CH(NH_3^+)\text{-}COO^- + OH^- \longrightarrow R\text{-}CH(NH_2)\text{-}COO^- + H_2O$$

You can, of course, reverse the whole process by adding an acid to the ion we've just finished up with.

That ion contains two basic groups - the $-NH_2$ group and the $-COO^-$ group. The $-NH_2$ group is the stronger base, and so picks up hydrogen ions first. That leads you back to the zwitterion again.

$$R\text{-}CH(NH_2)\text{-}COO^- + H^+_{(aq)} \longrightarrow R\text{-}CH(NH_3^+)\text{-}COO^-$$

Why isn't the Isoelectric Point of an Amino Acid at pH 7?

When an amino acid dissolves in water, the situation is a little bit more complicated than we tend to pretend at this level. The zwitterion interacts with water molecules - acting as both an acid and a base.

As an acid:

$$\underset{\text{R-CH-COO}^-}{\overset{\text{NH}_3^+}{|}} + H_2O \rightleftharpoons \underset{\text{R-CH-COO}^-}{\overset{\text{NH}_2}{|}} + H_3O^+$$

The $-NH_3^+$ group is a weak acid and donates a hydrogen ion to a water molecule. Because it is only a weak acid, the position of equilibrium will lie to the left.

As a base:

$$\underset{\text{R-CH-COO}^-}{\overset{\text{NH}_3^+}{|}} + H_2O \rightleftharpoons \underset{\text{R-CH-COOH}}{\overset{\text{NH}_2^+}{|}} + OH^-$$

The $-COO^-$ group is a weak base and takes a hydrogen ion from a water molecule. Again, the equilibrium lies to the left.

When you dissolve an amino acid in water, both of these reactions are happening.

But, the positions of the two equilibria aren't identical - they vary depending on the influence of the "R" group. In practice, for the simple amino acids we have been talking about, the position of the first equilibrium lies a bit further to the right than the second one. That means that there will be rather more of the negative ion from the amino acid in the solution than the positive one. In those circumstances, if you carried out electrophoresis on the unmodified solution, there would be a slight drift of amino acid towards the positive electrode (the anode). To stop that, you need to cut down the amount of the negative ion so that the concentrations of the two ions are identical. You can do that by adding a very small amount of acid to the solution, moving the position of the first equilibrium further to the left. Typically, the pH has to be lowered to about 6 to achieve this. For glycine, for example, the isoelectric point is pH 6.07; for alanine, 6.11; and for serine, 5.68.

GEL FITRATION

Gel Filtration

Gel Fitration encompasses a variety of techniques that involve a mobile phase flowing through a bed of stationary phase. A mixture of solutes is resolved or separated as a result of such differences in "affinity" of the solutions for the mobile phase relative to the stationary phase. A solute that has more "affinity" for the mobile phase moves forward with the mobile phase whereas a solute with more "affinity" for the stationary phase is retarded. Thus, solutes are separated into bands based on their distribution between phases.

Molecules of biological interest are separated or resolved by a variety of chromatography techniques including adsorption chromatography, ion-exchange chromatography, partition chromatography (paper or thin layer chromatography), gas-liquid chromatography, and molecular exclusion chromatography (molecular sieve or gel filtration).

Chromatography involves a "mobile" phase (liquid or gas) flowing over a "stationary" phase (solid or liquid; Figure 2.51). A mixture of solutes is separated based on the solute components' degree of affinity for the mobile and the stationary phases. If Component #1 has a stronger affinity for the stationary phase than Component #2, then #2 will migrate through the column more rapidly than #1.

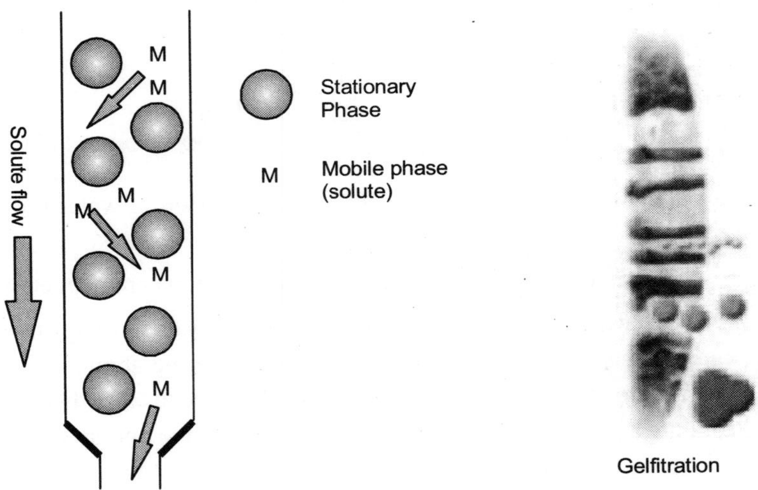

Fig. 2.51 & 2.51a. The Gelfitration diagram

Nutshell Overview of Gel Filtration

The most important factors to understand about gel filtration are:

- Gel filtration separates molecules based solely on their size.
- Larger molecules elute first from the column.
- Smaller molecules elute later from the column.
- Gel filtration chromatography is used primarily to purify proteins or other molecules of interest.
- Gel filtration chromatography can be used to estimate the size of unknown proteins.

Principles of Gel Filtration

In gel filtration or molecular exclusion process molecules in solution are separated by size as they pass through a column of cross-linked beads that form a three-dimensional network. These polymer beads (frequently made of dextran, agarose, or acrylamide) comprise the stationary phase. The liquid phase is the solvent that is found both around the beads and in the pores of the stationary phase matrix.

As a sample passes through the column, there are two routes that molecules can take through the column, depending on their size and the size of the pores in the beads. Molecules that are larger than the pores will not enter the stationary phase, staying in the solution (mobile phase) surrounding the beads. Hence they elute first from the column. Smaller molecules will enter the pores in the beads and so move more slowly through the column (Figure 2.52 & 2.53). Molecules of intermediate size will enter the stationary phase to some extent, but will not spend as much time there as do the smaller molecules.

To summarize, larger molecules will elute from the column first, and the smallest molecules will elute last, with intermediate molecules strung out in between.

Putting gel filtration in terms that fit the general description of chromatography given above, the beads comprise the stationary phase and the solution is the mobile phase. Larger molecules have a greater affinity for the mobile phase than the stationary phase, so they migrate through the column more rapidly than smaller molecules, which have a greater affinity for the stationary phase.

As the samples elute from the bottom of the column, they are collected in tubes as fractions. Usually, fractions of a particular volume are collected. In most cases, however, the eluted fractions must be tested to determine both what fractions contain the samples (usually proteins) and how much of the sample (protein) is in the fraction. Several methods that are commonly used are: (1) spectrophotometric examination of the fractions; (2) SDS-PAGE of the fractions; and (3) assaying the fractions for a particular enzymatic activity.

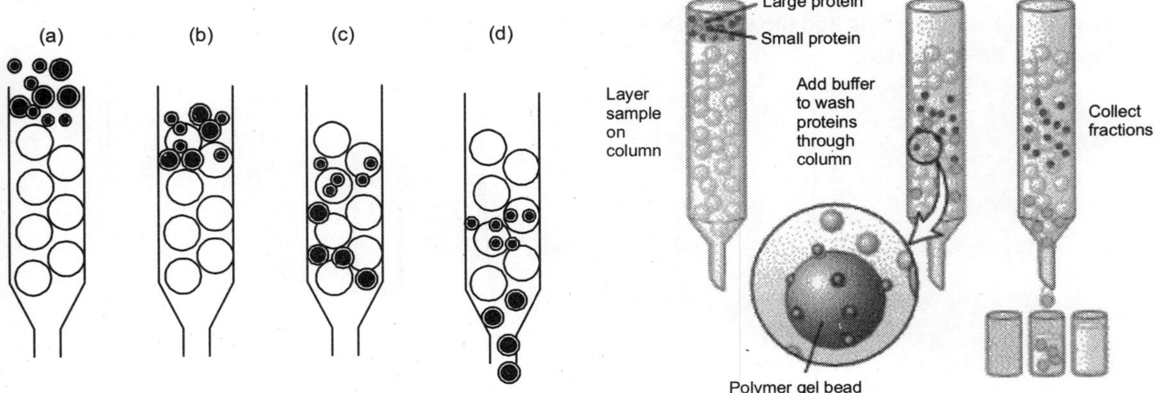

Fig. 2.52 & 2.53. The Protein Separation

Once the sample concentrations in the fractions have been determined, the concentrations can be graphed against the elution volume to create an elution profile (Figure 2.55). The first volume that elutes from the column before any fractions is called the void volume (V_0) of the column. Any proteins or other molecules that are too large to enter the pores of the beads will elute immediately after the void volume.

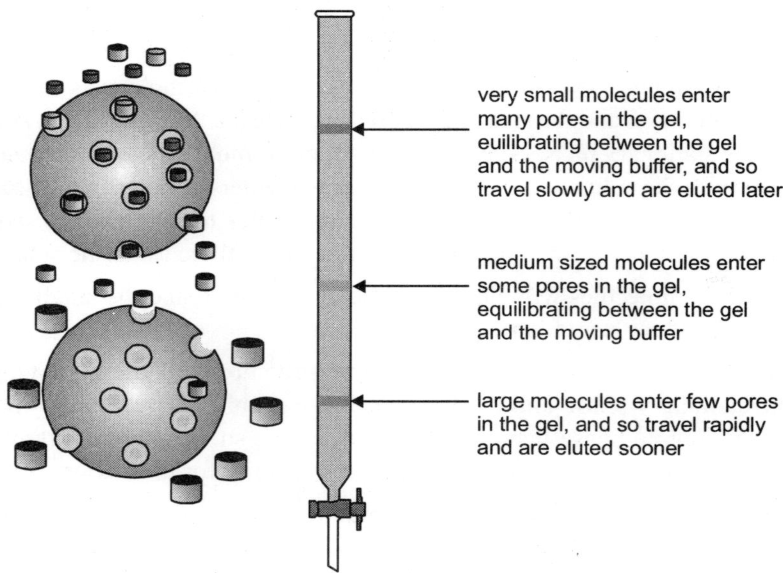

Fig. 2.54. The Details of Gel fitration in column

If the molecular weights of the proteins are already known, then these data can be used to create a standard curve to calculate the molecular weight of an unknown protein. The elution volume is plotted against the known molecular weights of the proteins on semilog graph paper (Figure 2.56) and a line derived. From this line and the elution volume of an unknown protein, the molecular weight of the unknown can be estimated.

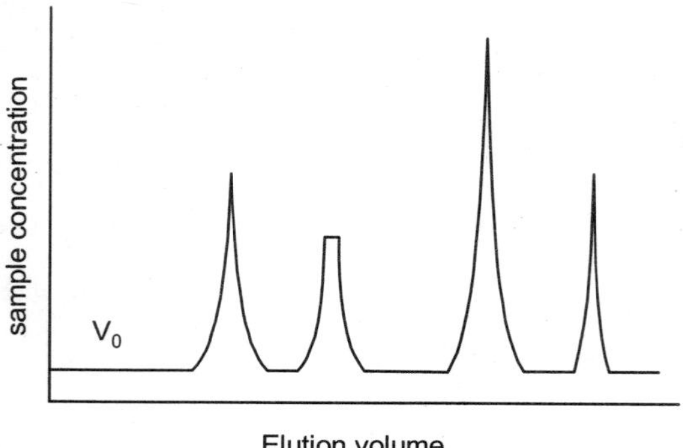

Fig. 2.55 : Elution profile from filtration chromatography. Plot of elution volume against concentration for 4 proteins. V_0 is the void volume of the column.

There are many types of gel filtration materials. The choice of matrix depends on the range of size of molecules to be separated and the goal of the separation. Different bead types have pores of different sizes. Consult a reference book such as those listed in the below for more details on factors to be considered in choosing a matrix.

The gel filtration material that will be used in the experiment below is called Sephadex G-75 and it will separated molecules with molecular weights from 3,000 to 70,000. Molecules with molecular weights larger than 70,000 will be excluded from entering the beads.

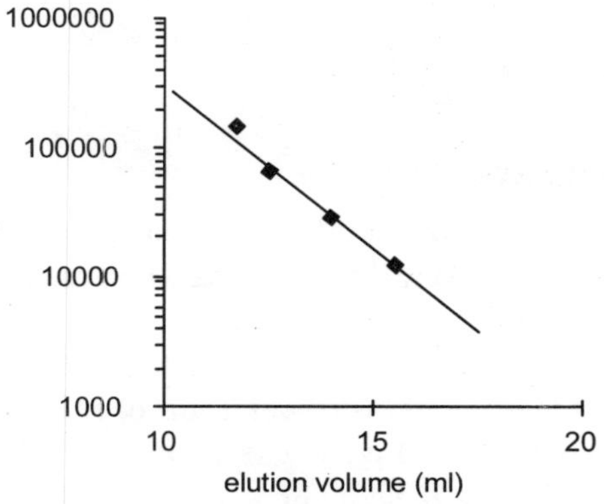

Notes on Use of Gel Filtration

- The choice of matrix depends on the range of size of molecules to be separated and the goal of the separation. Different bead types have pores of different sizes. Consult a reference book such as those listed below for more details on factors to be considered in choosing a matrix.
- The matrix beads normally come in dry form and must be swollen before use. It is important not to use a magnetic stirrer when preparing the beads, or the beads can be fragmented. It takes several days to swell beads like the Sephadex that you will use today. One short cut, however, is to autoclave the solution. This causes the beads to swell more rapidly without damaging them.
- Never allow a gel filtration column to dry out. If it dries out, the column must be re-poured. It is crucial for good separation that the column be consistent from top to bottom.

Fig. 2.56 : Molecular weight vs. elution volume. Plot of molecular weight of proteins against their elution volume from a gel filtration column. Proteins are yeast alcohol dehydrogenase [150 kilodaltons(kD)], bovine serum albumin (66 kD), carbonic anhydrase (29 kD), and cytochrome c (12.4 kD).

In Part 1 of this experiment you will pour your own gel filtration column (G-100 fast flow). You will then calibrate your column using four separate molecules: Blue Dextran, yeast alcohol dehydrogenase, BSA, and cytochrome-c.

In Part 2, you will use your calibrated gel filtration column to determine the molecular weight of your selected protein.

Part I: Column Pouring and Calibration

Materials

1 chromatographic columns, 1.0 cm (diameter) X 20 cm (length)

Sephadex G-150 -superfine resin equilibrated with 0.1 M phosphate buffer, pH 7.0

0.1 M phosphate buffer, pH 7.0

standards: solution containing blue dextran yeast alcohol dehydrogenase, BSA, and cytochrome-c.

test tubes and rack, ring stand, clamps, pinch clamps

selected protein sample: 1 mg/mL sample of your selected protein

96-well plate

BCA reagents

Procedure

Label the column (G-150) with your names. Use the pinch clamp to close the effluent tubes of the column. Pour several mL phosphate buffer into the column. Gently open the pinch clamp to allow some of the buffer to flow out of the column. The object of this step is to eliminate air bubbles from the bottom of the column and the effluent tubes. Once the air bubbles are eliminated, close the pinch clamp. There should be several mL of buffer remaining in the column at this point. Gently swirl the Sephadex G-50 and pour into the column labelled "G-50". After the resin as settled, open the pinch clamp and allow the column to drip until the settled chromatographic bed reaches a height of about 18 cm. *Do not allow the Column to Run Dry*!! Add phosphate buffer to the column to keep the fluid level above the resin bed.

Once an 18 cm resin bed has been established, close the column effluent tube with the pinch clamp. Measure the actual height of the resin bed. Remove most of the buffer from the top of the column. Carefully open the pinch clamp and allow the column to drip until the liquid level reaches the top if the resin bed. Immediately add 0.2 mL of the sample to the top of the resin by allowing the solution to gently run down the wall of the column. Place the effluent tube in the first test tube in the test tube rack (this will be fraction 1) and open the pinch clamp. Do not disturb the top of the resin! Allow the sample to enter the resin and then gently add a few drops of the phosphate buffer. Allow this buffer to penetrate the column and then gently add phosphate buffer to fill the column. Collect fractions until all the coloured material has eluted from the column. Close the patch clamp. Collect 1.0 mL of effluent in each tube. After 1 mL has been collected in the first tube (fraction 1), switch to the second tube (fraction 2) and collect the next 1 mL, etc. Perform a BCA assay on selected (every other, every third) tube to determine the amount of protein present. Note that blue dextran is not a protein; it will not be detected by the BCA assay. However, the molecule has a blue colour. The fraction containing the most intense blue colour is the should be taken as the Ve for the blue dextran (which is also the Vo for the column). Graphically represent the "elution profile" for your columns by plotting absorbance (y-axis) versus elution volume, mL (x-axis). You should have three series of data, one for each molecule/ color present.

Problems

1. Describe and explain your results.
2. From your data, calculate vt (total column volume), vo (void volume), and ve (elution volume for each solute).

   $vt = vo + vi + vm$

   where vm is the volume of the gel itself and vi is the volume included in the gel.

   Since vm is very small, it is assumed that

   $vt = vi + vo.$

3. Also be sure to include answers to the assigned problems.

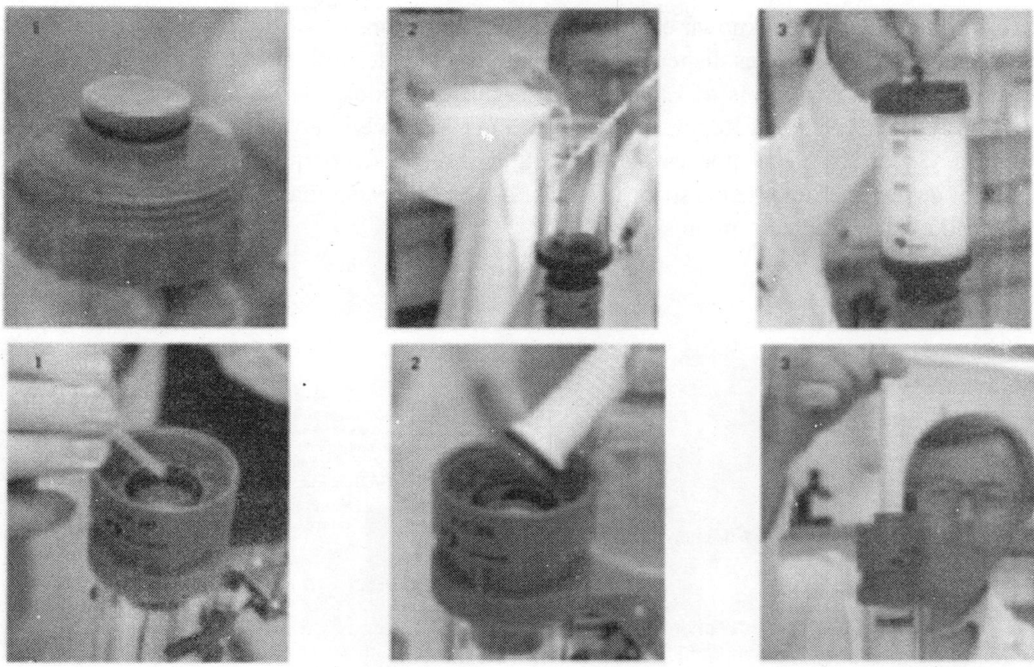

Fig. 2.57 : The Different photographs of gel Fitration Experiment

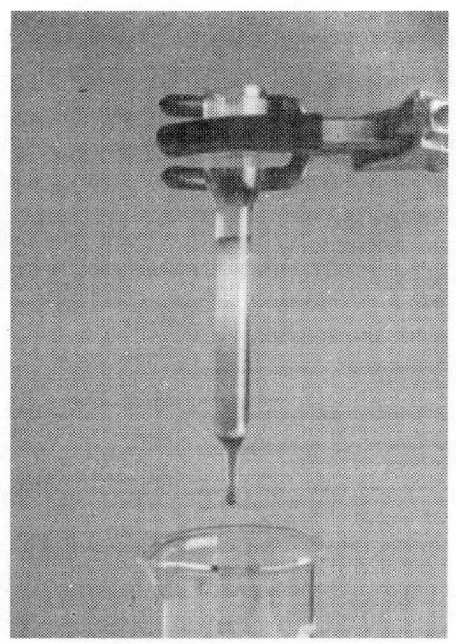

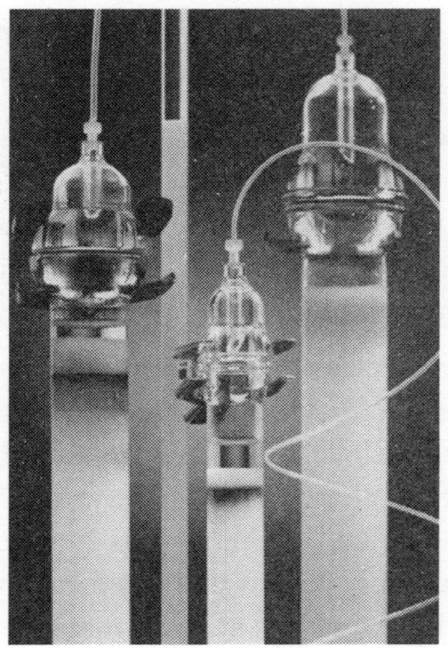

Fig. 2.57A & 2.57B. The Actual Gel Fitration colun in action

ELECTROPHORESIS

Electrophoresis is the best-known electrokinetic phenomenon. It was discovered by Reuss in 1807. He observed that clay particles dispersed in water migrate under influence of an applied electric field. There are detailed descriptions of Electrophoresis in many books on Colloid and Interface Science. There is an IUPAC Technical Report prepared by a group of well known experts on the electrokinetic phenomena. Generally, electrophoresis is the motion of dispersed particles relative to a fluid under the influence of an electric field that is space uniform. Alternatively, similar motion in a space non-uniform electric field is called dielectrophoresis.

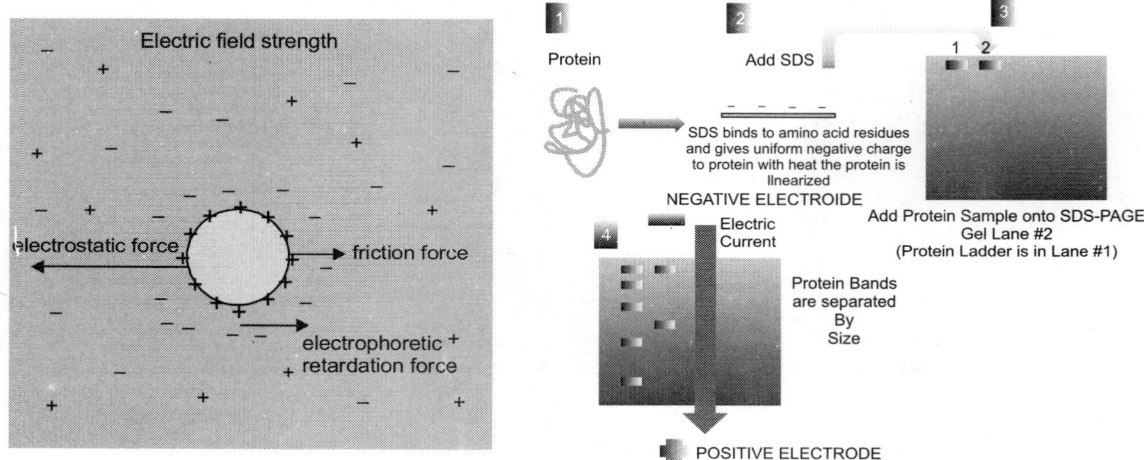

Fig. 2.58 and 2.58A: The Pattern of the electrophoresis system (diagramatic representation)

Electrophoresis occurs because particles dispersed in a fluid almost always carry an electric surface charge. An electric field exerts electrostatic Coulomb force on the particles through these charges. Recent molecular dynamics simulations, though, suggest that surface charge is not always necessary for electrophoresis and that even neutral particles can show electrophoresis due to the specific molecular structure of water at the interface.

The electrostatic Coulomb force exerted on a surface charge is reduced by an opposing force which is electrostatic as well. According to double layer theory, all surface charges in fluids are screened by a diffuse layer. This diffuse layer has the same absolute charge value, but with opposite sign from the surface charge. The electric field induces force on the diffuse layer, as well as on the surface charge. The total value of this force equals to the first mentioned force, but it is oppositely directed. However, only part of this force is applied to the particle. It is actually applied to the ions in the diffuse layer. These ions are at some distance from the particle surface. They transfer part of this electrostatic force to the particle surface through viscous stress. This part of the force that is applied to the particle body is called **electrophoretic retardation force**.

There is one more electric force, which is associated with deviation of the double layer from spherical symmetry and surface conductivity due to the excess ions in the diffuse layer. This force is called the **electrophoretic relaxation force**.

All these forces are balanced with hydrodynamic friction, which affects all bodies moving in viscous fluids with low Reynolds number. The speed of this motion v is proportional to the electric field strength E if the field is not too strong. Using this assumption makes possible the introduction of electrophoretic mobility μ_e as coefficient of proportionality between particle speed and electric field strength:

$$\mu_e = \frac{\upsilon}{E}$$

Multiple theories were developed during 20th century for calculating this parameter. Ref. 2 provides an overview.

Theory

The most known and widely used theory of electrophoresis was developed by Smoluchowski in 1903

$$\mu_e = \frac{\varepsilon\varepsilon_0\varsigma}{\eta}$$

where ε is the dielectric constant of the dispersion medium, ε_0 is the permitivity of free space ($C^2\ N^{-1}\ m^{-2}$), η is dynamic viscosity of the dispersion medium (Pa s), and ζ is zeta potential (i.e., the electrokinetic potential of the slipping plane in the double layer).

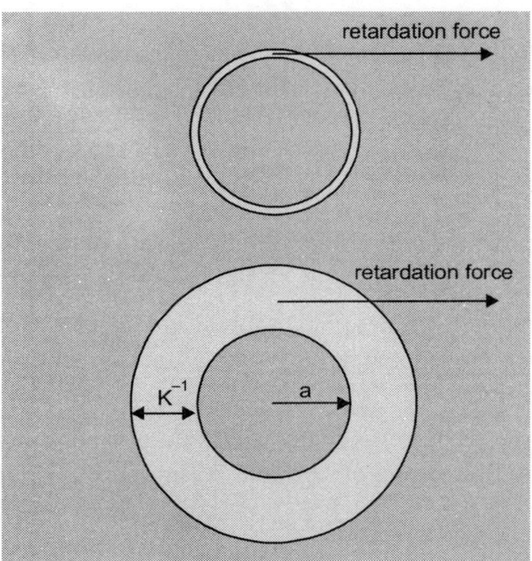

Fig. 2.59 : The Force Orientation of the protein molecules in the electrophoresis system

Smoluchowski theory is very powerful because it works for dispersed particles of any shape and any concentration, when it is valid. Unfortunately, it has limitations of its validity. It follows, for instance, from the fact that it does not include Debye length κ^{-1}. However, Debye length must be important for electrophoresis, as follows immediately from the Figure on the right. Increasing thickness of the DL leads to removing point of retardation force further from the particle surface. The thicker DL, the smaller retardation force must be.

Detailed theoretical analysis proved that Smoluchowski theory is valid only for sufficiently thin DL, when Debye length is much smaller than particle radius *a*:

$$\kappa a >> 1$$

This model of "thin Double Layer" offers tremendous simplifications not only for electrophoresis theory but for many other electrokinetic theories. This model is valid for most aqueous systems because the Debye length is only a few nanometers there. It breaks only for nano-colloids in solution with ionic strength close to water

Smoluchowski theory also neglects contribution of surface conductivity. This is expressed in modern theory as condition of small Dukhin number

$$Du << 1$$

Creation of electrophoretic theory with wider range of validity was a purpose of many studies during 20th century.

One of the most known considers an opposite asymptotic case when Debye length is larger than particle radius:

$$\kappa a < 1$$

It is called the "thick Double Layer" model. Corresponding electrophoretic theory was created by Huckel in 1924. It yields the following equation for electrophoretic mobility:

$$\mu_e = \frac{2\varepsilon\varepsilon_0\varsigma}{3\eta}$$

This model can be useful for some nano-colloids and non-polar fluids, where Debye length is much larger.

There are several analytical theories that incorporate surface conductivity and eliminate restriction of the small Dukhin number. Early pioneering work in that direction dates back to Overbeek and Booth.

Modern, rigorous theories that are valid for any Zeta potential and often any κa, stem mostly from the Ukrainian (Dukhin, Shilov and others) and Australian (O'Brien, White, Hunter and others) Schools.

Historically the first one was Dukhin-Semenikhin theory. Similar theory was created 10 years later by O'Brien and Hunter. Assuming **thin Double Layer**, these theories would yield results that are very close to the numerical solution provided by O'Brien and White.

The Electrophoresis System: The Electrophoresis Chamber and the Complete Unit

Electrophoresis chamber is quite easy, inexpensive and fun. After you have made your chamber, try it out using the Colourful Electrophoresis experiments.

SAFETY: Remember to wear safety glasses at all times while constructing this chamber.

TIP: Leave the paper on each Plexiglas piece until you are ready to use it. Write the name of each piece of Plexiglas® on the paper covering the surface before beginning construction. Remove the paper as you work with each piece.

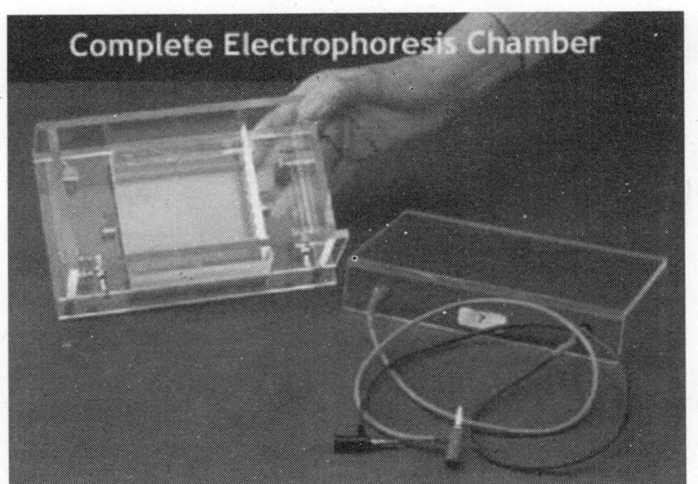

Fig. 2.60 and 2.60A : The Electrophoresis chamber and the complete unit (this complete product is also available with tarson)

Agarose gel electrophoresis is a method used in biochemistry and molecular biology to separate DNA, or RNA molecules by size. This is achieved by moving negatively charged nucleic acid molecules through an agarose matrix with an electric field (electrophoresis). Shorter molecules move faster and migrate farther than longer ones.

Applications

- Estimation of the size of DNA molecules following restriction enzyme digestion, e.g. in restriction mapping of cloned DNA.
- Analysis of PCR products, e.g. in molecular genetic diagnosis or genetic fingerprinting
- Separation of restricted genomic DNA prior to Southern transfer, or of RNA prior to Northern transfer.

---

if any student wants to make his or her's own electrophoresis unit (as it is very costly to purchase) consult the author's Architecture of Chemistry an instrumental approach.

The advantages are that the gel is easily poured, does not denature the samples. The samples can also be recovered. The disadvantages are that gels can melt during electrophoresis, the buffer can become exhausted, and different forms of genetic material may run in unpredictable forms. After the experiment is finished, the resulting gel can be stored in a plastic bag in refrigerator.

Factors Affecting Migration

The most important factor is the length of the DNA molecule, smaller molecules travel farther. But conformation of the DNA molecule is also a factor. To avoid this problem linear molecules are usually separated, usually DNA fragments from a restriction digest, linear DNA PCR products, or RNAs.

Increasing the agarose concentration of a gel reduces the migration speed and enables separation of smaller DNA molecules. The higher the voltage, the faster the DNA moves. But voltage is limited by the fact that it heats and ultimately causes the gel to melt. High voltages also decrease the resolution (above about 5 to 8 V/cm).

Conformations of a DNA plasmid that has not been cut with a restriction enzyme will move with different speeds (slowest to fastest): nicked or open circular, linearised, or supercoiled plasmid.

Visualisation: Ethidium Bromide (EtBr) and Dyes

The most common dye used to make DNA or RNA bands visible for agarose gel electrophoresis is ethidium bromide, usually abbreviated as EtBr. It fluoresces under UV light when intercalated into DNA (or RNA). By running DNA through an EtBr-treated gel and visualizing it with UV light, any band containing more than ~20ng DNA becomes distinctly visible. EtBr is a known carcinogen, however, and safer alternatives are available.

SYBR Green I is another dsDNA stain, produced by Invitrogen. It is more expensive, but 25 times more sensitive, and possibly safer than EtBr, though there is no data addressing its mutagenicity or toxicity in humans. SYBR Safe is a variant of SYBR Green that has been shown to have low enough levels of mutagenicity and toxicity to be deemed nonhazardous waste under U.S. Federal regulations. It has similar sensitivity levels to EtBr, but, like SYBR Green, is significantly more expensive.

Since EtBr stained DNA is not visible in natural light, scientists mix DNA with negatively charged **loading buffers** before adding the mixture to the gel. Loading buffers are useful because they are visible in natural light (as opposed to UV light for EtBr stained DNA), and they co-sediment with DNA (meaning they move at the same speed as DNA of a certain length). Xylene cyanol and Bromophenol blue are common loading buffers; they run about the same speed as DNA fragments that are 5000 bp and 300 bp in length respectively, but the precise position varies with percentage of the gel. Other less frequently used progress markers are Cresol Red and Orange G which run at about 125 bp and 50 bp.

Percent Agarose and Resolution Limits

Agarose gel electrophoresis can be used for the separation of DNA fragments ranging from 50 base pair to several megabases (millions of bases) using specialized apparatus. The distance between DNA bands of a given length is determined by the per cent agarose in the gel. In general lower concentrations of agarose are better for larger molecules because they result in greater separation between bands that are close in size. The disadvantage of higher concentrations is the long run times (sometimes days). Instead high percentage agarose gels should be run with a pulsed field electrophoresis (PFE), or field inversion electrophoresis.

Most agarose gels are made with between 0.7% (good separation or resolution of large 5–10kb DNA fragments) and 2% (good resolution for small 0.2–1kb fragments) agarose dissolved in electrophoresis buffer. Up to 3% can be used for separating very tiny fragments but a vertical polyacrylamide gel is more appropriate in this case. Low percentage gels are very weak and may break when you try to lift them. High percentage gels are often brittle and do not set evenly. 1% gels are common for many applications.

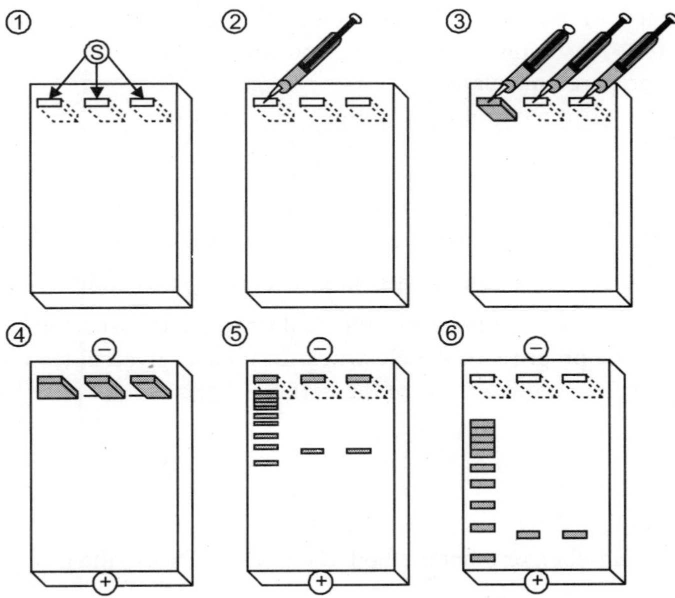

Fig. 2.61 : The Diagramatic representation of the agros gel electrophoresis

Buffers

There are a number of buffers used for agarose electrophoresis. The most common being: tris acetate EDTA (TAE), Tris/Borate/EDTA (TBE) and Sodium borate (SB). TAE has the lowest buffering capacity but provides the best resolution for larger DNA. This means a lower voltage and more time, but a better product. SB is relatively new and is ineffective in resolving fragments larger than 5 kbp; However, with its low conductivity, a much higher voltage could be used (up to 35 V/cm), which means a shorter analysis time for routine electrophoresis. As low as one base pair size difference could be resolved in 3% agarose gel with an extremely low conductivity medium (1 mM Lithium borate).

Analysis

After electrophoresis the gel is illuminated with an ultraviolet lamp (usually by placing it on a light box, while using protective gear to limit exposure to ultraviolet radiation) to view the DNA bands. The ethidium bromide fluoresces reddish-orange in the presence of DNA. The DNA band can also be cut out of the gel, and can then be dissolved to retrieve the purified DNA. The gel can then be photographed usually with a digital or polaroid camera. Although the stained nucleic acid fluoresces reddish-orange, images are usually shown in black and white.

Gel electrophoresis research often takes advantage of software-based image analysis tools, such as ImageJ.

Materials and Method

Typically 10-30 µl/sample of the DNA fragments to separate are obtained, as well as a mixture of DNA fragments (usually 10-20) of known size (after processing with DNA size markers either from a commercial source or prepared manually).

- Buffer solution, usually TBE buffer or TAE 1.0x, pH 8.0
- Agarose
- An ultraviolet-fluorescent dye, ethidium bromide, (5.25 mg/ml in H_2O). The stock solution be careful handling this. Alternative dyes may be used, such as SYBR Green.
- Nitrile rubber gloves. Latex gloves do not protect well from ethidium bromide
- A colour marker dye containing a low molecular weight dye such as "bromophenol blue" (to enable tracking the progress of the electrophoresis) and glycerol (to make the DNA solution denser so it will sink into the wells of the gel).
- A gel rack
- A "comb"
- Power Supply
- UV lamp or UV lightbox or other method to visualize DNA in the gel

Preparation

There are several methods for preparing gels. A common example is shown here. Other methods might differ in the buffering system used, the sample size to be loaded, the total volume of the gel (typically thickness is kept to a constant amount while length and breadth are varied as needed). Most agarose gels used in modern biochemistry and molecular biology are prepared and run horizontally.

1. Make a 1% agarose solution in 100ml TAE, for typical DNA fragments (see figures). A solution of up to 2-4% can be used if you analyze small DNA molecules, and for large molecules, a solution as low as 0.7% can be used.

> Wear gloves from here on, ethidium bromide is a mutagen

2. Carefully bring the solution just to the boil to dissolve the agarose, preferably in a microwave oven.

3. Let the solution cool down to about 60 °C at room temperature, or water bath. Stir or swirl the solution while cooling.
    1. Add 5 μl ethidium bromide stock (10 mg/ml) per 100 ml gel solution for a final concentration of 0.5 ug/ml. Be very careful when handling the concentrated stock. Some researchers prefer not to add ethidium bromide to the gel itself, instead soaking the gel in an ethidium bromide solution after running.
    2. Stir the solution to disperse the ethidium bromide, then pour it into the gel rack.
    3. Insert the comb at one side of the gel, about 5-10 mm from the end of the gel.
    4. When the gel has cooled down and become solid, carefully remove the comb. The holes that remain in the gel are the wells or slots.
    5. Put the gel, together with the rack, into a tank with TAE. Ethidium bromide at the same concentration can be added to the buffer. The gel must be completely covered with TAE, with the slots at the end electrode that will have the negative current.

Procedure

After the gel has been prepared, use a micropipette to inject about 2.5 μl of stained DNA (a DNA ladder is also highly recommended). Close the lid of the electrophoresis chamber and apply current (typically 100 V for 30 minutes with 15 ml of gel). The colored dye in the DNA ladder and DNA samples acts as a "front wave" that runs faster than the DNA itself. When the "front wave" approaches the end of the gel, the current is stopped. The DNA is stained with ethidium bromide, and is then visible under ultraviolet light.

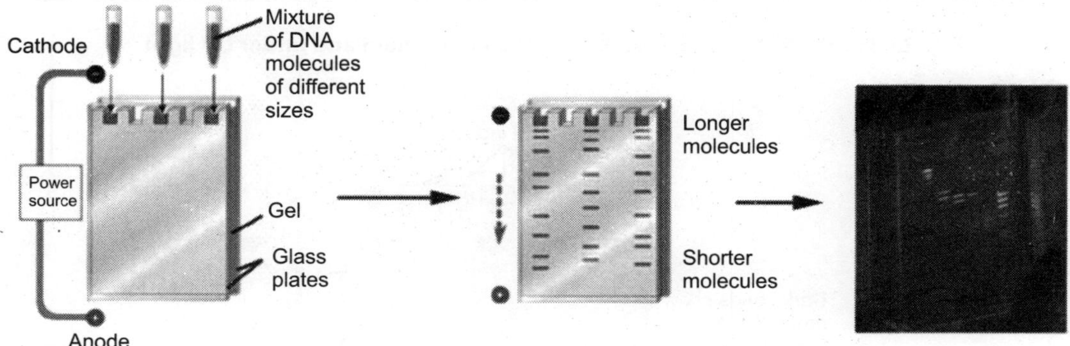

Restriction Map—a type of physical "map" of the banding pattern seen in gel electropherograms made from treating chromosomal DNA with restriction enzymes and then electrophoresing the fragments
...a kind of "fingerprint" of the DNA fragments from one restriction enzyme.

Fig. 2.62: The method before the interpretation of the electrophoresis pattern

1. The agarose gel with three slots/wells (S).
2. Injection of DNA ladder (molecular weight markers) into the first slot.
3. DNA ladder injected. Injection of samples into the second and third slot.

4. A current is applied. The DNA moves toward the positive anode due to the negative charges on its phosphate backbone.
5. Small DNA strands move fast, large DNA strands move slowly through the gel. The DNA is not normally visible during this process, so the marker dye is added to the DNA to avoid the DNA being run entirely off the gel. The marker dye has a low molecular weight, and migrates faster than the DNA, so as long as the marker has not run past the end of the gel, the DNA will still be in the gel.
6. Add the colour marker dye to the DNA ladder.

Fig. 2.63 and 2.63A: The gel Slab after the experiment and under UV light

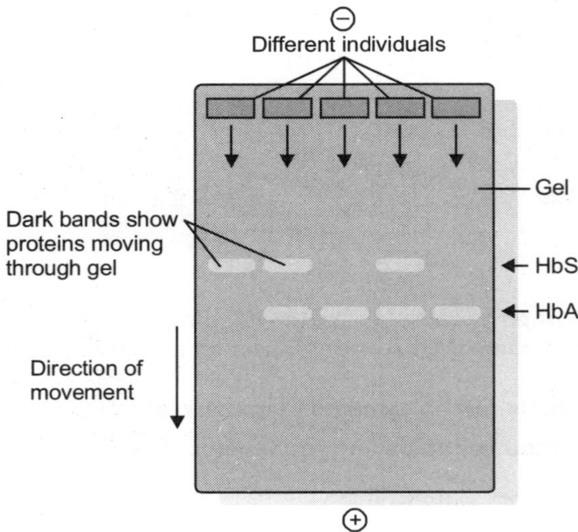

Fig. 2.63B: The digaramatic interpretration of a gel slab after electrophoresis

ULTRACENTRIFUGATION

When you put a particle in a centrifugal field, it is acted upon by the centrifugal force, which is proportional to the molecular weight (M), to the square of the speed (angular velocity, rpm) of the rotor (w^2) and to the distance of the solution from the center of rotation (r): Cent. Force ~ Mw^2r. Acting in the opposite direction to particle motion is friction. This frictional force is proportional to the velocity of the particle (V) and to a coefficient of friction (f) that depends on the shape of the particle: Friction force ~ fV. The particle will accelerate until a velocity is achieved such that these two opposing forces are equal, after which the particle will continue to sediment, but at a constant velocity. Setting the 2 forces equal to each other we have: $Mw^2r = fV$, and so:

$$V \sim Mw^2r/f.$$

In general, this means that velocity increases as mass (molecular weight) increase, and velocity decreases as the cross-sectional size (diameter) increases (*f* increases with cross-sectional size).

V depends on the particular machine and on the speed of rotation. So you can compare results irregardless of machine and speed, you calculate the Svedberg constant or S value instead of V. $S = V/w^2r$. Macromolecules usually have S values between 1 and 100×10^{-13} sec. A value of 1×10^{-13} sec is called one Svedberg unit or S. Many molecules are known by their S values, for example 23S ribosomal RNA. Its S value was known before its function or molecular weight, and so has become its name.

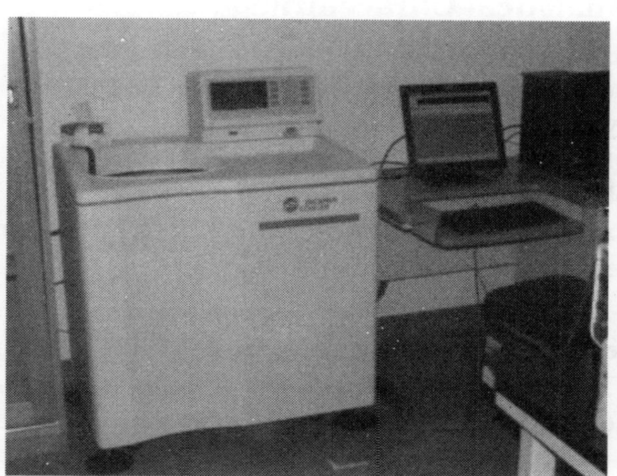

Fig. 2.64 : The Ultracentrifuge

Molecules of different S values are usually separated by layering them on top of a dense solution (such as sucrose) and then running the centrifuge to force the molecules through the solution. The molecules of larger S value go farther because of either larger molecular weight or less frictional drag or both. You stop the centrifuge when the different molecules have travelled different distances down the tube and then remove the solution by puncturing the bottom of the tube or pumping the solution off the top. The invention of the ultracentrifuge (a super fast centrifuge) by Svedberg was what originally made the purification of macromolecules possible.

The **ultracentrifuge** is a centrifuge optimized for spinning a rotor at very high speeds, capable of generating acceleration as high as 1,000,000 g (9,800 km/s²). There are two kinds of ultracentrifuges, the preparative and the analytical ultracentrifuge. Both classes of instruments find important uses in molecular biology, biochemistry and polymer science. Theodor Svedberg invented the analytical ultracentrifuge in 1923, and won the Nobel Prize in Chemistry in 1926 for his research on colloids and proteins using the ultracentrifuge.

The vacuum ultracentrifuge was invented by Edward Greydon Pickels. It was his contribution of the vacuum which allowed a reduction in friction generated at high speeds. Vacuum systems also enabled the maintenance of constant temperature.

In 1946, Pickels cofounded Spinco (Specialized Instruments Corp.) and marketed an ultracentrifuge based on his design. Pickels, however, considered his design to be complicated and developed a more "foolproof" version. But even with the enhanced design, sales of the technology remained low, and Spinco almost went bankrupt. The company survived and was the first to commercially manufacture ultracentrifuges, in 1947. In 1949, Spinco introduced the Model L, the first preparative ultracentrifuge to reach a maximum speed of 40,000 rpm. In 1954, Beckman Instruments (now Beckman Coulter) purchased the company, forming the basis of its Spinco centrifuge division.

Analytical Ultracentrifuge

In an analytical ultracentrifuge, a sample being spun can be monitored in real time through an optical detection system, using ultraviolet light absorption and/or interference optical refractive index sensitive system. This allows the operator to observe the evolution of the sample concentration versus the axis of rotation profile as a result of the applied centrifugal field. With modern instrumentation, these observations are electronically digitized and stored for further mathematical analysis. Two kinds of experiments are commonly performed on these instruments: sedimentation velocity experiments and sedimentation equilibrium experiments.

Sedimentation velocity experiments aim to interpret the entire time-course of sedimentation, and report on the shape and molar mass of the dissolved macromolecules, as well as their size-distribution. The size resolution of this method scales approximately with the square of the particle radii, and by adjusting the rotor speed of the experiment size-ranges from 100 Da to 10 GDa can be covered. Sedimentation velocity experiments can also be used to study reversible chemical equilibria between macromolecular species, by either monitoring the number and molar mass of macromolecular complexes, by gaining information about the complex composition from multi-signal analysis exploiting differences in each components spectroscopic signal, or by following the composition dependence of the sedimentation rates of the macromolecular system, as described in Gilbert-Jenkins theory.

Sedimentation equilibrium experiments are concerned only with the final steady-state of the experiment, where sedimentation is balanced by diffusion opposing the concentration gradients, resulting in a time-independent concentration profile. Sedimentation equilibrium distributions in the centrifugal field are characterized by Boltzmann distributions. This experiment is insensitive to the shape of the macromolecule, and directly reports on the molar mass of the macromolecules and, for chemically reacting mixtures, on chemical equilibrium constants.

The kinds of information that can be obtained from an analytical ultracentrifuge include the gross shape of macromolecules, the conformational changes in macromolecules, and size distributions of macromolecular samples. For macromolecules, such as proteins, that exist in chemical equilibrium with different non-covalent complexes, the number and subunit stoichiometry of the complexes and equilibrium constant constants can be studied.

Preparative Ultracentrifuge

Preparative ultracentrifuges are available with a wide variety of rotors suitable for a great range of experiments. Most rotors are designed to hold tubes that contain the samples. *Swinging bucket rotors* allow the tubes to hang on hinges so the tubes reorient to the horizontal as the rotor initially accelerates. *Fixed angle rotors* are made of a single block of metal and hold the tubes in cavities bored at a predetermined angle. *Zonal rotors* are designed to contain a large volume of sample in a single central cavity rather than in tubes. Some zonal rotors are capable of dynamic loading and unloading of samples while the rotor is spinning at high speed.

Preparative rotors are used in biology for pelleting of fine particulate fractions, such as cellular organelles (mitochondria, microsomes, ribosomes) and viruses. They can also be used for gradient separations, in which the tubes are filled from top to bottom with an increasing concentration of a dense substance in solution. Sucrose gradients are typically used for separation of cellular organelles. Gradients of caesium salts are used for separation of nucleic acids. After the sample has spun at high speed for sufficient time to produce the separation, the rotor is allowed to come to a smooth stop and the gradient is gently pumped out of each tube to isolate the separated components.

Hazards

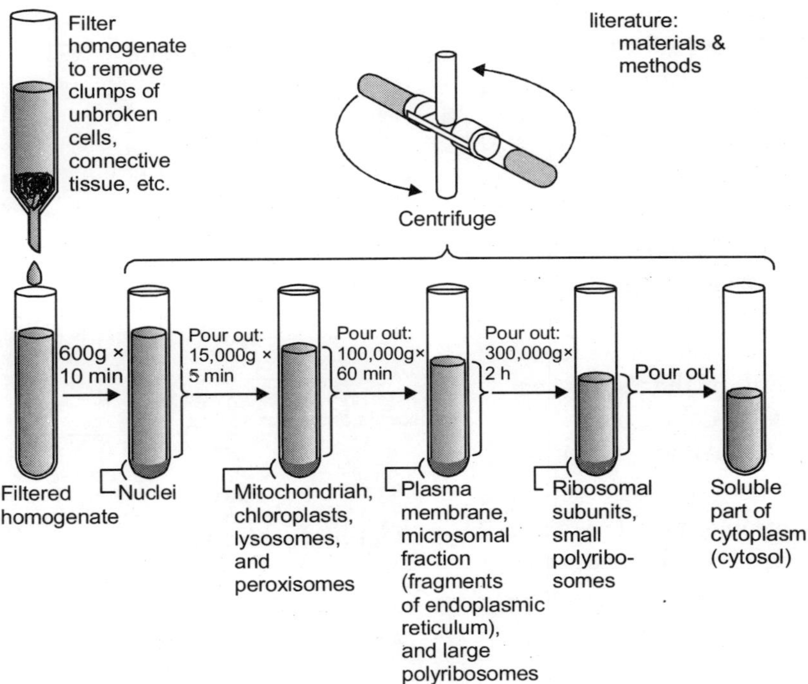

Fig. 2.65 : The Different biological components getting separated by means of differencial centrifugation

The tremendous rotational kinetic energy of the rotor in an operating ultracentrifuge makes the catastrophic failure of a spinning rotor a serious concern. The stresses of routine use and harsh chemical solutions eventually cause rotors to deteriorate. Proper use of the instrument and rotors within recommended limits and careful maintenance of rotors to prevent corrosion and to detect deterioration are necessary to avoid this hazard.

Differential Centrifugation. Generally, the cellular homogenate is first filtered or centrifuged at relatively low speeds to remove unbroken cells. Then centrifugation of the homogenate at a slightly faster speed or for a longer duration will selectively pellet the nucleus—the largest organelle (usually 5–10 μm in diameter). A centrifugal force of 600 *g* (600 times the force of gravity) is necessary to sediment nuclei; this is generated by a typical centrifuge rotor operating at 500 revolutions per minute (rpm). The undeposited material (the supernatant) is next centrifuged at a higher speed (15,000*g* × 5 min), which deposits the mitochondria, chloroplasts, lysosomes,and peroxisomes. A subsequent centrifugation in the ultracentrifuge (100,000 *g* × 60 min) results in deposition of the plasma membrane, fragments of the endoplasmic reticulum, and large polyribosomes. A force of 100,000 *g* requires about 50,000 rpm in an ultracentrifuge; at this speed, the rotor chamber is kept in a high vacuum to reduce heating due to friction between air and the spinning rotor. The recovery of ribosomal subunits, small polyribosomes, and particles such as complexes of enzymes requires additional centrifugation at still higher speeds. Only the cytosol—the soluble aqueous portion of the cytoplasm—remains undeposited after centrifugation at 300,000 *g* for 2 hours.

Differential centrifugation schemes involve stepwise increases in the speed of centrifugation. At each step, more dense particles are separated from less dense particles, and the successive speed of centrifugation is increased until the target particle is pelleted out. The final supernatant is removed, the pellet is resuspended, and further study or purification can be done on it. The fractionation of rat liver is an example of how this process works:

An important thing to note is that there is cross contamination between the second and third pellets. Mitochondria show up in Pellet 3 and lysosomes show up in Pellet 2. This shows that the separations made by this technique aren't absolute purifications, but relative enrichments of organelles.

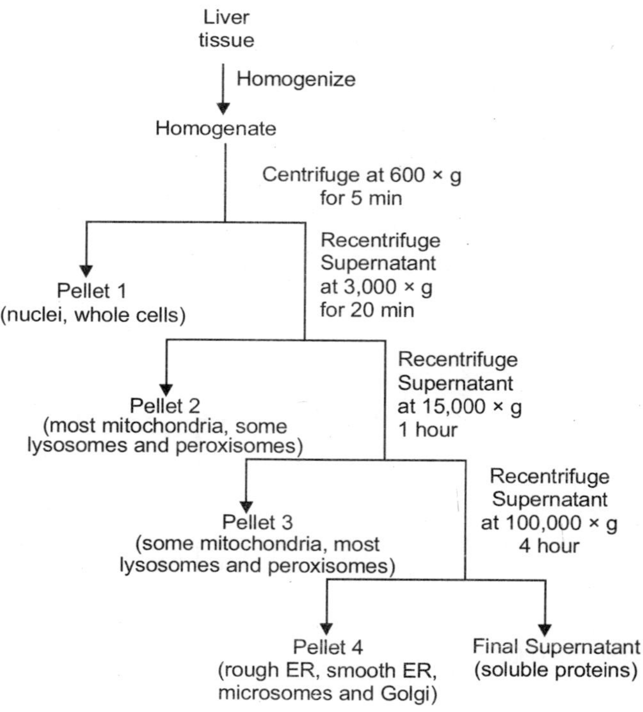

Fig. 2.66 : The flow chart for the protein separation from liver tissue

NINHYDRIN REACTION

Ninhydrin (2,2-Dihydroxyindane-1,3-dione) is a chemical used to detect ammonia or primary and secondary amines. When reacting with these free amines, a deep blue or purple colour known as Ruhemann's purple is evolved. Ninhydrin is most commonly used to detect fingerprints, as amines left over from peptides and proteins (terminal amines or lysine residues) sloughed off in fingerprints react with ninhydrin.

Uses

Ninhydrin can also be used to monitor deprotection in solid phase peptide synthesis (Kaiser Test). When the growing peptide chain is deprotected, a ninhydrin test yields blue. If the next peptide residue is coupled then the test is colourless or yellow.

Ninhydrin is also used in amino acid analysis of proteins: Most of the amino acids are hydrolyzed and reacted with ninhydrin except proline; Also, certain amino acid chains are degraded. Therefore, separate analysis is required for identifying such amino acids that either react differently or don't react at all with ninhydrin. The rest of the amino acids are then quantified colourimetrically after separation by chromatography.

A solution suspected of containing the ammonium ion can be tested by ninhydrin by dotting it onto a solid support (such as silica gel); treatment with ninhydrin should result in a dramatic purple colour if the solution contains this species. In the analysis of a chemical reaction by thin layer chromatography (TLC), the reagent can also be used. It will detect, on the TLC plate, virtually all amines, carbamates and also, after vigorous heating, amides.

Reactivity

The carbon atom of a carbonyl bears a partial positive charge, so the central carbon of a 1,2,3-tricarbonyl is less stable and more electrophilic than a simple ketone. In most compounds, a carbonyl is more stable than the dihydroxy (hydrate) form. However, ninhydrin is a stable hydrate of the central carbon because this form does not have the destabilizing effect of adjacent carbonyl partial-positive centers. Indane-1,2,3-trione reacts readily with nucleophiles.

Note that in order to generate the ninhydrin chromophore, the amine is condensed with a molecule of ninhydrin to give a Schiff base. Thus only ammonia and primary amines can proceed past this step. At this step, there must also be an alpha proton (H* in the diagram) for Schiff base transfer, so an amine adjacent to a tertiary carbon cannot be detected by the ninhydrin test. The reaction of ninhydrin with secondary amines gives an iminium salt, which is also coloured, and this is generally yellow-orange in color.

DETERMINATION FO PROTEIN STRUCTURE WITH NINHYDRIN

Spectrophotometric method is quantitative in contrast to previous ones (1, 3, 4), because exclusion of oxygen during the reactions prevents fading of the red and blue colours of hydrindantin. Control of the oxygen content and of the pH provides for a clear distinction between hydrindantin reactions and those of ninhydrin. These controls are necessary, for, on the one hand, oxidation of hydrindantin yields 2 moles of ninhydrin, and, on the other hand, ninhydrin at certain pH values can form hydrin-dantin by way of o-carboxyphenylglyoxal. The probable structure of the red-coloured derivative of hydrindantin is indanone-enediol and not, as Ruhemann believed, the monovalent salt of intact hydrindantin. In the reaction of amino acids and hydrindantin, 1 mole of indanone-enediol is used up for each mole of Ruhemann's purple formed. The reaction with amines summarized in formulae I could be either a simultaneous or a sequential condensation with indanone-enediol and ninhydrin. The alternatives will be considered in a separate ,paper on the order of reaction, which the present method has made possible.

Appartus—

1. Spectrophotometers. The Beckman model DU instrument, with quartz prism, and the Coleman clinical instrument, model 6, were used.
2. Cuvettes and reaction vessels. For the Beckman instrument silica for the measurement of ultraviolet absorption, Corex for visible light.

The length of the light path was 1 cm. For the Coleman instrument,

Hamilton vessels were calibrated for length of the light path, which averaged 1.89 cm, and were used as reaction vessels as well.

Preparation of Hydrindantin- A solution of 0.5 gm. of ascorbic acid and 1 gm. of ninhydrin in 200 ml. of McIlvaine's buffer (0.1 M) at pH 3 washeated to 90° and the crystals allowed settling at room temperature. Recrystallization from hot acetone yielded 350 mg. of colourless anhydrous hydrindantin. The purity of the product was checked by elementary analysis and by its melting point. Preparation of Oxygen-Free Reaction Solutions. Since at room temperature water does not dissolve hydrindantin, it was dissolved in acetone in a concentration of 1 mg. per ml. 1 ml. was delivered to each Hamilton vessel and dried by passing a stream of nitrogen above, not in, the solution. 10 ml. of buffer solution, either alone as a control or containing another reagent under test, oxygen-free after a stream of nitrogen was passed through it for 5 minutes, were delivered to each vessel, which was lubricated and quickly closed so as to be air-tight. The gas was then removed by suction from a motor pump until the pressure was constant at about 2 mm. of Hg. Each vessel in the control and test group was again made air-tight and was immersed upright in a frame in a boiling, distilled water bath for a known time interval. Spectrophotometric Readings-The optical density was recorded at wave-lengths of 490 and 570 ml.c in the Coleman instrument. Sometimes the measurements were made as quickly as possible after removal from the boiling water bath, in order to record the optical density at a temperature close to 100". In such a case, it was found that three measurements could be taken comfortably in from 1 to 2 minutes. For the most part, they were taken at room temperature after cooling in a water bath, and again after passing air through the solutions for 3 minutes. Absorption Spectra of, Red and Blue Colours Derived from Hydrindantin Different from Ruhemann's Purple .Solutions of hydrindantineither in Sorensen's NaOH-borate buffer at pH 9.2 or in 0.4 N NaOH were prepared to be oxygen-free in cuvettes. The visible colors were constant for at least 48 hours, but ultraviolet absorption, without change in position of the maxima, slowly lessened until a constant value was reached at the end of 48 hours. The result is explained by hydrolysis of hydrindantin into diketohydrindol and ninhydrin, followed by irreversible transformation of ninhydrin into either o-carboxyphenylglyoxal or o-carboxymandelic acid, depending on the pH; either process can be detected by change in the ultraviolet absorption spectrum of ninhydrin (see MacFadyen.The spectrum, Curve 1, Fig. 2.66, for the blue colour is markedly different from the curve for Ruhemann's purple with respect to position and intensity of maxima. These differences invalidate Retinger's concept. Formation of Ruhemann's Purple from ol-Alanine and Hydrindantin Independent of NH_3 Pathway (Table below)-At pH 5.3 at 100", and anaerobically, the intensity of purple colour in the reaction of hydrindantin, 0.3 mM, with the amino acid was 5 times that for ammonium salts in the same concentration, 0.56 mM. Therefore, the concepts of Ruhemann (1) and Harding et al. (3, 4), necessitating an NH_3 pathway, are inadequate. The of the reaction.Hydrindantin Red Colour Disappears As Ruhemann's Purple is Formed, Mole for Mole (Table I)-Difficulties of quantitative estimation of concentrations of Ruhemann's purple and the red colour from.hydrindantin, together in reaction mixtures, were obviated in the following ways.

Table 2.27 : Formation of Ruhemann's Purple from α-Alaninne and Hydrindantin with Disappearance of Hydrindantin Red Color

Reaction time in boiling water bath	Observed optical densities, units × 10^3					Hydrindantin disappe-arance	Formation$ of Ruhe-mann's purpose
	a-Alanine			Present	Absent		
	P_{570}	D^{-O}_{490}	D^{+O}_{490}	R'_{490}	R''_{490}		
	a*	b	c	d = b – c	c	f = 687 (c-d) × 10^{-3}	g = af_1f_2×10^{-3} (average
min.						*mM per l.*	*mM per l.*
10	360	283$$	123	160	175	4.8-13.7	12.1
	360	290	122	168	180		
20	620	365	200	165	187	15.1-20.6	22.7
	640	363	199	164	194		
30	860	441	278	163	205	25.4–31.6	35.5
	880	440	281	159	200		
40	1010	465	324	141	212	39.2–52.9	46.4
	1000	498	338	160	218		
50	1150	492	370	122	232	74.8–77.0	63.5
	1280	541	418	123	234		
60	1290	546	434	112	230	77.0–81.0	76.4
	1360	566	450	116	228		

*For explanation of P_{570}, R_{490}, computation of the factor 687, and for f1 and f2 see the text.

$For $(NH_4)_2SO_4$ the results, in chronological order, were 2.2, 5.0, 5.7, 9.5, 14.3 and 15.7.

$$The third figure was estimated by interpolation.

Ruhemann's purple is decolorized at pH 5.3, even under anaerobic conditions, when its solutions are heated to boiling, but not significantly at room temperature. Therefore, estimates of the amount formed in a given time interval, in contrast to the amount present, required measurements of the intensity of the purple color remaining in solution of known amounts of the sodium salt of diketohydrindamine-diketohydrindyli(lcne) under the conditions of the cu-alaninc reaction with hydrindanbin including all reagents except oc-alanine. Such tests provided us with factors, fl, by which the optical density at X = 570 rnp observed at a given time interval of boiling was converted into the initial optical density before heating. Furthermore, while it is true that Ruhemann's purple obeys Beer's law, it was necessary to correct for spectrophotometric conditions of the Hamilton vessels and the Coleman instrument with respect to deviations from Beer's law, but not in the case of the red colour. For this purpose the sodium salt of diketohydrindamine-diketohydrin-dylidene was dissolved in buffer at pH 9, and the optical densities at varied concentration determined in the Hamilton vessels in the Coleman instrument. From these results factors, fi, were computed which when multiplied by the optical density at X = 570 mp, in so far as it is referable only to Ruhemann's purple, yielded the concentration of the sodium salt in micromoles per liter.

Application of the factors, depending on a clear distinction between the purple colour due to the sodium salt and the red colour due to hydrindantin,was made by taking advantage of two facts. Whereas Ruhemann's purple is stable when oxygenated at pH 10 to 11 at room temperature, the red colour is discharged in 3 minutes. Therefore, the optical density after oxygenation, measured at X = 570 m,u, D 72, when corrected in the usual manner for variation in length' of the light path and translucency from vessel to vessel, is representative of Ruhemann's purple and is denoted as P670, whereas the optical density difference due to oxygenation, measured at h = 490 rnp, DG: - D, when similarly corrected, is representative of the hydrindantin red color and is denoted by ROW.

The experiments were carried out with 0.5 mg. of cr-alanine (or 0.37 mg. of (NH&Sod) and 1 mg. of hydrindantin in 10 ml. of 0.1 M acetate buffer at pH 5.3. The solutions, in duplicate, were heated to boiling, anaerobically, for a given time interval of 10, 20, 30, 40, 50, or 60 minutes. Then they were cooled in ice water for 4 minutes and brought to room temperature at 27", about 25 minutes later. The optical densities in the absence of oxygen, D-O, were recorded. The vessels were opened, the pH of the solutions changed to 10 to 11 by adding about 0.02 ml. of 40 per cent NaOH, and the red colour was discharged by bubbling air through the solutions for 3 minutes. Then the optical densities, D+O, were recorded.The observed data were converted to micromolar concentrations (X) of substance responsible for the red or purple colors as follows: In the case of the red color, (X) = Rw/(E X 10m6 X 1 X a), where E is the molar absorption coefficient at X = 490 rnp and at pH 9.2, assuming complete hydrolysis of hydrindantin into diketohydrindol, 1 is the length of the light path in .cm., and CY is the ratio of optical density at pH 5.3 to that at pH 9.2. The numerical values were 1400, 1.89, and 0.55 respectively. The equation simplifies to (X) = 687R. In the case of the purple color, (X) = Psm X fl X f2, previously described. For the time intervals 10,20, 30, 40, 50, and 60 minutes the numerical values of fl were 1.31, 1.44, 1.58, 1.74, 1.91, and 2.10, respectively, and for f~ were 24.8, 25.2, 25.8, 26.5, 27.4, and 27.5, respectively.

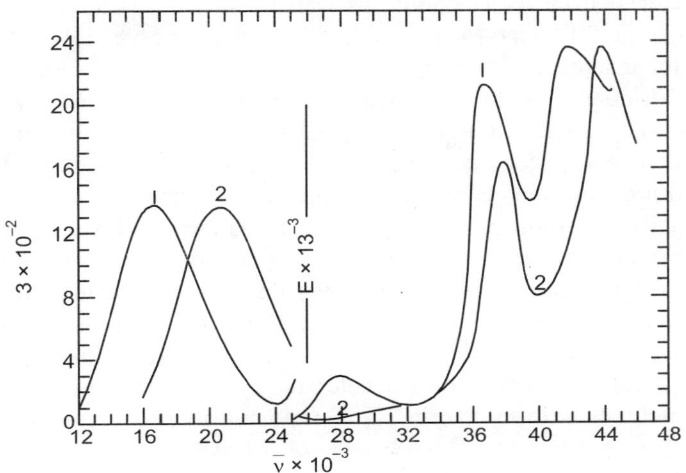

Fig. 2.67 : Absorption curves of the red and blue colours derived from hydrindantin. Curve 1, for the blue colour in 0.5 N NaOH; Curve 2, for the red colour at pH 9.2. The ordinates refer to the molar absorfition coefficients; the abscissae to the wave number per cm

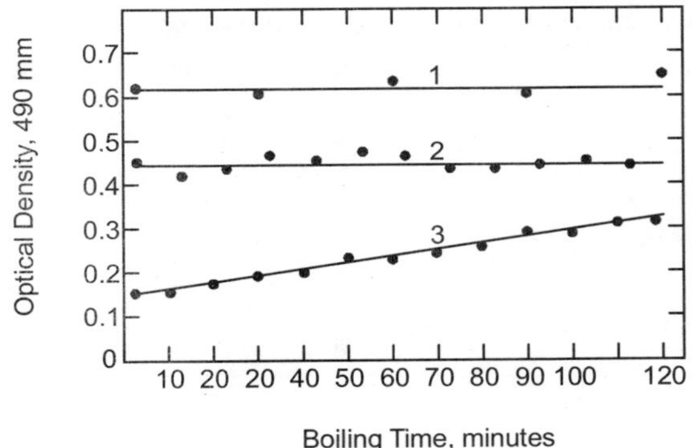

Fig. 2.68 : Hydrolysis of hydrindantin (0.31 mM) at pH 5.3 in boiling aqueous anaerobic solution with and without dimethyldihydroresorcinol. Curve 1, optical density at 100", with or without dimethyldihydroresorcinol; Curve 2, optical density after cooling to 25", in the presence of dimethyldihydroresorcinol; Curve 3, optical density after cooling to 25', without dimethyldihydroresorcinol

The disappearance of red color associated with formation of Ruhemann's purple was calculated from R&O, obtained when the reaction mixture contained cr-alanine, and from R, obtained when all reagents were present except cr-alanine. The disappearance, in terms of micromoles of hydrindantin per liter, = 687(R" - R')490.In the case of a-alanine, the results show that the disappearance of hydrindantin red color is proportional to the formation of Ruhemann's purple, within the limits of error of the method.Red and Blue Colors from Hydrindczntin Due to Dilcetohydrindol (Fig. 2.68). Our claim that the red colour evolved from hydrindantin, as well as the undisputed blue colour (l), is due to diketohydrindol rests on the following facts. (a) The blue colour is reversibly changed into the red by acid-base titration under anaerobic conditions, pK' = 12.3 at 25'. When the blue colour is formed, the other component of hydrindantin, namely ninhydrin, is changed into o-carboxymandelic acid (1) by irreversible internal oxidation-reduction. This change is complete in a few minutes, but the colour change is quantitatively reversible for days. Therefore, the change from blue to red does not necessitate the reformation of hydrindantin; the claim (10) that the red colour is due to the monovalent anion of hydrindantin is invalid. (b) The same play of colours with change in pH was observed by Hassall (12) in connection with the hydrolysis of acetoxyindan-dione to diketohydrindol, which was identified by adding ninhydrin to acidified solutions from which hydrindantin was obtained. (c) In acid solutions of hydrindantin dimethyldihydroresorcinol accelerates the formation of the red colour, which attains a constant intensity for a given concentration of hydrindantin. The explanation offered is hydrolysis of hydrindantin into the red colour and ninhydrin, accelerated by combinationwith the resorcinol (see Fig. 68). The resorcinol combines with ninhydrin.The compound, inert to oxygenation, can be detected spectrophoto-metrically in the solutions after oxygenation. (d) Above pH 7, there is no difference in red colour caused by dimethyldihydroresorcinol or by cooling the solutions from 100" to room temperature. If a red alkaline solution is acidified to pH 5.3, anaerobically, the colour fades as hydrindantin is precipitated. The data could be interpreted, as Ruhemann concluded (lo), to show that intact hydrindantin is responsible for the red colour. However, at a pH, temperature, and time interval (pH 10, 25°C, 24 hours) insuring complete irreversible transformation of ninhydrin to o-carboxyphenylglyoxal, acidification no longer causes reformation of hydrindantin, the fading of the red colour being what would be expected from its titration curve, pK' = 5.2. In this case, addition of ninhydrin causes and is necessary for precipitation

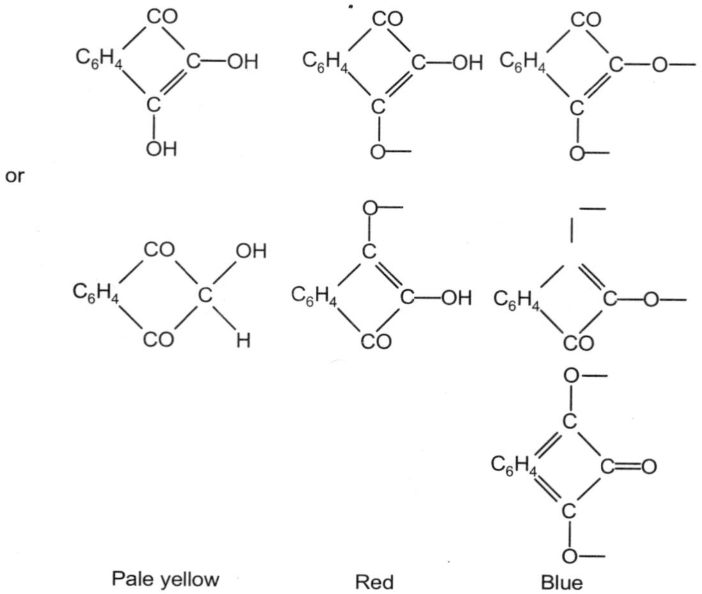

Pale yellow Red Blue

(II) Colored forms of indanone-enediol

of hydrindantin. (e) By careful adjustment of the concentration of added hydrosulfite, the intensity of the red colour from hydrindantin can be doubled. Having shown that the red and blue colours of hydrindantin solutions are derivatives of diketohydrindol readily convertible under anaerobic conditions one to the other and to diketohydrindol, simply by change in pH (pK' = 5.2 and 12.3), we conclude that the colours are due to the anions of diketohydrindol. Enolization of diketohydrindol would allow for two ionizable groups. Therefore, the red colour is attributable to the monovalent anion, the blue colour to the divalent anion. The ease of oxidation on exposure to air is consistent with the indene structure formable by enolization. The enediol formulae shown in II provide for resonance which would explain the difference in color and the chemical behavior of the substances.

2,2-dihydroxy-1H-indene-1,3(2H)-dione + 2-aminopropanoic acid → 2-hydroxy-1H-indene-1,3(2H)-dione + H_3C–CHO + CO_2 + NH_3

+ 2,2-dihydroxy-1H-indene-1,3(2H)-dione + NH_3 → 3 H_2O +

2-[(3-hydroxy-1-oxo-1H-inden-2-yl)imino]-1H-indene-1,3(2H)-dione

On reduction with hydrosulphite the red and blue colours disappear but careful oxygenation restores their original intensity. On the other hand, the effect of oxygenation is not reversible in the case of the blue colour and is only reversible in the case of the red colour if ninhydrin is present in, solution. Irreversibility is explained by oxidation of the red colour to o-carboxyphenylglyoxal and of the blue colour to o-carboxymandelic acid.

Mechanism of Reaction of a-Alanine with Hydrindantin-The disappearance of the red colour in this reaction may now be reconsidered. It is ascribed to conversion of indanone-enediol into Ruhemann's purple, the anion of diketohydrindamine-diketohydrindylidene. The alternative, reduction of Ruhemann's purple which would become colourless while the enediol was oxidized to colourless ninhydrin, is untenable, because the chromogenic reaction of a-alanine is faster with 1 mole of hydrindantin than with 2 moles of ninhydrin ($C_9H_6O_4$). The non-enolic component of Ruhemann's purple must be supplied by ninhydrin, also from hydrolysis of hydrindantin, for only 1 mole of indanone-enediol isused up for each mole of Ruhemann's purple formed. The details of the mechanism, summarized in formula, will be discussed in a paper on the order of the reactions of amines with hydrindantin and ninhydrin.

SEPARATIONS OF AMINO ACIDS

Chromatography is a word used to encompass a range of techniques in which mixtures of pure substances are separated into the individual substances by using a mobile phase (usually a liquid or gas) to push the mixture along a stationary phase (usually a solid or liquid coated on a solid). Because the individual substances have different molecular structures, they interact differently with both the stationary and mobile phases, and consequently are "pushed" at different rates by the mobile phase.

PAPER CHROMATOGRAPHY

A solution of the mixture to be separated is "spotted", usually from a micropipet, near one edge of a piece of filter paper, and the solvent is evaporated. Usually several sample and standard spots are placed along the edge. Then the chromatogram is "developed" by immersing that edge of the paper in a solvent that migrates through the paper as the mobile phase. The solvent often has two, three or four components, one of which is usually water. Development is normally done in a chamber that is saturated with solvent vapor. The water from the solvent, in particular, is adsorbed and tightly held on the paper fibers, so the sample components partition between the migrating mobile phase and the tightly-held water. If the bottom edge of the paper is immersed in the solvent, the process is "ascending" chromatography. Alternatively, the upper edge can be immersed in a trough of solvent and hung down for "descending" chromatography. Usually the process is stopped before the solvent front reaches the opposite edge of the paper, and the solvent is evaporated. The ratio of the distance moved by a particular component to the distance moved by the solvent front is the "R_f", or "retardation factor", which is characteristic of that component under the conditions used. After the separation, any strongly coloured spots are visible on the "chromatogram". Colourless materials can be visualized by heating the paper in an oven so that substances (but not the paper) char and leave black spots. Sometimes the paper is first sprayed with a solution of sulfuric acid for better charring. Fluorescent materials can be visualized with ultraviolet light. Reagents specific for certain components may be sprayed on to make colored spots. Radioactive spots can be located with a detector, or the chromatogram can be pressed against X-ray film for minutes or hours to expose the film. Sample spots can be tentatively identified if they have the same R_f values as known standard spots.

Sometimes spots, once located, are cut out so that the material in the spot can be recovered. There are also instruments that (more or less accurately) quantitate the material in the spot by measuring light absorbance.

To know more about these instrumental techniques please see The Architecture of Chemistry An Instrumental Approach by the same author.

Variant: Circular chromatography: the sample is applied to the center of a circle of filter paper. Then the mobile phase is slowly dripped onto the spot and the chromatogram develops as a circle.

Variant: Electrochromatography: as the mobile phase flows downward, an electrical field is applied transversely (electrophoresis) to get a two-dimensional separation of ionized molecules.

Variant: Two-dimensional chromatography: One spot is applied near a corner of a sheet of paper, and is developed with one solvent system. Then a second solvent system is applied from another edge so that mixed spots that did not separate with the first system may separate with the second one. These sheets may be as large as 75 cm × 75 cm.

Lab reports or procedures will usually specify the solvent mixture (e.g., methyl isobutyl ketone:isopropyl alcohol:ethyl acetate:water:ammonium hydroxide (50:45:35:18:3), the paper used and the means by which the spots are visualized.

THIN LAYER CHROMATOGRAPHY

Thin-Layer Chromatography (TLC) is a simple and inexpensive technique that is often used to judge the purity of a synthesized compound or to indicate the extent of progress of a chemical reaction. In this technique, a small quantity of a solution of the mixture to be analyzed is deposited as a small spot on a TLC plate, which consists of a thin layer of silica gel (SiO_2) or alumina (Al_2O_3) coated on a glass or plastic sheet. The plate constitutes the stationary phase. The sheet is then placed in a chamber containing a small amount of solvent, which is the mobile phase. The solvent gradually moves up the plate via capillary action, and it carries the deposited substances along with it at different rates. The desired result is that each component of the deposited mixture is moved a different distance up the plate by the solvent. The components then appear as a series of spots at different locations up the plate. Substances can be identified from their so-called R_f values. The R_f value for a substance is the ratio of the distance that the substance travels to the distance that the solvent travels up the plate. For example, an R_f value of 0.5 means that the spot corresponding to the substance travels exactly half as far as the solvent travels along the plate.

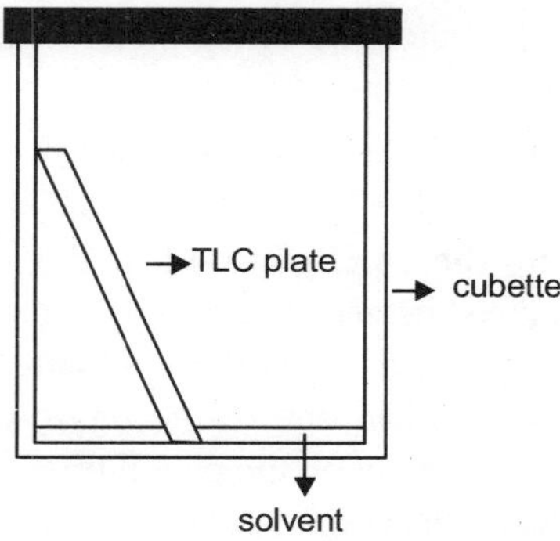

Fig. 2.69 : The Diagram of the TLC set up for experment

Fig. 2.70 : The Actual TLC set-up

The Process of TLC. Performing a TLC analysis consists of a number of steps: preparing a spotting capillary; marking the TLC plate; spotting the TLC plate; developing the TLC plate; drying the plate; visualizing the substance spots, and measuring the R_f values. We consider these steps in turn.

Preparing a spotting capillary. Glass capillaries used for spotting TLC plates are commercially available. However, it is occasionally necessary to make your own capillaries. To accomplish this, light a Bunsen burner and adjust for a medium flame. Hold a melting point capillary in the flame until it just begins to soften, then quickly pull the two ends of the capillary in opposite directions. The central, soft part of the glass will elongate and thin down to a capillary with very small diameter. Break the two pieces apart at the center of the thin portion to obtain two TLC spotting capillaries.

Marking the TLC Plate. Obtain a silica gel TLC plate that is approximately 2 cm wide and 5 cm long. Mark the TLC plate as follows using a pencil (pencil must be used rather than pen because inks are moved by many developing solvents). First, LIGHTLY draw a straight line parallel to the short dimension of the plate, about 1 cm from one end of the plate. Don't gouge the silica gel or make a trough with the pencil. Second, LIGHTLY make two small marks perpendicular to this line to divide the line into thirds. This subdivided line will serve as a guide for placing the substance spots, and as a point from which to measure R_f values. Third, LIGHTLY draw a second line parallel to the first line and about 1 cm from the other end of the plate. When you develop the plate, you will allow the solvent front to rise to this second line.

Activating the TLC plate. Place your marked TLC plate in an oven at 50-60°C for 15-20 minutes to "activate" it. Activation involves driving off water molecules that bond to the polar sites on the plate.

Spotting the TLC plate. Place the narrow end of one of your capillaries into a vial containing a solution of the substance to be analyzed, and allow the solution to rise in the capillary; this will happen spontaneously. Once the capillary is loaded, hold it vertically just above the subdivided pencil line on the plate and centered in one of the three sections. Lower it until the narrow end of the capillary just touches the plate right on the pencil line. You will observe that some of the solution leaves the capillary and deposits on the plate. Leave the capillary in contact with the plate only briefly so that the spot is no larger than 1 mm in diameter, then raise the capillary. Allow the solvent to completely evaporate from the spot (if the solvent is water, evaporation will be slow; you can hasten it by putting the plate in the oven for 5 minutes). Then, if desired, make a second deposit on the same spot with the capillary. Again allow the solvent to completely evaporate.

Developing the TLC plate. Pour the desired developing solvent into a small wide-mouth glass or plastic bottle to a depth of about 4-5 mm. Using tweezers, pick up the TLC plate at the top, which is the end opposite where the subdivided pencil line is drawn. Place it carefully in the developing bottle so that it stands somewhat but not excessively tilted—that is, the bottom of the slide should be somewhat away from the wall of the bottle, while the top of the slide rests against the wall of the bottle. It is important that you not allow the TLC slide to tilt too much when in the developing bottle. If the slide is excessively tilted, solvent will not advance uniformly along the plate and development will not take place properly. Similarly, if the bottom of the slide is against the wall of the bottle, solvent will advance more rapidly up the edges of the slide than in the middle, causing the substances to be pushed toward the center of the slide as they move up. Leave the slide in the chamber until solvent has advanced to the top pencil line on the slide. Development normally requires at least 30 minutes. When the solvent front has advanced to the top pencil line, use the tweezers to withdraw the slide from the chamber.

Drying the Plate. Place the plate flat on a clean dry surface and allow the solvent to completely evaporate. If the solvent is not highly volatile, this can be facilitated by placing the slide on a flat surface in an oven at a temperature of 50-60 °C (higher temperatures will melt the plastic substrate material). When the plate is completely dry, it is ready for visualization.

> Note: Ninhydrin Stains the Skin Purple, too, Due to the Presence of Protein (Amino Acids!). To avoid this, Wear Gloves When Handling It.

Visualization of the TLC Plate. If the substances being separated are coloured, the spots can be seen without any further effort. Using a pencil, draw a boundary around each spot that matches the shape of the spot. Many substances are colourless (white) and do not show up on the white silica gel unless steps are taken to make them visible. There are a number of techniques for doing this. *First* is the technique of iodination. The dry plate is placed in a chamber containing a few crystals of iodine. The iodine vapor in the chamber oxidizes the substances in the various spots, making them visible to the eye. Once the spots are visible, they may be outlined with a pencil before the iodine coloration fades. *Second* is the ninhydrin technique, which is particularly effective for visualizing amino acid spots. In this method, a solution containing 0.2% ninhydrin in ethanol is sprayed on the dry plate. Alternately, the plate can be dipped in ninhydrin solution. In contact with an amino acid, ninhydrin displays a purple coloration that is easily seen. It usually takes a few minutes for this color to develop, so after the plate is sprayed, it is allowed to sit for several minutes. Placing it in a 50 °C oven will hasten the appearance of the purple colour. Once this appears, the spots may be circled in pencil to permanently mark their positions. *Third*, one may use TLC plates that have been loaded with a fluorescent substance that is uniformly distributed in the silica gel. Substances moving up the plate block this fluorescence at their locations. When the dried plate is viewed using a special UV light, the substance spots are visible for their lack of fluorescence on an otherwise uniformly fluorescing field. The spots may be outlined with pencil while being viewed in the light.

Measurement of R_f. The distance between the 2 horizontal pencil lines is the distance of solvent advance. The distance from the bottom pencil line to the center of a substance spot is the distance of

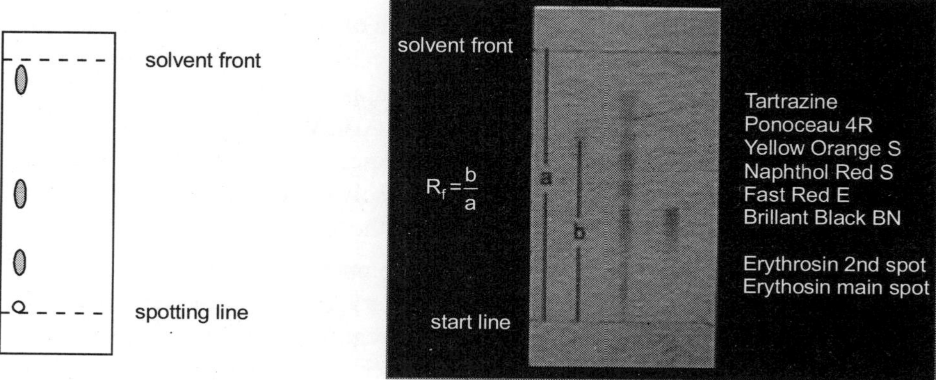

Fig. 2.71 and 2.71A. The TLC plate for interpretation

advance of the substance. The ratio of substance advance distance to solvent advance distance is the R_f value for the substance. The figure shows a developed plate for a mixture consisting of 4 components with Rf values of about 0.05, 0.2, 0.5, and 0.9.

TLC of Inks. Before attempting to apply TLC to the challenging problem of separating and identifying amino acids, it is advisable to learn and practice the technique by applying it to mixtures that are easily visualized and separated. Inks provide an ideal practice vehicle for TLC because they normally contain several coloured components that separate nicely in common solvents such as ethanol, acetone, or chloroform. Spotting the plate is also easy: it may be done simply by making a VERY small mark on the plate with the tip of a pen, just above the pencil line drawn across the bottom edge of the plate. Inks are of different types and colours, of course. Some are washable (water-soluble), others are permanent. Different types require different solvents for development. Common sources of ink are ball point pens, felt-tip markers, and roller ball pens. Bottled ink is still available for people using fountain pens. Particularly good ink sources are bottled inks by Parker, Sheaffer, and Mont Blanc; Sharpie marker inks (black, red, blue, orange, brown, yellow, green); Marks-a-Lot Stay sharp markers; and Sanford calligraphy pens.You should select inks from at least 2 different sources available in the lab and should find a solvent or mixture of solvents that separates each ink into its component colours.

Experimental Procedure. You will be assigned two ink samples to examine. Obtain four TLC plates, four TLC spotting capillaries, and 3 developing tanks (snap-top plastic vials). Mark your TLC plates with a narrow pencil line about 1cm from the bottom. Prepare spotting solutions of your ink samples in 1-dram vials by diluting 1 drop of ink with 9 drops of ethanol (this is called a 1-to-10 dilution). Insert one end of a spotting capillary into the first ink solution. You should see the ink solution rise in the capillary. Withdraw the capillary when the liquid has risen to a height of 1-2 cm. Hold the capillary vertically and briefly touch the filled end to the pencil line, about 5 mm from the left edge of the TLC plate. Liquid should flow from the capillary to the plate to form a spot. THIS HAPPENS RAPIDLY. If possible, lift the capillary before the spot gets any bigger than 1 mm in diameter. Allow the ethanol to evaporate. Touch the capillary to a Kimwipe to empty it of ink solution. Use the other end of the same capillary to draw up your second ink solution, and spot the plate about 5 mm from the right edge. Discard the capillary in the broken glass receptacle. Allow the ethanol to evaporate from the plate.

Into your first developing tank, pour ethanol to a depth of about 5 mm. Using tweezers to hold the top end of the spotted plate, lower the plate into the tank. Allow the bottom of the plate to rest on the bottom of the tank, with the top of the plate leaning against a tank sidewall. NOTE: IT IS VERY IMPORTANT THAT THE START LINE AND SPOTS BE ABOVE THE LEVEL OF SOLVENT. Close the tank and monitor the movement of solvent up the plate. When solvent has advanced to the top pencil line, remove the plate from the tank, and allow all solvent to evaporate. Use a pencil to outline each observed spot on the plate, preserving the shape of the spot.

If ethanol did not effect a separation of the ink into components, try another solvent. If you found that the ink moved along with or close to the solvent front, try a less polar solvent. If you found that the ink did not move at all with ethanol, try a more polar solvent (e.g., methanol). Your goal is to cleanly separate each of your ink samples into its constituents, each constituent producing a single TLC spot. Be aware that some inks contain only one component!

TLC of Amino Acids. TLC of amino acids is more difficult than TLC of inks, because amino acids are colourless. Therefore, not only can you not monitor their progress up the plate, but you cannot see the spots with the naked eye once the plate is fully developed and dried. To see the spots, it is necessary to use either the ninhydrin or the black-light visualization techniques. Of course, the latter works only if you use fluorescent TLC plates, which are available in the lab. Until you see the spots, you will not know whether or not a chosen solvent system has been effective in moving an amino acid or in separating a mixture. Therefore the process of finding an effective solvent system can be long and painstaking. For this reason, we will specify the solvent system that you are to use. As points of general information, amino acids are quite polar and tend to move on silica gel plates with polar solvents. They have R_f values close to 1 when water or concentrated ammonia is used as the developing solvent, probably because of their high solubility in water. Diluting a polar solvent with a less polar one results in smaller R_f values, roughly in proportion to the amount of less polar solvent used, Thus, alanine, glycine, threonine, and proline all have R_f values of around 0.60 when developed with a 50/50 mixture of water and n-propanol, and around 0.40 when developed with a 30/70 mixture of concentrated NH_3 and n-propanol. The following procedure assumes the use of 50/50 water/n-propanol as the developing solvent, but you are free to try other polar/non-polar combinations.

Experimental Procedure. In the hood, prepare 10 mL of a mixture consisting of 50% 1-propanol and 50% water by volume, and pour about half of this into a clean developing tank. Make sure that the level of liquid in the tank is no higher than 5 mm, and close the lid. In a 1-dram vial, prepare a solution of about 0.001 g of your amino acid in 0.2 mL of water. Dissolve the acid, then draw some solution up in a spotting capillary and double-spot a properly marked and activated TLC plate. Allow the plate to dry for 5 minutes, then use tweezers to carefully lower the plate into the developing tank so that its bottom is submerged in the developing solvent. NOTE: IT IS VERY IMPORTANT THAT THE START LINE AND SPOTS BE ABOVE THE LEVEL OF SOLVENT. Close the lid, and allow the plate to develop until solvent has risen to the pencil line at the top of the plate. Remove the plate from the tank and place it in an oven at 50 °C to dry. When the plate is dry, visualize it using ninhydrin spray or iodination. Circle the amino acid spots with pencil, and calculate R_f values. Compare the measured R_f values with the values posted for the amino acids.

On this basis, you can narrow down the identity of your amino acid. In combination with other data that you obtain, this information will help you unambiguously identify your amino acid. Suppose that your amino acid has an R_f value similar to that of, say, alanine. You should then prepare a small amount of alanine solution and spot it alongside your amino acid on a new TLC plate. Develop, dry, and visualize the plate to confirm that your amino acid indeed has exactly the same R_f value as alanine, and that the spot is the same shape and color.

Finally, it is very important to be observant of detail in doing TLC. In addition to the R_f value for a substance, the shape of the spot produced by a particular developing solvent and the shade of color produced by iodine or ninhydrin can be characteristic of the substance. Please note all of these things. For example, when alanine, glycine, threonine, and proline are spotted side-by-side on a plate and developed with 70% n-propanol/30% conc NH_3, the following observations can be seen in the table below:

Amino Acid	Solvent	Spot Color after Iodination	Spot Color with Ninhydrin	R_f Value	Spot Shape
alanine	30/70 conc NH_3/n-propanol	white on brown bkgrnd	purple		elongated oval
alanine	50/50 water/n-propanol	white on brown bkgrnd	purple	0.65	circle
glycine	30/70 conc NH_3/n-propanol	white on brown bkgrnd	pink	0.25	elongated oval
glycine	50/50 water/n-propanol	white on brown bkgrnd	pink	0.55	circle
threonine	30/70 conc NH_3/n-propanol	white on brown bkgrnd	purple		elongated oval
threonine	50/50 water/n-propanol	white on brown bkgrnd	purple	0.57	circle
proline	30/70 conc NH_3/n-propanol	dark brown on brown	yellow with pink border		elongated oval
proline	50/50 water/n-propanol	white on brown bkgrnd	yellow with pink border	0.65	circle

Problems in TLC

Over-large Spots. Sample spots made using TLC capillaries should be no larger than 1-2 mm in diameter, because component spots in the developed plate will be no smaller than, and will usually be larger than, the size of the initial spot. If the initial spot is larger than 2 mm in diameter, then components with similar R_f values may not be resolved because their spots will be so large that they will overlap considerably and may appear to be one large spot. Small initial spots, on the other hand, maximize the potential of complete separation of components.

Uneven Advance of Solvent Front. A common problem in TLC is uneven advance of solvent along the plate. Instead of a straight line, the solvent front may appear to bow either up or down in the center. Uneven advance of solvent leads to uneven advance of substance spots, and inaccurate R_f values result. A frequent cause of uneven solvent advance is the use of a developing chamber that does not have a flat bottom. Glass bottles usually have bottoms that curve upward from the edges to the center. If the bottom of the TLC plate is placed on this curved surface, the shape of the solvent advance line may mirror the shape of the container bottom. It is therefore important to use flat-bottomed developing tanks in TLC. A bowed solvent front may also result if too little developing solvent is placed in the chamber; if the plate is cut improperly, so that the sides are not exactly perpendicular to the bottom edge; and if the slide is excessively tilted in the chamber. Care in choosing and using a developing chamber is the best defense against curved solvent fronts.

Water is seldom used as a developing solvent because it has a tendency to produce a dramatically curved front. This may be due to its unusually high surface tension.

Streaking. Sometimes a substance will move along a TLC plate as a long streak, rather than as a single discrete spot. This is the result of spotting the plate with too much substance, more than the moving solvent can handle. The solvent moves as much substance as it can, but a substantial amount of substance is left behind. The substance is dragged along by the solvent leaving a trail of substance that may sometimes span the entire distance between the starting line and the solvent front. Streaking can be eliminated by systematically diluting the spotting solution until development and visualization show the substances moving as single spots, rather than elongated streaks.

HIGH PERFORMANCE LIQUID CHROMATOGRAPHY

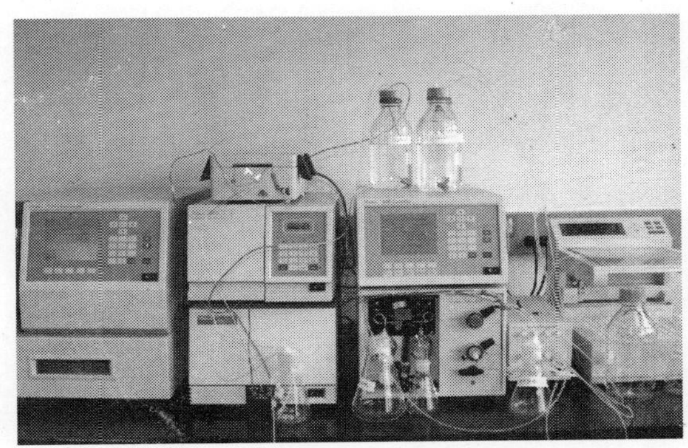

The Analytical HPLC system

Research with thin-layer and column chromatography showed that separations are much more effective when the stationary phase is a very thin layer on the surface of very small and very uniform spherical beads. However, resistance to flow of the mobile phase is very much higher, and in order to get a useful flow of a liquid mobile phase, e. g., 1 - 3 milliliters/minute, pressures of around 15 Mpa (about 2,000 psi) must be applied to the mobile phase. It is possible to apply such pressure from a cylinder of compressed gas, but most systems use a reciprocating piston pump or diaphragm pump with some means of damping the pressure fluctuations from the piston. The sample is usually dissolved in the mobile phase before injection. Columns are typically 4.6 mm ID (6 mm OD) stainless steel tubing 250 mm long. A typical packing will have octadecylsilyl (C_{18}-Si-) (ODS) groups bonded to 5 µm silica beads. The packing is held inside the column by "frits", discs with pores about 0.5 µm in diameter.

Liquid-liquid chromatography began with samples dissolved in organic solvents and a stationary phase of water adsorbed on particles or fibers of the solid support. More generally, the stationary phase was more "polar" than the mobile phase. That is the so-called "normal phase" chromatography. But stationary phases such as ODS have been particularly useful for separating samples dissolved in water (and most HPLC is now done with bonded phases). Liquid chromatography with the stationary phase less polar than the mobile phase is called "reverse phase", but is now the common situation. The mobile phase is very often not just water but a mixture of water with methanol (CH_3OH) or acetonitrile (CH_3CN). "Solvent programming", a stepwise or continuous change (gradient elution) of the mobile phase composition, is used to speed up separations, like temperature programming in gas chromatography.

"Chiral"columns have been developed relatively recently to separate optical isomers. This separation is important because many pharmaceuticals are active in only one chiral form. For instance, natural Vitamin E is D-a-tocopherol, while half of synthetic Vitamin E is the less active L- isomer.

As in gas chromatography, a few microliters of the solution are measured into a "sample loop" on an "injector". When the injector is operated, the sample loop is suddenly switched into the flow of mobile phase just before it reaches the column. The mobile phase leaving the column passes immediately into a "detector" which is used to determine the presence and concentration of a solute.

Because the pressure on the mobile phase drops rapidly as it leaves the column, bubbles may form from dissolved gases. Chromatographers "degas" the mobile phase before use by boiling it, ultrasonic treatment, bubbling helium through it to flush out other gases, or applying a vacuum.

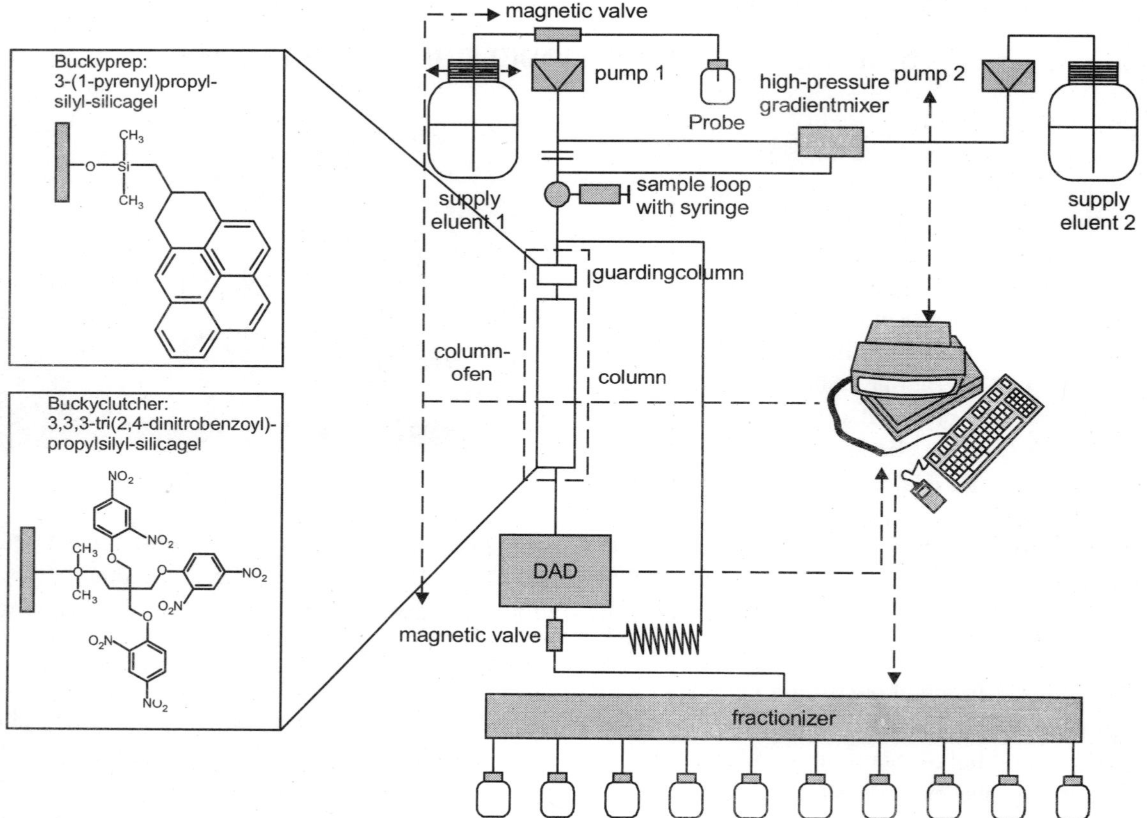

Fig. 2.72 : The Diagramatic representation of the HPLC system

There are no relatively inexpensive HPLC detectors as sensitive and broadly useful as the flame ionization and electron capture detectors used in gas chromatography. A refractive index (RI) detector responds to most components, but is not very sensitive. An ultraviolet (UV) detector is quite sensitive for molecules which absorb ultraviolet light, and a variable wavelength UV detector can be set to the absorption maximum for a particular molecule of interest, or to a short wavelength where most molecules absorb. A diode array detector (DAD) disperses the transmitted light into a spectrum, providing an absorption spectrum of each component that absorbs ultraviolet light. Still more sensitive, and still less general, is the fluorescence detector which measures fluorescence emitted from components which have absorbed ultraviolet light. There are various special-purpose detectors. Amperometric systems measure electron flow which oxidizes or reduces certain components (sugars, for instance), and polarimetric detectors, generally not very sensitive, detect components are optically active. Mass spectrometric detectors are now used, but were late to arrive because it was difficult to separate the mobile phase molecules so as to maintain adequate vacuum in the mass spectrometer.

Special case: Supercritical fluid chromatography (SFC). Liquids can vaporize, and gases can condense to the liquid phase. Both changes depend on the pressure and temperature. Every gas has a critical temperature above which it cannot be condensed at any pressure. The critical pressure is the pressure required at the critical temperature. If a liquid (such as a condensed gas) is warmed, under high pressure, to the critical temperature and beyond, strange things happen. The boundary between gas and liquid vanishes, and it is best not to ask physical chemists too many questions about the remaining phase, the "supercritical fluid". Whatever their exact nature, supercritical fluids can be very useful solvents for HPLC, which runs under high pressure anyway. The most commonly used of these fluids is carbon dioxide. (Supercritical fluids are also useful for extracting things from solids, and, when the solids are clothing, for dry cleaning.)Reports and procedures will specify the mobile phase composition, flow rate and perhaps the pressure; the column dimensions; the column packing (particle type and size; coating); the detector and its operating conditions (e. g., wavelengths for UV and fluorescence detectors), integrator settings, and, somewhere, retention times.There are some kinds of liquid chromatography that do not fit the partition or adsorption models very well. They are often done in simple vertical columns.

POLYACRYLAMIDE GEL ELECTROPHORESIS

Sodium dodecyl sulphate (SDS) is an anionic detergent which denatures proteins by "wrapping around" the polypeptide backbone - and SDS binds to proteins fairly specifically in a mass ratio of 1.4:1. In so doing, SDS confers a **negative charge** to the polypeptide in proportion to its length - ie: the denatured polypeptides become **"rods" of negative charge cloud** with equal charge or charge densities per unit length. It is usually necessary to reduce disulphide bridges in proteins before they adopt the random-coil configuration necessary for separation by size: this is done with **2-mercaptoethanol or dithiothreitol**. In denaturing SDS-PAGE separations therefore, migration is determined not by intrinsic electrical charge of the polypeptide, but by **molecular weight**.

Determination of Molecular Weight

This is done by SDS-PAGE of proteins - or PAGE or agarose gel electrophoresis of nucleic acids - of known molecular weight along with the protein or nucleic acid to be characterised. A linear relationship exists between the logarithm of the molecular weight of an SDS-denatured polypeptide, or native nucleic acid, and its ***Rf***. The Rf is calculated as the **ratio of the distance migrated by the molecule to that migrated by a marker dye-front**. A simple way of determining relative molecular weight by electrophoresis *(M*r) is to plot a standard curve of distance migrated vs. log10MW for known samples, and read off the log*Mr* of the sample after measuring distance migrated on the same gel.

Continuous and Discontinuous Buffer Systems

There are two types of buffer systems in electrophoresis, **continuous and discontinuous**. A **continuous** system has only a single separating gel and uses the same buffer in the tanks and the gel. In a **discontinuous** system, a non-restrictive large pore gel, called a stacking gel, is layered on top of a separating gel called a resolving gel. Each gel is made with a different buffer, and the tank buffers are different from the gel buffers. The resolution obtained in a discontinuous system is much greater than that obtained with a continuous system.

Protocol

NB: Acrylamide Monomer is a Potent Cumulative Neurotoxin. Do not Mouth Pipette Acrylamide Solutions, and Wear Gloves when Handling Unpolymerised Solutions.

Assembling Gel Apparatus

Assemble two glass plates (one notched) with two side spacers, clamps, grease, etc. as shown by demonstrators. Stand assembly upright using clamps as supports, on glass plate. Pour some **pre-heated 1% agarose** onto glass plate, place assembly in pool of agarose: this seals the bottom of the assembly.

Resolving Gels

Gel concentration of 12.5% in 0.25 M Tris-HCl pH 8.8

	Volume Reagent: (ml: TO MAKE 30 ML)	Volume (ml: TO MAKE 10 ML)
40% Acrylamide stock*:	9.4	3.1
water (distilled)	12.3	3.8
1 M Tris-HCl pH 8.8	7.5	2.5
10% SDS	0.3	0.1
Peroxydisulphate 1%	0.5	0.5
TEMED (added last)	20 ul	20 ul

* = 19:1 - 38:1 w:w ratio of acrylamide to N,N'-methylene bis-acrylamide

Mix ingredients **GENTLY!** in the order shown above, ensuring no air bubbles form. Pour into glass plate assembly **CAREFULLY**. **Overlay gel with isopropanol** to ensure a flat surface and to exclude air. Wash off isopropanol with water after gel has set (+15 min).

Stacking Gels

Gel concentration of 4.5% in 0.125 M Tris-HCl pH 6.8

* (Read about this in Architecture of chemistry An instrumental approach textbook).

	Volume Reagent: (ml: TO MAKE 30 ML)	Volume (ml: TO MAKE 10 ML)
40% Acrylamide stock	1.7	1.1
water	10.8	7.1
1 M Tris-HCl pH 6.8	1.9	1.25
10% SDS	0.15	0.1
Peroxydisulphate 1%	0.5	0.5
TEMED (stir quickly)	20 ul	20 ul

Mix as before, then **pour onto top of set resolving gel**, insert comb, allow to set, remove comb, fill with electrophoresis buffer. Assemble top tank onto glass plate assembly. Fill with electrophoresis buffer.

Electrophoresis Buffer

The final **TANK** buffer composition is **196mM glycine / 0.1% SDS / 50mM Tris-HCl pH 8.3**, made by diluting a **10x stock** solution. This goes in both top and bottom tanks.

Samples

Grind a little leaf material (eg. 2 grams) in a mortar. Centrifuge in an Eppendorf tube for 3 min. Take supernatant and **mix 100ul 1:1 (v:v) with SDS-PAGE disruption mix**: this is **125mM Tris-HCl pH 6.8 / 10% 2-mercaptoethanol / 10% SDS / 10% glycerol**, containing a little bromophenol blue. **BE CAREFUL WITH THIS AS IT SMELLS AWFUL and is poisonous to boot!!**

For liquid / purified samples, take eg. 100 ul and add 50 - 100 ul of disruption mix.

Heat sample Eppendorfs for 5 min at 95oC in a "float" in a waterbath. Layer samples under buffer on stacking gels. Connect up apparatus and electrophorese as shown.

Staining of Gels

1. Coomassie Brilliant Blue/Page-Blue 83

Make up stain: **0.2% CBB in 45:45:10 % methanol:water:acetic acid**. Cover gel with staining solution, seal in plastic box and leave overnight on shaker (RT) or for 2 to 3 hours at 37 c also with agitation. Destain with 25% 65% 10% methanol water acetic acid mix, with agitation.

2. Copper Chloride (0.3M CuCl2

Rinse gel in distilled water, immerse in copper chloride solution with agitation for about 20 minutes (RT), rinse with distilled water and immerse in sufficient fresh distilled water to cover the gel (this acts as the destaining step). Seal in a plastic box.

CLASSIFICATION AND STRUCTURE OF IMMUNOGLOBINS AND PLASMA PROTEINS

Antibodies (also known as **immunoglobulins**, abbreviated **Ig**) are gamma globulin proteins that are found in blood or other bodily fluids of vertebrates, and are used by the immune system to identify and neutralize foreign objects, such as bacteria and viruses. They are typically made of basic structural units—each with two large heavy chains and two small light chains—to form, for example, monomers with one unit, dimers with two units or pentamers with five units. Antibodies are produced by a kind of white blood cell called a B cell. There are several different types of antibody heavy chains, and several different kinds of antibodies, which are grouped into different *isotypes* based on which heavy chain they possess. Five different antibody isotypes are known in mammals, which perform different roles, and help direct the appropriate immune response for each different type of foreign object they encounter.

Although the general structure of all antibodies is very similar, a small region at the tip of the protein is extremely variable, allowing millions of antibodies with slightly different tip structures to exist. This region is known as the hypervariable region. Each of these variants can bind to a different target, known as an antigen. This huge diversity of antibodies allows the immune system to recognize an equally wide diversity of antigens. The unique part of the antigen recognized by an antibody is called an epitope. These epitopes bind with their antibody in a highly specific interaction, called induced fit, that allows antibodies to identify and bind only their unique antigen in the midst of the millions of different molecules that make up an organism. Recognition of an antigen by an antibody *tags* it for attack by other parts of the immune system. Antibodies can also neutralize targets directly by, for example, binding to a part of a pathogen that it needs to cause an infection.

The large and diverse population of antibodies is generated by random combinations of a set of gene segments that encode different antigen binding sites (or *paratopes*), followed by random mutations in this area of the antibody gene, which create further diversity. Antibody genes also re-organize in a process called class switching that changes the base of the heavy chain to another, creating a different isotype of the antibody that retains the antigen specific variable region. This allows a single antibody to be used by several different parts of the immune system. Production of antibodies is the main function of the humoral immune system.

Antibody forms

Activated B cells differentiate into either antibody-producing cells called plasma cells that secrete soluble antibody or memory cells that survive in the body for years afterward in order to allow the immune system to remember an antigen and respond faster upon future exposures. Antibodies are, therefore, an essential product of the adaptive immune system that learns and remembers responses to invading pathogens. Antibodies occur in two forms: a soluble form secreted into the blood and other fluids in the body, and a membrane-bound form that is attached to the surface of a B cell.

Soluble antibodies that are secreted from an activated B cell (in its plasma cell form) bind to foreign substances and signal for their destruction by the rest of the immune system. They may also be called *free antibodies* (until they bind an antigen and become part of an *immune complex*) or *secreted antibodies*.

The membrane-bound form of an antibody may be called a *surface immunoglobulin* (sIg) or a *membrane immunoglobulin* (mIg). It is part of the *B cell receptor* (BCR), which allows a B cell to detect when a specific antigen is present in the body and triggers B cell activation. The BCR is composed of surface-bound IgD or IgM antibodies and associated Ig-α and Ig-β heterodimers, which are capable of signal transduction. A typical human B cell will have 50,000 to 100,000 antibodies bound to its surface. Upon antigen binding, they cluster in large patches, which can exceed 1 micrometer in diameter, on lipid rafts that isolate the BCRs from most other cell signaling receptors. These patches may improve the efficiency of the cellular immune response. In humans, the cell surface is bare around the B cell receptors for several thousand Ångstroms, which further isolates the BCRs from competing influences.

Isotypes

Antibody isotypes of mammals

Name	Types	Description	Antibody Complexes
IgA	2	Found in mucosal areas, such as the gut, respiratory tract and urogenital tract, and prevents colonization by pathogens. Also found in saliva, tears, and breast milk.	Monomer IgD, IgE, IgG Dimer IgA Pentamer IgM
IgD	1	Functions mainly as an antigen receptor on B cells that have not been exposed to antigens. Its function is less defined than other isotypes.	
IgE	1	Binds to allergens and triggers histamine release from mast cells and basophils, and is involved in allergy. Also protects against parasitic worms.	
IgG	4	In its four forms, provides the majority of antibody-based immunity against invading pathogens. The only antibody capable of crossing the placenta to give passive immunity to fetus.	
IgM	1	Expressed on the surface of B cells and in a secreted form with very high avidity. Eliminates pathogens in the early stages of B cell mediated (humoral) immunity before there is sufficient IgG.	

Antibodies can come in different varieties known as isotypes or classes. In placental mammals there are five antibody isotypes known as IgA, IgD, IgE,IgG and IgM. They are each named with an "Ig" prefix that stands for immunoglobulin, another name for antibody, and differ in their biological properties, functional locations and ability to deal with different antigens, as depicted in the above table.

The antibody isotype of a B cell changes during cell development and activation. Immature B cells, which have never been exposed to an antigen, are known as naïve B cells and express only the IgM isotype in a cell surface bound form. B cells begin to express both IgM and IgD when they reach maturity—the co-expression of both these immunoglobulin isotypes renders the B cell 'mature' and ready to respond to antigen. B cell activation follows engagement of the cell bound antibody molecule with an antigen, causing the cell to divide and differentiate into an antibody producing cell called a plasma cell. In this activated form, the B cell starts to produce antibody in a secreted form rather than a membrane-bound form. Some daughter cells of the activated B cells undergo isotype switching, a mechanism that causes the production of antibodies to change from IgM or IgD to the other antibody isotypes, IgE, IgA or IgG, that have defined roles in the immune system.

Structure

Antibodies are heavy (~150kDa) globular plasma proteins that are also known as immunoglobulins. They have sugar chains added to some of their amino acid residues. In other words, antibodies are *glycoproteins*. The basic functional unit of each antibody is an immunoglobulin (Ig) monomer (containing only one Ig unit); secreted antibodies can also be dimeric with two Ig units as with IgA, tetrameric with four Ig units like teleost fish IgM, or pentameric with five Ig units, like mammalian IgM.Several immunoglobulin domains make up the two heavy chains (red and blue) and the two light chains (green and yellow) of an antibody. The immunoglobulin domains are composed of between 7 (IgC) and 9 (IgV) β-strands.

Immunoglobulin Domains

The Ig monomer is a "Y"-shaped molecule that consists of four polypeptide chains; two identical *heavy chains* and two identical *light chains* connected by disulfide bonds. Each chain is composed of structural domains called Ig domains. These domains contain about 70-110 amino acids and are classified into different categories (for example, variable or IgV, and constant or IgC) according to their size and function. They have a characteristic immunoglobulin fold in which two beta sheets create a "sandwich" shape, held together by interactions between conserved cysteines and other charged amino acids.

Heavy Chain

There are five types of mammalian Ig heavy chain denoted by the Greek letters: α, δ, ε, γ, and μ. The type of heavy chain present defines the *class* of antibody; these chains are found in IgA, IgD, IgE, IgG, and IgM antibodies, respectively. Distinct heavy chains differ in size and composition; α and γ contain approximately 450 amino acids, while μ and ε have approximately 550 amino acids.

Each heavy chain has two regions, the *constant region* and the *variable region*. The constant region is identical in all antibodies of the same isotype, but differs in antibodies of different isotypes. Heavy chains γ, α and δ have a constant region composed of *three* tandem (in a line) Ig domains, and a hinge region for added flexibility; heavy chains μ and ε have a constant region composed of *four* immunoglobulin domains. The variable region of the heavy chain differs in antibodies produced by different B cells, but is the same for all antibodies produced by a single B cell or B cell clone. The variable region of each heavy chain is approximately 110 amino acids long and is composed of a single Ig domain.

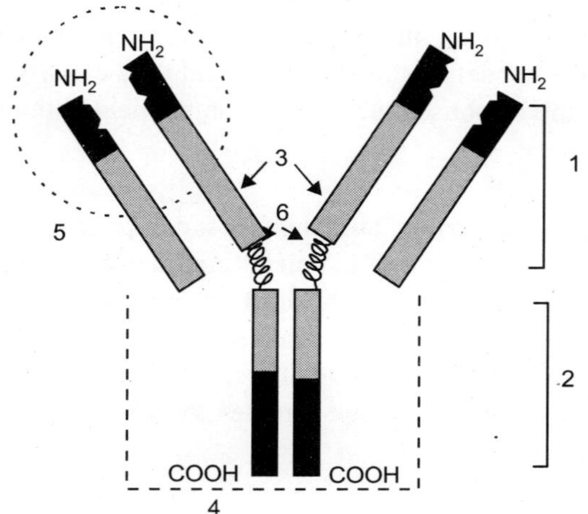

1. **Fab region**
2. **Fc region**
3. **Heavy chain with one variable (V_H) domain followed by a constant domain (C_H1), a hinge region, and two more constant (C_H2 and C_H3) domains.**
4. **Light chain with one variable (V_L) and one constant (C_L) domain**
5. **Antigen binding site (paratope)**
6. **Hinge regions.**

Fig. 2.73 : The Immunoglobin Structure

Light Chain

In mammals there are two types of light chain, which are called lambda (λ) and kappa (κ). A light chain has two successive domains: one constant domain and one variable domain. The approximate length of a light chain is 211 to 217 amino acids. Each antibody contains two light chains that are always identical; only one type of light chain, κ or λ, is present per antibody in mammals. Other types of light chains, such as the iota (ι) chain, are found in lower vertebrates like Chondrichthyes and Teleostei.

Fab and Fc Regions

Some parts of an antibody have unique functions. The tips of the Y, for example, contain the site that bind antigen and, therefore, recognize specific foreign objects. This region of the antibody is called the *Fab (fragment, antigen binding) region*. It is composed of one constant and one variable domain from each heavy and light chain of the antibody. The paratope is shaped at the amino terminal end of the antibody monomer by the variable domains from the heavy and light chains.

The base of the Y plays a role in modulating immune cell activity. This region is called the *Fc (Fragment, crystallizable) region*, and is composed of two heavy chains that contribute two or three constant domains depending on the class of the antibody. By binding to specific proteins the Fc region ensures that each antibody generates an appropriate immune response for a given antigen. The Fc region also binds to various cell receptors, such as Fc receptors, and other immune molecules, such as complement proteins. By doing this, it mediates different physiological effects including opsonization, cell lysis, and degranulation of mast cells, basophils and eosinophils.

Function

Since antibodies exist freely in the bloodstream, they are said to be part of the humoral immune system. Circulating antibodies are produced by clonal B cells that specifically respond to only one antigen (an example is a virus capsid protein fragment) . Antibodies contribute to immunity in three main ways: they can prevent pathogens from entering or damaging cells by binding to them; they can stimulate removal of a pathogen by macrophages and other cells by coating the pathogen; and they can trigger direct pathogen destruction by stimulating other immune responses such as the complement pathway.

Activation of Complement

Antibodies that bind to surface antigens on, for example a bacterium, attract the first component of the complement cascade with their Fc region and initiate activation of the "classical" complement system. This results in the killing of bacteria in two ways. First, the binding of the antibody and complement molecules marks the microbe for ingestion by phagocytes in a process called opsonization; these phagocytes are attracted by certain complement molecules generated in the complement cascade. Secondly, some complement system components form a membrane attack complex to assist antibodies to kill the bacterium directly.

Activation of Effector Cells

To combat pathogens that replicate outside cells, antibodies bind to pathogens to link them together, causing them to agglutinate. Since an antibody has at least two paratopes it can bind more than one antigen by binding identical epitopes carried on the surfaces of these antigens. By coating the pathogen, antibodies stimulate effector functions against the pathogen in cells that recognize their Fc region.

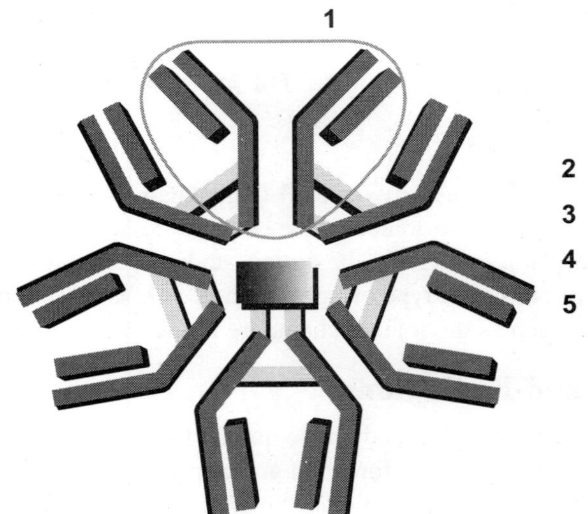

Fig. 2.74 : The secreted mammalian IgM has five Ig units. Each Ig unit (labelled 1) has two epitope binding Fab regions, so IgM is capable of binding up to 10 epitopes.

Those cells which recognize coated pathogens have Fc receptors which, as the name suggests, interacts with the Fc region of IgA, IgG, and IgE antibodies. The engagement of a particular antibody with the Fc receptor on a particular cell triggers an effector function of that cell; phagocytes will phagocytose, mast cells and neutrophils will degranulate, natural killer cells will release cytokines and cytotoxic molecules; that will ultimately result in destruction of the invading microbe. The Fc receptors are isotype-specific, which gives greater flexibility to the immune system, invoking only the appropriate immune mechanisms for distinct pathogens.

Immunoglobulin Diversity

Virtually all microbes can trigger an antibody response. Successful recognition and eradication of many different types of microbes requires diversity among antibodies; their amino acid composition varies allowing them to interact with many different antigens. It has been estimated that humans generate about 10 billion different antibodies, each capable of binding a distinct epitope of an antigen. Although a huge repertoire of different antibodies is generated in a single individual, the number of genes available to make these proteins is limited. Several complex genetic mechanisms have evolved that allow vertebrate B cells to generate a diverse pool of antibodies from a relatively small number of antibody genes.

Domain Variability

The region (locus) of a chromosome that encodes an antibody is large and contains several distinct genes for each domain of the antibody—the locus containing heavy chain genes (IGH) is found on chromosome 14, and the loci containing lambda and kappa light chain genes (IGL and IGK) are found on chromosomes 22 and 2 in humans. One of these domains is called the variable domain, which is present in each heavy and light chain of every antibody, but can differ in different antibodies generated from distinct B cells. Differences, between the variable domains, are located on three loops known as hypervariable regions (HV-1, HV-2 and HV-3) or complementarity determining regions (CDR1, CDR2 and CDR3). CDRs are supported within the variable domains by conserved framework regions. The heavy chain locus contains about 65 different variable domain genes that all differ in their CDRs. Combining these genes with an array of genes for other domains of the antibody generates a large cavalry of antibodies with a high degree of variability. This combination is called V(D)J recombination discussed below.

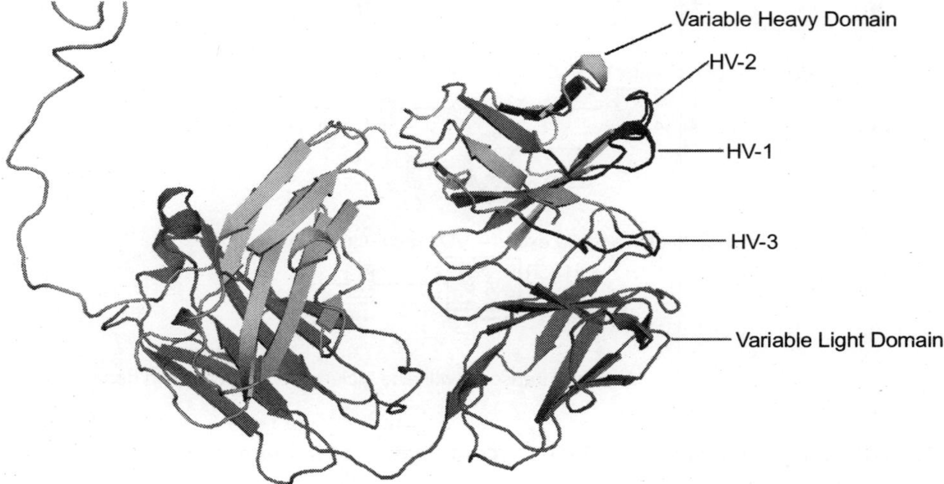

Fig. 2.75 : The hypervariable regions of the heavy chain is shown in red, PDB 1IGT

V(D)J Recombination

Somatic recombination of immunoglobulins, also known as *V(D)J recombination*, involves the generation of a unique immunoglobulin variable region. The variable region of each immunoglobulin heavy or light chain is encoded in several pieces—known as gene segments. These segments are called variable (V), diversity (D) and joining (J) segments. V, D and J segments are found in Ig heavy chains, but only

V and J segments are found in Ig light chains. Multiple copies of the V, D and J gene segments exist, and are tandemly arranged in the genomes of mammals. In the bone marrow, each developing B cell will assemble an immunoglobulin variable region by randomly selecting and combining one V, one D and one J gene segment (or one V and one J segment in the light chain). As there are multiple copies of each type of gene segment, and different combinations of gene segments can be used to generate each immunoglobulin variable region, this process generates a huge number of antibodies, each with different paratopes, and thus different antigen specificities.

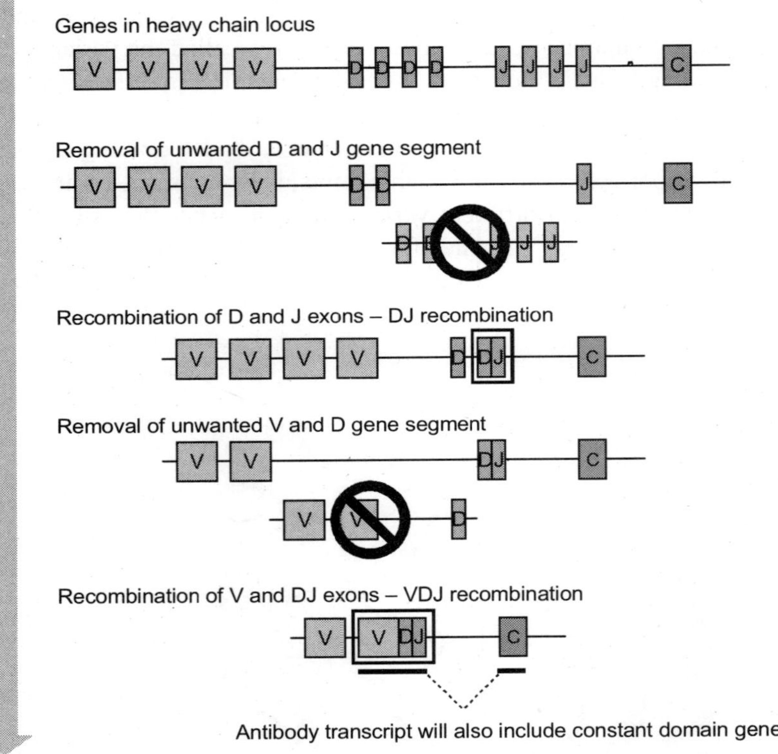

Fig. 2.76 : The Simplistic overview of V(D)J recombination of immunoglobulin heavy chains

After a B cell produces a functional immunoglobulin gene during V(D)J recombination, it cannot express any other variable region (a process known as allelic exclusion) thus each B cell can produce antibodies containing only one kind of variable chain.

Somatic Hypermutation and Affinity Maturation

Following activation with antigen, B cells begin to proliferate rapidly. In these rapidly dividing cells, the genes encoding the variable domains of the heavy and light chains undergo a high rate of point mutation, by a process called *somatic hypermutation* (SHM). SHM results in approximately one nucleotide change per variable gene, per cell division. As a consequence, any daughter B cells will acquire slight amino acid differences in the variable domains of their antibody chains.This serves to increase the diversity of the antibody pool and impacts the antibody's antigen-binding affinity. Some point mutations will result in the production of antibodies that have a weaker interaction (low affinity) with their antigen than the original antibody, and some mutations will generate antibodies with a stronger interaction (high affinity). B cells that express high affinity antibodies on their surface will receive a strong survival signal during interactions with other cells, whereas those with low affinity antibodies will not, and will die by apoptosis. Thus, B cells expressing antibodies with a higher affinity for the antigen will outcompete those with weaker affinities for function and survival. The process of generating antibodies with increased binding affinities is called *affinity maturation*. Affinity maturation occurs in mature B cells after V(D)J recombination, and is dependent on help from helper T cells.

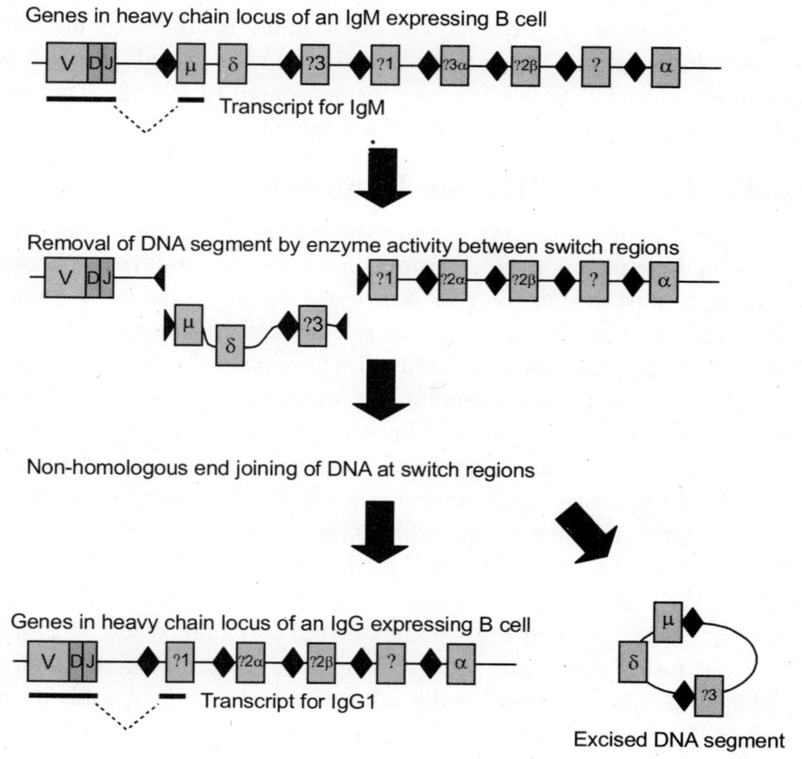

Fig. 2.77 : The Mechanism of class switch recombination that allows isotype switching in activated B cells

Class Switching

Isotype or class switching is a biological process occurring after activation of the B cell, which allows the cell to produce different classes of antibody (IgA, IgE, or IgG). The different classes of antibody, and thus effector functions, are defined by the constant (C) regions of the immunoglobulin heavy chain. Initially, naïve B cells express only cell-surface IgM and IgD with identical antigen binding regions. Each isotype is adapted for a distinct function, therefore, after activation, an antibody with a IgG, IgA, or IgE effector function might be required to effectively eliminate an antigen. Class switching allows different daughter cells from the same activated B cell to produce antibodies of different isotypes. Only the constant region of the antibody heavy chain changes during class switching; the variable regions, and therefore antigen specificity, remain unchanged. Thus the progeny of a single B cell can produce antibodies, all specific for the same antigen, but with the ability to produce the effector function appropriate for each antigenic challenge. Class switching is triggered by cytokines; the isotype generated depends on which cytokines are present in the B cell environment. Class switching occurs in the heavy chain gene locus by a mechanism called class switch recombination (CSR). This mechanism relies on conserved nucleotide motifs, called *switch (S) regions*, found in DNA upstream of each constant region gene (except in the ä-chain). The DNA strand is broken by the activity of a series of enzymes at two selected S-regions. The variable domain exon is rejoined through a process called non-homologous end joining (NHEJ) to the desired constant region (γ, α or ε). This process results in an immunoglobulin gene that encodes an antibody of a different isotype.

Medical Applications and Disease Diagnosis

Detection of particular antibodies is a very common form of medical diagnostics, and applications such as serology depend on these methods. For example, in biochemical assays for disease diagnosis, a titer of antibodies directed against Epstein-Barr virus or Lyme disease is estimated from the blood. If those antibodies are not present, either the person is not infected, or the infection occurred a *very* long time ago, and the B cells generating these specific antibodies have naturally decayed. In clinical immunology, levels of individual classes of immunoglobulins are measured by nephelometry (or turbidimetry) to characterize the antibody profile of patient. Elevations in different classes of immunoglobulins are sometimes useful in determining the cause of liver damage in patients whom the diagnosis is unclear. For example, elevated IgA indicates alcoholic cirrhosis, elevated IgM indicates viral hepatitis and primary biliary cirrhosis, while IgG is elevated in viral hepatitis, autoimmune hepatitis and cirrhosis. Autoimmune disorders can often be traced to antibodies that bind the body's own epitopes; many can be detected through blood tests. Antibodies directed against red blood cell surface antigens in immune mediated hemolytic anemia are detected with the Coombs test. The Coombs test is also used for antibody screening in blood transfusion preparation and also for antibody screening in antenatal women. Practically, several immunodiagnostic methods based on detection of complex antigen-antibody are used to diagnose infectious diseases, for example ELISA, immunofluorescence, Western blot, immunodiffusion, immunoelectrophoresis, and Magnetic immunoassay.

Disease Therapy

"Targeted" monoclonal antibody therapy is employed to treat diseases such as rheumatoid arthritis, multiple sclerosis, psoriasis, and many forms of cancer including non-Hodgkin's lymphoma, colorectal cancer, head and neck cancer and breast cancer. Some immune deficiencies, such as X-linked agammaglobulinemia and hypogammaglobulinemia, result in partial or complete lack of antibodies.

These diseases are often treated by inducing a short-term form of immunity called passive immunity. Passive immunity is achieved through the transfer of ready-made antibodies in the form of human or animal serum, pooled immunoglobulin or monoclonal antibodies, into the affected individual.

Prenatal Therapy

Rho(D) Immune Globulin antibodies are specific for human Rhesus D (RhD) antigen, also known as Rhesus factor. These anti-RhD antibodies are known under several brand names, including RhoGAM, BayRHo-D, Gamulin Rh, HypRho-D, and WinRho SDF. Rhesus factor is an antigen found on red blood cells; individuals that are Rhesus-positive (Rh+) have this antigen on their red blood cells and individuals that are Rhesus-negative (Rh–) do not. During normal childbirth, delivery trauma or complications during pregnancy, blood from a fetus can enter the mother's system. In the case of an Rh-incompatible mother and child, consequential blood mixing may sensitize an Rh- mother to the Rh antigen on the blood cells of the Rh+ child, putting the remainder of the pregnancy, and any subsequent pregnancies, at risk for hemolytic disease of the newborn. Anti-RhD antibodies are administered as part of a prenatal treatment regimen to prevent sensitization that may occur when a Rhesus-negative mother has a Rhesus-positive fetus. Treatment of a mother with Anti-RhD antibodies prior to and immediately after trauma and delivery destroys Rh antigen in the mother's system from the fetus. Importantly, this occurs before the antigen can stimulate maternal B cells to "remember" Rh antigen by generating memory B cells. Therefore, her humoral immune system will not make anti-Rh antibodies, and will not attack the Rhesus antigens of the current or subsequent baby. Rho(D) Immune Globulin treatment prevents sensitization that can lead to Rh disease, but does not prevent or treat the underlying disease itself.

Blood proteins, also called *serum proteins and Plasma Proteins*, are proteins found in blood plasma. Serum total protein in blood is 7g/dl, which in total makes 7% of total body weight. They serve many different functions, including

- circulatory transport molecules for lipids, hormones, vitamins and metals
- enzymes, complement components, protease inhibitors, and kinin precursors
- regulation of acellular activity and functioning and in the immune system.

Separating serum proteins by electrophoresis is a valuable diagnostic tool as well as a way to monitor clinical progress.

Often Mentioned Blood Proteins

Blood protein	Normal level	%	Function
Albumins	3.5-5.0 g/dl	60%	create osmotic pressure and transports other molecules
immunoglobulins	1.0-1.5 g/dl	37%	participate in immune system
Fibrinogens	0.2-0.45 g/dl	4%	blood coagulation
α1-antitrypsin			neutralize trypsin that has leaked from the digestive system
Regulatory proteins		<1%	Regulation of gene expression

Other types of blood proteins include: Prealbumin α1 antitrypsin, α1 acid glycoprotein α1 fetoprotein Haptoglobin, α2 macroglobulin Ceruloplasmin Transferring C3/C4 β 2 microglobulin β lipoprotein γ globulin proteins C-reactive protein (CRP)

- α2-macroglobulin
- Other globulins, which are of three types-α, β and γ.
- Lipoproteins (chylomicrons, VLDL, LDL, HDL)
- Transferrin
- Prothrombin

All the plasma proteins are synthesized in liver except γ globulins. We shall dicuss this topic in detain in the Hæmato-chemistry chapter.

60% of plasma proteins are made up of the protein albumin, which are major contributors to osmotic pressure of plasma which assists in the transport of lipids and steroid hormones. Globulins make up 35% of plasma proteins and are used in the transport of ions, hormones and lipids assisting in immune function. 4% is fibrinogen and this is essential in the clotting of blood and can be converted into insoluble fibrin. Regulatory proteins which make up less than 1% of plasma proteins are proteins such as enzymes, proenzymes and hormones. Current research regarding blood plasma proteins is centered on performing proteomics analyses of serum/plasma in the search for biomarkers. These efforts started with two-dimensional gel electrophoresisefforts in the 1970s and in more recent times this research has been performed using LC-tandem MS based proteomics.

HEAMOGLOBIN AND MYOGLOBIN

Hemoglobin (also spelled **haemoglobin** and abbreviated **Hb** or **Hgb**) is the iron-containing oxygen-transport metalloprotein in the red blood cells of vertebrates.

In mammals, the protein makes up about 97% of the red blood cell's dry content, and around 35% of the total content (including water). Hemoglobin transports oxygen from the lungs or gills to the rest of the body, such as to the muscles, where it releases the oxygen for cell use. It also has a variety of other roles of gas transport and effect-modulation which vary from species to species, and are quite diverse in some invertebrates.

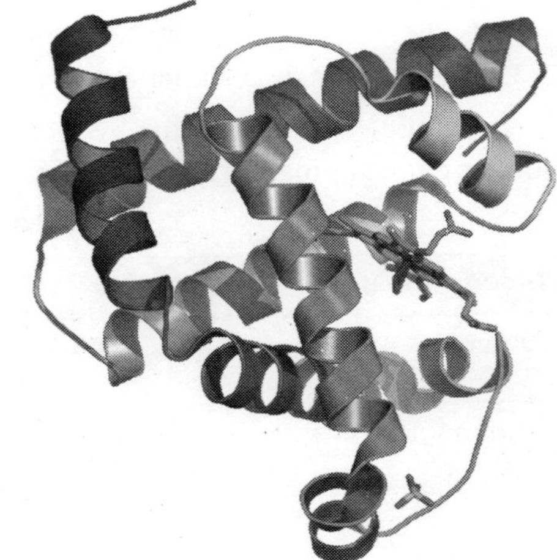

Fig. 2.78 : The Myoglobin protein Structure

Myoglobin is a single-chain globular protein of 153 amino acids, containing a heme (iron-containing porphyrin) prosthetic group in the center around which the remaining apoprotein folds. It has eight alpha helices and a hydrophobic core. It has a molecular weight of 16,700 daltons, and is the primary oxygen-carrying pigment of muscle tissues. Unlike the blood-borne hemoglobin, to which it is structurally related, this protein does not exhibit cooperative binding of oxygen, since positive cooperativity is a property of multimeric/oligomeric proteins only. Instead, the binding of oxygen by myoglobin is unaffected by the oxygen pressure in the surrounding tissue. Myoglobin is often cited as having an "instant binding tenacity" to oxygen given its hyperbolic oxygen dissociation curve. High concentrations of myoglobin in muscle cells allow organisms to hold their breaths longer. In 1958, John Kendrew and associates successfully determined the structure of myoglobin by high-resolution X-ray crystallography. For this discovery, John Kendrew shared the 1962 Nobel Prize in chemistry with Max Perutz. The human version of this gene is MB. Despite being one of the most studied proteins in biology, its true physiological function is not yet conclusively established: mice genetically engineered to lack myoglobin are viable and show no obvious defects.

Meat Colour

An X-ray diffraction image for the protein myoglobin.

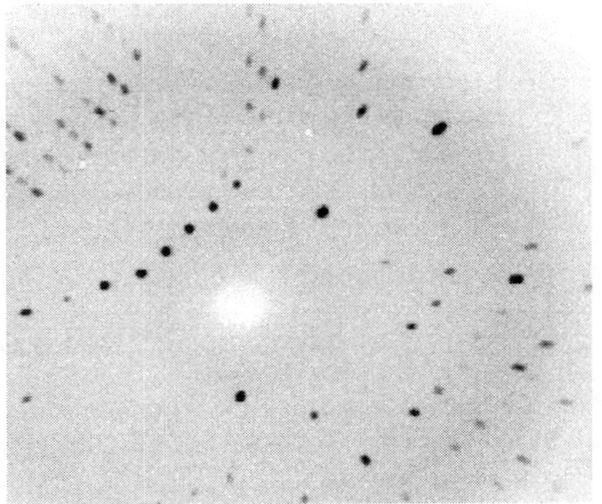

Fig. 2.79 : The X-Ray diffraction picture of the Myoglobin

Myoglobin forms pigments responsible for making meat red. The colour that meat takes is partly determined by the charge of the iron atom in myoglobin and the oxygen attached to it. When meat is in its raw state, the iron atom is in the +2 oxidation state, and is bound to a dioxygen molecule (O_2). Meat cooked well done is brown because the iron atom is now in the +3 oxidation state. Having lost an electron, and is now coordinated by a water molecule. Under some conditions, meat can also remain pink all through cooking, despite being heated to high temperatures. If meat has been exposed to nitrites, it will remain pink because the iron atom is bound to NO, nitric oxide (true of, e.g., corned beef or cured hams). Grilled meats can also take on a pink "smoke ring" that comes from the iron binding a molecule of carbon monoxide to give metmyoglobin. Raw meat packed in a carbon monoxide atmosphere also shows this same pink "smoke ring" due to the same coordination chemistry. Notably, the surface of the raw meat also displays the pink color, which is usually associated in consumers' minds with fresh meat. This artificially-induced pink color can persist in the meat for a very long time, reportedly up to one year. Hormel and Cargill are both reported to use this meat-packing process, and meat treated this way has been in the consumer market since 2003. Myoglobin is found in Type I muscle, Type II A and Type II B, but most texts consider myoglobin not to be found in smooth muscle.

Role in Disease

Myoglobin is released from damaged muscle tissue (rhabdomyolysis), which has very high concentrations of myoglobin. The released myoglobin is filtered by the kidneys but is toxic to the renal tubular epithelium and so may cause acute renal failure.

Myoglobin is a sensitive marker for muscle injury, making it a potential marker for heart attack in patients with chest pain. CK-MB and cTnT is used in combination with ECG, and the clinical signs to diagnose Acute Myocardial Infarction (AMI).

Structure and Bonding

Myoglobin contains a porphyrin ring with an iron center. There is a *proximal histidine* group attached directly to the iron center, and a *distal histidine* group on the opposite face, not bonded to the iron.

Many functional models of myoglobin have been studied. One of the most important is that of *picket fence porphyrin* by James Collman. This model was used to show the importance of the distal prosthetic group. It serves three functions:

1. To form hydrogen bonds with the dioxygen moiety, increasing the O_2 binding constant
2. To prevent the binding of carbon monoxide, whether from within or without the body. Carbon monoxide binds to iron in an end-on fashion, and is hindered by the presence of the distal histidine, which forces it into a bent conformation. CO binds to heme 23,000 times better than O_2, but only 200 times better in hemoglobin and myoglobin. Oxygen binds in a bent fashion, which can fit with the distal histidine.
3. To prevent irreversible dimerization of the oxymyoglobin with another deoxymyoglobin species

ROLE OF 2, 3 DPG

3-BPG is formed from 1,3-BPG by an enzyme called bisphosphoglycerate mutase. It is broken down by a phosphatase to form 3-phosphoglycerate. Its synthesis and breakdown are therefore a way around a step of glycolysis.

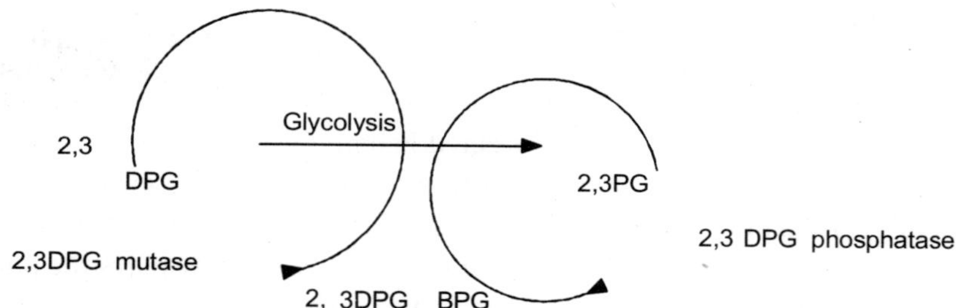

Effects of Binding

When 2,3-BPG binds deoxyhemoglobin, it acts to stabilize the low oxygen affinity state (T state) of the oxygen carrier. It fits neatly into the cavity of the deoxy- conformation, exploiting the molecular symmetry and positive polarity by forming salt bridges with lysine and histidine residues in the four subunits of hemoglobin. The R state, with oxygen bound to a heme group, has a different conformation and does not allow this interaction.

By selectively binding to deoxyhemoglobin, 2,3-BPG stabilizes the T state conformation, making it harder for oxygen to bind hemoglobin and more likely to be released to adjacent tissues. 2,3-BPG is part of a feedback loop that can help prevent tissue hypoxia in conditions where it is most likely to occur. Conditions of low tissue oxygen concentration such as high altitude (2,3-BPG levels are higher in those acclimated to high altitudes), airway obstruction, or congestive heart failure will tend to cause RBCs to generate more 2,3-BPG in their effort to generate energy by allowing more oxygen to be released in tissues deprived of oxygen. Ultimately, this mechanism increases oxygen release from RBCs under circumstances where it is needed most. This release is potentiated by the Bohr effect in tissues with high energetic demands.

Fetal Hemoglobin

Interestingly, fetal hemoglobin (HbF) exhibits a low affinity for 2,3-BPG, resulting in a higher binding affinity for oxygen. This increased oxygen binding affinity relative to that of adult hemoglobin (HbA) is due to HbF having two α/γ dimers as opposed to the two α/β dimers of HbA. The positive histidine residues of HbA β-subunits that are essential for forming the 2,3-BPG binding pocket are replaced by serine residues in HbF γ-subunits.

THALASSEMIA

The fundamental abnormality in thalassemia is impaired production of either the α or β hemoglobin chain. Thalassemia is a difficult subject to explain, since the condition is not a single disorder, but a group of defects with similar clinical effects. More confusion comes from the fact that the clinical descriptions of thalassemia were coined before the molecular basis of the thalassemias were uncovered. As a result, the organizational structure is somewhat disorderly. Review of thalassemia is best approached by separately examining its genetic basis and clinical expression.

Genetic Classification of the Thalassemias

Thalassemia includes disorders affecting the α hemoglobin chain genes and the β hemoglobin chain gene (see Hemoglobin Overview for explanation of α and β chains).

Alpha Thalassemia

Alpha thalassemia occurs when one or more of the four alpha chain genes fails to function. Alpha chain protein production, for practical purposes, is evenly divided among the four genes. With alpha thalassemia, the "failed" genes are almost invariably lost from the cell due to a genetic accident.

(*i*) The loss of one gene diminishes the production of the alpha protein only slightly. This condition is so close to normal that it can be detected only by specialized laboratory techniques that, until recently, were confined to research laboratories. A person with this condition is called a "silent carrier" because of the difficulty in detection.

(*ii*) The loss of two genes (two-gene deletion alpha thalassemia) produces a condition with small red blood cells, and at most a mild anemia. People with this condition look and feel normal. The condition can be detected by routine blood testing, however.

(*iii*) The loss of three α genes produces a serious hematological problem (three-gene deletion alpha thalassemia). Patients with this condition have a severe anemia, and often require blood transfusions to survive. The severe imbalance between the α chain production (now powered by one gene, instead of four) and beta chain production (which is normal) causes an accumulation of beta chains inside the red blood cells. Normally, beta chains pair only with alpha chains. With three-gene deletion alpha thalassemia, however, beta chains begin to associate in groups of four, producing an abnormal hemoglobin, called "hemoglobin H". The condition is called "hemoglobin H disease". Hemoglobin H has two problems. First it does not carry oxygen properly, making it functionally useless to the cell. Second, hemoglobin H protein damages the membrane that surrounds the red cell, accelerating cell destruction. The combination of the very low production of α chains and destruction of red cells in hemoglobin H disease produces a severe, life-threatening anemia. Untreated, most patients die in childhood or early adolescence.

(*iv*) The loss of all four α genes produces a condition that is incompatible with life. The γ chains produced during fetal life associate in groups of four to form an abnormal hemoglobin called "hemoglobin Barts". Most people with four-gene deletion α thalassemia die *in utero* or shortly after birth. Rarely, four gene deletion α thalassemia has been detected *in utero*, usually in a family where the disorder occured in an earlier child. *In utero* blood transfusions have saved some of these children. These patients require life-long transfusions and other medical support.

Beta Thalassemia

The fact that there are only two genes for the beta chain of hemoglobin makes beta thalassemia a bit simpler to understand than α thalassemia (2). Unlike α-thalassemia, β thalassemia rarely arises from the complete loss of a β globin gene. The beta globin gene is present, but produces little beta globin protein. The degree of suppression varies. Many causes of suppressed β globin gene expression have been found. In some cases, the affected gene makes essentially no β globin protein (β-0-thalassemia). In other cases, the production of beta chain protein is lower than normal, but not zero (β-(+)-thalassemia). The severity of β thalassemia depends in part on the type of beta thalassemic genes that a person has inherited.

(*i*) one-gene β thalassemia has one β globin gene that is normal, and a second, affected gene with a variably reduced production of β globin. The degree of imbalance with the α globin depends on the residual production capacity of the defective β globin gene. Even when the affected gene produces no β chain, the condition is mild since one β gene functions normally. The red cells are small and a mild anemia may exist. People with the condition generally have no symptoms. The condition can be detected by a routine laboratory blood evaluation. (Note that in many ways, the one-gene β thalassemia and the two-gene α thalassemia are very similar, from a clinical point of view. Each results in small red cells and a mild anemia).

(*ii*) two-gene β thalassemia produces a severe anemia and a potentially life-threatening condition. The severity of the disorder depends in part on the combination of genes that have been inherited: β-0-thal/β-0-thal; β-0-thal/β-(+)-thal; β-(+)-thal/β-(+)-thal. The β-(+)-thalassemia genes vary greatly in their ability to produce normal hemoglobin. Consequently, the clinical picture is more complex than might otherwise be the case for three genetic possibilities outlined.

Clinical Classification of the Thalassemias

Alphathalassemia

Alpha thalassemia has four manifestations, that correlate with the number of defective genes. Since the gene defect is almost invariably a loss of the gene, there are no "shades of function" to obscure the matter as occurs in beta thalassemia.

(*i*) Silent carrier state. This is the one-gene deletion alpha thalassemia condition. People with this condition are hematologically normal. They are detected only by sophisticated laboratory methods.

(*ii*) Mild alpha-thalassemia. These patients have lost two alpha globin genes.They have small red cells and a mild anemia. These people are usually asymptomatic. Often, physicians mistakenly diagnose people with mild alpha-thalassemia as having iron deficiency anemia. Iron therapy, of course, does not correct the anemia.

(*iii*) Hemoglobin H disease. These patients have lost three α globin genes. The result is a severe anemia, with small, misshapen red cells and red cell fragments. These patients typically have enlarged spleens. Bony abnormalities particularly involving the cheeks and forehead are often striking. The bone marrow works at an extraordinary pace in an attempt to compensate for the anemia. As a result, the marrow cavity within the bones is stuffed with red cell precursors. These cells gradually cause the bone to "mold" and flair out. Patients with hemoglobin H disease also develop large spleens. The spleen has blood forming cells, the same as the bone marrow. These cells become hyperactive and overexpand, just as those of the bone marrow. The result is a spleen that is often ten-times larger than normal. Patients with hemoglobin H disease often are small and appear malnourished, despite good food intake. This feature results from the tremendous amount of energy that goes into the production of new red cells at an extremely accelerated pace. The constant burning of energy by these patients mimics intense aerobic exercise; exercise that goes on for every minute of every day.

(*iv*) Hydrops fetalis. This condition results from the loss of all four α globin genes. The affected individual usually succumbs to the severe anemia and complications before birth.

Betathalassemia

(*i*) Thalassemia minor, or thalassemia trait. These terms are used interchangeably for people who have small red cells and mild (or no) anemia due to thalassemia. These patients are clinically well, and are usually only detected through routine blood testing. Physicians often mistakenly diagnose iron deficiency in people with thalassemia trait. Iron replacement does not correct the condition. The primary caution for people with beta-thalassemia trait involves the possible problems that their children could inherit if their partner also has β-thalassemia trait. These more severe forms of β-thalassemia trait are outlined below.

(*ii*) Thalassemia intermedia. Thalassemia intermedia is a confusing concept. The most important fact to remember is that thalassemia intermedia is a description, and not a pathological or genetic diagnosis. Patients with thalassemia intermedia have significant anemia, but are able to survive without blood transfusions. The factors that go into the diagnosis are:

- The degree to which the patient tolerates the anemia.
- The threshold of the physician to transfuse patients with thalassemia.

With regard to the tolerance of the anemia, most patients with thalassemia have substantial symptoms with a Hb of much below 7 or 8 gm/dl. With hemoglobins of this level, excess energy consumption due to the profound hemolysis can produce small stature, poor weight gain, poor energy levels, and susceptibility to infection. Further, the extreme activity of the bone marrow produces bone deformities of the face and other areas, along with enlargement of the spleen. The long bones of the arms and legs are weak and fracture easily. Patients with this clinical condition usually do better with regular transfusions. The need for regular transfusions would then place them under the heading of thalassemia major. On the other hand, some patients with marked thalassemia can maintain a hemoglobin of about 9 to 10 gm/dl. The exercise tolerance of these patients is significantly better. The question then becomes whether the accelerated bone marrow activity needed to maintain this level of hemoglobin causes unacceptable side-effects such as bone abnormalities or enlarged spleen. This is a judgment decision. A given patient at the critical borderline would be transfused by some physicians to prevent these problems, even if they are slight. The patient then would be *clinically* classified as having thalassemia major. Another physician might choose to avoid the complications of chronic transfusion. The same patient then would be *clinically* classified as thalassemia intermedia. The patient has thalassemia that is more severe than thalassemia trait, but not so severe as to require chronic transfusion as do the patients with thalassemia major. A patient can change status. The spleen is enlarged in these patients. The spleen plays a role in clearing damaged red cells from the blood stream. Since all of the red cells in patients with severe thalassemia have some degree of damage, clearance by the spleen accelerates the rate of cell loss. Therefore the bone marrow has to work harder to replace these cells. In some patients, removal of the spleen slows the rate of red cell destruction just enough, that they can manage without transfusion, and still not have the unacceptable side-effects. In this case, the patient converts *clinically* from thalassemia major to thalassemia intermedia.

(*iii*) Thalassemia major. This is the condition of severe thalassemia in which chronic blood transfusions are needed (3). In some patients the anemia is so severe, that death occurs without transfusions. Other patients could survive without transfusions, for a while, but would have terrible deformities. While transfusions are life-saving in patients with thalassemia major, transfusions ultimately produce iron overload. Chelation therapy, usually with the iron-binding agent, desferrioxamine (Desferal), is needed to prevent death from iron-mediated organ injury.

Relationship of the Genetic and Clinical Classifications of Thalassemia

The advent of modern molecular biology permits the genetic classification of thalassemias, outlined earlier in this document. A rough correlation exists between the clinical and genetic classifications. The relationship between genetics and clinical state is not absolute, however:

- thalassemia trait (minor)- normal β gene/ thalassemia gene (β zero or +)
- thalassemia intermedia- often two β-(+)-genes
- thalassemia major- two beta-(+)-genes (where the plus is not substantial); β-(+)-gene/β-0-gene; β-0-gene/β-0-gene

COLLAGEN

Collagen is the main protein of connective tissue in animals and the most abundant protein in mammals, making up about 25% to 35% of the whole-body protein content.

Uses

Collagen is one of the long, fibrous structural proteins whose functions are quite different from those of globular proteins such as enzymes. Tough bundles of collagen called *collagen fibers* are a major component of the extracellular matrix that supports most tissues and gives cells structure from the outside, but collagen is also found inside certain cells. Collagen has great tensile strength, and is the main component of fascia, cartilage, ligaments, tendons, bone and skin. Along with soft keratin, it is responsible for skin strength and elasticity, and its degradation leads to wrinkles that accompany aging. It strengthens blood vessels and plays a role in tissue development. It is present in the cornea and lens of the eye in crystalline form. It is also used in cosmetic surgery and burns surgery. Hydrolyzed collagen can play an important role in weight management. As a protein, it can be advantageously used for its satiating power.

Industrial Uses

If collagen is partially hydrolyzed, the three tropocollagen strands separate into globular, random coils, producing gelatin, which is used in many foods, including flavoured gelatin desserts. Besides food, gelatin has been used in pharmaceutical, cosmetic, and photography industries. Collagen and gelatin are poor-quality protein since they do not contain all the essential amino acids that the human body requires—they are not complete proteins. Manufacturers of collagen-based dietary supplements claim that their products can improve skin and fingernail quality as well as joint health. However, mainstream scientific research has not shown any evidence to support these claims. Individuals with problems in these areas are more likely to be suffering from some other underlying condition rather than protein deficiency.

From the Greek for glue, *kolla*, the word collagen means "glue producer" and refers to the early process of boiling the skin and sinews of horses and other animals to obtain glue. Collagen adhesive was used by Egyptians about 4,000 years ago, and Native Americans used it in bows about 1,500 years ago. The oldest glue in the world, carbon-dated as more than 8,000 years old, was found to be collagen—

used as a protective lining on rope baskets and embroidered fabrics, and to hold utensils together; also in crisscross decorations on human skulls. Collagen normally converts to gelatin, but survived due to the dry conditions. Animal glues are thermoplastic, softening again upon reheating, and so they are still used in making musical instruments such as fine violins and guitars, which may have to be reopened for repairs—an application incompatible with tough, synthetic plastic adhesives, which are permanent. Animal sinews and skins, including leather, have been used to make useful articles for millennia.

Gelatin-resorcinol-formaldehyde glue (and with formaldehyde replaced by less-toxic pentanedial and ethanedial) has been used to repair experimental incisions in rabbit lungs.

Medical Uses

Collagen has been widely used in cosmetic surgery, as a healing aid for burn patients for reconstruction of bone and a wide variety of dental, orthopedic and surgical purposes. Some points of interest are:

1. when used cosmetically, there is a chance of allergic reactions causing prolonged redness; however, this can be virtually eliminated by simple and inconspicuous patch testing prior to cosmetic use;
2. most medical collagen is derived from young beef cattle (bovine) from certified BSE (Bovine spongiform encephalopathy) free animals. Most manufacturers use donor animals from either "closed herds", or from countries which have never had a reported case of BSE such as Australia, Brazil and New Zealand;
3. porcine (pig) tissue is also widely used for producing collagen sheet for a variety of surgical purposes; and
4. alternatives using the patient's own fat, hyaluronic acid or polyacrylamide gel are readily available.

Collagens are widely employed in the construction of artificial skin substitutes used in the management of severe burns. These collagens may be derived from bovine, equine or porcine, and even human, sources and are sometimes used in combination with silicones, glycosaminoglycans, fibroblasts, growth factors and other substances.

Collagen is also sold commercially as a joint mobility supplement. This lacks supportive research as the proteins would just be broken down into its base amino acids during digestion, and could go to a variety of places besides the joints depending upon need and DNA orders.

Recently an alternative to animal-derived collagen has become available. Although expensive, this human collagen, derived from donor cadavers, placentas and aborted fetuses, may minimize the possibility of immune reactions.

Collagen is now being used as a main ingredient for some cosmetic makeup.

Conformation and Structure

Collagen structure is complex. Its conformation can be considered at the monomeric level (individual) collagen molecules and/or at its aggregate level, how the monomers are arranged i.e. their packing structure (fibrils, networks, etc.).

History and Background

The molecular and packing structures of collagen have eluded scientists for decades; the first evidence that it possess a regular structure at the molecular level was presented in the mid-1930s. Since that time many prominent scholars, including (but not limited to) Nobel laureate Crick, and Pauling, Rich, Yonath, Brodsky, Berman and Ramachandran concentrated on the conformation of the collagen monomoer. Several competing models although correctly dealing with the conformation of each individual peptide chain, gave way to the triple-helical "Madras" model which provided an essentially correct model of the molecules quaternary structure although this model still required some refinement. The packing structure of collagen has not been defined to the same degree outside of the fibrillar collagen types, although it has been long known to be hexagonal or quasi-hexagonal.As with its monomeric structure, several conflicting models alleged that either the packing arrangement of collagen molecules is 'sheet-like' or microfibrillar. Recently it was confirmed that the microfibrillar structure as described by Fraser, Miller, Wess (amongst others) was closest to the observed structure, although it over-simplified the topological progression of neighboring collagen molecules and hence did not predict the correct conformation of the discontinuous D-periodic pentameric arrangement termed simply: the microfibril.

Molecular Structure

The *tropocollagen* or "collagen molecule" is a subunit of larger collagen aggregates such as fibrils. It is approximately 300 nm long and 1.5 nm in diameter, made up of three polypeptide strands (called α peptides), each possessing the conformation of a left-handed helix (its name is not to be confused with the commonly occurring α helix, a right-handed structure). These three left-handed helices are twisted together into a right-handed coiled coil, a triple helix or "super helix", a cooperative quaternary structure stabilized by numerous hydrogen bonds. With type I collagen and possibly all fibrillar collagens if not all collagens, each triple-helix associates into a right-handed super-super-coil that is referred to as the collagen microfibril. Each microfibril is interdigitated with its neighbouring microfibrils.

Collagen Fibre

A distinctive feature of collagen is the regular arrangement of amino acids in each of the three chains of these collagen subunits. The sequence often follows the pattern Gly-Pro-Y or Gly-X-Hyp, where X and Y may be any of various other amino acid residues. Proline or hydroxyproline constitute about 1/6 of the total sequence. With Glycine accounting for the 1/3 of the sequence, this means that approximately half of the collagen sequence is not glycine or proline, a fact often missed due to the distraction of the unusual GXY character of collagen α-peptides. This kind of regular repetition and high glycine content is found in only a few other fibrous proteins, such as silk fibroin. 75-80% of silk is (approximately) -Gly-Ala-Gly-Ala- with 10% serine—and elastin is rich in glycine, proline, and alanine (Ala), whose side group is a small, inert methyl group. Such high glycine and regular repetitions are never found in globular proteins save for very short sections of their sequence. Chemically-reactive side groups are not needed in structural proteins as they are in enzymes and transport proteins, however collagen is not quite just a structural protein. Due to its key role in the determination of cell phenotype, cell adhesion, tissue regulation and infrastructure, many sections of its non-proline rich regions have cell or matrix association regulation roles. The relatively high content of Proline and Hydroxyproline rings, with their geometrically constrained carboxyl and (secondary) amino groups, along with the rich abundance of glycine, accounts for the tendency of the individual polypeptide strands to form left-handed helices spontaneously, without any intrachain hydrogen bonding.

Because glycine is the smallest amino acid with no side-chain, it plays a unique role in fibrous structural proteins. In collagen, Gly is required at every third position because the assembly of the triple helix puts this residue at the interior (axis) of the helix, where there is no space for a larger side group than glycine's single hydrogen atom. For the same reason, the rings of the Pro and Hyp must point outward. These two amino acids help stabilize the triple helix—Hyp even more so than Pro—a lower concentration of them is required in animals such as fish, whose body temperatures are lower than most warm-blooded animals.

Fibrillar Structure

The tropocollagen subunits spontaneously self-assemble, with regularly staggered ends, into even larger arrays in the extracellular spaces of tissues. In the fibrillar collagens, the molecules are staggered from each other by about 67nm (a unit that is referred to as 'D' and changes depending upon the hydration state of the aggregate). Each D-period contains 4 and a fraction collagen molecules. This is because 300 nm divided by 67 nm does not give an integer (the length of the collagen molecule divided by the stagger distance D). Therefore in each D-period repeat of the microfibril, there is a part containing 5 molecules in cross-section – called the "overlap" and a part containing only 4 molecules. The triple-helices are also arranged in a hexagonal or quasi-hexagonal array in cross-section, in both the gap and overlap regions.

There is some covalent crosslinking within the triple helices, and a variable amount of covalent crosslinking between tropocollagen helices forming well organized aggregates (such as fibrils). Larger fibrillar bundles are formed with the aid of several different classes of proteins (including different collagen types), glycoproteins and proteoglycans to form the different types of mature tissues from alternate combinations of the same key players. Collagen's insolubility was a barrier to the study of monomeric collagen until it was found that tropocollagen from young animals can be extracted because it is not yet fully crosslinked. However, advances in microscopy techniques (Electron Microscopy - EM and Atomic Force Microscopy -AFM) and X-ray diffraction have enabled researchers to obtain increasingly detailed images of collagen structure *in situ*. These later advances are particularly important to better understanding the way in which collagen structure affects cell-cell and cell-matrix communication and how tissues are constructed in growth and repair, and changed in development and disease.

Collagen fibrils are collagen molecules packed into an organized overlapping bundle. Collagen fibers are bundles of fibrils.

Collagen fibrils / aggregates are arranged in different combinations and concentrations in various tissues to provide varying tissue properties. In bone, entire collagen triple helices lie in a parallel, staggered array. 40 nm gaps between the ends of the tropocollagen subunits probably serve as nucleation sites for the deposition of long, hard, fine crystals of the mineral component, which is (approximately) hydroxyapatite, $Ca_{10}(PO_4)_6\,(OH)_2$with some phosphate. It is in this way that certain kinds of cartilage turn into bone. Type I collagen gives bone its tensile strength.

Types and Associated Disorders

Collagen occurs in many places throughout the body. There are more then 28 types of collagen described in literature. Over 90% of the collagen in the body, however, are of type-I, II, III, and IV.

- Collagen One - skin, tendon, vascular, ligature, organs, bone (main component of bone)
- Collagen Two - cartilage (main component of cartilage)
- Collagen Three - reticulate (main component of reticular fibers), commonly found alongside type-I.
- Collagen Four - forms bases of cell basement membrane

Collagen diseases commonly arise from genetic defects that affect the biosynthesis, assembly, postranslational modification, secretion, or other processes in the normal production of collagen.

Type	Notes	Gene(s)	Disorders
1	2	3	4
I	This is the most abundant collagen of the human body. It is present in scar tissue, the end product when tissue heals by repair. It is found in tendons, skin, artery walls, the endomysium of myofibrils, fibrocartilage, and the organic part of bones and teeth	COL1A1, COL1A2	osteogenesis imperfecta, Ehlers-Danlos Syndrome
II	Hyaline cartilage, makes up 50% of all cartilage protein. Vitreous humour of the eye	COL2A1	Collagenopathy, types II and XI
III	This is the collagen of granulation tissue, and is produced quickly by young fibroblasts before the tougher type I collagen is synthesized. Reticular fiber. Also found in artery walls, skin, intestines and the uterus	COL3A1	Ehlers-Danlos Syndrome
IV	basal lamina; eye lens. Also serves as part of the filtration system in capillaries	COL4A1, COL4A2,	Alport syndrome

...(Contd.)

1	2	3	4
	and the glomeruli of nephron in the kidney	COL4A3, COL4A4, COL4A5, COL4A6	
V	most interstitial tissue, assoc. with type-I, associated with placenta	COL5A1, COL5A2, COL5A3	Ehlers-Danlos syndrome (Classical)
VI	most interstitial tissue, assoc. with type-I	COL6A1, COL6A2, COL6A3	Ulrich myopathy and Bethlem myopathy
VII	forms anchoring fibrils in dermal epidermal junctions	COL7A1	epidermolysis bullosa
VIII	some endothelial cells	COL8A1, COL8A2	—
IX	FACIT collagen, cartilage, assoc. with type-II and XI fibrils	COL9A1, COL9A2, COL9A3	- EDM2 and EDM3
X	hypertrophic and mineralizing cartilage	COL10A1	—
XI	cartilage	COL11A1, COL11A2	Collagenopathy, types II and XI
XII	FACIT collagen, interacts with type I containing fibrils, decorin and glycosaminoglycans	COL12A1	—
XIII	transmembrane collagen, interacts with integrin a1b1, fibronectin and components of basement membranes like nidogen and perlecan.	COL13A1	—
XIV	FACIT collagen	COL14A1	—
XV	—	COL15A1	—
XVI	—	COL16A1	—
XVII	transmembrane collagen, also known as BP180, a 180 kDa protein	COL17A1	Bullous Pemphigoid and certain forms of junctional epidermolysis bullosa
XVIII	source of endostatin	COL18A1	—
XIX	FACIT collagen	COL19A1	—
XX	—	COL20A1	—

1	2	3	4
XXI	FACIT collagen	COL21A1	—
XXII	—	COL22A1	—
XXIII	MACIT collagen -	COL23A1	—
XXIV	—	COL24A1	—
XXV	—	COL25A1	—
XXVI	—	EMID2	—
XXVII	—	COL27A1	—
XXVIII	—	COL28A1	—
XXIX	epidermal collagen	COL29A1	Atopic Dermatitis

In addition to the above mentioned disorders, excessive deposition of collagen occurs in Scleroderma.

Staining

In histology, collagen is brightly eosinophilic (pink) in standard H&E slides. The dye methyl violet may be used to stain the collagen in tissue samples.

The dye methyl blue can also be used to stain collagen and immunohistochemical stains are available if required.

The best stain for use in differentiating collagen from other fibers is Masson's trichrome stain.

Synthesis

Amino Acids

Collagen has an unusual amino acid composition and sequence:

- Glycine (Gly) is found at almost every third residue
- Proline (Pro) makes up about 9% of collagen
- Collagen contains two uncommon derivative amino acids not directly inserted during translation. These amino acids are found at specific locations relative to glycine and are modified post-translationally by different enzymes, both of which require vitamin C as a cofactor.
  - — Hydroxyproline (Hyp), derived from proline.
  - — Hydroxylysine, derived from lysine. Depending on the type of collagen, varying numbers of hydroxylysines have disaccharides attached to them.

Collagen I Formation

Most collagen forms in a similar manner, but the following process is typical for type-I:

1. Inside the cell:
    1. Three peptide chains are formed (2 α-1 and 1 α-2 chain) in ribosomes along the Rough Endoplasmic Reticulum (RER). These peptide chains (known as preprocollagen) have registration peptides on each end; and a signal peptide is also attached to each
    2. Peptide chains are sent into the lumen of the RER
    3. Signal Peptides are cleaved inside the RER and the chains are now known as procollagen
    4. Hydroxylation of lysine and proline amino acids occurs inside the lumen. This process is dependent on Ascorbic Acid (Vitamin C) as a cofactor
    5. Glycosylation of specific hydroxylated amino acid occurs
    6. Triple helical structure is formed inside the RER
    7. Procollagen is shipped to the golgi apparatus, where it is packaged and secreted by exocytosis
2. Outside the cell:
    1. Registration peptides are cleaved and tropocollagen is formed by procollagen peptidase.
    2. Multiple tropocollagen molecules form collagen fibrils, and multiple collagen fibrils form into collagen fibers
    3. Collagen is attached to cell membranes via several types of protein, including fibronectin and integrin.

Synthetic Pathogenesis

Vitamin C deficiency causes scurvy, a serious and painful disease in which defective collagen prevents the formation of strong connective tissue. Gums deteriorate and bleed, with loss of teeth; skin discolors, and wounds do not heal. Prior to the eighteenth century, this condition was notorious among long duration military, particularly naval, expeditions during which participants were deprived of foods containing Vitamin C. In the human body, a malfunction of the immune system, called an autoimmune disease, results in an immune response in which healthy collagen fibers are systematically destroyed with inflammation of surrounding tissues. The resulting disease processes are called Lupus erythematosus, and rheumatoid arthritis, or collagen tissue disorders.

Many bacteria and viruses have virulence factors which destroy collagen or interfere with its production.

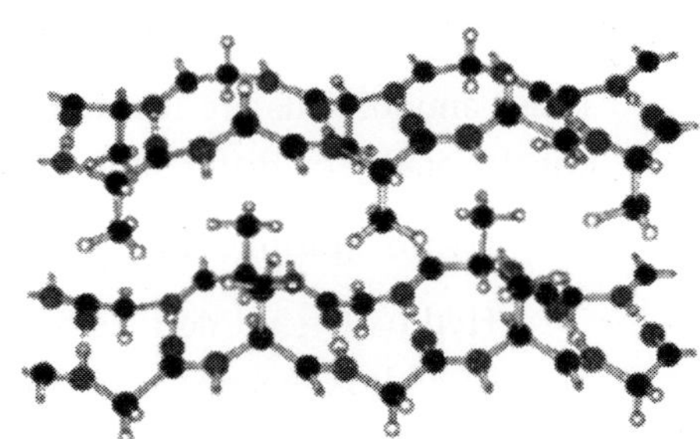

Keratins are a family of fibrous structural proteins; tough and insoluble, they form the hard but non-mineralized structures found in reptiles, birds, amphibians and mammals. They are rivaled as biological materials in toughness only by chitin.

There are various types of keratins within a single animal.

Variety of Animal Uses

Keratins are the main constituent of structures that grow from the skin:

The *α-keratins* in the hair (including wool), horns, nails, claws and hooves of mammals the harder *β-keratins* found in nails and in the scales and claws of reptiles, their shells (chelonians, such as tortoise, turtle, terrapin), and in the feathers, beaks, claws of birds and quills of porcupines. (These keratins are formed primarily in beta sheets. However, beta sheets are also found in α-keratins.)

Arthropods such as crustaceans often have parts of their armor or exoskeleton made of keratin, sometimes in combination with chitin.

The baleen plates of filter-feeding whales are made of keratin.

They can be integrated in the chitinophosphatic material that makes up the shell and setae in many brachiopods.

Keratins are also found in the gastrointestinal tracts of many animals, including roundworms (which also have an outer layer made of keratin).

Although it is now difficult to be certain, the scales, claws, some protective armour and the beaks of dinosaurs would, almost certainly, have been composed of a type of keratin.

In Crossopterygian fish, the outer layer of cosmoid scales was keratin.

Cornification

It used to be thought that keratins were separable into 'hard' and 'soft,' or 'cytokeratins' and 'other keratins', but those designations are now understood to be incorrect. In 2006, a new nomenclature was adopted for describing keratins which takes this into account. Keratins are intermediate filaments. Like all intermediate filaments, keratin proteins form filamentous polymers in a series of assembly steps beginning with dimerization; dimers assemble into tetramers and octamers and eventually, the current hypothesis holds, into unit-length-filaments (ULF) capable of annealing end-to-end into long filaments. During the process of epithelial differentiation, cells become cornified as keratin protein is incorporated into longer keratin intermediate filaments. Eventually the nucleus and cytoplasmic organelles disappear, metabolism ceases and cells undergo a programmed death as they become fully keratinized. In many other cells types, such as dermis, keratin functions in the cytoplasm to mechanically stabilize the cell against physical stress. It does this through connections to desmosomes, cell-cell junctional plaques, and hemidesmosomes, cell-basement membrane adhesive structures. Cells in the epidermis contain a structural matrix of keratin which makes this outermost layer of the skin almost waterproof, and along with collagen and elastin, gives skin its strength. Rubbing and pressure cause keratin to proliferate with the formation of protective calluses—useful for athletes and on the fingertips of musicians who play stringed instruments. Keratinized epidermal cells are constantly shed and replaced.

These hard, integumentary structures are formed by intercellular cementing of fibers formed from the dead, cornified cells generated by specialized beds deep within the skin. Hair grows continuously and feathers moult and regenerate. The constituent proteins may be phylogenetically homologous but differ somewhat in chemical structure and supermolecular organization. The evolutionary relationships are complex and only partially known. Multiple genes have been identified for the β-keratins in feathers, and this is probably characteristic of all keratins.

Molecular Biology and Biochemistry

The properties which make structural proteins like keratins useful depend on their supermolecular aggregation. These depend on the properties of the individual polypeptide strands, which depend in turn on their amino acid composition and sequence. The α-helix and β-sheet motifs, and disulfide bridges, are crucial to the conformations of globular, functional proteins like enzymes, many of which operate semi-independently, but they take on a completely dominant role in the architecture and aggregation of keratins.

Glycine and Alanine

Keratins contain a high proportion of the smallest of the 20 amino acids, glycine, whose "side group" is a single hydrogen atom; also the next smallest, alanine, with a small and noncharged methyl group. In the case of β-sheets, this allows sterically-unhindered hydrogen bonding between the amino and carboxyl groups of peptide bonds on adjacent protein chains, facilitating their close alignment and strong binding. Fibrous keratin molecules can twist around each other to form helical intermediate filaments.

Limited interior space is the reason why the triple helix of the (unrelated) structural protein collagen, found in skin, cartilage and bone, likewise has a high percentage of glycine. The connective tissue protein elastin also has a high percentage of both glycine and alanine. Silk fibroin, considered a β-keratin, can have these two as 75–80% of the total, with 10–15% serine, with the rest having bulky side groups. The chains are antiparallel, with an alternating C '! N orientation. A preponderance of amino acids with small, nonreactive side groups is characteristic of structural proteins, for which H-bonded close packing is more important than chemical specificity.

Disulphide Bridges

In addition to intra- and intermolecular hydrogen bonds, keratins have large amounts of the sulfur-containing amino acid cysteine, required for the disulfide bridges that confer additional strength and rigidity by permanent, thermally-stable crosslinking—a role sulfur bridges also play in vulcanized rubber. Human hair is approximately 14% cysteine. The pungent smells of burning hair and rubber are due to the sulfur compounds formed. Extensive disulfide bonding contributes to the insolubility of keratins, except in dissociating or reducing agents.

The more flexible and elastic keratins of hair have fewer interchain disulfide bridges than the keratins in mammalian fingernails, hooves and claws (homologous structures), which are harder and more like their analogs in other vertebrate classes. Hair and other α-keratins consist of α-helically-coiled single protein strands (with regular intra-chain H-bonding), which are then further twisted into superhelical ropes that may be further coiled. The β-keratins of reptiles and birds have β-pleated sheets twisted together, then stabilized and hardened by disulfide bridges.

Silk

The silk fibroins produced by insects and spiders are often classified as keratins, though it is unclear whether they are phylogenetically related to vertebrate keratins.

Silk found in insect pupae, and in spider webs and egg casings, also has twisted β-pleated sheets incorporated into fibers wound into larger supermolecular aggregates. The structure of the spinnerets on spiders' tails, and the contributions of their interior glands, provide remarkable control of fast extrusion. Spider silk is typically about 1 to 2 micrometres (μm) thick, compared with about 60 μm for human hair, and more for some mammals. (Hair, or fur, occurs only in mammals.) The biologically and commercially useful properties of silk fibers depend on the organization of multiple adjacent protein chains into hard, crystalline regions of varying size, alternating with flexible, amorphous regions where the chains are randomly coiled. A somewhat analogous situation occurs with synthetic polymers such as nylon, developed as a silk substitute. Silk from the hornet cocoon contains doublets about 10 μm across, with cores and coating, and may be arranged in up to 10 layers; also in plaques of variable shape. Adult hornets also use silk as a glue, as do spiders.

Pairing

A (neutral-basic)	B (acidic)	Occurrence
keratin 1, keratin 2	keratin 9, keratin 10	stratum corneum, keratinocytes
keratin 3	keratin 12	cornea
keratin 4	keratin 13	stratified epithelium
keratin 5	keratin 14, keratin 15	stratified epithelium
keratin 6	keratin 16, keratin 17	squamous epithelium
keratin 7	keratin 19	ductal epithelia
keratin 8	keratin 18, keratin 20	simple epithelium

The entries KRT23, KRT24, KRT25, KRT26, KRT27, KRT28, KRT31, KRT32, KRT33, KRT33A, KRT34, KRT35, KRT36, KRT37, KRT38, KRT39, KRT40, KRT71, KRT72, KRT73, KRT74, KRT75, KRT76, KRT77, KRT78, KRT79, KRT8, KRT80, KRT81, KRT82, KRT83, KRT84, KRT85 and KRT86 have been used to describe keratins past 20.

Clinical Significance

Some infectious fungi, such as those which cause athlete's foot, ringworm or the *Batrachochytrium dendrobatidis* (Chytrid fungus) which is killing amphibians all over the world, feed on keratin.

Diseases caused by mutations in the keratin genes include—

- Epidermolysis bullosa simplex
- Ichthyosis bullosa of Siemens
- Epidermolytic hyperkeratosis

- Steatocystoma multiplex
- Keratosis pharyngis

Elastin is a protein in connective tissue that is elastic and allows many tissues in the body to resume their shape after stretching or contracting. Elastin helps skin to return to its original position when it is poked or pinched. Elastin is also an important load-bearing tissue in the bodies of mammals and used in places where mechanical energy is required to be stored.

Composition

Elastin is primarily composed of the amino acids glycine, valine, alanine, and proline. It is a specialized protein with a molecular weight of 64 to 66 kDa, and an irregular or random coil conformation made up of 830 amino acids.

Elastin is made by linking many soluble tropoelastin protein molecules, in a reaction catalyzed by lysyl oxidase, to make a massive insoluble, durable cross-linked array. The amino acid responsible for these cross-links is lysine.

Desmosine and isodesmosine are both found in elastin.

Locations in Body

Elastin serves an important function in arteries and is particularly abundant in large elastic blood vessels such as the aorta. Elastin is also very important in the lungs, elastic ligaments, the skin, the bladder, elastic cartilage, and the intervertebral disc above the sacroiliac. It is present in all vertebrates above the jawless fish.

PURINES

Purines

- Adenine = 6-amino purine
- Guanine = 2-amino-6-oxy purine

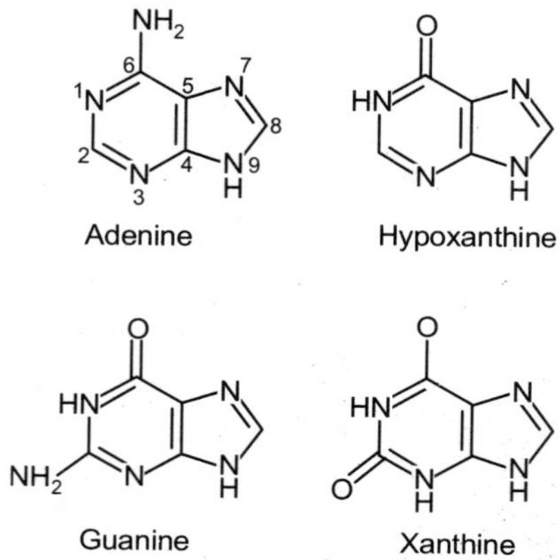

Adenine

Hypoxanthine

Guanine

Xanthine

- Hypoxanthine = 6-oxy purine
- Xanthine = 2,6-dioxy purine

Adenine and guanine are found in both DNA and RNA. Hypoxanthine and xanthine are not incorporated into the nucleic acids as they are being synthesized but are important intermediates in the synthesis and degradation of the purine nucleotides.

PYRIMIDINES

- Uracil = 2,4-dioxy pyrimidine
- Thymine = 2,4-dioxy-5-methyl pyrimidine
- Cytosine = 2-oxy-4-amino pyrimidine
- Orotic acid = 2,4-dioxy-6-carboxy pyrimidine

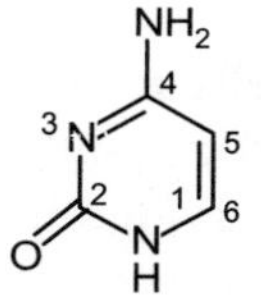

Cytosine

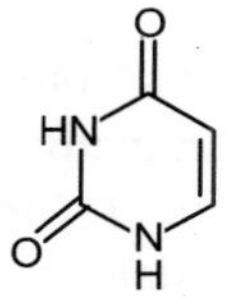

Uracil

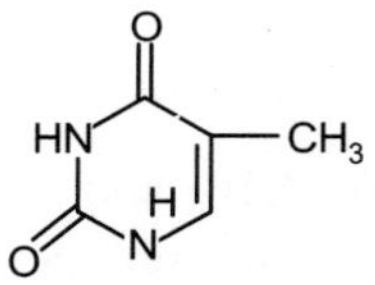

Thymine

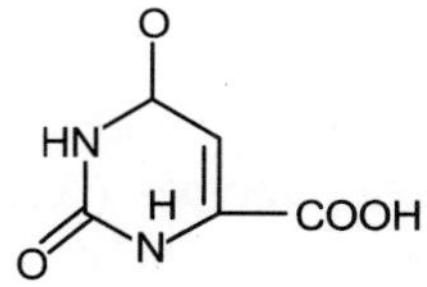

Orotic Acid

Cytosine is found in both DNA and RNA. Uracil is found only in RNA. Thymine is normally found in DNA. Sometimes tRNA will contain some thymine as well as uracil.

Nucleosides

If a sugar, either **ribose** or **2-deoxyribose**, is added to a nitrogen base, the resulting compound is called a **nucleoside**. Carbon 1 of the sugar is attached to nitrogen 9 of a purine base or to nitrogen 1 of a pyrimidine base. The names of **purine** nucleosides end in **-osine** and the names of **pyrimidine** nucleosides end in **-idine.** The convention is to number the ring atoms of the base normally and to use l', etc. to distinguish the ring atoms of the sugar. Unless otherwise specificed, the sugar is assumed to be ribose. To indicate that the sugar is 2'-deoxyribose, a **d-** is placed before the name.

- Adenosine
- Guanosine
- Inosine - the base in inosine is **hypoxanthine**
- Uridine

- Thymidine
- Cytidine

Adenosine is a nucleoside composed of a molecule of adenine attached to a ribose sugar molecule (ribofuranose) moiety via a β-N-glycosidic bond.

Adenosine plays an important role in biochemical processes, such as energy transfer - as adenosine triphosphate (ATP) and adenosine diphosphate (ADP) - as well as in signal transduction as cyclic adenosine monophosphate, cAMP. It is also an inhibitory neurotransmitter, believed to play a role in promoting sleep and suppressing arousal, with levels increasing with each hour an organism is awake.

Pharmacological Effects

Adenosine is an endogenous purine nucleoside that modulates many physiologic processes. Cellular signaling by adenosine occurs through four known adenosine receptor subtypes (A1, A2A, A2B, and A3).

Extracellular adenosine concentrations from normal cells are approximately 300 nM; however, in response to cellular damage (e.g. in inflammatory or ischemic tissue), these concentrations are quickly elevated (600-1,200 nM). Thus, in regards to stress or injury, the function of adenosine is primarily that of cytoprotection preventing tissue damage during instances of hypoxia, ischemia, and seizure activity. Activation of A2A receptors produces a constellation of responses that in general can be classified as anti-inflammatory.

Adenosine Receptors

The different adenosine receptor subtypes (A1, A2A, A2B, and A3) are all seven transmembrane spanning G-protein coupled receptors. These four receptor subtypes are further classified based on their ability to either stimulate or inhibit adenylate cyclase activity. The A2A and A2B receptors couple to Gαs and mediate the stimulation of adenylate cyclase, while the A1 and A3 adenosine receptors couple to GÜi which inhibits adenylate cyclase activity. Additionally, A1 receptors couple to Gαo, which has been reported to mediate adenosine inhibition of Ca2+ conductance, whereas A2B and A3 receptors also couple to Gαq and stimulate phospholipase activity.

Anti-inflammatory Properties

Adenosine is a potent anti-inflammatory agent, acting at its four G-protein coupled receptors. Topical treatment of adenosine to foot wounds in diabetes mellitus has been shown in lab animals to drastically increase tissue repair and reconstruction. Topical administration of adenosine for use in wound healing deficiencies and diabetes mellitus in humans is currently under clinical investigation.

Action on the Heart

When administered intravenously, adenosine causes transient heart block in the AV node. This is mediated via the A1 receptor, inhibiting adenyl cyclase, reducing cAMP and so causing cell hyperpolarization by increasing outward K+ flux. It also causes endothelial dependent relaxation of smooth muscle as is found inside the artery walls. This causes dilatation of the "normal" segments of arteries, i.e. where the endothelium is not separated from the tunica media by atherosclerotic plaque. This feature allows physicians to use adenosine to test for blockages in the coronary arteries, by exaggerating the difference between the normal and abnormal segments.

In individuals suspected of suffering from a supraventricular tachycardia (SVT), adenosine is used to help identify the rhythm. Certain SVTs can be successfully terminated with adenosine. This includes any re-entrant arrhythmias that require the AV node for the re-entry (e.g., AV reentrant tachycardia (AVRT), AV nodal reentrant tachycardia (AVNRT). In addition, atrial tachycardia can sometimes be terminated with adenosine.

Adenosine has an indirect effect on atrial tissue causing a shortening of the refractory period. When administered via a central lumen catheter, adenosine has been shown to initiate atrial fibrillation because of its effect on atrial tissue. In individuals with accessory pathways, the onset of atrial fibrillation can lead to a life threatening ventricular fibrillation.

Fast rhythms of the heart that are confined to the atria (e.g., atrial fibrillation, atrial flutter) or ventricles (e.g., monomorphic ventricular tachycardia) and do not involve the AV node as part of the re-entrant circuit are not typically converted by adenosine. However, the ventricular response rate is temporarily slowed with adenosine in such cases.Because of the effects of adenosine on AV node-dependent SVTs, adenosine is considered a class V antiarrhythmic agent. When adenosine is used to cardiovert an abnormal rhythm, it is normal for the heart to enter ventricular asystole for a few seconds. This can be disconcerting to a normally conscious patient, and is associated with angina-like sensations in the chest.

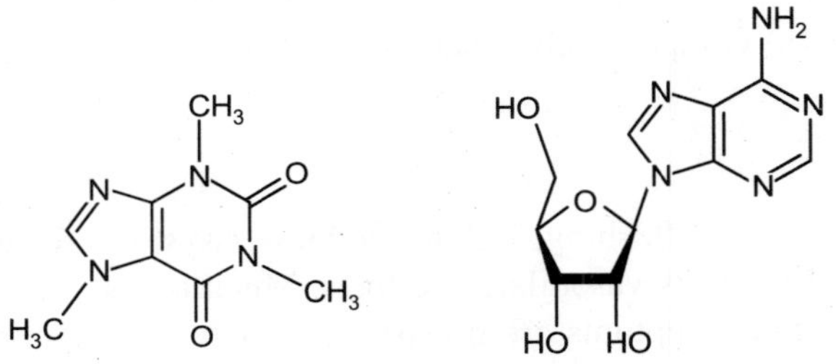

Caffeine Adenosine

Caffeine's principal mode of action is as an antagonist of adenosine receptors in the brain. They are presented here side by side for comparison.

By nature of caffeine's purine structure it binds to some of the same receptors as adenosine. The pharmacological effects of adenosine may therefore be blunted in individuals who are taking large quantities of methylxanthines (e.g., caffeine, found in coffee and tea, or theobromine, as found in chocolate).

Action in CNS

Generalized, adenosine has an inhibitory effect in the central nervous system (CNS). Caffeine's stimulatory effects, on the other hand, are primarily (although not entirely) credited to its inhibition of adenosine by binding to the same receptors, and therefore effectively blocking adenosine receptors in the CNS. This reduction in adenosine activity leads to increased activity of the neurotransmitters dopamine and glutamate.

Dosage

When given for the evaluation or treatment of an SVT, the initial dose is 6 mg, given as a fast IV/IO push. Due to adenosine's extremely short half-life, the IV line is started as proximal to the heart as possible, such as the antecubital fossa. The IV push is often followed with an immediate flush of 5-10ccs of saline. If this has no effect (e.g., no evidence of transient AV block), a 12mg dose can be given 1-2 minutes after the first dose. If the 12mg dose has no effect, a second 12mg dose can be administered 1-2 minutes after the previous dose. Some clinicians may prefer to administer a higher dose (typically 18 mg), rather than repeat a dose that apparently had no effect. When given to dilate the arteries, such as in a "stress test", the dosage is typically 0.14 mg/kg/min, administered for 4 or 6 minutes, depending on the protocol.

The recommended dose may be increased in patients on theophylline since methylxanthines prevent binding of adenosine at receptor sites. The dose is often decreased in patients on dipyridamole (Persantine) and diazepam (Valium) because adenosine potentiates the effects of these drugs. The recommended dose is also reduced by half in patients who are presenting Congestive Heart Failure, Myocardial Infarction, shock, hypoxia, and/or hepatic or renal insufficiency, and in elderly patients.

Drug Interactions

Dopamine may precipitate toxicity in the patient. Carbamazepine may increase heart block. Theophylline and caffeine (methylxanthines) competitively antagonize adenosine's effects; may require increased dose of adenosine.

Side Effects

Many individuals experience facial flushing, lightheadedness, asystole, diaphoresis, or nausea after administration of adenosine due to its vasodilatory effects. Metallic taste is a hallmark side effect of adenosine administration. These symptoms are transitory, usually lasting less than one minute. It is classically associated with a sense of "impending doom", more prosaically described as apprehension. This lasts a few seconds after administration intravenously.

Metabolism

When adenosine enters the circulation, it is broken down by adenosine deaminase, which is present in red cells and the vessel wall.

Dipyridamole, an inhibitor of adenosine deaminase, allows adenosine to accumulate in the blood stream. This causes an increase in coronary vasodilatation.

Guanosine is a nucleoside comprising guanine attached to a ribose (ribofuranose) ring via a β-N_9-glycosidic bond.

Guanosine can be phosphorylated to become GMP (guanosine monophosphate), cGMP (cyclic guanosine monophosphate), GDP (guanosine diphosphate) and GTP (guanosine triphosphate).

When guanine is attached to a deoxyribose ring, it is known as a deoxyguanosine.

Uridine is a molecule (known as a nucleoside) that is formed when uracil is attached to a ribose ring (also known as a ribofuranose) via a β-N_1-glycosidic bond.

If uracil is attached to a deoxyribose ring, it is known as a deoxyuridine.

Harvard researchers report that supplementation in rats with a combination of uridine and EPA/DHA omega-3 fatty acids has antidepressant activity equivalent to that of commonly prescribed antidepressant medications, such as Prozac and other SSRIs.

Dietary Sources of Uridine

Uridine is found is a variety of foods. Sugarcane extract is rich in nucleosides, especially uridine.Tomatoes also (about 500-1000 mg. of uridine per kilogram of dry matter). Brewer's yeast is also a good source of uridine, as yeast is high in RNA (ribonucleic acid), which after digestion is broken down into ribosyl pyrimidines (uridine and cytidine), which are absorbed intact . About 3 percent of yeast (dry weight) results in digestion uridine products. This assumes the usual 9% RNA content found in Brewer's yeast. Alternatively, drinking beer also results in increased plasma uridine. The ingestion of one liter of beer results in increased plasma uridine at a level that is comparable to those reached after ingestion of CDP-choline (citicoline) (as in, the increase is measured as a percent change relative to baseline plasma uridine). Alternative uridine/cytidine sources include other high RNA foods such as organ meats (liver, pancreas, etc) or broccoli. High RNA foods may result in high blood purine levels, which may increase uric acid production in humans, which may aggravate conditions such as gout. Because of this, it has been suggested that the RNA content of yeast products should be chemically reduced if these products are to be consumed in high amounts as a source of protein (50 grams or more per day). However, such processing is expensive, and as of today (2008), it seems that commonly available Brewer's yeast products are not RNA-reduced. Consumption of moderate amounts of yeast (5 grams per day) should provide enough uridine for improved health, while minimizing possible side effects such as increased uric acid

Cytidine is a nucleoside molecule that is formed when cytosine is attached to a ribose ring (also known as a ribofuranose) via a β-N_1-glycosidic bond. Cytidine is a component of RNA.

If cytosine is attached to a deoxyribose ring, it is known as a deoxycytidine.

Cytidine has been shown to exhibit "antidepressant-like effects in rats.".

Dietary Sources of Cytidine

Dietary sources of cytidine include foods with high RNA (ribonucleic acid) content, such as organ meats, Brewer's yeast, as well as pyrimide-rich foods such as beer. During digestion, RNA-rich foods are broken-down into ribosyl pyrimidines (cytidine and uridine), which are absorbed intact. In humans, dietary cytidine is converted into uridine, which is probably the compound behind cytidine's metabolic effects

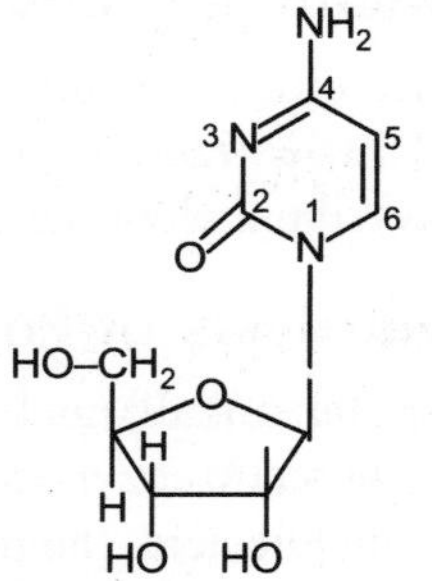

Cytidine

Thymine is one of the four bases in the nucleic acid of DNA that make up the letters ATGC. The others are adenine, guanine, and cytosine. Thymine (T)

always pairs with adenine. Thymine is also known as **5-methyluracil,** a pyrimidine nucleobase. As the name suggests, thymine may be derived by methylation of uracil at the 5th carbon. In RNA, thymine is replaced with uracil in most cases. In DNA, thymine(T) binds to adenine (A) via two hydrogen bonds to assist in stabilizing the nucleic acid structures.

Thymine combined with deoxyribose creates the nucleoside deoxythymidine, which is synonymous with the term thymidine. Thymidine can be phosphorylated with one, two, or three phosphoric acid groups, creating, respectively, TMP, TDP, or TTP (thymidine mono-, di-, or triphosphate).

One of the common mutations of DNA involves two adjacent thymines or cytosine, which, in presence of ultraviolet light, may form thymine dimers, causing "kinks" in the DNA molecule that inhibit normal function.

Thymine could also be a target for actions of 5-fluorouracil (5-FU) in cancer treatment. 5-FU can be a metabolic analog of thymine (in DNA synthesis) or uracil (in RNA synthesis). Substitution of this analog inhibits DNA synthesis in actively-dividing cells.

Thymine bases are frequently oxidized to hydantoins over time after the death of an organism.

Nucleotides

Adding one or more phosphates to the sugar portion of a nucleoside results in a **nucleotide**. Generally, the phosphate is in ester linkage to carbon 5' of the sugar. If more than one phosphate is present, they are generally in acid anhydride linkages to each other. If such is the case, no position designation in the name is required. If the phosphate is in any other position, however, the position must be designated. For example, 3'-5' cAMP indicates that a phosphate is in ester linkage to both the 3' and 5' hydroxyl groups of an adenosine molecule and forms a cyclic structure. 2'-GMP would indicate that a phosphate is in ester linkage to the 2' hydroxyl group of a guanosine.

Some representative names are:

- AMP = adenosine monophosphate = adenylic acid
- CDP = cytidine diphosphate
- dGTP = deoxy guanosine triphosphate
- dTTP = deoxy thymidine triphosphate (more commonly designated TTP)
- cAMP = 3'-5' cyclic adenosine monophosphate

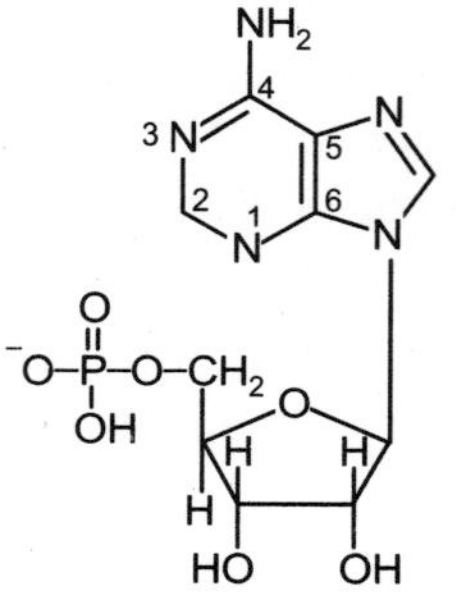

Adenosine Monophosphate
AMP

Polynucleotides

Nucleotides are joined together by 3'-5' phosphodiester bonds to form polynucleotides. Polymerization of ribonucleotides will produce an RNA while polymerization of deoxyribonucleotides leads to DNA.

Hydrolysis of Polynucleotides

Most, but not all, nucleic acids in the cell are associated with protein. Dietary nucleoprotein is degraded by pancreatic enzymes and tissue nucleoprotein by lysosomal enzymes. After dissociation of the protein and nucleic acid, the protein is metabolized like any other protein.

The nucleic acids are hydrolyzed randomly by **nucleases** to yield a mixture of polynucleotides. These are further cleaved by **phosphodiesterases** (exonucleases) to a mixture of the mononucleotides. The specificity of the pancreatic nucleotidases gives the 3'-nucleotides and that of the lysosomal nucleotidases gives the biologically important 5'-nucleotides.

The nucleotides are hydrolyzed by **nucleotidases** to give the nucleosides and P_i. This is probably the end product in the intestine with the nucleosides being the primary form absorbed. In at least some tissues, the nucleosides undergo phosphorolysis with **nucleoside phosphorylases** to yield the base and ribose 1-P (or deoxyribose 1-P). Since R 1-P and R 5-P are in equilibrium, the sugar phosphate can either be reincorporated into nucleotides or metabolized via the Hexose Monophosphate Pathway. The purine and pyrimidine bases released are either degraded or salvaged for reincorporation into nucleotides. There is significant turnover of all kinds of RNA as well as the nucleotide pool. DNA doesn't turnover but portions of the molecule are excised as part of a repair process.

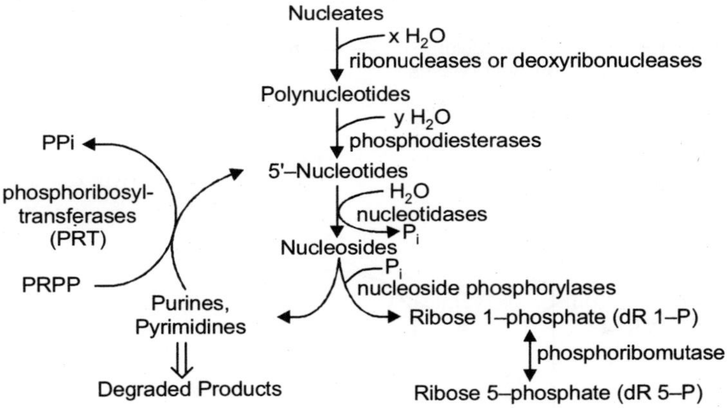

Nucleates
x H_2O
ribonucleases or deoxyribonucleases
Polynucleotides
y H_2O
phosphodiesterases
5'–Nucleotides
H_2O
nucleotidases
P_i
Nucleosides
P_i
nucleoside phosphorylases
Ribose 1–phosphate (dR 1–P)
phosphoribomutase
Ribose 5–phosphate (dR 5–P)
PPi
phosphoribosyl-transferases (PRT)
PRPP
Purines, Pyrimidines
Degraded Products

Purine and pyrimidines from tissue turnover which are not salvaged are catabolized and excreted. Little dietary purine is used and that which is absorbed is largely catabolized as well. Catabolism of purines and pyrimidines occurs in a less useful fashion than did the catabolism of amino acids in that we do not derive any significant amount of energy from the catabolism of purines and pyrimidines. Pyrimidine catabolism, however, does produce β-alanine, and the endproduct of purine catabolism, which is uric acid in man, may serve as a scavenger of reactive oxygen species.

Purine Catabolism

The end product of purine catabolism in man is **uric acid**. Other mammals have the enzyme urate oxidase and excrete the more soluble allantoin as the end product. Man does not have this enzyme so urate is the end product for us. Uric acid is formed primarily in the liver and excreted by the kidney into the urine.

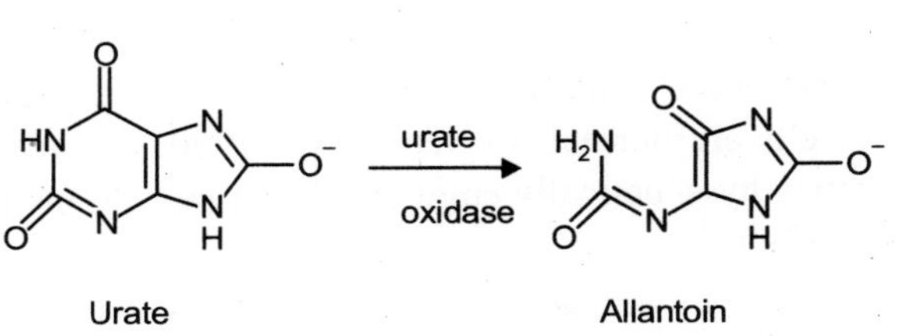

urate oxidase
Urate
Allantoin

Nucleotides to Bases

Guanine nucleotides are hydrolyzed to the nucleoside **guanosine** which undergoes phosphorolysis to **guanine** and **ribose 1-P**. Man's intracellular nucleotidases are not very active toward AMP, however. Rather, **AMP is deaminated by the enzyme adenylate (AMP) deaminase to IMP**. In the catobilsm of purine nucleotides, IMP is further degraded by hydrolysis with nucleotidase to **inosine** and then phosphorolysis to **hypoxanthine**.

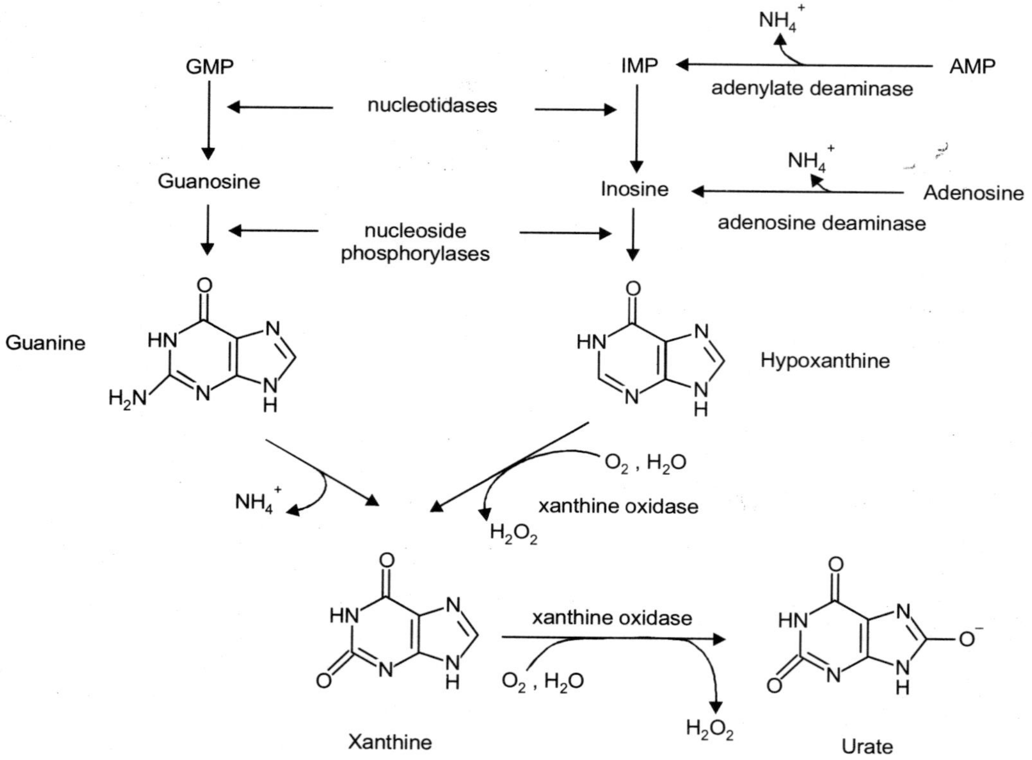

NH_4^+
GMP
IMP
AMP
adenylate deaminase
nucleotidases
NH_4^+
Guanosine
Inosine
Adenosine
adenosine deaminase
nucleoside phosphorylases
Guanine
Hypoxanthine
O_2, H_2O
NH_4^+
xanthine oxidase
H_2O_2
xanthine oxidase
O_2, H_2O
H_2O_2
Xanthine
Urate

Adenosine does occur but usually arises from S-Adenosylmethionine during the course of transmethylation reactions. Adenosine is deaminated to inosine by an adenosine deaminase. Deficiencies in either **adenosine deaminase** or in the **purine nucleoside phosphorylase** lead to two different immunodeficiency diseases by mechanisms that are not clearly understood. With **adenosine deaminase deficiency**, both T and B-cell immunity is affected. The **phosphorylase deficiency** affects the T cells but B cells are normal. In September, 1990, a 4 year old girl was treated for adenosine deaminase deficiency by genetically engineering her cells to incorporate the gene. The treatment,so far, seems to be successful.

Whether or not **methylated** purines are catabolized depends upon the location of the methyl group. If the methyl is on an $-NH_2$, it is removed along with the $-NH_2$ and the core is metabolized in the usual fashion. If the methyl is on a ring nitrogen, the compound is excreted unchanged in the urine.

Bases to Uric Acid

Both adenine and guanine nucleotides converge at the common intermediate **xanthine**. Hypoxanthine, representing the original adenine, is oxidized to xanthine by the enzyme **xanthine oxidase**. Guanine is deaminated, with the amino group released as ammonia, to xanthine. If this process is occurring in tissues other than liver, most of the ammonia will be transported to the liver as glutamine for ultimate excretion as urea.

Xanthine, like hypoxanthine, is oxidized by oxygen and xanthine oxidase with the production of hydrogen peroxide. In man, the urate is excreted and the hydrogen peroxide is degraded by catalase. Xanthine oxidase is present in significant concentration only in liver and intestine. The pathway to the nucleosides, possibly to the free bases, is present in many tissues.

Gouts and Hyperuricemia

Both undissociated uric acid and the monosodium salt (primary form in blood) are only sparingly soluble. The limited solubility is not ordinarily a problem in urine unless the urine is very acid or has high [Ca^{2+}]. [Urate salts coprecipitate with calcium salts and can form stones in kidney or bladder.] A very high concentration of urate in the blood leads to a fairly common group of diseases referred to as gout. The incidence of gout in this country is about 3/1000.

Gout is a group of pathological conditions associated with markedly elevated levels of urate in the blood (3-7 mg/dl normal). **Hyperuricemia** is not always symptomatic, but, in certain individuals, something triggers the deposition of sodium urate crystals in joints and tissues. In addition to the extreme pain accompanying acute attacks, repeated attacks lead to destruction of tissues and severe arthritic-like malformations. The term gout should be restricted to hyperuricemia with the presence of these tophaceous deposits.

Urate in the blood could accumulate either through an overproduction and/or an underexcretion of uric acid. In gouts caused by an **overproduction** of uric acid, the defects are in the control mechanisms governing the production of - not uric acid itself - but of the nucleotide precursors. The only major control of urate production that we know so far is the availability of substrates (nucleotides, nucleosides or free bases).

One approach to the treatment of gout is the drug **allopurinol**, an isomer of hypoxanthine.

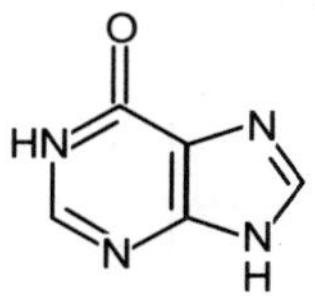

O
HN N
N N
H
Hypoxanthine

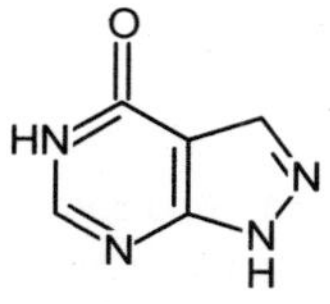

O
HN N
N N
H
Allopurionol

Allopurinol is a substrate for xanthine oxidase, but the product binds so tightly that the enzyme is now unable to oxidized its normal substrate. Uric acid production is diminished and xanthine and hypoxanthine levels in the blood rise. These are more soluble than urate and are less likely to deposit as crystals in the joints. Another approach is to stimulate the secretion of urate in the urine.

Summary

In summary, all, except ring-methylated, purines are deaminated (with the amino group contributing to the general ammonia pool) and the rings oxidized to uric acid for excretion. Since the purine ring is excreted intact, no energy benefit accrues to man from these carbons.

Pyrimidine Catabolism

In contrast to purines, pyrimidines undergo ring cleavage and the usual end products of catabolism are beta-amino acids plus ammonia and carbon dioxide. Pyrimidines from nucleic acids or the energy pool are acted upon by nucleotidases and pyrimidine nucleoside phosphorylase to yield the free bases. The 4-amino group of both cytosine and 5-methyl cytosine is released as ammonia.

Ring Cleavage

In order for the rings to be cleaved, they must first be **reduced by NADPH**. Atoms 2 and 3 of both rings are released as ammonia and carbon dioxide. The rest of the ring is left as a **beta-amino acid**. Beta-amino isobutyrate from thymine or 5-methyl cytosine is largely excreted. Beta-alanine from cytosine or uracil may either be excreted or incorporated into the brain and muscle dipeptides, **carnosine** (his-beta-ala) or **anserine**.

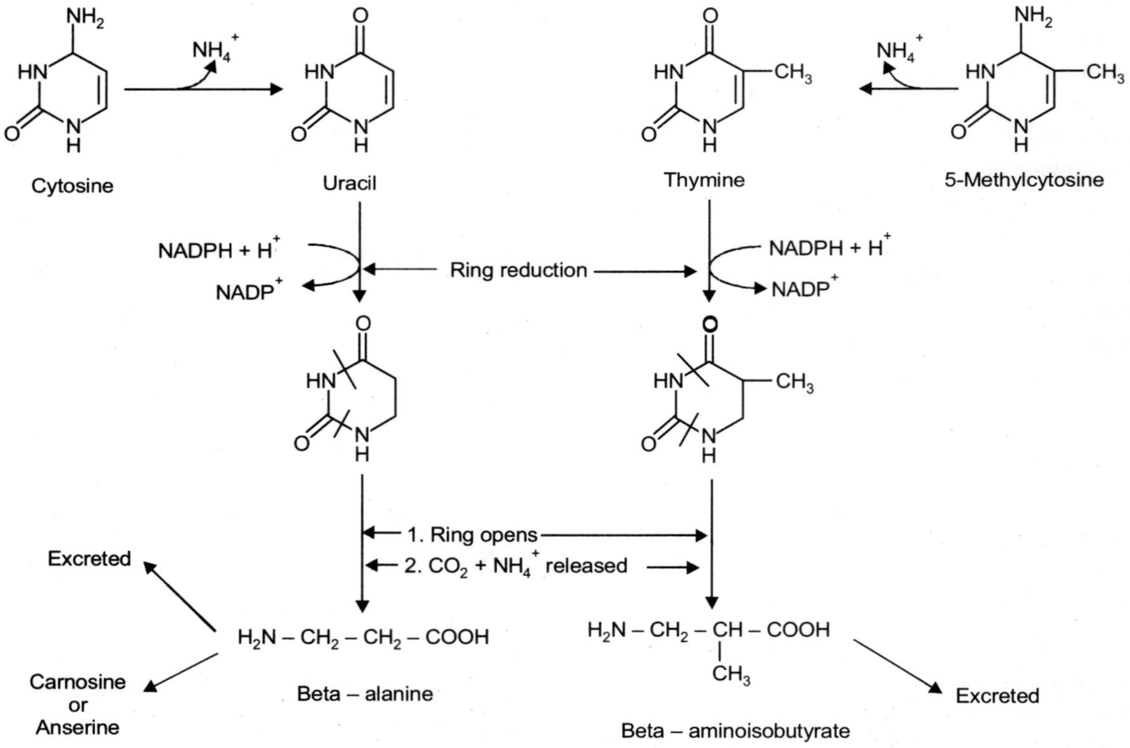

General Comments

Purine and pyrimidine bases which are not degraded are recycled - ***i.e.*** reincorporated into nucleotides. This recycling, however, is not sufficient to meet total body requirements and so some *de novo* synthesis is essential. There are definite tissue differences in the ability to carry out *de novo* synthesis. *De novo* synthesis of purines is most active in liver. Non-hepatic tissues generally have limited or even no *de novo* synthesis. Pyrimidine synthesis occurs in a variety of tissues. For purines, especially, non-hepatic tissues rely heavily on preformed bases - those salvaged from their own intracellular turnover supplemented by bases synthesized in the liver and delivered to tissues via the blood.

"Salvage" of purines is reasonable in most cells because xanthine oxidase, the key enzyme in taking the purines all of the way to uric acid, is significantly active only in liver and intestine. The bases generated by turnover in non-hepatic tissues are not readily degraded to uric acid in those tissues and, therefore, are available for salvage. The liver probably does less salvage but is very active in *de novo* synthesis - not so much for itself but to help supply the peripheral tissues.

De novo synthesis of both purine and pyrimidine nucleotides occurs from readily available components.

De Novo Synthesis of Purine Nucleotides

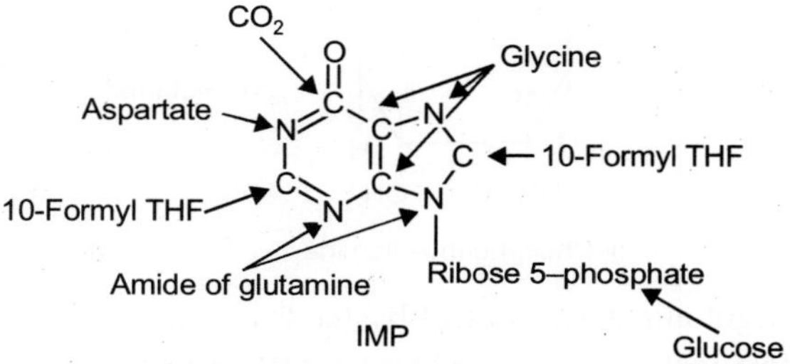

CO_2 · Glycine · Aspartate · 10-Formyl THF · 10-Formyl THF · Amide of glutamine · Ribose 5–phosphate · Glucose

IMP

We use for purine nucleotides the entire glycine molecule (atoms 4, 5,7), the amino nitrogen of aspartate (atom 1), amide nitrogen of glutamine (atoms 3, 9), components of the folate-one-carbon pool (atoms 2, 8), carbon dioxide, ribose 5-P from glucose and a great deal of energy in the form of ATP. In de novo synthesis, IMP is the first nucleotide formed. It is then converted to either AMP or GMP.

PRPP

Since the purines are synthesized as the ribonucleotides, (not as the free bases) a necessary prerequisite is the synthesis of the activated form of ribose 5-phosphate. Ribose 5-phosphate reacts with ATP to form **5-Phosphoribosyl-1-pyrophosphate (PRPP)**.

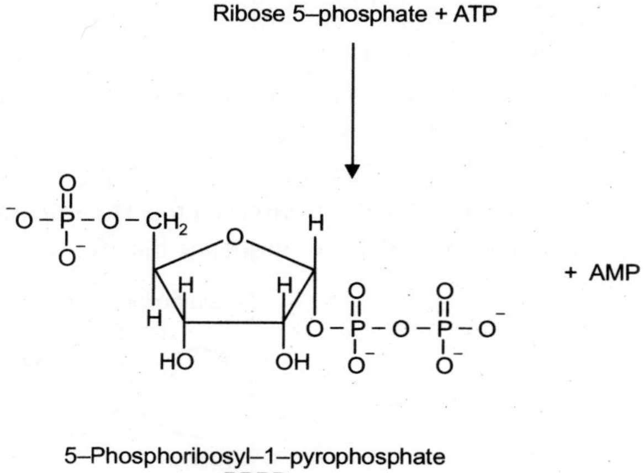

Ribose 5–phosphate + ATP

→ + AMP

5–Phosphoribosyl–1–pyrophosphate
PRPP

This reaction occurs in many tissues because PRPP has a number of roles - purine and pyrimidine nucleotide synthesis, salvage pathways, NAD and NADP formation. The enzyme is heavily controlled by a variety of compounds (di- and tri-phosphates, 2,3-DPG), presumably to try to match the synthesis of PRPP to a need for the products in which it ultimately appears.

Commitment Step

De novo purine nucleotide synthesis occurs actively in the cytosol of the liver where all of the necessary enzymes are present as a macro-molecular aggregate. The first step is a replacement of the pyrophosphate of PRPP by the amide group of glutamine. The product of this reaction is **5-Phosphoribosylamine**. The amine group that has been placed on carbon 1 of the sugar becomes nitrogen 9 of the ultimate purine ring. This is the commitment and rate-limiting step of the pathway.

Amide nitrogen of glutamine becomes N–9 of purine

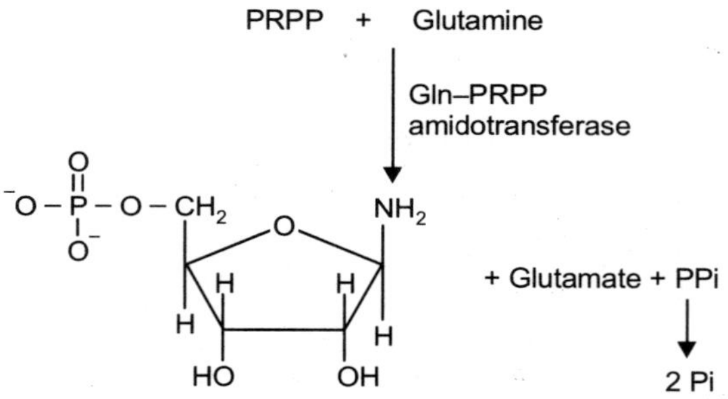

PRPP + Glutamine

Gln–PRPP amidotransferase

$^{-}O-P(=O)(O^{-})-O-CH_2$ ... NH_2

H H H H HO OH

+ Glutamate + PPi → 2 Pi

5–Phosphoribosylamine

The enzyme is under tight allosteric control by feedback inhibition. Either **AMP, GMP, or IMP** alone will **inhibit the amidotransferase** while **AMP + GMP or AMP + IMP** together act **synergistically**. This is a fine control and probably the major factor in minute by minute regulation of the enzyme. The nucleotides inhibit the enzyme by causing the small active molecules to aggregate to larger inactive molecules.

[PRPP] also can play a role in regulating the rate. Normal intracellular concentrations of PRPP (which can and do fluctuate) are below the KM of the enzyme for PRPP so there is great potential for increasing the rate of the reaction by increasing the substrate concentration. The kinetics are sigmoidal. The enzyme is not particularly sensitive to changes in [Gln] (Kinetics are hyperbolic and [gln] approximates KM). Very **high [PRPP]** also **overcomes the normal nucleotide feedback inhibition** by causing the large, inactive aggregates to dissociate back to the small active molecules.

Glutamine–PRPP Amidotransferase

2 Active ⇄ Inactive (AMP, GMP, IMP →; ← PRPP)

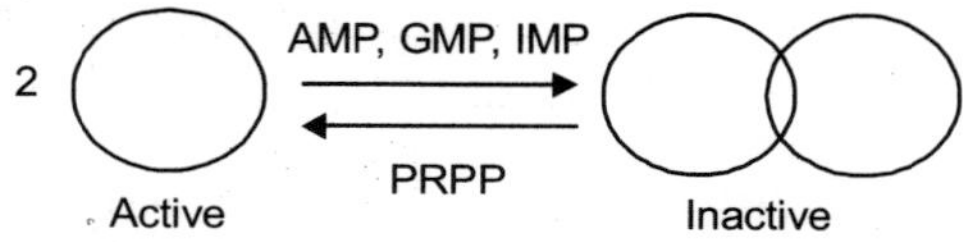

Purine *de novo* synthesis is a complex, energy-expensive pathway. It should be, and is, carefully controlled.

Formation of IMP

Once the commitment step has produced the 5-phosphoribosyl amine, the rest of the molecule is formed by a series of additions to make first the 5- and then the 6-membered ring. (Note: the numbers given to the atoms are those of the completed purine ring and names, etc. of the intermediate compounds are not given.) The whole glycine molecule, at the expense of ATP adds to the amino group to provide what will eventually be atoms 4, 5, and 7 of the purine ring (The amino group of 5-phosphoribosyl amine becomes nitrogen N of the purine ring.) One more atom is needed to complete the five-membered ring portion and that is supplied as 5, 10-Methenyl tetrahydrofolate.

Before ring closure occurs, however, the amide of glutamine adds to carbon 4 to start the six-membered ring portion (becomes nitrogen 3). This addition requires ATP. Another ATP is required to join carbon 8 and nitrogen 9 to form the five-membered ring.

The next step is the addition of carbon dioxide (as a carboxyl group) to form carbon 6 of the ring. The amine group of aspartate adds to the carboxyl group with a subsequent removal of fumarate. The amino group is now nitrogen 1 of the final ring. This process, which is typical for the use of the amino

Schematic Representation of Purine Nucleotide Synthesis

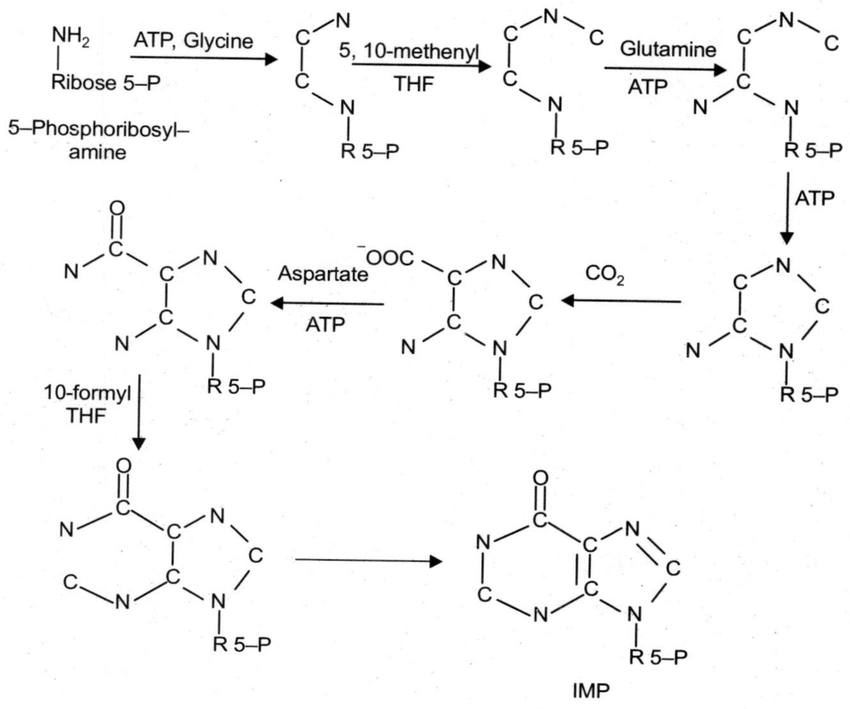

group of aspartate, requires ATP. The final atom of the purine ring, carbon 2, is supplied by 10-Formyl tetrahydrofolate. Ring closure produces the purine nucleotide, IMP.

Note that at least 4 ATPs are required in this part of the process. At no time do we have either a free base or a nucleotide.

Formation of AMP and GMP

IMP can then become **either** AMP or GMP. **GMP** formation requires that IMP be first oxidized to XMP using NAD. The oxygen at position 2 is substituted by the amide N of glutamine at the expense of ATP. Similarly, GTP provides the energy to convert IMP to **AMP**. The amino group is provided by aspartate in a mechanism similar to that used in forming nitrogen 1 of the ring. Removal of the carbons of aspartate as fumarate leaves the nitrigen behind as the 6-amino group of the adenine ring. The monophosphates are readily converted to the di- and tri-phosphates.

Conversion of IMP to either AMP or GMP

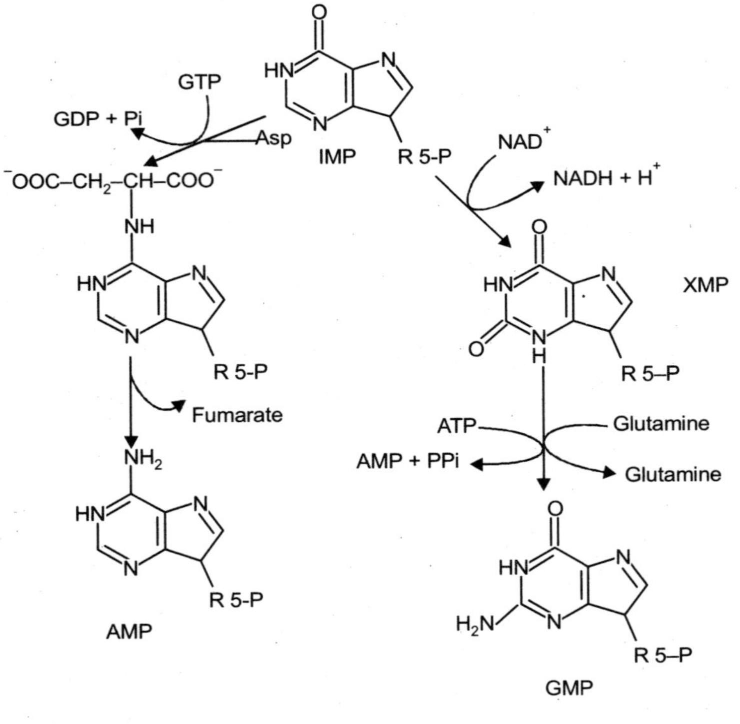

Control of De Novo Synthesis

Control of purine nucleotide synthesis has two phases. Control of the **synthesis as a whole** occurs at the amidotransferase step by nucleotide inhibition and/or [PRPP]. The second phase of control is involved with **maintaining an appropriate balance (not equality) between ATP and GTP**. Each one stimulates the synthesis of the other by providing the energy. Feedback inhibition also controls the branched portion as GMP inhibits the conversion of IMP to XMP and AMP inhibits the conversion of IMP to adenylosuccinate.

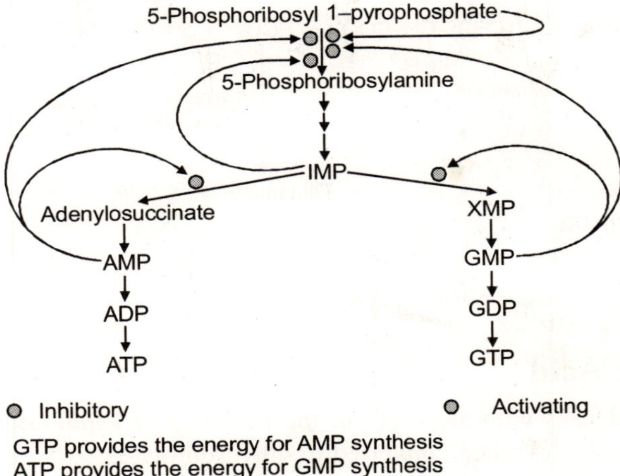

Possible Scenario

One could imagine the controls operating in such a way that if only one of the two nucleotides were required, there would be a partial inhibition of de novo synthesis because of high levels of the other and the IMP synthesized would be directed toward the synthesis of the required nucleotide. If both nucleotides were present in adequate amounts, their synergistic effect on the amidotransferase would result in almost complete inhibition of de novo synthesis.

De Novo Synthesis of Pyrimidine Nucleotides

Since pyrimidine molecules are simpler than purines, so is their synthesis simpler but is still from readily available components. Glutamine's amide nitrogen and carbon dioxide provide atoms 2 and 3 or the pyrimidine ring. They do so, however, after first being converted to carbamoyl phosphate. The other four atoms of the ring are supplied by aspartate. As is true with purine nucleotides, the sugar phosphate portion of the molecule is supplied by PRPP.

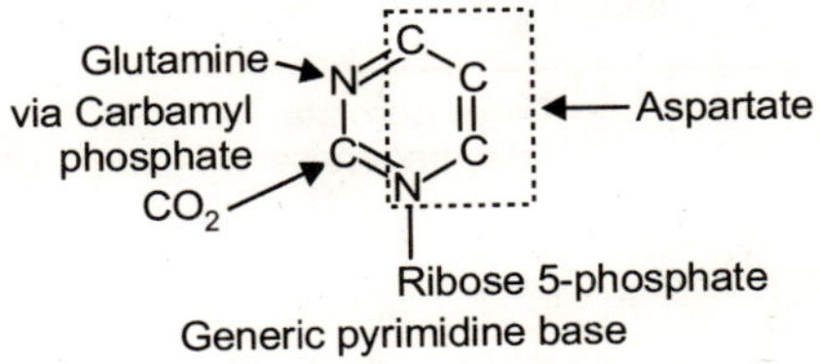

Glutamine via Carbamyl phosphate → N

CO_2 → C

Aspartate → C, C, C, N

Ribose 5-phosphate

Generic pyrimidine base

Carbamoyl Phosphate

Pyrimidine synthesis begins with **carbamoyl phosphate** synthesized in the cytosol of those tissues capable of making pyrimidines (highest in spleen, thymus, GItract and testes). This uses a different enzyme than the one involved in urea synthesis. **Carbamoyl phosphate synthetase II (CPS II)** prefers glutamine to free ammonia and has no requirement for N-Acetylglutamate.

$$H_2N-CO-(CH_2)_2-CH(NH_3^+)-COO^- + HCO_3^- + 2\,ATP \xrightarrow{\text{Carbamoyl phosphate synthetase II}} H_2N-C(=O)-OPO_3^{-2} + \text{Glutamate} + 2\,ADP + P_i$$

Glutamine → Carbamoyl phosphate

Formation of Orotic Acid

Carbamoyl phosphate condenses with aspartate in the presence of **aspartate transcarbamylase** to yield N-carbamylaspartate which is then converted to dihydroorotate.

In man, **CPSII, asp-transcarbamylase, and dihydroorotase activities** are part of a **multifunctional protein.**

Oxidation of the ring by a complex, poorly understood enzyme produces the free pyrimidine, orotic acid. This enzyme is located on the outer face of the inner mitochondrial membrane, in contrast to the other enzymes which are cytosolic. Note the contrast with purine synthesis in which a nucleotide is formed first while pyrimidines are first synthesized as the **free base.**

Carbamoyl phosphate + Aspartate → (Aspartate transcarbamylase +; releases P_i) → N-Carbamoylaspartate

N-Carbamoylaspartate → (Dihydroorotase +; releases H_2O) → Dihydroorotate

Dihydroorotate → (Dihydroorotate dehydrogenase; NAD^+ → $NADH + H^+$) → Orotic Acid

+ Part of the same multi–functional protein

Formation of the Nucleotides

Orotic acid is converted to its nucleotide with PRPP. **OMP** is then **converted sequentially** - not in a branched pathway - to the other pyrimidine nucleotides. Decarboxylation of OMP gives **UMP. O-PRT**

and OMP decarboxylase are also a **multifunctional protein**. After conversion of UMP to the triphosphate, the amide of glutamine is added, at the expense of ATP, to yield **CTP**.

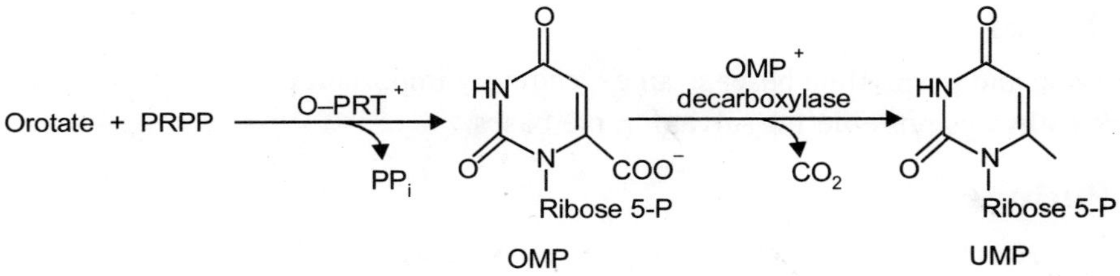

Orotate + PRPP —(O–PRT⁺; PP_i)→ OMP (Ribose 5-P; COO^-) —(OMP decarboxylase⁺; CO_2)→ UMP (Ribose 5-P)

UMP ⟶ UDP ⟶ UTP

UTP (Ribose 5-P-P-P) —(Glutamine → Glutamate; ATP → ADP + P_i)→ CTP (NH_2; Ribose 5-P-P-P)

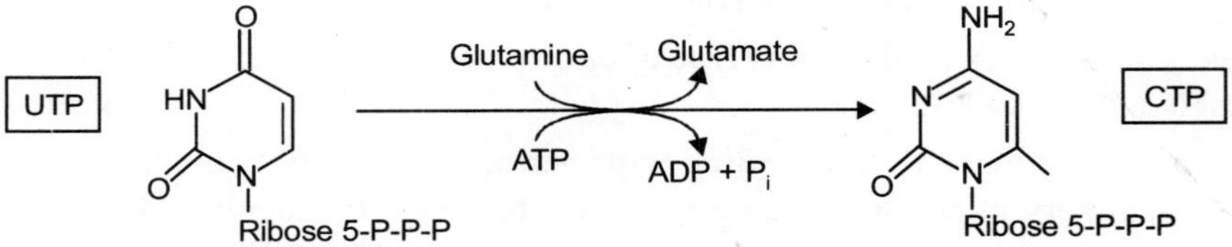

+ Part of the same multi–functional protein

Control

The control of pyrimidine nucleotide synthesis in man is exerted primarily at the level of **cytoplasmic CPS II**. **UTP inhibits** the enzyme, competitively with ATP. **PRPP activates** it. Other secondary sites of control also exist (e.g. OMP decarboxylase is inhibited by UMP and CMP). These are probably not very important under normal circumstances.

In bacteria, aspartate transcarbamylase is the control enzyme. There is only one carbamoyl phosphate synthetase in bacteria since they do not have mitochondria. Carbamoyl phosphate, thus, participates in a branched pathway in these organisms that leads to either pyrimidine nucleotides or arginine.

Interconversion of Nucleotides

The monophosphates are the forms synthesized *de novo* although the triphosphates are the most commonly used forms. But, of course, the three forms are in equilibrium. There are several enzymes classified as **nucleoside monophosphate kinases** which catalyze the general reaction:(= represents a reversible reaction)

Base-monophosphate + ATP = Base-diphosphate + ADP

e.g. Adenylate kinase: AMP + ATP = 2 ADP

There is a different enzyme for GMP, one for pyrimidines and also enzymes that recognize the deoxy forms.

Similarly, the diphosphates are converted to the triphosphates by **nucleoside diphosphate kinase**:

BDP + ATP = BTP + ADP

There may be only one nucleoside diphosphate kinase with broad specificity. One can legitimately speak of a pool of nucleotides in equilibrium with each other.

Salvage of Bases

Salvaging of purine and pyrimidine bases is an exceedingly important process for most tissues. There are two distinct pathways possible for salvaging the bases.

Salvaging Purines

The more important of the pathways for **salvaging purines** uses enzymes called **phosphoribosyltransferases (PRT)**:

PRTs catalyze the addition of ribose 5-phosphate to the base from PRPP to yield a nucleotide.:

Base + PRPP = Base-ribose-phosphate (BMP) + PPi

We gave already seen one example of this type of enzyme as a normal part of *de novo* synthesis of the pyrimidine nucleotides, - O-PRT.

As a salvage process though, we are dealing with purines. There are two enzymes, A-PRT and HG-PRT. **A-PRT** is not very important because we generate very little adenine. (Remember that the catabolism of adenine nucleotides and nucleosides is through inosine). **HG-PRT**, though, is exceptionally important and it is inhibited by both IMP and GMP. This enzyme salvages guanine directly and adenine indirectly. Remember that AMP is generated primarily from IMP, not from free adenine.

Lesch-Nyhan Syndrome

HG-PRT is deficient in the disease called **Lesch-Nyhan Syndrome**, a severe neurological disorder whose most blatant clinical manifestation is an uncontrollable self-mutilation. Lesch-Nyhan patients have very **high blood uric acid** levels because of an essentially **uncontrolled *de novo* synthesis**. (It can be as much as 20 times the normal rate). There is a significant increase in PRPP levels in various cells and an inability to maintain levels of IMP and GMP via salvage pathways. Both of these factors could lead to an increase in the activity of the amidotransferase.

Salvaging Pyrimidines

A second type of salvage pathway involves two steps and is the major pathway for the pyrimidines, uracil and thymine.

Base + Ribose 1-phosphate = Nucleoside + Pi (nucleoside phosphorylase)

Nucleoside + ATP - Nucleotide + ADP (nucleoside kinase - irreversible)

There is a uridine phosphorylase and kinase and a deoxythymidine phosphorylase and a thymidine kinase which can salvage some thymine in the presence of dR 1-P.

Formation of Deoxyribonucleotides

De novo synthesis and most of the salvage pathways involve the ribonucleotides. (Exception is the small amount of salvage of thymine indicated above.) Deoxyribonucleotides for DNA synthesis are formed from the ribonucleotide diphosphates (in mammals and *E. coli*).

A base diphosphate (BDP) is reduced at the 2' position of the ribose portion using the protein, **thioredoxin** and the enzyme **nucleoside diphosphate reductase**. Thioredoxin has two sulfhydryl groups which are oxidized to a disulfide bond during the process. In order to restore the thioredoxin to its reduced for so that it can be reused, **thioredoxin reductase and NADPH** are required.

*PPO – CH_2 Base (OH OH) BDP + Thioredoxin (SH, SH) —nucleoside diphosphate reductase→ * PPO – CH_2 Base (OH H) dBDP + Thioredoxin (S–S)

Thioredoxin (S–S) + NADPH + H^+ —thioredoxin reductase→ Thioredoxin (SH, SH) + $NADP^+$

* PPO represents the diphosphate group $^{-}O-\overset{O}{\underset{O}{\overset{||}{\underset{|}{P}}}}-O-\overset{O}{\underset{O^{-}}{\overset{||}{\underset{|}{P}}}}-O-$

This system is very tightly controlled by a variety of allosteric effectors. dATP is a general inhibitor for all substrates and ATP an activator. Each substrate then has a specific positive effector (a BTP or dBTP). The result is a maintenance of an appropriate balance of the deoxynucleotides for DNA synthesis.

Synthesis of dTMP

DNA synthesis also requires dTMP (dTTP). This is not synthesized in the *de novo* pathway and salvage is not adequate to maintain the necessary amount. dTMP is generated from dUMP using the folate-dependent one-carbon pool.

CDP —Thioredoxin system→ dCDP —kinase (ADP → ATP)→ dCMP —dCMP deaminase (NH_4^+)→ dUMP

dUMP (dRibose 5 – P) + 5, 10 – methylene THF ($-N-CH_2-$, CH_2) —Thymidylate synthetase→ dTMP (CH_3, dRibose 5–P) + Dihydrofolate

Dihydrofolate must be subsequently reduced to the tetrahydro form

Since the nucleoside diphosphate reductase is not very active toward UDP, CDP is reduced to dCDP which is converted to dCMP. This is then deaminated to form dUMP. In the presence of **5,10-Methylene tetrahydrofolate** and the enzyme **thymidylate synthetase**, the carbon group is both transferred to the pyrimidine ring and further reduced to a methyl group. The other product is **dihydrofolate** which is subsequently reduced to the tetrahydrofolate by dihydrofolate reductase.

Chemotherapeutic Agents

Thymidylate synthetase is particularly sensitive to availability of the folate one-carbon pool. Some of the cancer chemotherapeutic agents interfere with this process as well as with the steps in purine nucleotide synthesis involving the pool.

Cancer chemotherapeutic agents like **methotrexate** (4-amino, 10-methyl folic acid) and **aminopterin** (4-amino, folic acid) are structural analogs of folic acid and inhibit dihydrofolate reductase. This interferes with maintenance of the folate pool and thus of *de novo* synthesis of purine nucleotides and of dTMP synthesis. Such agents are highly toxic and administered under careful control.

STRUCTURES AND PROPERTIES OF DNA IN DIFFERENT FORMS

Deoxyribonucleic acid (DNA) is a nucleic acid that contains the genetic instructions used in the development and functioning of all known living organisms and some viruses. The main role of DNA molecules is the long-term storage of information. DNA is often compared to a set of blueprints or a recipe, or a code, since it contains the instructions needed to construct other components of cells, such as proteins and RNA molecules. The DNA segments that carry this genetic information are called genes, but other DNA sequences have structural purposes, or are involved in regulating the use of this genetic information.

Chemically, DNA consists of two long polymers of simple units called nucleotides, with backbones made of sugars and phosphate groups joined by ester bonds. These two strands run in opposite directions to each other and are therefore anti-parallel. Attached to each sugar is one of four types of molecules called bases. It is the sequence of these four bases along the backbone that encodes information. This information is read using the genetic code, which specifies the sequence of the amino acids within proteins. The code is read by copying stretches of DNA into the related nucleic acid RNA, in a process called transcription.

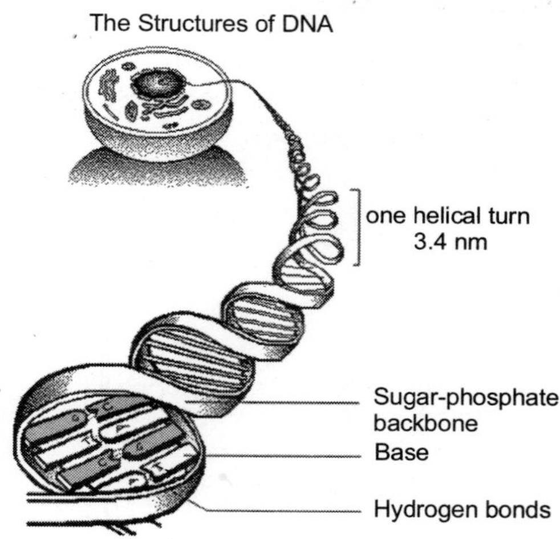

Fig. 2.79A. The Digramatic representation of the structure of DNA

Within cells, DNA is organized into structures called chromosomes. These chromosomes are duplicated before cells divide, in a process called DNA replication. Eukaryotic organisms (animals, plants, fungi, and protists) store their DNA inside the cell nucleus, while in prokaryotes (bacteria and archae) it is found in the cell's cytoplasm. Within the chromosomes, chromatin proteins such as histones compact and organize DNA. These compact structures guide the interactions between DNA and other proteins, helping control which parts of the DNA are transcribed.

Properties

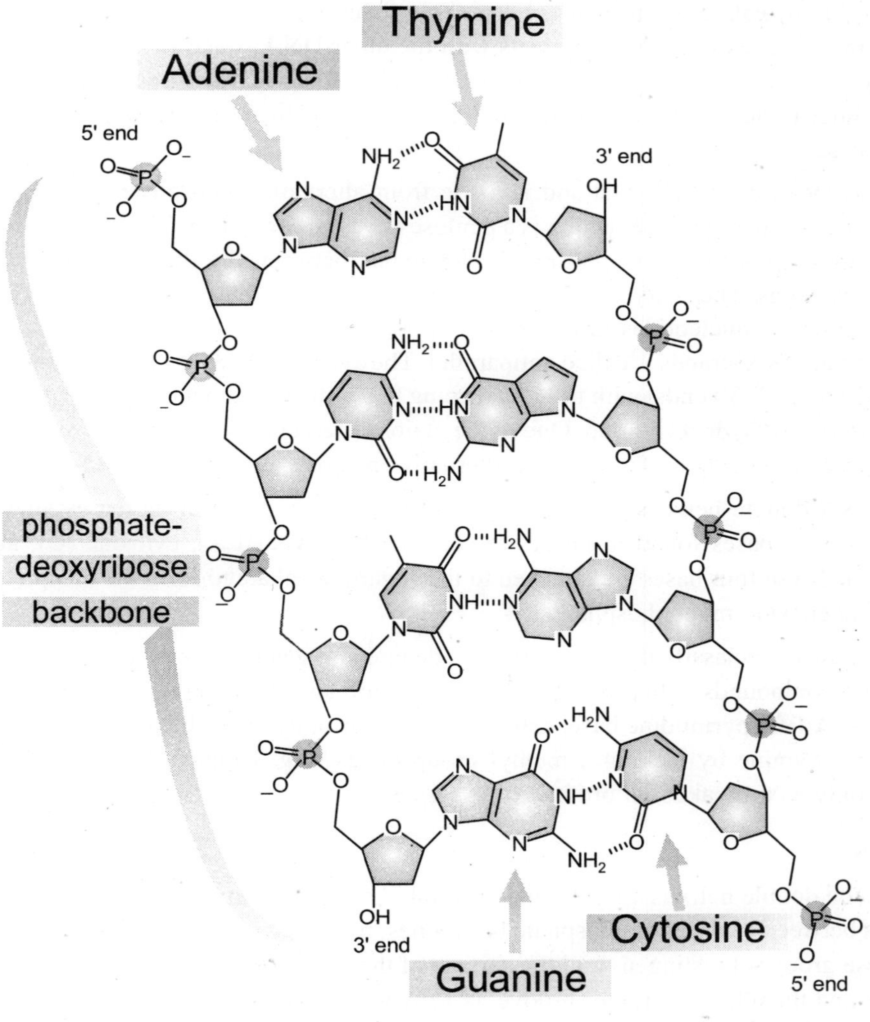

Fig. 2.80 : The chemical structure of DNA. Hydrogen bonds are shown as dotted lines

DNA is a long polymer made from repeating units called nucleotides. The DNA chain is 22 to 26 Å wide (2.2 to 2.6 nanometres), and one nucleotide unit is 3.3 Å (0.33 nm) long. Although each individual repeating unit is very small, DNA polymers can be very large molecules containing millions of nucleotides. For instance, the largest human chromosome, chromosome number 1, is approximately 220 million base pairs long.

In living organisms, DNA does not usually exist as a single molecule, but instead as a tightly-associated pair of molecules. These two long strands entwine like vines, in the shape of a double helix. The nucleotide repeats contain both the segment of the backbone of the molecule, which holds the chain together, and a base, which interacts with the other DNA strand in the helix. In general, a base linked to a sugar is called a nucleoside and a base linked to a sugar and one or more phospnate groups is called a nucleotide. If multiple nucleotides are linked together, as in DNA, this polymer is called a polynucleotide.

The backbone of the DNA strand is made from alternating phosphate and sugar residues. The sugar in DNA is 2-deoxyribose, which is a pentose (five-carbon) sugar. The sugars are joined together by phosphate groups that form phosphodiester bonds between the third and fifth carbon atoms of adjacent sugar rings. These asymmetric bonds mean a strand of DNA has a direction. In a double helix the direction of the nucleotides in one strand is opposite to their direction in the other strand. This arrangement of DNA strands is called antiparallel. The asymmetric ends of DNA strands are referred to as the 52 (5') and 32 (3') ends, with the 5' end being that with a terminal phosphate group and the 3' end that with a terminal hydroxyl group. One of the major differences between DNA and RNA is the sugar, with 2-deoxyribose being replaced by the alternative pentose sugar ribose in RNA.

The DNA double helix is stabilized by hydrogen bonds between the bases attached to the two strands. The four bases found in DNA are adenine (abbreviated A), cytosine (C), guanine (G) and thymine (T). These four bases are attached to the sugar/phosphate to form the complete nucleotide, as shown for adenosine monophosphate.

These bases are classified into two types; adenine and guanine are fused five- and six-membered heterocyclic compounds called purines, while cytosine and thymine are six-membered rings called pyrimidines. A fifth pyrimidine base, called uracil (U), usually takes the place of thymine in RNA and differs from thymine by lacking a methyl group on its ring. Uracil is not usually found in DNA, occurring only as a breakdown product of cytosine.

Grooves

Normally, the double helix is a right-handed spiral. As the DNA strands wind around each other, they leave gaps between each set of phosphate backbones, revealing the sides of the bases inside. There are two of these grooves twisting around the surface of the double helix: one groove, the major groove, is 22 Å wide and the other, the minor groove, is 12 Å wide. The narrowness of the minor groove means that the edges of the bases are more accessible in the major groove. As a result, proteins like transcription factors that can bind to specific sequences in double-stranded DNA usually make contacts to the sides of the bases exposed in the major groove. This situation varies in unusual conformations of DNA within the cell *(see Fig. 2.81)*, but the major and minor grooves are always named to reflect the differences in size that would be seen if the DNA is twisted back into the ordinary B form.

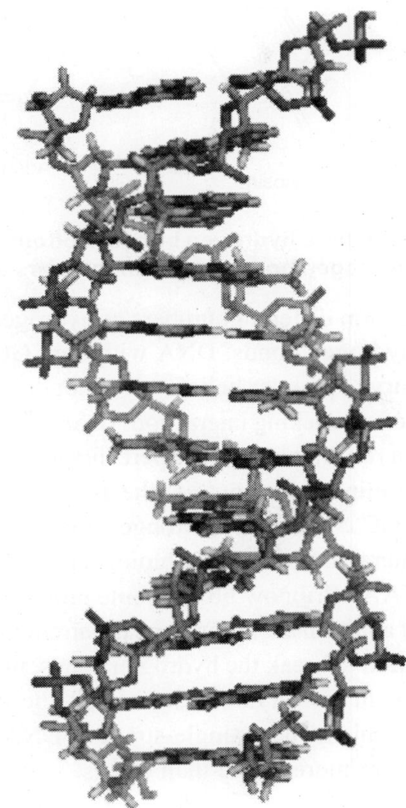

Fig. 2.81. Animation of the structure of a section of DNA. The bases lie horizontally between the two spiraling strands

Base Pairing

Each type of base on one strand forms a bond with just one type of base on the other strand. This is called complementary base pairing. Here, purines form hydrogen bonds to pyrimidines, with A bonding only to T, and C bonding only to G. This arrangement of two nucleotides binding together across the double helix is called a base pair. As hydrogen bonds are not covalent, they can be broken and rejoined relatively easily. The two strands of DNA in a double helix can therefore be pulled apart like a zipper, either by a mechanical force or high temperature. As a result of this complementarity, all the information in the double-stranded sequence of a DNA helix is duplicated on each strand, which is vital in DNA replication. Indeed, this reversible and specific interaction between complementary base pairs is critical for all the functions of DNA in living organisms.

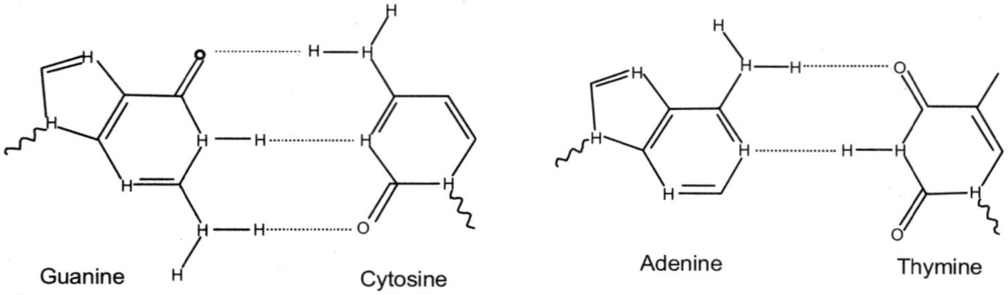

Guanine Cytosine Adenine Thymine

Fig. 2.82 : Top, a GC base pair with three hydrogen bonds. Bottom, an AT base pair with two hydrogen bonds. Non-covalent hydrogen bonds between the pairs are shown as dashed lines

The two types of base pairs form different numbers of hydrogen bonds, AT forming two hydrogen bonds, and GC forming three hydrogen bonds. DNA with high GC-content is more stable than DNA with low GC-content, but contrary to popular belief, this is not due to the extra hydrogen bond of a GC basepair but rather the contribution of stacking interactions (hydrogen bonding merely provides specificity of the pairing, not stability). As a result, it is both the percentage of GC base pairs and the overall length of a DNA double helix that determine the strength of the association between the two strands of DNA. Long DNA helices with a high GC content have stronger-interacting strands, while short helices with high AT content have weaker-interacting strands. In biology, parts of the DNA double helix that need to separate easily, such as the TATAAT Pribnow box in some promoters, tend to have a high AT content, making the strands easier to pull apart. In the laboratory, the strength of this interaction can be measured by finding the temperature required to break the hydrogen bonds, their melting temperature (also called T_m value). When all the base pairs in a DNA double helix melt, the strands separate and exist in solution as two entirely independent molecules. These single-stranded DNA molecules have no single common shape, but some conformations are more stable than others.

Sense and Antisense

A DNA sequence is called "sense" if its sequence is the same as that of a messenger RNA copy that is translated into protein. The sequence on the opposite strand is called the "antisense" sequence. Both sense and antisense sequences can exist on different parts of the same strand of DNA (i.e. both strands contain both sense and antisense sequences). In both prokaryotes and eukaryotes, antisense RNA sequences are produced, but the functions of these RNAs are not entirely clear. One proposal is that antisense RNAs are involved in regulating gene expression through RNA-RNA base pairing.

A few DNA sequences in prokaryotes and eukaryotes, and more in plasmids and viruses, blur the distinction between sense and antisense strands by having overlapping genes. In these cases, some DNA sequences do double duty, encoding one protein when read along one strand, and a second protein when read in the opposite direction along the other strand. In bacteria, this overlap may be involved in the regulation of gene transcription, while in viruses, overlapping genes increase the amount of information that can be encoded within the small viral genome.

Supercoiling

DNA can be twisted like a rope in a process called DNA supercoiling. With DNA in its "relaxed" state, a strand usually circles the axis of the double helix once every 10.4 base pairs, but if the DNA is twisted the strands become more tightly or more loosely wound. If the DNA is twisted in the direction of the helix, this is positive supercoiling, and the bases are held more tightly together. If they are twisted in the opposite direction, this is negative supercoiling, and the bases come apart more easily. In nature, most DNA has slight negative supercoiling that is introduced by enzymes called topoisomerases. These enzymes are also needed to relieve the twisting stresses introduced into DNA strands during processes such as transcription and DNA replication.

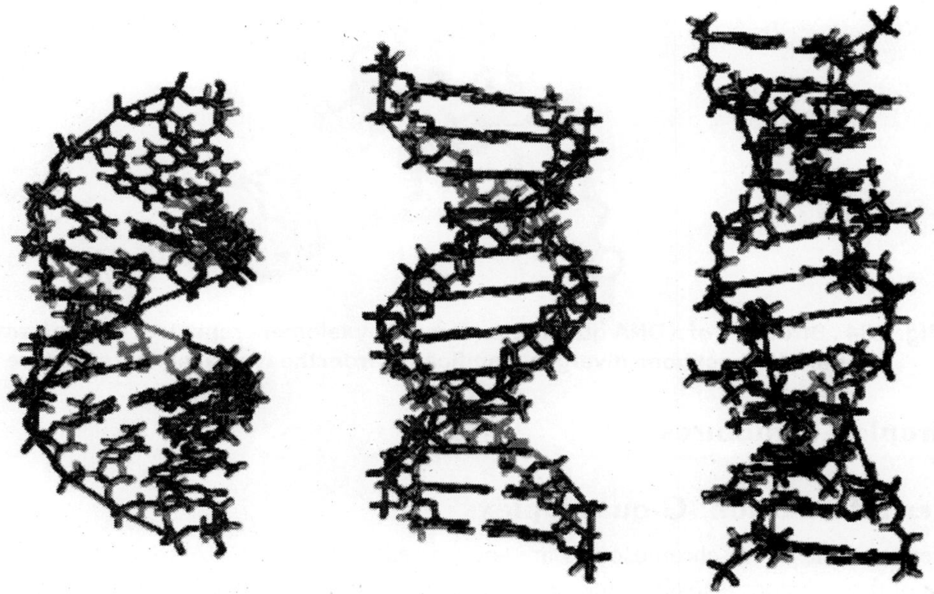

Fig. 2.83 : From left to right, the structures of A, B and Z DNA

Alternative Structures

DNA exists in many possible conformations. However, only A-DNA, B-DNA, and Z-DNA have been observed in organisms. Which conformation DNA adopts depends on the sequence of the DNA, the amount and direction of supercoiling, chemical modifications of the bases and also solution conditions, such as the concentration of metal ions and polyamines. Of these three conformations, the "B" form described above is most common under the conditions found in cells. The two alternative double-helical forms of DNA differ in their geometry and dimensions.

The A form is a wider right-handed spiral, with a shallow, wide minor groove and a narrower, deeper major groove. The A form occurs under non-physiological conditions in dehydrated samples of DNA, while in the cell it may be produced in hybrid pairings of DNA and RNA strands, as well as in enzyme-DNA complexes. Segments of DNA where the bases have been chemically-modified by methylation may undergo a larger change in conformation and adopt the Z form. Here, the strands turn about the helical axis in a left-handed spiral, the opposite of the more common B form. These unusual structures can be recognized by specific Z-DNA binding proteins and may be involved in the regulation of transcription.

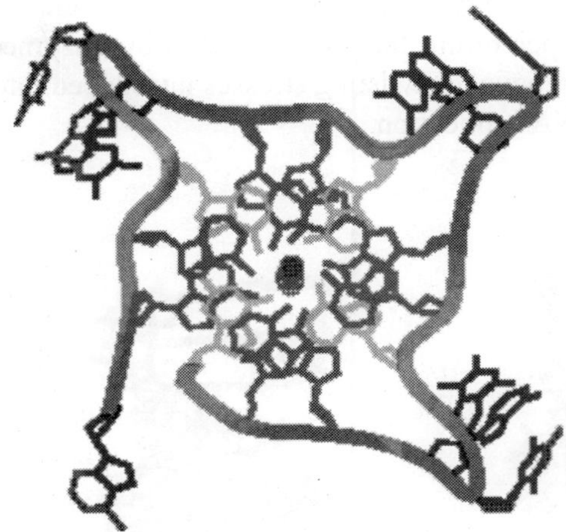

Fig. 2.84 : Structure of a DNA quadruplex formed by telomere repeats. The conformation of the DNA backbone diverges significantly from the typical helical structure

Quadruplex Structures

Further Information: G-quadruplex

At the ends of the linear chromosomes are specialized regions of DNA called telomeres. The main function of these regions is to allow the cell to replicate chromosome ends using the enzyme telomerase, as the enzymes that normally replicate DNA cannot copy the extreme 32 ends of chromosomes. These specialized chromosome caps also help protect the DNA ends, and stop the DNA repair systems in the cell from treating them as damage to be corrected. In human cells, telomeres are usually lengths of single-stranded DNA containing several thousand repeats of a simple TTAGGG sequence.

These guanine-rich sequences may stabilize chromosome ends by forming structures of stacked sets of four-base units, rather than the usual base pairs found in other DNA molecules. Here, four guanine bases form a flat plate and these flat four-base units then stack on top of each other, to form a stable *G-quadruplex* structure. These structures are stabilized by hydrogen bonding between the edges of the bases and chelation of a metal ion in the centre of each four-base unit. Other structures can also be formed, with the central set of four bases coming from either a single strand folded around the bases, or several different parallel strands, each contributing one base to the central structure.

In addition to these stacked structures, telomeres also form large loop structures called telomere loops, or T-loops. Here, the single-stranded DNA curls around in a long circle stabilized by telomere-binding proteins. At the very end of the T-loop, the single-stranded telomere DNA is held onto a region of double-stranded DNA by the telomere strand disrupting the double-helical DNA and base pairing to one of the two strands. This triple-stranded structure is called a displacement loop or D-loop.

Branched DNA

In DNA fraying occurs when non-complimentary regions exist at the end of an otherwise complimentary double-strand of DNA. However, branched DNA can occur if a third strand of DNA is introduced and contains adjoining regions able to hybridize with the frayed regions of the pre-existing double-strand. Although the simplest example of branched DNA involves only three strands of DNA, complexes involving additional strands and multiple branches are also possible.

Fig. 2.85 and 2.85A: A DNA structure with a single branching point

A DNA structure with multiple branches

Chemical Modifications

Structure of cytosine with and without the 5-methyl group. After deamination the 5-methylcytosine has the same structure as thymine.

Base Modifications

The expression of genes is influenced by how the DNA is packaged in chromosomes, in a structure called chromatin. Base modifications can be involved in packaging, with regions that have low or no gene expression usually containing high levels of methylation of cytosine bases. For example, cytosine methylation, produces 5-methylcytosine, which is important for X-chromosome inactivation. The average level of methylation varies between organisms - the worm *Caenorhabditis elegans* lacks cytosine methylation, while vertebrates have higher levels, with up to 1% of their DNA containing 5-methylcytosine. Despite the importance of 5-methylcytosine, it can deaminate to leave a thymine base, methylated cytosines are therefore particularly prone to mutations. Other base modifications include adenine methylation in bacteria and the glycosylation of uracil to produce the "J-base" in kinetoplastids.

Damage

DNA can be damaged by many different sorts of mutagens, which change the DNA sequence. Mutagens include oxidizing agents, alkylating agents and also high-energy electromagnetic radiation such as ultraviolet light and X-rays. The type of DNA damage produced depends on the type of mutagen. For example, UV light can damage DNA by producing thymine dimers, which are cross-links between pyrimidine bases. On the other hand, oxidants such as free radicals or hydrogen peroxide produce multiple forms of damage, including base modifications, particularly of guanosine, and double-strand breaks. In each human cell, about 500 bases suffer oxidative damage per day. Of these oxidative lesions, the most dangerous are double-strand breaks, as these are difficult to repair and can produce point mutations, insertions and deletions from the DNA sequence, as well as chromosomal translocations.

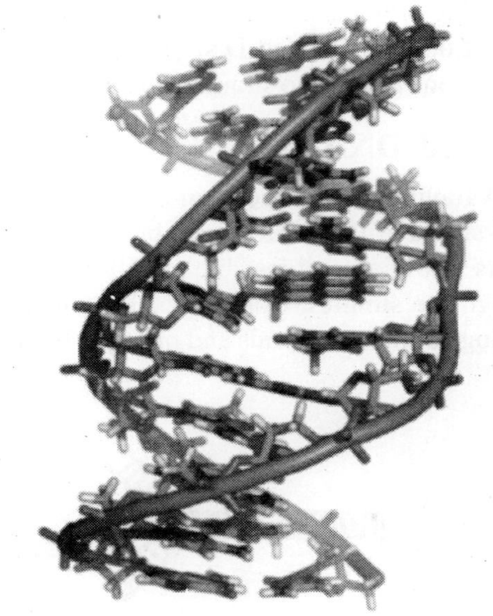

Fig. 2.86 : Benzopyrene, the major mutagen in tobacco smoke, in an adduct to DNA

Many mutagens fit into the space between two adjacent base pairs, this is called *intercalating*. Most intercalators are aromatic and planar molecules, and include Ethidium bromide, daunomycin, and doxorubicin. In order for an intercalator to fit between base pairs, the bases must separate, distorting the DNA strands by unwinding of the double helix. This inhibits both transcription and DNA replication, causing toxicity and mutations. As a result, DNA intercalators are often carcinogens, and benzopyrene diol epoxide, acridines, aflatoxin and ethidium bromide are well-known examples. Nevertheless, due to their ability to inhibit DNA transcription and replication, these toxins are also used in chemotherapy to inhibit rapidly-growing cancer cells.

Biological Functions

DNA usually occurs as linear chromosomes in eukaryotes, and circular chromosomes in prokaryotes. The set of chromosomes in a cell makes up its genome; the human genome has approximately 3 billion base pairs of DNA arranged into 46 chromosomes. The information carried by DNA is held in the sequence of pieces of DNA called genes. Transmission of genetic information in genes is achieved via complementary base pairing. For example, in transcription, when a cell uses the information in a gene, the DNA sequence is copied into a complementary RNA sequence through the attraction between the DNA and the correct RNA nucleotides. Usually, this RNA copy is then used to make a matching protein sequence in a process called translation which depends on the same interaction between RNA nucleotides. Alternatively, a cell may simply copy its genetic information in a process called DNA replication. The details of these functions are covered in other articles; here we focus on the interactions between DNA and other molecules that mediate the function of the genome.

Genes and Genomes

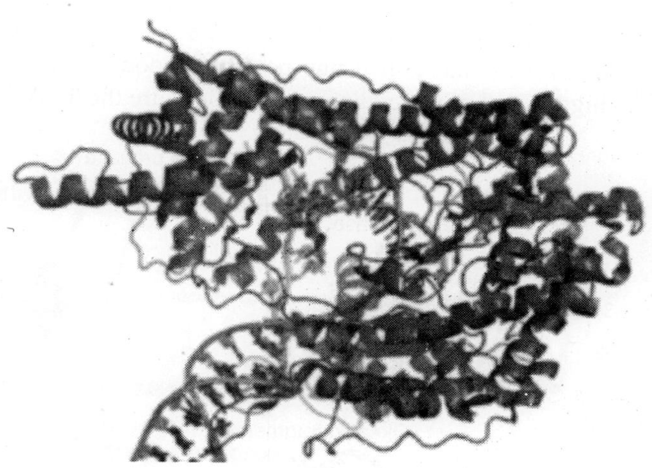

Fig. 2.87: T7 RNA polymerase (blue) producing a mRNA (green) from a DNA template (orange)

Genomic DNA is located in the cell nucleus of eukaryotes, as well as small amounts in mitochondria and chloroplasts. In prokaryotes, the DNA is held within an irregularly shaped body in the cytoplasm called the nucleoid. The genetic information in a genome is held within genes, and the complete set of this information in an organism is called its genotype. A gene is a unit of heredity and is a region of DNA that influences a particular characteristic in an organism. Genes contain an open reading frame that can be transcribed, as well as regulatory sequences such as promoters and enhancers, which control the transcription of the open reading frame.

In many species, only a small fraction of the total sequence of the genome encodes protein. For example, only about 1.5% of the human genome consists of protein-coding exons, with over 50% of human DNA consisting of non-coding repetitive sequences. The reasons for the presence of so much non-coding DNA in eukaryotic genomes and the extraordinary differences in genome size, or *C-value*, among species represent a long-standing puzzle known as the "C-value enigma." However, DNA sequences that do not code protein may still encode functional non-coding RNA molecules, which are involved in the regulation of gene expression.

Some non-coding DNA sequences play structural roles in chromosomes. Telomeres and centromeres typically contain few genes, but are important for the function and stability of chromosomes. An abundant form of non-coding DNA in humans are pseudogenes, which are copies of genes that have been disabled by mutation. These sequences are usually just molecular fossils, although they can occasionally serve as raw genetic material for the creation of new genes through the process of gene duplication and divergence.

Transcription and Translation

A gene is a sequence of DNA that contains genetic information and can influence the phenotype of an organism. Within a gene, the sequence of bases along a DNA strand defines a messenger RNA sequence, which then defines one or more protein sequences. The relationship between the nucleotide sequences of genes and the amino-acid sequences of proteins is determined by the rules of translation, known collectively as the genetic code. The genetic code consists of three-letter 'words' called *codons* formed from a sequence of three nucleotides (e.g. ACT, CAG, TTT).

In transcription, the codons of a gene are copied into messenger RNA by RNA polymerase. This RNA copy is then decoded by a ribosome that reads the RNA sequence by base-pairing the messenger

RNA to transfer RNA, which carries amino acids. Since there are 4 bases in 3-letter combinations, there are 64 possible codons (4^3 combinations). These encode the twenty standard amino acids, giving most amino acids more than one possible codon. There are also three 'stop' or 'nonsense' codons signifying the end of the coding region; these are the TAA, TGA and TAG codons.

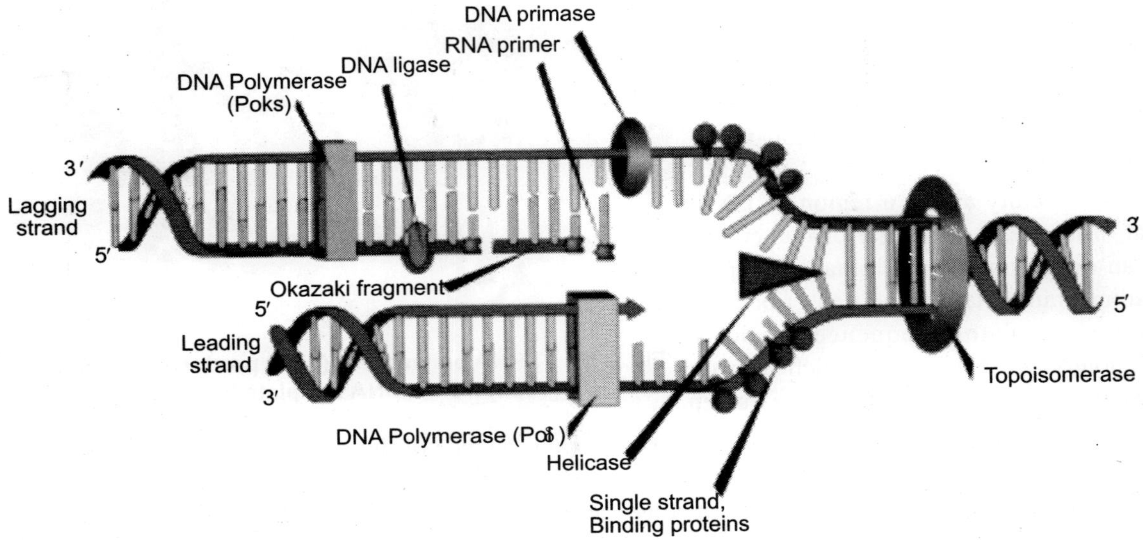

Fig. 2.88: DNA replication. The double helix is unwound by a helicase and topoisomerase. Next, one DNA polymerase produces the leading strand copy. Another DNA polymerase binds to the lagging strand. This enzyme makes discontinuous segments (called Okazaki fragments) before DNA ligase joins them together

Replication

Cell division is essential for an organism to grow, but when a cell divides it must replicate the DNA in its genome so that the two daughter cells have the same genetic information as their parent. The double-stranded structure of DNA provides a simple mechanism for DNA replication. Here, the two strands are separated and then each strand's complementary DNA sequence is recreated by an enzyme called DNA polymerase. This enzyme makes the complementary strand by finding the correct base through complementary base pairing, and bonding it onto the original strand. As DNA polymerases can only extend a DNA strand in a 52 to 32 direction, different mechanisms are used to copy the antiparallel strands of the double helix. In this way, the base on the old strand dictates which base appears on the new strand, and the cell ends up with a perfect copy of its DNA.

Interactions with Proteins

All the functions of DNA depend on interactions with proteins. These protein interactions can be non-specific, or the protein can bind specifically to a single DNA sequence. Enzymes can also bind to DNA and of these, the polymerases that copy the DNA base sequence in transcription and DNA replication are particularly important.

DNA-binding Proteins

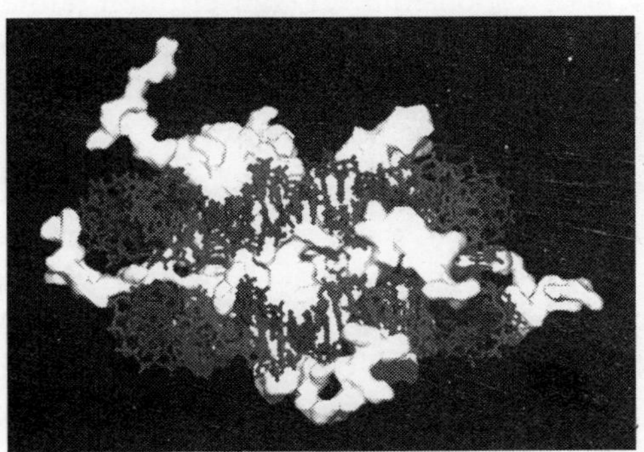

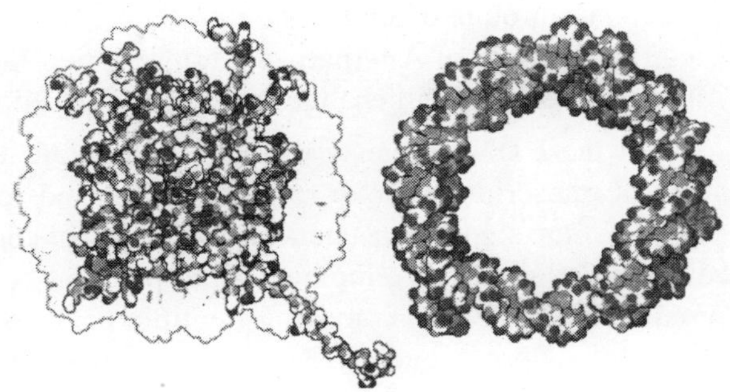

Fig. 2.89 and 2.89A: AInteraction of DNA with histones (shown in white, top). These proteins' basic amino acids (below left, blue) bind to the acidic phosphate groups on DNA (below right, red).

Structural proteins that bind DNA are well-understood examples of non-specific DNA-protein interactions. Within chromosomes, DNA is held in complexes with structural proteins. These proteins organize the DNA into a compact structure called chromatin. In eukaryotes this structure involves DNA binding to a complex of small basic proteins called histones, while in prokaryotes multiple types of proteins are involved. The histones form a disk-shaped complex called a nucleosome, which contains two complete turns of double-stranded DNA wrapped around its surface. These non-specific interactions are formed through basic residues in the histones making ionic bonds to the acidic sugar-phosphate backbone of the DNA, and are therefore largely independent of the base sequence. Chemical modifications of these basic amino acid residues include methylation, phosphorylation and acetylation. These chemical changes alter the strength of the interaction between the DNA and the histones, making the DNA more or less accessible to transcription factors and changing the rate of transcription. Other non-specific DNA-binding proteins in chromatin include the high-mobility group proteins, which bind to bent or distorted DNA. These proteins are important in bending arrays of nucleosomes and arranging them into the larger structures that make up chromosomes.

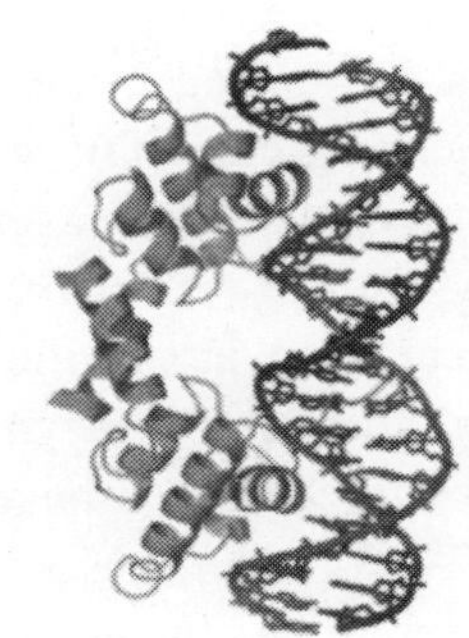

Fig. 2.90: The lambda repressor helix-turn-helix transcription factor bound to its DNA target

A distinct group of DNA-binding proteins are the DNA-binding proteins that specifically bind single-stranded DNA. In humans, replication protein A is the best-understood member of this family and is used in processes where the double helix is separated, including DNA replication, recombination and DNA repair. These binding proteins seem to stabilize single-stranded DNA and protect it from forming stem-loops or being degraded by nucleases.

In contrast, other proteins have evolved to bind to particular DNA sequences. The most intensively-studied of these are the various transcription factors, which are proteins that regulate transcription. Each transcription factor binds to one particular set of DNA sequences and activates or inhibits the transcription of genes that have these sequences close to their promoters. The transcription factors do this in two ways. Firstly, they can bind the RNA polymerase responsible for transcription, either directly or through other mediator proteins; this locates the polymerase at the promoter and allows it to begin transcription. Alternatively, transcription factors can bind enzymes that modify the histones at the promoter; this will change the accessibility of the DNA template to the polymerase.

As these DNA targets can occur throughout an organism's genome, changes in the activity of one type of transcription factor can affect thousands of genes. Consequently, these proteins are often the targets of the signal transduction processes that control responses to environmental changes or cellular differentiation and development. The specificity of these transcription factors' interactions with DNA come from the proteins making multiple contacts to the edges of the DNA bases, allowing them to "read" the DNA sequence. Most of these base-interactions are made in the major groove, where the bases are most accessible.

DNA-modifying Enzymes

Nucleases and Ligases

Nucleases are enzymes that cut DNA strands by catalyzing the hydrolysis of the phosphodiester bonds. Nucleases that hydrolyse nucleotides from the ends of DNA strands are called exonucleases, while endonucleases cut within strands. The most frequently-used nucleases in molecular biology are the restriction endonucleases, which cut DNA at specific sequences. For instance, the EcoRV enzyme shown to the left recognizes the 6-base sequence 5?-GAT|ATC-3? and makes a cut at the vertical line. In nature, these enzymes protect bacteria against phage infection by digesting the phage DNA when it enters the bacterial cell, acting as part of the restriction modification system. In technology, these sequence-specific nucleases are used in molecular cloning and DNA fingerprinting.

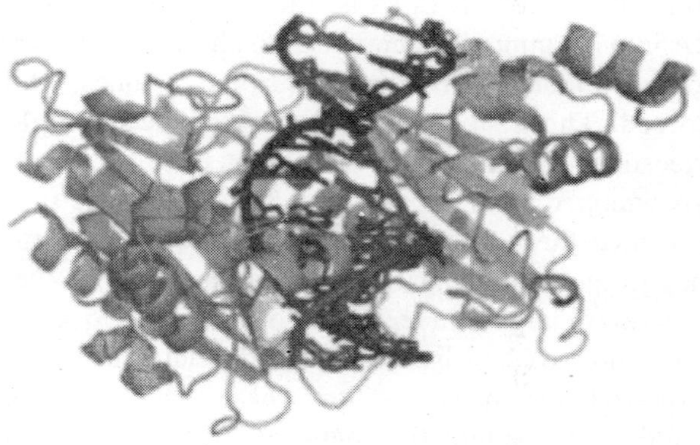

Fig. 2.91 : The restriction enzyme EcoRV (green) in a complex with its substrate DNA

Enzymes called DNA ligases can rejoin cut or broken DNA strands. Ligases are particularly important in lagging strand DNA replication, as they join together the short segments of DNA produced at the replication fork into a complete copy of the DNA template. They are also used in DNA repair and genetic recombination.

Topoisomerases and Helicases

Topoisomerases are enzymes with both nuclease and ligase activity. These proteins change the amount of supercoiling in DNA. Some of these enzyme work by cutting the DNA helix and allowing one section to rotate, thereby reducing its level of supercoiling; the enzyme then seals the DNA break. Other types of these enzymes are capable of cutting one DNA helix and then passing a second strand of DNA through this break, before rejoining the helix. Topoisomerases are required for many processes involving DNA, such as DNA replication and transcription.

Helicases are proteins that are a type of molecular motor. They use the chemical energy in nucleoside triphosphates, predominantly ATP, to break hydrogen bonds between bases and unwind the DNA double helix into single strands.These enzymes are essential for most processes where enzymes need to access the DNA bases.

Polymerases

Polymerases are enzymes that synthesize polynucleotide chains from nucleoside triphosphates. The sequence of their products are copies of existing polynucleotide chains - which are called templates. These enzymes function by adding nucleotides onto the 3′ hydroxyl group of the previous nucleotide in a DNA strand. Consequently, all polymerases work in a 5′ to 3′ direction. In the active site of these enzymes, the incoming nucleoside triphosphate base-pairs to the template: this allows polymerases to accurately synthesize the complementary strand of their template. Polymerases are classified according to the type of template that they use.

In DNA replication, a DNA-dependent DNA polymerase makes a copy of a DNA sequence. Accuracy is vital in this process, so many of these polymerases have a proofreading activity. Here, the polymerase recognizes the occasional mistakes in the synthesis reaction by the lack of base pairing between the mismatched nucleotides. If a mismatch is detected, a 3′ to 5′ exonuclease activity is activated and the incorrect base removed. In most organisms DNA polymerases function in a large complex called the replisome that contains multiple accessory subunits, such as the DNA clamp or helicases.

RNA-dependent DNA polymerases are a specialized class of polymerases that copy the sequence of an RNA strand into DNA. They include reverse transcriptase, which is a viral enzyme involved in the infection of cells by retroviruses, and telomerase, which is required for the replication of telomeres. Telomerase is an unusual polymerase because it contains its own RNA template as part of its structure.

Transcription is carried out by a DNA-dependent RNA polymerase that copies the sequence of a DNA strand into RNA. To begin transcribing a gene, the RNA polymerase binds to a sequence of DNA called a promoter and separates the DNA strands. It then copies the gene sequence into a messenger RNA transcript until it reaches a region of DNA called the terminator, where it halts and detaches from the DNA. As with human DNA-dependent DNA polymerases, RNA polymerase II, the enzyme that transcribes most of the genes in the human genome, operates as part of a large protein complex with multiple regulatory and accessory subunits.

Genetic Recombination

A DNA helix usually does not interact with other segments of DNA, and in human cells the different chromosómes even occupy separate areas in the nucleus called "chromosome territories".This physical separation of different chromosomes is important for the ability of DNA to function as a stable repository for information, as one of the few times chromosomes interact is during chromosomal crossover when they recombine. Chromosomal crossover is when two DNA helices break, swap a section and then rejoin.

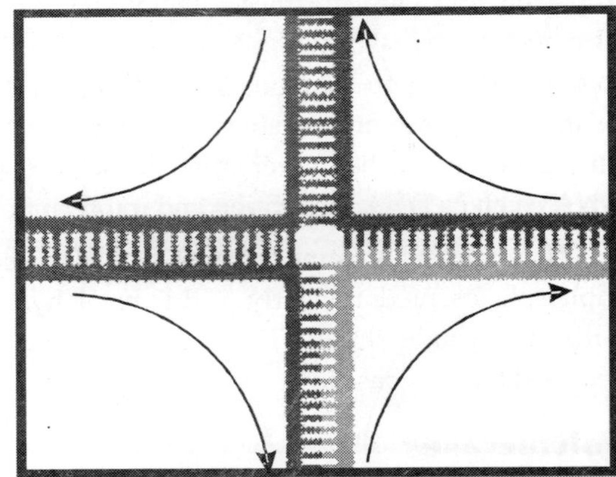

Fig. 2.92: A Structure of the Holiday junction intermediate in genetic recombination. The four separate DNA strands are coloured red, blue, green and yellow

Recombination allows chromosomes to exchange genetic information and produces new combinations of genes, which increases the efficiency of natural selection and can be important in the rapid evolution of new proteins. Genetic recombination can also be involved in DNA repair, particularly in the cell's response to double-strand breaks.

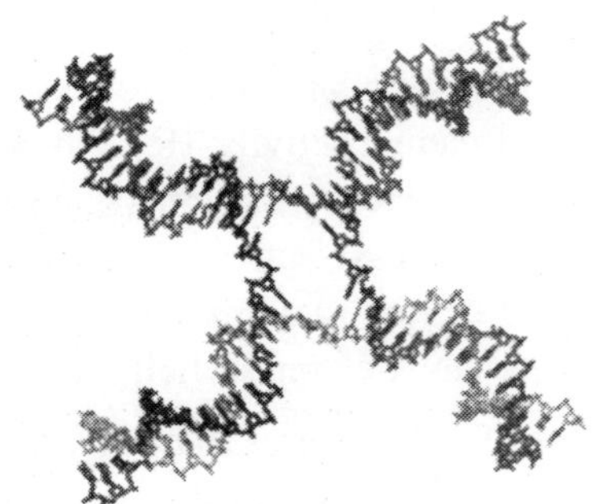

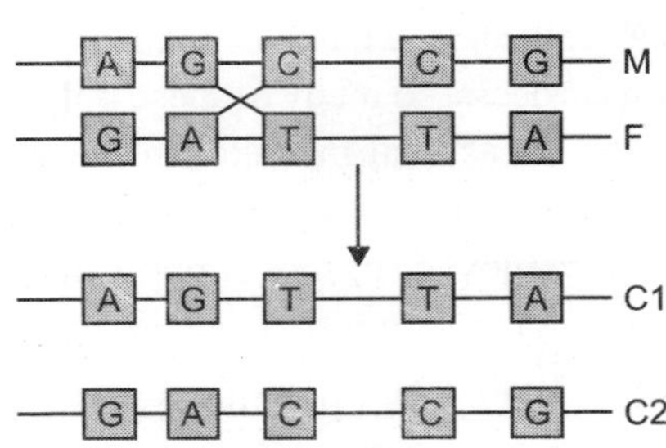

Fig. 2.93 : Recombination involves the breakage and rejoining of two chromosomes (M and F) to produce two re-arranged chromosomes (C1 and C2).

The most common form of chromosomal crossover is homologous recombination, where the two chromosomes involved share very similar sequences. Non-homologous recombination can be damaging to cells, as it can produce chromosomal translocations and genetic abnormalities. The recombination reaction is catalyzed by enzymes known as recombinases, such as RAD51. The first step in recombination is a double-stranded break either caused by an endonuclease or damage to the DNA. A series of steps catalyzed in part by the recombinase then leads to joining of the two helices by at least one Holiday junction, in which a segment of a single strand in each helix is annealled to the complementary strand in the other helix. The Holiday junction is a tetrahedral junction structure that can be moved along the pair of chromosomes, swapping one strand for another. The recombination reaction is then halted by cleavage of the junction and re-ligation of the released DNA.

Evolution

DNA contains the genetic information that allows all modern living things to function, grow and reproduce. However, it is unclear how long in the 4-billion-year history of life DNA has performed this function, as it has been proposed that the earliest forms of life may have used RNA as their genetic material. RNA may have acted as the central part of early cell metabolism as it can both transmit genetic information and carry out catalysis as part of ribozymes. This ancient RNA world where nucleic acid would have been used for both catalysis and genetics may have influenced the evolution of the current genetic code based on four nucleotide bases. This would occur since the number of unique bases in such an organism is a trade-off between a small number of bases increasing replication accuracy and a large number of bases increasing the catalytic efficiency of ribozymes.

Unfortunately, there is no direct evidence of ancient genetic systems, as recovery of DNA from most fossils is impossible. This is because DNA will survive in the environment for less than one million years and slowly degrades into short fragments in solution. Claims for older DNA have been made, most notably a report of the isolation of a viable bacterium from a salt crystal 250-million years old, but these claims are controversial.

Uses in Technology

Genetic Engineering

Methods have been developed to purify DNA from organisms, such as phenol-chloroform extraction and manipulate it in the laboratory, such as restriction digests and the polymerase chain reaction. Modern biology and biochemistry make intensive use of these techniques in recombinant DNA technology. Recombinant DNA is a man-made DNA sequence that has been assembled from other DNA sequences. They can be transformed into organisms in the form of plasmids or in the appropriate format, by using a viral vector. The genetically modified organisms produced can be used to produce products such as recombinant proteins, used in medical research, or be grown in agriculture.

Forensics

Forensic scientists can use DNA in blood, semen, skin, saliva or hair found at a crime scene to identify a matching DNA of an individual, such as a perpetrator. This process is called genetic fingerprinting, or more accurately, DNA profiling. In DNA profiling, the lengths of variable sections of repetitive DNA, such as short tandem repeats and minisatellites, are compared between people. This method is usually an extremely reliable technique for identifying a matching DNA. However, identification can be complicated if the scene is contaminated with DNA from several people. DNA profiling was developed in 1984 by British geneticist Sir Alec Jeffreys, and first used in forensic science to convict Colin Pitchfork in the 1988 Enderby murders case.

People convicted of certain types of crimes may be required to provide a sample of DNA for a database. This has helped investigators solve old cases where only a DNA sample was obtained from the scene. DNA profiling can also be used to identify victims of mass casualty incidents. On the other hand, many convicted people have been released from prison on the basis of DNA techniques, which were not available when a crime had originally been committed.

Bioinformatics

Bioinformatics involves the manipulation, searching, and data mining of DNA sequence data. The development of techniques to store and search DNA sequences have led to widely-applied advances in computer science, especially string searching algorithms, machine learning and database theory. String searching or matching algorithms, which find an occurrence of a sequence of letters inside a larger sequence of letters, were developed to search for specific sequences of nucleotides. In other applications such as text editors, even simple algorithms for this problem usually suffice, but DNA sequences cause these algorithms to exhibit near-worst-case behaviour due to their small number of distinct characters. The related problem of sequence alignment aims to identify homologous sequences and locate the specific mutations that make them distinct. These techniques, especially multiple sequence alignment, are used in studying phylogenetic relationships and protein function. Data sets representing entire genomes' worth of DNA sequences, such as those produced by the Human Genome Project, are difficult to use without annotations, which label the locations of genes and regulatory elements on each chromosome. Regions of DNA sequence that have the characteristic patterns associated with protein- or RNA-coding genes can be identified by gene finding algorithms, which allow researchers to predict the presence of particular gene products in an organism even before they have been isolated experimentally.

DNA Nanotechnology

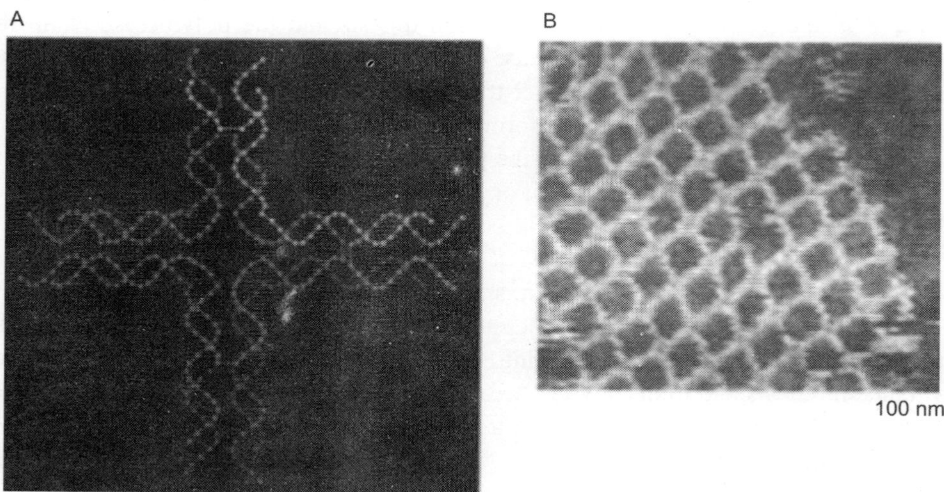

Fig. 2.94: The DNA structure at left (schematic shown) will self-assemble into the structure visualized by atomic force microscopy at right. DNA nanotechnology is the field which seeks to design nanoscale structures using the molecular recognition properties of DNA molecules. Image from Strong, 2004

DNA nanotechnology uses the unique molecular recognition properties of DNA and other nucleic acids to create self-assembling branched DNA complexes with useful properties. DNA is thus used as a structural material rather than as a carrier of biological information. This has led to the creation of two-dimensional periodic lattices (both tile-based as well as using the "DNA origami" method) as well as three-dimensional structures in the shapes of polyhedra. Nanomechanical devices and algorithmic self-assembly have also been demonstrated, and these DNA structures have been used to template the arrangement of other molecules such as gold nanoparticles and streptavidin proteins.

History and Anthropology

Because DNA collects mutations over time, which are then inherited, it contains historical information and by comparing DNA sequences, geneticists can infer the evolutionary history of organisms, their phylogeny. This field of phylogenetics is a powerful tool in evolutionary biology. If DNA sequences within a species are compared, population geneticists can learn the history of particular populations. This can be used in studies ranging from ecological genetics to anthropology; for example, DNA evidence is being used to try to identify the Ten Lost Tribes of Israel.

DNA has also been used to look at modern family relationships, such as establishing family relationships between the descendants of Sally Hemings and Thomas Jefferson. This usage is closely related to the use of DNA in criminal investigations detailed above. Indeed, some criminal investigations have been solved when DNA from crime scenes has matched relatives of the guilty individual.

Primary DNA Molecular Structure

Basic Properties of Primary DNA Molecular Structure

Fundamental Properties of DNA

- Two antiparallel and complimentary strands of deoxyribonucleic acid
- Hydrophillic polar external sugar-phosphate backbone
- Hydrophobic core of bases: Adenine, Thymine, Guanine, Cytosine
- DNA has significant secondary structure.

Sugar-Phosphate Backbone

The basic structure of DNA can be divided into two portions: the external sugar-phosphate backbone, and the internal bases. The sugar-phosphate backbone, as its name implies, is the major structural component of the DNA molecule. The backbone is constructed from alternating ribose sugar and phosphate molecules which are highly polar. Because the backbone is polar, it is hydrophillic which means that it likes to be immersed in water.

Molecular Structure of DNA

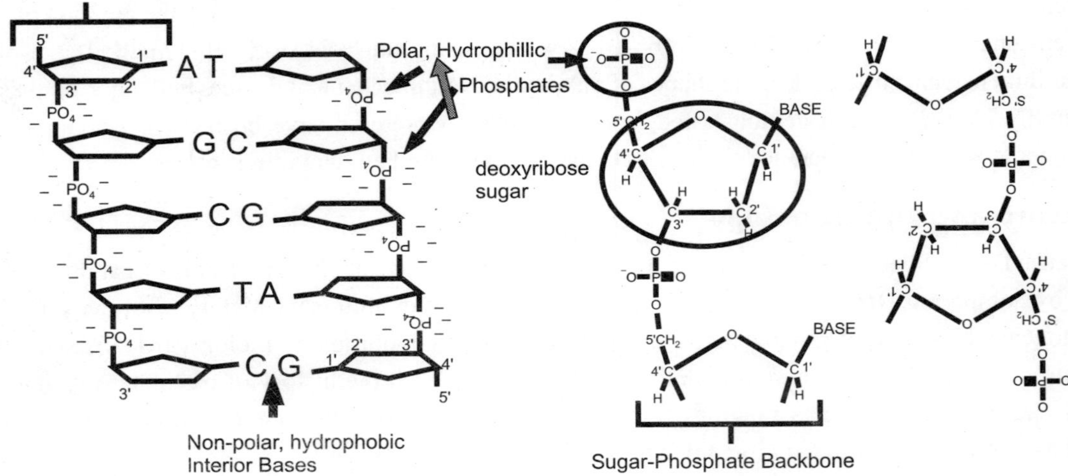

Fig. 2.95: The Molecular structure of DNA

Complimentary Core of Bases

The interior portion of a DNA molecule is composed of a series of 4 nitrogenous bases: adenine (A), guanine (G), thymine (T), and cytosine (C). These bases are non-polar and therefore hyrdophobic (they don't like water). Inside a DNA molecule these bases pair up, A to T and C to G, forming hydrogen bonds that stabilize the DNA molecule. Because the interior bases pair up in this manner, we say the DNA double helix is complimentary. It is this sequence of bases inside the DNA double helix that we refer to as the genetic code.

Hydrostatic Interactions

Hydrostatic forces are very important to the molecular structure of DNA. Hydrostatic forces arise because of hydrogen bonding between the hydrogen and oxygen atoms in water. Polar molecules, because of thier charge, can interact with water without disrupting the ubiquitous latice of hydrogen bonds that the water molecules naturally form. This allows polar molecules to easily dissolve into water. Therefore we call them hydrophilic. Non-polar molecules, however, cannot form electrostatic bonds with the hydrogen and oxygen atoms in water. Non-polar molecules, to be immersed in water, must break potential electrostatic hydrogen bonds between water molecules. Breaking potential bonds represents a net increase in the free energy of the system of water molecules and this has a destabilizing effect the result of which is that non-polar molecules tend to get pushed out of water. Thus we call them hydrophobic. If you have ever seen a water and oil mixture shook up and then slowly settle apart, you have seen these forces in action. The desktop wave boxes are an example of this. Actually, you see these forces in action constantly; you wouldn't exist without them: your cells would lyse apart, your proteins wouldn't fold properly, and your would die a horribly messy death if these hydrostatic forces quit working:)

The sugar-phosphate backbone of DNA is polar, and therefore hydrophillic; thus it likes to be proximal to water. The interior portion of DNA, the bases, are relatively non-polar and therefore hydrophobic. This duality has a very stabilizing effect on the overall structure of the DNA double helix: the hydrophobic core of the DNA molecule 'wants' to be hidden inside the sugar-phosphate backbone which acts to isolate it from the polar water molecules. Due to these hydrostatic forces there is a strong pressure gluing the two strands of DNA together.

Antiparallel—Direction and Polarity in DNA

DNA is Antiparallel

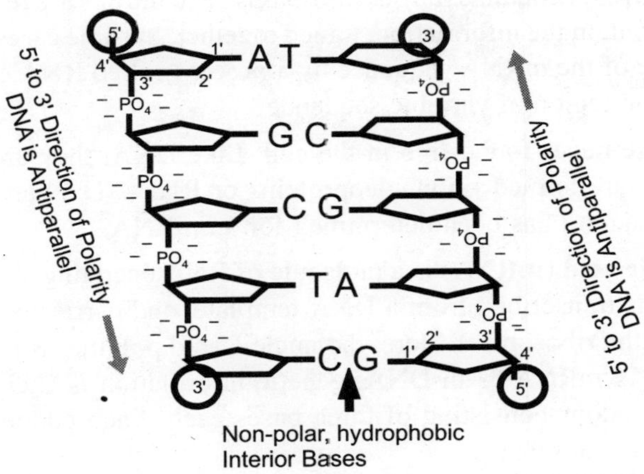

Fig. 2.96: The Direction polarity of DNA

That DNA is antiparallel means that the two strands of DNA have opposite chemical polarity, or, stated another way, their sugar-phosphate backbones run in opposite directions. Direction in nucleic acids is specified by referring to the carbons of the ribose ring in the sugar-phosphate backbone of DNA. 5' specifies the 5th carbon in the ribose ring, counting clockwise from the oxygen molecule, and 3' specifies the 3rd carbon in the ring. Direction of, and in reference to, DNA molecules is then specified relative to these carbons. For example, transcription, the act of transcribing DNA to RNA for eventual expression, always occurs in the 5' to 3' direction. Nucleic acid polymerization cannot occur in the opposite direction, 3' to 5', because of the difference in chemical properties between the 5' methyl group and the 3' ring-carbon with an attached hydroxyl group.

STRUCTURE AND PROPERTIES OF m-RNA, t-RNA, r-RNA AND hn-RNA

Messenger or mRNA is a copy of the information carried by a gene on the DNA. The role of mRNA is to move the information contained in DNA to the translation machinery.

mRNA is heterogeneous in size and sequence. It always has a 5' cap composed of a 5' to 5' triphosphate linkage between two modified nucleotides: a 7-methylguanosine and a 2' O-methyl purine. This cap serves to identify this RNA molecule as an mRNA to the translational machinery. In addition, most mRNA molecules contain a poly-Adenosine tail at the 3' end. Both the 5' cap and the 3' tail are added after the RNA is transcribed and contribute to the stability of the mRNA in the cell.

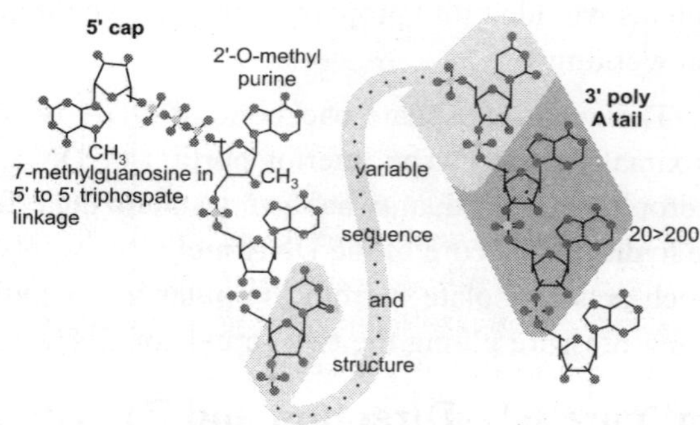

Fig. 2.97 : The m-RNA and its operation

mRNA is not made directly in a eukaryotic cell. It is transcribed as heterogeneous nuclear RNA (hnRNA) in the nucleus. hnRNA contains introns and exons. The introns are removed by RNA splicing leaving the exons, which contain the information, joined together. In some cases, individual nucleotides can be added in the middle of the mRNA sequence by a process called RNA editing. In the figure the exons are represented as the region of variable sequence.

hnRNA and mRNA are never found free in the cell. Like DNA, they are bound by cations and proteins. These complexes are termed ribonucleoproteins or RNPs. The variability in sequence and structure means that no structure has been determined for an mRNA.

Messenger ribonucleic acid (mRNA) is a molecule of RNA encoding a chemical "blueprint" for a protein product. mRNA is transcribed from a DNA template, and carries coding information to the sites of protein synthesis: (the ribosomes). Here, the nucleic acid polymer is translated into a polymer of amino acids: a protein. In mRNA as in DNA, genetic information is encoded in the sequence of nucleotides arranged into codons consisting of three bases each. Each codon encodes for a specific amino acid, except the stop codons that terminate protein synthesis. This process requires two other types of RNA:— transfer RNA (tRNA) mediates recognition of the codon and provides the corresponding amino acid, while ribosomal RNA (rRNA) is the central component of the ribosome's protein manufacturing machinery.

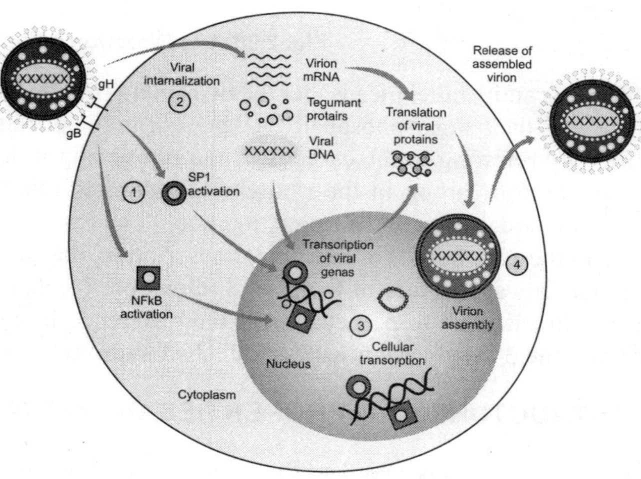

Fig. 2.98 : The Life Cycle and Action of m-RNA

Synthesis, Processing, and Function

The brief existence of an mRNA molecule begins with transcription and ultimately ends in degradation. During its life, an mRNA molecule may also be processed, edited, and transported prior to translation. Eukaryotic mRNA molecules often require extensive processing and transport, while prokaryotic molecules do not.

Transcription

During transcription, RNA polymerase makes a copy of a gene from the DNA to mRNA as needed. This process is similar in eukaryotes and prokaryotes. One notable difference, however, is that eukaryotic RNA polymerase associates with mRNA processing enzymes during transcription so that processing can proceed quickly after the start of transcription. The short-lived, unprocessed or partially processed, product is termed *pre-mRNA*; once completely processed, it is termed *mature mRNA*.

Eukaryotic pre-mRNA Processing

Processing of mRNA differs greatly among eukaryotes, bacteria and archea. Non-eukaryotic mRNA is essentially mature upon transcription and requires no processing, except in rare cases. Eukaryotic pre-mRNA, however, requires extensive processing.

5' cap Addition

A *5' cap* (also termed an RNA cap, an RNA 7-methylguanosine cap or an RNA m^7G cap) is a modified guanine nucleotide that has been added to the "front" or 5' end of a eukaryotic messenger RNA shortly after the start of transcription. The 5' cap consists of a terminal 7-methylguanosine residue which is linked through a 5'-5'-triphosphate bond to the first transcribed nucleotide. Its presence is critical for recognition by the ribosome and protection from RNases.

Cap addition is coupled to transcription, and occurs co-transcriptionally, such that each influences the other. Shortly after the start of transcription, the 5' end of the mRNA being synthesized is bound by a cap-synthesizing complex associated with RNA polymerase. This enzymatic complex catalyzes the chemical reactions that are required for mRNA capping. Synthesis proceeds as a multi-step biochemical reaction.

Splicing

Splicing is the process by which pre-mRNA is modified to remove certain stretches of non-coding sequences called introns; the stretches that remain include protein-coding sequences and are called exons. Sometimes pre-mRNA messages may be spliced in several different ways, allowing a single gene to encode multiple proteins. This process is called alternative splicing. Splicing is usually performed by an RNA-protein complex called the spliceosome, but some RNA molecules are also capable of catalyzing their own splicing.

Editing

In some instances, an mRNA will be edited, changing the nucleotide composition of that mRNA. An example in humans is the apolipoprotein B mRNA, which is edited in some tissues, but not others. The editing creates an early stop codon, which upon translation, produces a shorter protein.

Polyadenylation

Polyadenylation is the covalent linkage of a polyadenylyl moiety to a messenger RNA molecule. In eukaryotic organisms, most messenger RNA (mRNA) molecules are polyadenylated at the 3' end. The poly(A) tail and the protein bound to it aid in protecting mRNA from degradation by exonucleases. Polyadenylation is also important for transcription termination, export of the mRNA from the nucleus, and translation. mRNA can also be polyadenylated in prokaryotic organisms, where poly(A) tails act to facilitate, rather than impede, exonucleolytic degradation.

Polyadenylation occurs during and immediately after transcription of DNA into RNA. After transcription has been terminated, the mRNA chain is cleaved through the action of an endonuclease complex associated with RNA polymerase. After the mRNA has been cleaved, around 250 adenosine residues are added to the free 3' end at the cleavage site. This reaction is catalyzed by polyadenylate polymerase. Just as in alternative splicing, there can be more than one polyadenylation variant of a mRNA.

Transport

Another difference between eukaryotes and prokaryotes is mRNA transport. Because eukaryotic transcription and translation is compartmentally separated, eukaryotic mRNAs must be exported from the nucleus to the cytoplasm. Mature mRNAs are recognized by their processed modifications and then exported through the nuclear pore. In neurons mRNA must be transported from the soma to the dendrites where local translation occurs in response to external stimuli. Many messages are marked with so-called "zip codes" which targets their transport to a specific location.

Translation

Because prokaryotic mRNA does not need to be processed or transported, translation by the ribosome can begin immediately after the end of transcription. Therefore, it can be said that prokaryotic translation is *coupled* to transcription and occurs *co-transcriptionally*.

Eukaryotic mRNA that has been processed and transported to the cytoplasm (i.e. mature mRNA) can then be translated by the ribosome. Translation may occur at ribosomes free-floating in the cytoplasm, or directed to the endoplasmic reticulum by the signal recognition particle. Therefore, unlike prokaryotes, eukaryotic translation *is not* directly coupled to transcription.

Structure

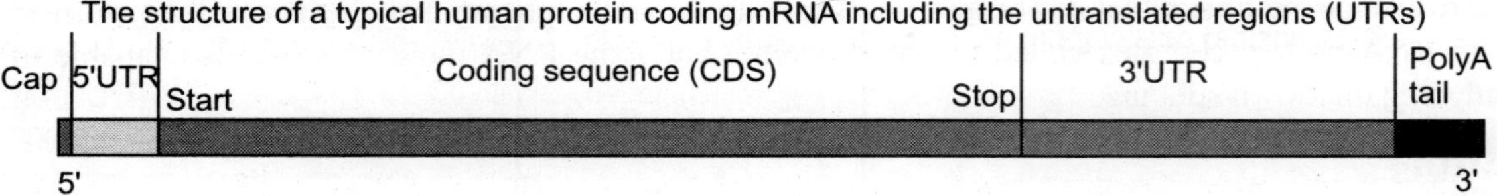

Fig. 2.99 : The structure of a mature eukaryotic mRNA. A fully processed mRNA includes a 5' cap, 5' UTR, coding region, 3' UTR, and poly(A) tail

5' Cap

The *5' cap* is a modified guanine nucleotide added to the "front" (5' end) of the pre-mRNA using a 5'-5'-triphosphate linkage. This modification is critical for recognition and proper attachment of mRNA to the ribosome, as well as protection from 5' exonucleases. It may also be important for other essential processes, such as splicing and transport.

Coding Regions

Coding regions are composed of codons, which are decoded and translated into one (mostly eukaryotes) or several (mostly prokaryotes) proteins by the ribosome. Coding regions begin with the start codon and end with the a stop codons. Generally, the start codon is an AUG triplet and the stop codon is UAA, UAG, or UGA. The coding regions tend to be stabilised by internal base pairs, this impedes degradation. In addition to being protein-coding, portions of coding regions may serve as regulatory sequences in the pre-mRNA as exonic splicing enhancers or exonic splicing silencers.

Untranslated Regions

Untranslated regions (UTRs) are sections of the mRNA before the start codon and after the stop codon that are not translated, termed the five prime untranslated region (5' UTR) and three prime untranslated region (3' UTR), respectively. These regions are transcribed with the coding region and thus are exonic as they are present in the mature mRNA. Several roles in gene expression have been attributed to the untranslated regions, including mRNA stability, mRNA localization, and translational efficiency. The ability of a UTR to perform these functions depends on the sequence of the UTR and can differ between mRNAs.

The stability of mRNAs may be controlled by the 5' UTR and/or 3' UTR due to varying affinity for RNA degrading enzymes called ribonucleases and for ancillary proteins that can promote or inhibit RNA degradation.

Translational efficiency, including sometimes the complete inhibition of translation, can be controlled by UTRs. Proteins that bind to either the 3' or 5' UTR may affect translation by influencing the ribosome's ability to bind to the mRNA. MicroRNAs bound to the 3' UTR also may affect translational efficiency or mRNA stability. Cytoplasmic localization of mRNA is thought to be a function of the 3' UTR. Proteins that are needed in a particular region of the cell can actually be translated there; in such a case, the 3' UTR may contain sequences that allow the transcript to be localized to this region for translation.

Some of the elements contained in untranslated regions form a characteristic secondary structure when transcribed into RNA. These structural mRNA elements are involved in regulating the mRNA. Some, such as the SECIS element, are targets for proteins to bind. One class of mRNA element, the riboswitches, directly bind small molecules, changing their fold to modify levels of transcription or translation. In these cases, the mRNA regulates itself.

Poly(A) Tail

The 3' poly(A) tail is a long sequence of adenine nucleotides (often several hundred) at the 3' end of the pre-mRNA. This tail promotes export from the nucleus and translation, and protects the mRNA from degradation.

Monocistronic versus Polycistronic mRNA

An mRNA molecule is said to be monocistronic when it contains the genetic information to translate only a single protein. This is the case for most of the eukaryotic mRNAs. On the other hand, polycistronic mRNA carries the information of several genes, which are translated into several proteins. These proteins usually have a related function and are grouped and regulated together in an operon. Most of the mRNA found in bacteria and archea are polycistronic. Dicistronic is the term used to describe a mRNA that encodes only two proteins.

mRNA Circularization

In eukaryotes it is thought that mRNA molecules form circular structures due to an interaction between the cap binding complex and poly(A)-binding protein. Circularization is thought to promote recycling of ribosomes on the same message leading to efficient translation.

Degradation

Different mRNAs within the same cell have distinct lifetimes (stabilities). In bacterial cells, individual mRNAs can survive from seconds to more than an hour; in mammalian cells, mRNA lifetimes range from several minutes to days. The greater the stability of an mRNA, the more protein may be produced from that mRNA. The limited lifetime of mRNA enables a cell to alter protein synthesis rapidly in response to its changing needs. There are many mechanisms which lead to the destruction of a message, some are described below:

Prokaryotic mRNA degradation

In prokaryotes the lifetime of mRNA is generally much shorter than in eukaryotes. The regulation of mRNA degradation in prokaryotes is much simpler than in eukaryotes. Prokaryotes have numerous RNases which degrade messages rapidly regardless of the sequence of the mRNA. Alternatively, small RNA molecules (sRNA) of tens to hundreds of nucleotides long can recognize specific mRNAs and stimulate their degredation. Complementary sequences in the sRNA bind to the mRNA creating a double-stranded RNA molecule which is a substrate for certain classes of RNAses. It was recently shown that bacteria also have a sort of 5' cap consisting of a triphosphate on the 5' end. Removal of two of the phosphates leaves a 5' monophosphate causing the message to be destroyed by the exonuclease RNAse E.

Eukaryotic mRNA Turnover

Inside eukaryotic cells there is a balance between the processes of translation and mRNA decay. Messages which are being actively translated are bound by polysomes, the eukaryotic initiation factors eIF-4E and eIF-4G, and poly(A)-binding protein. eIF-4E and eIF-4G block the decapping enzyme (DCP2), and poly(A)-binding protein blocks the exosome complex, protecting the message. In nutrient-starvation conditions or during viral infection translation may be compromised and decay is stimulated. The balance between translation and decay is reflected in the size and abundance of the cytoplasmic structures known as P-bodies During rounds of translation the poly-A tail of the mRNA is shortened by exonucleases. This is thought to disrupt the circular structure of the message and destabilize the cap

The MAINE LOBSTER Book

The MAINE LOBSTER Book

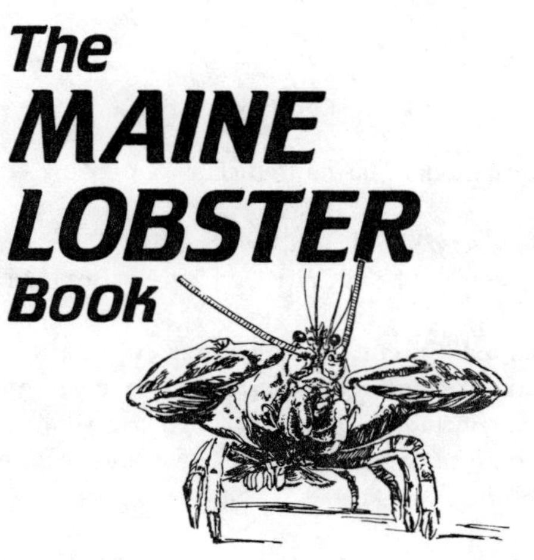

Mike Brown

Illustrations by Jim Sollers

INTERNATIONAL MARINE
PUBLISHING COMPANY
Camden, Maine

Typeset by The Key Word, Inc., Belchertown, Massachusetts
Printed and bound by BookCrafters, Inc., Chelsea, Michigan

Published by International Marine Publishing Company
21 Elm Street, Camden, Maine 04843
(207) 236-4342

Library of Congress Cataloging in Publication Data

Brown, Mike, 1929–
The Maine lobster book.

1. Lobster fisheries—Maine. I. Title.
SH380.2.U6B76 1986 639'.541'09741 86-2760
ISBN 0-87742-216-8

Contents

Publisher's Preface

In 1985, International Marine published a no-holds-barred look at the Maine lobster fishery. *The Great Lobster Chase: The Real Story of Maine Lobsters and the Men Who Catch Them* was written by Mike Brown, a lobsterman and newspaper editor who also happens to be a very good writer. The book has created a stir. A lobster co-op manager in Marblehead, Massachusetts bought copies for each of his fishermen. Maine legislators on the Marine Resources Committee have read it while debating lobster legislation. Lobstermen have been reading it, and lovers of the Maine coast from all around the country have been reading it. We reprinted the book in February 1986, and hope to reprint many times more.

At just over 200 pages in hardcover, however, *The Great Lobster Chase* costs $22.95, and it seems probable that many people who should read it or would enjoy reading it will hesitate at the book's size and price. We knew that if we could publish a shortened version, we could bring the story of a people and the fishery that sustains them to a wider audience. So we chiseled away at *The Great Lobster Chase*, struggling to preserve its character and intent while reducing its dimensions, and *The Maine Lobster Book* is the result. The carefully developed arguments about the political issues surrounding this fishery are gone, but the wit, affection, and descriptive power remain. *The Maine Lobster Book* is, we think, the best *little* book about this rich way of life. We hope you like it.

1

The Lobsterman

It is cold, damn cold, one of those hanger-on late March teen-degree mornings when lobsterman Harold Martin slams the door of his pickup truck at the cooperative wharf. There is still snow on the ground, dirty gravel snow like butterscotch pudding that has been frozen and thawed too many times, finally to be left in the dog dish in the hope that it will disappear. Dirty snow, thinks Harold, as he walks down the planked dock to the co-op office. This is my fortieth year of dirty snow.

Already, at 5 o'clock in the morning, there are other lobstermen at the co-op. The double 55-gallon-drum wood stove is humping, as the co-op boy Stevie likes to say. The warmth spreads around the small office like flowing honey, and Harold turns his body rotisserie-fashion, making morning small talk. The starting price of lobsters. No-good wire traps. The $2-a-hundred increase in oak laths. Heading twine is up three bucks a ball. Jesus, Harold thinks, as he has thought every March morning for the past five years, this is my last year. Forty years of hauling. Forty years of stinking bait barrels, stinking boats, and leaky boots. Forty years of dirty snow.

Last night on high water Harold put 20 traps aboard his Jonesport lobsterboat, some new, some patched, some on their last voyage secretly riddled with yards of tunnels bored by invisible marine worms. Give me one more year, Harold

At the co-op

silently muttered to the tender traps as he loaded them aboard *Bobby & Jean* under a cirrus sky. One more year, the story of my life.

The cleats on the gangway to the lobster float are clogged with ice. They fill with water and slush during the day and freeze at night, making a ladder of ice wedges. Twenty years ago Harold slid the whole length of the gangway breaking two fingers on his left hand. Forty years of ice wedges . . . .

His punt painter is on the bottom of the float as usual, and Harold has to uncleat four others before coming to his own.

The ropes are semifrozen from salt vapor and water and wind. He thinks about when he was a kid and they chose up for scrub baseball. Somebody threw the bat and a kid grabbed it and everybody started piling fist upon fist. The top fist was up first. Harold was always up last. His fist on the bottom, like his punt painter.

There is only one oar in Harold's punt and it is lying in the white bottom slush. Just an old broken orange oar that Harold found floating one day. He has never rowed his punt. There are no oarlocks aboard. Harold, like all lobstermen, just stands up and paddles back and forth to his moorings. A warden stopped him one afternoon and gave him a warning because he didn't have a life preserver aboard. Now he carries one. It is frozen up under the stern seat of the punt. It has been that way since Thanksgiving. An iceball of kapok and straps and buckles. An official league ball to play the game with.

Harold thumps against *Bobby & Jean* and notices her short skirt of ice at the waterline. Tying the punt to the ringbolt on the stern, he climbs into the cockpit and walks forward and unlocks the sliding door to the pilothouse. Never locked it 20 years ago, never even thought about it. But lobstering ain't the same nowadays, for sure. Lifting the engine-box cover, Harold notices the bilge water is where it should be, the Rule automatic switch sitting high and dry like a fat, red bullfrog. He hits the starter switch and the Chevy growls its required seconds of protest, like a sleeping hound dog kicked out from under the kitchen table for a goodnight watering.

Harold is one to idle his engine. Warm her up slowly like a woman. They appreciate it and last longer is Harold's engine/woman philosophy. The bare stainless steel exhaust stack is giving out heat now, and Harold molds a pair of soggy blue dunker gloves around the pipe to dry them. They sizzle and sear and fall to the floor. Harold switches on his Radio Shack CB. Channel 14. Fifteen years ago when he installed the radio he set the dial at channel 14. It has never moved from that number. Harold has thought about getting a new CB, one of those 40-channel jobs. Everybody has them. But, what the hell, he's giving up lobstering this year.

Taking the mike from the overhead where it lives between two galvanized trap nails, Harold presses the talk button. "You on, Effiejean?"

He waits a couple seconds and then calls again. "You on there yet, Effiejean? C'mon back."

"I'm here, Harold."

"Christ, I've been calling you f'half an hour. Where the hell you been?"

"Just here. Where would I go, Harold?"

"Well, I'm on. Leaving the mooring now. Give you a call later on this morning. Ten, maybe half past."

Harold returns the mike to its double nail rest and climbs forward and lets go his mooring chain and ball after tying the punt painter through the chain loop.

"Take the day off, you little bastard," Harold says to his one-oared watchdog.

He clears the harbor buoy and heads sou'east. He'll set a 20-trap string south of Seal Island and then haul nor'west the 150 traps he set out last week.

"A little chop," Harold says to *Bobby & Jean*, "but a pretty good chance, old girl, considering March." He talks to his boat all day long. He is always calling her old girl.

"C'mon old girl, let's find that trap."

"C'mon old girl, what's a little fog."

"Well, old girl, let's eat lunch."

"That's it old girl, let's head f'the barn."

Running sou'west, Harold sees the other lobsterboats now. If he hadn't been an idler, he'd be hauling like the rest of them. Funny how they all look the same, haulers tugging on pot warp like a puppy on a rope—circling, tugging, and the trap jumping out of water to bite the rail. Yellow midgets throwing black specks back overboard. Splash and splash all day long. Tug, throw, splash. What damn fools we are, Harold thinks. I mean it, this is my last year.

Harold switches on his paper recorder, and the stylus offers him a soft, wide line at 30 fathoms. But he already knew he was on his grounds. The marks are there. Ledge, island spruce, water tower.

"Chore time, old girl."

For the next hour Harold does what 40 years of lobstering make him do. The ritual is pat. Harold likens lobstering to cowboying. All day in the saddle. Good horse under you knowing what to do. Roping in those doggies. He thinks he could go West and be cowboying in two days. Effiejean would like cowboys.

The twentieth trap buoy hits the water and Harold turns *Bobby & Jean* due north. The slanting morning sun shows Harold's track, his string of white and red buoys a mile long, straight as a string of popcorn, claiming that piece of ground he has claimed for 40 years. Rangeland for a cowboy, Harold thinks to himself, watching his colors dance in the chop.

"Not a bad chance out here this morning, old girl," he says as he picks up the mountain-range marks for his first string of traps. . . .

* * * *

The evolution of the lobsterman, the breaking away and specialization from a wide-ranging fishing industry, was gradual. Pot fishing for lobsters was not everybody's cup of tea and still isn't. Of the 9,000 lobster licenses issued in 1985, only about a third are full-time-work-at-it-52-weeks-a-year fishermen like Harold Martin. Not all flyers are jet fighter pilots.

The Maine lobsterman was born when the demand for lobsters increased to a single voice in the fishmonger's song. Early Maine fishermen used their boats and homemade gear to harvest the sea and all that lived in it. Their crop was fish—all creatures from the ocean were fish. The necessity of fish for protein lessened as more affluence and the permanent settlements and structures of a society emerged. The diet changed, and palates, tastes, and desires created demands for particular foods. The lobster rose above his fodder-only seashore colleagues and became a special being. And so, to catch a taste, a unique harvester emerged.

Lobster fishing was an attractive occupation, considering some other tangents of fishing. A large boat was not needed because the gear was small and a large harvest volume was not required to earn a respectable living. Deep, long, and expensive nets could be ignored. The lobstermen turned to the abundant woods, where all the materials for a lobster trap could be taken. Boats could even be rowed or sailed. And the work was nearshore and comparatively safe from the winds and waves and dangerous elements the offshore fishermen faced.

But perhaps the most compelling attraction of lobstering was the lonely independence of the life. The early lobstermen were not far from the traits and mores and instincts that had been needed by the early settlers to survive in a new country. Lobstering was not a new course but the sustaining of an old one. The lobsterman and the creature he pursued were remarkedly alike in their different worlds. There was, and is, an uncanny compatibility in their coexistence. Just as the lobster has survived the aeons by his solitary habits and independent senses, so has the lobsterman. They were made for each other.

* * * *

A lobsterman sitting on McLoon's dock at Rockland was asked by another what he was going to do that day.

"Take a tuck in my gumption, Junior, them lobsters are on t'me. Let 'em wait a day."

What he was saying was maybe he didn't feel like working that day. Maybe he didn't like the wind backing around to the east. Or he had had a fight with his wife. He had found out he needed a valve job on his old pickup. Whatever his attitude, he related it to lobsters—let 'em wait.

As the lobster profession progressed, the reasons for joining it became more varied. Demand, price, family, environment. The father-son heritage was especially strong. Perhaps no other profession, unless it is farming, retains such family bondage and values. A lobsterman's son is expected to assume the birthright of his father and his father before him, much as a monarchy needs heirs to survive. A majority of the lobstermen today come from family genes.

* * * *

The dangers of the sea are inherent in its being there. The marine elements are forever challenging, luring, deceptive, subtle, and raging. And it seems quite paradoxical that a person surrounded by water has no fear of it. That is the strange case of lobstermen—and fishermen in general. The reason is the awesome respect fishermen have for the water, which completely inhabits any void for fear. They work a step away from death every day, and yet, there is no thought of drowning. There is a fatalistic demeanor acquired by fishermen over the centuries, but few believe that God or Neptune will spirit you away if you maintain one hand for the boat and one for yourself. Fatalism surrounds the few unexplained lost lobstermen. The rest were stupid, is the saying.

One key to his survival on a demanding sea is his self-sufficiency. The lobsterman often builds his own boat. He surely maintains it—changes the oil and filter, waters the batteries, paints the bottom, repacks the stuffing boxes, inspects the plugs and points. A hundred first steps into the drink are smothered by the inherent preventative maintenance schedule programed into the fisherman from birth. No calling for appointments; he alone is the beneficiary of the timely execution of his responsibilities. For better, or for dead.

But the best of lobstermen can be lost at sea. By far the most common cause of accidental drowning is being hauled overboard in a tangle of pot warp as the lobster trap descends to the bottom. The combination of the weight of trap and warp and the forward speed of the boat means that a misplaced foot in a turn of rope will result in an overboard situation, or a maiming, or both. The breaking strength of today's nylon pot warp is about 2,600 pounds. Even if a lobsterman manages to untangle himself from a clinging mass of rope before being dragged under, the boat is probably in gear and moving away. Rubber boots, a set of oilskins, and 40-degree water are no comfort. If the boat happened to be in neutral and the lobsterman could reach it, he probably could not get aboard. Few lobsterboats have side ladders, and the rail is a forever 3½ feet away. And all this assumes the lobsterman can swim—which he often cannot.

There are sad stories and funny sad stories and tragic stories of drownings and near drownings. They are usually taken in the context of their severity. Nobody talks of the young fisherman lost and of the tearful wife and two small children who must overcome last images of their dad and husband every time they watch a lobsterboat leave the harbor. But let a lobsterman fall overboard from his punt while paddling ashore and then porpoise the rest of the distance with a half-learned dog paddle, and the whole town talks for years about the time old Charlie walked ashore from his mooring because he drank the harbor dry.

Drunks are woven through lobster stories like shards of blue mussel shells in shore sand. The public's conception of hard-drinking fishermen, however, is grossly overrated. Drink they do, but a daily hangover is not an enjoyable mate all day long on a bouncing boat engulfed in continual roar from an eight-cylinder mechanical bull. When a fishing community member is forced to explain a genuine town drunk, there is always a disclaimer so that the inquisitor knows that the man is not a derelict, or a bum, or in the same league with the common stupor. He simply has his gear on the bank. Old Bill, they will explain, used to be a highliner, had the biggest, fastest boat in the harbor, fine wife. He, errrr, put his gear on the bank 10 years ago.

Squirt Carter had stacked his gear on the bank 20 years ago. He did odd jobs, changed oil in boats at the co-op, lived in a trailer, and talked incessantly about the good old days. There was not a man in the village who would not trust Squirt with anything he owned. Except a can of beer. Sitting around on the wharf drinking a six-pack after coming in from hauling, the lobstermen would see Squirt coming down the road and say, "Oh, oh, Squirt-alert, Squirt-alert." The beer would go under a bait bucket. When Squirt joined the group they would give him two bucks to go to the store for a quart of Narragansett beer for the group. Squirt would leave, the fishermen would leave, and Squirt knew that when he returned the lobstermen would not and he and the quart would go home to his trailer and watch a soap opera through a tall sudsy glass. The lobstermen treat their community life with dignity for all.

* * * *

The matter of territorial imperative is important. Lobster wars, gear disappearing, shots in the night, boat rammings—all are true. They have persisted since the beginning of lobstering. The water is a public body, and once a person has purchased a state license he should be able to set his lobster traps anyplace he desires, but in the real lobster world, it doesn't work that way. Lobster grounds and fishing rights are a ball of string wound with various-length pieces of family, integrity, geography, and fear. Territories are staked between communities as well as fishermen. There is, for instance, an imaginary line drawn down eastern Penobscot Bay, with Saddleback Ledge and Eagle Island the landmarks. The Deer Isle-Stonington-Isle au Haut fishermen lobster on the eastern side, and the North Haven-Vinalhaven lobstermen set and haul their gear on the western side. First-time trespassers will be warned in the traditional way, a half hitch around the buoy spindle. Sometimes a second warning of two half hitches is issued, sometimes not. If the foreign trap is still over the line in a week, the trap is cut off. There is usually retaliation by the loser even if this means a raid into enemy territory. A few skirmishes of this magnitude and a full-scale lobster war can erupt.

Lobster wars involve all the tactics of general battle from Biblical times, from the relatively mundane warning shot across the bow to actual cases of bottling up an enemy lobsterboat fleet in harbor with armed marine picket boats. A single blockaded lobsterman, pinned in a cove, unable to use his radio because of jamming, and the light and tide falling, has one of two choices. He can either try to ram one of the pickets and risk losing either his boat or life, and maybe both, or he can repent in a truce raft-up whereby he is released from bondage with the promise that before God, Jesus Christ, Mother's honor, Neptune, and suicide, in no particular order, his line-jumping days are over.

Territories within communities are more subtle. The dimensions diminish in scale and relate to particular sides of the harbor, ledges, coves, west and east ends of islands, a channel, or certain compass courses. If the Smith family lobstermen have fished Herring Cove for centuries, the family name is branded upon that lobster ground as surely as granite monoliths at the Roman Colosseum. It is in trust, however, a life tenancy, and Smith family fishermen must work their heritage. It cannot be sustained by marital infusion. A brother-in-law has no more right to Herring Cove than the next guy. If the island queens do not conceive boy heirs, the lobster throne begins to crumble.

Territorial imperative is above judicial law, and in-community violators are rarely turned over to uniformed marine law enforcement people. Brothers-in-law, maybe. But a neighbor trap molester is reckoned with much differently than the foreign-fleet pirates. Warnings start with anonymous phone calls. Then there is the Citizens Band weapon. Everybody, just everybody, listens to CB, and subtle innuendo messages dart around the waterfront ether like arrows shot into the air, and they know where, just where, they will fall. Ignored prescriptions and continual lobster trap illness will bring on more drastic medicine. Morning sunrises have revealed white lobsterboats with "Trap Thief" spray-painted from stem to stern. The silent treatment at the store, the co-op, and basketball games sometimes gets a point home. Violence is the last refuge. Boats, pickups, and even trailers and houses have been flamed by irate vigilantes.

In a small community near Deer Isle there was a rash of thefts from lobsterboats moored in the harbor. Eventually, the fishermen traced the stolen gear to a young lobsterman who lived in a trailer with his wife and two small children. Six fishermen drove up to the trailer one night, knocked on the door, and walked right past the startled young man. He was given a trial right there in front of the 6 o'clock TV news with wife huddled on the turf-bare sofa holding two frightened children. Confronted by the prosecutor, judge, and jury, the young man confessed and admitted he had stolen and sold the gear. The fishermen laid ten $100 bills on the dirty Formica kitchen table—full payment, they said, for the young fisherman's fiberglass skiff, outboard, and 50 traps. The family was to load all their trailer belongings in their pickup and be out of town by the next day noon. They had no choice. The laws of the open range.

The Maine lobsterman is a unique creature in an animal world. He possesses instincts for survival much like those of the chosen adversary he pursues. But the lobsterman has learned the parameters of life and death and does not dwell on either. The middleground of life to him is his house of cards, and he builds it with as many aces as he can muster. He will convert tragedy to humor, abstract to reality, stars to kitchen matches, and his goal, which he will not discuss candidly with anyone, is something he calls the simple good life.

2

The Gear

Once upon a time, so goes the tale, if people wanted lobsters they merely walked down to the shore, kicked aside the seaweed, and picked them up. The evolution from fetching to catching has taken over 300 years of trial and error with assorted paraphernalia known by lobstermen as simply—gear.

Gear can be anything from pot warp (rope) to bait (cuttings, brim) to the officially sanctioned brass gauge, which the lobsterman, in his homely way, calls a measure.

The Maine lobsterman today uses two styles of pots or traps. (The words "pot" and "trap" are interchangeable, but generally the fishermen west of Penobscot Bay say pot and the eastern fishermen use the term trap.) The two types are known as square and round. Neither is the case. The lobsterman sees no purpose in being superfluous in word or description. For the record, the round trap is half-round and the square trap is rectangular. In the beginning, about 1800, American wooden traps were almost always round style, probably from the heritage of the European creel, inkwell, or eel pots. Square traps gradually replaced round traps in Massachusetts and then showed up in the southern and western areas of Maine. The Bay State coast, like western Maine, is an exposed fishery ground with vast areas of sandy bottom. Strong tides and storms raise hell with lobster gear, and the square trap does not have the tendency to roll along the bottom in heavy weather that the round trap has. With increasing use of larger boats, the square trap stacked much better. Amazing numbers of square traps, up to 100, are carried on 32-foot lobsterboats when fishermen transfer traps at the beginning and closing of the season.

The standard wooden parlor trap is constructed of oak laths, bows, sills, and runners. Trap mills are specialty sawmills dealing only with trap stock. Not only do they saw the laths and other straight stock, but they also steam-bend the bows. Probably 90 percent of wooden-trap fishermen build their own after buying trap stock. There are, however, a number of increasingly "big" fishermen, full-time types, who fish up to 4,000 traps and have them factory made because they simply do not have the time to build them.

The wooden parlor trap is a relatively simple box divided into two sections—kitchen and parlor. There are three knit heads shaped like funnels. Two of the heads, one either side of the kitchen, allow the lobsters to enter the trap to get bait, which is positioned in the middle of the trap at the inner end of the kitchen. Once the lobster has eaten or ripped off a piece of bait, he tries to escape. The most convenient egress is up the apron of the third head, which opens into the parlor section. Of course, some of the smart-ass lobsters refuse to accept this cunning route and crawl, no room to swim, back out the ring openings of the side heads. But, considering that the trap design has been around and little changed for two centuries, something about the catchability must be right. Lobster trap design and effectiveness have been much studied by scientists, who generally conclude that a better mousetrap would be hard to build.

The heads, as mentioned, are knit of nylon twine from two-pound "balls" purchased at marine-supply stores and fishery cooperatives. It's fair to assume that almost all lobster-trap heads are hand-knit at home. The lobsterman knits in the evening by the black iron kitchen stove where he has conveniently screwed a hook into the wall to function as a holdfast for his knitting. The wife knits. Grampa and Grandma knit. And head knitting, along with making bait

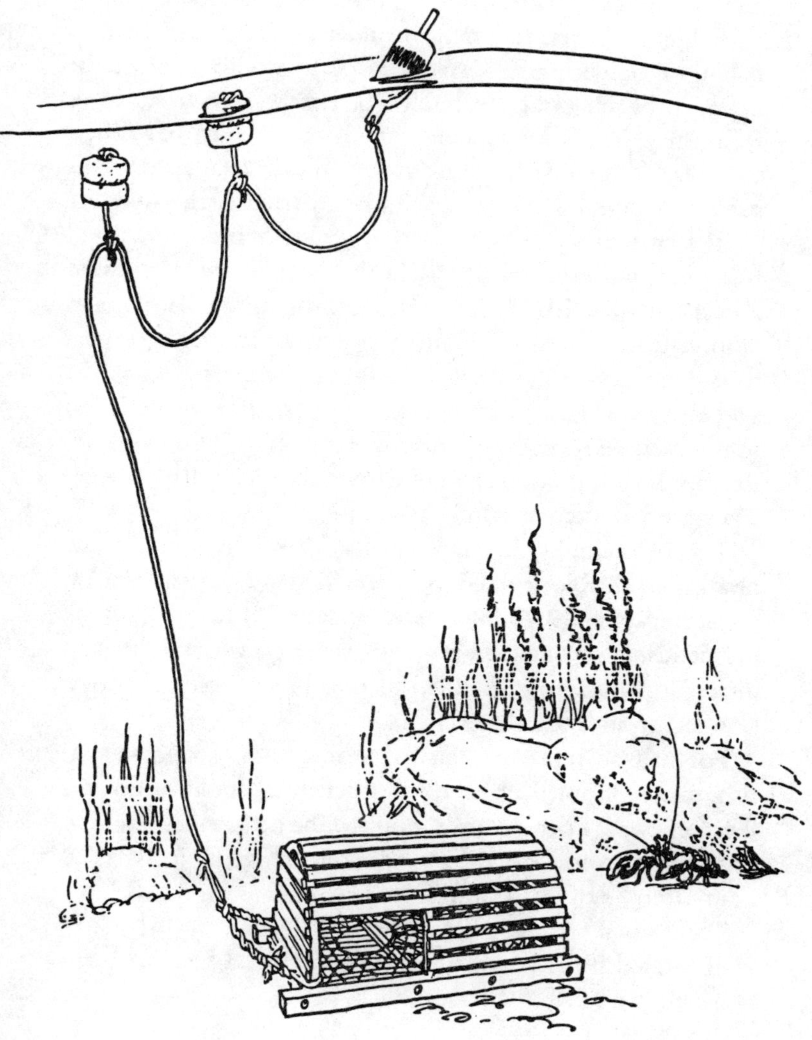

bags, is one of the first teachings of children in lobster families.

There are many variations on the standard wooden parlor trap. Size is one. The old faithful measured 36 inches long, 26 inches wide, and 18 inches high. Traps just naturally got bigger with bigger boats, hydraulic haulers, and offshore winter fishing. Double-headers have two parlors with two sets of heads to better hold the catch when traps aren't hauled as frequently. There are center-headed traps, which have their entrances in the middle with parlors on each end. Some traps have entrances on each end with one parlor in the middle.

It seems every fishing community also has quirks and traditions in trap building. The lobsterman-built traps of Kittery, Bailey Island, Vinalhaven, Stonington, Beals, and Winter Harbor all have slight variations. Even the big-timers have their traps built to order. And the old-timers will go over a strange trap like a cat around a bird's nest. They will tell you sometimes, and sometimes not, that this trap is a Dyer trap, that one a McInnis trap or a Lot Brown trap. They have names. They are patented as surely as Coca-Cola.

The old standby, however, has some competition in the new kid on the block—the wire trap. When wire first came on the scene about 1970, some lobstermen at Al Roberts's dock in Friendship spent an afternoon inspecting a wire trap that an intrepid manufacturer had left alongside a couple of pilings. (He left it and ran, that is.)

For half an hour the trap was ignored, although several fishermen continually passed it by. Finally a couple sat down a few feet away. Then a couple more on the other side. The trap was eventually bracketed but not acknowledged. Nearly a full hour later, the first fisherman spoke.

"What is it, Alfie?"

"Damned if'n I know, Junior. Looks like a bird cage but there ain't no water dish inside."

The farthest lobsterman put his whittling knife down, strolled over to the trap, peered down like an eagle on a rabbit, and said, "Boys, either this is a goddamn lobster trap or somebody's funning us."

Today, Alfie, Junior, and most of the Friendship lobstermen have given up their wooden ways and are fishing wire traps.

Wooden traps have their assets, such as availability of building stock, ease of construction with familiar tools, and relatively low cost. In 1982, a standard wooden trap could be built for under $15. But wooden traps have problems. They are heavy and cumbersome to handle, and a lobsterman moves his gear about as much as he does his woodpile. When they are first put overboard in the spring, they have to be ballasted with extra rocks and bricks to sink them, because the wood has dried out during the winter. Wire traps start fishing right away, and no extra ballast is needed. Wooden traps also have a predator—the wood borer. It's a rare trap that lasts three years. Sometimes, one year. The advantages of wire traps are many, and the number of these in use grows each year.

The line running from the trap to the surface is called pot warp, sometimes trap warp. When the word "warp" was first substituted for rope or line is a good question, but it is a short, terse, rather elegant word, perfectly suited to the lobsterman's vocabulary.

The warp is made fast to the trap by means of a small (about 4-foot) bridle half-hitched from corner sill to corner sill on the kitchen end of the trap. The bridle arrangement is used so that the warp can be easily removed from the trap when it is brought ashore for winter storage or when the warp needs replacement. The bridle also gives two contact points in case one end chafes away. Positioning the bridle on the kitchen end is an interesting practice. Presumably, because the entrance heads are on either side at the kitchen end, the trap is hauled from that end so that any lobsters will be thrown backward, away from the entrances. Also, when the trap comes aboard kitchen end first, it is swung forward onto the boat's washrail with the trap door facing inboard. But, as we should have learned by now, lobstermen are individuals, and some fish left-handed boats—that is, with steering and hauling gear on the port side. They are called queer fishermen by others. Well, anyway, queer fishermen have the doors made on the opposite side of the traps.

Both straight and queer fishermen occasionally tie their trap bridles on the parlor end. Ass backwards, they call it. As an explanation they usually say, "always have" or "like it that way." How about "none of your goddamn business." All responses are valid, but very likely these lobstermen feel that the pot warp moving around in the tide in front of the head entrances has a tendency to spook lobsters. They tie the spooker on the other end. Why not?

Buoys are the visible brands of the range. Cutting, stealing, or molesting a lobster buoy is worse than cattle rustling. Bullets have been fired in anger, and boats of known buoy rustlers have settled slowly to the bottom under a no-telling starry northern smile from the heavens.

The buoys of yesteryear were simply pieces of wood—usually pine or cedar, but anything would do. As time passed, fishermen became more fanciful and began to shape their buoys so that they torpedoed into the tide. The pieces of wood were shaped to individual likes and whims. Names or initials were carved into the buoys when some started to look alike. When the licensing system came into being, the individual fisherman's numbers were carved or branded into his buoys. Now it's the law, and each trap, crate, car, or lobster holding device must also be branded. Wooden lobster buoys are rare-ish today. Plastic prevails.

By law, buoys also have individual colors. With 9,000 lobstermen and half a dozen basic colors, there can be a few duplications. But again, individuality and originality prevail. Such combinations as hot pink and cool green, Jonesport yellow and North Haven red show up on application forms. They are dutifully recorded by state department clerks in the serious vein in which they were conceived.

Lobstermen select their colors in strange ways. Inquiries will reap a non-answer harvest. Yellow is the favorite color, and red is next, both because of logical visibility reasons. Old-timers usually have single colors. Why go to all that trouble of a combination paint job? As other fishermen came along they found all the good, visible, single colors taken and had to double up, sometimes triple up, on a paint scheme. So, it's somewhat safe to assume that a solid yellow buoy floating in a maze of confetti buoys belongs to a harbor mentor.

Eccentricity in hues is known. Yellow buoys with black polka dots obviously belong to a part-timer who has nothing better to do than paint buoys. Solid black is an interesting color, if you can call black a color. One skiff lobsterman fishing black said he rarely gets hauled after dark with his midnight shade. Ocean blue and sea green are beautiful colors but don't survive too well in traffic. One fledgling fisherman started out with pearly white buoys, and on his first haul-back the wind was blowing and the waves were cresting. He chased down 40 miles of whitecaps before finding his first trap. Those Day-glo colors are definitely in. They can be seen for miles. A little gaudy, but visible.

One veteran all-yellow fisherman in Penobscot Bay always advises newcomers to paint their buoys like his—all yellow. "So's I can keep an eye on 'em," he says. The "old yellars" don't get that way letting Irish moss grow on their thinking.

One of the strictest marine laws is that governing the molesting of lobster gear, which includes buoys. Fines up to $500 are applicable. But if there is one souvenir that each Maine tourist covets, it's a lobster buoy. Most of them are authentic, once-fished, genuine buoys. And legally, they are still the property of those individual fishermen. Perhaps a million pilfered Maine lobster buoys adorn recreation rooms, bar walls, and dens all over the world. Bringing these out-of-state thieves to justice would wipe out the Maine debt.

There are a couple of other small pieces of gear between the buoy and the lobster trap. They are called toggles. They, like the buoys, were originally small pieces of cork or wood, and their purpose was to keep the warp from fouling on the bottom. Two toggles are used, one about three fathoms from the trap and the other about the same distance from the buoy. Sometimes the lower toggle is replaced by floating polypropylene split-warp. With the introduction of motorboats into the industry, the top toggle became quite important. Propellers foul buoys and cut them off, but with the upper toggle still visible at low tide, the trap can be saved.

The cork or wood toggle evolved into bottle toggles. Before the days of returnables, the most popular toggle was the Narragansett pint beer bottle. Picwik Ale was a close second. These two brands carried more line than the United States Navy. Solid rubber stoppers sealed the bottles, and they were made onto the warps by a couple of half hitches. Sometimes short gangions were used.

As anyone who has purchased a live lobster can attest, the business ends of the creature have been neutralized. The claws are made inoperative by either plugging or banding. This practice involves another piece of lobster gear.

The "plugging" of lobsters is not for the benefit of the consumer, really. The lobsterman could care less if the buyer gets nipped. And this might be the time to digress a moment from plugs, bands, and the sort, to explain how to pick up a lobster. Orienting a lobster into automobile terminology, the creature should be picked up by the body, from the rear, with thumb and forefinger on either side close behind the right and left front tires (claws, of course). Disregard those numerous little walking legs with pincer ends, they merely tickle. It's important to hold the lobster firmly and close behind his claws because the reflex action of the lobster is to swim away. In the water, which he still thinks he is, this is accomplished by vigorously snapping his tail, which has a very strong muscle. This is a rather violent movement, and if the grasp is loose, the lobster will end up on the floor. Another reason for holding the lobster high on the body is that when the lobster flexes his tail shut, the sharp, pointy ends of the tail sections, called somites by experts, will pierce the fingers like needles if they are not out of the way.

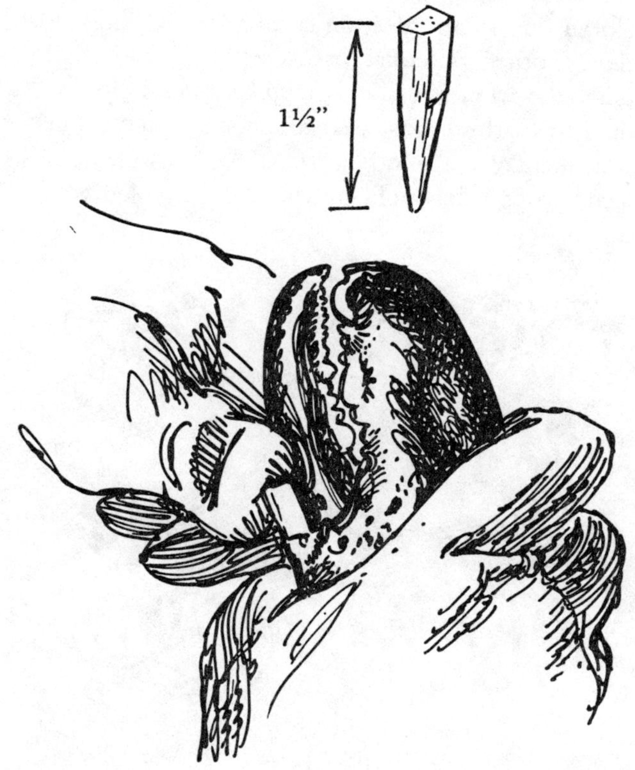

Plugging a claw

Bands and plugs are used by lobstermen to prevent lobsters from harming each other in the confinement of lobster pounds. In Maine, the plugging of lobsters began with string tying but quickly evolved to small wooden pegs hand-whittled from woods such as pine and cedar. Inserted in the claw joint, these keep the claw from opening. The industrial age gave Maine lobstermen a machine-made plug from basswood. Plastic, naturally, entered the scene in the 1950s, mimicking the wooden plugs, and for a period of several years these plugs were the accepted method of immobilizing lobsters.

In 1951, rubber bands were introduced. Wholesalers and retailers were getting consumer flak about claw meat that was discolored due to an infection caused by the plugs. Also, a couple of other potential problems with plugging were worrying the wholesalers and pound keepers. One was the bleeding to death of lobsters through the punctures in their external skeletons. The other was a blood disease caused by the same wound. It's unlikely that lobsters bled to death,

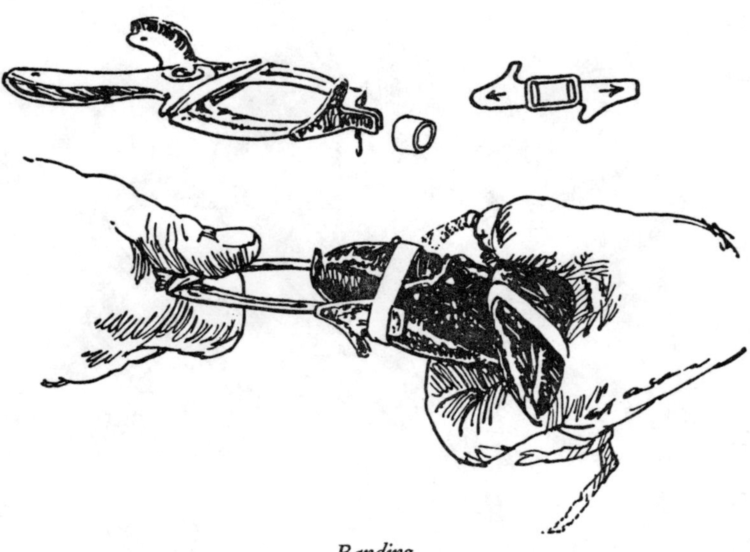

Banding

although the belief still exists today. They routinely "throw" (or discard) their claws when excited, stressed, or when fighting. The fresh wound is quickly covered with a crust of coagulated blood, which prevents further bleeding until a skin is formed. But blood disease is another matter. This infection from a bacterium classified as *Gaffkya homari* exists in the natural environment, and when lobsters are confined closely in holding pounds, the disease can spread rapidly and fatally, causing losses of thousands of pounds of lobsters. The disease, commonly called "redtail," is transmitted in several ways. Healthy lobsters feeding on dead, diseased lobsters is one way. Scientific data and common sense both indicate that redtail can more easily infect a lobster through any break in its exoskeleton. Plugging a lobster claw creates a most decided wound.

For a decade, Maine lobstermen avoided rubber bands like a split oar. Trying to slip slippery, stretchy, snappy rubber bands on lobster claws aboard a pitching, broaching sea platform in a nasty November breeze was just another burr of progress under the fisherman's saddle of lifestyle.

The original rubber bands had tabs, completely useless, as an aid to stretching the band over the claw. L-shaped hooks strung with bands helped, but hardly. The savior of the rubber band was the invention of rubber band pliers, the action of which is opposite that of regular pliers. A band is placed on the plier end tabs. Closing the handles then stretches the band, which is inserted over the claw quite neatly. A simple device. The pliér invention and some real hardball pressure from dealers has brought banding into general acceptance. Many dealers will not even accept lobsters with plugs. In most banding practices, only the larger crusher claw is plugged.

* * * *

Lobster bait is a science, no doubt about it. Fish heads and fish guts are not thrown haphazardly into traps and lobsters come running. The annals of lobstering are replete with the myth and lore of lobster bait.

Fish are the staple bait. But a fish is not a fish is a fish.

Herring. The bread and butter of lobster bait. More herring and parts thereof are used than all other fish combined. The reasons are availability, price, and good catching qualities. Herring and sardines are the same thing, although to be Maine proper, sardines are made from herring and never the other way around. The last sardine plants in the United States are in Maine. The procedure of making a sardine from a herring is quite exact and simple. Cut the head off and stick it in a can. Herring are individually hand-trimmed to can size, one at a time, by experienced ladies wielding sharp scissors. One herring head may be snipped right behind the ears and the next halfway down its belly. It's the fit that counts, not the fish.

Anyway, the headless fish travel their pathways to eventual saltine snacks and the heads are sloshed down a chute into an offal or bait truck parked outside the sardine factory and underneath endless conveyor belts. Lobstermen in sawed-off 1956 Buick sedan pickups fitted with rough-sawn wood platforms pull up to these bait trucks every day "fish are in." Coal-like chutes divert the herring heads, which at this point assume the name "cuttings," into wooden bait tubs, plastic garbage cans, 55-gallon oil drums, and every other kind of container, including the bare hermaphrodite pickup body itself. The price is interesting. The factory pays about $2.50 a bushel for the whole fish, snips off the heads, sells the sardines for $30 and then the heads to lobstermen for $3.50 a bushel. This has to be Yankee business ingenuity at its apex. The factory also sells the lobsterman a couple of 80-pound bags of salt for $4 each, so each 200-trap fisherman baiting up at the factory every week drops 50 bucks into the plant till for the privilege of hauling away its offal and gurry.

The off-islanders get their fish from bait boats. These are old sardine carriers usually run by owners and one-man crews. They purchase fish from herring stop seiners, who fish at night and tie up by early morning in lobster harbors. Fishermen come alongside and take on whatever they need for that day's lobstering. It's a fine arrangement because a fisherman doesn't have to salt, store, and transport bait from home to harbor to boat. Bait boats like Joe Upton's *Amaretto* and Pete Collins' *Maine* have serviced large island communities such as Vinalhaven and Stonington in the summer months. The islanders also store bait for rainy days or when the herring aren't running. Fishery cooperatives buy cuttings and redfish racks or brim in hundreds-of-bushel lots for use by members.

The cuttings not carried away from the sardine plant are trucked to commercial bait dealers, and lastly, to reduction plants that convert the offal to fish meal, fish oil, and by-products. The large plant in Rockland draws its share of public

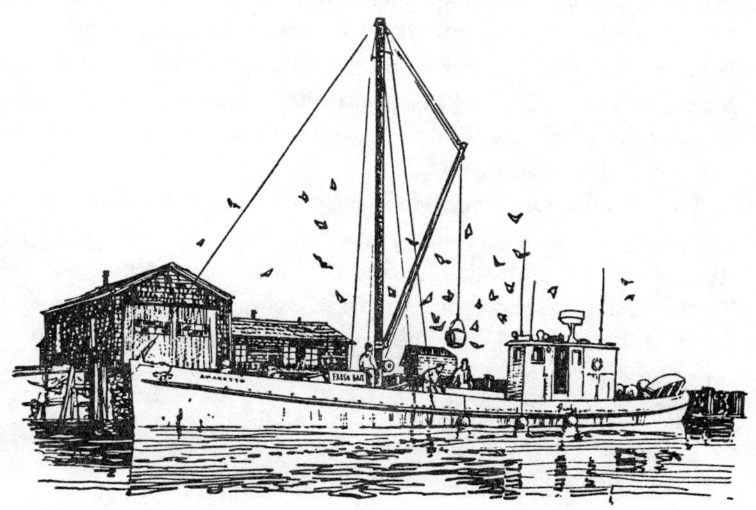

comment on hot, muggy days in summer when all that is left of the princely herring, the odor, settles on residential surroundings like a blanket of Liederkranz cheese. A similar plant in South Portland was closed by court order in 1984.

Redfish. Not to be outdone by the herring canners, and with considerable one-upmanship, the redfish plants run their rosy ocean perch through mechanical filleting (a true fisherman pronounces fillet as "fill-it," not "fill-laý," for heaven's sake) machines, leaving only the head and backbone, which they sell to lobstermen for $7.50 a bushel. When the remains hit the bait truck they are called racks or "brim." The precise meaning of the latter term, taken from a fat dictionary, is "the upper edge of anything hollow." It's doubtful that a more precise description could be found for a piece of fish with an intact head and long-gone loins.

The redfish bait industry is a large one. Dealers usually contract with the companies fishing redfish and groundfish, and individual fishermen find it hard to buy direct from those primary catchers. The bait dealer wholesales to co-ops and lobster buyers, who supply those lobstermen selling to them. Brim is also stored in frozen blocks, which negates the need for expensive salting at the wholesale level. Redfish stored for any length of time must be preserved with salt or frozen or it will rot most magnificently.

The Vinalhaven Fisherman's Cooperative, a heady band of independent fishermen, once had a new manager who spared the salt and ruined his life. A bait shed is no rose garden, and Tommy Hildings had seen bait come and go with the best of noses. But one day Tommy was sitting in the office munching on a Hershey bar, without nuts, when his ears heard a strange thing.

"What's that chirping noise?" asks Tom.

"Ain't nothing I hear," answers brother Dicky.

"Jesus Christ, I know a strange sound when I hear it, Dicky."

"Probably the wind or somebody's fartin' down on the dock."

Tom got up and went into the bait shed, which had about the same decor as the office, except for the telephone. Then he saw it. The noise. The top layer of redfish was rising up and down like a bellows. Tom hauled back a layer of brim and then jumped like he'd been stabbed. The bait was a solid mass of black, munching, clicking maggots. A couple of co-op meetings later a secret landsite was selected, a convoy of pickups was loosely collected into a passive posse, and maggots et al were given a midnight pauper's burial.

Fishing brim has its advantages. First of all, the lobsterman doesn't have to fill "those goddamn bait bags." Brim is "strung on" with a baiting iron. The redfish racks last longer than bait bags, and brim fishermen will tell all to a man that lobsters crawl to bones. But redfish is not an oily fish like

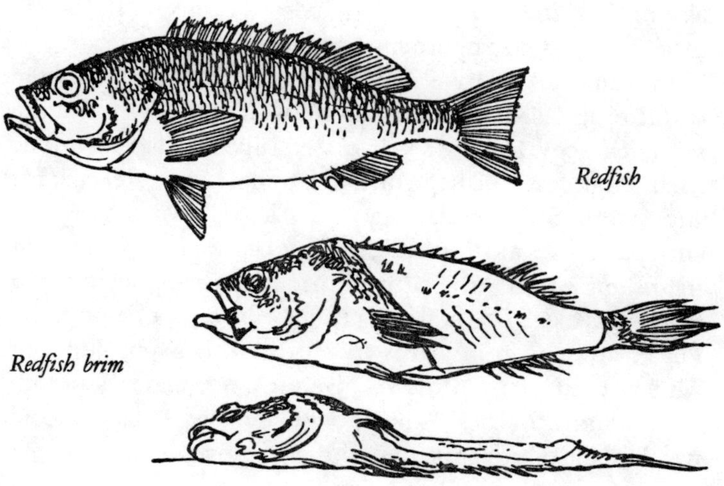

Redfish

Redfish brim

herring and menhaden, so the popularity of the bait is more in the ease of handling than the chemical attraction of the carcass.

Menhaden. This is the pogy, the mossbunker, the fat back. In Maine everybody calls them pogies. This pelagic fish feeds chiefly on microscopic plankton, primarily diatoms, and migrates to Maine waters in the summer months. Surface schools acres wide infiltrate shallow-water bays and travel far up the estuaries and rivers. In recent years they have been very abundant in July, August, and September.

Pogy boats from as far away as Virginia come to the Maine coast in summer. The pogies are caught by large purse seines and pumped aboard vessels that transport them to reduction plants in Rockland, Maine, and Gloucester, Massachusetts. And talk about an oily fish! Almost 20 gallons of oil can be pressed from 1,000 pogies. The menhaden was once one of the most commercially important fish swimming on the Atlantic Coast of the United States. They were used not only for oil but also for fertilizer and increasingly, post-World War II, for fish meal, which is an important ingredient in animal feed.

In Maine, pogies have always been used for lobster bait, the oil factor, in this case, being viewed as a decided asset. But the saturated pogy, like the sophisticated lady, may have outdone itself. Pogies are one huge dilemma to hold in a bait stage for any length of time. If they are salted down whole, the entrapped gases bloat the fish, and one of the eight wonders of the fishing world is the odor from a punctured pogy that is a few weeks old. "T'would gag a maggot" is an apt description. Pogies are cut up in pieces to prevent this happening, but whole or cut, the profuse oil gets on, over, under, and into everything it touches. When pogies are used as bait in cold weather, the oil congeals and hands and gloves look as though they have been dipped in a Crisco can.

Alewives. The alewife is a fish similar to the herring, but it has a greater depth of body. Other names heaped upon this fish are sawbelly and kyack. The sawbelly moniker comes naturally, for the fish has a decidedly sharp midline down its belly, and the way to handle alewives is carefully. The label of "kyack" is one of those fisherman metaphors that is hard to

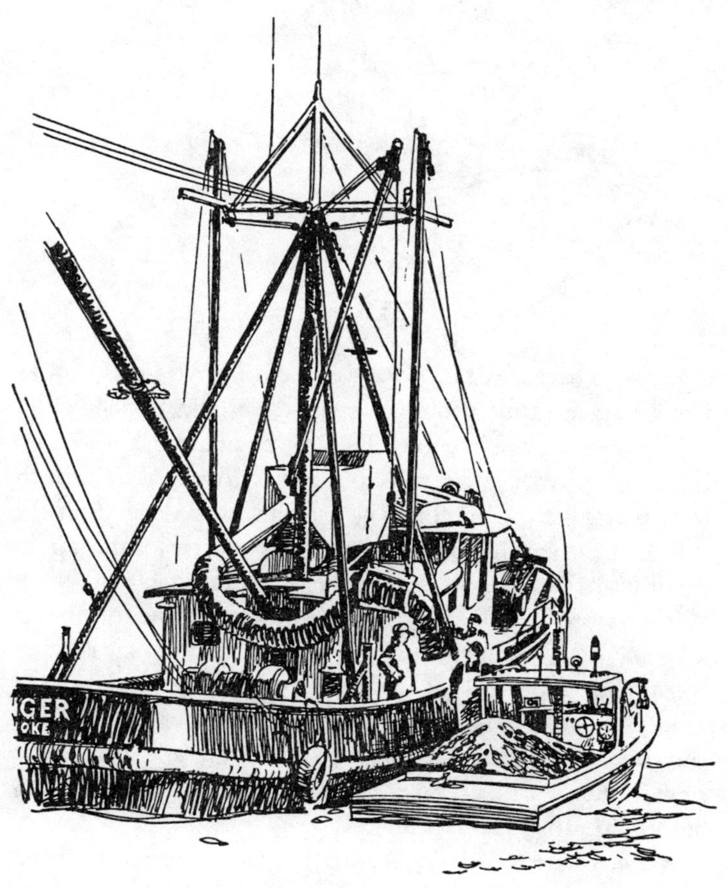

Bait from a pogy boat

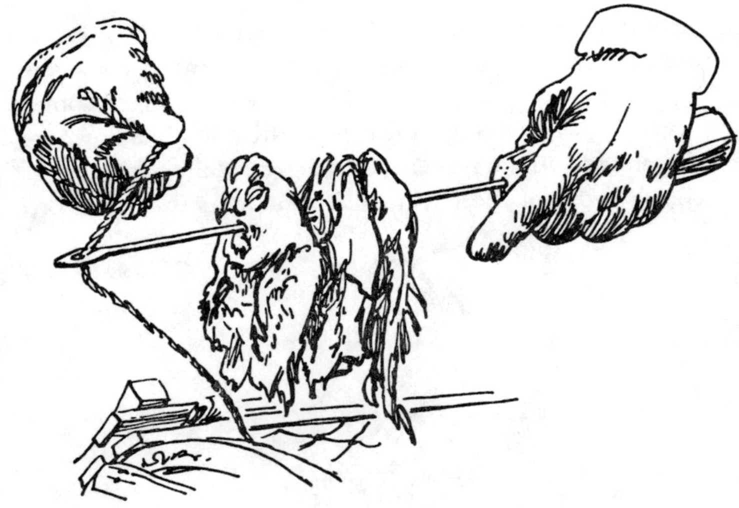

The baiting iron

argue with because of its primitive accuracy. An alewife looks like a kayack (same pronunciation, less efficient spelling).

Alewives are anadromous. They are born in fresh water, go to sea for growth, and return to the exact rivers of their birth to spawn. Like the salmon. The arrival of the alewives in the spring of the year has been the salvation of early lobstering. In April, the first schools of herring are lurking far offshore. Winter-stored bait is sparse. The pickup truck line for brim is long and discouraging. Scurrying for any kind of bait is time-consuming.

Lobstermen in pickups drive long distances to get alewives, and there is no assurance that they will come home from the alewife trap with bait. Sometimes it takes a week of driving before "the fish ran last night."

Alewives are salted like herring but are also used fresh and strung on whole. When cut up, kyacks are stuffed in bait bags a la herring cuttings. The lobster jury verdict on alewives is unanimous—finestkind. Price per bushel to lobstermen in the spring of 1982 was an average of $5, from the traps.

Artificial bait. Fake bait, it's called, and it comes off the wall at times. There are more weird fake baits than places in Maine you can't get to from here. The most serious of the fake baits is a meal, or paste, manufactured from fish waste. It is essentially a fish meal. To this are added ingredients that practically eliminate odor and stabilizers that give the bait an almost unlimited shelf life. The paste does not freeze in the winter nor rot in the summer. The advantages are many. None of the bait-fetching rituals are needed—no trips to the sardine factory or salting and storing headaches. The meal comes in convenient 5-gallon plastic tubs that can be easily stored aboard a lobsterboat.

Because it is a mealy paste, the method of using it involves some kind of a container other than bait bags. These are half-pint round plastic cups with screw-on covers. Molded into the bottom of the bait cup is a small tab with a hole in it, so that the cup can be strung on a bait string. The cup lids come perforated with several holes to allow the release of bait scent. The frugal fisherman, never one to buy new when an adaptation will do, borrows the family ice-cream scoop, which is the ideal tool for filling bait cups. The consistency of the paste is remarkably on the order of chocolate whorl.

There may be a social stigma attached to the use of artificial bait. Something about going lobstering with a 5-gallon pail of mush, a bushel of sundae cups, and an ice-cream scoop. One hundred tubs of complimentary artificial bait sat gathering weeks of mildew and backward glances at the Vinalhaven cooperative before retreating to the mainland aboard the ferryboat with its perplexed manufacturer.

Still, a few lobstermen use the stuff. They are mostly skiff fishermen, part-timers. The 5-gallon tub gives 150 baitings, and the cost is in the neighborhood of $30.

If lobster bait is a science, then salting it is an art. The best definition of a finestkind lobster bait is, "It'll draw a crow but drive a skunk." The betweens is the catch. There are lobstermen who use salt sparingly, and they and their families are preceded by their custom wherever they wander. Lobster bait permeates like penetrating oil. A fisherman who has filled bait bags for 50 years has veins of half blood and half fish gurry. The gurry ratio increases in direct proportion to the amount of salt he uses to put down his bait.

Bait used within three to four days can usually be treated lightly, but anything stored longer than that needs some sodium-chloride attention. The best bait put-downers salt liberally (say, 80 pounds to five bushel), and after a few days, when the bait turns to pickle (leakproof containers must be used), the bait is re-salted. This procedure produces a sweet lobster bait that will last for years.

Lobstermen who don't salt heavily claim that lobsters like fresh rather than salty bait. Others claim the opposite. A majority will admit that crabs like fresh bait better, a plus for those keeping and selling crabs and a minus for lobstermen who find crabs a pest. It is also claimed that lobsters will not so readily enter a trap full of scampering crabs.

* * * *

A lobsterman "ain't particular" about his wearing apparel. A wool shirt bought is a wool shirt worn through thick to thin. He will wear what is nearest when he rises in the morning. He cares not a whit about style, only that a piece of clothing is held together in some semblance of order and that specific garb for a selected site is usually worn there. A hat, for example, is usually worn on the head. Lobstermen are not eccentric in dress, they just don't give a damn.

Lobstermen rarely wore anything on their hands until recent years. Inshore summer fishing wasn't particularly cold work, and mittens and gloves were cumbersome. Also, either a man had "fisherman's hands" or he did not. If a man's hands got cold in the summer working around the water, he might just as well go into milking cows for a living. For years the traditional cold-weather hand covering was the fisherman mitten. These were home-knit using raw wool, real raw, with all the oils and even burdock bits and pieces. They were knit thick and oversize and prewashed in hot water to shrink them. The mittens were knit long for a purpose. They were obviously wet all the time (fishermen wet them before putting them on, in fact), and the water accumulated in the tip of the mitten. The fishermen then simply slapped the tip of the mitten on the boat rail or some other hard surface, and the water was knocked from the end. The wet wool would stay warm from the friction of the hand inside. Fishermen still wear them, but most all are home-knit by relatives.

The newer breed of lobstermen wear cotton gloves summer and winter. A box of "dunkers," which are blue cotton with an elastic gathering band across the back, will usually last a fisherman a season. Twelve pair, a pair every couple of weeks. More gloves are lost overboard than wear out.

Jackets, sweaters, and shirts are cotton flannel and wool. The lobsterman dresses in layers. The mornings are cool, and as the day progresses, a layer is shed according to the whims of weather. Dark green wool pants, like the old Johnson brand, are favorites. All lightweight cotton pants are called khakis, no matter what color they are. They are never called chinos. Matching cotton shirts are favored. Khaki pants and shirts in green are the most popular.

The lobsterman's dress sock is white and cotton. If the country ever awakened to its senses and elected a lobsterman president, he would be inaugurated in black tie, tails, and white socks. A lobsterman can be singled out in any crowd, no matter his upper sartorial splendor, by his white socks. A

Rockland haberdasher who sold white socks to island lobstermen for decades said the reason related to him was that colored socks "poisoned the feet." In any case, fishermen wear white socks even under wool socks in the cold months. Raw wool homemade socks are the best, and with white liners, they last for years.

Fishermen's boots have come a long way. They were once all-leather and kept half waterproof and half flexible with lard, turpentine, linseed oil, and chicken fat "tried out" on the back of the kitchen wood stove. The legendary stories of lost feet from wet and frozen leather hi-cuts are frightful. Vulcanized rubber came to the rescue of suffering feet quite recently. In 1929, only nine percent of boots *and shoes* were made of rubber, but today, rubber boots are the first piece of clothing a fisherman purchases. Wet feet, unlike wet hands, are not tolerated.

There are several styles of rubber boots, and they repeat themselves like the cyclic nature of the sea. The first popular boots were knee length. This style faded in favor of three-quarter length or "over-the-knee" boots. Nowadays, the knee style is back. The three-quarter boot is favored by fishermen who hate to wear oilskin pants, since the added length gives some protection against wet thighs and legs. The shorter boots are lighter. In the three-quarter boot, there is a cinch strap at the top that is cut off immediately after purchase. No upstanding fisherman would be caught dead with boots sporting functioning cinch straps. After removal, sometimes a marline loop is tied into the ex-cinch hole to hang the boots.

The rubber boots don't necessarily come off when the lobsterman comes ashore. He wears his boots to the beer store, co-op meetings, movies, and to town when he needs to go to town. He paddles around his house in stocking feet much like the Japanese fisherman. The dress shoe, when the wife absolutely, positively puts her foot down against wearing rubber boots to weddings and funerals, is the Romeo slipper.

This is an elf-like half-slipper type of footwear with no laces, stretchy side pieces, leather sole, and rubber heel. It's the perfect shoe that is not a shoe. When a lobsterman is forced into conventional shoes because he doesn't want to spring for a new pair of Romeo's, he will reluctantly wear a pair of his discarded wicked-lace shoes. However, once the laces are tied, usually in a square knot, the fisherman never unties them. He just wiggles in over a crunched-down heel until there is no resistance left in the leather, and out of shape and out maneuvered, the shoe surrenders to the inevitable power of the white-socked foot.

Technology and plastics hung the oilskins on an attic peg with hoop skirts, and the emerging raingear must be considered a welcome improvement. Raingear or "foul weather gear" is a misnomer, because a fisherman uses his raingear as working gear. Sunday supplement pictures of lobstermen in full battle dress on sunny days are accurate . . . . Well, the sou'wester is a bit much. A lobsterman wears rain pants, or sometimes just an apron, to stay dry from the water coming up, not down. Aprons are worn in the summer because they are cooler and provide the same amount of protection. And there are those lobstermen who just *hate* pants.

Gear such as life rafts, survival suits, and even an anchor, for heaven's sake, are not mandated by law, and the fisherman on the smallish lobsterboat operated nearshore rarely has such equipment aboard.

There are several reasons. One is the old story that a lobsterman can't swim and if he fell overboard he'd drown anyway. True, most of the old-timers could not swim. Taking dips in year-round 40-degree water was not a pastime of lobstermen. Even the sometimes Saturday night bath in tepid water was a chore. But the majority of contemporary fishermen can swim. They are not accomplished Buster Crabbes but can keep their heads above water for respectable lengths of time doing the Friendship crawl or the Beals Island dog paddle. Fishermen have an uncanny ability to hang onto things, which relieves them of learning to swim. Lobstermen under the influence of sinful spirits have been falling off wharves and docks for hundreds of years. Although positively incapable of fundamental decisions on land, once in the water some rational and spiritual sixth sense kicks into gear and they grasp and cling to anything solid or floating. Lobster lore is pregnant with tales of fishermen clinging to wharf spiling for days, unable to speak but hugging those slippery poles through tide and time. Fishermen falling overboard are often found floating accompanied by the smallest chip of wood. Lobstermen have divine faith that when they do go overboard, something will be there to grasp. It's simply a calculated risk with the odds on their side. So why learn to swim? The new breed swims because of outside social contacts with the schools and "Y" and not because they believe mastering a few arm strokes will save their skins. Of course, lobstermen are born fatalists, and learning to swim only serves up a devil's advocate to their doctrine and who needs another quandary?

The lobsterman learned from his father, and his father before him, that once in 40-degree water, a person has precious few minutes to disassociate himself from it. In the winter, falling overboard can even shock a fisherman to death. Water is a wonderful conductor of body heat and there is little use in trying to prolong the inevitable, the lobsterman's logic goes.

There are tragic stories of lost and drowned fishermen. If a lobsterboat is found with engine running but nobody aboard, fellow fishermen search the immediate area very quickly. The finding of the living is paramount. But, as time passes, there is the inevitable hauling back of the fisherman's string of traps to look for a body fouled in twisted pot warp. A life taken by the sea reinforces the fatalism of those who trespass upon it.

There is no question that safety equipment would save lives, fisherman logic aside. The documented evidence cannot be refuted that chances of living through an overboard ordeal are much higher wearing survival suits and life preservers. But the arguments for lifesaving gear must take into account working modes of fishermen. It is next to impossible, for instance, to tend traps wearing an approved life preserver. And although life preservers must legally be aboard a boat, they are thrown forward to the fo'c's'le, which is the very worst place they could be stored. An engine fire cuts off access, and water coming aboard a boat heads almost always like a torrent into the forward below-deck areas.

Pit Ginnis was a lobsterman who fished around Penobscot Bay in the 1940s. One day his motor up and quit. The old Chevy had seen better days. The points had carbon warts on them bigger than a bullfrog. You could throw a fixed tomcat through the gap in the spark plugs. Pit was never one for preventative maintenance.

Anyway, on this day Pit swore and yelled and fumed and ground his battery nearly to death. The Chevy wouldn't even cough. Pit eased a little gas into the carburetor. A little more. A sputter. More gas, and more . . . .

Then the whole boat exploded into a mass of wood splinters, engine block, wholesome lobster bait—and Pit. Luckily, it was summertime, the water was tolerable, and several other lobstermen were in the vicinity of the disaster. Easing through the debris, another fisherman saw Pit about midmess hanging onto his half-submerged bait barrel.

"Christ a'mighty, Pit!" yelled the rescuer, "what the hell happened to your boat?"

"Sonofabitch blew up. Backfired, the bastard," sputtered Pit.

He was hauled aboard, bait barrel and all, because Pit wasn't about to let go of anything solid unless he could swap it for land.

"Pit," said the savior, "if'n you hadn't found that barrel, you'd drowned f'sure. Ain't you got any life jackets aboard?"

"Naw," Pit said, sitting high on his bait barrel and dumping seawater out of his high-top sneakers, "them things are useless as tits on a clam."

3

The Boat

Maine lobsterboats were traditionally built of oak frames and timbers and planked with cedar. Wood, wood, wood. And then in the early 1960s an upstart, a new kid on the block, came to town, and the repercussions are still vibrating throughout the length and breadth of coastal Maine.

One of the first fiberglass boats came from Cy Cousins's Webber's Cove Boat Yard in South Blue Hill. The yard had built a few of the 40-footers for the U.S. Navy for use as personnel carrier boats. Cy decided that Maine lobstermen could use a few.

He took a bare hull to an Ellsworth shopping center and set the hull on a cradle for display. In due time, the lobstermen passing by would stop, take a turn around the hull, and get back in their pickups and drive off. One day, a contingent of lobstermen arrived in a beat-up old Buick. Four creaky doors opened and out stepped six rubber-booted fishermen, each holding a warm pint of Ballantine Ale. It was obviously a planned from-afar inspection party with all the credentials needed to tell the folks back home just what the hell this plastic boat thing was all about.

For at least 10 minutes the fishermen patted, climbed, stroked, kicked, and eyed Cy Cousins's alien from outer space. They huddled amidships. Then the first and only words were spoken. Pointing to the scribed waterline, the inspection team leader said, "Well, boys, that's where they put that sonofabitch together and that's where she'll come apart." They en-Buicked and sped away toward parts downeast.

Despite this unpromising beginning, fiberglass lobsterboats have become the norm on the Maine coast in the last few years. The wood-fiberglass argument can be heard today in about every harbor and fish house. The once thriving wooden-boat shops on Beals Island are still there, turning out magnificent craft made of wood, shaped and formed by experienced boatbuilders who learned their trade from fathers and grandfathers. But they are fewer. And fiberglass-boat shops are sprouting like lawn mushrooms after a week's rain.

Whatever the construction of a lobsterboat, the gearing out of the craft has progressed pretty much to a standard list of mechanics and electronics.

Nothing since the internal combustion engine has altered lobsterfishing as much as the hauler—first, the mechanical, and then, hydraulic. Hand hauling of traps is still a common occurrence in Maine, especially with skiff and part-time fishermen, but when strings of gear reach the hundreds, other means of lifting are required. Power takeoffs are used. In early versions, pulleys rigged to the engine shaft and then to V-belts driving other pulleys or gear linkages powered a winch known as a niggerhead. Two or three turns of pot warp were taken on the turning head, and the leverage ratio multiplied many fold, greatly easing the aching backs of lobstermen. These early haulers usually did not have throw-out clutches and ran constantly when the engine was turning. This caused some nasty accidents. Loose clothing got wound on the heads, pulling fishermen into the whirling dervish like moths to a flame. The most common malady, the "riding turn," reached epidemic proportions when niggerheads first came into the fishery. A fisherman's hand would be caught between the winch head and the warp, drawing the arm into wrenching distortions and usually breaking a few bones before the engine stalled or the turns slipped. The riding turn was especially prevalent during the winter, when fishermen wore gloves or mittens. Many old-timers have fingers and hands permanently formed like the Coriolis effect. Another problem was the exposed linkage of the hauler, with shafts and pulleys and gears strung along under the rail of the boat. One lobsterman likes to tell of the time one of his pant legs got caught in the hauler shaft and ripped his pants, shirt, and jacket right off his body. "I was left ballicky bare-assed," he says, "and that's a sight."

Amazingly, there are still lobsterboats with pants-stealing and riding-turn mechanical haulers operating on the coast. But in the 1950s, the hydraulic sheave hauler was born. This device literally revolutionized lobsterfishing. It is quite simple. The boat's engine shaft drives a small hydraulic motor. Flexible hoses connected to the motor, with a small oil-expansion tank and speed-control valve, drive a hauler. This hauler is essentially two disks bolted together with a small V-space between them. The pot warp is placed between the two disks, which the fishermen call "shivs," and as the hauler starts spinning, via the position of the hand control, the warp is drawn tight between the two disks. The action retrieves the line at any speed desired by the operator. Hydraulic fluid is compressed oil, and it offers amazing power for the space it occupies. One of the very first hydraulic haulers in Maine was the Hydro-Slave, and thousands have been sold. The haulers are relatively simple to install and maintain, with lobstermen doing both.

The lobsterboat engine can be one of nearly a hundred makes or models. The early boats were powered with 6-cylinder Chevy engines. The 4-cylinder Jeep engine was popular right after World War II. But like the boy with a toy, fishermen began trading and installing bigger and bigger "mills." Buick and Oldsmobile engines became the rage in the 1960s. The 325 h.p. Buick gas engine with 2:1 Warner "Velvet" drive was considered the golden dream of a boy

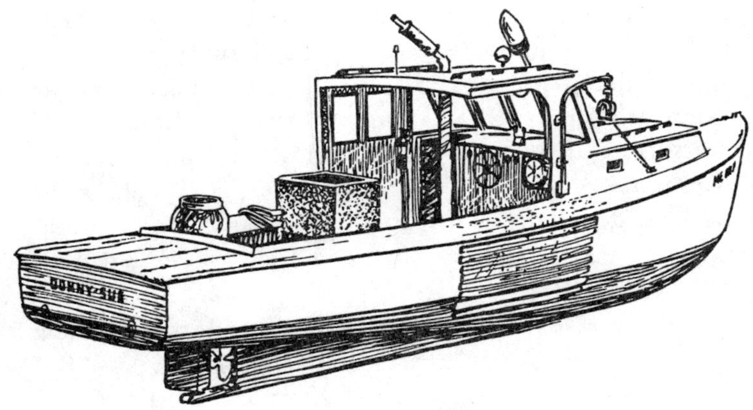

Today's lobsterboat

fisherman. He had to someday race his buddy Billy, who wanted a 403 Oldsmobile. Big gas engines were definitely "in" for many years. But the fishermen who didn't care to race began buying diesels, and today probably three-quarters of the lobsterboats are diesel powered.

A lobsterman fishes his ground by familiarity. He knows the ledges and rock piles and bottom configurations as he knows the path home. But few boats are without electronics in some measure, including radar (sometimes) and loran (increasingly). Depth sounders are in common use. The lobsterman will use his sounder most frequently when fishing in shoal water during the shedding season and when "setting off" in the deeper water during the fall, when lobsters begin to move out of the bays and inshore depths. The fisherman will set strings in different depth contours to determine just where the lobsters are traveling. At these times, his sounder is extremely helpful.

Lobstermen use two kinds of radios, the ever-faithful and chatty citizens band and the Coast Guard–monitored VHF "marine service" radio. The FCC regulates both, but in the case of CB, the government has pretty much thrown in the towel.

With inexpensive radio communications, the lobsterman (and his wife) embrace the means of breaking the day's monotony. And, well, the radios just might be handy in case of emergency, shore or waterside. The FCC found out early in the big-brother monitoring business that provincial language patterns were not about to be altered no matter the damn rules. But for a while, mobile FCC vans zeroed in on racy highway CBers and closed them down via license revocation. That is, if the operator had a license. It was a little harder to nail boats on the water. Today, the CBers and the FCC are adversaries only in the sense that the FCC pays very little attention to CB activity. A constant XXX-rated station will probably get a call-warning from the FCC, but little else. The Coast Guard adamantly refuses to monitor the so-called CB emergency standby station, Channel 9, and what regulations come in the box are discarded with the wrapping paper. CB is the old phone party line, the backyard tattler, the smoke signal, the two-cans-on-a-string method of water communication along the coast of Maine. Lobstermen do not use call letters. They are "Dirty Water Harry" and "The Preacher" and "Dockside Dink" and "Snot Sleeve" and "Wherefuckarewe." Each fishing community will have a different standby channel, and wives will leave their sets tuned to that station all day long. Their lobstermen husbands are a lot more chatty than their stoic, reticent Maine reputations lead one to believe. It's no secret that thousands of CB radios are purchased in Maine by people who will never push the talk button. It's pure, unforgettable downeast entertainment.

Lobstermen have no use whatever for a chart. Charts are for boatmen who don't know where they are going, a condition not compatible with lobsterfishing. If a lobsterman doesn't need a chart, why would he need such things as parallel rules, plotters, protractors, and dividers? He doesn't. Now, a good leadline is another basket of fish.

Maine lobstermen have an individualistic custom in naming their boats. More often than not they are named for sons, even though the fishermen follow the normal sea tradition of calling boats a "her" or "she." A lobsterman might refer to his boat as "she's a good old girl" when the name on the bow is *Donny M.* The Donnies and Bobbies and Billies outnumber the Susies and Betties and Barbaras. There are compromises such as *Donny-Sue* and even acronyms concocted from the first

The pilothouse

names of the family—the Billy, Elizabeth, Sam, and Steve becomes the *BESS*. Lobstermen also take considered pains in having the boat-name paint job a first-class application. Usually there is a sign painter or talented artist in a community who does the names on all boats. Block letters, shaded letters, Old English, and other styles surrounded in filigree make for handsome hailing ports.

The Maine lobsterboat is the fisherman's working world, and he will maintain that structure oftentimes with more diligence than he devotes to his domicile. After all, his house has a co-captain, and his boat does not tolerate that shared responsibility. The model development of the lobsterboat came through compatibility with the seas upon which the lobsterman works. But design had to accommodate the whims and idiosyncrasies of that independent fisherman. The lobsterman mixes mores and structured ideologies like no other artisan. And so perhaps the evolution of the Maine lobsterboat came about through a handful of pragmatism, a pinch of eccentricity, a dose of science, a bit of wit, and a headful of respect for the sea, which is after all the final reviewing stand.

Whatever, the Maine lobsterboat is an elegant lady.

4

The Lobster

In small pockets along the Maine coast lobsters are traditionally called "spiders." And that they are, more or less.

Animals without a backbone and having external skeletons are classified as arthropods—"arthro" meaning jointed and "pod" meaning leg. Included among the arthropods are crustaceans, insects, and arachnids, which are the spiders. Crustaceans include decapods, or 10-legged critters, which are lobsters, crayfish, crabs, and shrimp. (There are others that don't taste very good and won't be mentioned.) The scientists who have zeroed in on the American lobster have labeled the genus *Homarus* and the species *americanus*. These same gentlemen and ladies claim the "Maine" lobster is the only true lobster, and those who have taste-tested southern spiny lobster and Maine lobster side by side surely would have to agree.

The "Maine" lobster ranges along the Atlantic seaboard from Labrador to North Carolina and is found in waters a foot deep nearshore to over 1,000 feet deep off the edge of the continental shelf. The offshore lobster fishery in the canyons of southern Georges Bank, Stellwagon Bank, and Tillies Bank operates at great depths for the same lobster that might be found under the seaweed at low tide on Vinalhaven.

The lobster has 21 segments, 14 of them in the head and chest area (cephalothorax) and seven around the stomach or

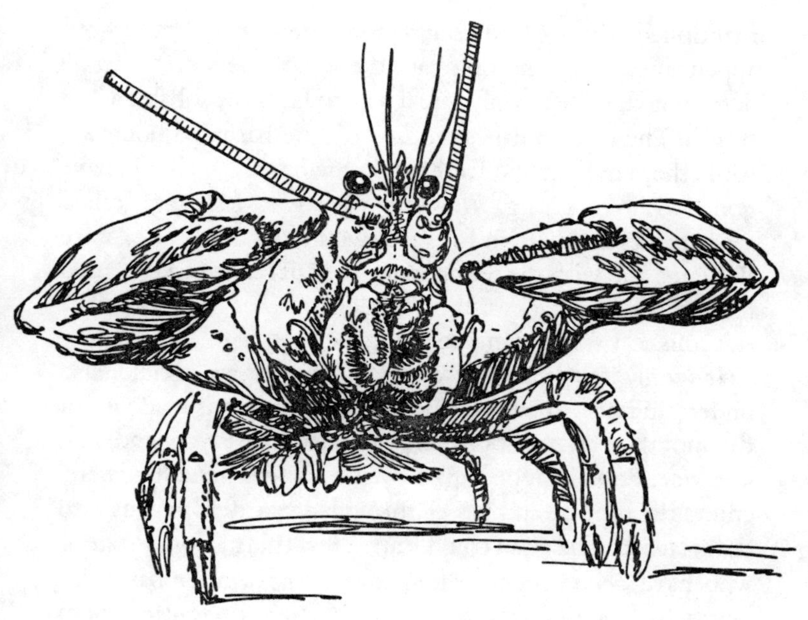

abdomen. The top and sides of the upper body are covered by the carapace, and this is what is used for the measurement of the lobster for legal and scientific purposes. The sharp, bony spur jutting forward from between the eyes is a rostrum. Shrimp have a similar appendage. The eyes are stalked, movable, and compound. The lobster has two sets of antennae, short and long, which are used as sensory organs. The mouth is a mess to look at but very efficient. One would assume that a lobster grabs a hunk of food in a manner reminiscent of King Henry VIII and stuffs it in his mouth with his big fat fist. Nope, the lobster eats daintily with mandibles and things called maxillae, which pass food to the jaws for crushing.

The lobster has claws, of course. Those two big ones out front are called chelipeds. Behind those are four pairs of walking legs; the first two pairs are hinged. The abdomen, we mentioned, has seven segments, and each has paired appendages. From the first pair, the sex of the lobster can be determined. It is the only visual way to label the lobster a boy or girl. These appendages on the male are hard and bonelike, while they are light and feathery on the female. It should have gone without saying. Anyway, the next four are called swimmerets and help in forward swimming. The last pair are flattened and with the telson make the tail fan. The fan is used for swimming backward. The lobster has a four-speed transmission stuck right under its belly.

It's only fair that if we eat a lobster, we should also understand how they eat. The digestive system is made up of the mouth, a short esophagus, stomach, digestive gland, and intestine. Food is ripped apart by all that front-end hardware, enters the mouth, and passes through a short esophagus into the stomach. The cows could learn a few things from lobsters, who have been around a lot longer. The lobster has three stomachs. The first stomach is sort of a how-do waiting room from which the food passes to number two, which is a gastric

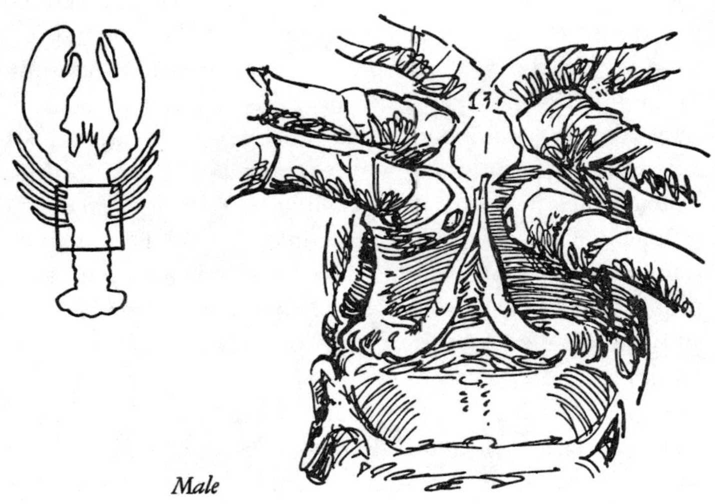

Male

mill where the food is ground very fine and then passed to the third, or posterior stomach. Here, the juices from the digestive gland (tomalley, and now nobody will eat tomalley) start digestion, and the food in solution is distributed throughout the body. Waste material is extracted from the blood by the "green gland" and excreted through openings at the base of each second antenna. Other waste continues through the digestive tract and is excreted through the anus, which is in the tail.

The circulatory system of the lobster is an open one. That is, the blood mills around a lot, rather than traveling narrowly defined routes through closed veins and arteries. The blood cavities, called sinuses, surround the important organs. In these spaces the blood sloshes like the wash water in a Bendix, bathing and feeding the organs. The heart has one chamber and is located forward of the thorax. The blood of mammals and lobsters differs. Oxygenated mammal blood is bright red and the nonoxygenated is dull red. But in the lobster, the oxygenated blood is bluish and the non-oxygenated blood is colorless. Sawing a lobster in half will not create a gory red mess all over the floor.

Pit Ginnis was giving a group of tourists a biology lesson on lobsters and was asked if lobsters had blood.

"Not a goddamn drop," Pit stammered. "They pump salt water through a flapper valve."

"Amazing, just amazing," said a tourist.

"Tell ya another goddamn thing," Pit said, waving his hands around like he was knitting bait bags. "They ain't got tongues, Adam's apples, or ears, but they do have scales."

When another tourist told Pit that he had never seen scales, Pit said, "Hell no, you never saw 'em 'cause you ain't no lobsterman."

Getting back to some more scientific stuff. On each side of the body, enclosed by the curved lower edges of the carapace, are the gills. There are 20 on each side, each with numerous filaments deployed around the central axis in an arrangement resembling a bottle brush. There are two openings into the gill chamber. Water enters in back and leaves from the front. It is fanned out of the chamber by the second maxillae, which are also called gill bailers. Once every few minutes the gill bailers reverse their beat for a few strokes to wash silt or sand that has settled on the gills back out the gill chamber entrance. Since respiration occurs as blood passes through the gills, the lobster would smother if the gill bailers should stop beating. Carbon dioxide diffuses from the blood through the thin walls of the gill filaments, and oxygen is removed from the air dissolved in the water.

Scientific literature states that reproduction in the lobster is of the sexual type. Pit Ginnis would have loved that. Lobster sex means that the male lobster mates with the female, usually when she is in a soft condition after shedding. He does this by rolling the female on her back and inserting his sperm into her seminal receptacle using his modified first pair of swimmerets.

The mating behavior of lobsters has been studied at some length. Biologists suggest that the most successful matings occur less than three hours after the female sheds her old shell. As she sheds, she releases a hormonal substance that attracts the male. And apparently this hormone attraction decreases quite rapidly with time. If a male is transferred to a tank with a female that has molted within the last 12 hours, mating will usually occur. But mating seldom takes place if more than two days have elapsed since the female's molt.

After mating, according to conventional wisdom, the sperm cells are held in the seminal sac until they are needed. This could mean 9 to 12 months later, when the female extrudes (lays) her eggs. She lies on her back, and the eggs pass through the opening at the base of the second pair of walking legs and over the sac, where fertilization takes place. If the female did not mate, and the eggs therefore cannot be fertilized, she may nevertheless extrude her eggs but may drop them after a few weeks rather than carrying them "full term."

Just before mother lobster lays her eggs, she grooms her swimmerets very carefully of all dirt and sand. Then, as the eggs emerge, they are covered with a gluey substance. After fertilization, they are guided along her stomach area by her swimmerets, the glue causing them to stick to one another and to the swimmerets. The eggs are greenish-black, and their number depends upon the size of the lobster. A female lobster with a carapace length of 3¼ inches lays an average of 10,000 eggs; a 4½-inch lobster produces, on average, 38,000.

Mother lobster carries her eggs around on her belly about as long as mothers ashore—10 months. Come birthing time, she arches her tail upward and vigorously agitates her swimmerets. Off come the eggs, which now have become transparent, revealing the eyes of the embryos within. The eggs hatch from the female over a one- to two-day period. The

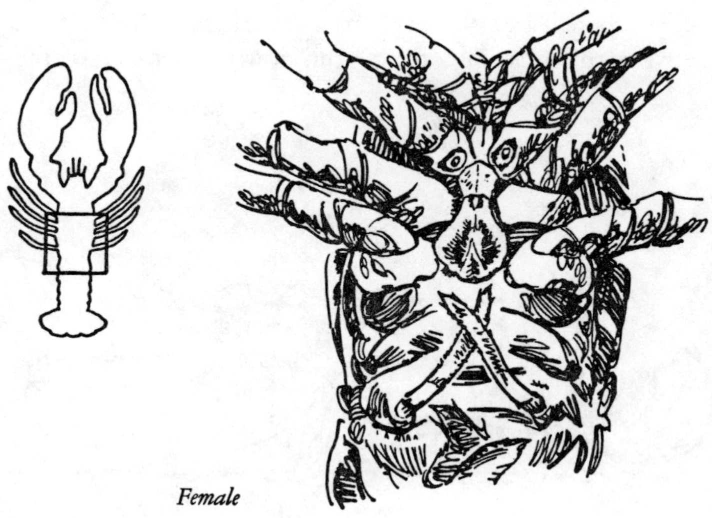

Female

newly hatched lobster does not resemble its mother. It is about a third of an inch long with very large eyes, and its double walking legs have hairlike projections that allow the larval lobsters to swim near the surface. During the free-swimming period, which lasts 15 to 25 days, the larvae shed their shells three times, changing their appearance somewhat each time. Each between-molt period is called a stage. At the fifth stage, the lobster settles to the bottom. The young lobsters are extremely defenseless and vulnerable until they sink to the bottom and find home in a sand or mud burrow or under a rock. More than 99 percent of them die during the free-swimming phase.

Because larval lobsters are at or near the surface for as long as a month or more, a major oil spill during this period in the summer could literally wipe out the total lobster year class for that area. And because small lobsters feed exclusively on plankton, which also would be killed by the metric ton, the oil spill scenerio is a scary one.

"You ain't got the brains God gave a lobster," is an old chestnut heard along the shore. The lobster ain't too brainy. What's there is a small whitish mass at the base of the rostrum, between the eyes. Two nerves run down either side of the

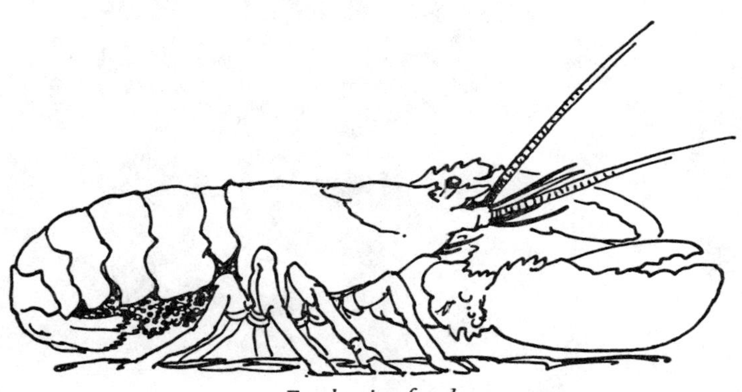

Egg-bearing female

lobster from the brain to the tail, with branches in every tail segment.

The sense organs are the eyes and sensory bristles. The eyes are compound, like those of the housefly, but probably aren't much use because the lobster is active mostly at times (night) or in places (great ocean depths) where there is little if any light. Lobsters cannot see in the dark, although some lobstermen can. It is the bristles that are most important. There are two types, chemical and sensory (touch). The chemical bristles that detect taste are located around the mouth. Those that pick odor from the water are on the antennae. Remove these and the lobster has to stumble over breakfast before he finds it.

The sensory bristles are distributed all over the body. One amazing group of these functions as a balancing organ. Seems that at the base of the first antennae there is a tiny pore opening into a water-filled sac called a statocyst. On the floor of this sac is a ridge of sensory bristles, and interspersed among them are extremely fine grains of sand. Any movement of the lobster causes the bristles to sway and the grains of sand to roll over them, which stimulates the nervous system to make the proper balancing response. Humans got their idea for ears from the lobster.

While on the subject of amazing lobster stunts, how about a neat trick called autotomy, or reflex amputation? If a lobster is caught by an appendage—or perhaps merely shocked by rapid changes in temperature or by rough handling—it can "throw" that appendage, which could be a claw, walking leg, or antenna. Tricky enough—indeed, brave—cutting off your own hand, but the real act starts at the next molt when the lobster begins replacing that which was lost. With each succeeding molt the appendage will grow larger, until it "catches up" with the rest of the critter.

The lobster reaches full growth by increments, and from its hatching size of 3/16 inch (length of carapace, also called gauge measure) the lobster must shed 25 old shells and grow 25 new ones to reach a legal length. All this house refurbishing takes place within a time span of 5 to 7 years. Lobsters have to molt because they get too big for their britches. Those lobsters just below and within the legal size range increase about 14 percent in length and about 40 percent in weight with each molt. That means that a lobster right at the legal minimum gauge length of 3³⁄₁₆ inches, weighing about one pound, will increase to 3⅝ inches and approximately 1⅜ pounds. Given enough time (and no one really knows how much time is enough) he can grow to between 30 and 40 pounds. Lobster lifespan is the stuff of myth, and supposedly responsible people have said a lobster can live a hundred years or more.

There are some general and accepted ideas on where lobsters live. The preferred environment is a rocky bottom, though lobsters do move around throughout the seasons, crossing large areas of mud bottom in the process. But for every concept there is a disclaimer. Lobsters do sometimes live on sandy and muddy sediments, especially in deep water. They dig burrows when no rock formations are available and live quite comfortably.

In 1982 some scuba divers made a night dive onto a bed of mahogany quahogs in the Penobscot Bay area. The site had been investigated a couple years before, and photos had been taken. The surface of the bed, about 60 feet in depth, was shown by the photos to be hard-packed sand with numerous old shells and many "pit" marks. The small depressions were a mystery until the night dive, when hundreds of lobsters were observed apparently digging for the quahogs, which were only an inch or so beneath the sand. The lobsters had come out of a nearby area that was riddled with burrows.

A suitable habitat must include a source of food, of course, which brings us to the question of a lobster's diet. Outside the trap, that is. In two separate research investigations, the

First-stage larva

Fourth-stage larva

stomach contents of lobsters taken from the wild by scuba divers were examined. In Newfoundland, rock crabs and spider crabs were found to make up half the stomach contents. Starfish comprised 10 percent of the diet with sea urchins next at 7 percent and then jellyfish, clams, mussels, periwinkles, and worms, in that order. Fish was a mere 3 percent of the diet.

A study in Connecticut also identified crabs as 50 percent of the diet, followed by snails, worms, mussels, fish, eel grass, sea squirts, and sea urchins.

* * * *

Scientists admit to very little usable knowledge of the migration patterns of the northern lobster. Research indicates that what migration exists along the Maine coast is to the southwest. In early studies of tagged and released lobsters, some of the recaptures were made 16 miles from point of release. In a 1957–58 study, four large lobsters liberated from Penobscot Bay traveled 75 or more nautical miles. Another, a 4½-inch (carapace) sprightly female was tagged near Tom Rock, Sheepscot Bay, and was recaptured seven months later near Race Point Light off Provincetown, Massachusetts. Those four Penobscot Bay travelers showed no common destination. One released near Little Green Island was recaptured near Timber Island, Cape Porpoise, Maine two months later. Another tagged at Little Green Island was recaptured 14 months later near Gloucester, Massachusetts. Another starting out from Little Green in September 1957 was recaptured 19 months later near Cape Ann Light, Massachusetts. The champion long-distance crawler was a 5-inch (unidentified sex) lobster released in Penobscot Bay and recaptured 13 months later at Nauset Light, Massachusetts—an estimated straight-line distance of 138 nautical miles. The smallest lobster recaptured traveled less than six nautical miles.

These and more recent migration studies in Maine, together with studies conducted in other states and in offshore waters, weave an interesting mosaic of the life and travels of *Homarus* over his briny north-south range. There is a south-to-southwesterly movement of predominantly large, mature lobsters along the Maine coast. There is a directed movement of egg-bearing lobsters along the eastern shore of

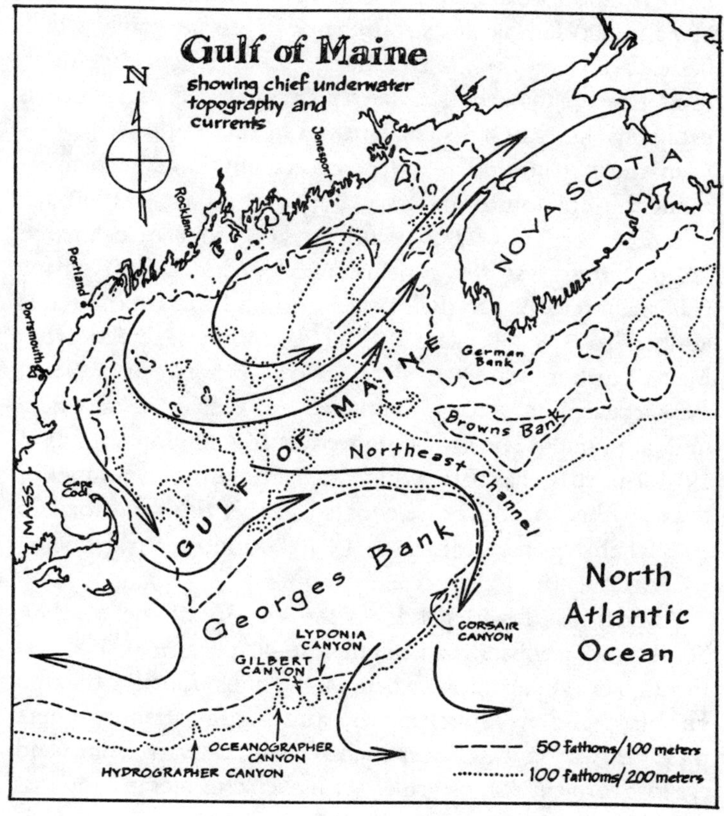

Cape Cod and into Cape Cod Bay during summer months. There is a seasonal inshore-offshore migration of lobsters inhabiting Corsair Canyon southwest to Baltimore Canyon. Lobsters intermix between offshore canyons. And there is a seasonal movement of some lobsters from eastern Long Island Sound and southern Rhode Island to the middle and outer ridges of the Continental Shelf, while there are few, if any, extensive migrations of lobsters living in Virginia and North Carolina waters.

(The offshore canyons—canyons such as Corsair, Lyndonia, Gilbert, and Oceanographer—are steep valleys between the 100 and 500 fathom lines southeast of Georges Bank, which lies east of Cape Cod, Massachusetts. Like fingers, they reach their tips up into the shoaler waters of the bank, which is probably the richest fishing ground in the world.)

In the last 40-odd years there certainly have been many changes in the lobster industry. Lobster landings in 1940, for instance, were 7,644,735 pounds, taken from an estimated 222,000 traps. At the conclusion of World War II many fishermen returned to their profession, and traps in the water jumped to 378,000, with 19,132,785 pounds landed from Maine's inshore waters. In all but two years between 1949 and 1982 the landings have fluctuated between 17.5 and 22.6 million pounds. The lowest figure was 16.4 million pounds in 1974. But there has been a dramatic increase in the number of traps in the water. The scientists call this fishing effort. In 1982 there were an estimated 1,846,000 lobster traps being fished along the Maine coast.

Annual trap loss depends on the severity of the weather. Some years a lobsterman might lose 40 percent of his gear through heavy storms, which move the traps along the bottom until they end up broken on the shore or far offshore where they "drown" in water deeper than their pot warp length and are lost forever. A lobsterman with even a moderate trap loss of 15 percent from his total of 400 will be building or buying 60 traps to replace his gear. But many overbuild, anticipating a further loss the coming year. More traps enter the water via this philosophy. And then there are the new fishermen coming into the business. In 1962, there were 5,658 lobster licenses issued. That year, 22 million pounds of lobsters were landed. In 1982, there were 8,500 licensed lobstermen and just about the same poundage caught. Using these and other statistics, the lobster fishery managers scream about the "overcapitalization of the fishery." They define this as a condition whereby the input of vessels, fishermen, and technology is greater than that necessary to harvest the maximum sustainable yield—the surplus yield or maximum yield per lobster recruit.

Yet Maine still leads the United States in lobster landings. The ferocious-looking creature seems to be holding his own despite many doomsayers. Today there is a thriving offshore lobster fishery to impose further pressure on the resource, and yet the inshore annual landings seem to be holding near or over 20 million pounds.

Clearly the American lobster has a mind of his own, and just as clearly the people who know the lobster are no mind readers.

5

The Buyers

The lobsterman in this century has never had a problem selling his catch. The old tales of using lobsters for fertilizer under hills of corn and five-pounders going for a nickel are exaggerated without question. After all, lobsters are a tasty food with an easily shucked exoskeleton, and what housewife wouldn't prefer a simply made lobster stew or salad to the tedious task of preparing those other bony things that come from the sea?

Our warped retrospect of lobster fertilizer and such arises partly because we forget that prices were commensurate with the times. We should remember, but rarely do, that these were the "good old days." The City Fish Market in Belfast, Maine, catered to a fish-loving populace and did a thriving business. E.F. Bramhall, proprietor, was an ardent advertiser in the city's weekly newspaper, *The Republican Journal*. In 1907, the prices for seafood were "quite reasonable except for salmon," according to the historians of that day. The advertisements read: "flounder, 15¢ a dozen; halibut, 22¢ lb.; clams (shucked) 10¢ pint; cod, 9¢ lb.; scallops, 15¢ pint; haddock, 10¢ lb.; fresh Atlantic salmon, 52¢ lb." And those locally caught Penobscot Bay lobsters were selling for 22¢ a pound no matter the size, but the preferred were two-pounders. Mr. Bramhall's account ledger indicates that lobster sales were equal to sales of other seafood, although prices were almost double that of fish. Of course, lobster could never compete with flounder at 15¢ a dozen. And flounder, incidentally, is still one of the best-selling fish in the world, but the draggerman is receiving about 85¢ a pound for his efforts while Mr. Bramhall's modern-day counterpart displays "sole" fillets between rows of lemon halves for $3.50 per pound. And if one wants to get shocked into fish-price perspectives, those 20¢-per-pint retail scallops of Mr. Bramhall were bringing an ex-vessel price to scallop boats in October 1983 of $7.65 per pound in New Bedford, Massachusetts. The retail ticket was outasight.

The Maine lobsterman doesn't have much truck with fish prices except when he wants to build a fish chowder and checks out the price of whole haddock, which he reluctantly buys with a few choice remarks about just who in the fishing industry is getting rich. Lobstermen prefer to exclude the middleman wherever possible, selling direct to the retail customer. This is for obvious reasons. The wholesaler's profit, or a part of it, goes into his pocket. Most fishermen selling direct will "take off" a few nickels from their price, so that persons buying direct from the lobstermen claim they are getting a helluva deal. Actually the price is about the same. The lobsterman didn't get off the turnip boat yesterday. The part-timer and small lobsterman is usually the one selling direct to consumers, however; it's a matter of logistics. The little guys can store their smaller catches in floating sunken cars, using their home as a place of business. After all, mom or the kids are always around to answer the phone.

Selling direct has other advantages besides price. These are "casual sales" in the reasoning of the lobsterman, requiring a cash transaction only and the IRS can go fish. The small and part-time lobsterman usually builds a summer clientele over the years or inherits one from his father, and those summer folk can eat a lot of lobsters, especially when they are bought from the local lobsterman and are flipping fresh from the cold, salty waters of the Maine coast. The purchase transaction is a continuing source of summer talk among the folks from away.

"Well, everybody, we are having lobsters tonight. Hope you can eat a few selects I buy from Harold over in the cove. I get them wholesale, of course, have been for years, pay just what it costs Harold to catch 'em. Harold always saves me the few selects he gets each hauling. You have never tasted anything like them, right, Martha? Say, why don't we all go down to the wharf and watch Harold bring them in from his car. Wouldn't that be the old school tie?"

Following a phone call, all meet Harold at the wharf and watch as he rows, backwards, out to his buoy in the cove and hauls a brown scummy bunch of boxy slats into his punt. Filling a plastic 5-gallon pail, Harold rows back, backwards, to the dock, where he is immediately surrounded by mid-cocktail hour giggles and gaffs from the folks from away.

"Jesus Christ, Harold, these are beauties," says the host.

The lobster car

"Oh, just average, Albert," says the old salt shyly, hitching up his beltless green wool pants.

The ladies giggle and then chorus an "oooohh" as the host replies, "Average, my ass, Harold, you saved them special for me, didn't you?"

"Now that could be, Albert," says Harold, putting the host into the inextricable position of paying top dollar or facing humiliation at the lobster dinner boil.

"Tell you what," says the host, "how about I pay you a bonus, Harold, on top of the regular price?" Albert folds three crisp twenty-dollar bills into Harold's salty, scummy hands.

"Aw, Albert, now you didn't have t'go do that," says the lobster catcher as he transfers the bills to his shirt pocket bank, the one with the button padlock.

Happy as clams on a coming tide, the folks from away return to their interrupted cocktail hour to hear the host once again expound on the what-a-goddamn deal he gets from Harold over in the cove on the price of lobsters.

Meanwhile, back home, Harold throws another alder stick in the kitchen Clarion stove and smiles at the finestkind of prices the summer folks are paying this year.

The home-seller lobsterman is faced with a problem, however, when fall comes and the summer folk have shuttered all their shingle cottages and returned to city ways. This takes place, unfortunately, just when the lobstering is the best—the fall months. Now the entrepreneur is faced with a surplus of perishable goods. He must take sou'wester in hand and visit the wholesaler and accept middleman prices. Some lobster dealers are outright belligerent with these guys and refuse to buy their lobsters. But most could care less and purchase whatever from whomever on their terms. A lobster is a lobster is a lobster.

By far, most Maine lobster poundage traffics through the lobster dealer, and he comes in all sizes. The dealer is the one who buys direct from lobstermen, or from other "buyers," and then sells to various levels of the trade from the retail restaurant to another wholesaler. Some dealers offer bait, supplies, and even cash up front to their fishermen to keep traditional and dependable sources of product. Other dealers will have nothing to do with bait, believing that lobsters alone are headache enough to handle. Obviously, only the shore-based dealers can supply full services at their own docks and wharves. Many dealers operate from inland or at another middleman level requiring little more than a truck. It should be evident that "middlemen" in the lobster business have an indefinable geographic and economic location, and they seek their own individual levels according to market supply and demand.

Sometimes lobstermen band together to take marketing into their own hands. In 1976 there were about 20 cooperatives along the Maine coast. They were almost all lobster co-ops. The oldest was the Pemaquid Fisherman's Cooperative Association organized in 1947, followed by the Stonington Co-op in 1949. Stonington was the largest with 164 members. The smallest was in Winter Harbor with 19 members. There has been a considerable increase in membership in all Maine co-ops during the early 1980s.

The larger co-ops sometimes operate more than one buying station and market their own lobsters, sometimes out of state via their own trucks. As a business corporation, the fishing co-ops buy in quantity, which includes just about everything marine from trap-building supplies to gloves. Merchandise is sold to members only at enough of a markup to cover expenses. The lobsterman is morally obligated to sell all his lobsters to the co-op, which pays, in cash, near the going price. At the end of the year, however, after business expenses are paid, the co-op members receive a bonus of so much per pound that they sold to the co-op. It can be as small as 2¢ a pound and occasionally more. In bad years, or with unforeseen capital expenditures, there is no bonus.

The co-op

A fisherman is not obliged to join a co-op, and more do not than do. They prefer their independent marketing freedoms and sell to private buyers, which are usually set up right in the same harbors as the co-ops. The private buyers usually pay more per pound, but there is, obviously, no year-end bonus.

It is, claim most of the lobster dealers, supply and demand that sets the lobster price at the fisherman and dealer levels. This fails to wash with lobstermen, who believe that the price is set by dealers, perhaps in collusion, as "low as they dare." It is a fact that there is nowhere the lobstermen can turn for authoritative price guidelines. It is the dealers who determine the price of lobsters.

That price is set on a number of factors, one of which is the ageless supply-and-demand equation. Through the winter and spring months lobsters are scarce because there are few offshore lobstermen, and the price goes up. The bay or nearcoast lobstering nearly ceases during the period from December to April. There is another hiatus during the July-through-August period when a majority of the Maine lobsters are burrowed beneath rocks acquiring new shells. When "new shells" start moving into traps, the price gradually falls to the lowest point in the year, during September to November, when the large catches occur. This is the supply-and-demand argument used by most all dealers.

The Canadian lobster industry is tied very closely with the U.S. industry. In fact, dealers claim to a man that the Canadian lobsters "enhance" the price paid to U.S. fishermen. The Yankees have a hard time swallowing that wisdom, but it is a fact that without Canadian lobsters at certain times of the year, many dealers would be out of business. Imports of live Canadian lobster in 1982 accounted for about 30 percent of the U.S. market. Although Canadian buying by Maine dealers varies considerably, those that do admit the 30 percent figure is quite accurate. The bulk of this foreign buying is in the winter months, when the American lobstermen couldn't begin to supply the market. During the Christmas season, for instance, there is a heavy demand for lobsters, right at the precise time that Maine has few to sell. The Maritime Provinces of Nova Scotia and New Brunswick know their way around the lobster business and play a key opportunist's role in the U.S. lobster market at this time. In the fall months, when the U.S. harvest is the greatest, the Canadians close their seasons.

The Maine lobster does not immediately crawl from lobster trap to table. The logistics of handling a fluctuating supply of perishable live product were solved nearly a hundred years ago with tidal lobster pounds. These are enclosures within small coves and other protected shore indentations, all with some natural impoundment characteristics but with help from manmade additions. They are, in effect, dams with slatted gate openings allowing the ingress and egress of the changing tides. The pounds, however, are deep enough so that lobsters are not exposed at low tide. With each flood, new salt water and accompanying nutrients are brought into the pound, and on each ebb the wastes are flushed to sea.

There are approximately 56 tidal lobster pounds in Maine with a combined capacity of about 4,225,000 pounds of live lobster. The pounds come and go and change ownership, but Maine has a nearly constant capacity to store 4 million pounds of lobster in any year. The figure is actually considerably higher, because there are many live-lobster pounds with considerable capacity that are nontidal; they must pump seawater through their systems because their locations will not allow them to use the free tidal energy.

When the market price is at the season perigee, owners begin filling their pounds with "cheap lobsters." The larger pounds start first—some can store 200,000 pounds of lobster—and the smaller ones begin in November. The length of time the lobsters are held is the name of the lobster game. Fortunes have been made and lost on lobster speculation. Having a pound full of live lobsters in the winter is like driving a truck full of old, unstable dynamite down Pikes Peak without brakes.

The risk is quite easy to comprehend. Assume that pound owner Salty Dog began filling his 100,000-pound-capacity facility in September and by hook or crook (lobstermen presume the latter) he paid an average price of $2.10 per

The lobster pound

pound. One might innocently assume that Salty could unload these on the Christmas market for $3.25 per pound and make himself a neat little profit of $115,000. Disregarding such things as shrinkage for a moment, probably Salty is a true lobster entrepreneur and plays his hunch, gamble, or whatever *Homarus* speculators play. Salty waits. The price goes to $3.50. Next week it goes up again. For every 25¢ the lobster price goes up, Salty makes another 25 grand. It's not hard to see Salty's gambling blood beginning to boil. At around $4 per pound, Salty decides to wait just one more day to unload and then spend the rest of the winter in West Palm Beach. Well, overnight, some Canadians and Newfoundlanders plus a half dozen countrymen decide to unload, too. The price drops $1 and Salty Dog has just watched $100,000 go down a clam hole. His one big shot of the year fizzled like a wet firecracker from Formosa.

That was the simple scenario. Salty obviously had a few more problems and expenses. He and most of his colleagues need ready cash to play the lobster game, for lobstermen don't play with wooden matches. So, bank loans at 14% enter the profit-and-loss ledger. Salty must feed his lobsters while they are confined, at the standard formula of one bushel of bait (fish cuttings, racks, brim, etc.) per 1,000 pounds of lobsters per week. Bait can run $5 per bushel. Then, there is the weather. Extremely cold temperatures can freeze the surface of lobster pounds, which are located next to the shore where freshwater runoff can reduce salinities, thus raising the freezing point of the water. Most lobster pounds operate in the same manner when removing the lobsters. They haul drags over the bottom to catch them. Cutting and removing the ice first is fairly cost intensive.

Theft (pounds are usually in isolated, remote areas of downeast Maine and subject to guerrilla tactics by unsavories) cannot be disregarded, and the ever-present red tail can literally wipe out a lobster pound in a couple of days. Dead lobsters, incidentally, are not a total loss, and they are not taken to the dump. Dead lobsters, whether in summer or winter, in small or large lobster pounds, are cooked and picked for the meat. However, the wholesale price of lobster meat is about $8 per pound, and the same lobsters live would sell for three times that. It takes about five live chicken lobsters (one to 1¼ pounds) to produce a pound of clear meat. (Lobster lovers purchase millions of gallons of water a year.) And so, dead lobsters are not a profitable item, as Salty Dog will find out if he continues to play the lobster game much beyond Christmas.

Lobster dealers supply a variety of markets within and without Maine. These vary from other wholesalers to retail restaurants to supermarkets. Restaurants in Maine do not sell the quantity of lobsters one would think, although during the tourist season lobsters are a must item on Maine menus. Handling live lobsters presents problems. Unless the restaurant has one of those expensive display tanks, the lobsters must be kept alive in refrigerators, and therefore buying cannot be in any volume. Restaurants close to dealers can run over and pick up a few pounds when that party of 12 arrives and all order baked stuffed you-know-what, but very few live lobsters are kept on the premises. Another concern of the restaurateur is the price fluctuation, and just about every eating place will leave the lobster price on the menu a blank space. The markup on a lobster dinner, even in Maine, is considerable. A 1½-pound lobster cost the Maine native or visitor an average of $13 in 1983. Restaurants usually purchase lobsters from their regular seafood distributors, which are high in the middleman ranks, and they therefore pay more than, say, a primary wholesaler. Lobster meat is another matter. This product can be frozen quite well and has a considerable shelf life. Restaurants can also get "deals" on lobster meat, and the meat is a familiar product that meshes comfortably into the portioned logistics of such menu items as stews, rolls, and salads. The live lobster is the only live product handled by restaurants and is therefore an "outsider." But restaurant sales of *Homarus* account for hundreds of thousands of pounds a year.

The lobster buyer, whether in a moored sailboat in a secluded cove or the seated guest around the fancy restaurant table, certainly needs little urging to dine.

6

The Government

Anything, be it wardens or weather, impeding a lobsterman in his work is the fault of the "goddamn gum-mint." The "gum-mint" can be a coastal warden checking for a valid license, a judge handing down a stiff fine for short lobsters, or even the legislature passing a law. They are all lumped into the pot called "gum-mint."

Lobstermen have plenty of allies in their adversary relationship with the government—just about the whole fishing community. One time four clam diggers were returning from a day of island clamming in their 16-foot skiff with the usual big-horse Mercury outboard bolted to the stern and running wide open. The body weight plus 10 bushel of clams plus pet hound dog (barks when wardens approach) plus a couple of cases of empty beer cans (returnable) left little that any prudent person could call freeboard. The just-awash apparition was sighted by a marine patrol boat, which eventually overtook the clammers. Coming alongside amidst blue flashing light and wailing siren, the patrol boat's chief badge threw a line, which was ignored by the clam diggers.

"Hey, you, you in the boat there, let me see your registration," shouted the warden, not taking the prima facie valid boat decal's display word for it.

The throttle handler dug around in his muddy pants and eventually came up with an equally soggy piece of paper that passed as a current boat license. The other clam diggers sat looking like a pack of dogs denied their supper.

"Your license is OK," said the warden, passing the wad back to the operator, "but you're overloaded. I'll have to terminate you right here," added the warden in learned but unfortunate terminology.

"Jesus Christ, he's gonna shoot us!" said the bow ballast.

"No, sir," said the warden. "However, I cannot allow you to proceed in your precarious position."

"We been clammin' like this for forty years," said a forward thwartsman.

"I must terminate you at once," replied the warden.

"You go first, Willy," said the bow, "you ain't got kids."

The warden explained the terminal operation. Half the skiff would get out on that there ledge while the boat operator took the remainder ashore. Then he would return and pick up the rest of the motley crew. This would prevent the dangerous overloading condition, which would otherwise without a doubt cause loss of life.

The aft thwartsman had a question—make it an observation—for the warden. "That there ledge is a half-tide rock, cap, and before Willy here can get t'the harbor n'back we ain't gonna be standing on no rock, we gonna be treading water, that is, them's that can tread."

There was a large huddle in the patrol boat cabin over an area chart, which took some 10 minutes. When the final play was called the chief badge emerged and said rather sourly that all the clam diggers should come aboard and they and the clam skiff would be escorted to the town dock. When the patrol boat approached the dock, the pile sitters and sunny lee slouchers all yelled and applauded. The clammers waved and yelled back like returning heroes.

The fisherman concept of the government shooting first

and making it a law later persists, but beating the "gum-mint" is so sweet.

* * * *

Some fishermen claim that there are more laws on lobsters than on humans in the state. Not quite, but fishing laws are extensive, so much so that at one public hearing a fisherman suggested that DMR bind the laws in Styrofoam and issue them as life jackets.

State "gum-mint" regulations say that a lobster license (including crabs) can be issued only to a resident. The residency requirements over the years have diminished. At first there were none, because nonresident lobstermen were unheard of. In 1939, a person applying for a license was required to have lived in Maine for 10 years. In 1970, that applicant must have been a resident of Maine for eight months of the year for a period of three years. In the late 1970s this residency law was challenged, and the law courts determined that a requirement of "longevity" was unconstitutional but that the state (under the banner of conservation, again) could regulate the lobster industry by issuing a resident lobster license only (or so the decision was interpreted). That is the position of the State of Maine today. The applicant, if he has not held a license before, must show proof of residency by property tax receipts, driver's license, or other supporting evidence.

The lobster and crab fishing license is also only issued to an individual, and this has effectively maintained the individual status of the industry. Corporations, companies, co-ops, and any other groups are not allowed permits. Although it is true that most full-time fishermen "fish for" a co-op or dealer (which can supply capital, among other things), these organizations operate under a dealer or wholesaler license, another breed of cat. One could easily imagine the chaos if the law were changed and General Motors was allowed to purchase 5,000 licenses and institute the sharecropper plan.

When lobstermen run upon the ledges of the law, they appear before a bar of justice that is in general quite

sympathetic to the plight of fishermen, much to the consternation of perplexed wardens. Although this story does not involve a lobsterman, it could have, and it illustrates the sympathy of the court and the reputation and mystique of fishermen—or maybe a smidgeon of all three.

A clam digger appeared in a coastal community's district court charged with digging clams in a closed area. It was not the first time, and the judge and the digger were not total strangers.

"Aubrey," said the voice from the bench, "you are charged with digging clams in a closed area. Do you have anything to say for yourself?"

"Yes, your honor, I was not digging clams."

"Aubrey, the warden states that he observed you on the flats with a clam hoe, digging here and there. Is that correct, warden?"

"Yessir, your honor, and Aubrey hadn't had time to find a good bed of clams and I thought I'd better stop him before he did," answered the badge man in green.

"Well, Aubrey?" asked the judge.

"Mr. Judge, you see it was this way like I tried t'tell the warden here, fall's coming on, ya' know, damn, this mornin' I had t'build up a good kindlin' fire for the missus, well, anyways, I ain't got the money f'high price firewood, judge. Christly, they want all getout f'a cord nowadays. I 'member when slab edgings was $5 a dump body. Well, they ain't now f'sure. Like I said these nip mornings turned my thoughts t'my wood plight n' I 'membered all that driftwood in the flats, been hunkered in f'years, all water-logged, so's I took a piece home week or two back n' dried her out n' goddamn if'n it wa'nt the best firewood I ever burnt. So's come yesterday on low water I headed f'the flats t'stock up on winter firewood. I wa'nt moseying around more than a few minutes when Warden Norm there comes a frothin' down over the bank n' pinches me f'clamming. Hell, judge, I was woodin'."

"Warden?" came the one-word query from the bench.

"Your honor," said the warden, gaining his composure, "that's the most preposterous story I have ever heard in my life."

"I'll be the judge of absurdity, officer, just tell me if it was possible that Aubrey could have been collecting firewood."

"I would say, your honor, that the odds would be somewhere around one in a million."

"But there is that possibility, isn't there, warden?"

"But sir, Aubrey had a clam hoe, a clam hod, and was poking around where clams live on an illegal clam flat."

"I get your point, warden, but it seems to me there is reasonable doubt in this case that Aubrey was, in fact, digging clams. It is coming on fall, and just this morning I said to the missus, 'Martha, we have to get a couple cords of firewood.' Anyway, I am dismissing this case with a warning to you, Aubrey. If you do any more fire-wooding, make it a high-tide venture."

"Thank you, Mr. Judge," said the clam-flat woodsman.

"Jesus H. Christ," muttered the warden, ripping a page from his summons book.

The courts were so lenient in dealing with fishermen that the state eventually had mandatory sentences passed, which puts a little more starch in the game. In keeping with modern-day justice, the fines are on a sliding scale. The court "shall" impose a base fine of $25 for each offense involving illegal-size lobsters, with another $10 per short lobster up to five and then $30 for each short in excess of five. A bushel of short lobsters could run into a fairly long court cash-register tape. But probably the penalty that does more to curtail short lobsters is that upon conviction a lobsterman can lose his license for three years. It rarely happens on first offense, but the option is there. A few license losers tried to circumvent

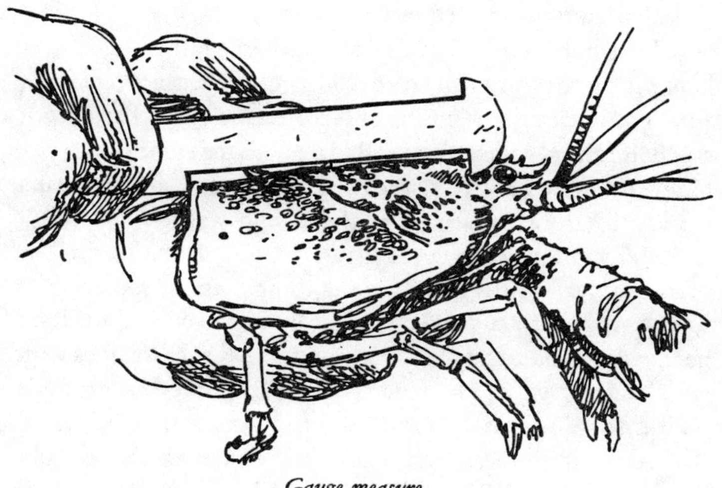

Gauge measure

this penalty by transferring their gear to a wife and fishing in an "advisory" role, meaning the lobsterman would do all the work until a warden appeared and then he would sit back and read the paper while his wife fiddled and faddled and did a lot of make-work until the warden sped off down the bay. Alas, the loophole was closed, and license losers must now take their gear out of the water.

Full-time and thoughtful lobstermen rarely take shorts anyway. Just ask them.

What happens when a fisherman is caught red-handed with the little fellas? Well, again, this stimulates the undercover psyche that all lobstermen are born with. Some of the more common responses after being caught with hands in the short pail are:

"Well officer, I lost my measure overboard and I was taking these ashore to gauge 'em but sure as hell would throw the illegal ones off the wharf."

"These ain't mine."

"Warden, you won't believe this, but . . . ."

"Short huh? Well, that's what I thought but old Bill here kept saying they was legal n'all. Bill's getting along you know, ain't got a family or nothing. No money. Ain't hardly got much life ahead of him. I take old Bill with me some days n'pay him off in lobsters so's he can at least get one square meal a week. Bill's lobsters are really short, huh warden?"

"Goddamn, warden, am I glad t'see you. I was settin' out t'look for 'ya. OK, now here's the thing. I think something's wrong with my gauge. Can't put my finger on it, but, look here, would you mind measuring up these few close ones with your gauge just t'set my mind at ease? Just put the shorts aside n'I'll throw 'em back. Sure glad I run into you, warden."

"Judge, sir, this warden has been after me ever since the day he ran outa gas n'I towed him home but made five passes 'cause there was so many people on the dock. He thought I was hangin' him high on purpose, judge, but I wa'nt. Wind wa'nt right t'land."

"Ya'honor, we Eatons are God-fearing folk. My granddaddy was a peace officer. An Eaton don't go 'round breaking the law on purpose. No sir, I really didn't think them were shorts. Yes sir, I was in here a couple weeks ago."

There are other lobster violations as serious as the taking of shorts. Possession of an egg-bearing lobster brings a fine of $30 a mama, and if a lobsterman scrubs the eggs from mama's tummy to be cute, then, as mentioned, the fine is $100 for each scrubbed lobster plus a mandatory one-year loss of license. It would seem rather difficult to prove that a clean female lobster's tummy was laundered, but the biologists say they can tell—but are not telling how. Mutilated lobsters and those with a V-notch in the right tail flipper bring a $30-apiece fine. The most severe punishment is reserved for "molesting" another person's lobster gear. That is defined as fooling around with a lobster trap, warp, buoy, or car not one's own.

Conviction means a mandatory loss of license—if the violator is a lobsterman, of course—for a period of three years. In addition, the commissioner "may" actually suspend the license or certificates of anyone convicted of violating any marine-resource law. Duration of the suspension is six months for the first offense, nine months for the second, and one year for the third collar. In the real world, the commissioner rarely exercises this option, since to do so could be poor politics.

There are other nonregulatory happenings that the lobstermen call "gum-mint" and that do have short- and long-range implications to their livelihoods. Certain science workshops can formulate policy. An example was the Canada-United States Workshop on the Status of Assessment Science for Northwest Atlantic Lobster. Although less than a handful of working lobstermen in Maine ever heard of the meeting, their prey and subsistence was the star of the show. Attending were 19 representatives from the National Marine Fisheries Service, state scientists from 10 states, 26 Canadian federal and provincial biologists, and two European lobster experts. The symposium consisted of 23 papers discussing everything from statistical reporting systems to larval studies to lobster growth rates. Since these were key people within the world lobster industry, there was bound to be some regulation spinoff, and indeed, much of this science did manifest itself in what would become the federal American Lobster Fishery Management Plan. A well-known lobstercrat was asked once why such professional meetings never included a bona fide lobsterman who just might offer some different perspectives and probably some revealing eyeball data.

"My God," he said, "we'd never get anything accomplished. They just don't understand the language."

Money sources, whether private or public, are also considered "gum-mint." Unless one inherits a lobsterfishing business by working through it from age five, the entry fee for full-scale give-her-hell lobstering can be frightening. The sum of $100,000 for a boat and gear "to make a living" isn't too far out of the ball park. There are several sources that underwrite lobstermen, or say they do, but when it comes to fish or cut bait, the prospective lobsterman is usually cutting bait. The federal programs are not geared to small fishermen, even when there is money available. The Maine Guarantee Authority statutes and intent state that the program permits a loan guarantee for any documented fishing vessel, but the red tape surrounding such a loan makes the route torturous, with little light at the end of the wharf. Local banks seem to be the prime moneylenders.

Outside of a loan here and there, the lobsterman has always operated on a cash-only basis, and when it came time to pay his income tax, he figured up what he had for income and what it cost to get that income, and then he paid the "gum-mint" what he thought was a fair share. All those IRS tables and rules were for other people, he figured, those nine-to-fivers. Not that the lobsterman paid no income tax at all; he merely changed the rules in that he decided "fair share," not the IRS, which didn't understand the fishing game. As one lobsterman expressed it, "The IRS goes about collecting taxes like a fart in a mitten."

Up until the mid-1970s, there were some feeble attempts by the IRS to audit lobstermen. The federal auditors came away from these sessions as if they had had a lobotomy— just staring off into space. After one encounter in Rockland, a young IRS accountant said, "These guys keep their records on the head of a match."

Some say it began with complaints from fishermen themselves. Others claim the record-keeping practices were so arrogantly flaunted that the IRS was bound to take some action—if only to save face. Known variously as the "Great Gum-mint Strike Force," the "Invasion," or the "Gestapo," among other unflattering titles, a surprisingly large IRS contingent did, indeed, show the flag in 1975. Gathering information as only the IRS can, they called lobstermen to audits all along the coast, mailing out notices to come and bring "seafood receipts for the past three years." A refusal to honor the invitation, some lobstermen discovered, produced a "G-man" at the door in the form of a United States Marshal—shades of the Old West.

A part-timer with 10 traps and a very common last name got notice to appear for an audit. He took along his wife, who did what bookkeeping was necessary in the household. The auditor proceeded on prescribed course with duplicate license, seafood receipts. . . . "Whoa, there, how many pounds did you say I sold last year, mister?"

The young accountant flipped through four sheets of figures and calmly said, "It looks like, er, right here on the bottom line, $34,500."

"Jesus Christ, I never sold a one. Shit, I only got ten traps."

"Ten?"

"Five and five—ten."

"Well, it doesn't seem that you could catch that many, does it?" asked the auditor. "But just how many lobsters do you catch a week?"

The wife, who had taken a no-pay day off from work to attend her husband's honor, had had it by now. She jumped up, grabbed her shoebox of "receipts" from the auditor's desk, and said in a loud voice, "He gets about two and we fight over them!"

"Guess there's been a slight mistake here," said the auditor to himself as the highliner and his wife slammed the door.

It was not long before the IRS discovered what they probably had known all along—that it would be impossible to change 300-year-old habits, and throwing any Maine lobsterman in the slammer would be political suicide. When it was over in about 18 months, a handful of civil fines went out to individual lobstermen who admitted to "a couple of mistakes," and there was just one criminal indictment, which was later changed to a civil offense when the fisherman agreed to pay taxes on $22,000 rather than the $367 income he had previously reported. "An oversight," he called it. The IRS left seemingly happy that its integrity, honor, and watchdog image had been preserved.

Like the kid with freckles, the lobsterman knows he has the "gum-mint" and doesn't like it but can't do much about it. It's an affliction he can live with and sometimes takes pleasure in, as when confrontation leads to victory. But most of the time he ignores regulation if it doesn't suit his own fashion.

Old Pit Ginnis once said that "gum-mint people are like a good case of the shits. They don't know when t'stop."

7
The Community

Maine lobstermen, for all their fiercely independent image, do not walk alone. Like the rest of us less fortunate mortals, they have families and live in communities. They exist within normal social mores, and except for the hours spent upon the water, lobstermen and their brother fishermen are ordinary citizens—contrary to many rumors. Whatever one does in these small villages, which most lobster ports are, it is almost immediately known by everybody. And just about everything is tolerated. Extramarital affairs, drinking, even an occasional theft. But the rules and the laws of the sea are inviolable. Gear molestation is not tolerated, and invasions of historic territories are scorned. Trespasses are severely dealt with.

Boy fishermen turn into father and grandfather fishermen as surely as the tides turn. In the early years of establishing lobstering as an occupation distinct from handlining and seining and trawling, it was expected of a son that he would become a lobsterman, too, that he would continue the tradition, the profession of his father. With that mandate, small boys started very young to learn their trade.

The teaching method then and now is a show-and-learn affair with the father showing his son, quite methodically, the ways "things" are to be done. The boy is then on his own, and father will rarely offer advice unless the lesson is hopeless.

Knitting bait bags

Boys, for instance, are never shown twice how to knit a bait bag. "Is this it, Dad?" Or, "I got it right this time, Pop?" is invariably answered with a "nope" or "yep." Knitting 50 wrong-way bait bags can do something for the powers of concentration when the next lesson rolls around.

Who can say who laid down the rules and duties of a fishing family? Whoever sealed that package did most certainly leave the daughter's role caught in the seam. Poor missy, sometimes the daughter feels that she was predestined by the gods of the sea forever to sweep floors, do dishes, and knead a million loaves of bread into food for the men in her life. Unfair, unfair, she says in her prayers, when dad takes brother to the island for clamming and to town to buy new pot warp. Modern missy is not quite so jilted as her grieving predecessors, but it's tough being a girl in the lobstering world.

Daughters can also be the object of much love from their father lobstermen. Being out in the boat all day with Junior doesn't leave much for man-to-man conversation when the supper dishes are cleared. That's when dad takes a keen interest in his daughter's activities of the day, although father fisherman doesn't know much about the women's world of things like knitting and kneading dough. Sometimes, daughters demand to enter the lobsterhood. And succeed. They can knit lobster heads, lath a trap, and fill bait bags with the best of brothers.

The teen lobsterman begins with the inevitable skiff and small outboard, but the starting gate is jammed with all manner of craft. A few young men jump into lobstering with complete financial abandon, going for the big-time inboard boat and lobster gear to match. Dad is usually the backer or at least the cosigner at the community bank. Other, less reckless young fishermen buy used boats and gear or "buy out" a fisherman who is "putting it on the bank" either because of age, lagging enthusiasm, or money woes.

Whatever his entry level, the young fisherman, because of his community status and having spent a few years with the old man, does not have the territorial problems associated with a stranger. Waters and ground fished by the father are inherited by the son, who may or may not expand his fishing territory to include some ledges and kelp beds and "ground" of his own. Once on his own, however, he is treated with the same competitive tolerance as any other lobsterman, even by his father. Setting traps too close, dumping excessive gear on good common ground, or, heaven forbid, molesting gear, is dealt with as if the family did not exist. "Every man for himself" is hardly a meaningless cliche in lobstering. Lobstering anecdotes are common concerning father-son friendships turning downright ugly and even causing lifelong intrafamily splits. And in most cases, it is the mother who suffers disproportionately from the harsh, unbending rules of lobstering. Not that the rules are absolute, and many an evening ashore is consumed in the father-to-son lectures on lobstering responsibilities.

Young lobstermen in recent years have joined the moving-out generation when it comes to speed on the water. Outboards up to 200 h.p. are used by teenage fishermen on seemingly ridiculous boats. Imagine those power plants bolted to the stern of a 19-foot boat. With the throttle only half open, nothing but the prop touches the water. You could set up camp under the hull, there is so much room.

"Sufferin' Jaysus," said one old-timer watching the kids return to the harbor after a day's lobstering. "They're gettin' back 'fore they're goin' out."

Fathers rarely interfere with a son's gear purchases once they start fishing on their own. Dad may wince when Junior bolts that Mercury mill on his transom, but there is no "take that goddamn thing back or else" attitude. One reason is that fathers have almost absolute faith in the seamanship of their sons. Two hundred horsepower or three, father lobstermen have a pragmatic mind set that sons have the inherent knowledge to recognize the moods of the sea and handle their boats accordingly. After all, who taught them?

Although the young lobsterman is an integral part of the

family life until he finishes school or is old enough, usually 18, to drop out, his finances are his own. Incidentally, many young islanders are faced with a schooling problem when they reach the secondary school level. It happens about the same time they are beginning to lobster alone. A few islands on which lobstering is the only income of consequence have only primary grade schools. No high school. Three examples are Frenchboro (Long Island), Matinicus, and the Cranberry Isles. The state subsidizes these students to attend mainland high schools, but they must board and live on the mainland. There is usually a traumatic shock within the family when a teenager leaves for the mainland to live during his academic months. These are the first times away from home for close-knit, sensitive kids, and even though some reside with relatives, the parting is difficult. Also, the spring and fall months are good lobstering times as well as school-keeping times. These sons of fishermen forfeit a head start and are confined to the vacation lobstering of June through August.

Through the period up to, perhaps, about 1956, a young fisherman contributed his share of earnings to the family financial pool. And a lot of them still do. But increasingly, the beginning lobsterman is an enterprising entity with a bank number of his own. He is an unquestionable saver—saving for a boat, saving for a pickup, saving for a 140 h.p. Merc. It would seem that the cash-only nature of the lobstering industry would be too great a temptation for so young a mind. But frugality is also taught by fathers along with knitting bait bags. They are not selfish, these emerging lobstermen, and the industry is one of the very few where teenage boys—young, developing professionals engaged in hard, manual, proletarian labor—regularly give their mothers new washing machines or living room sofas and their fathers new TVs and sets of snow tires and their sisters new sleds and fluffy Snoopy toy dogs for bedroom companions.

Bank managers in small fishing communities will confirm

that there is hardly any risk in giving a young fisherman reasonable credit. One said, "I remember cases where youngsters took days off from fishing to come in and see me about one problem or another. That's unusual. Even adults find that difficult."

The last real socioeconomic study of the Maine lobsterman was done by the University of Maine in 1970. Because of money restrictions and the time allotted (six weeks), it was decided not to tackle the industry as a whole but to concentrate on three lobster fishing communities. These were Corea, representing a highly specialized, isolated economy where lobstering was the predominant activity; Beals Island, again because it is a highly specialized lobster community, although less isolated than Corea; and lastly, Phippsburg, a lobstering port with a more diversified economy and a close proximity to alternative job opportunities. Each of the areas had one thing in common—the lobster fishery was the major economic activity.

The survey results indicated that the average age was 42.6 years and the average income was $6,213. Two-thirds of the fishermen said they went lobstering for the money. The rest wanted lobsters for home consumption only, or they lobstered out of a preference for the particular way of life. There was a scattering of other reasons such as "the old man made me." The average years of education was 9.8, and approximately 40 percent had less than a ninth year of education. However, 41 of the 131 fishermen indicated that they received some type of formal vocational training in such areas as carpentry, metalworking, and mechanics. When the questionnaire asked about preference for receiving vocational training, only a small fraction expressed any interest. Perhaps the fishermen thought they were being asked whether they could, in fact, "do" carpentry and mechanics and woodworking. Most lobstermen are jacks-of-all-trades.

There has been no socioeconomic survey comparable to the 1970 study done since then. The current lobster license application does not require the applicant to declare completed years of formal education, but in nearly a decade and a half the average school years most certainly have risen above a high school sophomore.

There aren't many fishermen around nowadays like Pit Ginnis. Pit was asked one time how far he went in school.

"Far's I could, sonny, far's I could."

"Whatcha mean by that, Pit, mean you went right up through college n'everything?"

"Goddamn right, sonny. Tell ya how it was. Began in the grammar school of spankings, woodbox fillings n'dory cleaning. Learned good. Went right on up t'the high school of fetchin' f'father, punt building, trap heading n'staying dry. Took an extra course there, gettin' wet. From there went right up t'the college of hard knocks. Been knocking around ever since. Get my sheepskin every night I make it home safe and sound."

And there are some still, who like Pit, live by their wits.

A lobstering community is comprised of both full- and part-time lobstermen. There is no absolute handle on who's who. There are some educated and uneducated guesses as to the percentage makeup, and the spread is considerable. One study found that only about 17 percent of commercial lobstermen surveyed in four Maine communities fished over 150 days a year. According to statistics compiled by the National Marine Fisheries Service in 1983 for inclusion in the American Lobster Fishery Management Plan, Maine had 3,409 full-time and 4,952 part-time or casual lobstermen. In other words, 41 percent of lobstermen are full-time. Whether NMFS defines "full-time" as fishing 150 days or more a year, or just what criterion does designate that category, is not revealed. In Maine, however, inshore lobstermen who do not work outside their profession, at any other job, can hardly support that 41 percent figure. The very nature of the inshore fishery predetermines that a lobsterman could not physically fish year-round. The lobsters are not there. The weather is. And so shore jobs are sought.

One of the coveted jobs along the Maine coast is that of caretaker for the "summer people." Hundreds of sou'westers have been permanently hung up on the hallway peg when an opportunity for a caretaker job comes up. They are coveted for many reasons. Although the jobs are not particularly well paying, neither is the work that strenuous. It consists primarily of opening the "cottage" in the spring, maintaining the dwelling(s) and grounds in the summer, and closing in the fall. Summer folk are usually in residence for about a month only. The rest of the time the caretakers work on pretty much puttering-around schedules.

In the communities that are made up of local and summer folk, there is a distinct and accepted dividing line. There is a peaceful, passive, respectful cleavage between the two types of dwellers. Lobstermen do frequently take out summer folk on day lobstering trips, but the association does not extend to any social friendships or even infrequent visits between "those from away" and those that work and stay. There are two defined sanctuaries, both privileged and honored. The lobsterman and his wife do not attend the summer folk cocktail party, nor do the vacationers sit and drink coffee by the native kitchen Clarion.

Children from both side of the fence are taught this at an early age. Home and away loves that do manage to bloom among the rich downeast humus generally fail to consummate in anything but tearful love letters going on until Thanksgiving.

There are occasional case stories from the lobstering circuit involving intersocial sexualizing between home and away adults, and the consequences are far more disastrous. It is rumored, just that of course, that a rather massive shingled summer cottage burned to the ground in less than 10 minutes when the summer owner-gentleman began to pay considerably more attention to a lobsterman's wife than was called for in the rules of summertime fair play. The volunteer fire department, in fact, never responded, despite the fire chief having been ashore resting for the day.

There is common and acceptable cruising ground for the home folk and the aways. One is the public supper. All Maine communities, be they marine or meadow, have public suppers. Some are "put on" apparently for no good reason. Others are a benefit to prop up the financial ruins of a member of the community who may have been burned out, drowned out, or plain tuckered out. There isn't much in the way of advance notice of a public supper except word of mouth, and that's probably just fine with everybody.

Summer folk plainly love local public suppers, although they are never, but never, included in the solicitation of food. And although there may appear to be chaos in the public-supper spectacle, it is in fact run with the precision of a Marine brigade.

The person soliciting the food and handling general operations is the Commander-in-Chief, a position that is inherited. This victual General will know each local woman's capacity for cookery. She does not call on a sour-apple-pie maker for pastries or a runny-casserole concocter for scalloped dishes. Each housewife is tapped for her best. If a housewife had a mealy potato harvest in her garden, the General knows it, and mashed potatoes do not come from that quarter. And a housewife would not be asked, for instance, to provide a scalloped clam dish if her husband dug water-belly clams.

ASSOC'N

The General also commands the kitchen corps, and there is status associated with this seemingly routine KP operation. It is organized along the lines of an army platoon. There is the Top Sergeant, whose duty is to wield a big spoon. She nibbles the meat, slurps the soup, pinches the rolls, and dispatches condiments to her personal satisfaction.

The turnip masher and potato whipper are long-standing areas of experience. Not everybody could mash a turnip for a public supper. Left-handed mashers are more in demand than right for some strange reason.

Head-of-Pies is almost a commissioned officer. Her performance is often the critical stroke that saves a drowning supper. She is a sculptress possessed with an uncanny eye for measurement. Unassisted by any foolishness such as mechanical pie cutters, her miter-box sight cleaves wedges to the sixteenth of an inch. Allocation is within her authority also. The kids at the end of the table do not get chocolate cream, nor does the first selectman or prominent summer folk get even a slight glance at a crumbling graham cracker or sagging squash.

The Meat Slicer is just below Head-of-Pies in rank. She must possess many of Pie's assets plus more or less of a blacksmith's knowledge of stone and steel.

The waitresses are the privates of the public supper. They come under the jurisdiction of the Chief in the table of organization. Often recruited haphazardly, they are treated with tolerance by the Chief. Only when gross bobbles are committed will the Chief intervene. A patron receiving a lap full of gravy from a private waitress will not usually have time to utter a syllable before the Chief is by his side with a four-foot Turkish towel and a select wedge of Boston cream pie.

The Ticket Seller comprises the Chief's entire financial section. Not especially a position of importance, the Ticket Seller is usually a busted Meat Slicer who left the string on the roast beef. She sits with countenance demure behind a shaky card table at the portal to the public supper, dispensing reusable hand-printed meal tickets.

Meshed among the positions of trust, the Commander-in-Chief has numerous fledglings such as napkin runners, coffee refillers, sugarers, creamers, dropped-fork replacers, and pie fetchers. All these proletarian categories are filled exclusively from the native labor pool. There has not been a documented case of a summer folk ever regularly staffing at local suppers. A couple of guest appearances, maybe.

When the last supper patron has loosened his belt a peg or two and departed, the Chief supervises cleanup maneuvers. Pot wallopers swing into sudsy action. Floor sweepers and table cleaners know their appointed tasks. The hall now shipshape, the Chief allots leftover grub to the worthy, pulls the puckering strings on the canvas sack containing the "proceeds," and follows her troops out the door to their respective billets.

Summer folk absolutely marvel at the local efficiency talent displayed at public suppers.

Only one other home-and-away confrontation of any importance occurs. This is the annual volunteer firemen's fund drive, which conveniently coincides with the time in summer when summer folk occupancy is greatest. There are two types of solicitation. One is cold cash, and summer folk fall all over themselves writing out checks, usually generous. The other is goods for the firemen's auction. There have been goods donated to local firemen's auctions by summer folk that would make an antique dealer turn forever green. Local fishermen families are not much taken with such collectibles as period wicker furniture, for instance, and what would bring $500 on the New York market goes for $5 at the local fire department sale.

For any coastal Maine lobsterman there is a hiatus. Even the hardy few that fish offshore take a break. The inshore

Rigged for scalloping

fisherman usually has his gear on the bank by December, and setting out does not commence until April, sometimes May. When scallops were fairly abundant, the lobsterman would re-rig his boat with winch, wire, and scallop dredge. Throughout the winter months—scalloping is legal from November 1 to April 15—inshore scalloping could be quite lucrative. That industry started to decline in 1970 and has gotten progressively worse since.

Lobstermen also turn to shrimping in the winter. The Northern shrimp, or *Pandalus borealis*, comes inshore to colder waters to spawn starting about December, and lobstermen convert to dragging with small otter trawls. In recent years there has developed a substantial shrimp fishery using wire traps that are somewhat similar to the lobster trap. The boat need not be converted because the traps are hauled with the same hydraulic hauler. Lines, or pot warps, are taken from the lobster gear and used on the shrimp traps. Shrimp-trap fishermen were receiving about 70¢ a pound in the winter of 1984.

One problem with shrimping, however, is that it is either offshore or confined primarily to the midcoastal area. The Boothbay-Sheepscot River area is the center of the shrimping industry, probably because of the large river systems that feed fresh water to these coastal waters, giving them a lower sea temperature than many other parts of the coast. Whatever the reason, there is little shrimping in the Washington or Hancock County areas, which could use an alternative winter fishery.

The lobstermen, except those in the mosquito fleet, rarely turn to clam digging or worming. It is somewhat beneath the status of a lobsterman to dig in the mudflats. Besides, clammers and wormers are a breed apart and they probably wouldn't allow lobstermen into their exclusive brotherhood.

If a lobsterman doesn't switch fish, then he either gets a shore job or works his gear. Most work their gear. Because of the weather and outside winter storage, boats are usually left for springtime. But there are traps to be built. Fish houses become brotherhood lodges. "Fish house" is a generic term meaning any place with a stove, workbench, and just enough room to store some trap-building materials. Fish houses can be on wharves, in the barn, in the garage, or even the kitchen.

Coastal communities in which fishing is the primary economic factor are "quite religious," a downeast designation meaning that the fundamentalist denominations—Nazarene, Church of God, Pentecostal, Baptist, and Seventh Day Adventist—are the dominant congregations. But there are Roman Catholics, Episcopalians, Congregationalists, Unitarians, Mormons, Christian Scientists, Jews, Lutherans, Greek Orthodox, and members of some sects peculiar to Maine such as the Bullockites, a society of primitive Baptists.

The widely known Seacoast Mission headquarters in Bar Harbor operates the vessel *Sunbeam*, which brings religious, educational, hospital, and recreational facilities to the coastal communities, especially those on the offshore islands. There are 1,500 churches representing 26 denominations in Maine. And it's probably fair to assume that fishermen and their families are represented in all of them.

Fishermen are also in the mainstream of service clubs in the state. There are about 200 of these, including the familiar Rotary, Lions, Kiwanis, Elks, Knights of Columbus, Odd Fellows, VFW, and American Legion. The VFW and Legion dances, usually held on a Saturday night, are the liveliest game in town in many places. Another surprising focus of community participation is the Granges, which one usually associates with farming families. However, there are 600 local Granges throughout the state and no dearth of these organizations along the coast.

The lobsterfishing ports are as different in character as the men who sustain the industry and the families who live on their shores. Yet the people of lobsterfishing communities have a common bond, although their independence belies that fact. They like where they live and defend those roots with uncommon tenacity when their life is challenged by encroaching civilization or "from away" meddlers. Their idiorhythmic hamlets, large or small, are their business, and nobody should pry from without or prattle from within. And yet, their communities are too rich in love to ignore.

ME

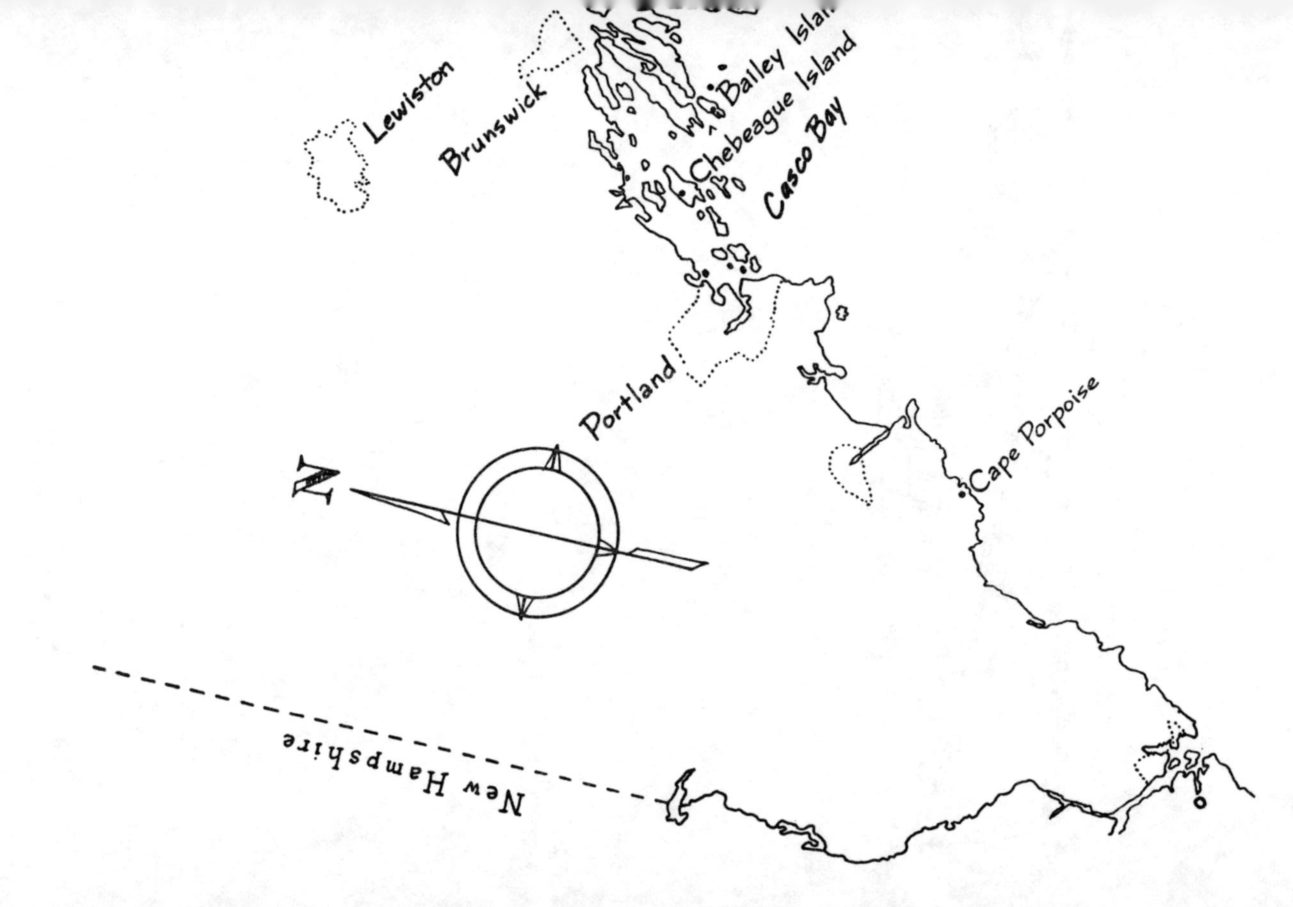

Lewiston
Brunswick
Bailey Isla
Chebeague Island
Casco Bay
Portland
N
Cape Porpoise
New Hampshire

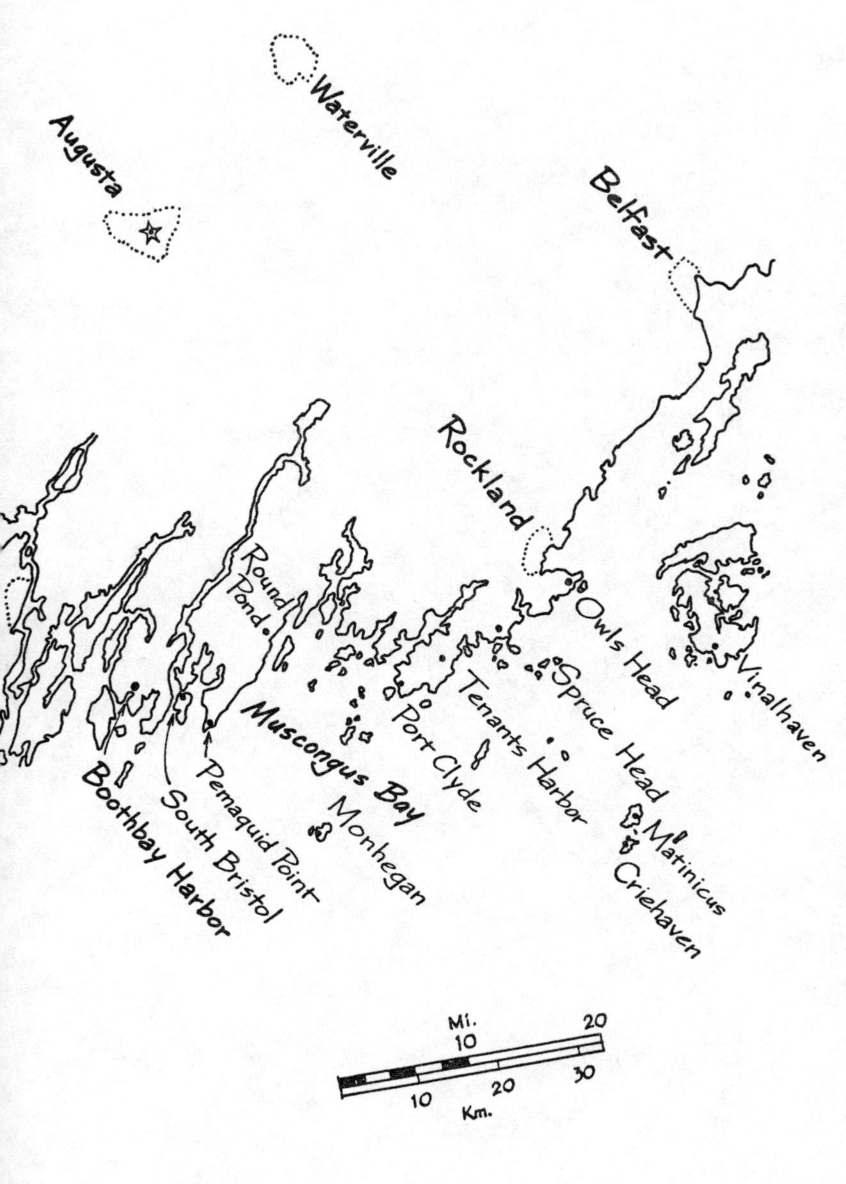
Augusta
Waterville
Belfast
Rockland
Round Pond
Owls Head
Vinalhaven
Spruce Head
Tenants Harbor
Port Clyde
Muscongus Bay
Monhegan
Pemaquid Point
South Bristol
Boothbay Harbor
Matinicus
Criehaven
Mi.
10
20
30
Km.

Core
Prospect Ha
Winter Harbor
Cranberry Isles
Bangor
Ellsworth
Tremont
Bass Harbor
Frenchboro
Swans Island
Deer Isle
Isle Au Haut
Penobscot Bay
Vinalhaven
Belfast
Islesboro
Mi.
10
20
10
20
30
Km.

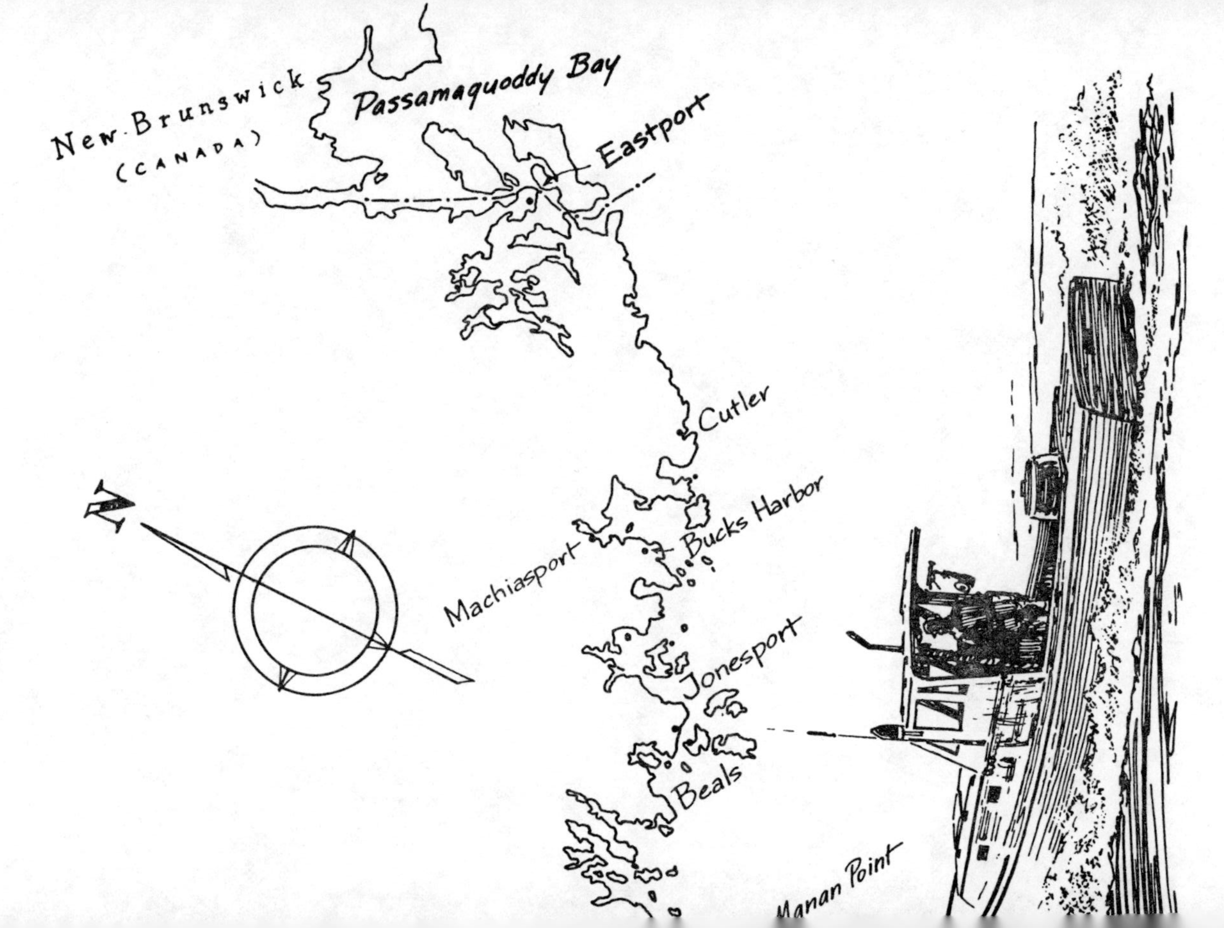
New Brunswick
(CANADA)
Passamaquoddy Bay
Eastport
Cutler
Bucks Harbor
Machiasport
Jonesport
Beals
Manan Point
N